Gems

Gems

Their Sources, Descriptions and Identification

Sixth Edition

Edited by
Michael O'Donoghue

AMSTERDAM • BOSTON • HEIDELBERG • LONDON
NEW YORK • OXFORD • PARIS • SAN DIEGO
SAN FRANCISCO • SINGAPORE • SYDNEY • TOKYO

ELSEVIER

Butterworth-Heinemann is an imprint of Elsevier

Butterworth-Heinemann is an imprint of Elsevier
Linacre House, Jordan Hill, Oxford OX2 8DP, UK
30 Corporate Drive, Suite 400, Burlington MA 01803, USA

First published 1962
Second edition 1970
Third edition 1975
Fourth edition 1983
Fifth edition 1994
Sixth edition 2006

British Library Cataloguing in Publication Data

A catalogue record for this book is available from the British Library

Library of Congress Cataloguing in Publication Data
A catalogue record for this book is available from the Library of Congress

ISBN 13: 978-0-75-065856-0
ISBN 10: 0-75-065856-8

For information on all Butterworth-Heinemann publications
visit our website at www.books.elsevier.com

Typeset by Integra Software Services Pvt. Ltd, Pondicherry, India
www.integra-india.com
Printed and bound in Great Britain

Working together to grow
libraries in developing countries

www.elsevier.com | www.bookaid.org | www.sabre.org

ELSEVIER   BOOK AID International   Sabre Foundation

Contents

Contents

For Annie Satsuma

Preface to the Sixth Edition

This text is intended as an exhaustive study of those substances, both organic and inorganic, natural and man-made, which have at some time been used as ornament. A number of authors have helped me and are acknowledged in the appropriate places and some historical material has been edited from the Fifth edition of *Gems*, by the late Robert Webster. Some material from this work, revised to a greater and sometimes a lesser extent over five editions, has been retained, though a few chapters which appeared largely untouched from the 1962 original have been scrapped out of respect for the materials described.

I, as editor and major contributor, am ultimately responsible for those omissions and inaccuracies which will inevitably have crept in. This book is not intended to be a bench manual; see references in many places below to *The Identification of Gemstones* by O'Donoghue and Joyner (2003, ISBN 0750655127), *Synthetic, Imitation and Treated Gemstones* by O'Donoghue (1997, ISBN 0750631732) and *Artificial Gemstones* (2005, ISBN 071980311X).

Readers should know that the only way to keep up with the work of determinative mineralogy is to consult the relevant journal literature. Though they are just what I am about to recommend, monographs by their nature cannot be so up-to-date; the ones I am using at the time of writing are the *Handbook of Mineralogy* (Anthony, Bideaux, Bladh and Nichols, five parts in six volumes, Mineralogical Society of America, 1990–2003). The best and pretty well the only easily available guide to mineral species and groups (outside journal literature) is *Fleischer's Glossary of Mineral Species*, latest edition 2004 (ISBN 094500530X). This is updated regularly in *The Mineralogical Record* and new issues seem to come out every four to five years. A number of minerals lose their status as valid species as investigation grows more sophisticated and the books recommended above are the best guides. Those who wish for the prolongation of outdated and discredited names can take the matter up with the International Mineralogical Association, Commission on New Minerals and Mineral Names (CNMMN).

Figures for dispersion are given only in cases where the property is significant: lustre is assumed to be vitreous in the majority of cases, exceptions are cited. Pleochroism of light and dark of the same colour is also assumed for anisotropic coloured specimens and is only described where different colours are seen.

Locations are mentioned only when they provide significant sources of ornamental-quality material. For many of the better-known species there are far too many locations for all to be mentioned, let alone described in the detail that would be pleasing to all! I have consulted a very large number of authorities and cited full details in the text; I am grateful to my very large personal library for virtually all sources.

Derivations of gem and mineral names are not given; they can be found in almost all cases and old ones more conveniently in Bayliss's *Glossary of Obsolete Mineral Names* (2000, ISBN 0930259041) who conveniently collects those names used for mineral substances throughout history, which are not now considered valid or current.

While many gem collectors collect mineral specimens too, the text does not cater for them so that many famous sites for crystals which are never likely to be fashioned, probably fortunately, have not been entered. They can be located in issues of *The Mineralogical Record* and in Anthony et al. cited above. The history of mineral collecting is excellently summarized by a single-topic issue of *The Mineralogical Record* (Vol. 25, No. 6, 1994), edited by Wendell E. Wilson. This paper includes a bibliography of collection catalogues and a general bibliography. I have not mentioned in each instance some properties which ought to be obvious: one such is the propensity of carbonates to effervesce in some acids.

Robert Webster's *Gems* was first published in 1962 and subsequently achieved further editions in 1970, 1975, 1983 and 1994. I was one of the major revisers of the text for the 1994 edition. In this new edition, the gem testing section has been omitted; this information is more easily available in Peter Read's *Gemmology* (2005, ISBN 0750664495). Though previous editions of *Gems* provide a great deal of information, it is not always efficiently sorted or really useful; in addition, some of the revisions over the years have added very little to the 1962 text! Chapters in this category have been re-written.

This is particularly the case with organic materials where I have retained text from *Gems* for the sake of the history, while modifying the rather individual didactic and flowery style to suit modern standards. I am strengthened in this decision by the welcome appearance of such books as Maggie Campbell Pedersen's *Gems and Ornamental Materials of Organic Origin* (2005, ISBN 0750658525). This text is the most up-to-date monographic account of these materials so far and gives all the details necessary for testing (though this should be carried out by a laboratory).

In the case of some of the 'classic' species, I have deliberately omitted continual comparisons with specimens of similar appearance but of a different species. Space forbids the presentation of more than a descriptive coverage of as many species, synthetics and imitations as possible.

I am heartened in this decision by the publication of Peter Read's *Gemmology* (third edition, 2005, ISBN 0750664495) and by the recent appearance of the gem testing survey *The Identification of Gemstones* (2003) by O'Donoghue and Joyner. "Sinkankas" refers to *Gemology: an annotated bibliography*, 1993.

I have cited in the text as many as possible of the papers from which information has been taken instead of building up a large block in a single place. I have found that this arrangement is preferable. Some of the contributors have also added short lists of references.

Michael O'Donoghue

Preface to the Fifth Edition

When, after three reprints of the 1983 fourth edition of Webster's *Gems*, the publishers decided it was time to embark on a fifth edition, they contacted several gemmologists to canvas nominations for the task. I was only just recovering from the effort of producing *Gemmology*, a vastly expanded update of my original paperback *Beginner's Guide to Gemmology*, and offered several possible names. When none of these avenues proved fruitful, the publishers again contacted me, this time with the proposal that I consider attempting such a revision myself. I countered this with the comment that while I would be very happy to act as the overall editor of a fifth edition (as with the Butterworth-Heinemann Gem Book series), the revision work involved with so large a book would be more sensibly shared out among a group of authors/gemmologists, each member tackling his or her own speciality subjects. This was agreed to by the publishers and led to the approach subsequently adopted.

Because of the high regard and affection in which Webster's great work is held, I had little difficulty in finding willing contributors. As a result, the bulk of the thirty-nine chapters, identification tables and appendices were revised by the fifteen specialists named in this edition. It is interesting to note that these include contributors from Australia, Canada, Germany, Thailand, Vietnam, the UK and the USA, making this a truly international collaboration. Driven by my own interests in gem testing equipment and techniques, I have taken advantage of my position as editor to revise a few of the relevant chapters in Part 2 of this volume.

In addition to the necessary updating of chapters by the contributing authors, the contents of three chapters have been rearranged, 'Composite and Artifically Coloured Stones' becoming 'Composite Gemstones', 'Colour in Gemstones' becoming 'Colour and Colour Enhancement', and 'Opal' being extracted from 'Gems of the Silica Group' to form a separate chapter. I have also reintroduced Table 1 from the third edition, which contains a comprehensive list of gemstone constants.

I have been privileged to know both Robert Webster and Basil Anderson, and I trust that this fifth edition of *Gems* will live up to the high standards set in Robert Webster's three earlier editions, as well as those of the fourth edition, which was revised by Basil Anderson. In keeping with the tradition set by the very first two-volume publication, which included contributions and advice from Webster's colleagues in the London Chamber of Commerce's Gem Testing Laboratory, this edition is enhanced by contributions from colleagues in the Gemmological Association and Gem Testing Laboratory of Great Britain.

Peter G Read

Preface to the Fourth Edition

When approached by the publishers with the suggestion that I should undertake the revision, which had clearly become needed, of Robert Webster's great book, I was under no illusion that the task would be an easy one if it were to be carried out with the care and completeness that the work demanded. Nevertheless, I felt compelled to accept the heavy assignment, partly as a memorial tribute to an old and valued colleague, and partly also because the seed from which *Gems* eventually grew to such an imposing stature was planted as long ago as 1953 as a joint project from the London Gem Testing Laboratory, in which not only Robert Webster but CJ Payne and myself were to be actively involved. In my diary for that year the 36 intended chapters of this ambitious proposition were listed and allocated to what seemed the most appropriate authors. CJ Payne and I actually got so far as to write a chapter apiece, but our will to continue gradually dwindled, partly because the removal of the Laboratory from its old gloomy quarters at 55 Hatton Garden to new accommodation in itself provided a distraction and also resulted in an increase in routine work and better facilities for research, and partly because the excellent revision carried out by FC Phillips on Herbert Smith's classic *Gemstones* had diminished the urgent need for an up-to-date work on gemmology. Robert Webster, however, quietly persisted in his determination to compile a book on a more comprehensive scale, and to that end contributed long serial articles to *The Gemmologist* on the microscope, X-rays, and luminescence suitable for eventual incorporation in the projected work. As his colleagues and friends we were quite content to supply him, where needed, with data from our records and occasional specialised advice. But the sheer slogging hard work, the painstaking gathering of facts, and eventually the unusual ordering of the book with its description of species preceding that of the gemmological techniques needed for their study, were all Robert Webster's work, and in 1962 the first edition appeared in two volumes, a form which had obvious disadvantages for the reader.

In earlier and less scientifically inventive days the lapse of some six or seven years between editions of a standard work of this kind could be expected to entail no more than minor trimming and a few new facts or descriptions of fresh techniques. Not so today. The period which I have tried to cover for this fourth edition has seen numerous changes of far-reaching consequence for the science of gemmology. It is a strange but true fact that a large proportion of these changes have been brought about as a direct consequence of the appearance in quantity of cubic zirconia, an extremely successful simulant of diamond, which has virtually 'taken over' from the succession of previously acclaimed imitations of that most important of all gemstones. It is an encouraging fact that the precious stone trade (after a brief flirtation with meaningless fancy names for the product) has been accepting and selling 'CZ' under its correct and unequivocal description. Fifty years ago, if one may judge by the absurd 'scare' caused by the skilful promotion of synthetic white spinel as a diamond substitute, both the trade and the public might have shown some degree of panic, and there might even have been a temporary fall in the price of diamond. Thanks to the world-wide spread of gemmological knowledge, the outcome of the appearance of CZ took the much more practical form of an outburst of invention and the production of new forms of testing apparatus (reflectivity meters, heat-probes, high-index refractometers and the like) which served to enable not only the trained gemmologist but the scientifically untrained working jeweller to distinguish quickly between diamond and the new product, even when the stones were mounted in jewellery and therefore unavailable for a simple density test. A full description of CZ and of these new instruments was obviously a foremost concern of this new edition.

There have also been a number of additions to the range of synthetic gemstones produced by Pierre Gilson including, most importantly, synthetic opal, both white and black, and a corresponding ingenuity in the production of 'Slocum stone', a special type of glass closely resembling opal in appearance. Quartz, which for many years has been manufactured in large, clear, twin-free crystals for industrial purposes, has now entered the field of synthetic coloured stones not only in shades of blue and green, but in stones virtually indistinguishable from natural citrine and amethyst, which are now produced in Russia and Japan on a commercial scale.

Alteration in the colour of gemstones by heat-treatment, irradiation or bombardment with sub-atomic particles has been practised for a considerable period, but there have been advances in such treatments quite recently which have caused some perturbation in the trade. These are described and discussed at some length in this edition.

Diamond grading has now become so embedded in trade practice for stones of any importance that it seemed necessary to describe in some detail the current procedures followed. Though the diamond market has recently been through a difficult period, new and unexpected sources for the mineral continue to make their appearance. Discoveries of other extensive fields in the prophetically named Kimberley district of Western Australia promise to be of world importance when production is fully advanced in a few years time, while diamonds have also been discovered in China in some quantity.

Several new gem minerals, or minerals recently found in gem quality, have seemed of sufficient interest to be included in the list of gemstones described briefly in these pages. Of these, charoite is probably the most important and certainly the most spectacular. Further, two of the rarest of all gemstones, taaffeite and painite, have been re-analysed using modern techniques and new formulae assigned to them.

To compensate for the space needed for these and many other necessary additions to the text in a book which was already amply long for a single volume, a number of cuts have been made in certain portions of the text, which, I am confident, will not be seriously missed.

In gathering the necessary information concerning all these advances in gemmology I have relied very heavily on the papers and abstracts published in the specialised journals with which our science is now so splendidly endowed. I have been particularly indebted to Britain's own *Journal of Gemmology* whose editor, Mr John Chisholm, has been successful in providing in each quarterly issue a rich harvest not only of original papers but also of abstracts from papers published in similar journals from all over the world. Wherever my own brief résumé of certain papers has been inadequate for reasons of space I have inserted the appropriate reference in the text – a practice which I have myself found helpful when adopted by other authors. One book which I found outstandingly valuable has been *Gems Made by Man* by Kurt Nassau, in which the talented author provides gemmologists with detailed and accurate information which goes far beyond the scope suggested by the title. For this and for his many recent articles in various journals I feel that all gemmologists should be grateful.

I have received help in smaller but nonetheless valuble ways from a large number of my gemmologist friends, and I hope they recognise that to mention them all by name would be a somewhat tedious and perhaps an invidious procedure, as knowing the frailty of my memory, some equally worthy names would be unintentionally omitted. I am confident that I have personally thanked all those who gave me help.

I should, however, like to say how glad I was, before embarking on this lengthy project, to receive a letter from Jean Smith, Robert Webster's eldest daughter, expressing her pleasure that I should be undertaking the preparation of a new edition of her father's book. She also kindly sent me a small note-book containing his jottings of possible value in a future edition.

B W Anderson

Preface to the Third Edition

Since the second edition of *Gems: Their Sources, Descriptions and Identification* of 1970, there has been a considerable increase in the number of man-made gem materials now available as gemstones or having a potential in that direction. This has made necessary a completely rewritten chapter on synthetic gems, and this has been necessary, too, with the text on composite stones of which there are now a number of new types.

The chapters on diamond have been revised in order to bring the discussion more in line with modern knowledge. In the cases of emerald and ruby and sapphire new sources of supply have necessitated additional text to the chapters on these gemstones. Minor adjustments have been made to other parts of the text and some new and unused gem materials have been mentioned where it is known that cut stones are in the hands of collectors.

The alteration of place names, particularly those of the new republics of Africa and Asia, has caused problems. Where possible these new names are put in the section headings, but they may not be continued throughout the text, for the older names are still in current circulation and are much better known. Thus 'Siam rubies' are understood by all, but 'Thai rubies' may well cause raised eyebrows, and even more so with 'Sri Lanka rubies and sapphires' in place of 'Ceylon rubies and sapphires'.

New designs of testing instruments are described and something is told of the more unusual instruments and techniques used by highly equipped laboratories coupled to universities, museums and large industrial concerns which have these types of apparatus for research and control. These are often made available when normal gem-testing methods fail to provide an answer. The author tenders his thanks to Mr Alan Jobbins and Mr Brian Young of the Institute of Geological Sciences for their assistance in making these additions.

Again I am indebted to those many friends who have sent criticisms and fresh information, and, rather belatedly, my thanks to Mr Ove Dragsted of Copenhagen for allowing me to use the colour plate of *tugtupite* at a critical moment in the production of the second edition. Thanks, too, to De Beers Consolidated Mines Ltd, for the information on the Botswana diamond mining; Messrs Eickhorst & Co of Hamburg for the illustrations of the diamond colorimeter and dark field illuminators, and to Gemmological Instruments Ltd and P W Allen for illustrations of new types of instruments. I also acknowledge the help given by Mr Charles Schiffman of Lucerne in making certain criticisms and providing a new picture of the Gübelin spectroscope, Mr Monnickendam for advising on diamond classification, and Professor Hermann Bank for his helpful advice.

Robert Webster

Preface to the Second Edition

Experience with the two volumes of the first edition has shown that a single volume would be more convenient to the user. Despite the production of a book having a greater weight, *Gems: Their Sources, Descriptions and Identification* is now produced complete within a single cover.

The arrangement of the text has not been altered except that additions have been made to various chapters so that new information could be incorporated, in particular that on synthetic stones. New tables on plastics and their identification, drawn up by Mr H Lee, will provide more help in dealing with the identification of these difficult materials, and separate tables on refractive indices and specific gravities have been included to supplement the comprehensive main table of constants. The new blue zoisite gemstone, Tanzanite, has called for an alteration in the text. The section 'Thulite', which was previously the only gem material of the zoisite species, has now been entitled 'Zoisite' in order to cover more conveniently the new variety.

My thanks go to the many friends all over the world who have sent me items of information, or pointed out errors of omissions in the first edition. Dr Kurt Nassau has been most helpful in advising on the newer synthetic stones and Mr Craig C Smith gave valuable information on the new chrome chalcedony found in Rhodesia. The staff of both western and eastern headquarters of the Gemological Institute of America gave helpful advice, and I must give my thanks to Mr B W Anderson for his ever-ready assistance, advice and criticism. I thank Mr H Lee and Mr Dennis Smith for their help in proof reading this second edition.

New coloured plates, with a somewhat different outlook, replace those of Hallwag used in the first edition, as these earlier plates are no longer available. The new plates have been provided by the courtesy of De Beers

Consolidated Mines Ltd; Messrs Garrards Ltd (The Crown Jewellers); Messrs Christie, Manson, and Woods; and The Institute of Geological Sciences through the staff of the Geological and Survey Museum, who prepared the coloured plates of the stones specially for this edition.

Robert Webster

Preface to the First Edition

For two decades, an earlier work of mine – *Practical Gemmology* – has proved a useful elementary textbook on the subject of gem materials and their testing, a study now known as 'gemmology'. During this period, new materials have been found and testing methods greatly improved, so that the simple expositions given in *Practical Gemmology* do not now cover the subject adequately and a new book has become a necessity.

The majority of books written on the subject of gems start with a detailed account of the physical and optical properties of gem materials. They then go on to the theory and use of instruments used in gem-testing before describing the gems themselves. In this work, the usual scheme is reversed and a more practical approach is made. A short introduction to the formation of such minerals in the earth leads to descriptions of the various gem materials.

Following this, there are chapters on synthetic stones, composite and imitation stones, and how gemstones are fashioned for the market. The first volume concludes with descriptions of pearl, coral, jet and amber, and other materials used in ornamentation which owe their genesis to organic processes. The second volume deals with the technical aspects of gem materials and is followed by descriptions of the various methods used in gem identification. The book is completed by a section containing tables and useful data.

The arrangement of describing the gemstones themselves before their technical aspects may be considered open to objection as some technical data must be included in the description of the stones. This is a minor point against the value of an arrangement which introduces the subject in the logical sequence of the finding in the rough state of the natural gems, the production of synthetic and imitation stones, and the fashioning of these various natural and man-made materials into a finished gemstone. Where possible, some simple explanation of the meaning of the technical references is brought out as the gem story unfolds.

One problem which besets any writer on gemstones is to select the order in which the gem materials are discussed. Whatever method is used, some criticism is inevitable. In this work the better-known jewellery stones are placed first, and are followed by the lesser-known stones in alphabetical order, unless, for some reason, the stones are better placed in a group forming a small subchapter.

In the technical section an endeavour is made to tell something of the history of gem-testing which has led to the present high standard of what has with some truth become known as 'scientific gemmology'. In this section, too, are details of the use and working of a number of special instruments not usually included in such books.

Throughout the compilation of this work considerable assistance has been received from many members of the Gemmological Association and of the jewellery trade. A special debt of gratitude is owed to Mr G F Andrews and Mr H Lee, who read and criticised the original manuscript.

I owe much to my colleagues Mr B W Anderson and Mr C J Payne for their encouragement and for unstintingly supplying me with much of their data accumulated in the archives of the Laboratory of the London Chamber of Commerce. Further, Mr Anderson freely handed over to me much of his unpublished work on the causes of colour, the hardness and the methods of determining the density of gemstones. These notes have been duly incorporated in the text. Likewise, Mr Payne allowed the incorporation of his articles on the method of refractive index determination by minimum deviation, and that on interference figures, both of which were published in *The Gemmologist*.

Mr Lee, with his technical knowledge of chemistry, gave considerable help on the chemical aspects of plastics, and the chapter on chemistry is based mainly on his information.

My thanks go to Dr W E Smith of Chelsea College of Science and Technology for his help with the chapter on the geology and the formation of minerals which forms the first part of the book. Dr W Stern advised on the part dealing with the marketing and industrial uses of diamonds, and Herr G O Wild of Idar-Oberstein gave information on a number of topics, particularly on the quartz gems.

Mr J Asscher, Jnr, of Amsterdam checked the notes on diamond polishing, and Mr C L Arnold and Mr G E Bull-Diamond likewise checked the part on lapidary working of gemstones.

A number of line drawings, particularly of crystals, were kindly prepared for me by Mr G A White of Norwich. Dr E Gübelin allowed the use of some of his excellent photomicrographs to illustrate the chapter on gemstone inclusions, which was mainly compiled from his published researches. It is appropriate here to mention that Plates I–XVI appeared in Dr E Gübelin's *Edelsteine*, and are reproduced by courtesy of Hallwag, Berne. Mrs V G Hinton kindly took the pearl surface photomicrograph specially for me, and Dr E H Rutland, Mr B W Anderson, Mr R K Mitchell and Mr H Lee assisted in the task of reading the proofs.

The Gemmological Association kindly allowed the incorporation in this work of my articles on emerald, ruby and sapphire, and marble, as well as a number of illustrations which had previously been published in the *Journal of Gemmology*. The chapters on the microscope, X-rays and luminescence are largely made up from my articles published in *The Gemmologist* and are reprinted, together with the illustrations, by courtesy of NAG Press Ltd. Finally, the pictures of the pearl cultivation in Japan were supplied by Shell Photographic Unit and the Cultured Pearl Company, and those of diamond mining by the Anglo-American Corporation.

Robert Webster

Acknowledgements

I am indebted to my contributors, and to Gwyn Green for a keen scrutiny of the manuscript, and to Louise Joyner for her help and support in many ways.

I gratefully acknowledge the use of photographs from the late Professor Dr Edward Gübelin, from Alan Jobbins (Chapter 1), from Kenneth Scarratt, from the contributors, and from GemA, London.

Stephani Havard and Jackie Holding of Architectural Press have been the embodiment of kindness and patience.

Contributors

Brown, Grahame (Chapters 28, 29, 31)
Collins, Alan T. (Chapter 4)
Dominy, Geoffrey (Chapter 30)
Harding, Roger (Chapter 1)
Hughes, Richard W. (Chapter 5)
Jackson, Brian (Chapters 11, 12)
Mercer, Ian (Chapter 2)
Middleton, Andrew (Chapter 20)
Pedersen, Maggie Campbell (Chapter 27)
Rouse, J. (Chapter 25)
Sanderson, Robin (Chapter 22)
Woodward, Christine (Identification Tables)

All other chapters were written and/or revised by Michael O'Donoghue

Part I

1

The Geological Sources of Gems

Roger Harding

Introduction

Gem materials are rare, and the finding of a single stone can be either the result of a happy accident or the culmination of a determined effort by perhaps thousands of people. Gems are recovered at or beneath the surface of the Earth, and in order to understand how or why they form, some appreciation of the wider aspects of Earth science or geology is necessary.

Geology is the study of the planet Earth – its rocks, minerals and fluids, its systems and its processes. Earth's *systems* comprise the solid earth, the hydrosphere, the atmosphere and the biosphere, and each interacts with the other through a wide range of *processes*. The different gem materials come from many different sources in the Earth: most are from the solid earth, but some owe their existence also to the biosphere (organic gems) and processes operating in all four systems or may contribute to gem formation.

With appropriate illustrations this chapter outlines the different geological contexts in which gems may be found, introduces geological terms which may be referred to in the chapters on gem species, places gems in the context of Earth's history, and briefly outlines some methods of recovering gems.

Structure of the Earth

The Earth's interior consists of three main components (*Figure 1.1*). The outermost layer is the crust, beneath which is the mantle, and in the centre is the core. The core has a diameter of just under 7000 km and probably consists of two parts, an inner solid region of diameter 2440 km and density about 8 g·cm^{-3}, and an outer region about 2260 km thick and

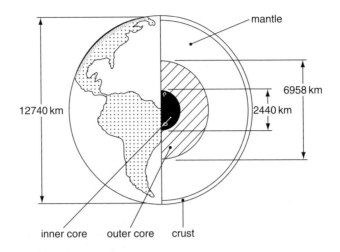

	Depths (average) km	Thicknesses (average) km
Crust	0–40	40
Mantle	40–2891	2851
Liquid iron outer core	2891–5150	2259
Solid iron inner core	5150–6370	1220 (radius)

Figure 1.1   Schematic diagram of the division of the Earth into the inner core, outer core, mantle and crust

density about 5.5 g·cm^{-3}, which behaves as a liquid. Around the core is the mantle, about 2850 km thick, which, with a density about 3.5 g·cm^{-3}, is much less dense than the core. Around the mantle lies the Earth's crust, a 'thin' skin of rock, less dense than the mantle, and derived from it by geological processes which have operated for millions of years.

The boundaries between the core, mantle and crust have been deduced from studies of earthquakes, heat flow measurements and behaviour of the Earth's magnetic field. The change, for example, from crust to mantle is marked by an abrupt increase in the speeds of seismic primary and secondary waves (P and S waves) generated by earthquakes. Such properties also enable geologists to subdivide the mantle and distinguish a zone beneath the crust which is rigid and strong – the lithosphere. Beneath the lithosphere is a zone which is partially molten and weak – the asthenosphere; below that in the solid mantle, there are two abrupt changes of increasing density before the outer core is reached.

At the surface of the Earth, seven-tenths of the crust is covered by water and this points to a fundamental geological distinction between two kinds of crust: oceanic and continental. The oceanic crust averages 5 km in thickness and about 3 $g \cdot cm^{-3}$ in density, and nowhere is it older than 200 million years. The continental crust averages 2.7 $g \cdot cm^{-3}$ in density and 25–40 km in thickness, but under mountain chains such as the Himalayas it may reach 65 km.

Large areas of the continental crust have been relatively stable for at least 1500 million years and are called shield areas; some areas within the shields consist of rocks as old as 3900 million years.

The oceanic crust has a simple layered structure and relatively consistent composition; by contrast, the continental crust contains a wide variety of rocks found in an equally wide range of simple to complex rock structures. This range of rock types and their inter-relationships are the records of geological processes, and by studying their compositions and properties, their histories can be interpreted.

Rocks and the Processes That Formed Them

Rocks are composed of minerals, and although over 4000 minerals are known, only about 100 are common enough to be considered as rock-forming. Some rocks consist of one kind of mineral only, and the white ornamental marble, composed entirely of calcite, is a good example (*Figure 1.2*). Most rocks, however, are composed of two or more minerals.

Figure 1.2   Fantiscritti quarry, Carrara, Italy where marble is mined

A mineral can be defined as a naturally occurring solid, generally inorganic, with a specific chemical composition and crystal structure. Minerals with the qualities of beauty, durability or rarity and considered suitable for use as gems number about 20, but up to 200 could be considered as 'lesser-known' gem or ornamental materials.

In detail, the compositions of rocks in the Earth's crust are many and varied, but traditionally it has been convenient to consider them under three broad headings which indicate the processes that formed them: igneous, sedimentary and metamorphic. An outline of the main rocks in these categories will be followed by a section, particularly important for gems, on rocks derived from water-rich fluids.

Igneous Rocks

Some regions of the Earth's crust and mantle are so hot that parts may melt to form molten rock or magma. When magma cools, it forms igneous rocks, and the rate of such cooling has a major influence on the appearance and texture of the rock. Deep underground, slow cooling allows time for crystals to grow and leads to coarsely grained textures; these rocks are described as plutonic or intrusive. If the magma reaches the surface of the Earth and is extruded from a volcano (*Figure 1.3*), cooling is rapid and leads to fine-grained textures or even glassy rocks (*Figure 1.4*); these are volcanic or extrusive rocks.

Figure 1.3   Basaltic lava flow of 1944 in a valley leading down to Naples, Italy. The Monte Somma crater wall is in the background

Appearance and texture are the most obvious features to use in classifying igneous rocks, but rocks of similar texture, whether coarse or fine, may have very different compositions, different origins and different histories. One of the earliest compositional classifications of

Figure 1.4   Basaltic lava containing white skeletal crystals of feldspar about 0.2 mm across in a groundmass of pale brown glass and tiny grains of magnetite. Island of Mull, Scotland

igneous rocks was based on their silica (SiO_2) content; most igneous rocks contain between 40 and 70% silica, and this range gives considerable scope for discrimination. Modern classification, however, puts the emphasis more on silicate minerals than on simple silica content. The most important minerals for classification purposes are olivine, the pyroxenes, the amphiboles, the micas, the feldspars and quartz. Broadly, these minerals represent the range from *mafic* minerals, those rich in magnesium and iron, to those rich in silica – the *felsic* minerals, a name derived from feldspar and silica. Since mafic and felsic are compositional terms, they are applied to rocks as well as minerals, and typical mafic rocks are basalts (extrusive) and gabbros (intrusive), while typical felsic rocks are rhyolite and obsidian (extrusive) and granite (intrusive) (*Figure 1.5*). Between the mafic and felsic rocks are the intermediate rocks: the andesites, which are extrusive and the intrusive diorites. Outside this range are rocks such as peridotite, which are particularly rich in magnesium, and a number of much rarer rock types which may be high in carbonates, sodium and potassium, or compositionally extreme in some way.

The processes involved in forming igneous rocks can take place over millions of years and can affect wide areas of country rock. Such processes have resulted in large volmes of granitic rocks underlying mountain ranges like the Andes and the Rockies. Some of the granites may be homogeneous and consistent in texture, but other bodies may consist of multiple intrusive phases, with different rock types being emplaced at different times. Around the margins of granite intrusions there may be evidence that the country rock has been pushed apart to create tension cracks that are then filled with the intruding granite. These

(a)

(b)

(c)

Figure 1.5 (a) Gabbro on the left in contact with basalt on the right. The twinned feldspar crystals in the gabbro are about 1 mm across. St Kilda, Scotland, (b) Bear carved in obsidian by Fabergé craftsmen. Obsidian is a natural glass with the composition of granite, (c) Granite texture shown in thin section and consisting of quartz (clear), feldspar (pale brown, 'dusty') and magnetite (black). Field of view: 4 mm across. St Kilda, Scotland

sheet-like bodies of granite are known as dykes if they are steeply dipping or sills if they are flat-lying. Similar features are shown by mafic rocks such as gabbros and basalts, and in some areas such as north-west Scotland, the parallel or radial arrays of basaltic dykes are so dense as to form a 'dyke swarm' (*Figure 1.6*).

When magma cools slowly, individual minerals start to crystallize around local centres. The first to do so are those that form at the highest temperatures – generally the mafic minerals. These develop relatively good crystal outlines extracting the magnesium and iron from the magma and leaving a liquid which becomes increasing rich in silica, alumina, alkalis and the rarer elements. If the region is quiescent and all the magma solidifies in the one intrusion, an igneous rock consisting of a mixture of well-formed minerals and later interstitial minerals will be formed. But if about half way through crystallization, the region becomes

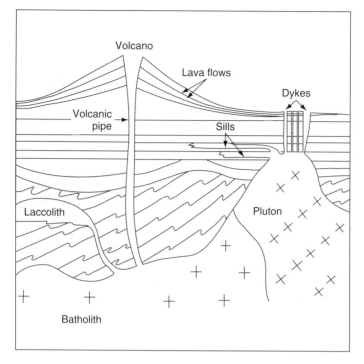

Figure 1.6   Schematic section of the Earth's crust showing igneous rock structures, their names and relationships with metamorphic rocks (wavy line pattern) and sedimentary rocks (horizontal lines)

Figure 1.7   Sheets of basaltic rock intrusive into diorite, on the island of St Kilda, Scotland. The sheet, lower right, splits in mid-picture to form a dyke towards the top and a sill to the left

active, then liquids may be squeezed out of the crystal mush and migrate elsewhere.

Towards the end of crystallization of a granite – say when about 80% of the mass has solidified – the residual liquids are very often water- or gas-rich, and contain many of the rarer elements that could not find a place in the earlier-formed minerals. These residual liquids and gases may remain in the granite and cool to form miarolitic cavities (*Figure 1.8a*) lined with such crystals as beryl and topaz, or they may coalesce into pools or sheets and cool to form pegmatities (*Figure 1.8b*).

Pegmatites are very coarse-grained rocks which occur in pockets in their host igneous rock or as sheets or veins at its edges – both within the

(a)

(b)

Figure 1.8   (a) Miarolitic cavity in granite from the Mourne Mountains, Northern Ireland. A topaz with good crystal form projects into a cavity bordered with white feldspar and slightly smoky quartz, (b) Very coarse pegmatite at least 8 m (26 ft) thick at Urubu, near Araçuai, Minas Gerais, Brazil. Fine pink and green gem-quality tourmaline has been mined here

intrusion and in the surrounding country rock. They represent a late stage in the crystallization history of a rock melt and can contain crystals of extraordinary dimensions. Giant crystals weighing some tons and measured in metres occur in a number of localities, and although the largest crystals are opaque some gem-quality aquamarine and topaz crystals can weigh many kilograms.

In geological usage, veins are generally understood to be less regular in form than sheets – which are parallel-sided. They are more or less irregular cracks and fractures, filled with minerals which can range from millimetres to metres in width and from metres to kilometres in length. Quartz veins in parts of Australia, California and elsewhere are famous for carrying gold, and pegmatite veins may contain minerals with rare elements such as lithium, beryllium, tin, tungsten and uranium.

When igneous bodies and pegmatites penetrate country rock they may well activate waters held in these rocks, and these in turn could influence which minerals form on cooling. Some veins contain minerals with significant chemically bound water and comparison with the results from appropriate laboratory experiments indicates that many such minerals could form from hot water solutions at temperatures between 200 and 400 °C. The results of such hydrothermal activity are common around the margins of large intrusions and because the liquids and gases can be in contact with both igneous and sedimentary sources they may scavenge and distill a range of elements and deposit these as economic concentrations of minerals or gems. Mobile liquid or gaseous activity may leave well-formed crystals projecting into cavities in veins or pockets. Or again, if the fluids have been particularly pervasive they may penetrate a rock completely and cause a transformation to a massive gem material such as the lapis lazuli in northern Afghanistan. Such processes can equally be considered as metamorphic as they involve changing both the form and content of a rock, but before the sedimentary and metamorphic categories are described, some important sources of gems linked to volcanic rocks should be mentioned.

Significant areas of the Earth's surface are covered by volcanic rocks and by far the greater part of these consists of basalt, a dark, fine-grained rock extruded as lava from volcanic pipes (*Figures 1.3b, 1.4* and *1.6*) or through parallel fissures (cf. dyke swarms described before). Basalts lie at the mafic end of the composition range. Intermediate volcanic rocks are mostly andesite and these too can locally be extruded in considerable volume, one example being Mount St Helens in the Rocky Mountains. At the felsic end of the range are rhyolites and trachytes.

Gems associated with volcanic rocks include, for example, the rubies, sapphires and zircons found close to basaltic flows in Thailand, Australia and Nigeria, the amethysts occurring in spectacular cavities in intermediate and acid members of the volcanic province in southern Brazil and Uruguay, and the fire opals occurring in small cavities in the rhyolites of Mexico (*Figure 1.9*). At first sight, it may appear that volcanic magmas

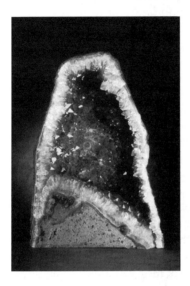

Figure 1.9 A geode containing amethyst from volcanic rocks in Rio Grande do Sul, Brazil. The amethyst crystals are more strongly coloured at their tips and are up to 3 cm long

over the whole compositional range may be important as sources of gems, but it is very probable that none of the above examples (with the possible exception of some sapphires) crystallized from the magma of its host rock. Both the fire opal and the amethyst have formed from hydrothermal fluids (see p. 27) *subsequent* to the main emplacement of the lavas, and the rubies, zircons and most of the sapphires formed *prior* to the eruption of their host basalts.

These basalts are important not as the hosts from which gems crystallized but as the means of transport by which they were brought to the surface. The zircons, sapphires, rubies and spinels were already present in pegmatites or metamorphic rocks, deep in the Earth's crust, before volcanic activity from below fractured the rocks and carried fragments to the surface. Such foreign materials which have their origins outside the volcanic fluid and which have been accidentally incorporated are known as *xenocrysts* if they are single crystals, or *xenoliths* if they are rocks (*Figure 1.10*).

Another gem which owes its presence at the Earth's surface to volcanic transport is diamond. Like the rubies and sapphires of Australia and

Figure 1.10 Xenocryst of sapphire in basalt from Nigeria. A loose xenocryst from the same locality has reflective rounded margins which indicate that the sapphire was affected by the heat of the basaltic magma

Thailand, diamonds are xenocrysts accidentally incorporated by volcanic fluids at considerable depth. The latest estimates place the sources of gem diamonds between 100 and 700 km beneath the surface. The fluids generated at these depths are mixtures of gases, liquids and solids, and on their turbulent ascent they may incorporate a wide range of rock and mineral fragments from the fractured wall rocks. Consequently the rocks at the surface are mixtures of magma and variable quantities of xenoliths and xenocrysts – the most important of which are diamond. Such rocks were first found in Kimberley, South Africa, and because they did not resemble any currently known rock, they were given the name *kimberlite*. Kimberlites have been found subsequently in other parts of Africa, Siberia (*Figure 1.11*), India,

Figure 1.11 Diamond in kimberlite from the Mir kimberlite pipe, Yakutia

China, Canada and elsewhere; they have only been found in areas of ancient crustal rocks which are described as *shield areas* or *cratons*.

Although gem diamonds did not crystallize from kimberlites, their host rocks do contain two kinds of xenolith which may contain diamonds. One is *peridotite*, a coarse-grained ultramafic rock rich in magnesium and poor in silica, and the other is *eclogite* (*Figure 1.12*), a coarse-grained metamorphic rock transformed at high pressure and temperature from a basalt or rock of similar composition.

Figure 1.12   Octahedral diamond crystal (top right) in eclogite, which also contains garnet (red) and pyroxene (black)

One other kind of volcanic rock is the source of diamonds at Argyle in Western Australia. This is lamproite and, like kimberlite, it is rich in olivine and mica; unlike kimberlite it contains negligible carbonate minerals.

Sediments and Sedimentary Rocks

Rocks of the second major category owe their formation to the actions of air, water, organisms, and biological processes on rocks already in existence.

Sediments result from the breaking down of rock masses by physical and chemical weathering, the erosion and transportation of fragments and solutions, and the deposition as particles or precipitates downstream from the source (*Figure 1.13*). As the layers of sediment accumulate, the older material beneath is compacted and buried. The burial process merges imperceptibly into '*diagenesis*' which encompasses the physical and chemical changes which take place during the transformation from sediment to sedimentary rock: these include decrease of the spaces between grains – **pore spaces** – by compaction and growth of new minerals into the pore

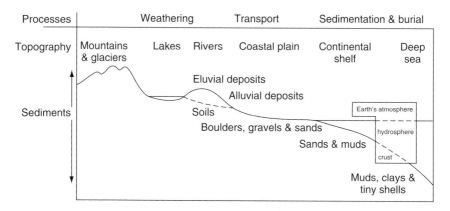

Figure 1.13 Schematic diagram indicating the broad relationships between Earth's processes, topography and sediments

spaces that remain. At first, sediments are very loosely packed and may be saturated with groundwater but as compaction and pressure increase, the water is squeezed away and the detrital minerals may break down in the presence of water to form new minerals filling the pore spaces and cementing the rock. During these cementation or *lithification* processes, sands become sandstones, gravels become conglomerates, muds and clays become mudstones and shales, and rock screes become breccias.

In some places, the erosion processes may remove only the 'lighter' minerals and leave the 'heavier' or denser minerals – which may include any gems – almost *in situ*: these are called eluvial deposits. Where the whole rock is broken down and carried away by water, wind or ice and deposited perhaps in a river system, the resulting deposits are called alluvial.

In the headwaters of a river system, fast-flowing streams will transport quite large and dense particles, but when the flow rate decreases on flatter ground these denser fragments are deposited and separated from the finer particles which are held in suspension longer – to be deposited even farther away from the source. Sorting of the debris by water will depend greatly on the density of the minerals being carried. The more important gem minerals, being relatively dense, tend to fall into any depression in the river bed where water velocity is low, and there they form placer deposits (which are also known as gem gravels or alluvials) (*Figure 1.14*). Gems may be found not only in modern river beds but also in the beds of ancient rivers, which may now be covered by more recent sediments and soils. It is in such old river courses in countries like Sri Lanka and Thailand that the water-worn pebbles of many gem materials are found.

*Figure 1.14   Alluvial sands containing diamonds near Sopa,
Minas Gerais, Brazil*

The above discussion has concentrated on clastic sedimentary rocks – those consisting of fragmentary material; now those formed by evaporation of water and precipitation (chemical sediments), by the actions of organisms (biochemical sediments) or of mixed origin will be outlined. Gypsum, halite and other salt deposits crystallize from evaporating seawater and from lakes in desert regions that have no river outlets. During evaporation, the dissolved salts become more concentrated and eventually the liquid becomes saturated and then supersaturated. The minerals then crystallize in the following sequence: calcite ($CaCO_3$), dolomite ($CaMg(CO_3)_2$), gypsum ($CaSO_4 \cdot 2H_2O$), halite (NaCl), and then magnesium and potassium chlorides and sulphates. The relatively pure, fine-grained and compact form of gypsum is the ornamental material alabaster (*Figure 1.15*).

Calcite is the major constituent of limestones and although it is the first to precipitate from evaporating seawater, rocks formed from this single

*Figure 1.15   White nodules of alabaster in grey marls
(calcic mudstones), Marmolaio quarry, near Volterra, Italy*

16

chemical purely inorganic process are quite subordinate to those of mixed origin. Most limestones consist of the shells and skeletons of marine organisms living near the surface or on the sea bed. These include a wide range of seashells and coral reefs, all of which have formed as a result of the biochemical actions of living organisms. At the end of their life cycle, the organisms decay and the shells either sink to the sea bed or are transported by currents to become bioclastic sediments. The seashells commonly contain two forms of calcium carbonate – calcite and aragonite – and since aragonite is generally responsible for the attractive aspects of shells such as the iridescence of mother of pearl, one might expect limestones to contain areas of similar interest, but during burial, aragonite in the shells progressively recrystallizes to the more stable calcite. During this process the calcite will tend to grow in the pore spaces, cementing and strengthening the rock.

Chert, which is popularly known as flint, is also of mixed chemical, biochemical and clastic origin. It consists largely of silica derived from siliceous shells of tiny, single-celled marine animals. When these animals die, the shells accumulate as layers of silica sediment. Most of the silica is in the form of quartz, but some is non-crystalline opal which may be partly or wholly crystallized to quartz during diagenesis. The resulting chert or flint is much harder than limestone and was worked by early man for tools and weapons.

Despite the processes of diagenesis, limestones and other sedimentary rocks may preserve useful or even spectacular reminders of past life. For example, when mixed with sands or clays and then compacted, many plants and animals remain recognizable as fossils and some are of importance as gem materials. In Australia, shells and fragments of wood have been replaced by precious opal; in the USA, the woody cell structures of quite large trees have been replaced and preserved in detail by silica, the range of reds, yellows and browns of the petrified forest in Arizona being caused by traces of iron and other metals. Fossil shells, corals or crinoids (sea lilies) occur in an abundance of shapes and colours, perhaps best seen in polished marble surfaces. Plant remains in some regions form peat deposits and, on compaction, these become lignites or coals; in the north of England at Whitby, jet is a particularly fine-grained, hard and lustrous variety of lignite.

Metamorphic Rocks

Heat and pressure from intrusions of igneous rocks, or form the compression and folding of the crust during mountain-building periods may so alter the country rocks that new ones are created. Such an alteration is termed metamorphism, meaning literally a change of shape. In practice, this means a change in rock texture and usually a change in mineral content. In contrast to the formation of an igneous rock which involves melting, during metamorphism, the rock remains solid, the changes taking place at an atomic scale.

There are two main kinds of metamorphism: regional and contact. Regional metamorphism occurs over wide areas and incorporates both thermal changes and deformation. These conditions are characteristic of the interiors of fold mountains where heat flow may be high over a wide area and the stresses and strains of mountain building create different degrees of compression, shearing and faulting in different parts of the fold belt. Contact metamorphism takes place locally around hot, if not molten, igneous bodies intruded into relatively cool country rock. The zone of alteration in the country rock is called a contact aureole with the maximum alteration closest to the intrusion.

In addition to the two main regimes of metamorphism, rocks may be mechanically crushed or sheared in faults and joints (deformational metamorphism); they may be subject to percolation and alteration by fluids from igneous intrusions (hydrothermal metamorphism); or they may suffer slowly increasing temperatures and pressures in a sinking basin of sediments (burial metamorphism). The temperatures and pressures in burial metamorphism are not as high as in regional metamorphism, and the rocks are described as low grade; high-grade rocks have been subject to high pressure, high temperature or both.

All these kinds of metamorphism cause changes in rock textures, and the sizes, shapes and mutual relations of the mineral grains can be radically altered. In regionally metamorphosed rocks, a very common change is development of a foliation where the minerals grow in parallel bands. Heat and pressure commonly transform sedimentary minerals such as clays into micas and because these are flat platy minerals, the surrounding pressure commonly forces them to grow in parallel. The platy micas are very reflective and when a hand specimen is turned, it sparkles, and the direction of foliation will be indicated by the greatest number of reflections.

At low grade, the foliation in **slates** and **phyllites** may be difficult to see, but at higher grade, the foliation becomes more conspicuous as the micas and other minerals are coarser and easier to see. If such rocks contain more than 50% mica, they are called **schists**. With less than 50% mica, these medium and high-grade rocks tend to be more massive and less fissile and are called **gneiss**. Even though they are less fissile, gneisses may contain black amphiboles and creamy white feldspars in contrasting bands and show strong foliation.

The platy crystal habit of mica is the most important factor in forming foliated rocks, but where the elements for creating mica are absent, regional metamorphism produces rather more massive, non-foliated rocks. These include marbles, quartzites, amphibolites and greenstones.

18

In contrast to the new minerals and obvious banding in schists and gneisses, the changes in non-foliated rocks can be more subtle. Development of the Carrara marble in Tuscany, for example, involved recrystallization of a limestone consisting of calcite to a marble consisting of calcite (*Figure 1.16*); the particular quality of the marble is due to the calcite crystals being of even, uniform size, closely intergrown and relatively free of impurities.

Figure 1.16 Thin section of Carrara marble showing interlocking grains of calcite in cross-polarized light. Some grains show bright interference colours, others are at extinction indicating random orientation of the grains. This makes the rock stronger. Field of view: 4 mm across

Sandstones, consisting mostly of quartz grains, can behave similarly under increases of temperature and pressure. Pure varieties will transform to quartzites, which, like the Carrara marble, consist of a mosaic of intergrown grains and are white. If a sandstone is impure, the nature of the impurity will determine whether an attractively coloured rock will be produced on metamorphism. The best-known example in the gem world is green aventurine quartz which consists of quartzite with tiny flecks of emerald-green mica (*Figure 1.17*).

Of widespread occurrence in many metamorphic rock sequences is **amphibolite**, a medium- to high-grade metamorphic rock consisting largely of amphibole group minerals and feldspars. In geology, this understanding of what the term *amphibolite* means is quite specific; confusingly for gemmologists, it cannot appropriately be applied to the one gem material consisting entirely of amphibole that is important in the gem world – nephrite. Nephrite jade consists of a mass of intergrown fibres of tremolite amphibole (*Figure 1.18*) which, in many localities, was formed at low temperature and pressure, and it would be misleading to use the term *amphibolite* for a rock so different in origin

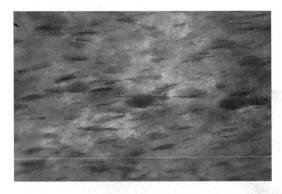

Figure 1.17   Aventurine quartz: Green to dark green flakes of mica in a quartz matrix. The flakes are up to 2 mm long and show strong alignment

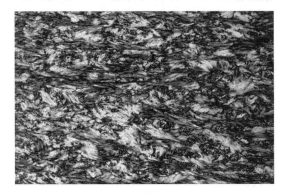

Figure 1.18   Thin section of nephrite jade from Taiwan. The variation in interference colours indicates that the fibres of amphibole vary in orientation and that they have intergrown to form a very tough rock. Field of view: 4 mm across

from those formed at medium or high grades. Nephrite is discussed further below.

Some rocks formed at low grade, generally in the presence of fluids at 150–300 °C, contain an abundance of the green mineral chlorite, and this feature is the reason for their name – greenstones. In the ancient shield areas of the Earth's crust, they form 'greenstone belts' and are important as the source of chromium for emerald in these regions.

In the formation of some schists, the metamorphic conditions may favour faster growth of one mineral more than the others, and, for example, some almandine garnet crystals may grow to several centimetres in diameter while the matrix micas and quartz grains are only a few millimetres across (*Figure 1.19*). These large crystals are porphyroblasts,

Figure 1.19   Garnet

resembling texturally but quite different in origin from phenocrysts and xenocrysts in igneous rocks.

Igneous rocks are the cause of contact metamorphism but the effects an igneous intrusion or lava flow may have on the local country rocks depend on a number of factors. These include the temperature of the country rock, the size of the intrusion and its temperature, and whether or not there are accompanying gases and water-rich solutions. For example, a large granite at 700 °C intruded deep under a mountain chain will have a more extensive effect on the warm rocks under those mountains than, perhaps, a small body of basalt at 1000 °C would have on cold rocks at the Earth's surface. The effect on the country rock of the basalt close to its margin would be quite striking, but at a distance, the effects would be small. Between these limits and if the rocks are not pure limestones or pure quartz sandstones, it is generally possible to define zones in which different metamorphic minerals have developed.

If the country rocks were sandstone or shale, metamorphism would cause new silicate minerals to be formed, but if, for example, a magma intruded limestone with some quartz and clay minerals, the zone closest to the intrusion may contain calcic minerals such as grossular garnet and diopside. The cooler zone further away from the intrusion may contain calcite, serpentine and chlorite – indicating their formation at lower temperatures in the presence of water. If sufficient alumina were present (in, say, the clay minerals), and temperatures were high enough, corundum and spinel may form, and this may be how ruby and spinel formed in

Figure 1.20   Ruby crystals and ruby in coarsely crystalline marble from Jegdalek, Afghanistan. Marble specimen 6 cm across

the marbles of Mogok in Myanmar (formerly known as Burma) and of Jegdalek in Afghanistan (*Figure 1.20*). Both of these marbles belong to sequences of gneiss and schist, and if the regional metamorphism were intense enough, this could also account for the presence of these gem minerals. The third possibility, a suggestion made more recently for the Mogok deposits, is that a combination of regional and contact metamorphism could be responsible. Whatever the detailed origin, metamorphic rocks like these could be the real source of the rubies found as xenocrysts in the volcanic rocks of Australia and Thailand. Another gem formed as a result of the effects of granitic or pegmatitic fluids on gneiss, schist and marble is the lapis lazuli at Badakhshan in Afghanistan (*Figure 1.21*).

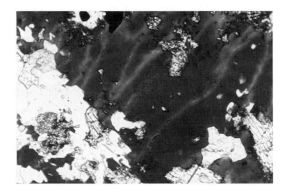

Figure 1.21   Thin section of lapis lazuli which consists of lazurite (blue) accompanied on the left by mica (clear) and pyroxene (dusty brown), at centre top by opaque pyrite, and at lower right by mica and quartz. The assemblage is similar to that at Badakhshan, Afghanistan

There the lapis lazuli has been mined intermittently for up to 6500 years from large lenses of rock rich in lazurite and pyrite which occur in a sequence of marbles up to 400 m thick.

Like ruby, diamond is another gemstone whose true geological source has been obscured somewhat by volcanic activity, in this instance, by that of kimberlites. But although gem diamonds did not crystallize directly from the kimberlites we see, they are present in eclogite and peridotite xenoliths also carried to the surface by the volcanism. Eclogite is a high-grade metamorphic rock consisting mostly of garnet and pyroxene (*Figure 1.21*) which has formed at depths of more than 35 km (equivalent to more than 10 000 atmospheres pressure) and at temperatures between 400 and 800 °C. The other major source of diamond in these temperature and pressure conditions is peridotite, which is an ultramafic igneous rock consisting mostly of olivine and pyroxene, with only minor quantities of diamond, pyrope garnet or other minerals.

That diamonds crystallized and grew in eclogite and periodotite is not in doubt because inclusions in the diamonds such as garnet, pyroxene or olivine have the same compositions as those in the eclogite or peridotite. Further, these minerals are sufficiently distinctive to provide evidence of a diamond's origin. Garnets from peridotites, for example, are rich in pyrope, while those from eclogites have a significant content of grossular, so if the garnet inclusions in a diamond are analysed, a strong indication of its origin can be gained.

Two categories of metamorphism remain to be noted: deformational and burial metamorphism. Deformational metamorphism commonly breaks up rocks along narrow zones, and the resulting mix of angular fragments can be recemented to form coherent rocks. These are termed *breccias*, and decorative examples include some of the impure marbles from Italy, some serpentines from the Lizard in south-west England (*Figure 1.22*), and fragments of agate and jasper cemented to form 'brecciated agate'.

In burial metamorphism, sediments undergo cementation and become increasingly rock-like (lithified) as they sink and as more sediments accumulate above them. For each 4.4 m of sediment above, the pressure rises 1 atmosphere, and for each kilometre of depth, the temperature rises about 30 °C. A sandstone in a typical sedimentary basin at 10–12 km depth could be under about three thousand atmospheres of pressure and at 300–350 °C, and in these conditions, burial metamorphism can be considered to have passed into low grade regional metamorphism.

Minerals from Water-Rich Fluids

Having outlined the three major categories of rocks, there remain a number of rocks, important as sources of gems, which are difficult to

Figure 1.22   Three varieties of decorative serpentine from the Lizard, south-west England. Grey serpentine supports a round plinth to red serpentine on which rests the carving of a mythical beast in serpentine breccia

assign simply as sedimentary, igneous or metamorphic. These are minerals and rocks formed by the chemical actions of one or more of the following:

- magmas rich in water and gases
- hydrothermal solutions
- groundwaters
- seawater.

The process of changing the composition of a rock by introduction and removal of chemical elements is called **metasomatism**. It is replacement at an atomic scale and the process is brought about through chemical or electrical gradients, or by the more easily imaginable movement and percolation of fluids. At high grades of metamorphism, metasomatism causes changes that are usually compositional and textural, i.e. new minerals are created in new textures, but at low grades, mineral textures may survive during chemical change and even the cellular textures of plant fossils can be preserved during hydrothermal action.

Hydrothermal activity was referred to briefly under both igneous rocks (as a stage following pegmatite formation where hydrothermal minerals were formed in veins) and metamorphic rocks, where reference was made to alteration by fluids from igneous intrusions. Such activity can give rise at one extreme to a rock of relatively consistent grain size and composition, and at the other extreme to a network of veins with a range of minerals of different sizes. Examples of the former include the massive deposits of nephrite jade, and of the latter, the vein deposits of amethyst, emerald and tanzanite.

Gem minerals such as amethyst, aquamarine, morganite, topaz and tourmaline may be found both in pegmatites and in hydrothermal veins, and determination of their precise origins will depend on careful local studies. Similar care has to be taken for minerals that may have formed either solely from hydrothermal fluids or from mixtures of such fluids with waters from sedimentary rocks, with groundwaters or with seawater. Mixtures of hydrothermal fluids and seawater, for example, were involved in the formation of the jadeites of Myanmar and Guatemala, and mixtures of sedimentary pore-fluids and groundwaters are thought to be a major agent in the formation of opal in Australia. Groundwater alone is considered to be the major agent in turquoise formation – particularly in the Yun Yang deposit in Hubei Province, China.

For many years on the basis of experimental work, jadeite jade was thought to have formed at depths of 35 km or more beneath the Earth's surface. More recently, it has been found that where a region is saturated with water, jadeite can form in veins and metasomatically at depths of 16–20 km. The host rocks are commonly serpentinite and these are transformed into jadeite by hydrothermal fluids derived from local sediments or igneous rocks (*Figure 1.23*). The evidence that the jadeite formed also in the presence of seawater and at temperatures of 200–400 °C comes from studies on the fluid inclusions.

Figure 1.23   Jadeite boulder section showing interlocking platy and fibrous crystals, the different orientations of which, catch the light in different ways. The brown rim is caused by weathering at the boulder edges

Nephrite jade is also formed by hydrothermal activity at temperatures similar to those found for jadeite. But in contrast, the pressures are much lower and a different range of fluid–rock combinations has generated tremolite-actinolite amphiboles rather than jadeite. Such combinations include: the action of granitic or hydrothermal fluids on dolomite, the action of serpentine fluids on silica-rich rocks, or the action of hydrous fluids on serpentinite.

In the Merelani region of Tanzania, veins containing tanzanite and tsa-vorite garnet have resulted from pegmatitic and hydrothermal fluids reacting with medium- to high-grade metamorphic rocks along the crests of folds. The good crystal form of many of the tanzanites is a result of their growth into cavities, fed by mineralizing solutions and free of inter-ference from other minerals (*Figure 1.24*). The zone of mineralization is only a few metres wide but extends for about 9 km and includes other spectacular gems such as chrome tourmaline and chrome diopside.

Amethyst occurs in many localities worldwide but some of the most important are in Brazil, where the gem-quality deposits are associated

Figure 1.24   Tanzanite crystal and cut stone of 21.3 ct

either with pegmatites or with volcanic lavas. In the states of Minas Gerais and Bahia, pegmatites and hydrothermal veins with amethyst were emplaced in metamorphic rocks, while in Rio Grande do Sul and across the border in Uruguay, amethyst crystallized from hydrothermal fluids in fissures and cavities in volcanic lavas. In some places, the cavi-ties in the lavas (geodes) reach 3 m in length and the walls may be lined with crystals up to 10 cm long (see *Figure 1.8b*).

North and east from where amethysts are found in Brazil are probably the best-known gems from South America: the emeralds from Colombia. Emeralds occur at a number of sites north of the capital, Bogotá, where they have been and are recovered from calcite and pyrite or albite and pyrite veins which cut across a sequence of black shales (*Figure 1.25*). Traditionally, emeralds were thought to need the proximity of granite (as a source of beryllium) for their formation, but none could be found sufficiently close in Colombia for this idea to be viable. Now it is thought that fluids from the regional sedimentary succession (comprising detrital and chemical deposits – sediments and evaporites) have reacted with organic matter in the black shales to release beryllium, chromium and vanadium. These rare

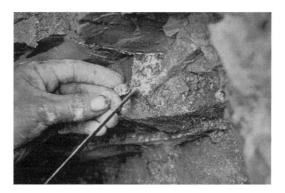

*Figure 1.25   Emerald being prized from a vein in black shale
at Chivor, Colombia*

elements have then joined other elements in the fluids, including silicon
and aluminium derived from the shales, to form emerald. Studies of the
fluid inclusions in the emeralds indicate that they formed in the veins at
temperatures between 200 and 400 °C and at depths of about 4 km.

Fluids of mixed hydrothermal and groundwater origin could also
be responsible for the precious opal found in volcanic rocks. In the
Querétaro area of Mexico, fire opal has formed in cavities in rhyolite lava
flows from fluids saturated in silica. Again, studies of the fluid inclusions in
the opal have indicated its temperature of formation, and this is estimated to
be about 160 °C. In Australia, volcanic rocks have not been found near any
of the major opal deposits in South Australia, New South Wales or
Queensland. In these localities, precious opal occurs in sedimentary rocks
both parallel to the bedding planes of the sandstones and claystones, and in
faults which cut across these beds. The origin of this opal is still subject to
debate, and although hydrothermal fluids may have played a part, the present consensus is that opal precipitated from hot waters saturated in silica.
Such hot waters could consist of pore waters from the surrounding sediments and groundwaters enriched in silica from weathering and leaching
cycles mixed with alkaline waters from the basement which have some
exotic trace elements. The disposition of the various opal deposits suggests
that these solutions were then squeezed along faults and porous sedimentary horizons by gentle earth movements, depositing precious opal on
cooling, on change of alkalinity (pH) and in the presence of suitable nuclei.

The movements of hydrothermal fluids, pore fluids from sediments
and groundwaters derived from weathering (meteoric waters) combine to
provide a powerful mechanism to concentrate elements and deposit minerals. In the mineral world, gold in quartz veins, copper sulphides in igneous
rocks and lead sulphides in sedimentary rocks are examples of economically important hydrothermal deposits. Such orebodies can themselves

subsequently be re-worked by groundwater, and around copper deposits, turquoise may be formed. In the famous deposits at Nishapur, Iran, and in the extensive deposits in the south-western USA, copper sulphides in trachytes (alkaline volcanic rocks) have been dissolved and the copper precipitated in turquoise; in Egypt, porphyry is the source of the copper; and in Yunyang, China, groundwaters have dissolved the copper sulphides and phosphates from sedimentary slates to form extensive turquoise deposits and a range of other phosphate minerals in veins and fissures.

When groundwaters rich in minerals flow into cavities as large as caves, they are likely to drip from the roof, evaporate and precipitate crystals at both top and bottom, forming stalactites and stalagmites respectively. With sufficient time, the two structures may join and form a complete pillar between roof and floor. Further deposition of minerals may well occur in irregular sheets on pillar and floor and is commonly banded in structure. It is from such formations that the banded calcites, the 'onyx marbles' and the banded pink and white manganese carbonate rhodochrosite are obtained.

Rock Systems

In classifying a flat-lying sequence in sedimentary rocks, the geologist starts with two principal assumptions: first, that the sediments were deposited in horizontal layers, and second, that each layer (in an undisturbed sequence) is younger that the one beneath it. Fossils and distinctive rock types are used to compare and correlate sequences from different areas and allow the relative ages of rocks to be established – this is part of the science of stratigraphy.

Using these methods over the past 200 years, geologists have assembled stratigraphic and faunal or fossil successions and constructed a time scale for rocks over most of the Earth. However, the time scale established by stratigraphic methods is a relative one – it does not enable one to say in terms of years when a particular rock was formed. The opportunity to determine the ages in years only came after the discovery of radioactivity in uranium in 1896.

Uranium does not consist of a single kind of atom, but of atoms with a range of masses (isotopes), some of which are radioactive. Two of the most important isotopes are uranium-235 and uranium-238, and the latter is useful in determining ages of rocks and minerals. Radioactivity describes an unstable state of an element in which the 'parent' isotopes spontaneously and randomly 'decay' by loss of atomic particles to 'daughter' isotopes, and although the process on an atomic scale is random, the rate of decay overall for a mass of the element is constant. This decay rate is commonly described in terms of the 'half-life' of the isotope – the time taken for a quantity of the isotope to reduce to half of its mass. For uranium-238, this period is about 4500 million years. Isotopes of other elements such as potassium-40, rubidium-87 and carbon-14 have different decay rates.

By measuring the relative amounts of 'parent' and 'daughter' isotopes in a mineral – for example, uranium in zircon, rubidium in beryl, or potassium in feldspar – the time taken for the 'daughter' element to form can be calculated. This gives a direct estimate in years of the age of the mineral. Such geochronological studies started seriously in the 1950s and have been invaluable in clarifying the histories of a wide range of rocks and particularly those igneous and metamorphic rocks which lack fossils and an apparent stratigraphic position.

The second great benefit of these studies is that they have enabled ages in years to be assigned to the periods established from stratigraphy. And again, the work of the past half century has led to a continual refinement of the classification. This is summarized in *Table 1.1*.

Table 1.1
The relative ages of Eras, Systems and Series of Rocks

Eon	Era	Period or system		Epoch or series	Age in Millions of Years (Ma)
		Quaternary		Holocene	0–0.01
				Pleistocene	0.01–1.8
	Cainozoic		Neogene	Pliocene	1.8–5
		Tertiary		Miocene	5–24
				Oligocene	24–35
			Paleogene	Eocene	35–57
				Paleocene	57–65
Phanerozoic		Cretaceous			65–144
	Mesozoic	Jurassic			144–206
		Triassic			206–251
		Permian			251–290
		Carboniferous			300–354
	Paleozoic	Devonian			354–409
		Silurian			409–439
		Ordovician			439–510
		Cambrian			510–543
	Proterozoic	Late			543–1000
		Middle			1000–1700
Precambrian		Early			1700–2500
	Archean				>2500

The aspect of the sequence shown in *Table 1.1* that is most difficult to absorb is the sheer scale of geological time. The period of recorded history from about 4000 years ago is minute compared with that of the 65 million years since the dinosaurs died out, or the 543 million years since complex organisms first appeared. But what this span of time does make conceivable are some of the extremely slow processes described for the mantle and crust which are responsible for the diversity of rocks and minerals and which drive their evolution.

The Ages of Gems

How do gems fit into the geological framework? Are all emeralds of the same age? Are diamonds the same age as other gems, and what is the oldest diamond?

To answer such questions, gems need to be considered as part of the rocks and processes described above. Like them, gems have formed throughout a significant proportion of the Earth's history, and some idea of the diversity shown by different gems can be gained from the following examples of emerald, diamond and amber.

Emerald

Among the oldest emeralds that have been cut as gems are those from Sandawana in Zimbabwe, where the pegmatite veins containing the emeralds (*Figure 1.26*) have been dated to about 2600 million years (Ma), which is the late Archaean (see *Table 1.1*). Emeralds from a very similar geological setting also occur at the Menzies mine in Western Australia and these rocks also have been dated to about 2600 Ma.

Figure 1.26   Emerald in pegmatite from Sandawana, Zimbabwe

About 2565 million years later, and in completely different geological conditions, the emeralds of present-day Colombia were being formed in hydrothermal veins in the mountains now known as the Andes.

Between the Tertiary (35 Ma) deposits in Colombia and the Archaean (2600 Ma) deposits in Zimbabwe, emeralds have formed at many different times and in many different places. In the Kaduna-Plateau states of Nigeria, for example, emeralds were formed in two distinct episodes: the first was associated with the 'Older granites' between 450 and 600 Ma ago, and the second associated with the 'Younger granites' between 140 and 190 Ma ago.

From these examples, two general observations can be made:

- Emerald can form in similar geological conditions at the same time in different parts of the Earth's crust.
- Emerald can form in the same region of the crust at different times.

It is important to note that the ages assigned to emeralds, for example at Sandawana, are based on analyses of associated minerals in the host rocks, and the reliability of any dates depends on correct interpretation of the rock textures. This way of estimating ages can be extended to many other gem minerals occurring in pegmatites but not to diamond in kimberlite.

Diamond

Unlike emerald in pegmatite, one cannot say that a diamond grew in its host rock, kimberlite. Most, if not all, gem-quality diamonds have an origin deep in the Earth's crust or mantle and have been transported to the surface as crystals or rock fragments by fluid, volcanic kimberlites. Therefore diamonds are older that their host kimberlites and dating them has to rely on microanalysis of their inclusions (*Figure 1.27*).

Figure 1.27 Cut and polished diamond with an inclusion of red pyrope garnet

Such work indicates that the oldest diamonds found so far are about 3300 Ma old, and stones of this age occur in the Kimberley and Finsch kimberlite pipes in South Africa. Slightly younger diamonds, 2900 Ma old, occur in the Orapa kimberlite in Botswana, accompanied by even younger stones 'only' 990 Ma old. The age of the kimberlite host rock at Orapa has also been determined and is significantly younger than its diamonds at 93 Ma. So in many kimberlites, minerals and rocks of vastly different ages may exist side by side.

The very old ages for diamonds obtained from isotope analyses are given some support by stratigraphic evidence for diamonds at least 2500 Ma old in South Africa. The Witwatersrand quartzites and conglomerates are ancient sediments which contain alluvial diamonds. The sediments were folded and metamorphosed 2500 Ma ago, so the diamonds must have come from an even older source (*Figure 1.28*).

Figure 1.28 Diamond-bearing conglomerate at Sopa, near Diamantina, Minas Gerais, Brazil. The conglomerate is Precambrian, probably 1500 million years old, so the diamonds must have come from a source even older

Amber

The third example chosen to illustrate the many ages of gems is the organic gem, amber. Amber is a resin, derived from trees, which has been preserved and fossilized in sedimentary rocks. In terms of Earth's history, the first resin-producing trees appeared relatively recently, and so the range of amber ages is correspondingly small. Even so, the oldest amber dates from the Coal Measures in the Upper Carboniferous period about 320 Ma ago. However, this amber is not of gem quality, and most of the amber fashioned for jewellery is Tertiary in age. Amber recovered from the coastal regions of the Baltic Sea is 35–40 Ma old, and amber from the other main source for the jewellery trade, the Dominican Republic, comes from a wider range of horizons 15–40 Ma old. The oldest amber-containing insects occurs in sediments of Cretaceous age.

Discussion

Amber ages are based on fossils found in the amber-bearing sedimentary rocks, i.e. they are dates derived by association. Similar reasoning applies to many dates derived for emeralds. Truer ages are potentially obtainable by analysing the gems themselves or their inclusions, and this is the approach used to date diamonds. However, the limitations of such methods have so far restricted their application, and more resources would be needed to extend the methods used for diamonds to other gem species – but the results could be surprising.

Meanwhile, until more information is available, the oldest gems dated so far are diamonds. But it is clear from the above discussion that without the fortuitous occurrence of the right inclusions and knowledge of their composition one cannot say what the age of a particular stone will be. So the diamonds in a shop window are likely to be of a range of ages.

It is equally difficult to pick out the youngest gemstone in a shop window, and although there may be 'young' emeralds, amethysts or ambers on show, the youngest gems are likely to be the cultured pearls grown in the last few years, or synthetic stones, perhaps only a few weeks old.

Gem Regions

Gemstones are rarities in the geological world and are not evenly distributed in crustal rocks. But despite their overall scarcity certain areas do seem to be favoured with local concentrations of a number of gem species while others are barren. Among the better endowed are the very old shield areas of the major continents, and of these, Brazil, East Africa, South India and Sri Lanka are exceptional. The shield areas are also the locations for kimberlites and gem diamonds, the richest regions being southern Africa, Siberia and, more recently, north-western Canada.

Other favoured localities containing gem minerals occur in major fold belts, with emeralds coming from the Andes in Colombia, the Urals in Russia and the Himalayas in Pakistan; rubies also from the Himalayas in

Pakistan and Nepal and from their extension into Myanmar (formerly known as Burma); jades come from the Rockies in western Canada, the mountainous South Island of New Zealand and, again, from Myanmar.

Pegmatites can provide a wide range of gem minerals including beryl, garnet, kunzite, topaz and tourmaline, and examples can be found both in shield areas such as Minas Gerais (Brazil) and in fold belts such as in California.

Large areas of New South Wales and Queensland, central Thailand and central Nigeria are covered with basaltic rocks which are associated with sapphires, zircons and spinels. Also in parts of Australia, but far removed from mountain building or volcanic action, precious opal was formed in the tranquil conditions of near-surface groundwaters.

Many other gem regions or districts could be cited, but reference is more appropriately made to them where individual gem species are described.

Gem Recovery Methods

Gem gravels are one of the most productive sources of fine gems. In many parts of the world, gem minerals are weathered out of surface rock and washed downhill and downstream to be deposited some distance from their source. Flawed minerals tend to be destroyed during this transport and the ultimate deposits consequently contain high-quality gem material. Many deposits, however, are patchy and in difficult terrain, making large-scale mining uneconomic. Consequently, much exploitation has been small-scale and simple.

Some gem gravels occur in the beds of ancient rivers, long since vanished and covered by more recent sediments. The method of working such gem gravel is to dig a small pit down to the gems and, if conditions are favourable, drive short horizontal galleries from the base of the pit. Usually the gem-bearing gravels need to be washed in sluices, a process which floats away the less dense minerals and leaves the concentrate which will contain any gems.

In the running water of streams and rivers, the gem fragments are recovered by 'panning'. A shallow dish containing the gem gravel and water is swirled and rotated in such a manner as to wash the lighter mineral over the edge. The gems are left in the bottom of the pan and then hand-picked.

By these straightforward methods applied in countries such as Brazil (*Figure 1.29*), Madagascar and Sri Lanka, almandine garnet, aquamarine, chrysoberyl, ruby, sapphire, spinel, topaz, zircon and many other rarer gems have been recovered.

*Figure 1.29   Washing gravel and hand-picking gems from the
alluvial deposits at Marambaia, Minas Gerais, Brazil. The valley
is well known as a source of fine aquamarine and chrysoberyl*

The first mechanized mining of gem gravels containing ruby and
sapphire began in the Australian sapphire fields in New England
in the 1970s. Large earth-moving equipment removes up to 20 m of
overburden before extracting the gravels which contain the sapphires
(*Figure 1.30a* and *b*). The gravels are broken up in a rotating drum (trom-
mel) and sorted into size fractions before passing onto pulsating jigs.
The jigs concentrate the heavy minerals from which the gems are picked
by hand.

Emeralds present a different problem. They are neither as hard nor as
tough as rubies and sapphires and, to minimize cracking and breaking,
have traditionally been mined straight from the veins in country rock by
pick, hammer and chisel. In the Colombian mountains, the black shales
which contain the emerald veins are relatively soft and amenable to
working in this way. In recent years 'individual' approach has evolved
and grown in scale to 'terrace mining' where the topsoil and shale is cut
away in steps, enabling easier and quicker access to the productive veins.
In other countries where the emerald occurs in hard, unweathered pegmatite, more difficulty may be experienced in following the veins, and
controlled blasting is used to loosen large blocks which are then broken
down by individuals working with drills or hammers. Where prospects
are good, it may be worth investing in shafts and tunnels underground,
and good gem material has rewarded this kind of investment in both
Sandawana (Zimbabwe) and Itabira (Brazil). The mining of emerald
focuses on individual veins and this approach is also applied to extract
many other gems: these include lapis lazuli, peridot, opal, tanzanite and
other rarer gems.

In each of the methods outlined above, the object is to recover
stones of gem quality; those which lack good colour and durability

(a)

(b)

Figure 1.30 (a) Sapphires, rubies and zircons are recovered from sediments at the base of the over-burden which is being removed by the digger; New England, Australia, (b) Sediment containing abundant sapphires

are rejected. But with diamond, the approach is different because a use can be found for all diamonds. Whether they are large or small, of gem or industrial quality, recovery is designed to be as complete as possible.

Diamonds are extracted either from pipe-like bodies of kimberlite or lamproite, from alluvial deposits in rivers or from shoreline beach deposits. In recovering diamonds from kimberlite, the first stage is to mine the upper part of the pipe as an open pit to a maximum depth of about 300 m (*Figure 1.31*). Below this, it is more economic to continue by

*Figure 1.31   Mining Kimberlite at the Majhgawan deposit,
Panna District, Madhya Pradesh, India*

one of a variety of underground methods. The mined ore is then crushed and processed to concentrate the heavy minerals. Historically these were then passed over grease tables – which retained the diamonds but let other minerals flow through – but more modern methods rely on the diamonds fluorescing under X-rays.

Alluvial diamonds have been found in Brazil since about 1725 and a few large dredgers continue to process thousands of tons of gravel in the bigger rivers to extract diamonds, gold and platinum. In the beach deposits on the western shores of Namibia and South Africa, diamonds are recovered from huge open pits behind banks of sand constructed to prevent flooding from the sea. Offshore, mining is carried out by ships with large-scale vacuum equipment. Sea-bed sand is sucked up to the ship where the diamonds are extracted and the sand returned to the sea bed.

Note on the Word 'Origin'

Over hundreds of years, a reputation for quality and excellence has been acquired by gems from certain countries and, for example, rubies from Burma, sapphires from Kashmir and emeralds from Colombia are considered by the 'market' to be the most desirable of their kind.

Despite the geological evidence that these countries can also produce stones of inferior quality and that stones of high quality and rather similar characteristics may come from East Africa (ruby), Madagascar (sapphire) and Nigeria or Pakistan (emerald), the market desire for top quality stones from the traditional countries remains undiminished. Indeed, demand has grown for certificates to state their country of origin and the attachment of one of these to a particular stone can significantly enhance its value.

Most origin certificates are issued by reputable independent gem laboratories, but some stones do not contain sufficiently distinctive characteristics to determine where they come from, and the question of their country of origin is debatable. Such a stone may pass through less than scrupulous hands and emerge with a probable 'origin'. This has had the effect of discouraging use of the term 'origin' in any objective discussion about a gem in favour of more specific terms such as geological source or geographical location. For these reasons 'geological sources' was preferred to the term 'origins' in the title to this chapter.

Further Reading

Aracic, S., 1999. *Rediscover opals in Australia*. Stephen and Mary Aracic. Lightning Ridge, New South Wales. ISBN 0-9595830-2-5

Bowersox, G.W. and Chamberlin, B.E., 1995. *Gemstones of Afghanistan*. Geoscience Press Inc. Tucson, Arizona. ISBN 0-945005-19-9

Deer, W.A., Howie, R.A. and Zussman, J., 1992. *An introduction to the rock-forming minerals*. 2nd ed. Longman Group Ltd. Harlow, Essex. ISBN 0-582-30094-0

Deveson, B., 2004. The origin of precious opal: a new model. *Australian Gemmologist*, **22**, 50–8

Fritsch, E., Rondeau, B., Ostrooumov, M., Lasnier, B., Marie, A.-M., Barrault, A., Wery, J., Counoué, J. and Lefrant, S., 1999. Découvertes récentes sur l'opale. *Revue de Gemmologie*, **138/139**, 34–40

Fuhrbach, J.R., 1992. Kilbourne Hole peridot. *Gems and Gemology*, **28**(1), 16–27

Gaines, D., Skinner, H.C.W., Foord, E.E., Mason, B. and Rosenzweig, A., 1997. *Dana's new mineralogy*. Eighth edition. John Wiley & Sons Inc. New York. ISBN 0-471-19310-0

Giard, D., Giuliani, G., Cheilletz, A., Fritsch, E. and Gonthier, E. (Ed. and co-eds), 1998. *L'émeraude: connaissances actuelles et prospectives*. Association Française de Gemmologie, Paris

Gübelin, E. and Erni, F.-X., 2000. *Gemstones symbols of beauty and power*. Geoscience Press Inc. Tucson, Arizona. ISBN 0-945005-36-9

Gübelin, E.J. and Koivula, J.I., 1986. *Photoatlas of inclusions in gemstones*. ABC Edition. Zurich. ISBN 3-85504-095-8

Harlow, G.E. (Ed.), 1997. *The nature of diamonds*. Cambridge University Press, Cambridge. ISBN 0-521-62935-7

Harlow, G.E. and Sorenson, S.S., 2001. Jade: occurrence and metasomatic origin. *Australian Gemmologist*, **21**, 7–10

Hoover, D.B., 1992. *Topaz*. Butterworth-Heinemann, Oxford, ISBN

Hughes, R.W., 1997. *Ruby and sapphire*. RWH Publishing, Boulder, Colorado. ISBN 0-9645097-6-8

Keller, P.C., 1990. *Gemstones and their origins*. Van Nostrand Reinhold, New York. ISBN 0-442-31945-2

Keller, P.C., 1992. *Gemstones of East Africa*. Geoscience Press Inc. Tucson, Arizona. ISBN 0-945005-08-3

Legrand, J., 1980. *Diamonds myth, magic and reality*. Crown Publishers, Inc., New York. ISBN 0-517-53981-0

Pearson, D.G. and Shirey, S.B., 1999. *Isotopic dating of diamonds*: Economic Geology special publication: Society of Economic Geology Reviews in Economic Geology vol. 9; Applications of radiogenic isotopes to ore deposit research. eds J. Ruiz and D.D. Lambert, 143–71

Poinar, Jr, G.O., 1992. *Life in amber*. Stanford University Press. Stanford, California. ISBN 0-8047-2001-0

Press, F. and Siever, R., 2001. *Understanding earth*. 3rd ed. W.H. Freeman and Company, New York ISBN 0-7167-3504-0

Qi Lijian, Yan Weixuan, and Yang Mingxin, 1998. Turquoise from Hubei Province, China. *Journal of Gemmology*, **26**(1), 1–12

Sevdermish, M. and Mashiah, A., 1996. *The dealer's book of gems and diamonds*. Volumes I and II. Mada Avanim Yekarot Ltd, Israel. ISBN 965-90072-0-5

Sinkankas, J., 1981. *Emerald and other beryls*. Chilton Book Company. Radnor, Pennsylvania. ISBN 0-8019-7114-4

Gemmological Journals

Australian Gemmologist. Australia
Canadian Gemmologist. Canada
Gemmologie. Germany
Gemmology. Japan
Gems and Gemology. USA
Journal of Gemmology. Great Britain
Journal of Gems and Gemmology. China
Revue de gemmologie. France

2

Crystalline Gem Materials

Ian Mercer

The subject of crystals is a source of both pleasure and consternation within the study of gem materials. Crystal perfection and the variety and beauty of crystal form are a delight to gem and mineral collectors, designers, photographers and many others. Yet the fear of scientific complexity in the study of crystal structure and crystallography has turned many people away from a deeper and more fascinating study of gems. This need not be so.

Most solid materials beneath, around and within us are crystalline or at least partially so; nearly all gem and ornamental materials are crystalline. The realm of the crystal is fundamental to gemmology. However, it would be a mistake to think that knowledge of classical crystallography is fundamental to gemmology. It is not. A deeper appreciation of gem materials starts simply with having a reasonable idea of what is meant by the words 'crystalline' and 'crystallinity'.

People who have an interest in crystals, who might indeed need to appreciate crystallinity in order to continue their professions and pursuits, include the collector locating and describing significant gem crystals for an exhibition; a retail jeweller checking a piece of gem-set jewellery brought in for repair; the lapidary faced with an unfamiliar gem material; a jewellery designer working with a group of cut and uncut gems; an artist producing a wall plaque made of various ornamental materials; an earth scientist analysing soil samples in prospecting for gems; a technologist developing new crystal X-ray analysis methods; and all those gemmology students who seek to develop their ability to check gems and ornamental materials in their ambition to advance their future business or hobby. During the study and observation of crystallinity, the significance of *non*-crystallinity also is important in its own right: the lack of crystallinity and the properties of non-crystalline materials are extremely important in gem identification and detection.

Internal Structure

For any material to be considered as crystalline, it has to be made up of atoms arranged with some degree of orderliness. It need not display any external crystal form, let alone any degree of 'perfection'. It may indeed be a fashioned gem, with all external crystal evidence removed: cut, ground or carved and polished away. It is still crystalline inside: nothing has changed there.

A crystalline gem material can be thought of as a three-dimensional arrangement made up of units of natural, electronically linked, atomic 'design', normally in a regularly repeating pattern. One wonderful consideration in the study and identification of gems as crystalline materials is that many of the gems in our jewellery are made up from *single*, transparent arrangements of such natural atomic design: they are 'single crystals'. This consideration is of great significance to all gemmologists, as is apparent in this chapter and in other sections of this book. Some appreciation of the orderly, crystalline world of gem materials is of fundamental importance to all who study and enjoy gems and jewellery. Ever since the first half of the twentieth century, the inner reasons for the nature of crystallinity have been appreciated by those who have striven to succeed in the development of countless types of crystalline materials throughout industry, and in high-technology or for further research. By the early 1930s, Sir Lawrence Bragg, Nobel physicist, President of the Gemmological Association of Great Britain from 1954 until 1972 and one of the great insightful clarifiers of science for ordinary people, was able to state:

> In the past, the study of external crystalline form has been of the greatest importance, because it has been the principal means of identification and classification. - - - The fundamental property of the crystal, however, is its atomic pattern, and the external form is only one result of this pattern – a relatively unimportant feature which depends in a complex way on external factors. - - - The new methods probe deeply into the crystal structure, telling us how the atoms are arranged, and throwing light on the nature of the molecule and the relation between structure and physical properties.
>
> From Sir Lawrence Bragg, *The Crystalline State*
> (London: Bell, 2nd ed, 1949)

The use of the study of crystallinity, apart from help with the answers to the most basic and frequent questions about gems and jewellery, is broadly in the areas of observation, testing, detection, appreciation and enjoyment of all types and conditions of gem materials, in and out of jewellery. It is well known that crystal structures are classified in *crystal systems*, based on a characterization of 3-D abstract patterns and their symmetry. Such a classification may best be used after, rather than before, crystalline material has been studied far enough to be able to use the classification. That study is truly an inside story. The descriptions that follow are given hopefully to enable a deeper insight into the significance and uses of the knowledge of the crystalline state for a working approach to practical gemmology. Although not an educational course in crystallinity for gemmology students, the descriptions could be used greatly to their benefit and may act as useful supplement and examples for courses such as those provided by the Gemmological Association of Great Britain.

The Occurrence of Crystallinity

The word *crystal* derives from an ancient conviction that its occurrence followed from a cryogenic origin: almost two thousand years ago, Pliny the Elder wrote, 'a violently contracting coldness forms the rock crystal in the same way as ice', in the belief that clear quartz is made from ice so deeply frozen that it will not thaw. The word is derived from the Greek, κρύσταλλoζ *krystallos*, from the word *kryos*, 'deep cold'. However, philosophers before and since Pliny's time have concluded that most crystals are unlikely to be made of deeply frozen ice. Now we try to understand the crystalline occurrence of materials in terms of their constituent atoms and molecules and the conditions and patterns in which these have become assembled.

Conditions for the crystalline assembly of atoms and molecular units are found within, upon and far out from this planet. Different conditions allow different assemblies of atoms, so long as the ingredients for assembly are available. The planet Earth contains a mix of ingredients that have been provided by the natural processes of star destruction and planetary birth more than 4000 million years ago. Crystalline order has been the rule rather than the exception for the Earth ever since. Gems are a relatively rare consequence of an apparently universal and predominant natural process. Most of the natural crystalline materials in and upon this planet are inorganic in origin, and most gems are composed of inorganic crystals. However, a few crystalline gems are biogenic in origin, and a very few gems are made of extraterrestrial crystals. By far the majority of gem materials are crystalline solids of natural inorganic origin within the Earth; on that basis, those materials are defined as *minerals*.

There is a constant drive for new techniques of growing and using ever more perfect crystals for optics, electronics and other industrial processes and research. A huge variety of materials depend on carefully controlled crystallization to function, including crystals of semiconductors and electronic materials, oxides, synthetic materials, magnetic and optical crystals, organic and biological crystals, metals and thin-film crystal growth. This work involves physicists, chemists, materials scientists, electronic engineers and metallurgists as well as specialists in crystal growth. Amid this accelerating pace of development, it is likely that more and more artificial gem materials, both single-crystal and polycrystalline, will be developed, both by chance and by design. Gemmologists have to cope with an ever-increasing range of synthetic and other artificial crystalline materials, some of which are very important in the gem trade: cubic zirconia has now the greatest sales by weight of all faceted gems, while there is a very large output of synthetic sapphire single crystals of various colours, a large proportion produced in China for use in graduation jewellery in the USA. Future gem developments may derive from research into biogenic crystals and composites, thin-film and new inorganic polycrystalline and composite materials. Together with the increasing occurrence of gem treatments, such developments are likely to change gemmological thinking and gem education radically, and ultimately, to change the way the jewellery buyer is likely to think about gems.

Despite all this interest and consequent rapid developments, the mechanics of crystallization still remain largely mysterious, which is an important and fundamental problem. The essential difficulty is that the processes taking place at the atomic scale are still barely or not at all accessible to current experimental methods. *Crystallinity* and the word *crystal* are concepts that are based on a submicroscopic scale of pattern. The crystalline world of atoms, with its natural, three-dimensional designs, is not particularly familiar to us. However, certain relevant and fascinating aspects of design and pattern are well within our everyday experience. Crystallinity, the state of being crystalline, is a state of lowered energy, down to which atoms and their linking mechanisms naturally sink, given the right conditions. The right conditions, in our surroundings, exist very frequently for the many different combinations of the many types of atoms making up our world. It is 'easy' – that is, energetically favourable – for salt or sugar to crystallize in our kitchens; it is energetically 'easy' for the crystallinity of bone and fingernails to develop in the womb; technologists make furnace conditions energetically favourable for the natural coming together of a crystalline pattern of atoms in a synthesis of a beautiful ruby, to copy what the Earth has long produced. Yet nature, the Earth, is still producing ruby, and many other gems, regardless of the existence and understanding of humans on the surface.

All this happens in a regularly repeating state of order. To gain some idea of the regularity of crystallinity and of the significance of energy, consider the analogy of equal-sized ball-bearings being shaken energetically in a shallow saucer; as the shaking diminishes, the balls begin to aggregate in an orderly way, in a hexagonal pattern reminiscent of the top view of a honeycomb. The underlying arrangement of the balls changes from a high-energy random, 'liquid' mass into a lower-energy, hexagonally patterned, regularly repeating arrangement. The mass of ball-bearings has become 'crystalline': a regularly repeating solid pattern. You can visualize this as a 'crystal structure'. The boundary to the whole orderly mass of ball-bearings is analogous to a single 'crystal': notice how the orderly pattern within governs the angles of the external outline, analogous to the crystal 'shape'. These angles are between the 'crystal faces', the planes which exist where the atoms were freely aggregating without external interference from other solid materials; the planes where the mass stopped growing as a crystal (*Figure 2.1*).

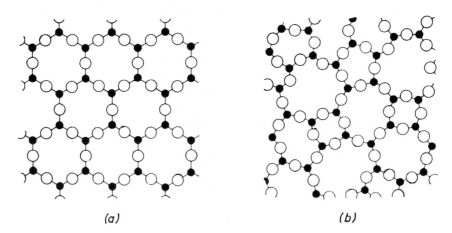

(a)   *(b)*

Figure 2.1   (a) A two-dimensional representation of the internal structure of a crystal, showing the high degree of geometric order, (b) a glass of the same composition, with its less orderly arrangement of atoms (after W. H. Zachariasen)

Crystalline materials, whether they are natural, artificial, inorganic or have an organic origin, can be imagined as *naturally formed 3-D designs of atoms and atomic links in regular, repeating arrangements*. This natural action is controlled artificially in the manufacture of artificial crystalline gem materials, including the various synthetic gems: the energy and action in the rearrangement of atoms amount to a natural process of crystallization.

There are many thousands of different regular atomic arrangements, or *crystal structures*, yet they all conform to only 14 different types of 3-D *pattern*. Such patterns are mathematical *lattices* and they are used as an aid to the classification of the huge variety of crystal structures. Later, in describing the orderliness of these lattices in terms of the concept of symmetry, which we can apply to the regularity of the repeat directions in each lattice, we can draw the description of all crystals into an even tighter classification called the 'seven systems of symmetry' or, more usually, the *crystal systems*.

An atom may be considered as consisting of a nucleus, small and compact, surrounded by a cloud of tiny, orbiting electrons. It is the electrons which are involved in the joining together of atoms to make up all kinds of materials, including gemstones. The chemical properties of an element and the formation of chemical compounds and crystals depend a great deal upon the arrangement of the electrons. In each atom, the cloud of orbiting electrons is arranged in a system of 'orbitals' which represent the most likely positions of each electron in the cloud. Orbitals of electrons are grouped in various energy levels. Atoms with the same number of electrons in their outer orbitals have similar chemical properties. The linkage of atoms to produce a solid crystalline substance is essentially a regularly repeated pattern of electronically joined – *bonded* – electrons in three dimensions. The bonding is achieved in different ways. The strongest forms of bonding common within gem materials are ionic bonding and covalent bonding. Ionic bonding involves donation of electrons whereas covalent bonding involves sharing of electrons.

In an *ionic bond* the forces holding the structure together are due to electrical attraction between atoms. Each electron carries an electrical charge. By transfer of one or more outer electrons from one atom to another, each atom achieves a stable electronic configuration. However, by this transfer action the adjacent atoms become oppositely electrically charged and are strongly attracted to each other. The electrically charged atoms are termed *ions*; when linked in this way they are said to be ionically bonded. Another type of strong bonding is the *covalent bond*, in which atoms share electrons to form an electron pair: each atom shares its electron with the other and so achieves a stable electronic configuration without any electrical charge transfer to become an ion. Covalent bonds are prevalent in the bonding of carbon and silicon atoms; they are therefore important in the diamond and silicate structures. These two types of bonding, ionic and covalent, are not mutually exclusive. For example, the silicon–oxygen bond in silicates is about half-and-half ionic and covalent, whereas the sodium chloride bond is highly ionic and the carbon–carbon bond in diamond is highly covalent.

Isomorphism

When two ions are brought together there is a distance where a force of repulsion abruptly sets in and resists any closer approach. With this in mind it is convenient to consider the ions as spheres in contact, and the distance between their centres is taken as the sum of the radii of the two ions. The relative sizes of certain ions are shown in *Figure 2.2*. Certain ions, such as the oxygen ion, have large ionic radii, while the other ions have radii which are much smaller. Small ions which very frequently make up the structure of minerals are those metal ions which have

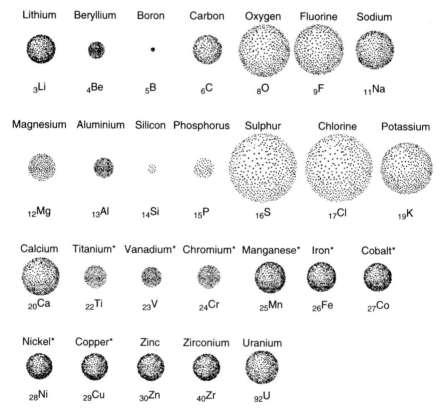

Figure 2.2   Some of the atom sizes of chemical elements which are important in gem materials; shown 50 million times actual size. The name of each element is shown, with its atomic number and chemical symbol; * indicates transition elements. The relative sizes of electron clouds are shown here for just one valence state or type of bonding for each element. However, the size depends upon the degree of ionization and type of bonding. The surroundings within a crystal structure also influence (and are influenced by) the shape of the electron cloud for certain elements, notably the transition elements and carbon: rather than the spherical shapes shown here, these will have lobes of electron density where orbital directions are exploited in crystal structure bonding arrangements

similar-sized radii and hence can replace each other in the crystal structure without unduly straining it. This accounts for the phenomenon of isomorphism, or isomorphous replacement, which is so common in many gem minerals, such as the garnets, the feldspars and tourmaline. As the phenomenon describes alternative ions in the same crystal structural positions, a more useful term might be 'isostructuralism'.

The Shapes of Crystals

The internal structure of any crystalline substance is constant and the outward shape of a crystal must have a definite relationship to this structure. The regularly repeating arrangement of the atoms and bonds which make up any crystal is the inward reason for the arrangement of exterior natural crystal faces. Many crystals show a symmetrical arrangement of crystal faces and such external evidence can be of use in the identification of rough gem material. Crystal faces are natural surfaces which reveal where layers of atoms and bonds in crystal structures have stopped growing or reforming. Crystal faces are relatively rare in crystalline materials because most crystals have grown together, leaving no chance for faces to develop. Faces on most gem materials are likewise rare: normally they are cut and polished away in fashioning.

Well-formed crystals may be made up of similar, symmetrically related faces to give a single *form*. A cube and an octahedron (*Figure 2.3*) are examples of single forms, while a crystal which consists of two or more forms is termed a combination of forms (*Figure 2.4*). If a crystal is made up entirely of one form, that form is termed a 'closed' form (*Figure 2.5*). Forms which do not enclose space are called 'open' forms. Thus the hexagonal prismatic crystal of emerald can have just two open forms, that of the six parallel faces of the hexagonal prism form, which is closed by the two parallel faces of the pinacoid form at each end of the crystal (*Figure 2.6*). The intersection of any two adjacent faces is called the edge.

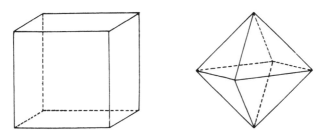

Figure 2.3   The cube and the octahedron

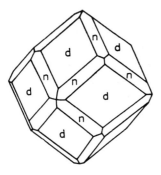

Figure 2.4 *Combination of two closed forms. The faces of the dodecahedron have their edges truncated by the faces of the icositetrahedron, a 24-faced form. Such a combination of forms is common in garnet*

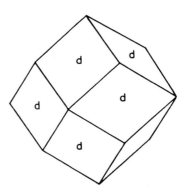

Figure 2.5 *The dodecahedron with 12 four-sided faces is a closed form. This is a common form in which garnet crystallizes*

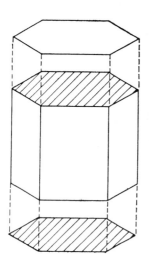

Figure 2.6 *Two types of faces are needed to enclose space in the hexagonal emerald crystal. They are the six faces which are all parallel to one direction (hexagonal prism), and the two parallel faces (pinacoid) which close the top and bottom. Both of these are open forms*

Crystals showing good faces are said to be euhedral, and those which do not are termed anhedral.

Symmetry

The array of points in space at which the pattern of atoms and bonds repeats itself is called a crystal *lattice*, of which there are only 14 variations possible. The lattice is a mathematical concept, represented by an imaginary and infinite array of points. However, the actual positions of all the atoms and the bonds in a crystalline material constitute the crystal *structure*, of which there are thousands of examples. All these structures can be thought of as natural designs, based on the 14 representative three-dimensional lattice patterns. The regular inner structures of crystals, that is their natural atomic designs, as distinct from the lattice patterns upon which they can be based, are arranged according to certain laws of symmetry which form the basis of a classification into 32 mathematically possible classes and, conveniently, into 7 crystal systems. Symmetry is a concept which is used to describe the orderliness of objects or arrangements in two or more dimensions; it can be imagined as operating in three ways, as a plane of symmetry, an axis of symmetry and a centre of symmetry. A plane of symmetry is an imaginary plane which divides a crystal structure into two parts, one of which is the mirror image of the other. The faces may sometimes conform to the symmetry of the crystal structure within, and at least some of the symmetry of the crystal might be discerned from the symmetry of the faces present. Thus in an orthorhombic crystal shaped like a brick, which has three pairs of faces (pinacoid forms), three planes of symmetry can be operated (*Figure 2.7*); and in a cube there are nine planes of symmetry.

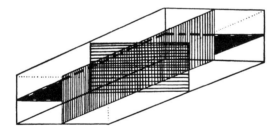

Figure 2.7   The three planes of symmetry in a brick, which represents a basic shape in the orthorhombic system, made up from three pinacoid forms (each a distinct pair of parallel faces) at mutual right angles

The orthorhombic symmetry shown in *Figure 2.7* reflects an underlying crystal structure which has lattice pattern repeats that are different in three directions at right angles. By contrast, cube forms conform to greater symmetry, as a consequence of underlying structures which all have lattice pattern repeats which are equal in three directions at right angles.

The second kind of symmetry is the symmetry axis, an axis about which a crystal structure, when rotating, comes to occupy the same configuration in space more than once during a complete revolution of

360 degrees. Because of this, the crystal faces of any one form will all be parallel to the same angular configuration more than once in a revolution. This will happen regardless of the shape or size of each face in a single form. Depending upon the symmetry a crystal structure may occupy the same configuration twice (twofold, diad or digonal axis), three times (threefold, triad or trigonal axis), four times (fourfold, tetrad or tetragonal axis) or six times (sixfold, hexad or hexagonal axis) during a complete rotation about a symmetry axis. Axes of symmetry of a cube are shown in *Figure 2.8*. The centre of symmetry, the third type of symmetry, is only

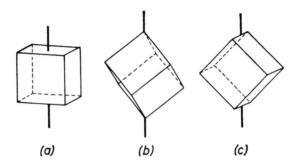

(a)   (b)   (c)

Figure 2.8 Axes of symmetry in a cube: (a) one of three fourfold axes, (b) one of four threefold axes, (c) one of six twofold axes

possible when like structural features, including faces and edges, occupy corresponding configurations on reflection to opposite sides of a point within the structure. A brick or cube has such a centre but a tetrahedron, a form in the cubic system with four triangular faces, cannot have a centre (*Figure 2.9*); it results from a cubic crystal structure with lower symmetry than that of most cubic crystals.

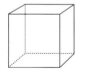

Figure 2.9 The crystal illustrated on the left has a centre of symmetry, but the tetrahedron on the right cannot have one

Crystal Systems

Crystals are further divided into seven systems based on the essential symmetry of their inner structures. In describing crystals it is convenient and necessary to assume certain lines passing through the crystal to indicate specific directions of reference. These imaginary directions, called crystallographic axes, are depicted as lines intersecting within the crystal at a point called the origin. The relative lengths of axes indicate the relative repeat distances of the crystal structure in those directions. For many crystal structures these axes are parallel to symmetry axes or normal to symmetry planes. Every face of a given crystal form has similar or related intercepts with the crystallographic axes; for example, each face of an octahedron intercepts each of the three crystallographic axes at an equal distance from the origin. The seven systems with their axes and maximum symmetry (each system has several 'classes' with differing degrees of symmetry) are as follows. Note that the crystallographic axes must not be confused with the symmetry axes.

Cubic System

Crystal structures of the cubic (or isometric) system are referred to three crystallographic axes, all of which are of equal length and at right angles to one another. These lengths represent the relative lattice repeat amounts for all cubic crystal structures: these are equal in the three directions regardless of their actual amount. There is a maximum of nine planes of symmetry and thirteen axes (three fourfold, four threefold and six twofold) and there is a centre of symmetry. The crystallographic axes in a cube and examples of crystals of the cubic system are shown in *Figure 2.10*. Diamond, garnet, spinel, fluorspar, zinc blende (sphalerite)

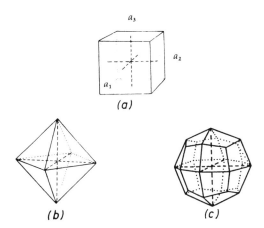

Figure 2.10   The crystallographic axes of the cubic system and examples of cubic crystals: (a) cube (pyrite, fluorspar), (b) octahedron (diamond, spinel) and (c) icositetrahedron (garnet)

and pyrite crystallize in the cubic system, but the two minerals mentioned last do not have the full symmetry of the system. The four threefold symmetry axes are essential to all structures of the cubic system regardless of the degree of overall symmetry.

Tetragonal System

Crystal structures of the tetragonal system are referred to three crystallographic axes, all of which are at right angles to one another. The two lateral axes are of equal length, but the principal axis is either longer or shorter than the lateral axes. Maximum symmetry shows five planes, and five axes of symmetry (one fourfold and four twofold) and there is a centre of symmetry. The crystallographic axes in a tetragonal prism and examples of crystals belonging to this system are shown in *Figure 2.11*. Zircon, idocrase, rutile and cassiterite form crystals having the full symmetry of this system, while the gem scapolite belongs to this system but has less than the full symmetry. The single fourfold axis is essential to all tetragonal crystals.

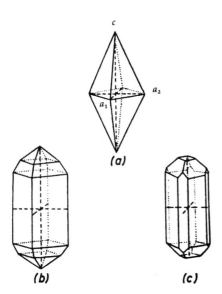

Figure 2.11 The crystallographic axes of the tetragonal system and examples of crystals of the system: (a) bipyramid, (b) prism and two bipyramids (zircon) and (c) two prisms, a bipyramid and a pinacoid (idocrase)

Orthorhombic System

Crystal structures of the orthorhombic system have three crystallographic axes which are all at right angles to one another but which all have different lengths. Maximum symmetry shows three planes, three twofold axes of symmetry (parallel to the crystallographic axes and perpendicular to each symmetry plane) and a centre of symmetry. *Figure 2.12* shows the axes in an orthorhombic crystal with three pinacoid forms, giving six faces in parallel pairs. Topaz, chrysoberyl, peridot (*Figure 2.13*), staurolite, tanzanite (zoisite), iolite, andalusite, danburite, kornerupine and enstatite are some of the minerals which crystallize in the orthorhombic system.

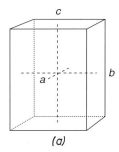

(a)

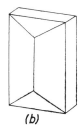

(b)

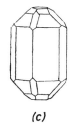

(c)

Figure 2.12   The crystallographic axes of the orthorhombic system and examples of crystals of this system: (a) three pinacoids with axes, (b) two prisms and two pinacoids (staurolite) and (c) four prisms, one pinacoid and two bipyramids (topaz)

Monoclinic System

Crystal structures of the monoclinic system have three axes of unequal length. Two (*a* and *c*) are inclined to each other (at an angle other than 90°), and the third (lateral) axis is at a right angle to the plane which contains the other two. The lateral axis is designated the *b*-axis. Such crystal structures have one twofold axis of symmetry (which is parallel to the *b*-axis) perpendicular to one plane of symmetry (which contains the *a*- and *c*-axes), and a centre of symmetry. *Figure 2.14* illustrates the crystallographic axes and examples of crystals of the system. Important gemstones which crystallize in the monoclinic system are kunzite, sphene, jadeite, nephrite, malachite, spodumene, serpentine, orthoclase feldspar, epidote, diopside, brazilianite and datolite.

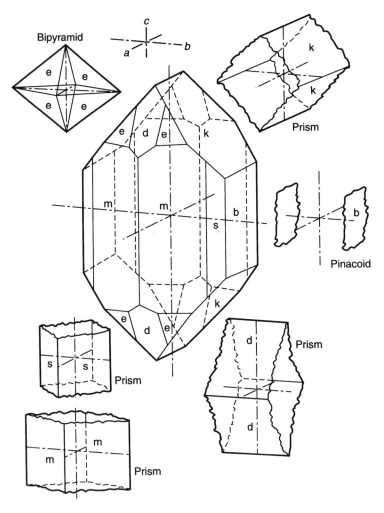

Figure 2.13   A crystal of peridot showing six combined crystal forms: two principal orthorhombic prisms, m and s; a side pinacoid, b; two lateral orthorhombic prisms, d and k, parallel to the two lateral crystallographic axes; and one orthorhombic bipyramid, e

Triclinic System

Crystal structures of the triclinic system have three crystallographic axes which are all of unequal length and which are all inclined to each other. There is a centre of symmetry but no symmetry planes or axes. *Figure 2.15* shows crystals of this system. Rhodonite, kyanite, the plagioclase feldspars, turquoise and axinite crystallize in the triclinic system, the last having lower symmetry.

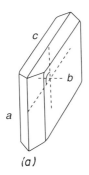

(a)

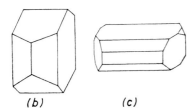

(b)   (c)

Figure 2.14   The crystallographic axes and examples of crystals of the monoclinic system: (a) one pinacoid and two prisms, with axes (gypsum, variety selenite), (b) three pinacoids and one prism (orthoclase feldspar) and (c) four pinacoids and two prisms (epidote)

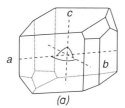

(a)

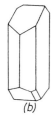

(b)

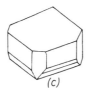

(c)

Figure 2.15   The crystallographic axes and examples of crystals of the triclinic system: (a) twelve pedions (single-faced forms) (axinite), (b) six pinacoids (albite) and (c) eight pinacoids (rhodonite)

Hexagonal System

Crystal structures of the hexagonal system have four crystallographic axes. Three are lateral axes of equal length which intersect at 120 degrees to each other, while the fourth, the principal axis, is either longer or shorter than the lateral axes and is at right angles to the plane containing them. Maximum symmetry shows seven planes and seven axes of symmetry (one sixfold and six twofold) and there is a centre of symmetry. *Figure 2.16* shows the axes of the hexagonal system and shows crystals of emerald and apatite which crystallize in the system, the latter having lower symmetry.

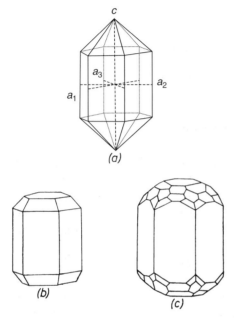

Figure 2.16   The crystallographic axes and examples of crystals of the hexagonal system: (a) one bipyramid and one prism, with axes, (b) one bipyramid, one prism and one pinacoid (apatite) and (c) one prism, one pinacoid and four bipyramids (beryl)

Trigonal System

The trigonal system has a maximum of only three planes of symmetry and four axes of symmetry (one threefold and three twofold axes). There is a centre of symmetry. In this system the principal crystallographic axis is paralleled by the axis of threefold symmetry and not sixfold as in the hexagonal system. For this reason, we describe the trigonal as a separate crystal symmetry system. However, as crystal structures of the trigonal system have the same crystallographic axial arrangement as that of the hexagonal system, some descriptions incorporate the trigonal as a subdivision of the hexagonal system. This difference in description presents no problem so long as the terms are clearly defined. *Figure 2.17* shows the crystallographic axes and examples of crystals of the trigonal system. The important gems ruby, sapphire, quartz and tourmaline, as well as dioptase, hematite, calcite and phenakite, crystallize in the trigonal system.

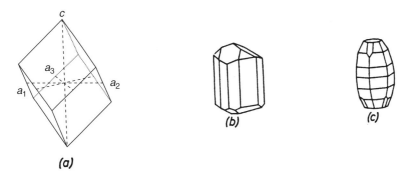

Figure 2.17   The crystallographic axes and examples of crystals of the trigonal system:
(a) rhombohedron with axes, (b) two prisms, four pyramids (tourmaline) and (c) one
rhombohedron, one pinacoid and three bipyramids (corundum)

Face Angles

Because the disposition of crystal faces is controlled by the underlying organization of regular crystal structure, the angle formed by the inclination of two corresponding faces on different crystals of the same substance and structure is constant. This fact may provide means of identifying crystals, even if the crystals are of widely different shapes and sizes, or are of uneven growth so common in natural crystals. In crystallography the interfacial angles are by convention the angles between the normals (perpendiculars) to the two faces and not the outside angle formed by them (*Figure 2.18*). For gemmologists, a visual appreciation of sets of symmetrically related angles and faces may allow recognition of forms present on crystals, of their crystal system and even of the identity of the material.

Most crystals which exhibit faces have grown with some departure from 'perfect' form owing to some faces having grown at the expense of

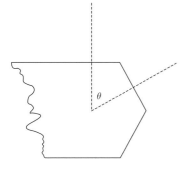

Figure 2.18   Interfacial angles are measured from the perpendiculars of the two faces and not the angle made by the two faces. The interfacial angle is θ

others. Their aspect may appear to be very different from that of an equivalent perfect crystal and can even appear to have the symmetry of another crystal system. *Figure 2.19* illustrates how quartz crystals can show such imperfect growth. However, the crystal structure within each of these crystals is the same regardless of the imperfection of the face shapes and sizes. Because of this the angles between like faces remain constant for all crystals of the same substance and structure. The atomic design and the bonding angles within always conform to the same pattern, so there can be no gross variation in face angle, but only in face size and shape. These 'imperfect' crystals are not distorted; the growth of the inner crystal structure was merely faster in some directions than in others, resulting in a less than 'perfect' shape. The constancy of equivalent face angles, which remain undistorted, provides strong evidence of the constancy and lack of distortion of the inner crystal structure.

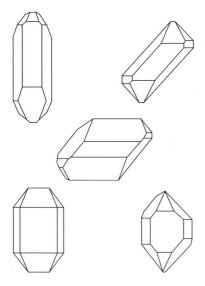

Figure 2.19   'Distorted' quartz crystals

Habit

The characteristic shape of a crystal is known as its habit (*Figure 2.20*). While the interfacial angles are always the same for crystals of the same mineral, the relative development of different crystal forms can differ greatly. A different habit can occur in crystals of the same mineral which come from different localities, or are of a different colour. This is because different forms are favoured by different conditions of growth. Sapphire crystals from Sri Lanka take the habit of a bipyramid, while others from

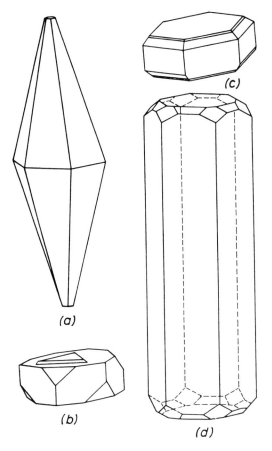

(a)

(b)

(c)

(d)

Figure 2.20   Habit in crystals: (a) Sri Lankan sapphire,
(b) Montana sapphire, (c) pink beryl, (d) aquamarine

Montana are short prismatic in habit. Aquamarine is usually found as long prismatic crystals, while pink beryl (morganite) has often formed as crystals which are short and tabular. Other types of habit are octahedral (diamond and spinel), dodecahedral (garnet), bladed as in kyanite and acicular, in slender needle-like crystals, such as are found as rutile inclusions in quartz. Synthetic crystals are very often grown with habits which differ markedly from those of their natural counterparts.

A crystal having the full symmetry of the system is termed a holohedral, or holosymmetric crystal. Some crystals, however, possess less than full symmetry and some of their forms have a restricted number of faces, or may have forms which are not repeated across the crystal. Crystals which exhibit different forms at either end of the crystal, notably in tourmaline, are termed hemimorphic or polar crystals.

Crystals may occur in groups which are commonly irregularly related to one another, but sometimes the faces and edges of the crystals are in parallel arrangement with other crystals in the group and usually joined to them. Such crystals are said to show parallel growth (*Figure 2.21*).

Figure 2.21   A tourmaline crystal showing parallel growth

Polymorphism

When a material has a chemical composition the same as another material but crystallizes in a different system or class, the effect is termed polymorphism. For example, diamond and graphite are both carbon but crystallize in the cubic and hexagonal system respectively, and the trigonal calcite has the same chemical make-up as the orthorhombic aragonite. A material will solidify or crystallize (or recrystallize) with a bonding structure which depends partly on the conditions under which it forms. Many substances can each exist in any of two or more alternative structural arrangements, depending on the particular conditions of growth or recrystallization. For two polymorphs to coexist, at least one would be in an unstable state; however, the energy required by the crystal structure to revert to a stable state in the new environment may be very great, for example in diamond at room temperature.

Twinned Crystals

Many crystals are twinned, that is they are composed of two or more individual parts such that one part of the structure is distinct from another and not related by the normal symmetry of the crystal, yet both are in structural continuity. Twinning is, in effect, a flip-over of crystal structure. In many types of twin there is an identifiable plane dividing the twinned parts of the crystal, called the twin plane, and an axis about which an imaginary rotation would produce an untwinned crystal. This is termed the twin axis and is frequently perpendicular to the twin plane. Twinning is quite distinct from parallel growth.

Many twin crystals have been twinned from the commencement of growth and physical movement of two parts has not occurred. Others have twinned during changes in physical conditions. Quartz and calcite often have twinned structures caused by temperature and stress changes respectively.

There are various types of twin crystals, such as the simple or contact twin, the most important of which is the spinel twin where one half of the octahedron is in reverse position producing the flattened star-shaped or triangular crystals which are called macles in diamond nomenclature. Some types of twinning, particularly in the tetragonal system, show geniculate (knee-shaped) forms. In interpenetrant twins the crystals are so interrelated that they cannot be divided into separate parts; such a twinning is common in fluorspar. Repeated twinning may occur: when the twins repeat three times they may be called 'trillings'. Such types of repeated twinning often cause the crystal to appear to have a higher symmetry than it actually has. A chrysoberyl example is the trilling of alexandrite (*Figure 2.22c*). Repeated twinning of thin lamellar crystals often shows as parallel striations on a face or cleavage. This polysynthetic or lamellar twinning is well shown by the plagioclase feldspars. Some twin crystals with faces are characterized by their having re-entrant angles. *Figure 2.22* illustrates some types of twin crystals. Twinning may often be revealed within a crystal or cut gem by observation of the materials between crossed polarizers.

Pits and Growth Marks

Many crystals exhibit small pits (etch marks) (*Figure 2.23*), growth marks or striations on their faces, and such marks are useful in showing whether the faces belong to the same crystal form. They may also give an indication of the symmetry: an equilateral triangular pit, for instance, indicates a threefold symmetry at right angles to the face on which it occurs. Striations are due to an oscillation between forms: thus the lateral striations on the prism faces of a quartz crystal are the result of oscillation between the prism form and the rhombohedral form. Likewise the striations in alternate directions on the faces of a pyrite cube, which reveal its lower symmetry, are due to similar oscillations between the cube form and the edges of the pentagonal dodecahedron form (*Figure 2.24*).

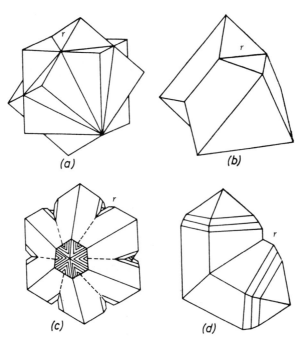

Figure 2.22 Some examples of twin crystals: (a) interpenetrant twin of fluorite, (b) contact twin of spinel, (c) pseudo-hexagonal trilling of alexandrite, (d) geniculate twin of zircon

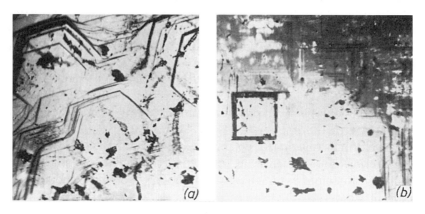

Figure 2.23 Etch marks on the faces of a beryl crystal: (a) hexagonal markings on the basal pinacoid, (b) rectangular markings on the prism face (by courtesy of B.W. Anderson)

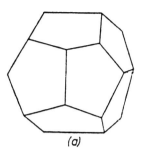 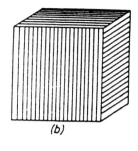

Figure 2.24 (a) The pentagonal dodecahedron (pyritohedron) and (b) the cube are two common forms in which pyrite crystallizes. The striations in alternate directions on adjacent faces of the cube are the traces of the edges of the faces of the pentagonal dodecahedron

Directional Properties

Crystallinity is fundamental to directional properties in solid materials. Directional properties influence how gemstones are fashioned and they can assist greatly in the observation, recognition and identification of nearly all gem materials. An important factor in gemmological observation and testing is that many gem material objects, such as a faceted ruby mounted in a ring, are of single crystals: the directional properties of single crystals are usually far more readily discernable than those of polycrystalline gem materials.

In gemmology, the term 'directional' normally refers to all variations in gem properties where the variation is controlled by the crystal structure of the material. This internal structure influences a material's physical, including optical, properties in particular ways by producing certain directional effects both on the surface and within the material. Directional physical properties are characteristic of *all* crystalline materials. These properties may be evident on the surface of the material, e.g. crystal form, face-markings (see 'pits and growth marks' above) and differential hardness, as well as within the material, e.g. oriented colour zones, internal crystal growth zones or 'grain', oriented inclusions and cleavage. Note that fracture is not a directional property, as it is not crystal-structure controlled: it is a property of solid materials in general and is not specific to those that are crystalline.

In lapidary, and in diamond manufacture ('cutting'), the existence of cleavage is significant. For instance, to a lapidary a cleavage direction is something that should be avoided in the choice of a main facet to be polished; to a designer in diamond manufacture cleavage may be exploited in dividing a rough stone prior to further fashioning. Incipient cleavage cracks within the rough stone may profoundly influence cutting decisions. Apart from cleavage, other directional properties influence cutting. For example, the shape of crystal rough, in terms of crystal form and habit, influence decisions on the shape of the cut, and the number of stones to be cut, from one piece of rough, in order to maximize yield and, therefore, value; also, colour zones can influence the orientation to reduce their conspicuousness, or to increase their impact on overall colour when viewed within jewellery. In various coloured gems, oriented inclusions are a particularly important directional consideration when fashioning star stones and cat's eyes; and good eyesight, together with good lighting, is crucial in making decisions on the choice of cut orientation for viewing the best possible colour in a number of coloured gemstone varieties such as aquamarine, tourmaline, ruby, sapphire and tanzanite.

In diamond manufacture and the use of diamond abrasive, the existence of the directional effect of differential hardness across each diamond crystal is one of the most important phenomena in the whole of modern industry. Diamond is the hardest known substance. If it possessed no differential hardness, then its surface would not have any direction along which it is slightly less resistant to scratching or abrasion. Without this slightly 'softer' direction, no diamond would be 'cuttable', not even by another diamond. Six directions along a diamond crystal structure, parallel to the cube-face diagonals, are the 'hard' directions: these directions present the hardest-known cutting agent for industrial manufacturing and for gem diamond cutting. No other agent can be used as an abrasive medium in gem diamond manufacture. The recognition of crystal twinning within diamond rough is also of enormous importance in making design decisions before fashioning, due to the change in orientation of the differential hardness directions across the twin plane. A saw blade may come up against a direction of greater hardness on the far side of a twin plane, making further work exceedingly slow or even impossible. An understanding of the crystallinity of rough diamond is central to all of these fashioning considerations.

In gemmology, we consider all of the useful directional properties that are 'physical'. These are caused by the interaction of a particular crystalline material with the physical forces acting through or upon it. However, of these physical properties the sub-group of *optical* properties is of particular importance in gem testing. Directional optical properties are possessed only by *optically anisotropic* materials due largely to their ability to polarize the transmitted light. Not all crystals are optically anisotropic; even though all crystals are physically anisotropic. Again, it is the inner structure of a crystalline material that exerts its control of the optical behaviour of the material. Structurally controlled optical behaviour may be detected in, and used in the testing of, uncut as well as cut and polished crystalline materials, including those which possess no crystal form as external evidence. The fundamental importance of crystal structure in optical gem testing manifests itself in, for instance, double refraction (DR) and birefringence measurement, pleochroism, optical extinction and in interference figures. These polarization phenomena are visible results of crystal structure effects, hence the term 'optical': we indeed see the effects of the orderly crystalline bonding directly. Crystalline materials in all crystal systems other than the cubic system are *optically* anisotropic. Cubic and non-crystalline materials are optically isotropic; however, many of these do exhibit a degree of optically anisotropic behaviour, because they are under strain or otherwise have some degree of order in their electronic bonding respectively. For this reason, interesting and sometimes characteristic anisotropic optical effects can often be displayed in paste (cut glass), plastics and certain cubic crystals, such as almandine garnet and Verneuil synthetic spinel. Crystalline materials of all crystal systems are physically anisotropic; whereas non-crystalline materials are physically isotropic and can never possess or display directional features such as structurally controlled growth faces or cleavage.

It should by now be apparent that in observation and testing, the directional properties of crystalline gem materials are of great importance. Simple, first-line observation of crystal form, face markings (e.g. striations and trigons) of colour, growth zones, 'grain' and oriented inclusions and DR, is a process that is rapid and consequently available to anyone equipped with a $10\times$ lens and simple flashlight in virtually any working circumstance. Of similar importance is the observation of cleavage cracks and surfaces in single crystals, as in a cut diamond or in a rough pebble of topaz; of crystal growth and colour-zones, as in many amethysts; of oriented inclusions, as in ruby and sapphire. Other observable directional effects include the undercutting in surfaces of ornamental materials. This is visible in reflected light, where differential hardness in individual grains (crystal particles making up the material) in different orientations results in different degrees of abrasion or polish affecting adjacent grains, leaving a mottled or 'orange peel' effect, seen notably in older jadeite carvings polished with mineral sand rather than diamond dust; similarly, observation of cleavage 'twinkles' in the grains within marble and jadeite can greatly aid identification and distinction from other, similar materials. With simple and portable instruments, observation can be extended, for instance, to the detection of extinction effects using a polariscope, including interference figures, and of pleochroism, e.g. in tourmaline or ruby using a dichroscope. The use of all these observations is possible with both cut and rough materials, in set stones and strung beads. Elsewhere in this book, there are many instances demonstrating the importance of crystal directional effects in gemmology, such as the significance of refractometer observations and readings, and the relationship of these to crystal anisotropy and orientation and system classification.

Added to all these benefits of the study of crystallinity in gem materials are the undoubted benefits and enjoyment of crystal form and beauty. The collecting of beautiful and unusual or particular crystals or groups of crystals has for long been a very important hobby, even contributing to the advent of the world's most important natural history museums; crystal collecting has a dedicated, sometimes fanatical, following, with an important specialist literature and full mineral and gem show calendar. The enjoyment of crystalline gem materials is open to all.

Part II

3

Diamond I

Diamond Genesis

Even today it is not completely established how diamond is formed though it is known to be the product of the deep-seated crystallization of ultrabasic igneous magmas which have intruded as dikes or pipes of kimberlite or lamproite (*Dana's New Mineralogy*, 1997). These primary deposits on erosion have concentrated diamond crystals in placers of different geological ages. Conglomerates and different stream and wave deposits are prospected. Readers are recommended, in view of the dated nature of most monographic literature on diamond genesis, to check the latest journals in a national or university library: gem textbooks cannot accommodate this type of information in addition to all the other topics they need to cover.

For any study of diamond genesis the study of inclusions is essential. While Koivula and Gübelin, *Photoatlas of Inclusions in Gemstones* (cited passim) is invaluable, Koivula has produced a study of inclusions in diamond alone, *The Micro World of Diamonds* (2000; ISBN 0964173352). Here all the solid inclusions are shown in colour where appropriate (there are no liquid inclusions in diamond) and there is a useful bibliography. Readers are advised to obtain this book.

Diamond Composition

Diamond, crystalline carbon, is polymorphous with graphite, chaoite and lonsdaleite. The hardness is 10 and the SG 3.515. The RI is 2.42 and the dispersion 0.044. Crystals are commonly octahedral, also cubic, dodecahedral or tetrahedral. The unique hardness arises from the three-dimensional covalent linkage of each carbon atom to the four neighbouring ones. The point group is $4/m\bar{3}2/m$, possibly bar $\bar{4}3\,m$ (for an explanation of point groups and related topics see *International Tables for Crystallography: Brief Teaching Edition of Space Group Symmetry*, second, revised edition, 1988, published at Dordrecht for the International Union of Crystallography by D Reidel Publishing Company, ISBN 9027725780). A more comprehensive survey for the International Union is Lima de Faria *et al.*, 1990 (ISBN 079230649X). Here the history of the terms used in crystal descriptions is given and there is a unique and excellent bibliography.

Diamond crystals present special problems in fashioning. Diamond crystals unless very small are not available to collectors; some are well illustrated from the standpoint of the polisher in two books written by Nizam Peters and published by the American Institute for Diamond Cutting, *Rough Diamonds: internal and external features*, 1998 (ISBN 0966585402). There are many studies of diamond cutting; one fairly recent one is Basil Watermeyer's *The Art of Diamond Cutting*, 1994 (ISBN 0412984113).

Diamond crystals have an adamantine to greasy lustre and sometimes show curved and striated faces; some crystals are spherical with an internal radial structure. Contact twinning with {111} as the twin plane is common and crystals are often flattened on {111}. For explanations of these symbols, which are given in this text only for diamond, see any text of mineralogy, for example *Dana's New Mineralogy* (1997) pages xviii–xxxii. The ISBN of this essential text for anyone needing to know about validated mineral species is 0471193100. Crystal faces are commonly marked with trigons whose outline does not echo the edges of the face but is reversed. Etch pits do follow the crystal edges.

Penetration twins, sometimes repeated, are also found. There is a perfect cleavage on {111} and a conchoidal fracture. Diamond shows fluorescence and phosphorescence and is triboelectric. It possesses the highest thermal conductivity of any known substance. Diamond is transparent to translucent and may be colourless, pale to deep yellow, brown, white and blue-white; less common colours reported are orange, pink, green, blue, red and black. Those diamonds with strong colours (not merely off-white rather than canary yellow) are known as fancies. Cell data: space group Fd3m. $a = 3.5595$ $Z = 8$. The X-ray powder patterns (given in this text only for diamond) are 2.06 (100), 1.261 (25), 1.0754 (16), 0.8182 (16), 0.8916 (8).

Diamond has an SG of 3.52 and varies little; the single RI is 2.42 and the dispersion (fire) 0.044. The ideal polished diamond is cut as a colourless round brilliant – if the ideal combination of brilliance (from the unique hardness) and dispersion is to be achieved. The flashes of colour from the small upper facets surrounding the table are irresistible and are best seen by distant single white spotlights – as in many cathedrals and museums. Direct sunlight is less effective (less subtle) in producing the best dispersion. Whatever the desired effect from a diamond, the polisher has to pay attention to the facet angles – in a round brilliant, light must not leak from the base or the effects of dispersion will be lost. The flat (spread) brilliants with their large tables may impress (and give the impression of greater weight) but in comparison with the deeper old-cut diamonds with their smaller tables and steeper facet angles their dispersion is low. Old-cut stones show their fire however small they are.

There are many books describing diamond polishing but I recommend, in particular, Herbert Tillander, *Diamond Cuts in Historic Jewellery, 1381–1910*, London, 1995; ISBN 1874044074. For the grading of polished colourless or near-colourless diamonds GemA have issued a useful text, *Diamond Grading Manual* (2000; no ISBN) and there are several editions of the commercially produced grading handbook by Verena Pagel-Thiesen (*Diamond Grading*, Antwerp, 1980). Gary A. Roskin, *Photo Masters for Diamond Grading*, 1994; ISBN 0964173301 will also be found useful.

A study of diamond crystals should begin with the early consultation of the English translation of Yuri L. Orlov, *Mineralogiia almaza* (*The Mineralogy of the Diamond*), New York, 1977; ISBN 0471018694. The original Russian edition was published in 1973 and one of its strengths is the presence of profuse illustrations of crystals and a very strong list of references. By the time of the English translation, Orlov had been able to add to his material.

Details of individual modes of occurrence are given below as are different responses to UV and X-rays.

The history of diamond mining is on the whole exhaustively written up, occurrences in Africa being especially well covered. South African diamond deposits and mining are described by Gardner F. Williams in *The Diamond Mines of South Africa*, second edition, New York 1905 [Sinkankas #7234]. Also on South African diamond is Alpheus Fuller Williams, *The Genesis of the Diamond*, 2 vols, London, 1932 [Sinkankas #7224]. Gardner Williams' book has been criticized for shortcomings in some details of the mineralogy and for omitting descriptions of mines not at the time under De Beers control. Much of the text discusses personalities (the second edition had undergone some revision). Nonetheless the text gives details of the discovery of the South African diamond deposits upon which many subsequent writers have drawn.

On the other hand, Alpheus Williams' book gives detailed studies of the petrology and mineralogy of the materials filling kimberlite pipes and should be consulted in the early stages of investigation into this topic.

General Surveys of Diamond

Some general surveys of gemstones have become cluttered with repetitive material whose authenticity cannot be established without access to such original sources as can be accessed (this field of study is open only to those who can deal with oriental and Slavonic languages; no single individual could attempt it – the literature would need to be compared with more modern records). The thought is a daunting one! For this reason I am recommending, as an excellent and up-to-date general survey of the history of diamond discovery, mining, use, testing and synthesis, George S. Harlow's *The Nature of Diamonds* (1998; ISBN 0521623957). This is as good a survey as can be found at the time of writing.

The synthesis of diamond is well described in Robert M. Hazen's *The Diamond Makers*, 1997 (ISBN 0521654742). Here we meet Hannay and other bold experimenters as well as the scientists who finally succeeded in manufacturing diamond. Hazen has also written *The New Alchemists* (1993; ISBN 0812922751) in which he describes, in layman's terms, the development of high-pressure technology, including its use in the manufacture of diamond. Amanda S. Barnard's *The Diamond Formula* (2000; ISBN 0750642440) is another first-rate overview of diamond at the present time; it includes an account of diamonds grown by chemical vapour deposition (CVD) (I was told at a conference in October 2004 that no CVD diamond had so far reached the trade).

Coloured diamonds, long known as fancies, now have their own large and beautifully presented book, Stephen C. Hofer's *Collecting and Classifying Coloured Diamonds*, 1998 (ISBN 0965941019). The book is a catalogue of the Aurora Collection of coloured diamonds which are illustrated in their entirety with many small features closely depicted. This is a particularly important study, as fancies now have their own grading system which is used by the major auction houses in their catalogue descriptions. Another beautifully produced study of coloured diamonds is *Fancy-colored Diamonds*, by Harvey Harris, published by the Fancoldi Registered Trust, Liechtenstein, 1994; ISBN 39520643).

Diamonds from India were probably the first to be described, the Koh-I-Noor and Jehangir diamonds being now assigned to India. The traveller Jean Baptiste Tavernier (1605–1680) wrote *Les six voyages de Jean Baptiste Tavernier*, Paris, 1676 [Sinkankas #6499] but also see many subsequent editions and translations of which the best is Valentine Ball's English version of 1889 and later. Ball includes a biography of Tavernier who travelled in the Near East during 1631–68; he then appears to have begun a fresh career as a jeweller with royal clients. He was buried near Moscow.

Tavernier's travels in India are agreed to be a significant record rather than a compilation of hackneyed legends; he visited the diamond mines there and the diamond-bearing rivers; one such visit was to the mines at Raolconda.

In the *Science and Technology of Diamond*, edited by G.S. Bhatnagar in 1999 (ISBN 1898326487), Bhatnagar and Murthy give a valuable introduction to the science of diamond in protohistoric India. As the book is not easy to find, a short resume is given below. Ancient diggings were along the Penner, Krishna and Godavari rivers, stones reaching the temples of southern India and later Europe. Valentine Ball's *Manual of the Geology of India*, part 3, 1881 and Ritter, *Erdkunde von Asien*, vol. 4 (2), 1836, are sources. Further information can be found in Murthy, *Gemmological Studies in Sanskrit Texts*, vol. 1, 1990 and in Sircar, *Studies in the Geography of Ancient and Medieval India*, Varanasi, 1971.

Diggings along the banks of the Penner, Krishna and Godavari rivers provided first the temples of South India and later Europe with fine diamonds. No really up-to-date account is available in English but Kautilya's *Arthashastra* (Mysore, Mysore Publishing and Printing House, 1967) and Samasatri, *Kautilyarthasastram*, Oriental Research Institute, University of Mysore, 1960, include considerable data. These sources date back to the third century BC and are especially valuable for the study of the geology, mineralogy and commerce of diamond. Varahamihira's *Brahat Samhita* from the fifth century AD is also of great value. Yet more information can be found in other Sanskrit texts. Bhatnagar (1999) gives an excellent bibliography.

The earliest record of diamond is in *Rig Veda* (see Murthy, *op. cit.*). The name Golconda, not always linked with a specific location, has often been used as an epithet for Indian diamonds in general, but does cover diamond-bearing ground in areas from which the Penner River discharges into the Bay of Bengal. Nonetheless it is difficult to link ancient names with present-day places though Bhatnagar proposes several connections: sabharastraka is identified as diamond from the Vidarbha area, madhyamarastraka from the present-day Madhya Pradesh, Kasmaka is believed to refer to diamond from near Baranas, Srikatanaka to the place named Vedotkata. Manimantaka is diamond from the mountain Maniman and Indravanaka is from Orissa. Vidharbha region is probably the area known for current diamond workings. These attributions are of course conjectural.

Sanskrit sources attribute many properties to diamond so its potential for ornament must have been realized from the earliest times.

The diamond sources on the Krishna in present day Andhra Pradesh are probably the mines which have produced, in present-day Andhra Pradesh, diamonds of the quality of the Koh-I-Noor, a celebrated diamond now in the English crown jewels; disputes over its re-cutting can be found on books devoted to diamond. It is possible that the slaty blue Hope diamond with the remarkable reddish fluorescence (in the National Museum of Natural History, Smithsonian Institution, Washington DC, USA) also comes from the same general area. Details of the major stones in the Crown Jewels can be found in the HMSO publications dealing with them.

In general, Indian diamonds are notably limpid though at the time of writing there is no way in which the place of origin of any diamond can be determined by normal gemmological testing. In the state of Andhra Pradesh some diamond pipes were discovered in the 1960s and additional pipes have also been found in the state of Madhya Pradesh. Not all pipes have been found to contain diamonds. The workings in Madhya Pradesh are in the Bundelkhand area; mining is sporadic. Alluvial, pipe mining and investigation of conglomerate rock have all taken place at least from time to time. Well-shaped green octahedra have been found at Majgawan and further investigation of the region has been carried out by government; there has been some recovery of diamond.

Extensive reports on diamonds in India were provided by Bauer in *Edelsteinkunde*, 1896 (English translation by Spencer, 1904; Dover Press reprint, 1968 [Sinkankas #471]). Those interested are recommended to read them with an eye on accuracy as no original material is cited.

Apart from the Majgawan pipe, the Indian diamonds are found in sandstone or conglomerate, or in the sands and gravels of old river beds. The associated minerals are usually confined to quartz pebbles and variously coloured jaspers. Panna diamonds have been known for 200 years.

Perhaps the best general surveys of Indian diamonds are contained in books by Valentine Ball, once of the Geological Survey of India and the translator of Tavernier's travels (q.v.). In 1880 he contributed a paper on the mode of occurrence and distribution of diamonds in India to Journal of the *Royal Geological Society of Ireland* 6, part 1, new series.

The paper [Sinkankas #334A] contains details of Indian diamond deposits, their history and geology. The paper can also be found in *Proceedings of the Royal Society of Dublin* 2 ns, 1880, pp. 551–89. Pages 69–70 in the first-mentioned journal contain notes on obscure localities, including Raolconda.

In another paper, published in *Journal of the Royal Asiatic Society of Calcutta*, 50, part 2, 1881, pp. 31–44 [Sinkankas #334B], Ball identifies those Indian diamond mines visited by Tavernier. Ball's 1881 book, *The Diamonds, Coal and Gold of India*, published in London [Sinkankas s#335], discusses, among other topics, the relationship between the Great Mogul diamond and the Koh-I-Noor. The book describes Indian diamond deposits and their geology and the prospect of success if the mines were taken over by Europeans.

Indonesia

Diamonds are found at a number of locations in present-day Indonesia but production appears to be sporadic. Deposits in the general area of Pontianak, a coast town on the west coast of Borneo have been mentioned in unsupported notes by some writers but current evidence is not forthcoming. The same can be said of reported deposits in other locations in Borneo. Information in Webster's *Gems*, various editions, appears to derive from Bauer, *Edelsteinkunde* (various German editions and Spencer's translation of 1904); for deposits in what is now Indonesia, Bauer cites Edmond Boutan's book *Le Diamant*, Paris, 1888 [Sinkankas #838].

Brazil

Diamonds were discovered in Brazil at least as far back as the early seventeenth century. Panning for gold brought diamond crystals to light, sites adjoining the Jequitinhonha River being particularly prolific. Remarks on Bauer's account of Indian diamond mines are equally applicable here as none of the admittedly profuse and interesting information is referred to authorities. Those wishing to investigate further into the history of Brazilian diamond deposits should read the second volume of Sir Richard Burton's *Explorations of the Highlands of Brazil*, London, 1869 [Sinkankas #1078]. This volume describes his journey to the area now known as Diamantina; there are also descriptions of alluvial diamond deposits in the state of Bahia. This is the major introduction to Brazilian diamond mining.

Another outstanding survey of Brazilian diamond is Wilhelm Ludvig Eschwege's *Pluto Brasiliensis* of 1833 [Sinkankas #1947]. Here the author describes gold as well as diamond and other gem minerals, giving, in passing, the regulations of the Portuguese crown for the control of the mining and marketing of diamonds. Gold and diamond smuggling are also covered. Brazil is relatively rich in diamond since in addition to the Diamantina area of Minas Gerais there are occurrences in several other states.

Diamonds have been found in possibly Precambrian rocks overlain by the Itacolomy series, diamond occurring in intrusions of an altered rock. Diamond has also been found in the Lavras series of conglomerates and other rocks and in alluvial deposits.

Most Brazilian diamonds are small but quality can be good. The Presidente Vargas diamond weighs 726 ct.

North America

The best accounts of any gemstone found in North America can be found in the three volumes of John Sinkankas's *Gemstones of North America* (1959, 1976, 1997). Occurrences except for those in Canada tend to be adventitious crystals turning up in California in 1849 during gold mining operations. In Merrill's *Handbook and Descriptive Catalog of the Collections of Gems and Precious Stones in the United States National Museum*, 1922, only one diamond (yellow, uncut but polished, weight 0.77 ct) is recorded, from Kentucky, and another from Rutherford county, North Carolina, a pale yellow flattened octahedron of 0.14 ct. In 1972 the Crater of Diamonds State Park was established in the area of Murfreesboro, Pike county, Arkansas, where three types of igneous rock are found, lamproite, lamproitic breccia and tuff, though only the breccia produces diamonds. A good account of the area and its geology can be found in *The Mineralogical Record* 21 (6) 1990. Diamonds have been found in Alaska, North and South Carolina (the latter at least possible), Alabama, Georgia, Louisiana, Kentucky, Michigan, Wisconsin, Montana, Iowa, Wyoming and California (Sinkankas 1997). One or two of the reports of diamond crystals (many pale yellow distorted octahedra in general) may be in error. However, some finds have been remarkable; the Uncle Sam diamond from the Arkansas deposit weighs 40.42 ct and the Star of Arkansas 15.31 ct.

The different state mineralogies give more detail and notes in journals can be found listed in the three Sinkankas volumes.

The diamond scene in North America was completely changed when, in the late 1970s kimberlite pipes were discovered in the north-west Territories of Canada when Charles E. Fipke identified diamonds in a kimberlite pipe situated beneath Point Lake in the Lac de Gras area. The story of some of Fipke's life and work can be found in his picaresque *Fire into Ice* (1999; ISBN 1551923343). Construction of the Ekati mine was completed in 1998. It is interesting to note that some exploration in Finland has shown the existence of similar pipes and small diamonds have been found.

Australia

Diamonds from New South Wales are well described in A.A. Macnevin's *Diamonds in New South Wales* (1972; Sinkankas #4115). The publishers are the Department of mines, Geological survey of New South Wales and the book forms #42 of the Mineral resources series. Diamond is reported to have been found in 1851 at Sutter's Bar on the Macquarie river near Bathurst. Most diamonds from New South Wales are small and yellowish or off-coloured. Diamonds have also been found in Queensland. In the second volume of *The Geology of the Commonwealth of Australia*, T. W. Edgeworth David (1950) briefly describes the deposits in New South Wales. The diamonds occur in stream deposits. Near Inverell, diamonds are found at Copeton, where many stones are recovered from a tertiary wash.

Western Australia

The Ashton Joint Venture's exploration for diamonds in the furthest north-east area of Western Australia has led to some of the most important developments in the diamond world for many years. Particularly notable are pink diamonds, some inclining to red.

Exploration of the Kimberley district began in 1972. Up to 28 pipes were found in Ellendale province and the quality of the diamonds recovered from the start made it likely that there would be a good proportion of gem-quality crystals. Later, deposits at Smoke Creek produced alluvial diamonds; a large lamproite pipe, AK-1, was found near Lake Argyle (man-made). Possible yield from Smoke Creek was reported to be as much as 470 ct for each tonne of ore.

The very large pipe AKA, said to be sixth in extent among world pipes, was expected to produce very large quantities of diamonds even though the proportion of gem-quality material might not be exceptional. Certainly the discovery of the coloured Argyle diamonds has led to a growing interest in fancy stones.

South Africa

The history of the discovery and mining of diamonds in southern Africa has been told many times and is not examined in detail here; several authoritative studies of mining in South Africa itself have already been cited (Gardner F. Williams, *The Diamond Mines of South Africa*, second edition, 1905; Alpheus F Williams, *The Genesis of the Diamond*, 1932 – A.F. Williams was the son of G.F. Williams and their combined experience of the area made both books essential for diamond studies). It may not be widely known that A.F. Williams also produced a book of essays concerning mining and miners in South Africa, *Some Dreams Come True*, 1948 [Sinkankas #7225]. The essays include brief biographies of a number of personalities, including Cecil Rhodes.

In 1873 Augustus F. Lindley published *Adamantia. The Truth about the South African Diamond Fields or, a Vindication of the Right of the Orange Free State to that Territory, etc.* This rare book was published in London and is Sinkankas #3984. As Sinkankas says ['the book is] an important historical document in which the seamy side of the acquisition of the diamond fields of South Africa is carefully examined'. The British Parliament and public is told about the actions of their government and the case is supported by considerable evidence.

The Anglo American Corporation of South Africa and the work of Sir Ernest Oppenheimer are described in Theodore Gregory's *Ernest Oppenheimer and the Economic Development of Southern Africa*, published in 1962 by Oxford University Press (Sinkankas #2525). Chapters 3–5, pp. 109–383, tell 'The diamond story' and the student can find a great deal of information on the exploitation of diamond resources. The establishment of the De Beers organization is described in 'The Story of De Beers', by Hedley A. Chilvers (London, 1939: Sinkankas #1287). The appendices, in particular, contain details of mine operations and statistics while the main text traces the history of the company with commentary on some of the politics involved. The diamond cutting industry in South Africa is also described.

A useful account of the diamond occurrence and mining in the Kimberley area can be found in Cairncross and Dixon, *Minerals of South Africa*, Geological Society of South Africa, 1995 (ISBN 0620193247). The book includes some useful maps of the mining areas as well as a comprehensive bibliography.

Zimbabwe

Alluvial diamonds were discovered during 1903 in the Somabula Forest west of Gwelo in Matabeleland, Zimbabwe. There seems to have been little development in recent years.

Namibia

In 1908 diamond was found near the settlement then called Lüderitz in present-day Namibia. Marine beach deposits were later found in separated areas extending around 100 km along the coast north of the mouth of the Orange River.

In 1927 extensive marine beach deposits were found in Namaqualand, south of the Orange river. Rich deposits were located some 25 m above sea level of a terraced beach. A minor diamond field was discovered in 1956 on the Skeleton Coast of Namibia. The area lies in the Kaokoveldt, south of the Kunene river, which forms the border with Angola.

These coastal fields are worked by excavators after overburden has been removed. Diamond-bearing gravel is removed by hand or by excavator brought to a central treatment plant. The origin of these marine alluvial diamonds is open to debate. Was the original source inland, or did the parent rocks lie submerged under the Atlantic Ocean? The size of the stones diminishes as the distance from the mouth of the Orange River increases. It is therefore thought that the source is inland, the stones having been carried by rivers of former geological ages to the sea to be later cast up on the beaches near the river mouths by ocean currents.

About 1964 the Marine Diamond Corporation commenced diamond recovery from the sea bed off the Namibia mainland, and in 1967 a specially equipped barge operated about a mile off the mouth of the Orange river. The sand of the sea bed, up to 30 m below, is raised by airlift and suction dredge. The diamondiferous sand is then screened, washed, concentrated by heavy media separation and finally sorted by hand. Shares in Sea Diamond, either the same or a similar venture, never really prospered.

Zaire, Congo and Angola

Diamonds from some areas of Africa are reported to have been sold for arms at various times and have become known as conflict diamonds. There is general agreement that trade in stones known to be conflict diamonds should not take place.

In 1910 diamonds were found in the south-western part of the Congo in alluvial gravels in the valleys of the tributaries of the Kasai river. Farther west, diamond deposits have been located in the Kwilu basin. A more productive area was discovered around the Bushimaie river to the east of the Kasai river and about 600 miles north-west of Lubumbashi in the district of Katanga. Near the Kundelungu plateau, just north of Lubumbashi, a number of kimberlite masses occur, a few of which have proved to contain diamonds. The town of Tshikapa is the centre of the diamond-mining operations in the Congo, which is the world's largest producer of diamond in terms of quality. Before the West African production came on the market, the Congo produced at least three-quarters of the crushing boart so widely used in industry.

During 1916 the Congo diamond fields were found to extend into the north-eastern part of Malange in Angola, particularly along the western bank of the Chuimbe river and both banks of the Luembe and Luana rivers. The Angola deposits are alluvial and are derived from an ancient drainage system which no longer exists.

Mining is carried out, in both Angola and the Congo, by removing the overburden to expose the diamondiferous gravel. This gravel is then excavated, taken to washing plants and treated in a heavy media separator. The concentrates are processed in a recovery section for the extraction of the diamonds.

Ghana

The discovery, in 1919, of diamond at Abomoso near the Birim river opened up diamond mining in Ghana. Later prospecting disclosed diamond in the district south-east of Kade and about 100 km north-west of Accra. Diamonds from Ghana are mostly of industrial quality, although gem crystals are also recovered.

The Ghana deposits consist of gravels from 0.5 to 2 m thick and are interesting in that the rock formations in the area are steeply dipping metamorphosed igneous and sedimentary rocks of Precambrian age. The mining for diamond is carried out by removing the overburden and depositing it in previously mined sections. The exposed diamond-bearing gravel is mechanically loaded into one-tonne trucks and taken to the washing and recovery plants where, for the smaller diamonds, a method of 'skin flotation' is used. Diamond are also found in the Ivory Coast.

Sierra Leone

The history of diamonds in Sierra Leone and the establishment of the diamond trade there is covered by H.L. Van der Laan in *The Sierra Leone Diamonds*, published by Oxford University Press in 1965 (Sinkankas #6808). The account covers 1952–61. The author does not omit the many political problems associated with mostly tribally organized and alluvial diamond mining; later troubles are described by other authors, and Matthew Hart, in *Diamond: The History of a Cold-Blooded Love Affair*, London, 2002; ISBN 184115279X, outlines what he sees as a threat to the long-term prospects of the De Beers organization.

From the geological standpoint, useful texts include Grantham and Allen, *Kimberlite in Sierra Leone*, London, 1960 (Sinkankas #2489) forming vol. 8 no. 1 of *Overseas Geology and Mineral Resources*. This review described kimberlite dikes with an outcrop map. Alluvial crystals are also described and there are useful crystal drawings.

The diamond fields of Sierra Leone were discovered in 1930, when a small diamond was found in the gravels of the Gboboro stream near the village of Fotingaia. Consolidated African Selection Trust Ltd sent a prospecting party from the Cold Coast to Sierra Leone. During 1932 they found diamonds north of the Bandafayi watershed in the Shongbo stream near Tongoma village. Small-scale mining operations showed the deposit to be worth investigating. The company was granted the exclusive right to prospect for diamond over an area of 10 800 km of the eastern district of Sierra Leone.

Further exploration showed that diamond occurred in the gravels of the Sewa river, an area not included in the licence granted to the Cold Coast company, and in 1934 a new subsidiary company, the Sierra Leone Selection Trust, was floated and was granted the sole right to prospect for diamond in the colony. Owing to the great expanse of diamond-bearing gravels, policing the area was difficult. During the early 1950s it was estimated that more diamonds left the country illegally than legally.

The diamonds produced in Sierra Leone are in the main of good quality and are often of large size. Many of the crystals show excellent octahedral form with bright faces. Crystals of 100 ct or more are occasionally found: The Woyie River Diamond recovered in 1945 weighed 770 ct, but this was surpassed by the finding of the Star of Sierra Leone at the Diminco mine at Yengema in February 1972. This crystal weighed 968.9 ct and was the third largest diamond ever found. Natural bottle-green-coloured diamonds are often found and such diamonds will turn light brown on heating.

Mining is carried on in a similar manner as in the other alluvial fields of Angola and Ghana. The Sierra Leone diamond fields are situated some 225 km east of Freetown.

West Africa

Among the first diamonds to be discovered in the former French West Africa were those in the Région Forestière of the Republic of Guinea. A minute diamond found in a river in the N'Zêrêkorê district on the border of Liberia was probably the first. A year later, several workable deposits of diamond were found in the Makona river district and in the rivers of Haute Guinée.

The main workings of this diamond field are located some 75 km north-north-east of Macenta. The diamonds are mostly found in alluvial deposits in the upper courses of the smaller streams and in old terraces associated with such minerals as ilmenite, magnetite, zircon, rutile, corundum, monazite, epidote, topaz, spinel, staurolite, tourmaline and almandine garnet. The diamond crystals are usually octahedra or dodecahedra and are frequently of a brown colour, or a clear yellow or blue colour. The stones often contain black inclusions. The majority of the crystals are industrial quality only.

Diamonds have been recovered form ancient river terraces along the region of the Mambere river which runs through the western part of the Oubangui-Chari district of the Central African Republic. An economic study of West African diamonds was published by Peter Greenhalgh, *West African diamonds*, 1919–83, 1985; ISBN 0719017637.

East Africa

In *Geology of East Africa* by Thomas Schluter (1997; *Beitrage zur regionalen Geologie der Erde* Bd 27; ISBN 3443110274) there is a brief note stating that alluvial diamonds have been found about 20 km east of the Kichwamba volcanic field in Uganda but that it was not possible to say whether or not the diamonds originated in kimberlites. The East African rift system has yet to be fully investigated.

Tanzania

Schluter (1997) describes the kimberlites of Tanzania, estimating that there are more than 200 examples. They are concentrated in the Lake Victoria area: John T. Williamson discovered the Mwadui kimberlite pipe in 1940; it was, at least at that time, the largest known pipe in the world (about 1500 × 1200 m in diameter). Water lain tuffs form a circumventing ring with a diamondiferous gravel halo. The kimberlite has an unusually high water content. Millions of carats of diamond have been obtained from this mine; the Williamson pink diamond, in the personal jewellery of Queen Elizabeth II, is a celebrated polished stone of 23.06 ct from the original crystal of 54.5 ct. In *Famous Diamonds*, fourth edition, 2000; ISBN 090343265X, Ian Balfour describes some of the events surrounding the finding and eventual polishing of the crystal, an account also to be found in Emily Hahn, *Diamond*, 1956 (Sinkankas #2655). The present writer (MO'D) has examined this stone (now set in a brooch) and can assure readers that it is a quite remarkable pink.

Mwadui diamond crystals are often twinned and have an average weight of 0.24 ct. Other diamond-producing kimberlites in Tanzania are the Uduhe mine in Maswa district and the Kahama pipe. Fifteen pipes were discovered in the Shinyanga region by aeromagnetic survey.

Lesotho

During 1958 diamonds were discovered some 2800 m in a valley in the Maluti Mountains of the Mokhotlong area west of the Drakensberg mountain range. Mining was carried on by primitive methods. Lesotho became famous when, in 1967, a miner's wife picked up a diamond weighing 601.25 ct. It was the largest diamond ever found by a woman. This crystal was cut into 18 stones. Two other large diamonds, one of 527 ct and the other of 338 ct, have also been found in Lesotho. The mine was closed when diamond prices slumped in the late 1970s.

Botswana

The search for diamonds in Botswana probably started in 1955 in the area west of the Limpopo river. A few diamonds were found along the dry Motloutse river at Foley in the Bamangwato tribal territory. Later searches were carried out to the headwaters of the dry river bed. Diamonds were found along the river course but the traces petered out and there was no sign of a pipe from which the diamonds could have come. The De Beers geologist, Dr Gavin Lamont, found traces of a large shallow valley which he connected with a suggestion made years earlier by Du Toit that an earth folding had occurred along this line and would have thrown the source of the Motloutse river much further west. This was proved when a group of pipes were discovered in the area of Letlhakane. One of these, found in 1967, is now the Orapa mine and is the second largest pipe mine, only being surpassed by the Williamson mine.

Sampling showed that diamonds were in payable quantity and in 1969 a pilot recovery plant was put in operation. Although Orapa was said to have come into production in July 1971, it was not officially opened until

1 May 1972. Mining is carried out by the open-cast methods. Since then production has largely increased in Botswana's diamond mines, with Orapa producing 15 000 ct a day in 1981, though only 15% of this was suitable for gems. In contrast 40% of the diamonds produced from the nearby Letlhakane mine are considered to be of gem quality. A further large diamond mine at Jwaneng has been producing a high proportion of gem-quality diamonds.

Guyana

Diamonds were discovered in the gold washings of the Puruni and Mazaruni basins of Guyana, during 1887. The diamonds, small in size, are found in a gravel of ferruginous clayey or sandy nature which is overgrown with tropical jungle. The nature of the terrain makes transport difficult and recovery of the stones uneconomic. Diamonds are also found in the Cuyuni, Potaro and Berbice rivers and their tributaries.

The diamonds of Guyana are commonly worn and pitted octahedra and are most common in the Mearnu-Apiqua district and in the Potaro deposits. The finest quality and the best shapes come from the kurupung and Eping districts. The commonest colour of Guyana diamond is white with a slight yellow tinge, followed by Capes, and light and dark browns. Unusual colours are rare as are blue-white stones, but occasionally a small aquamarine-blue or deep greenstone is found.

During 1948 a diamond deposit was found on the Ireng river which lies along the Guyana–Brazilian frontier, which led to a miniature rush by miners from both sides of the frontier. A cheerful account of the area and his adventures in exploring it for diamonds is given by Victor Norwood in his two books *Man Alone!* (1956) and *A Hand Full of Diamonds* (1960) [Sinkankas #4790 and 4791].

Bolivia

Diamonds have been found in the foothills of the Andes in Bolivia. The diamonds were discovered in the river gravels of the Rio Tuichi and have been reported from the Rio Tequeje, Rio Unduma and some other tributaries of the upper Rio Beni. The diamonds are found in gravel beds of 1–3 m thick overlying a bedrock of Permian age.

Venezuela

The diamond occurrences along the Cuyuni and Paragua rivers in the state of Bolivar in Venezuela may be said to be an extension of the diamondiferous fields of Guyana. The exploitation of these deposits is hindered by their inaccessibility but some mining and marketing of Venezuelan stones is carried out. Diamonds from the Paragua river are coated with iron oxide, but fine-quality stones can be cut from them. Green diamonds are frequently encountered at Icabaru. Fancy colours, such as yellow-green, brown, reddish brown, pink, light blue and black, are common in Venezuelan diamond deposits.

Russia

There is still a dearth of literature in Western languages on the diamond deposits of Siberia. In Russian, probably the all-time most important historical and geological account in A.P. Bobrievich *et al., Almaznye Mestorozhdeniya Yakutii*, 1959 [Sinkankas #732]. This is an account of the Yakutia diamond fields in Siberia, describing the area geology and the pipes as well as the properties of the diamonds themselves. A later study of the diamondiferous placers of the Malo-Botuobinsk region of western Yakutia. *Siberia* was published by I.S. Rozhkov *et al.* in 1963 [Sinkankas #5606]. Rock formations in the same area are particularly well described in the same author's 1967 publication, *Yakutskiy Filiala Sibirskogo Otdeleniya* [Sinkankas #5609]. A paper in *The Mining Magazine* 97(6), pages 329–38, 1957, reviews a number of USSR Russian-language publications on diamond deposits.

As in many places, the first diamond appears to have turned up as a pebble which attracted sufficient attention to make further investigation desirable. Again, the diamond crystal was apparently found when gold and platinum were being sought.

In *Mineralogy of the Diamond* (1997 *op. cit.*) Orlov reviews some of the occurrences of Russian diamonds. His opinion is that the sources of many diamonds in the Urals are Paleozoic clastic rocks and that diamonds may have entered these rocks on disintegration of the kimberlites. Some diamond-bearing pyrope serpentinites in one of the Yakutia pipes have given rise to the theory that they are fragments of completely serpentinized garnet peridotites or olivenites generally associated with a kimberlite magma originating in the upper mantle. A further theory, also originating from Russian sources, is that all diamonds in kimberlites are genetically associated with deep-seated pyrope periodotites. When these are fused they form kimberlite magma which acts as the transporting agent to the upper part of the crust. In general, the theories can be made to conclude that diamondiferous peridotite rocks must exist at depth.

In 1958 an eclogite xenolith was found in the Mir pipe at Yakutia. Orlov found that the diamond content was exceptionally high and the octahedra show stepped faces reminiscent of crystals from South Africa.

Other eclogitic sources of diamond have been found in Yakutia and in Africa.

Kimberlites are the only commercial diamond-bearing rocks found in the upper part of the crust. The first diamond-bearing kimberlite pipe in Yakutia was discovered in 1954 and given the name Zarnitsa. The Mir and Udachnaya pipes were located a year later, followed by Aikhal.

Up to 1938 the total diamond yield of placers in various parts of the Urals was 270–300 crystals. The placer deposits are extensively described by Fersman (1922), *Precious and Coloured Stones in Russia* vol. 1 (in Russian).

China

Occasional reports on the possible occurrence of diamond in China have been noted for more than two centuries but it was not until systematic prospecting began in 1953 in the provinces of Shantung and Hunan that recovery became organized. Orlov (1977) states that the finds in Shantung were in the basins of the rivers I Ho and Shu Ho; in Hunan finds were made in the basin of the Yuan Chiang river, a tributary of the Yangtze. Placer diamond deposits have been found in Yunnan province and on the island of Hainan.

Sorting and Grading

On recovery and after cleaning, diamond crystals, from whatever source, are sorted. Many crystals have no ornamental potential and are classed as industrials. Gem-quality crystals are sorted into categories of size, shape, clarity and colour at various stages before polishing. Details of the sorting process, carried out in different places, can be found in textbooks on the diamond though the closer the student gets to the process of pricing the fewer the details become! Common-sense dictates that the whole sorting process as far as gem-quality crystals are concerned is market-driven: this means that in some circumstances a crystal will achieve a higher grade (and price), sometimes a lower grade. The final classification, again according to colour, size, shape and clarity forms the basis of the pricing for the sights (sales) held by the De Beers orgainsation. Sight-holders are allowed a certain percentage of the type of goods they most want and have to accept some less acceptable material so that an unsaleable stockpile does not form. Not all rough diamonds are sold through De Beers – private enterprise of every kind continues! Previously the sight-holders' list has been shortened.

Readers are referred to textbooks on diamond for further information; the whole sorting and pricing process is never static.

Effects of Light on Diamond

Optical Properties

Absorption Spectra

The absorption spectra of most diamonds may be divided into two main groups. Diamonds of the Cape series, which fluoresce with a blue light, and have a body colour varying from colourless to yellow. In this group the strongest absorption line is at 415 nm, varying in strength with the depth of the body colour and often noted in colourless specimens. Associated with it are lines at 478 nm (often the only one visible), 465, 452, 435 and 423 nm.

Diamonds in the second group have a brown, greenish yellow or green body colour and show a green response to UV. Absorptions include a strong narrow line at 503 nm with weak lines at 537 and 495 nm. Blue and green fluorescing crystals may also show the 415 nm line.

Some colourless, bright yellow and brownish yellow diamonds, with a yellow fluorescent glow under UV, may show no discrete bands at room temperature except for, maybe, a weak line at 415 nm. Blue (type IIb) diamonds absorb slightly in the red which is difficult to observe with the hand spectroscope. Fluorescence is often seen to be banded or sectorized!

Ultraviolet

Under LW the response may be blue, green or yellow, sometimes a reddish glow. The strengths of the response varies from weak to a strong sky-blue. Fluorescence response has been the basis of a number of ingenious methods for recording individual diamonds so that they can be traced in the event of loss.

Diamonds fluorescing a bright blue do invariably, however, show yellow phosphorescence varying in strength with the strength of the blue glow. This combination has been recorded only in diamond. Some pink diamonds show an orange fluorescence with persistent orange phosphorescence. The spectroscope will show a bright line at 575 nm weaker bands at 586, 598 and 618 nm. The bright line at 575 nm is sometimes accompanied by an additional bright line at 537 nm. These bands have been reported from diamonds coloured by atomic bombardment.

Under short-wave ultraviolet light (SWUV) the fluorescent effects are in general similar to those observed in LW but the response is markedly weaker. Under X-rays most diamonds show a rather uniform bluish white glow; the exceptions are those diamonds which show a yellow glow under UV and show a similar glow under X-rays but this is not always so.

Robertson, Fox and Martin (1934) found that some diamonds were more transparent to UV than others. This was made the basis of a classification of diamonds into two types – type I and type II.

Type I diamonds are transparent down to about 300 nm while type II are much more transparent and pass UV down to about 225 nm. In 1952 Custers subdivided the type II diamonds into type IIa and type IIb. The type IIa are said not to phosphoresce when irradiated with SWUV, while the type IIb when similarly irradiated shows a bluish phosphorescence and will also conduct electricity. In these type IIb diamonds sometimes the afterglow is red, and this is so with the famous Hope blue diamond. The electrical effects shown by type IIb diamonds are due to the presence of boron atoms.

During 1959 Kaiser and Bond found the reason for the difference in the absorption of UV to be due to free nitrogen in type I diamonds, and this type has been divided into two sections by Dyer and others in 1965): type Ia in which the nitrogen has been shown to be in groups of two or more atoms, and type Ib in which the nitrogen is dispersed in substitutional sites in a paramagnetic form. Any nitrogen in type II diamonds is in extremely low concentrations. There is some evidence that well-formed diamond crystals depend to some extent on the presence of nitrogen; and, further, it is noticeable that large diamonds which are found do not show any good crystal form and are almost surely type II diamonds.

X-Rays

The transparency of a substance to X-rays is a function of the atomic mass of its constituent elements p. The higher the atomic mass of the elements the less transparent it is to X-rays. Diamond is pure carbon (atomic mass 12) and is notably transparent to X-rays, whilst zircon which has the heavier zirconium atom (atomic mass 91) and silicon (atomic mass 28) and indeed all colourless stone and glass which simulate diamond are much more opaque to X-rays.

Inclusions and Light

To be of the best quality, diamond must be clean (free from conspicuous inclusions). These imperfections may, however, be so small as to be undetectable under a 10× lens (the stone is then said to be flawless, internally flawless or loupe clean) or the flaws may be so obvious so that the beauty of the stone is affected. A single large imperfection in the wrong place in a cut diamond may be reflected from all the back facets so that to the eye the stone may appear full of inclusions.

A most important study of the internal features of diamond has been made by Gübelin who has identified in diamond inclusions of graphite, hematite, magnetite, garnet, diopside and enstatite among others. Diamond crystals can often be seen. Transparent inclusions may appear as black spots when light is totally reflected at their surfaces.

Artificial Colouration

The artificial colouration of diamond is a hot topic today!

History

The production of a green colour in diamond by bombardment with radioactive particles from radium compounds has been known since early in the twentieth century. Owing to the expense of the method and the ease with which such artificial colouration can be detected – for the stones themselves become radioactive from the treatment – 'radium-greened' diamonds are not commonly encountered. Other methods of colouring diamonds by particle bombardment, using high-voltage particle accelerators, have led to the sale of artificially coloured diamonds.

Cyclotrons using protons (a particle in the nucleus of all atoms) and deuterons (the nucleus of the heavy hydrogen atom) produce in diamond a green colour, provided the heat generated by the impact of the particles is dissipated (usually by the use of a jet of liquid helium), otherwise a brown colour is produced. In any case a brown or yellow colour may be produced by subsequently heating the greened stones at a controlled temperature.

A similar type of colouration has been produced by using neutrons (uncharged particles in the nucleus of most atoms) generated in an atomic reactor (atomic pile), and such artificial colouration of diamond is now carried out commercially both in Great Britain and in the USA. The diamonds leave the reactor a pale green in colour which can be altered to brown or yellow by subsequent heating under controlled conditions. Unlike cyclotroned diamonds in which the colour is little more than skin deep, these diamonds are coloured throughout. Electrons can produce a light green or an aquamarine-blue colour in diamonds.

A type of colour alteration sometimes practised, usually for fraudulent purposes, is to paint the rear facets of off-coloured yellowish diamonds in order to make them appear whiter. Scientifically it is well known that when two complementary colours mingle white is produced; it is this effect which is used in the whitening of diamonds. Experiments are not usually carried out on the finest diamonds – Cape stones are usually chosen – because something might go wrong!

Synthetic Moissanite

The synthetic form of the silicon carbide moissanite, SiC, has been manufactured for ornamental and gem use: in the 1960s some iridescent, though
opaque, crystal groups were around at gem and mineral shows but it
was only in recent years that a transparent variety was able to be synthesized, the manufacturer being located in North Carolina, USA. Most of
the properties of diamond are quite well imitated and the usual anxiety
associated with new diamond imitations was reported to be pervading the
trade. Such reports are usually exaggerated – the properties of moissanite
may be detected easily by a gemmologist with simple equipment.

Synthetic moissanite belongs to the hexagonal crystal system and shows
birefringence absent from diamond though doubling of back facet edges
cannot usually be seen through the table facet, which would have been easier to detect, but in a direction at right angles to this. Near-parallel needles
and stringers may be seen at right angles to the table. Some specimens
show rounded facet edges (those of diamond are exceptionally sharp)
which are not in themselves a vital clue. There are also some uni-directional
polishing lines on adjacent facets, which do not occur on diamond.

Gemmological properties are: hardness 9.25, RI 2.648–2.691 with a birefringence of 0.043, uniaxial positive and dispersion 0.104, which is more
than twice as great as diamond. The SG is 3.22 (diamond is 3.52).

These properties can all be tested with a little effort but with any diamond imitation it is usually worth devising a catch-all detector. Reflectivity
meters have usually been used to separate diamond-like transparent stones
from their more serious imitators YAG and CZ though they can only reach
a few stones in a piece of jewellery which contains many small ones in
hard-to-reach places. These lurking 'diamonds' can more often be reached
with the thermal conductivity tester which will very effectively separate
diamond from most of its simulants.

It has been suggested that manufacturers of synthetic moissanite
might be able to alter the RI of their product so that the reflectivity
meters might give a 'diamond' reading. [Reflectivity meters do not measure RI as such but RI does influence reflectivity.] In such a case it might
have been possible to repolish the specimen so that the original RI could
be assessed. Rumours of this kind often circulate through the trade and
those involved should ask themselves 'what might the manufacturer
gain from all this trouble?'

A synthetic near-colourless moissanite has been heat-treated, the treatment causing a brownish colour to develop across all the facets. Cleaning
and hand-polishing the samples, using cerium oxide on leather, restored
the reflectivity to 98% of the non-treated material. If heating forms part of
any testing experiment on a suspected moissanite, the gemmologist should

remember that surface oxidation could occur and keep the level of heating to a minimum. The colour of the surface might undergo alteration.

The thermal conductivity tester will give a 'diamond' reading for synthetic moissanite in any case so that its use may be confined to separating diamond and synthetic moissanite from other diamond simulants: a further test to separate the two could well be magnification.

Diamond will sink and moissanite float in di-iodomethane (SG 3.34).

Some coloured moissanites have been brown, green, yellow and blue but the colours are not very strong. Green specimens that I have seen are not like green diamond. A brown moissanite grown in Russia is reported to have been grown by CVD: the specimen described was opaque.

The first firm to produce synthetic moissanite and sell it (exclusively, at first) was C3 Inc. in North Carolina, USA. The same firm, now called Charles & Colvard, produced a tester, which they sold under the name Tester Model 590. Other instruments have followed. These separate moissanite from diamond by scanning the blue and near-visible UV areas of their absorption spectrum. In moissanite there is an intense region of absorption extending down from about 425 nm to the UV region. Colourless diamond, on the other hand, transmits well down into the UV. A halogen light source directs a beam on to the table facet which reflects it. If it transmits wavelengths from the blue to the UV region the tester gives a visual indication plus a bleep, indicating that the specimen is diamond. If no response is given the specimen will be moissanite, having absorbed this range of wavelengths.

Another instrument, the Presidium moissanite tester, detects the very small current passed by semiconducting materials. The operator receives a signal indicating diamond/not diamond. As most diamonds, apart from type IIb blue diamonds, are not semi-conductors, any current detected may be caused by impurities in the material.

In a paper in *Australian Gemmologist* 20, 483–85, 2000, the authors found that the testing of synthetic moissanite was made easier by the Presidium tester. As synthetic moissanite is a semiconductor, the instrument is able to sense a forward leakage of current. Other testers are based on the 'breakdown voltage' (which overcomes resistance to an insulator causing current to flow).

The authors found that all synthetic moissanite specimens were detected as such by the apparatus, indicating them by the illumination of a bright red window display and a sound alert. Other species caused a green 'Test' lamp to be illuminated. False positive synthetic moissanite readings occurred during the evaluation, particularly when the tip of the probe was in contact with metal (such as the setting) and also when a germanium transistor was touched. A synthetic moissanite response was also given by a black electrically conductive industrial-quality diamond.

The synthesis of diamond is excellently described by Amanda S. Barnard, *The Diamond Formula* (Oxford: Butterworth-Heinemann, 1999) and further notes can be found in two books by Robert M. Hazen *The New Alchemists*, New York, Times Books, 1993, ISBN 0812922751 and *The Diamond Makers*, Cambridge University Press, 1999, ISBN 0521654742.

Another trick is the application of a film of blue or violet dye on the back facets of the stone: the blue or violet and the yellow colour of the stone combine to give a whiter effect. The film is thin enough to escape notice except where it has caught up on the raw edge (the unpolished girdle or setting edge of the stone). As the treatment is usually carried out by using a water-soluble dye, a thorough washing in hot water will dissolve the colour. Should a coloured lacquer be used, solvents such as acetone or amyl acetate, or even acid, may be needed to remove the colour. Similarly pink diamonds have been imitated by painting the back facets with a pink dye or enamel.

Synthesis and Simulation

Synthetic diamond is described in the next chapter.

Imitations of Diamond – Natural Stones

The high lustre and dispersion of zircon would make it an excellent imitation of diamond though it is not very hard and its high birefringence would easily be detected. More difficulty can arise when a coloured natural stone is mistaken for diamond. A search of the descriptions in the relevant chapters may lead readers to find out for themselves which ones are most likely to cause confusion.

Imitations of Diamond – Man-Made Stones

Descriptions of synthetic rutile, strontium titanate, the synthetic garnets YAG, GGG and their analogues, cubic zirconia and synthetic moissanite are described in the next chapter. Composites are described in the next chapter.

Details of testing instruments and how they are used can be found in the 3rd edition of Peter Read's *Gemmology*, 2005 (ISBN 0750664495).

4

Diamond II

Alan T. Collins

Introduction

On average, approximately 1 tonne of rock from a diamond mine must be processed to recover each carat of diamond. In other words, the diamond concentration is about one part in 5 million. Even then, less than 25% of the material is of gem quality. However, once a gem diamond has been found, what a prize it represents! As the hardest known material, it can take a polish unsurpassed by any other gemstone; its transparency in the visible region, its high RI and high dispersion result in a brilliance and a fire that are truly distinctive.

Uniquely, diamond is the only gem material comprised of a single chemical element; pure diamond is made exclusively of carbon. Following the research in the late eighteenth century by Antoine Lavoisier, Smithson Tennant and Humphrey Davy, who, amongst them, first recognized that diamond was simply one of the possible forms of carbon, there followed many attempts to synthesize diamonds in the laboratory from one of the other, less valuable forms of carbon. Some of the researchers who attempted to manufacture diamond concluded that, because natural diamonds were probably produced under geological conditions of high pressure and high temperature (HPHT), such conditions would be necessary to manufacture diamonds in the laboratory. However, it was not until the middle of the twentieth century that the first undisputed synthetic diamonds were produced.

The original process for synthetic diamond produced grit-sized particles suitable for industrial applications, but developments of that process now allow HPHT synthetic diamonds weighing a few carats to be grown on a commercial basis.

Of far greater concern to the gem trade is the fact, made public in 1999, that the colour of some natural diamonds can be enhanced by HPHT processing of natural diamonds. In particular, certain brown diamonds can be converted to near-colourless, or occasionally pink or blue colours. If suitable starting material is available, it is far more lucrative to use HPHT equipment to enhance the colour of natural diamonds (which takes only a few minutes) than to tie it up for days growing large synthetic diamonds.

Within the last few years a new method has been developed for producing gem-quality diamond, using chemical vapour deposition (CVD). This does not require high-pressure presses or extremely high temperatures. Instead, a carbon-containing gas is decomposed at a pressure somewhat below atmospheric pressure in an energetic plasma, and the carbon is deposited as diamond.

This progress in diamond research therefore presents the diamond gem trade with a number of unwelcome developments:

- It is possible to produce gem-quality diamonds by HPHT synthesis, on a commercially viable basis. Depending on the growth process, and subsequent treatments, the diamonds can be grown in a wide range of colours.
- Equipment designed for HPHT synthesis can also be used to enhance the colour (and therefore the selling price) of natural brown diamonds; this appears to be more commercially attractive than using the equipment to grow synthetic diamonds.
- Gem-quality diamonds can be produced by CVD.

At present the quantities of HPHT synthetic diamonds, and of HPHT colour-enhanced diamonds, form a very small fraction of the total market. The production of CVD diamond gems is in its infancy, and the economics of the process are not yet well established. Despite the small numbers of these relatively new forms of diamond, gem testing laboratories will nevertheless be expected to detect such specimens when they turn up. In some cases, a simple visual inspection will suffice, in other cases sophisticated spectroscopic techniques will be required.

Once regarded simply as curiosities, during the last decade coloured diamonds have begun to come into fashion, and to create their own market. HPHT synthetic diamonds can be produced in a range of 'fancy' colours, the HPHT processing of natural brown diamonds brings about a dramatic change in the colour of the diamond and CVD diamond frequently has a brown colour which can be enhanced by HPHT processing. It is therefore vitally important that we have a reasonable understanding of why some diamonds are coloured. In fact, the vast majority of natural diamonds are coloured, which is why a top-quality near-colourless stone can command such a high selling price. In the following section of this chapter, we shall explore the current understanding of the origins of colour in diamond. The HPHT and CVD processes will then be considered, and we will conclude by examining methods by which naturally coloured natural diamonds can be differentiated from other diamonds which are increasingly finding their way into the market.

The Origin of Colour in Diamond

A perfect diamond, made up of a regular array of carbon atoms, and containing no impurities, would be completely colourless. In practice, no diamond is perfect and most diamonds are coloured. Brown and yellow are the most common colours encountered in natural diamonds, although other colours (red, orange, green, blue, violet) are found occasionally. These colours result from the presence of defects – either point defects which have dimensions comparable with the distance between carbon atoms in the diamond crystal or extended defects which may be large enough to see with an optical microscope. The scientific understanding of point defects is more advanced than that for extended defects, but, even so, is far from comprehensive. What is clear, however, is that nitrogen, which is a very common impurity in diamond, is involved in many of the point defects. Nitrogen can exist in several different forms in diamond, which we describe below. The properties of boron in diamond are also mentioned, as are details of one important extended defect. Vacancies can occur naturally in diamond, and can be introduced by radiation to produce 'treated colours'. A few vacancy-nitrogen complexes are also described which have particular relevance to the colour of diamond.

Nitrogen

Nitrogen can be present in diamond in at least three different forms. Each of these causes absorption in the infrared part of the spectrum at frequencies less than 1332 cm^{-1} (wavelengths longer than 7.5 μm). This is known as the 'one-phonon region'. Diamonds that contain sufficient nitrogen to produce absorption are generically known as 'type I'. If there is no easily detectable absorption in this region the diamond is classified as 'type IIa'.

It is believed that nitrogen is initially incorporated into diamonds on isolated substitutional sites. Very few natural diamonds are found with nitrogen in this form, but almost all manufactured diamond produced commercially by HPHT synthesis has most of the nitrogen in the isolated substitutional form. This sort of diamond is referred to as 'type Ib'. The presence of the nitrogen causes optical absorption in the visible region, starting at approximately 500 nm and increasing towards shorter wavelengths. Diamonds like this therefore have a distinctive yellow colour, which is sometimes described as a 'canary yellow'. At higher nitrogen concentrations, or in larger diamonds, the colour is yellow/brown.

Nitrogen as a Donor

To understand some of the colour phenomena in diamond we need to use a concept familiar in the study of semiconductors. Single substitutional nitrogen in diamond acts as an electrical donor. A nitrogen atom may therefore *donate* an electron to a certain defect if the nitrogen and the defect are sufficiently close together. This causes the defect to be in a negative charge state (because an electron has a negative charge), producing a different absorption than if the defect had no electrical charge and so was neutral.

Aggregated Nitrogen

The majority of natural diamonds have spent substantial periods of time (millions of years to perhaps 3000 million years) in the earth's mantle at geological temperatures up to 1200 °C. At this temperature, given sufficient time, the single nitrogen atoms come together to form aggregates. The first aggregate to be produced is the 'A aggregate' which comprises two nitrogen atoms on adjacent lattice sites. The next stage of aggregation leads to the 'B aggregate' in which four nitrogen atoms symmetrically surround a vacancy. (A vacancy is a position in the diamond crystal which would normally contain a carbon atom, but from which the carbon atom has been removed.) The formation of the B aggregate therefore requires a carbon atom to be ejected into an interstitial position (i.e. in between positions normally occupied by carbon atoms in the diamond structure). In the majority of cases these self-interstitials also aggregate to form extended defects called 'platelets'. The precise structure of these platelets is still not known, and they may also include some nitrogen. The platelets give rise to a distinctive absorption peak in the infrared spectrum, at approximately 1365 cm^{-1} (7.3 μm), but do not cause any absorption in the visible region.

Diamonds containing aggregated nitrogen are known generically as 'type Ia'. Those in which most of the nitrogen is present as A aggregates are classified as 'type IaA' and those that have the majority of the nitrogen present as B aggregates are classified as 'type IaB'. Diamonds containing significant concentrations of both the A form and the B form of nitrogen are termed 'type IaAB' or 'type IaA/B'. Neither the A aggregates nor the B aggregates produce any absorption in the visible region, and therefore do not affect the colour. However, when the B aggregate is formed there is another minor product produced which is referred to as the 'N3 centre'. This has three nitrogen atoms on a {111} plane, surrounding a common vacancy. Optical transitions at this centre give rise to an absorption band known as 'N3'. This has a sharp line at 415.2 nm and a structured band to shorter wavelengths. There is, at the N3 centre, another transition which produces the broad 'N2' peak at 478 nm with weaker peaks at shorter wavelengths. (The nomenclature of the N2 and N3 absorption peaks stems from early observations in the 1950s when the naturally occurring absorption lines in type Ia diamonds were listed in order of decreasing wavelength.) Because of the response of the eye, it is predominantly the N2 peak, and the peaks at slightly shorter wavelengths, that are responsible for the colour. Diamonds containing this absorption are known as 'cape stones'. At low concentrations of the N3 centre the colour of the diamond is just perceptibly yellow when compared with a colourless master stone; at progressively higher concentrations the yellow colour becomes increasingly obvious, eventually tending to a fancy colour.

The N3 absorption is one of the most common in diamond, and its presence underlies the system of colour grading. Colourless (D colour) diamonds command the highest price, and as the colour increases through the grades E, F, G, etc., a slightly tinted yellow colour is reached which is regarded as least desirable; then the selling price moves up again as the fancy yellow colour grades are approached.

If it is possible to take a diamond with a colour near the minimum point on this curve and increase the intensity of the colour by some technique, there is clearly a possibility of increasing the selling price. We shall see later two methods by which this may be achieved.

Boron

Natural diamonds in which substitutional boron is the major impurity are extremely rare. These are the type IIb diamonds, and the presence of boron results in optical absorption which starts at a wavelength of approximately 4 μm in the infrared region and extends into the red region of the visible spectrum. In favourable cases this absorption gives the diamond an attractive blue colour.

Extended Defects

We have already encountered platelets earlier in this chapter. Another type of extended defect in diamond is associated with plastic deformation; it appears that in certain stones an external shearing stress has caused some planes of carbon atoms to slip with respect to each other. Many of the brown and pink diamonds from the Argyle production are in this category.

If the crystals are examined under a microscope it is observed that the colour is not uniform, but is striated with the striations oriented in the direction of slip. These striations are referred to as 'coloured graining' in the gem trade. It is not known in detail why the plastic deformation produces coloured diamonds, or why many are brown and some are pink. One possibility is that a point defect is also involved, and this is trapped at, or 'decorates' the dislocation. An alternative explanation has been proposed by Tom Anthony of General Electric (GE): he has suggested that the pink colour is associated with small displacements of the planes of carbon atoms and that larger displacements result in a brown colour.

Type IIa brown diamonds simply exhibit a featureless absorption that rises continuously from the red end to the blue end of the visible spectrum. The brown type Ia diamonds from the Argyle production also have this increasing absorption towards short wavelengths. In addition they generally have some N3 absorption present, and a broad absorption band with a maximum near 560 nm. There may also be a weak absorption band known as 'H3' with a sharp line at 503 nm.

Vacancies

Vacancies can be produced in diamond by radiation. In summary an energetic particle, such as a high-energy electron or a fast neutron, displaces some of the carbon atoms, leaving a vacancy and placing the carbon atoms into interstitial positions. Absorption associated with the vacancy in its neutral charge state produces the GR absorption features. Vacancies are produced by all forms of sufficiently energetic radiation in all types of diamonds and the GR stands for **G**eneral **R**adiation. The absorption spectrum shows a sharp line at 741 nm (the GR1 line) and a band to shorter wavelengths. This absorption gives the diamond a green or blue-green colour. Many uncut natural diamonds have a green 'skin' which has been produced by alpha particles, but this is only a few μm deep. Diamonds with a naturally produced absorption due to vacancies throughout their bulk are extremely rare; the Dresden Green is one example.

Vacancies are also released when plastically deformed diamonds are subjected to HPHT processing, and this process will be considered in more detail in a later section.

Complexes Involving Nitrogen and the Vacancy

When diamonds containing vacancies are heated in the laboratory to 800 °C the vacancies become mobile, and after an hour or so the GR1 absorption band disappears.

In type I diamonds the vacancies are captured by the nitrogen to form new colour centres. The dominant defect in type Ib diamonds is the nitrogen-vacancy centre in the negative charge state $(N-V)^-$ and this produces an absorption system with a sharp line at 637 nm and a band to shorter wavelengths. This gives the diamond a pink colour or, with larger amounts of radiation, a pink/red colour. The reason why most of the $(N-V)$ centres are in the negative charge state is that type Ib diamonds have a high concentration of single-nitrogen atoms which act as donors. Some nitrogen-vacancy centres may be in the neutral charge state $(N-V)^0$ and this defect produces a sharp line at 575 nm. The relative absorption intensity of the 575 nm line to the 637 nm line is higher in diamonds with a low concentration of single nitrogen because the N–V centres are, on average, further away from the nitrogen donors and so remain neutral.

In type IaA diamonds vacancies are trapped by the A aggregates, and a rearrangement of the atoms occurs to produce the structure N–V–N. Research described below has shown that when all the nitrogen is in the A-aggregate form this centre is in the neutral charge state $(N-V-N)^0$, and produces absorption in the H3 system with a sharp line at 503 nm and a band to shorter wavelengths. This absorption produces a yellow, orange or reddish brown colour, depending on the amount of radiation to which the diamond was subjected. (The 'H' series of optical bands are those that are produced by radiation followed by Heat treatment.)

Following irradiation and annealing of type IaB diamond, the vacancies are trapped by the B form of nitrogen to produce H4 centres which give rise to an absorption band with a sharp line at 496 nm and a band to shorter wavelengths. Again, this results in a yellow, orange or reddish brown colour.

Other Colours

There are many other absorption bands in diamond for which there is, as yet, no adequate scientific explanation. These include blue and violet diamonds from the Argyle production, orange diamonds which owe their colour to a broad absorption band with a maximum at 480 nm, and 'chameleon diamonds' which change colour, depending on their environment and/or temperature. The blue and violet diamonds show evidence of a high hydrogen concentration which causes substantial absorption in the near-infrared region, but it has not yet been demonstrated whether the hydrogen is, in fact, responsible for these unusual colours.

Synthetic Diamonds

Many attempts to synthesize diamond were made in the first half of the last century, and by the 1940s the conditions of temperature and pressure under which diamond is the stable phase of carbon were being established experimentally. These results were extended by Berman and Simon. Interestingly, this shows that *graphite* is the stable phase at normal temperatures and pressures. Diamond, the ultimate gemstone, is only meta-stable at room temperature and standard atmospheric pressure! Fortunately, a large amount of energy must be given to a diamond to convert it to graphite and, in an inert atmosphere, diamond can safely be heated to at least 1500 °C without damage. However, once the temperature exceeds 1800 °C there is a rapid transformation to graphite.

Serious attempts at synthesizing diamond began in the 1930s following the availability of tungsten carbide. This extremely tough material allowed pressure cells to be developed which could generate pressures of 70 000 atmospheres at elevated temperature, and up to 400 000 atmospheres at room temperature. Early experiments were carried out by Percy Bridgman at Harvard University. No diamond was formed in any of the room-temperature experiments on graphite, and even when graphite was subjected to a pressure of 30 000 atmospheres and temperatures up to 3000 °C, diamonds still refused to form. Bridgman simply could not sustain sufficiently high pressures and temperatures simultaneously in order to directly convert graphite to diamond. As a result of these failures, Bridgman coined the phrase 'graphite is Nature's best spring'.

The problem of converting graphite to diamond was first solved by a team of scientists working at the Allmänna Svenska Elektriska Aktiebolaget (ASEA) laboratory in Stockholm in 1953. However, that early success was not publicized until after the announcement, on 15 February 1955, by GE that they had transformed graphite to diamond. In both cases, the secret of success was using a molten metal to dissolve the graphite. As more and more graphite is dissolved, the metal becomes saturated with carbon and so small crystals nucleate and begin to grow. Because the temperature and pressure are maintained in the region where diamond is the stable phase the crystals grow as diamonds. It is interesting that this approach had been tried 60 years earlier by the French chemist Frédéric-Henri Moissan, but he had been unable to reach the conditions where diamond is the stable phase of carbon. However, it is almost certain that the GE scientists had Moissan's ideas in mind when they tried, after many previous failures, the technique of dissolving the graphite in a molten metal. It is still a matter of debate whether the metal acts solely as a solvent, or whether there is also some catalytic action, and the description 'solvent-catalyst' is generally used. The transition metals cobalt, nickel and iron work well as solvent-catalysts, and these metals or their alloys are used in most commercial systems. Typical values of temperature and pressure used in commercial synthesis are 1400 °C and 55 000 atmospheres respectively.

Synthetic Diamond Grit

There are basically two different methods for growing synthetic diamond, depending on the final size required. To produce diamond grit, with grain sizes up to about 1 mm, graphite and the metal are mixed together and placed in a growth capsule, which has electrically insulating walls. The capsule is placed at the center of a high-pressure press, and an electric current is passed through the graphite/metal mixture to raise the temperature above the melting point of the metal (this is referred to as 'direct heating'). Once nucleation occurs, carbon flux is transported across a thin film of molten metal between the graphite and the growing diamond. After a few tens of minutes, a substantial fraction of the graphite is converted to diamond. The diamond grit is recovered by dissolving the metal in acid.

When GE first announced this process, it was an extraordinary technological breakthrough; today, 50 years later, approximately 200 tonnes (1000 million carats) of synthetic diamond grit is produced each year for industrial cutting and grinding applications.

Diamonds produced in the manner described have an attractive yellow colour because of the presence of single nitrogen atoms. The graphite, metal solvent-catalyst and the slightly porous material, from which the growth capsule is fabricated, all absorb nitrogen from the atmosphere, and the synthetic diamonds typically contain a few hundred parts per million of single nitrogen.

Research in the 1960s by GE showed that *colourless* synthetic diamond grit could be made if the nitrogen in the growth capsule were removed using 'nitrogen getters'. These are materials like aluminium or titanium which combine with the nitrogen and prevent it from being incorporated in the diamond. They also showed that blue diamonds could be grown if the nitrogen were removed and a small quantity of boron was added.

Gem-Quality Synthetic Diamonds

In 1970 GE announced a modified growth process which enabled them to produce large synthetic diamonds by the HPHT process. This technique is known as the 'temperature gradient method'. With this approach, the carbon source is placed near the middle of the growth capsule, and one or more synthetic diamond seeds are placed on a 'seed pad' at the bottom of the capsule. A substantial layer of the metal solvent-catalyst is placed between the carbon source and the seeds. The temperature of the reaction volume is raised by a graphite heater sleeve surrounding the growth capsule (this is referred to as 'indirect heating'). The bottom of the capsule is slightly cooler than the centre, and so there is a temperature gradient between the carbon source and the seeds. When the metal is melted, carbon dissolves at the central region of the capsule and moves down towards the seeds. Because the seed region is slightly cooler, diamond crystals begin to grow. As before, this growth capsule is at the centre of a high-pressure press with about 55 000 atmospheres pressure applied. The process is sometimes referred to as the 'reconstitution method' because often the carbon source is synthetic diamond grit, which is relatively inexpensive. If graphite is used, there is a substantial reduction in the volume as the graphite is converted to diamond; however, if diamond grit is converted to large diamond crystals there is no significant change in the volume of the growth capsule, and it is much easier to control and maintain the conditions necessary for high-quality growth.

If the growth process is extended to a few days, rather than a few minutes, large synthetic diamonds can be produced. The colour of the crystals is normally yellow to brown, again because of the presence of single nitrogen. However, as with the diamond grit, if the nitrogen is gettered, colourless crystals can be produced, and blue diamonds can be grown using nitrogen getters and adding boron.

The high-pressure presses (the so-called 'belt' presses) developed by GE, De Beers and others were very costly, and in the 1970s it was uneconomic to grow gem-quality diamonds by the HPHT method. More recently, Russian researchers have developed the BARS[1] apparatus, and this has been re-engineered by companies in the US, making it suitable for mass production. These presses have a small heated volume, and can only produce one diamond per growth run. Nevertheless, following these developments, a number of groups have set up, and are expanding, factories with dozens of BARS machines, each capable of turning out a gem-quality diamond every few days.

Colourless diamonds are difficult to produce; the presence of small concentrations of nitrogen improves the crystalline properties of the diamonds and also results in a faster growth rate. For that reason, most of the current production of gem-quality synthetic diamonds is fancy yellow. Such diamonds can also be given a pink colour, using radiation and annealing.

At present the annual production of synthetic diamond gemstones is only a few thousand carats, compared with 120 million carats of natural diamond mined each year. At first sight, the production rate of synthetic diamond gems may seem insignificant. However, amongst the natural diamonds there is perhaps only one fancy coloured diamond for every 10 000 near-colourless ones. Consequently, in future when a fancy yellow diamond, with a weight of up to 1.5 ct, or so, is encountered in the gem trade, there will be a high probability that it is synthetic.

Colour Enhancement by HPHT Annealing

The selling price of yellow diamonds varies with the depth of colouration. There is a similar curve for brown diamonds. For diamonds equal in all other respects, as the colour goes from very pale brown to a more noticeable brown the selling price drops, and for some attractive pronounced brown colours like 'champagne' and 'cognac', it may increase again. As before, the question arises: is it possible to take a cheap brown diamond, and improve its colour in some way, thereby enhancing its selling price? In particular, is it possible to take a brown diamond and remove the colour, converting the stone to a near-colourless D or E colour?

The answers to these questions again involved research by GE. In March 1999, the diamond gem trade was staggered by an announcement from Pegasus Overseas Ltd (POL) of Antwerp, Belgium, a subsidiary of Lazare Kaplan International (LKI), that they had begun marketing colour-enhanced diamonds. The diamonds had been processed by GE, and it was generally assumed, and subsequently confirmed, that the process involved treatment at high pressure and high temperature. Information circulating in the trade indicated that the starting material was brown type IIa diamond and that the GE process substantially reduced the brown colour.

The acronym GEPOL was soon attached to these colour-enhanced diamonds. Shortly after the initial announcement, the diamonds were marketed in the USA under the trade name 'Bellataire'. In March 2004, Littlejohn and Co., a private equity firm, announced the completion of its acquisition of the GE Superabrasives division from the GE Company. The company was renamed 'Diamond Innovations', and the transaction also included the gem technology behind the Bellataire diamond brand.

Prior to the revelation by GE in 1999, it had generally been thought that permanent colour changes to natural diamonds (produced by, for example, radiation and annealing) can only add to the pre-existing colour. Clearly, that is no longer the case. Later in the same year, Novatek, which manufactures high-pressure presses used in diamond synthesis, announced that they were also processing natural brown diamonds. Novatek set up a subsidiary company, NovaDiamond Inc., to produce and market these diamonds. NovaDiamond did not pre-select the diamond type, and many of the stones they processed changed from brown to yellow/green.

Although these announcements were made in 1999, there was evidence a year or so earlier that similar colour changes were being produced by laboratories in Russia. In retrospect, questions being raised in the mid-1990s about colour enhancement of natural brown diamonds probably indicate that these activities were already underway at that time.

In order to predict the colour change that will be produced in a brown diamond by HPHT processing, it is necessary to carefully characterize the stone. NovaDiamond did not have the facilities to do this, and are no longer producing significant quantities of colour-enhanced diamonds. However, in 2003, Sundance Diamonds, a division of U.S. Synthetic, announced the launch of its HPHT diamond-processing service for the diamond industry. (U.S. Synthetic is the world's largest manufacturer of synthetic diamond components for use in oil and gas drilling.) To complement the HPHT processing, Sundance has set up a characterization laboratory to determine the suitability of stones for HPHT processing and to make a preliminary analysis of the colour change to be expected. They have demonstrated the capability of changing brown and cape diamonds to fancy colours, including yellow, orange and green, and of changing type II diamonds from unmarketable colours to colourless or near-colourless.

For any company that has invested in HPHT equipment, the annealing of natural brown diamonds (which only takes a few minutes) can be much more profitable than growing gem-quality synthetic diamonds (which takes a few days). It is therefore not surprising that another manufacturer of synthetic diamond, Iljin of South Korea, announced in the summer of 2003 that they will be processing type I diamonds at HPHT conditions to produce fancy yellow, orange and greenstones. These diamonds are being marketed under the trade name 'Nouv'.

The Colour Change Processes

One of the problems in trying to understand why HPHT annealing reduces the brown colour of type IIa diamonds, and changes type I brown diamonds to yellow or green colours, is that the reason for the brown colour is still open to question. It certainly seems to be the case that, for naturally coloured stones, all brown diamonds (and, indeed, all pink diamonds) show evidence of plastic deformation. However, not all plastically deformed diamonds are brown or pink — some are near-colourless. During the last five years the colour change processes produced by HPHT annealing have been studied extensively. The findings have allowed at least a partial explanation of the phenomena to be put forward.

Plastically deformed diamonds, containing some nitrogen in the form of B aggregates, frequently exhibit yellow luminescence which originates from the slip planes. Following HPHT annealing the intensity of this luminescence is greatly reduced, showing that changes are taking place at the slip planes. This treatment also reduces the absorption causing the brown colour. In addition, there is evidence that vacancies are released, presumably from the vicinities of the slip planes, as these processes occur.

If the HPHT processing is carried out at approximately 1800 °C, the vacancies are trapped by the A and B aggregates of nitrogen to form H3 and H4 centres, respectively. Previous research has shown that H4 centres are less stable than H3 centres. If a diamond containing H4 centres is heated to 1700 °C, the H4 centres (which comprise 4 nitrogen atoms and 2 vacancies) dissociate, and some H3 centres (2 nitrogen atoms and 1 vacancy) are formed. Consequently, in the HPHT annealing process, the end result is to produce H3 centres, regardless of whether the diamond is type IaA or type IaB. Absorption by the H3 centres gives the diamonds a yellow colour. The overall effect of HPHT processing at 1800 °C is therefore to reduce the brown colour and to increase the yellow colour.

The aggregation of single nitrogen to produce nitrogen pairs (A aggregates) was discussed earlier. Research shows that this is, in fact, a reversible process:

$$N + N \rightleftharpoons A$$

If the HPHT annealing process is carried out at rather higher temperatures, say 2300 °C, on a plastically deformed diamond containing nitrogen in the A form, then some A aggregates will dissociate to form single nitrogen atoms. This affects the colour in two important ways. The absorption produced by the single nitrogen adds to the yellow colour; in addition, the presence of nitrogen donors results in some of the H3 centres acquiring a negative charge to become H2 centres, i.e.:

$$(N\text{–}V\text{–}N)^0 + N^0 \rightarrow (N\text{–}V\text{–}N)^- + N^+$$
$$\quad\;\; H3 \qquad\qquad\qquad H2$$

The H2 absorption begins with a sharp line at 986 nm and extends into the red end of the visible spectrum; together with the H3 absorption at the blue end of the spectrum, this results in the diamond having a green colour.

Many of the HPHT-processed diamonds appear to have a green component to their colour, even when no H2 absorption is present. In these cases the green colour is due to luminescence. Like many optical centres in diamond, H3 exhibits absorption at wavelengths less than that of its characteristic sharp line, and luminescence at wavelengths longer than that of the line. In the case of H3 the line is at 503 nm, and so the diamond absorbs blue wavelengths and emits green wavelengths. The luminescence is strongest in diamonds containing a low concentration of the A form of nitrogen, and many natural brown diamonds fall into that category.

Moving now to the type IIa diamonds which contain negligible concentrations of nitrogen, we may assume that the vacancies produced during the HPHT processing disappear by annihilation with the interstitials. Consequently, no new colour centres are formed and the end result is a diamond in which the brown colour is greatly reduced. This is the situation which GE and Lazare Kaplan first addressed, treating mainly brown type IIa diamonds. The Gemological Institute of America also examined 11 pink and 4 blue diamonds that had been colour-enhanced by GE. These results can be explained if we recognize that occasionally the starting material would have been pink or blue, if the brown colour been absent. Presumably, once such diamonds have been subjected to HPHT processing, the brown colour is greatly reduced and the underlying pink or blue colour is then dominant. It is certainly the case that some naturally coloured pink diamonds are type IIa, and we have noted earlier that type II diamonds that contain boron as the major impurity can be blue.

Finally, we consider the HPHT processing of cape diamonds which are not plastically deformed, and so do not have a brown component to their colour. If such diamonds are heated to quite high temperatures (up to 2300 °C), some of the A aggregates of nitrogen will dissociate to produce single nitrogen. However, because there are no dislocations, no vacancies are released, and so no H2 or H3 absorption is produced. With careful control of the heating time and temperature an unattractive tinted yellow diamond can therefore be transformed to a fancy canary yellow.

Artificial Colouring Using Combined Techniques

In an earlier section, we saw that the irradiation of a type Ia diamond, followed by heating at 800 °C, resulted in diamonds having a yellow or orange colour. This treatment is easily detected because an absorption line is also produced between 594 and 595 nm; this absorption is very rarely seen in a faceted natural fancy-coloured diamond. With a type Ib diamond, either natural or synthetic HPHT, irradiation followed by annealing produces a treated pink diamond.

The technology now exists to combine the traditional 'treatment' of diamonds, using radiation and annealing, with the HPHT annealing techniques described in the previous section. We will consider two examples, although other variations are doubtless being explored.

Radiation Followed by High Temperature (c.1400 °C) Annealing

It has been known for many years that if a treated yellow diamond, which shows absorption at 594 nm, is heated further to approximately 1100 °C, the 594 nm line is destroyed without significantly changing the colour of the diamond. The absorption spectrum, in the visible region, of a type IaA diamond would then be dominated by the H3 absorption band. Natural fancy-coloured diamonds are found occasionally which have a visible absorption spectrum which is very similar. It is therefore difficult to decide from the visible absorption spectrum alone whether such diamonds are naturally or artificially coloured. However, if the absorption measurements are extended into the infrared region, it is observed that, as the 594 nm absorption in a treated diamond is destroyed, an absorption line, known as H1b at 2024 nm, is produced. More recent research has shown that, if the annealing temperature is increased still further, to approximately 1400 °C, the H1b line is destroyed and, as this decreases in intensity there is a growth of the H2 band, mentioned previously. This combined treatment can give the diamond an attractive green colour.

HPHT Processing Followed by Radiation and Annealing

Earlier we saw that a tinted yellow (cape) diamond could be given a fancy canary yellow colour by HPHT processing. This colour is due to the production of single nitrogen atoms. Following this processing, it is possible to subject the diamond to irradiation and annealing at 800 °C. The single nitrogen atoms capture vacancies far more efficiently than the A and B forms of nitrogen, and consequently an appreciable concentration of $(N–V)^-$ centres is formed, giving the diamond a pink or mauve colour.

Chemical Vapour Deposition Diamonds

Earlier we examined the growth of gem-quality diamond using HPHT synthesis. Japanese researchers at the National Institute for Research in Inorganic Materials (NIRIM), Tsukuba, in an extension of earlier experiments carried out by John Angus in the USA and Boris Spitsyn in Russia, showed that it was possible to grow diamond at a pressure of less than 1 atmosphere by a process of CVD. In this procedure, a carbon-containing gas (usually methane), mixed with a large amount of hydrogen, was passed into a quartz tube, and a pump maintained the pressure at about 5% of atmospheric pressure. Microwave power fed into the tube created a plasma and also heated a substrate to approximately 800 °C. In the microwave plasma the hydrogen molecules (H_2) are dissociated into hydrogen atoms (H) which are extremely reactive, and the methane dissociates into carbon and hydrogen. The carbon is subsequently deposited on the substrate, mostly as graphite, but some carbon atoms bond together in the diamond crystal structure. The atomic hydrogen in the plasma removes graphite very readily but does not react with the diamond. The net result is that a thin film of diamond remains on the substrate.

Research on the growth of CVD diamond began at NIRIM in 1974, and an English-language publication from that institute in 1981 was the catalyst for a huge worldwide explosion of research activity in this field. In the pioneering work, the diamond films were typically no more than 25 μm thick, and, at that thickness, were generally brown or completely opaque to visible light. The diamond films were generally grown on substrates such as flat sheets of silicon or molybdenum, and were polycrystalline. The brown colour of these early films is generally attributed to non-diamond carbon at the grain boundaries. As research in this area continued, ways to speed up the growth rate, and to minimize the non-diamond carbon content of the CVD layers, were developed, and De Beers, for example, demonstrated the ability to produce transparent wafers of CVD diamond more than 100 mm in diameter and with thicknesses of at least 630 μm. This material was suitable for a number of industrial applications, but being polycrystalline, had a hazy transparency, because of light-scattering at the grain boundaries, and so was not suitable for the production of diamond gemstones.

The most recent research into the growth of CVD diamond has been concerned with thick layers deposited on a diamond substrate. If carefully controlled, this homo-epitaxial process can produce single-crystal specimens of CVD diamond several millimetres thick. Most of the detailed studies of CVD single-crystal diamonds have been carried out by the De Beers Diamond Trading Company. They have examined the optical properties of more than 1000 specimens grown by Element Six (formerly De Beers Industrial Diamond Division). Using high-purity gases and a well-designed growth chamber, it is possible to grow diamonds with no obvious colour and with electronic properties that exceed those of the best natural diamonds. As with HPHT synthetic diamond, the growth rate is somewhat higher in the presence of nitrogen. However, the addition of nitrogen to the growth chamber results in CVD diamond that has a brown colour and, for diamonds of the same thickness, the saturation of the brown colour is correlated with the concentration of the nitrogen in the process gases. This brown colour may be due to the presence of non-diamond carbon, and in some specimens can be removed by HPHT annealing. If, instead of nitrogen, a small concentration of a boron-containing gas (diborane B_2H_6) is added to the gas stream, the resulting CVD material is blue in colour.

Some material is lost when the CVD layer is separated from the diamond substrate, but the thickest layers have allowed round brilliants up to 0.82 ct to be fashioned. However, most CVD diamond gemstones are prepared using rather flat cuts to make better use of the full area of the material.

At present there is no commercial production of CVD diamond. However, Ricardo Sussmann, former head of the De Beers CVD diamond facility, believes that with the right know-how, and an adequate investment, it would be possible to produce CVD diamond gemstones on a commercial basis.

Detecting Treated, Synthetic, Colour-Enhanced and CVD Diamond

At the present time large numbers of natural diamonds are treated, using radiation, on its own, or followed by heating at approximately 800 °C to produce green, yellow and orange diamonds, and possibly a few pink stones. Small quantities of HPHT synthetic diamonds are appearing in the diamond trade; most of these are fancy yellow and some are irradiated and annealed to generate fancy pink colours. Colourless and blue HPHT synthetic diamonds are more difficult to manufacture, and are seen in only very small numbers. Significant numbers of natural brown diamonds are being colour-enhanced to produce near-colourless stones, from type IIa starting material, and fancy yellow and fancy yellow/green colours, from type Ia starting material. Because fancy colours are rare, the latter would immediately arouse suspicion, but of far greater concern is that any near-colourless diamond has potentially been colour-enhanced. At present there is no commercial production of CVD gem-quality diamond, but it is technically possible to produce near-colourless and fancy blue stones by this process. Perhaps in the future such material will be encountered in the diamond trade. In this section the range of colours that may be encountered is summarized, and methods are outlined for deciding whether a diamond is natural or synthetic and whether a colour is natural or artificial. In some cases reaching such a decision is straightforward, and in others it is extremely difficult.

Visual Observation

Important clues about the type of diamond, and whether the colour is natural can often be obtained by visual inspection with a 10× loupe. We consider two important examples in the sections below. Visual observation of the distribution of luminescence may also be valuable, and is considered later.

HPHT Synthetic Diamond

Impurities are incorporated in very different concentrations in the different growth sectors of HPHT synthetic diamond, and this is certainly the case with nitrogen and with boron which cause the yellow and blue colours, respectively. The uneven distribution of these colours is therefore a strong indicator that the diamond is not naturally coloured.

HPHT-Processed Natural Brown Diamonds

In the HPHT colour enhancement process, temperatures around 2300 °C may be used. Very few of the standard HPHT presses, designed for diamond synthesis at lower temperatures (c.1400 °C) are able to reach the necessary pressures where diamond is the stable phase of carbon at this temperature. Although the diamonds only experience these conditions for a few minutes, the areas of diamond around inclusions, cracks and feathers can exhibit severe graphitization. Observation of such regions of graphitization would reinforce the conclusion that the diamond had been subjected to HPHT processing. Surfaces of faceted stones are badly etched during processing, and any residual etching on the repolished stone would also suggest that it has been processed. However, it is possible that some facilities can heat diamonds to the required temperature and maintain the pressure in the diamond-stable region; the absence of graphitization does not therefore indicate that the colour is natural.

Red and Pink Diamonds

Red and pink diamonds can be produced by irradiation and annealing of type Ib diamond – either natural or synthetic. The red colour is an extreme version of the treated pink, and results from giving a larger dose of irradiation than would be used to produce a pink stone. The optical band responsible for the colour is that due to the $(N-V)^-$ centre with a sharp line at 637 nm and a structured band to shorter wavelengths. This absorption band is never seen in natural diamonds in sufficient strength to cause a pink or red colour, and so these treated pink stones are readily identified. In synthetic diamond the pink colour is uneven, because of the different nitrogen concentrations in the different growth sectors, but this is not easy to see in a strongly coloured stone. The 637 nm absorption can also be produced by HPHT processing of a type Ia diamond at sufficiently high temperatures to generate single nitrogen, followed by irradiation and annealing at 800 °C. It may be possible to see signs of the HPHT treatment from graphitized cracks and inclusions.

Orange Diamonds and Yellow Diamonds

Yellow diamonds can be produced by radiation and annealing to approximately 800 °C of type IaA diamonds, by HPHT processing of brown type Ia diamonds (to rather lower temperatures than those required to create yellow/green specimens) and by HPHT processing of near-colourless or pale cape yellow type Ia diamonds.

Radiation and Annealing

Radiation and annealing of type Ia diamonds produces the H3 and H4 absorption bands which are responsible for the yellow colour. If a heavy irradiation is used, the diamond becomes almost opaque at wavelengths less than approximately 500 nm, and may then have an orange or even reddish brown colour. The treatment also produces a sharp absorption line at approximately 595 nm. This line can be eliminated by annealing the diamond to approximately 1000 °C, but, as the 595 nm line anneals out, two further absorption lines in the near infrared spectrum are created, known as H1b (2024 nm) and H1c (1934 nm). It is exceedingly rare to find any of these three absorption lines, or the H4 line, occurring naturally in diamond, and it is sensible to regard as treated a stone in which the 595 nm, H1b or H1c lines are found. The H1b and H1c lines can be eliminated if the diamond is annealed further to approximately 1400 °C, but then the H2 absorption is produced, giving the diamond a green hue, and again providing evidence of treatment.

HPHT-Processing of Type Ia Brown Diamonds

The yellow diamonds produced by HPHT processing of brown type Ia diamonds generally still have some brown component remaining, and, in particular, the board band at approximately 560 nm may still be present. In addition, the stones exhibit the coloured graining characteristic of the starting material and may have graphitized inclusions. Intense absorption in the H3 band is extremely rare in naturally coloured diamonds, and it is this feature that would first arouse suspicion.

HPHT Processing of Type Ia Pale Yellow Diamonds

The absorption spectrum produced by HPHT processing of near-colourless or pale cape yellow stones is virtually identical to that observed in type Ib diamonds, and, if the processing conditions are optimized, a colour similar to that of natural 'canary yellow' diamond can be obtained. Unless a processed diamond contains other clues, such as graphitized cracks or inclusions, it is therefore difficult at present to be sure that the colour has been created artificially.

Green Diamonds

Green diamonds can be produced using radiation damage without annealing, radiation damage of type Ia diamonds followed by prolonged annealing at 1400 °C, and by the HPHT processing of brown type Ia diamonds using very high temperatures. In the first case the colour is caused by the GR1 absorption, and in the latter two cases the colour is caused by the combination of H2 and H3 absorption.

Irradiated Diamonds

Very few diamonds owe their green colour to naturally occurring GR1 absorption. The Dresden Green is one, but the intensity of the band is extremely low; at the peak of the band near 625 nm the absorption coefficient is no more than 0.2 cm^{-1} above the background. The diamond only appears to have a green tint by virtue of its size (40.7 ct). The shape of the absorption spectrum of the Dresden Green differs in no perceptible way from the absorption spectra of diamonds that have been artificially coloured using radiation. Consequently gem testing laboratories are reluctant to issue origin-of-colour certificates for green diamonds containing the GR1 absorption. However, the author is of the opinion that the Dresden Green, or some other diamond known to have GR1 absorption through natural processes, should be used as a benchmark, and that any diamond which exhibited a large GR1 absorption coefficient at 625 nm, compared with the benchmark, should be regarded as a 'treated green'.

HPHT-Processed Type Ia Brown Diamonds

To the author's knowledge, H2 absorption does not occur in natural diamonds to a sufficient extent to produce a green colouration. Any diamond that is coloured green or yellow/green because of the presence of H2 absorption should therefore be treated as suspect. The natural brown type Ia diamonds that successfully respond to HPHT processing frequently have low concentrations of nitrogen. Consequently, after processing, a green diamond will exhibit pronounced green (H3) luminescence on excitation with long-wave ultraviolet (LWUV); this is another indication. Before processing, such diamonds often exhibit brown graining, and this graining is evident in the green and yellow/green colouring of the end product, and in the distribution of the luminescence, providing further confirmation of colour enhancement.

Blue Diamonds

Irradiation of diamonds with electrons results in a blue or blue/green colour because of absorption in the GR1 band. Blue diamonds can also be grown by HPHT synthesis and by CVD. Blue colour in natural diamond is rarely, if ever, associated with GR1 absorption. The intense blue colours produced by boron doping of HPHT synthetic and CVD diamond are also unlike the relatively weak blue colour of natural type IIb diamond. In HPHT diamond the boron is taken up in different concentrations in the different growth sectors, and the uneven blue colour is easily noticed. It is therefore relatively straightforward to differentiate a naturally coloured natural blue diamond. In practice very few blue HPHT gem-quality diamonds are produced and there is currently no commercial production of gem-quality CVD diamond.

Colourless and Near-Colourless Diamonds

Following the GE/LKI revelation, every colourless or near-colourless diamond is potentially a type IIa diamond that has been colour-enhanced by HPHT processing of brown starting material. The first step in investigating such a stone must be to determine whether it is type II material. A simple procedure is to check the UV transmission at 254 nm. Diamonds that contain nitrogen in predominantly the B aggregate form are also transparent at this wavelength, and a specimen that is transparent at 254 nm must subsequently be examined spectroscopically in the defect-induced one-phonon region to check whether it is type IIa or type IaB. If the diamond is type IIa, further analysis will be required to assess whether it has been colour-enhanced. These analyses will involve both absorption and luminescence measurements. It is also possible to grow near-colourless diamonds by HPHT synthesis. The detection methods for this rarely encountered material are discussed in a later section.

Absorption Spectroscopy

Some nominally type IIa diamonds contain small concentrations of nitrogen, often predominantly in the B-form, which can be detected by very careful infrared spectroscopy. It may also be possible in some stones to detect an absorption feature known as N9, at 236.0 nm. If such a diamond is subjected to the HPHT conditions used to remove the brown colouration, some of the aggregated nitrogen decomposes to form single nitrogen, and it may be possible to detect this spectroscopically. The presence of single nitrogen gives a very faint yellow colour to the diamond, and therefore, to produce the best colour grades, groups producing colour-enhanced type IIa diamonds are selecting the starting material, using infrared spectroscopy, to have a very low nitrogen concentration. Absorption spectroscopy is therefore not likely to provide definitive results for such stones.

Luminescence Spectroscopy

Many type IIa diamonds contain defect centres involving nitrogen and vacancies, but at much too low a concentration to detect using absorption spectroscopy. The presence of these centres, in particular, N3 H4, H3, 575 nm [i.e. $(N–V)^0$] and 637 nm [i.e. $(N–V)^-$], can, however, be demonstrated using laser-excited luminescence spectroscopy. Significant luminescence is only observed at longer wavelengths than the wavelength of the laser, and the relative intensities of features in the luminescence spectrum will depend on the wavelength used for excitation. The laser also produces a 'Raman spectrum' which comprises a first-order line shifted by 1332.5 cm^{-1} from the laser line and a second-order band with shifts from the laser line of between 1900 and 2665 cm^{-1}.

A number of gem-testing laboratories have examined the luminescence spectra of brown type IIa diamonds before and after HPHT processing, and compared these with the spectra of natural untreated near-colourless type IIa diamonds. There are a number of indicators which, when taken with other evidence, can detect the majority of colour-enhanced specimens:

- In untreated diamonds the intensity of the 575 nm line is normally larger than the intensity of the 637 nm line, if these lines are present, whereas the converse is true in the colour-enhanced diamonds. The explanation is straightforward: the 575 nm transition comes from the $(N–V)^0$ centre, and the 637 nm transition comes from the $(N–V)^-$ centre. In the untreated diamonds there is a negligible concentration of single nitrogen (which acts as a donor) and so the neutral $(N–V)^0$ centre dominates. During the HPHT processing, some of the aggregated nitrogen (which is present in very small concentration) dissociates, providing single nitrogen donors which convert some of the $(N–V)^0$ to $(N–V)^-$. Occasionally an untreated near-colourless type IIa diamond is encountered in which the intensity of the 637 nm line exceeds that of the 575 nm line, so this criterion must be used with caution.
- In colour-enhanced diamonds the width of the 637 nm peak is frequently greater than the width of that peak in untreated diamonds (if present). The reason is that the colour-enhanced diamonds, which were originally brown, are plastically deformed. The random strain which results from this deformation can broaden the luminescence line to a greater extent than occurs in a naturally near-colourless diamond.

- Many sharp luminescence lines observed in some untreated type IIa diamonds disappear following HPHT annealing, but no new lines are produced. A few of the lines which disappear on HPHT processing are quite well defined, and consequently a diamond in which they *are* observed could, with reasonable confidence, be certificated as a natural colour. However, it would be dangerous to assume that a diamond in which they are absent has been colour-enhanced.
- Depending on the time and temperature used in the HPHT processing, some colour-enhanced diamonds show no luminescence when excited with a 514 nm laser. None of the untreated colourless diamonds examined to date shows a complete absence of luminescence. The absence of luminescence may therefore be an indicator of HPHT processing.

Although none of these tests is definitive, taken on their own, they do allow an accurate decision to be made in the majority of cases when combined with other observations (for example, the presence of graphitized inclusions, graining, etc.).

Instruments for Assessing Gem Diamonds

In order to examine and evaluate diamonds rapidly, DTC has developed three screening instruments – the DiamondSure™, the DiamondView™ and, more recently, the DiamondPLus™.

The DiamondSure

The DiamondSure measures the abosorption spectrum of a faceted gem diamond. The majority of natural near-colourless diamonds are type Ia, and absorb strongly in the UV spectrum at wavelengths less than approximately 330 nm. Many also contain sufficient N3 centres to be detectable in a room-temperature absorption measurement. Near-colourless type IIa diamonds, on the other hand, have neither of these absorption systems present. In the case of fancy yellow diamonds, a natural cape diamond would exhibit strong N3 absorption whereas a synthetic yellow diamond has a quite different absorption spectrum. Using these, and similar criteria, the DiamondSure will specifically identify about 98% of natural diamonds and refer all synthetic diamonds for further tests. About 2% of natural diamonds will also be referred for further tests. The instrument will explicitly identify type IIa diamonds (natural, synthetic HPHT and synthetic CVD) as well as synthetic Moissanite (a diamond simulant).

Type IIa diamonds are very rare, and, in a series of tests on 550 000 polished natural diamonds, DTC found that 98% 'passed' and could confidently be identified as natural without any further measurements being performed.

The DiamondView

The DiamondView instrument generates a luminescence image from the surface of a diamond by illuminating the stone with UV with wavelengths less than 225 nm. An image is also generated 0.1 s after the UV source has been turned off to check whether the diamond displays long-lived phosphorescence.

The standard screening sequence is that diamonds that have not been passed by the DiamondSure are examined using the DiamondView. Natural type IIa diamonds exhibit a blue luminescence that originates from a network of dislocations. HPHT synthetic diamonds that contain some nitrogen usually display green, yellow and blue luminescence, in distinct geometric growth patterns, from the characteristic growth sectors. Type IIa HPHT synthetic diamonds have predominantly blue emission, and show strong phosphorescence that is very rarely seen in natural specimens.

The majority of CVD diamond, although classified as type IIa, generally contains small concentrations of single substitutional nitrogen. Vacancies are introduced as part of the CVD growth process, and so these specimens contain sufficient concentrations of $(N–V)^-$ and $(N–V)^0$ centres to produce a strong orange luminescence, frequently with characteristic striations. It is technically possible to produce high-purity CVD diamond which shows very weak blue luminescence; however, this does not originate from dislocation networks as in natural type IIa diamond, but from dislocation bundles which have a different appearance. CVD diamond of this quality is, in any case, unlikely to find its way into the gem market; the absence of nitrogen makes the growth rate very low, and the resulting diamond is prohibitively expensive for gem applications.

The DiamondView, then, confirms the identification of yellow HPHT synthetic diamonds; it also allows differentiation of the HPHT synthetic type IIa diamond, the CVD synthetic type IIa diamond and the natural type IIa diamond. There is, today, a very real possibility that some near-colourless diamonds, identified as natural type IIa, were originally brown and have been colour-enhanced by HPHT annealing.

The DiamondPLus

To identify colour-enhanced type IIa diamonds, DTC have recently developed the DiamondPLus. This instrument measures the photoluminescence (PL) spectrum of the diamond, using excitation with more than one laser. In order that the PL lines are sufficiently sharp to be identified, it is necessary to carry out the measurement with the diamond cooled using liquid nitrogen. In tests carried out by DTC the DiamondPLus successfully identified all those type IIa diamonds that had been colour-enhanced, but it also 'referred' approximately 30% of natural type IIa diamonds that had not been treated. These referred diamonds have to be subjected to more sophisticated tests and careful visual inspection. Some of the criteria used in the operation of the DiamondPLus, and in any further measurements that need to be performed, have been outlined in a previous section, and that section also outlined the limitations of the analyses. These limitations explain why, at the conclusion of all the tests, there is a possibility that an occasional untreated diamond will be wrongly categorized as colour-enhanced.

The DiamondPLus has been specifically engineered by DTC to examine polished diamonds, and to give a result within about 30 seconds. There are other instruments on the market for measuring Raman and PL spectra which can yield the same information, but take rather longer to obtain a result.

Summary

During the last 15–20 years there has been increasing interest in the diamond gem trade in fancy coloured diamonds. This period has also seen the development of commercially viable production techniques for HPHT synthetic diamonds, and the introduction of the HPHT process for the colour enhancement of brown diamonds. The growth of gem-quality diamonds by CVD has also been demonstrated, and this process may become commercially viable in the future.

To retain consumer confidence in the diamond trade it is important to be able to differentiate between natural, naturally coloured diamond, natural treated-colour or colour-enhanced diamonds and diamonds produced by HPHT synthesis or CVD. This chapter has summarized recent developments, and covered the current understanding of many of the colour centres in diamond. This understanding has led to a methodology for examining diamonds spectroscopically and, in the majority of cases, it is possible to determine the diamond type and whether the colour is natural or artificial.

Note

1. Bespressovoi Apparature 'Razreznaya Sfera', meaning Press-less apparatus 'split sphere'.

5

Ruby and Sapphire

Michael O'Donoghue

with contributions by
Richard W. Hughes

Ruby and the different colours of sapphire are colour varieties of the mineral corundum, aluminium oxide (Al_2O_3). Entirely colourless corundum is not common in Nature as iron (the best example) is rarely far from the action when minerals are forming so that faceted colourless corundum must be tested for clues to its origin. Small amounts of chromium are the cause of the red colour in ruby and iron plays a similar part in the colouration of green and yellow sapphire. Blue sapphires contain iron and titanium, both being necessary for colouration as we shall see below. Chromium does not always give red; orange-pink or pink-orange specimens from Sri Lanka are called padparadscha and command high prices. Purple and mauve sapphires are known and highly desirable; brown sapphires can also be attractive if not too dark. Star rubies and blue sapphires at their best can be magnificent and named specimens are known. Titanium dioxide (rutile) is essential for star formation and as it is not found in sufficient quantities (if at all) in other colours of corundum the absence of asteriated yellow specimens, for example, is explained.

Cat's-eyes are not common but can form if the inclusions are appropriately distributed. Some sapphires may show a colour change though any specimen known to be corundum and showing a change from mauve to purple will, virtually invariably, turn out to be a vanadium-doped flame-fusion synthetic product.

The root *rub* suggests red in a number of languages but careful study is necessary if the derivation of today's names (for any natural product) needs to be traced back to Classical or other literatures. In his translation of Pliny's *Naturalis Historia*, Eichholz (1962) [*Loeb Classical library* no. 419] tells us that *sappirus* is not the sapphire of today; further efforts towards the investigation of corundum (or any other) varietal nomenclature should be directed towards the philologist rather than the mineralogist! The same can be said of the considerable amount of information available on the lore and legend of corundum. It is likely that some of the names have developed from Sanskrit roots. A translation of the thirteenth-century Ahmad ibn Yusuf Al Tifashi's *Best Thoughts on the Best of Stones*, incorporated in Samar Najm abul Huda's *Arab Roots of Gemology*, 1998 (ISBN 0810832941) is a good example of the type of study that may be needed for similar non-European MSS.

Named rubies include the De Long star ruby, in the American Museum of Natural History, New York, weighing 100 ct and the Rosser Reeves star ruby, in the National Museum of Natural History, Smithsonian Institution, Washington DC, USA, 138.7 ct (stone from Sri Lanka).

Named blue sapphires include the 423 ct Sri Lankan Logan sapphire, in the Smithsonian Institution, and the Myanmar Star of Asia, 330 ct, in the same museum. The Black Star of Queensland (733 ct, in a private collection) is the largest known black star sapphire. The Star of India, in the American Museum of Natural History, is a blue 536 ct sapphire. Other examples are quoted by Arem in the second edition of *Color Encyclopedia of Gemstones*, 1987, ISBN 0442208332.

Ruby

The lore and legend of ruby and sapphire are best left as received, attempts to disentangle them needing access to serious collections of material from Myanmar (in particular) if repetition of oft-quoted tales is to be avoided. The best sources to begin with are the oriental collections of The British Library, which include some, at least, of the reports of Burma Ruby Mines Ltd. The 5th and 6th editions of Edwin W. Streeter's *Precious Stones and Gems* (1892 and 1898) [Sinkankas, *Gemology, an Annotated Bibliography*, 1993, #6399, #6400] contain not only an early if not the first attempt in English to describe the ruby mines of the then Burma but also an account of Streeter's own short connection with the mines.

The most complete early geological description of the ruby mines of Myanmar is Barrington Brown and Judd, 1896, *The Rubies of Burma and Associated Minerals*. This appeared as *Philosophical transactions of the Royal Society of London*, series A for 1896, 187, 151–228 [Sinkankas 957]. The paper was published separately and gives a great deal of information on the geology and mineralogy of the area and (to quote Sinkankas, *op. cit.*) 'establishes the fact that alluvial rubies derive from decay of host rock (marble)'. Harbans Lal Chibber's *The Mineral Resources of Burma* (London, 1934) describes mining methods, production and fashioning and the ruby deposits themselves. Another book, with a promising title, is A.N. Iyer's *The Geology and Gemstones of the Mogok Stone Tract*, Burma, published in 1953 as *Memoirs of the Geological Survey of India*, vol. 82 [Sinkankas #3152]. Sinkankas regrets the comparative lack of space devoted to gemstones, despite the title; the book does provide a description of the crystalline Mogok marble, however, and some of the gem sources are discussed.

One of the earliest accounts of corundum as it was known in the eighteenth century is the Count de Bournon's *Description of the Corundum Stone and its Varieties*, commonly known by the names of oriental ruby, sapphire, etc. This account was published in *Philosophical transactions of the Royal Society of London* 92, 1802, pp. 233–326. Separate copies were published [Sinkankas #824]. Here we find notes on properties, crystallographical observations and nomenclature problems. It is the first paper in English to describe the contents of the Ceylon gem gravels.

Peter Bancroft's *Gem and Crystal Treasures* (1984; ISBN 0961346116) gives an illustrated description of ruby mining in the Mogok area, and a more recent account by Ted Themelis, *Mogok, Valley of Rubies and Sapphires* (2000; ISBN 0940965208) gives readers an excellent overview of the history and mining in the region with attractive photographs.

In 1945 D.N. Wadia and L.J.D. Ferrnando published *Gems and Precious Stones of Ceylon*, forming *Professional paper* no. 2 of *Records of the Department of Mineralogy of Ceylon*, pp. 13–44 with two tipped-in maps [Sinkankas #6952].

Until the publication in 1968 of E.J. Gübelin's *Die Edelsteine der Insel Ceylon* [Sinkankas #2575], surprisingly no other serious monograph on Sri Lankan gem minerals has been published.

Wadia and Fernando give a reasonable account of Sri Lankan gemstones with useful notes on native methods of cutting and on geology of gemstone occurrences. As expected, corundum locations are noted and the map at the end of the text shows where many deposits are. Some major specimens are mentioned. The Gübelin book which, despite its date, still cries out for an English translation from the German as the text gives a profusion of detail on mining, including prospecting, and on gemmological properties of the stones. This is a rare book, apparently published at the author's expense rather than through a commercial publisher.

Indian Precious Stones, by Iyer and Narayana, 1942 [Sinkankas #3148], *Records of the Geological Survey of India* 76, *Bulletins of economic minerals* no. 6, gives useful accounts of gem species found in India, Sri Lanka and Burma; a revised edition by Iyer and Thiagarajan, 1961 [Sinkankas #3150] gives considerable detail on Indian deposits alone. This formed no. 18 of the Series A, *Economic Geology*, of *Bulletins of the Geological Survey of India*. A recent study of Indian corundum is Panjikar, *Comparative Study of Gem Minerals, Beryl and Corundum from various Indian occurrences* (2004: 8190122208).

I have listed these works in some detail as more modern publications dealing with gem and especially corundum mining in south Asia draw from one another. One major source must have been the seventeenth-century traveller Jean-Baptiste Tavernier whose book '*Les six voyages de Jean-Baptiste Tavernier*' was first published at Paris in 1676. The best account of the history and content of this work is given by Sinkankas, 1993; as far as I know, writing in late 2004, there is no modern version in English. The book was usually re-published in two parts though the original edition was in three parts, the third volume containing no material on gemstones. In 1889 Valentine Ball of the Geological Survey of India made an English translation which has remained the best, perhaps the only, version published since that time.

While Tavernier remains the best account of Indian diamonds, other gemstones also find a place. In the second volume of Valentine Ball's translation (1889), an appendix describes the ruby mines of Burma. Sinkankas numbers for the major editions of Tavernier are: #6499 for the first edition 1676, #6533 for the Valentine Ball edition of 1889.

In 1996 Dr Nandkishor R. Barot published a two-volume reprint of *Mani-Mala: A Treatise on Gems* (Nairobi: no ISBN). The original work, by Raja Sourindro Mohun Tagore [Saurindramohana Thakura] was published in 1897 and is one of the rarest of all books on gemstones. By now the reprint will also be hard to find. The original is described, under #6475, by Sinkankas (1993). Much of the text is taken from the Puranas and other classics of Indian literature, and, to quote Sinkankas again, the major text of part 22 is an amazing pot-pourri of information, derived from both oriental and western sources, which includes sections on gemstone genesis according to ancient Sanskrit sources and many remarks on curious lore. It is now one of the rarest of all gemmological treatises.

In the first four editions of *Gems* (1962–1975) Webster presents some information both on early corundum mining and on lore and legend. None of it appears to be attributed and sadly no attempt seems to have been made to provide anything more than a book list rather than a bibliography. For this reason it is not easy to separate wheat from chaff but at this distance in time perhaps such work is best left to scholars. In *Corundum* (1990) Hughes draws more successfully on some of the older sources and, fortunately, cites them in end-of-chapter references.

Judging Quality of Rubies and Sapphires (RWH)

Next to red diamonds, rubies are probably the most expensive of gems in the world in sizes over 3 ct. The ideal for ruby is that of a red traffic signal, a highly fluorescent red of high intensity. Unlike diamond, small amounts of silk in a stone actually help the colour because they scatter light into areas it would not otherwise go. This helps cover up the extinction which would otherwise be found. Thai/Cambodian rubies suffer the double deficiency of too much iron, which cuts the fluorescence, and no silk to scatter light.

Top prices for blue sapphire are paid for stones of an intense blue verging on the violet. Large sapphires are more common than large rubies. Colour zoning can be a problem in sapphire: look for stones which have no major zoning problems. Many blue sapphires are too dark in colour, especially those from basalt sources such as Thailand and Australia. Such stones are generally inexpensive. Similarly, sapphires which are too light in colour are also inexpensive. The best stones are those which are well cut, eye clean and, most importantly, with intense blue colours. Orange sapphires should be rich in colour without blackish areas. In star stones, transparency is an important factor. Too much silk results in a lowering of colour intensity, because the length of light paths through the stone is reduced due to scattering from the silk. The ideal star stone contains just enough silk to show a star, but not enough to cause the colour's intensity to be reduced. A stone can be expensive if its colour is good, even if the star is poor. In contrast, stones with sharp stars but poor colour are not valuable.

Chemistry and Crystallography

Chemical Composition

Corundum is a crystalline form of aluminium oxide (Al_2O_3), traces of impurities giving the colour. Ruby varies in shade from near colourless through pink through all shades of red to a deep crimson. All other colours are called sapphire with a colour prefix.

Colouration of Ruby

The colour of ruby is due to a trace of chromic oxide (Cr_2O_3), which enters the crystal structure by small-scale isomorphous replacement of some of the alumimium atoms. The amount, about 1–3%, determines the depth of colour, but the presence of ferric iron also modifies the tint, reducing fluorescence and giving the rubies from Thailand a characteristic brownish

121

tinge. The finest rubies will be a strongly fluorescent red, resembling the colour of a red traffic signal. Such stones often contain extremely fine particles which serve to scatter light onto all facets and reduce extinction (facets which are dark). In the past in Myanmar this colour was termed 'pigeon's blood', but the term has little meaning today as so few people have seen the blood of this Burmese bird.

Colouration of Sapphire

While there is evidence of two or more possible causes for the blue colour of sapphire, the colour of most results from a combination of iron and titanium, where they have replaced aluminium in the corundum structure. Intervalence charge transfer involves $Fe^{2+} + Ti^{4+} > Ti^{3+} + Ti^{3+}$. Understanding this and related processes is essential for those intending to alter the colour of corundum by heating.

Yellow sapphire may be coloured by one of a number of operations and it is not always possible, in general testing at least, to identify which process has coloured a particular specimen. Nassau (1994) has identified seven different ways in which natural and synthetic yellow sapphires can be coloured; some operations produce unstable colour but only a fade test can establish which stones possess this! Purple and mauve sapphires usually contain some chromium. Black-appearing corundum is usually dark brown from profuse inclusions.

Crystallography

Corundum crystallizes in the trigonal system, but the habit varies with locality and/or colour. In many crystals the pinacoid is developed at the expense of the pyramid and rhombohedron; bipyramidal habit is also common. Clues to origin rather than diagnosis may be provided by some corundum crystals.

Combinations of any of these three main habits are also possible.

Ruby and sapphire crystals often show different habits. Rubies from Myanmar may occur as hexagonal prisms terminated at both ends by the pinacoid, with rhombohedral faces at alternate corners. The rhombohedral faces may be more or less developed or entirely absent, especially in the large, and usually opaque, crystals from Tanzania and Madagascar. In many ruby crystals the prisms are flattened and, although they may be of large diameter, are relatively thin. Such crystals often exhibit a stepped or platy appearance, as though the crystal was composed of a number of thin plates. Basal planes of many crystals are traversed in three directions by fine parallel striations, which take the form of hair-like lines crossing at angles of 60/120°, dividing the area into small triangles.

Sapphire (and some ruby) takes the form of a hexagonal bipyramid of twelve triangular faces, six above and six below, meeting at a girdle. This habit may occur in combinations of bipyramids of different inclinations, with the girdle sometimes formed by a narrow hexagonal prism. The ends of many bipyramidal crystals are capped by the flat basal face to give a barrel-shaped habit. The hexagonal bipyramidal habit is common in sapphire crystals from Sri Lanka, and in this form the faces are often deeply striated horizontally, owing to repeated oscillation between different pyramids or between the pyramid and the basal pinacoid. Sapphires from Yogo Gulch, Montana, as described by Clabaugh in *Corundum Deposits of Montana* (US Geological Survey Bulletin 983), 1952, showed a number of forms whose true nature was obscured by alteration of plane surface and original faces by etching. A basal pinacoid and rhombohedron were the two forms identified in a study by Pratt in *Corundum and Its Occurrence in the United States* (US Geological Survey Bulletin 269), 1906. In general, corundum crystals show predominating pinacoid and pyramid forms; tourmaline might be suspected when a red crystal with predominating prism forms is encountered and red spinel when triangular faces are all that can be seen.

Physical Properties

Specific Gravity

The SG of pure corundum, synthetic colourless sapphire, is 3.989, but natural ruby and sapphire usually have SG near 3.997: there is no significant variation for specimens from different localities though iron-rich yellow, green and blue-green sapphire from Australia or Thailand may reach 4.00.

Hardness, Fracture, Cleavage and Parting

Corundum is 9 on Mohs' scale. Ruby and sapphire may be brittle and should be carefully handled. The fracture varies from uneven to conchoidal.

There is no cleavage. Parting, from exsolved bohmite, may occur parallel to the basal plane or rhombohedral faces.

Effects of Light

Refraction

Anthony *et al*. (*Handbook of Mineralogy*, 1997, the most reliable source) assign the values 1.767–1.772 to the ordinary ray and 1.759–1.763 to the extraordinary ray. Corundum is uniaxial negative. Gem materials for the most part accord with these values and show a birefringence of 0.008. Both ruby and sapphire show distinct pleochroism: blue sapphire normally shows blue-green to yellow-green in the extraordinary ray and pale to deep blue in the ordinary ray, ruby showing orange-red in the extraordinary ray and bluish red in the ordinary ray. When possible, both rubies and sapphires should be cut with the c-axis at right angles to the table facet, to display the best colour.

Lustre

The lustre of ruby and sapphire varies from sub-adamantine to vitreous and is pearly on parting surfaces.

Dispersion

The dispersion of the stones is only 0.018 so little fire is perceptible.

Absorption Spectra

In describing absorption spectra throughout the book details have been checked against B.W. Anderson and C.J. Payne, *The Spectroscope and Gemmology*, edited by R Keith Mitchell, London, NAG Press, 1998; ISBN 071980261X. This is a compilation of papers by the authors in the journal *The Gemmologist* between 1954 and 1957, with additional notes by Mitchell who also re-drew some of the diagrams. The book should be available to all engaged in gem testing. Details and techniques of absorption spectrum investigation are also found in this book and in *Gemmology* (Read), 3rd edition, 2005 (ISBN 0750664495).

Ruby shows the classic chromium absorption spectrum though corundum of any colour which contains significant chromium may show some of its elements. Two lines close together in the red form a doublet which, depending upon the lighting conditions, may be coloured (emission doublet) or dark (absorption doublet). Emission lines occurring elsewhere in the spectrum will be found to have emanated from the room lighting, never from the specimen under examination. I have not recorded the fainter absorption lines but only those which are prominent in normal conditions of observation or which are significant in diagnosis.

The lines are situated at 694 nm and are followed by absorption lines in the orange at 668 and 659.2 nm. A broad absorption band extending from 610 to 500 nm obscures the green (this band, present though with slight differences of position in all materials coloured by chromium is the major cause of the residual colour (the colour seen by the unaided observer).

Three narrow lines in the blue, two of them equally strong, at 476.5 and 468.5 nm are accompanied by another weaker line at 475 nm and by a general absorption of the violet. Red spinel, easily confused with ruby when testing is undertaken in haste or under unsatisfactory conditions, does not show the strong, close lines in the blue. The absorption spectrum of synthetic ruby is identical though all elements may be deceptively easy to see. A strong emission doublet and other elements suggest that further tests may well be needed. Readers should also remember that elements of the absorption/emission spectrum may well be stronger in different directions and specimens should always be tested in several directions (back-to-front will give the same results as front-to-back!).

Anderson and Payne (1998) remind us that a chrome-rich ruby need not always be synthetic as some Myanmar and East African stones have a higher-than-usual chromium content. Furthermore, it has been shown that in deep red rubies the absorption beginning in the violet extends, in fact, not only close to the absorption line at 468.5 nm but also extends, after more transmission, into the near UV. The final absorption region begins in natural rubies between 300 and 290 nm: in synthetic rubies this region does not begin until about 270 nm (or not before).

Fluorescence

A red colour, seen through a red filter when a ruby is illuminated only by monochromatic blue light (the crossed filter effect), proves the presence of chromium. As might be expected, the effect is particularly spectacular with synthetic rubies since they contain no iron to weaken the effect. Iron is usually included in natural ruby and if present in sufficient quantities inhibits fluorescence, which is the transmission of energies of certain wavelengths when a specimen is irradiated by energies of a lower wavelength (but higher energy).

Despite the occasional claim, the response of ruby to UV and X-radiation should not be relied on either for identification, place or type of origin of a suspected ruby, though any effects seen may make very useful confirmatory tests. In ruby the excitation of a chromium ion to a higher energy level produces red light from the emission doublet described above. The effect produced by the crossed filter experiment, ruby's response to the ultraviolet radiations present in daylight and to other radiations of higher energy than these and the mechanism operating are all lucidly described by Kurt Nassau in the two editions of *The Physics and Chemistry of Colour* (second edition, 2000; ISBN 0471391069). Ruby fluoresces most effectively if no iron is present; this can only be guaranteed when the specimen is man-made so that a strong crimson response to UV or to X-rays requires the investigator to make additional tests. Rubies, whatever the origin, can be separated from red garnet or tourmaline if a UV source is available. While most Myanmar rubies commonly fluoresce more vividly than most (relatively iron-rich) rubies from Thailand, place of origin determination needs a study of inclusions as well as or more than fluorescence. Purple and mauve sapphires will usually respond with something of a red colour to activating sources.

Absorption Spectra of Other Colours of Corundum

Iron causes the colours of natural non-ruby corundum occurring, like chromium, as a replacement for aluminium. This is trivalent iron and less than 1% may be enough to cause colour. What might be thought of as the classic iron spectrum in corundum is that shown by green sapphire in which no other element influences the colour. Here there are three bands, at 471, 460 and 450 nm in the blue region of the spectrum. The band at 450 nm is the strongest and nearly coalesces with its neighbour at 460 nm, the next strongest. The 471 nm band is distinctly separate from the other two and is the weakest of the three. The same spectrum can be seen in natural but not synthetic flame-fusion blue sapphires but specimens from different locations show variations. In blue sapphires from Australia all three bands are strong and easily seen; on the other hand, Sri Lanka blue sapphires show only a faint 450 nm band. Anderson and Payne tell us that blue sapphires from Myanmar, Kashmir, Thailand and Montana show a clear 450 nm band accompanied by a smudge on the long-wave side – the only trace of the other two bands.

These bands belong to the ordinary ray and observations may be made more easily if the specimen is illuminated by light which has passed through a blue filter (copper sulphate solution is still used). If a Polaroid filter is available it can be useful in finding the strongest effect direction. Some deep blue natural sapphires show a rather vague broad absorption band near 585 nm.

Yellow sapphires may also show the three ferric iron bands. Sapphires from Australia, Montana and Thailand show them most clearly, while yellow sapphire from Sri Lanka may show only a faint 450 nm band or none at all (though they do give a characteristic apricot-yellow fluorescence under both types of UV). Generally speaking, synthetic flame-fusion sapphires neither fluoresce nor show any absorption bands – but exceptions have been recorded.

In some cases absorption bands in the UV may give a clue not only to the identity of an unknown yellow stone but also to its possible origin.

Photography needs to be used (a photograph will often show otherwise undetectable features in a synthetic stone, in particular) to find absorption bands in the near-UV at approximately 379 and 364 nm. These bands have been noted in Sri Lankan yellow sapphires when the 450 nm band is scarcely perceptible. The UV bands are usually obscured in sapphires which show the '450 complex' strongly. Synthetic sapphires in general, certainly the nearly ubiquitous flame-fusion products, show absorption neither in the visible nor in the UV regions. Anderson and Payne remark on the oddness of this since iron (with titanium) needs to be present in any sapphire to give a blue colour. Some observers have noted a weak 451 nm absorption band in some Verneuil and Chatham flux-grown sapphires.

Vanadium

The part played by vanadium in the colouration of gemstones often echoes that of chromium since trivalent vanadium absorbs in similar areas of the visible spectrum. The size and symmetry of the vanadium site is critical and this is one reason why vanadium-bearing minerals show a wide variety of colours (Schmetzer, *Absorption spectroscopy and colour of vanadium (3^+) – bearing natural oxides and silicates, N. Jb. Mineral. Abh.* 144, 73–106, 1982 deals with this question).

In corundum, vanadium causes the unmistakable (unless you have amethyst in mind) slate to purple colours of the material so often offered as, or at least confused with alexandrite. This is a flame-fusion product and shows a very sharp though not necessarily strong absorption band in the blue at 475 nm. This band is diagnostic for the material. colour change sapphire from Myanmar has been reported to show a vanadium spectrum.

A colour change sapphire from a recently reported source in the far south-west of Tanzania showed bluish green in daylight and reddish brown in incandescent light. Stones showed red though the Chelsea colour filter but did not respond to either type of UV. Both chromium and vanadium were found to be present. Nine out of ten specimens tested showed the true alexandrite effect, which is caused by the centering of the broad absorption band near 585 nm (550 nm in ruby). Narrow lines are also seen in the red, similar to a chromium spectrum. The vanadium spectrum is extremely rare in natural corundum. Mauve and purple sapphires will usually show elements of both chromium and iron absorption spectra.

Since blue sapphire needs iron for a blue colour to be produced, any response to UV or stronger radiations would be surprising: one or two examples of blue sapphire with significant chromium content have been recorded, these specimens showing some orange or reddish glow under LWUV, such examples are hardly typical. Similarly some heat-treated blue Sri Lankan sapphires have been found to respond with a weak blue to SWUV. These specimens have contained colourless areas which have given rise to the response, the areas corresponding with growth structures.

Most sapphires are inert under X-rays, except the Sri Lankan, Montana and some Indian (Kashmir) stones which may show a dull red or yellowish orange glow. It has been reported that under bombardment by cathode rays (fast-moving electrons), Kashmir sapphires show a greenish blue glow, Myanmar stones show a strong dark purple, Thai stones show a weak dull red and sapphires from Sri Lanka a vivid red.

The iron-rich green and yellow sapphires show no luminescence of any kind, but yellow and orange stones from Sri Lanka show a strong apricot-yellow glow under UV, X-rays and gamma rays. The cause of this particular luminescence is not known. Such stones when bombarded by X-rays turn to a rich topaz colour, however weakly yellow they were originally. This colour is not permanent and reverts on exposure to about 3.5 hours' sunlight or quickly when the stone is heated to about 23° T. Colourless sapphires may also suffer this change of hue after irradiation, but the shade of yellow attained is usually lighter; and further some blue sapphires will change to a dirty amber colour.

Inclusions

Any study of inclusions must turn first to Gübelin and Koivula, *Photoatlas of Inclusions in Gemstones*, second revised edition, 1992, ISBN 38555040958. Before exploring a suspected natural or synthetic ruby we might remember some useful general points: very high magnifications are not always necessary; 40–100× should identify most of the important inclusions. Polarized light is also useful but less easily achieved. Far more important are the mental pictures built up by the gemmologist as a 'working library' and a range of light sources (not necessarily integral parts of the microscope). This writer (MO'D) particularly likes the horizontal-tube microscope with its facility for specimen immersion. In dark-field illumination the specimen is lit from the side rather than by light transmitted through it. Details of microscope operating techniques are covered by Peter Read's *Gemmology* (passim). The term 'negative crystals' refers to included cavities which obey the crystallographic laws of the host crystal while not invariably showing the outline of the crystal forms. Negative crystals may have gas or liquid contents.

Schmetzer, in *Rubine, Eigenschaften und Bestimmung*, 1986; ISBN 3510651251, gives probably the best monographic account of the inclusions in ruby. The book includes an excellent bibliography.

Inclusions may be solid, liquid or gaseous. RWH has made a useful list of the major inclusions in corundum, reproduced, with some alterations, below. Further details will be found in the discussions of particular localities.

1. Straight angular growth lines following various crystal faces, often in a hexagonal pattern and often featuring associated minute exsolved needles or particles following these growth lines. The lines vary in thickness and spacing, and are never curved (if examined parallel to the face along which they grew), and always lie inside the stone. They are associated with crystal faces, not with polished facets. Sharp lines are seen best with dark-field illumination, or better, immersion with light-field shadowing illumination. Broad bands or ill-defined patches are best seen with immersion and diffused light-field illumination. In rubies, the colour often occurs in treacle-like swirls when looking in directions other than along the crystal faces.
2. Exsolved rutile needles and/or hematite plates (silk) forming parallel to the hexagonal prism (three directions, intersecting at 60/120° in the basal plane) often forming dense clouds. The rutile occurs as intergrown twins with re-entrant angles at the broad end while hematite tends to form plates. Sizes vary greatly, some being much longer than others, some appearing as mere dots, some broad, some narrow. Overhead fibre-optic illumination is often best, looking down the c-axis. Minute exsolved particles are often best seen with the fibre-optic light guide from below or to the side of the stone.
3. Crystals of different minerals of various types, including spinel, apatite, zircon, calcite, dolomite, uranpyrochlore [this name valid in Dana 8 though not in *Glossary of Mineral Species* 2004], mica group minerals plagioclase, pyrrhotite and other species, best viewed in dark-field illumination or via fibre-optic lighting.
4. Secondary liquid inclusions in patterns of infinite variety and thickness; often referred to as fingerprints or feathers. They are created when fractures are healed by post-formation geological activity. Their patterns may often be wispy or veil-like, and so are easily confused with flux inclusions in synthetic corundum. Their surfaces should be examined under high magnification with fibre-optic lighting to see if liquid (natural) or flux (synthetic) fills the small channels.

As natural stones healed over a much longer period of time, their healing patterns are often far more detailed. The higher viscosity of a flux also produces a coarser and less detailed healing in flux-grown synthetics.

5. Polysynthetic twinning along the rhombohedron (in three directions, but only two in any one plane) meeting at 86.1 and 93.9°. These lie about 30–60° off the *c*-axis. Growth twins may also be seen along other faces. Immersion and examination between crossed polars will separate true twinning from sharp colour zoning. True twinning planes will show interference fringes and appear light against a dark background.

6. Long white exsolved boehmite needles which form at the junctions of intersecting rhombohedral twinning planes. Thus their directions and angles are the same as that described in 5. Rhombohedral twinning with the boehmite needles has yet to be seen in the flux-grown synthetic corundum and so is extremely important for identification.

7. Rhombohedral parting (due to exsolved boehmite) and basal parting (due to exsolved hematite).

8. Wavy parallel cracks ('fire marks') near the facet junctions due to overly rapid polishing. These are more commonly seen on synthetic stones, as less care is taken in the polishing process, but may sometimes be seen in natural corundum, too.

Methods of Fashioning (RWH)

Faceting

Rubies and sapphires may be faceted in many different styles; mostly the mixed cut is used; the brilliant-cut crown is backed with a step-cut pavilion. For fine stones the step cut is often employed, and if the material is poor quality or much flawed it may be cut into beads or even carved. Pale stones are often mounted with a closed setting and the back of the stone is sometimes foiled with a suitable colour.

Star Stones

Star stones must be cut as cabochons in order to exhibit asterism. The cause of the asterism is exsolved rutile or hematite 'silk', multitudes of tiny oriented needles or plates. These inclusions form by exsolution – the unmixing of a solid solution. At high temperatures, the corundum may absorb more impurities, but as the crystal cools, the impurities can sometimes be forced out of solution. Since they are still in a solid, the impurity atoms only crystallize where there is space and this is determined by the host structure. In corundum, exsolved rutile and hematite crystallize in three directions in the basal plane, parallel to the faces of the hexagonal prism.

When a stone contains enough of such inclusions, it can display a six-rayed star effect if it is cut as a cabochon, with the base of the cab parallel to the basal pinacoid (parallel to the layers of silk). The hematite crystallizes parallel to the first-order hexagonal prism while the rutile follows the second-order hexagonal prism (30° off the first-order prism). If a stone contains enough of both hematite and rutile silk, it can display a twelve-rayed star. This is fairly common in black star sapphires, but is rare in other colours. To observe the star effect the stone should be examined in direct sunlight, or under a single, intense light, such as a fibre-optic light, pen light or spot light.

Synthesis and Simulation (RWH)

Details of gem crystal growth are comprehensively discussed in Chapter 2 but features characteristic of the types of synthetic corundum likely to be encountered are given here additionally.

Ruby was the first major gemstone to be synthesized, the first experiments taking place in the mid-nineteenth century. Today corundum of all colours is manufactured by a variety of processes, for both gem and industrial use. Star corundum has been synthesized since the late 1940s.

Melt-Growth Processes

Verneuil (Flame-Fusion) Process
Various manufacturers (all colours, including six-rayed stars).

1. Curved growth lines (thin striae or bands), seen best at roughly 90° to the boule's length. These curved lines are not concentric.
2. Gas bubbles, round or elongated at 90° to direction of growth lines, from pinpoints to large distorted doughnut-shaped spheres or highly irregular worm-like distortions. They are usually distributed in clouds which follow the curved growth structure of the boule. In blue stone, the gas bubbles may show concentrations of blue colour around them.
3. (a) Polysynthetic twinning along the *c*-axis ('Plato lines'), seen between crossed polars, most effectively when the specimen is immersed, in sets of one, two or three directions.
   (b) Polysynthetic twinning along the rhombohedron, sometimes with accompanying boehmite needles identical to natural corundum.
4. Induced fingerprints and feathers entirely similar to natural corundum.
5. Traces of the seed rod or seed crystal. Found at the base of the boule and featuring frosted surfaces at the seed junction.
6. Some varieties may show useful UV fluorescence. The V-doped colour change type shows a diagnostic line at 475 nm. The full Fe spectrum (451, 460, 470 nm) is not seen.

7. Irregular colour distribution and rounded facet junctions of blue varieties, in particular, may cause confusion with surface diffusion-treated corundums.
8. Dense clouds of extremely fine, exsolved rutile silk in star material. The clouds do not show the angular zoning patterns that are common to natural stones. Instead, they may display curving bands (which are sometimes concentric).

Czochralski (Pulling) Process

Inamori [Japan] (red, orange and red star), Novosibirsk [Russia] (red) and others (red, blue and colourless).

1. Extremely fine and narrow curved growth lines (curved striae), which may be concentric.
2. Gas bubbles of various sizes, shapes and orientations.
3. Faint 'smoke-like' or 'rain-like' wisps of tiny particles, probably representing remnants of a flux used to aid melting during growth.
4. Small unidentified black prismatic crystals in groups (Inamori).
5. Extremely fine exsolved clouds of what may be rutile silk in the red star material (Inamori).

Floating-Zone Process

Hattori Seiko (red, orange, blue), Novosibirsk (red).

1. Gas bubbles of various sizes, shapes and orientations.
2. Irregular colour swirls.
3. Secondary lamellar glide twinning along the rhombohedron faces. Twin planes may intersect at 87 and 93°, and may also show long white needles (boehmite?) at the junctions of intersecting twinning planes. Best seen between crossed polars while immersed.
4. Rectilinear parting.

Solution-Growth Processes

Flux Process
Chatham (red, orange, blue), Kashan (red), Ramaura (red), Knischka (red), Lechleitner (overgrowths in various colours), Novosibirsk (red), Douros (red).

1. Primary flux-filled negative crystals, often only partially filled (two-phase) and featuring a characteristic crazed surface appearance. Primary flux in Ramaura stones often has a yellow-orange colour

and may show distinct growth striation on flux surfaces, open cavities and crystal faces which mirror the colour zoning.

2. Secondary flux-filled fingerprints, feathers, etc.
3. Tiny flux particles, often arranged in streamer or comet-like patterns (such as 'rain' in Kashan stones).
4. Platinum plates, flakes, crystals, needles, etc. (Chatham, Knischka); black (platinum-rich?) growth planes (Chatham only, especially along the seed crystal).
5. Very small, oriented, exsolved silk-like needles and/or particles in zoned clouds (Chatham and Knischka).
6. Polysynthetic and growth twinning in various orientations, but without the boehmite needles often present in the natural, Verneuil and Seiko (floating zone) synthetics.
7. Straight growth lines running parallel to crystal faces and meeting at 11 specific angles. Unusual growth-line boundaries (Ramaura, Douros [Greece]).
8. Chatham: rounded transparent crystals of low relief (possibly chrysoberyl).
9. Seed crystal, generally with trapped flux on the boundary. The boundary may be difficult to see in the Lechleitner overgrowth.
10. Knischka: primary negative crystals which often display a two-phase filling. These may be bipyramidal or irregular in shape and sometimes are surrounded by irregular bluish white clouds.

Hydrothermal Process

Novosibirsk only (red).

1. Extremely strong growth zoning (graining) parallel to the basal plane.
2. Small, highly reflective crystals of a gold colour, consisting of copper alloys.
3. Secondary healed fractures (fingerprints) believed to be liquid filled.

Simulants and Imitations

Although there are a number of materials which can have a similar appearance to ruby and sapphire (such as certain spinels, garnets, benitoite, etc.), all are easily separated by reference to the properties, such as RI and SG.

More dangerous are assembled stones, generally consisting of one or more parts of natural corundum, attached to other materials (generally synthetic corundum). They may show natural inclusions and colours because they are part natural. The key to identifying any assembled stone is to locate the separation plane where the stone is joined together. It must be a distinct join, completely unbroken around the entire stone. It is usually on the girdle, but may be on the pavilion or even the crown. The glue layer often shows curved brush strokes and flat gas bubbles. Since the glue's RI is quite different from corundum, it will stand out in high relief when the gem is immersed in di-iodomethane.

Various kinds of assembled star stones are also possible. One of the most deceptive is simply taking a white star sapphire and coating it with red plastic.

Corundum from Different Areas (RWH)

Myanmar

Something of the history of ruby from the Mogok Stone Tract of Myanmar is described in the works cited at the beginning of the chapter. At this distance in time and without translations of major works in the vernacular the authors recommend the cited works. Readers may consult the previous editions of *Gems* (Webster, 1962–94) though the sources of his material are not given.

The most famous locality for fine rubies and sapphires is the district around Mogok in upper Myanmar. The Mogok Stone Tract is an area of some 1000 km². The dates of discovery are not known and the legends later associated with ruby finds can be discounted. The area produces fine ruby and blue sapphire today.

The Burma Ruby Mines Company

After the annexation of upper Burma by the British in 1886 (see A.T.Q. Stewart, *The Pagoda War*, London, Faber, 1972; ISBN 0571087221), Edwin Streeter, the London Bond St jeweller, obtained a concession to work the ruby mines though his position was disputed. On regularization of the position and flotation of the Burma Ruby Mines Company (with over-subscription) mining was able to begin.

As so often seems the case when a new gem mining venture actually begins work and miners look at the terrain without the assistance of rose-coloured spectacles, extraction of the rubies was far from easy. The alluvial deposits in which the ruby crystals were contained lay under the Mogok settlement or beneath waterlogged valleys. Nonetheless in time mining was able to commence. No underground mining was involved. Up to 1908 the Company was able to pay a dividend but depression in the US and the availability of cheap Verneuil synthetic rubies caused a price fall, affecting even the finest specimens. The Bombay market (where most lower-grade rubies were sold) suffered at least an equally drastic fall in activity. The First World War occupied attention from 1914 to 1918 and in 1925 the Company went into voluntary liquidation, surrendering its lease to the government.

Reasons for the Company's Failure

Many have speculated about the reasons for the Company's failure, most concluding that it was unavoidable, the difficulties being too great to surmount. But recent evidence uncovered by Richard Hughes suggests that the Company owed its failure less to the difficulty of the task and more to greed. In a confidential report written in 1925, the head of the Geological Survey of India, J. Coggin Brown, pointed the finger for the Company's failure straight at the De Beers diamond cartel. He stated:

> "at this juncture I cannot refrain from writing an opinion which I have already expressed verbally, that the influence of the De Beers diamond concern has had more to do with the present [1927] position of mining for coloured gems in Burma than appears on the surface. The reasons for this are obvious, and it is significant that there has always been a powerful representative of the great South African concern on the Board of the Burma Ruby Mines, Limited."

Brown went on to chronicle a number of bad decisions taken by the Board that resulted in missed opportunities. The report of 1927 is entitled *Gem Mining in the Mogok Stone Tract of Upper Burma from the Annexation to the Present Time with some Suggestions for Future Operations*. It was published by the Superintendent of Government Printing and Stationary, Burma and marked 'Confidential'.

At the time of this writing the company's equipment has lain fallow for over 60 years.

Native Mining

After the failure of the Company, mining in Burma reverted to native methods. Mining was allowed on payment of a monthly fee, which was higher if water and explosives were used. To obtain a mining licence in Burma, it was then necessary for one's name to be on arbitrary list of registered miners, but those on the list were often willing to lend their names to the less fortunate for a consideration.

Pit Mining

In the dry season a shaft (twinlon) is sunk to the gem-bearing alluvial gravel (byon). On reaching it tunnels are driven as necessary. Wet conditions force miners to construct a timbered shaft (lebin).

Hillside Mining

The mining method called hmyawdwin involves the cutting of holes into hillsides and is resorted to when wet conditions make pit mining too dangerous. Caverns (loodwin, loo) may be encountered and their contents are washed out. Fine specimens are often retrieved by this method of working. Dredging streams tends to produce less fine material.

Mode of Occurrence

When ruby is found *in situ* the host rock will be silica-deficient and most commonly a marble. This is the common mode of occurrence of ruby from Myanmar and Afghanistan. At Mogok ruby occurs in a metamorphosed limestone, traces of which provide calcite and apatite inclusions in the host crystals.

Sapphire from Myanmar is not apparently found in the marble and may have originated in syenites or in pegmatites though specimens are recovered from alluvial deposits as mentioned earlier.

Ruby has been found at other localities in Myanmar including the Sagyin Hills near Mandalay and Namseka, near Mainglon. The notable purple-cored rubies from Mong Hsu, 250 km east of Mandalay, are heavily fractured and the purple colour needs to be removed by heat treatment. The deposit was discovered in 1992.

Myanmar Today

The Mogok mines were nationalized in 1968. By 1991 the government began to allow limited access to some gem mines by foreigners. Little appeared to have changed since the days of the British Company. The government runs several mines in the Mogok area, but smuggling is persistent and little progress has been made. The mines are operated in a manner similar to that of the Burmese kings, as almost a private concession. What is not smuggled abroad is sold at the yearly emporium in Rangoon. While Myanmar still produces some of the finest sapphires, it has been eclipsed to a large degree by the recent ruby discoveries in

Vietnam. In 1992 Kane and Kammerling described, in the Fall 1992 issue of *Gems & Gemology*, the then status of the Mogok area. This paper includes very useful maps.

Thailand and Cambodia

Important sources for rubies and sapphires exist in Thailand and Cambodia. The ruby deposits occur along the Thai/Cambodia border in Chanthaburi and Trat province in Thailand, and neighbouring Battambang province in Cambodia. The rubies tend to be brownish red and somewhat dark in colour, while the sapphires are generally dark blue. Fine yellow sapphires occur at Khao Ploi Waen and Bang Ka Cha, while blue sapphires occur mainly at Khao Ploi Waen, Bang Ka Cha and Bo Ram in Thailand and Pailin in Cambodia. Rubies are found in a wide area including Bo Rai, Nong Bon, Bo Waen and Tok Prom in Thailand and Pailin in Cambodia. The centre for the gem trade in this area is the Thai town of Chanthaburi.

All gem deposits in this area are derived from iron-rich basalt. The high iron content of the mother rock carries over into the corundum, which is generally darker than the ideal. The stones are found in coarse yellow or brown sand, overlaying a bed of clay or basaltic rock. The beds are mostly within 2.5 m of the surface, but some of the mines are over 6 m deep. These deposits have only been worked to any extent since the late nineteenth century. Mining is by simple methods. Blue sapphires are also found in western Thailand at the town of Bo Ploi, in Kanchanaburi province. In the late 1980s mining at Bo Ploi was expanded dramatically, but has since declined.

Vietnam

Since 1983 Vietnam has become a major ruby producer: Vietnam rubies are striking in their similarity to those from Myanmar, as the geology of the two areas is similar; some experts have found Vietnam rubies to be at least the equal in clarity of Myanmar stones.

In a major survey published in the July 2004 issue (vol. 29(3)) of the *Journal of Gemmology*, Van Long *et al.* describe the geology of the ruby deposits in northern Vietnam. From 1991 blue, green and yellow (BGY) sapphires have been recovered from alluvial deposits associated with basaltic-type rocks in southern Vietnam and other types of corundum deposits have been found elsewhere in the country. An earlier paper by Kane *et al.* in *Gems & Gemology* 27(3), Fall 1991, also gives a useful account of ruby and sapphire from Vietnam. Pinkish purple sapphires are described; some rubies were found to contain distinct medium dark to dark blue colour zoning.

In northern Vietnam the main ruby-producing sites are in Yen Bai province (the Luc Yen and Yen Bai mining districts) and the Quy Chau deposit in Nghe An province. The southern deposits of BGY sapphires are located at Dark Nong and Binh Thuan in the Dek Lak and Lam Dong provinces respectively. Van Long *et al.* note a number of other sites.

It is pleasing to know that two exceptional ruby crystals weighing 2.58 and 1.96 kg have been designated as State Treasures and are inalienable. They were found in the placers of the Tan Huong and Truc Lac areas.

Van Long *et al.* describe the gemmological properties of corundum (ruby and blue sapphire) from the Quy Chau area. Rubies range from purplish red to purplish pink with varying saturations and sapphires are blue. As a contrast the authors describe ruby from the Yen Bai and Luc Yen deposits as showing all tones from red to pink; sapphires range from colourless through pale yellow to grey and blue. Rubies from Luc Yen are usually less saturated than those from Quy Chau. RI and DR are normal for corundum. A notably high SG of 4.08 is recorded for some BGY sapphires from southern Vietnam.

Straight and angular parallel growth features are characteristic of rubies and sapphires from Luc Yen and Quy Chau. Some rubies have been reported to show swirliness long noted for Myanmar ruby. Van Long *et al.* give a list of solid inclusions for both rubies and sapphires by location of deposit: the commonest mineral inclusions in Luc Yen and Quy Chau rubies are calcite, dolomite, rutile, diaspore and phlogopite. Rutile is found as short needles and also as twinned platy crystals. It may also appear as transparent or opaque orange-brown crystals trapped along growth zones. Zircon has been found in both types of deposit. Fluid inclusions may also contain euhedral negative crystals and flat, broad tubes, some capped by a mineral inclusion. All red and pink samples from Quy Chau and Luc Yen showed red under both types of UV and some blue sapphires from the Khoan Tthong area of Luc Yen also showed red under both types of UV.

Quy Chau produces some beautiful orange sapphires, the colour of papaya, as well as dark blue sapphires and yellow chrysoberyl. At Luc Yen some off-red and blue spinels have been reported and cobalt-blue spinels have also been reported in Vietnam, along with pink and green tourmaline and colourless topaz.

The trapiche effect seen in some Colombian emerald and described in the chapter on that species has also been noted in some ruby from Vietnam.

Kashmir (MO'D)

Blue sapphires from Kashmir are considered fine enough by the auction houses to advertise their place of origin (as with Myanmar blue sapphire and ruby, Colombian emerald). Accounts of the areas and its mines are very hard to find, the best being C.S. Middlemiss, Mineral Survey Reports, Jammu and Kashmir Government, *Precious and Semi-Precious Gemstones of Jammu and Kashmir*, 1931 [this description taken from the cover]. The report is followed by a photograph and six folding sketch maps and is #4464 in the Sinkankas bibliography mentioned passim in this book. This is the only geological report on the sapphires of Kashmir, the mines being located at Soomjam (Sumsam) Padar district, between 14 250 and 14 950 feet above sea level. Other gem minerals (tourmaline and pegmatite-hosted beryl) are also described from the region. Sapphires are covered in pages 1–50.

A later though much less detailed record, published in 1934 by the Government of India, is *A Sketch of the Geography and Geology of the Himalaya Mountains and Tibet*, part IV, The Geology of the Himalaya, pp. 277–359. This is a general overview – notes pertaining to sapphires begin on page 351. Here is an extract:

"some fifty years ago the beautiful azure blue sapphire of Sumjam in Padar, Zanskar, Kishtwar tahsil, was accidentally discovered at an altitude of 15 000 feet among rocks which have since been determined as feldspathic pegmatite veins in actinolite-tremolite lenticles in the marble bands of the area. The actinolite-tremolite schist appears to have been a modification of the marble. Large quantities of excellent stones were found and they yielded considerable revenue to the Kashmir government. . . . Subsequently the actual source of the rock appeared to be exhausted though the placer deposit continued to yield a diminishing output. Later on, new veins of corundum, sapphire and pink corundum (ruby) were discovered in the area and its neighbourhood . . . in colour, the sapphire is of a pale china-blue tint but more generally it is a rich sky blue which in the best stones becomes extremely vivid. Occasionally a more slaty blue tint appears. The colour is irregularly and unevenly distributed in the sapphire crystals, being found in stripes and patches of different dimensions among the milky grey and colourless corundum. A reddish tint is also found, though somewhat rare. Where found it varies from pale pink to rosy red and in a few cases to carmine with a slight blue tone in the red."

Gübelin and Koivula describe and illustrate some of the characteristic inclusions of Kashmir sapphire in *Photoatlas of Inclusions in Gemstones*, second edition, 1992. They describe a 'milky turbulence' (some have called this effect 'sleepiness'); this is thought to be caused by a profusion of cavities and exsolutions which diffuse the incident light. Mineral inclusions are rare but the authors illustrate a fine crystal of tourmaline magnified by 32×. Since blue sapphires from other places usually show a rather more defined inclusion pattern anyone testing a sapphire with no obvious inclusions should not rule out a possible Kashmir specimen.

India

Rubies are found in several Indian states. Facet-grade ruby has been reported from the Kangayam area of Tamil Nadu and from Karnataka. In general, the quality is not high and many stones are treated in one way or another; star rubies are fairly common.

Pakistan

In *Gemstones of Pakistan* (1990) Kazmi and O'Donoghue consider the attractive if frequently calcite-included rubies from the Hunza area in the north of Pakistan. The ruby deposits are associated with the Karakoram suture zone, the stones occurring in metamorphosed recrystallized dolomite marble in a narrow belt extending from the Hunza valley to Ishkoman, a distance of more than 100 km. Crystals are well formed and range from pink to a fine red. Included material is cut into cabochons but some of the finest reds are high facet-grade. The writer (MO'D) who visited, with Dr Kazmi, the gem-bearing areas of Pakistan for the United Nations Industrial Development Organization in late 1985 examined some beautiful specimens both *in situ* and after fashioning and found that despite the inclusions many of the pink stones were most attractive.

Other mineral inclusions were dolomite, the potassium mica phlogopite which occurs as reddish brown flakes and feather-like greenish crystals of the calcium mica, margarite. Pyrite, pyrrhotite and apatite were also found. However, no trace of rutile has yet been reported so that no Pakistan star rubies are likely to occur (so far). The absence of rutile is of course useful in the diagnostics of locality. No sapphires have yet been reported from Pakistan. Some Pakistan rubies show a distinct phosphorescence following X-ray irradiation, a phenomenon most commonly seen in some synthetic rubies. Readers are advised to consult Bender and Raza, *Geology of Pakistan* (1995; ISBN 3443110258).

Afghanistan

Bowersox and Chamerblin (1995) in *Gemstones of Afghanistan* give some details of the Jegdalek (spellings vary as with many gem-producing areas worldwide) ruby deposits which are, again as always, not easy to link with the mines described by previous authorities. Two papers by Griesbach in *Records of the Geological Survey of India* (Afghan and Persian field notes (*Records*, vol. 19, pp. 48–68, 1886) and Field notes to accompany a geological sketch map of Afghanistan and Khorassan (*Records*, vol. 20 no. 2, 93–103, 1887)) are quoted by Bowersox and Chamberlin as well as a book by Catelle (1903) – although omitted from the list of suggested reading this must be *Precious Stones, A Book of Reference* [Sinkankas #1190] which was well thought of in its day.

Bowersox and Chamberlin place the Jegdalek ruby deposit in the southern portion of the Sorobi district, 60 km south-east of Kabul in the Jegdalek river valley. They attribute ruby formation to the contact metasomatism of ultra-acid granites found in contact with Precambrian carbonate and magnesium rocks in the Pamir-Nuristan median mass. Ruby mineralization is irregularly developed. Gem ruby is found *in situ* in interstratified Proterozoic marble and gneisses intruded by granitic rocks from the Oligocene. The ruby-bearing marble is notably coarse-grained.

Ruby crystals are also found in a white dolomitic granular limestone. Mining, as attested by Gübelin in *Gems & Gemology*, 18(3) 123–29, 1982, is primitive. Bowersox in *Gems & Gemology*, 21(4) 192–204, 1985, notes that the most noticeable inclusions seem to be small spots or zones of a strong blue colour; their shape is mostly trigonal or hexagonal in outline. Some stones show parallel blue bands. A sharp division between red body colour and blue patch is highly characteristic of Jegdalek rubies. Crystals, apparently of calcite, are common.

A personal communication from Themelis to Bowersox states that the blue inclusions are effectively removed by heating, leaving the stone pink.

Sri Lanka

Corundum with a wide range of colours is found in south-west Sri Lanka. A good account may be found in Gübelin, *Die Edelsteine der Insel Ceylon* (1968 – no English translation, privately published but with apparently a very limited one-time commercial distribution, now very rare: Sinkankas #2575). Sinkankas (1993) makes the point that almost all the literature pertaining to the gemstone mining in Sri Lanka is in the form of papers in journals, there being no monographic survey. This is still the case in late 2004.

Sri Lankan ruby inclines to pink rather than crimson but stones are very bright and lively – this applies also to blue sapphire. Star rubies and blue sapphires can be very beautiful. Fine examples include the Rosser Reeves star ruby of 138.7 ct (Smithsonian Institution) and the 392 ct star sapphire owned by the State Gem Corporation of Sri Lanka is exceptional.

Mining is simple and any method of reaching the illam (gem gravels, perhaps 15 m down) is used, sites being chosen by chance finds of pebbles at the surface.

Virtually all Sri Lankan gems are cut and polished locally. Much blue sapphire shows the best or only blue in one part of the crystal; lapidaries manage to place the blue in the culet (bottom) area of the faceted stone so that the blue shows through the table though not in any other direction – such stones are intriguing and beautiful. Colour change (blue to purple) may be seen in some sapphires.

Sri Lankan ruby shows inclusions distinctive enough for the location to be determined. Flakes of 'biotite' (this name is now taken as a series name for dark lithium-free mica rather than as an individual species name) are very characteristic and rutile needles are generally longer and more slender than those seen in Myanmar stones. Grains of metamict zircon are accompanied by dark tension haloes and liquid inclusions form feather-like patterns. Sri Lankan blue sapphire shows distinctive 'arrowheads' of rutile which form V-shaped structures; the feldspar group mineral albite, apatite and hematite may also be found. Gübelin and Koivula (1992) illustrate albite crystals forming 'comet-tails' (an assemblage more familiar in synthetic corundum).

Sri Lankan blue sapphires on the market today may very well have begun life as milky and colourless. The name geuda has been used for both the colourless material and the blue crystals resulting when it is heated. The treatment of gemstones for colour improvement is a major concern in the gem trade today, difficulties arising as much over disclosure as identification. Scientific and practical matters in this context are excellently covered in monographs by Nassau (*Gemstone Enhancement*, second edition, 1994; ISBN 0750617977) and Themelis (*The Heat Treatment of Rubies and Sapphires*, 1992). Further details can be found in Chapter 30. Today it is not easy to find an untreated sapphire in Sri Lankan markets.

China

While some of the minerals of China are described in issue 26/27, 1994, of the monographic series ExtraLapis, a preceding paper in the Winter 1992 issue of *Gems & Gemology* deals with the blue sapphires found in alluvial deposits and, most unusually, *in situ*, near Wutu, Changle county, central

Shandong Province. The sapphires show a range of colours, including dark blue, blue, greenish blue and yellow. The most abundant trace elements found are iron, gallium, titanium, cobalt and vanadium. Gemmological properties accord with those of other basaltic-hosted sapphires; among the mineral inclusions are uranium- and thorium-rich zircon, titanium-rich columbite, sodium feldspar, apatite, ilmenite and magnesium-iron spinel (cf. sapphires from the Inverell-Glen Innes, Rubyvale and other areas in eastern Australia). The orange-red zircon inclusions have reached 1 mm and the columbite is black with a metallic lustre.

Australia

Although the three volumes (one of maps) of T.W. Edgeworth's David's *Geology of the Commonwealth of Australia* say little about sapphire, useful accounts of the corundum deposits can be found in I.A. Mumme's *The World of Sapphires*, 1988 (ISBN 0959306919). The discovery of sapphire in New South Wales dates from the middle nineteenth century but commercial production did not begin until the First World War had ended. Sapphires were found in the alluvium of streams on Frazer's Creek near the town of Inverell in the New England area and the name Sapphire was taken up by the settlement which grew up. Production tailed off in the 1930s as more stones were coming from deposits in South-east Asia. As prices for Australian sapphires improved the search for and marketing of New South Wales material took on new life. Deposits in the New England district which at first were accessible to the solitary miner produced less and mechanized operations took over. The export of geuda (see above) from Sri Lanka to Thailand for treatment also made it more difficult for the Australian sapphire industry.

Mumme gives considerable detail of the New England sapphire production and also describes the crystals. In the alluvial deposits crystals are found as fragments with the common form of tapering hexagonal prisms though some well-rounded crystals also occur. Colours are dark blue or dark green with intermediate shades of either colour; yellow crystals are rare. The dark blue colour is seen parallel to the ordinary ray and colour zoning is pronounced. Hollow cavities, rutile needles and cavities are common.

The Anakie sapphire deposit in central Queensland was proclaimed in 1902 as a recognized commericial field. While corundum fragments have been found in the basalt, crystals are recovered from alluvial sources ('wash'). The proportion of gem to non-gem-quality material is low but as well as blue and green which predominate, a wide range of colours has been found, including golden yellow and star stones. The very occasional ruby and colour change crystal has been reported. The name Rubyvale does not appear to indicate a plentiful supply of ruby. The official geological report on the Anakie field can be found in *Publications of the Geological Survey of Queensland*, 172, 1902.

In general, the Australian sapphire is either dark or heavily included, but often both. Nonetheless many stones may be considered the best of their colour, this applying in particular to green specimens. Heating can improve many examples. Green and yellow stones, in particular, show a clear absorption spectrum though the iron content inhibits the yellow stones from fluorescing (Sri Lankan yellow stones usually show a distinct fluorescence but faint absorption spectrum).

In 1978 flattish ruby crystals, rubies of poor cabochon grade, were recovered from a shallow subsurface deposit on the Harts Range, northeast of Alice Springs, Northern territory.

United States of America

To European gemmologists (this one [MO'D] at any rate), Americans seem seriously to under-value their gem minerals! Some of the sapphires found in North America rank with the most beautiful yet recovered. Two different types of occurrence, *in situ* (hard rock) and alluvial deposits, both produce specimens of high individuality.

Sapphires of Yogo, Montana

In 1900 Weed and Pirson produced *Geology of the Little Belt Mountains, Montana*, the study forming part 3 of the 20th *Annual Report of the United States Geological Survey*, 1898–99. Discussing the transition from analcite-basalt to minette [orthoclase-biotite lamprophyre consisting of biotite and subordinate diopside phenocrysts in a groundmass of orthoclase or sanidine]. Weed and Pirsson study the sapphire-bearing rock of Yogo Gulch, Judith Basin county. The first report of the occurrence was published in *American Journal of Science*, 4th series, vol. IV, p. 421. Stepen M. Voynick's *Yogo, the Great American Sapphire* (1990; ISBN 087842217X) gives an excellent account of the history of the area as do the three volumes of *Gemstones of North America* by John Sinkankas.

Highly characteristic sapphire crystals occur in a dark silica-poor lamprophyre dike consisting of biotite and pyroxene group minerals. The dike intrudes the Madison limestone and though only a few feet wide, extends for at least five miles.

The history of the deposit is well worth reading, especially as there are few important examples of gem corundum crystals being mined from their parent rock. At one time British and American companies worked different ends of the dike and a business venture, The New Mine Sapphire Syndicate, was set up and in 1914 produced a beautiful and now very rare advertising booklet, *A Royal Gem*, by Arthur Tremayne. The beauty of this booklet [Sinkankas #4741] is in its (blue) colour illustrations of Yogo sapphires in their cut and rough state and also as set in jewellery. For some time, sapphires from Yogo were handled by the London firm of Johnson, Walker and Tolhurst who retain illustrated catalogues, featuring Yogo sapphires, at the present time.

Colour illustrations of Yogo sapphires are also given in the fifth and sixth editions of Edwin Streeter's *Precious Stones and Gems* [Sinkankas #6399, #6400]. The specimens depicted are the characteristic flattened crystals in which a regular outline is hard to detect as prism faces are virtually absent. Sapphires are, in general, free from particularly characteristic inclusions but Koivula and Gübelin (1992) specify an inclusion pattern reminiscent of Thai rubies; analcime crystallites with haloes of fluid drops. Some analcime crystals show cube-icositetrahedron combinations.

Yogo blue sapphires are almost recognizable from their exceptionally bright but deep colour; as crystals are not usually large, Yogo sapphires make superb melee. Lilac and purple stones can also be beautiful; a very few rubies have been found. Without doubt Yogo sapphires are the most commerically important of all the coloured stones so far found in North America. Price guides for any gem species are notably unreliable owing to the market's inevitable volatility but, in *Standard Catalog of Gem Values*, second edition, 1984 (ISBN 0945005164), Miller and Sinkankas suggest US$ 1000–3500 for fine blue Yogo sapphires weighing 0.5–2 ct.

In *The 1922 Handbook and Descriptive Catalog of the Collections of Gems and Precious Stones in the United States National Museum* (1922) Merrill lists only two blue sapphires under the heading Yogo Gulch and in *Corundum Deposits of Montana* (*United States Geological Survey Bulletin* 983, 1952) Clabaugh mentions a price of $75 a carat for the larger Yogo sapphires when faceted. This is a vital report for anyone interested in Montana corundum in general.

In *Gemstones, Quality and Value*, vol. 1 (1993; ISBN 4418999027) Yasukazu Suwa says that Montana blue sapphires are a watery blue and do not command the highest prices for that reason. He does not say, however, that the sapphires described are Yogo rather than from one of the alluvial deposits – Yogo sapphires at their best are far from watery in appearance.

Crystals from Yogo are illustrated in *extraLapis* 15 (1998) the photographs showing the crystallographic nature of eight typical specimens.

Sapphires showing a wide range of colours are found in the (dry, overlain) gravels of the Missouri River in the general neighbourhood of Helena, Montana. While the colours can be very attractive, most of the production is routinely heated to give brighter colours. In general, crystals are barrel-shaped though some are flattened and characteristically striated across the flat pinacoid faces. Colours usually sold are blue, green and yellow. A fine 'hot pink' is achieved by heating and, in general, the treated stones are most attractive.

Details of the results obtained by heating can be found in papers by John L. Emmett and Troy R. Douthit, who have made a close study of heating methods and their results. Yogo sapphires are not heated.

The main working sites are Rock Creek, Dry Cottonwood Creek and Eldorado Bar. It is interesting and instructive to study the efforts made by a number of individuals and companies over the years to establish a serious and profitable trade in the heated sapphires. There has been limited and local success but the sapphires are not known in Europe or even very well in the US; though the position could improve if only the supply of sapphires from South and south-east Asia and Australia was not so well established.

Sapphires, apart from Yogo specimens, known to be from Montana are so routinely heated that disclosure is probably not needed though if there were to be a regular export of sapphires from North America some form of certification might be called for. When not heated the sapphires from the gem gravels may show negative crystals occupied in some instances by natural glass and one or more gas bubbles. Heating alters inclusions and obscures their outlines.

In North Carolina ruby of fair quality has been found in Macon County, at Cowee Creek. Ruby crystals are found in gravels with red garnet. Rubies with rutile inclusions have been cut to give star stones of good quality.

Africa

Kenya

In 1973 American geologists Tim Miller and John Saul discovered ruby in Tsavo West National Park, in the Mangari area. As so often happens, facet-grade crystals were much less plentiful than heavily included cabochon-only material. Crystals in some cases were well formed and would have made desirable collectors' specimens had many not evaded the cabochon cutters and perhaps exporters. The Kenya government of the time appropriated the mines: though ownership disputes took time to resolve private owners eventually regained control and good-quality cabochon-grade and a few facet-grade crystals are being recovered. Most crystals are sent to

Thailand for fashioning. Turkana and its vicinity in north-west Kenya produce dark, variously coloured sapphires.

Tanzania

Corundum from Tanzania occurs in a wide range of colours including orange, purple, mauve and yellow as well as red, green and blue. In general, the colours can be said to be pastel rather than well defined but are particularly attractive for this reason. Ruby inclusions in a bright green chromiferous rock make good carving material and have been given the vernacular name anyolite.

Of greater importance are the varieties of sapphire from the Umba river valley in north-eastern Tanzania, near the Kenya border where sapphires are mined from several pits in and around a greyish green serpentine pipe. The sapphires and rubies of Umba occur in various colours, often of a pastel shade which is very attractive. Purple and mauve and orange sapphires, in particular, are beautiful.

In the Fall 1991 issue of *Gems & Gemology*, Hänni and Schmetzer report the occurrence of ruby in parcels of red spinels said to be from the Morogoro area of Tanzania. The colour and inclusions were similar to some found in Myanmar ruby; inhomogenieties in growth zoning parallel to distinct faces and irregular swirls; sets of twin lamellae parallel to one, two or three rhombohedral faces; clouds of rutile crystals and coarse rutile; octahedral inclusions with slightly rounded solid material (spinel) or negative crystals looking similar. Zircon, apatite and pyrope have also been identified. The potential for asterism has been found in some specimens.

The mines are situated in the western Uluguru Mountains. An excellent account of ruby mining in Tanzania appeared in *extraLapis* 15, *Rubin*, *Saphir*, *Korund*, 1998. See also zwaan garnet, corundum and other gem minerals from Umba in *Scripta Geologica* 20, pp. 1–4, 1974.

Nigeria

Starting in the late 1970s, corundums of a blue, green and yellow colour began appearing in world markets. These are basalt-derived stones and tend to be dark in colour. The mines are located in the Kaduna district.

Malawi

While Malawi corundum has been known at least since 1958 when the Nyasaland Geological Survey reported on it, regular mining of the deposits at Chimwadzulu Hill is more recent. The area is 145 km south-east of the capital city, Lilongwe. The area comprises mainly metamorphosed ultramafic rocks with an exposed surface and an exposed surface diameter of about 1 km. The surrounding county rocks are Precambrian metasediments metamorphosed into schists and gneisses. The bedrock is deeply weathered on the hill with a surface layer of iron-rich red porous clay. Corundum is found in the topsoil and there is good evidence of bedrock resources. Minerals in aggregate found with the corundum include quartz, magnetite, hematite and chromite and separation work is necessary for extraction of corundum. David Hargreaves, Chairman and CEO of Minex (Pvt.) Ltd and the owner and operator of the mine, who has supplied me [MO'D] with details of its operation, reports that more than 50% of the corundum is magnetic and this property may be found useful in separation. Corundum-bearing soil is scraped off the bedrock and sent for processing.

When Minex began developing the mine in 1995 about 10% of the orebody had been worked. Recovery potential seems promising. The area had been described as a sapphire deposit in which the occasional ruby was found. Naturally coloured sapphire crystals of quality were rare and seemed excellent candidates for heat treatment. Conventional heat-treatment methods, however, did not achieve good colour. Fortunately the discovery of ruby in natural colours ranging from pink to purple with some orange padparadschah helped the operation to survive. Further experimental work on heating the sapphires has now produced some good blue, yellow and fancy colours though no red stones were achieved, nor could naturally red crystals be enhanced. Taking into account the high probability that the gem-buying public will turn to stones whose colour can be shown to be natural, the potential of Chimwadzulu corundum is good.

Chimwadzulu corundum can be described as ranging from a true pink to a very dark purplish blue. Stones weighing more than 1 ct sell on their individual merits though melee, high in volume and with sales vital, needs colour consistency. For this reason stones have been grouped into nine categories; ruby, padparadschah, purple, pink, yellow, teal, blue, cognac and fancy.

Columbia Gem House in the US has distributed the Malawi corundum and the profile is still expanding. Locality information is always useful when selling gemstones. A paper by Professor Andy Rankin in *The Journal of Gemmology* in 2003 described the Chimwadzulu material and the results of heating trails. The ratios (wt% oxides) of Cr/Ga vs Fe/Cr for Chimwadzulu and Thai and Kampuchean material shows only a minor overlap: there is no overlap with Mogok, Mong Hsu, Vatomandry and Longido.

Other African Sources

Sapphires of various colours have been reported from Namibia. Blue sapphire has been found in the Beraketa region of Madagascar where it occurs with feldspar; blue sapphire is also reported (*extra Lapis English* no. 1, 2001) from a skarn south-east of Andranondambo. Blue sapphire of moderate quality is found in some abundance in the Ilakaka river; the town of Ilakaka has become a major gemstone trading centre. Readers should consult Lapis for June 1999 for a paper by Karl Schmetzer on Madagascar corundum. Fancy sapphires (not ruby or blue sapphire) appear to be characteristic of African deposits. Crystals with a colour change (the colours varying) may become another African speciality.

Small deposits of corundum of various colours have been reported from the Somabulu forest in south-western Zimbabwe and a deposit of sapphire has been reported from a pegmatite and in alluvial gravels in north-eastern Zimbabwe.

Dark blue sapphires have been found near Jauru, Mato Grosso, Brazil, and both ruby and sapphire in the Rio Mayo, a tributary of the Patia, and in the sands of the Platayaco, in the Caqueta area of Colombia.

6

Beryl

The beryl group of silicates includes the important gem varieties emerald, blue aquamarine, pink morganite and red and yellow beryl which have no regular names. Other colours sometimes turn up and the commoner ones may be pale or strongly coloured. The formula usually given for the beryl group is $Be_3Al_2Si_6O_{18}$; however, the general formula may be expressed as $A_{2-3}B_2Si_5(Si,Al)O_{18}$, where A = beryllium, magnesium or iron and B = Aluminium scandium or iron. The mineral crystallizes in a hexagonal or in a closely related orthorhombic space group (*Dana's New Mineralogy,* 1997). Cordierite (iolite), see below, is also a member of the beryl group but in this section we are looking at the varieties of beryl itself. Crystals are usually columnar hexagonal prismatic with a basal pinacoid; some emeralds may show second-order prisms. Recently the new mineral pezzottaïte has been found in Madagascar and is described below.

Emerald is coloured by the trace element chromium though vanadium is often present; attractive green beryls which can closely resemble emerald may be coloured by traces of vanadium alone. Some alkalis may be present but do not usually affect beryl's ornamental properties. Beryl is a ring silicate in which the rings are centred on hexagonal axes; the rings consist of six silicon–oxygen tetrahedra. The colouration and properties of the other ornamental beryls are discussed below.

Emerald

Types of Deposit: General Note

In *Geochemistry of Beryllium* (1966), Beus describes a number of types of deposit, including those of emerald from Colombia, South Africa and Russia. Colombian emerald deposits are described in Moncaca and Quinn,

El maravilloso mundo de la Esmeralda colombiana (1995; ISBN 9589578004); for an in-depth study of the gem beryls the most important book is Sinkankas, *Emerald and Other Beryls* (1981; ISBN 0801971144). Another Colombian study of that country's emeralds was published in 1948. Munoz and Barriga Villalba present colonial and later history together with accounts of current mining, geology and mineralogy. This book, *esmeralda de Colombia*, published for the 25th anniversary of the Banco de la Republica, is rare [Sinkankas #4650].

The best survey (in Russian) of the Russian emerald deposits is by K.A. Vlasov and E.I. Kutukova, *Izumrudnye Kopi*, Moscow, Akademia Nauk SSR, 1960 [Sinkankas #6905]. This gives details of the history of the deposit which was located in 1830. Further details are given below during the discussion of occurrences. A translation was published in 1960 by the US Joint Publications Research Service with the serial number JPRS 5979 but Sinkankas considers this inferior to the original as the illustrations are poorly reproduced and in any case is an abridgement of the original text.

Beryl minerals occur in granites and granite pegmatites, in mafic metamorphic rocks and in high-temperature hydrothermal veins and in vugs in rhyolite. The Colombian emerald fields are located in the Cordillera Oriental, covering parts of the provinces of Boyaca, Cundinamarca and Santander. The important deposits at Muzo are Cretaceous (Sinkankas, 1981, *op. cit.*) and the mines lie in a series of black shales. It is interesting to note that while some varieties of beryl occur in granite pegmatites, this scarcely ever seems to be the case with emerald.

Properties

Emerald is one of the major gem species and one in which it is possible sometimes to identify the place and conditions of origin from a study of its inclusions. Chromium must be present if a green beryl is named as emerald; green beryl coloured by vanadium alone (these tend towards apple-green) should be termed green beryl. Gem-quality emerald normally has hardness in the range 6.5–8, and while there is no cleavage, specimens are notably brittle and should never be subjected to ultrasonic cleaning. The specific gravity of emerald as well as its refractive index varies with place of origin: Gübelin in *Gems & Gemology* 18(3) 1982 published a list of these figures.

Beryl is uniaxial negative and the low dispersion, at 0.014, does not enhance specimens. Pleochroism shows most strongly in darker-coloured stones, emerald showing yellowish green and blue-green. Chromium-rich specimens show red through the Chelsea filter, though with increasing iron content the effect diminishes and some specimens from South Africa and India, in particular, show little if any change. Synthetic emeralds glow a pronounced red.

Emerald Deposits

In this section I have presented what I consider the most interesting and useful feature of some of the best-known emerald deposits. I have not attempted to give much historical detail unless supported by references of repute and I have had to leave out some of the less authenticated discoveries. Sinkankas (1981) covers all omissions.

Egypt

The emerald deposits of Egypt are best described in Hume, *Geology of Egypt*, vol. 2, part 1, 1934, a publication of the Survey of Egypt [Sinkankas #3074]. The study describes metamorphic rocks and includes those forming the basement formations at the emerald-mining district of the Etbai desert. This account should be studied closely by anyone contemplating a detailed account of emerald locations in Egypt – the index can be a minefield (at least in my copy!). A useful account of the emerald mines, with some history, was published in *The Journal of Gemmology* for July 1990. In general the emeralds are not of the highest quality.

Austria

Emerald and other beryl varieties are found in Salzburg solely on the north slope of the Hohe Tauern range, which runs east to west about 80 km south of Salzburg. Emerald occurs on the side of the Legbach ravine; maps of the area and the deposits can be found in Sinkankas, *Emerald and Other Beryls*, 1981, ISBN 0801971144 Sinkankas #6110 (later abridgement by Peter Read). The deposit lies along the Central Gneiss of the Hohe Tauern and amphibolite rocks, where emerald crystals are found in a dark mica schist and in tremolite amphibolite schists (Leitmeier, *Das Smaragdvokommen im Habachtal in Salzburg und seine Mineralien, Tschermak's mineralogische und petrographische Mitteilungen* 49, pp. 303–32, 1937). Crystals show few clear areas though the colour can be very attractive. It is possible for collectors to purchase large masses of dark mica schist from which a few specks of emerald protrude; the author (MO'D) has retrieved a fine emerald crystal of about 4 cm from such a specimen.

Habachtal emeralds occur as characteristic prismatic crystals and have an RI in the range 1.585–1.597 for the ordinary ray and 1.578–1.590 for the extraordinary ray, with birefringence 0.007–0.008 and with yellowish green and bluish green dichroism. The absorption spectrum was in general fairly weak and characteristic for emerald (see p.161). The inclusions are the best clue to origin, showing a ready cleavage along basal planes and often profuse dark mica flakes. Dark green tremolite needles or platelets are often seen. Gas and gas-liquid inclusions form feathers and veils. A whitish effect is seen when these inclusions are prominent.

Colombia

A preliminary note on sources for the study of the detailed history of emerald mining in Colombia can be found at the beginning of this chapter.

We have already noted that the Colombian emerald deposits are situated in the Cordillera Oriental, the eastern range of the Andes. They lie north to north-east of Bogotá, the Colombian capital, in the departments of Boyacá, Cundinamarca and Santander. The formations in which the emeralds are found are extensive enough to provide a satisfactory supply for the foreseeable future. Sinkankas (1981) gives a chronology of the Colombian emerald. The best-known emerald-producing regions are to the north and east of Bogotá and can conveniently be described under the headings Muzo, Cosquez, Chivor, Gachala and Buenavista, following Sinkankas (1981).

Colombian emeralds are mined from two zones: the western zone that hosts, from north to south, the Peñas Blancas, Cosquez, Muzo and Yacopi mining regions, and the eastern zone that hosts the Chivor/Somondoco and Cachala mining regions.

The Chivor Mines

The mines are about 1000 m of conformable sediments made up of light grey calcareous shales with some carbonaceous lenses. The top member is hard grey fossiliferous limestone and the rarely exposed lower member is hard blue thin-bedded limestone or carbonaceous shale. In the late Cretaceous or early Tertiary the strata were tilted and folded with considerable distortion and fragmentation of the rock into angular pieces which are easily mined by hand (Sinkankas, 1981). During the faulting pyrite was formed and can be found as inclusions in emerald. No emeralds were found by the Spanish in the upper formations at Chivor but they were found to occur in albite-quartz-apatite veins in a hard bluish shale at two levels lower down, the lower level providing some large and fine-quality crystals. Crystals from Chivor occur predominantly as first-order prisms and as a rule are more elongated than emerald crystals from Muzo. The commoner inclusions are pyrite which forms well-formed cubes and pyritohedra and may be taken as diagnostic, albite and quartz, together with characteristic three-phase inclusions and goethite.

The emeralds of Chivor have an SG of 2.69, and median refractive indices of 1.577 for the ordinary ray and 1.571 for the extraordinary ray, with birefringences of 0.005–0.006. Chivor stones usually show a fairly strong red under the colour filter and a red fluorescence under UV.

Muzo and Cosquez Emeralds

The geology of the Muzo is similar to that of Chivor, but the emerald-hearing veins, containing calcite, quartz, dolomtite and pyrite, run through a black carbonaceous limestone and shale which form the country rock. Crystals are hexagonal prisms and often of a magnificent colour with a mean SG of 2.70 and mean RI for the ordinary and extraordinary rays respectively of 1.58 and 1.57 with DR 0.005–0.006. Three-phase inclusions are characteristic, containing a salt crystal, liquid (water or brine) and gas bubble (carbon dioxide). Calcite crystals have not been recorded as inclusions in Muzo emeralds. Brownish crystals of the cerium carbonate, parisite, are, however, characteristic.

The name trapiche emerald is given to crystals with a cog-wheel-like core and spokes of fine green emerald which occur both at Muzo and Chivor, the name deriving from a Spanish word for cane-crushing gear. Cores in Muzo trapiches tend to be darker than in Chivor specimens.

Buena Vista emeralds are found in an area with similar geology and mineralogy to that of Chivor; the group of mines is in the municipality of Ubala. There are other emerald-producing sites in Colombia. The Gachala area is south-east of Muzo and closer to Bogotá; in the 1970s emeralds of attractive colour were on the market in Germany for a short time. The colour is attractive with noticeable banding. The geology and mineralogy are similar to that of Chivor.

Large crystals of the finest emerald are almost certain to have come from Colombian sources; in *Smaragde unter dem Urwelt* (1941; Sinkankas #3456) Fritz Klein gives full details of the recovery of a large emerald crystal 'Patrizius' (sometimes 'Patricia'), which later went to the American Museum of Natural History, New York. A picaresque account of emerald mining in Colombia can be found in Peter Rainier's *Green Fire* (1942; Sinkankas #5313 and several later editions). The area described is Chivor and neighbouring Somondoco.

Brazil

First rumours and later finds of emerald encouraged colonial expeditions to the interior of Brazil and the royal house of Portugal received many specimens from the early seventeenth century onwards. Other varieties of beryl were found during prospecting for mica deposits. In

1812 John Mawe published the first edition of *Travels in the Interior of Brazil* [Sinkankas #4270] though gold and topaz feature as well as beryl varieties. The book became a classic for Brazil mining historians. This work, however, was not the first mention of emerald from Brazil, since in 1565 Conrad Gesner published *De rerum fossilium* [Sinkankas #2366], in which 'smaragdus Bresilicus' is mentioned, with a drawing which is clearly tourmaline; this was a mistake causing misconceptions for the future.

In the state of Bahia, emerald was found in a mica schist near Salininha and the deposits at Carnaiba also occur in mica schists. Crystals from Bom Jesus dos Meiras are notable for their fine forms; one crystal with 18 pyramidal faces is recorded [Sinkankas, 1981]. In Goias the only beryl variety found is emerald, and in the otherwise beryl-rich state of Minas Gerais relatively few specimens have been found. In general, Brazilian emeralds show two-phase inclusions with crystals of dolomite, biotite and talc. The most frequent inclusions, however, are liquid films. On the whole the colour does not equal that of the Colombian or the (much smaller) Pakistan crystals.

Pakistan

Emerald in Pakistan is found in deposits which occur along or close to the suture zones in which two plates meet, in this case the Indus suture zone characterized by the Main Mantle Thrust and the Karakoram suture zone characterized by the Main Karakoram thrust (Kazmi and O'Donoghue, *Gemstones of Pakistan* 1990). Emerald showings have been reported from a number of sites. The finest emerald crystals have been found east of Mingora town, where the Indian subcontinent sequence is composed of Precambrian crystalline schist, which is overlain by Alpurai talc-mica-garnet schist and the Saidu calc-graphitic schist possibly of Palaeozoic to early Mesozoic age. These rocks comprise an antiform in the core of which the Swat granite gneiss has been intruded. The Indus suture Melange Group (an assemblage of rocks from various origins) has been thrust over these crystalline schists and gneisses. One of the thrust sheets of the Melange Group is the Mingora ophiolitic melange and is composed of blocks of talc-dolomite schist, serpentinite and other rocks, all in a matrix of talc-chlorite-dolomite schist and calcquartz-mica-chlorite schist. Emerald mineralization in the suture zone is confined to the talc-chlorite-dolomite schist. Considerable work on the mineralization of emerald has been carried out by the Geological Survey of Pakistan and by the University of Peshawar. The most exhaustive study is A.H. Kazmi and L.W. Snee, *Emeralds of Pakistan, Geology, Gemology and Genesis*, New York, 1989.

Primary and secondary fluid inclusions are characteristic of Pakistan emeralds whose colour can equal that shown by the best Colombian stones. However, crystals are almost always small in the finest qualities. Inclusions of dolomite have been found as rhombohedra.

The Khaltaro emerald deposits are reported to be in pegmatites (and at high altitude) in the Gilgit Agency of the Northern areas of Pakistan. The best description is by Kazmi and Snee (1989).

Russia

Despite occasional claims that emeralds from the Urals were known and described in classical times the reports cannot be substantiated. It is known that emeralds were discovered in its biotite schist host rock in 1830 or 1831 by a small stream, the Takovaya. Versions of their finding and subsequent history are recounted in Sinkankas (1981). The best emeralds from the Urals are beautiful and some large crystals have been recorded. The emerald deposits of the Urals were one of the many mineral and gem sites to be investigated by the celebrated geochemist and mineralogist, A.E. Fersman, after whom the A.E. Fersman Mineralogical Museum in Moscow is named. Fersman wrote papers of particular value for those studying emerald: *Izumrudnye kopi na Urale*, in *Sbornik statei i materialov 46, Materialy dlya izucheniya estesyvennykh proizvoditelnykh sil Ossii*, Petrograd [St Petersburg] 1923 [Sinkankas #2044]. Readers should also consult Fersman's other works. The description of the emerald mines in the Urals by Vlasov and Kutakova has already been mentioned above. The darkest and finest emerald crystals occur in biotite schists, others are found in quartz and feldspar. Crystals, often jointed or broken, are long prismatic; frequently a colourless zone surrounds a green core. Quartz and brown mica were reported as inclusions in Ural emerald by Söchting in 1860 (*Die Einschlusse von Mineralien*, Freiberg, Germany [Sinkankas #6201]). Refractive indices tend to be higher than those shown by other emeralds.

Examination of 100 newly mined and 20 older Uralian emeralds, by Schmetzer *et al.* (*Gems and Gemology*, Summer 1991), indicated that the characteristic inclusions of Uralian emerald (healed fractures, tubes parallel to the c-axis, flat cavities at right angles to the c-axis, growth banding and biotite-phlogopite mica) are common to emeralds from other sources. Careful examination of emeralds presented from the Urals is therefore needed though confusion with other locations would have little commercial significance.

While the Takovaya emerald mines are the best-known in Russia, deposits of beryl are also found in the Chita area of Transbaikalia (south-central Siberia) and in the Ukraine. As far as emerald is concerned, only the deposits in the Ukraine have been reported from altered rocks enclosing a granite pegmatite, at an unpublished location.

Australia

Emerald was found in the Emmaville area of New South Wales during prospecting for cassiterite in 1890. The crystals are reported to occur in quartzose veins with cassiterite, topaz, fluorite and other minerals. Colour is often uneven. Better specimens are found in the Poona area of Western Australia where granitic pegmatites intrude into Archaean greenstones surrounded and invaded by granites. Well-developed prismatic crystals are found in biotite schists.

South Africa

In 1927 emerald crystals were found about 20 km east-north-east of Gravelotte station in the Leysdorp district of the Letaba area of north-eastern Transvaal. The deposits are associated with acid pegmatite inclusions in biotite, chlorite and actinolite schists belonging to the Swaziland system. Gem-quality crystals occur almost invariably in the biotite schists at or near contact with the pegmatite bodies. Associated with the emeralds are quartz, apatite, schorl, molybdenite, pyrite and feldspar.

The crystals have unusual prismatic form and range in size up to 50 mm or more in length, but most often they are variable in colour, cloudy, cracked and flawed. Silver-grey molybdenite inclusions suggest Gravelotte as a possible source. The South African emeralds rarely show red through the Chelsea colour filter.

Zimbabwe

During 1956 the prospectors Contat and Oosthuizen discovered emeralds in the Belingwe district in the southern part of the country. The occurrence is located in the Mweza Greenstone Belt, in the Sandawana Valley, on the south side of the Mweza Range, the crystals occurring in tremolite schist bordering pegmatite dykes. The bright and very attractive, though usually small, crystals are highly priced today. The name Sandawana emerald is invariably used.

Sandawana emeralds have an SG of near 2.75 and mean RI of 1.590 for the ordinary ray and 1.584 for the extraordinary ray, with a DR of 0.006. The inclusions are rather typical tremolite needles, occurring either as short rods or as fine, often curved, fibres, and garnets with a yellowish limonite halo. Some Zimbabwe stones have been seen to show colour zoning. Zimbabwe emeralds seem to be inert under ultra-violet light and show a weak reddish under the Chelsea colour filter. Emeralds are found over a fairly wide area but the stones from other mines do not seem to compare with the stones from Sandawana. Other mines are the adjacent Machingwe mine, the Belingwe mine and the Mustard, Flame Lily and Cohen's Luck claims which lie in the Filabusi Greenstone Belt to the north-west of Sandawana, and the Chikwanda, Novello, Twin Star and Mayfield claims in the Victoria Greenstone Belt to the north-east.

India

Emerald from India is widely but irregularly documented to the extent that almost any source compiled before the nineteenth century is of doubtful value though pleasing to read. References to Canjargum (Cangagem) probably refer to aquamarine rather than emerald as greenish blue beryl (the names may be a corruption of Kangayam, near Coimbatore, Tamil Nadu – formerly Madras). Emeralds were not found until 1943 when crystals turned up at a number of deposits in Rajasthan. However, there is no evidence that this area supplied the emeralds of antiquity as no sign of ancient workings has been found.

The Rajasthan deposits are described by Crookshank in Minerals of the Rajputana pegmatites in *Transactions, Mining, Geological and Metallurgical Institute of India* 42, 105–89, 1948. Green crystals were found near Kaliguman in south Rajasthan, occurring in schists and perhaps formed under pneumatolytic conditions. The emerald crystals are simple prisms which sometimes form aggregates. The best specimens are said to come from the area of Rajgarh and to show a colour equal to that of the Muzo emerald. None of the specimens shows red through the Chelsea filter; two-phase (liquid and gas) inclusions are characteristic. They resemble commas; biotite plates are also characteristic. Jaipur is the marketing centre.

Good illustrations of emerald (and other major stones) with useful commentary can be found in the large study, *Jewels of the Nizams*, by Usha R. Bala Krishnan, Government of India, 2001; ISBN 8175084069.

Tanzania

In 1970 emerald was found near the western shore of Lake Manyara at the base of the west escarpment of the Gregory Rift Valley in northern Tanzania (Schluter, *Geology of East Africa*, 1997, *Beitrage zur regionalen Geologie zur Erde*, Bd 27; ISBN 3443110274). Schluter's account suggests that the most productive era for the Mayoka mine was from 1970 to 1973. The emerald and alexandrites occur in biotite mica schists and also in the contact zone between biotite-actinolite schists and pegmatites. The schists appear to occur as bands or lenses in the gneisses making up the rocks of the Mozambique Belt. Pegmatites in the area are usually small.

The emeralds are a fine grass-green, sometimes with a yellowish or a bluish tinge, and the best-quality crystals may weigh up to about 8 ct. None of the emeralds, whatever their host, shows a deep red through the Chelsea filter. Apatite and mica flakes, with some orthoclase and quartz, are the main mineral inclusions. Properties are in the normal range for emerald.

Zambia

Emerald has been recovered from areas not far from the Zambian copper belt since about 1970. The Miku mine is about 56 km south-west of Kitwe. Other deposits in the Kafubu emerald field, SE of the Miku mine have been identified. Emeralds occur in contact zones within a talc-chlorite-amphibole-magnetite schist. The mode of occurrence echoes that of South African and Sandawan emerald; quality can be high. Properties are in the usual emerald range; solid inclusions reported include dravite, magnetite, two types of dark mica, orange-red rutile, chrysoberyl, hematite and apatite.

Nigeria

Possible gem-quality blue-green beryl, which Hänni suggests be termed *blue-green emeralds*, are being mined from near Jos in the Rafin Gabas Hills district of Nigeria's central Plateau state. Here, blue-green beryls occur with other coloured beryls, topaz and smoky quartz in narrow stringers and thin layers forming a stockwork over the roof zones of granitic hills. Moderately saturated blue-green Nigerian beryls, coloured by combinations of up to 0.08% chromium, 0.06% vanadium and 1.2% iron, have a mean SG of 2.67, RI which range from 1.570 to 1.574 for the ordinary ray and 1.564 to 1.568 for the extraordinary ray, a DR of 0.006, and an absorption spectrum that displays features of both aquamarine and emerald coloured by both chromium and vanadium. Inclusions include complex growth zoning, partly healed two-phase fractures and two-phase jagged cavities oriented parallel to the *c*-axis.

Madagascar

Emerald, with characteristics comparable to those of Zambian emerald, occurs in mica schists of the Ankadilalana mine, near Kianjavato, on the south-east coast of Madagascar. Blue-green emeralds, containing an emerald component due to chromium and an aquamarine component due to iron, have been mined for the past five years from mica schists in the Morafeno region, 30 km south of the central east coast city of Manjaray. These partly transparent emeralds have an SG of 2.68–2.71, refractive indices of 1.588–1.591 for the ordinary ray and 1.580–1.582 for the extraordinary ray, and a DR of 0.008–0.009. Inclusions include colour zoning, two-, three- and multiphase partly healed fractures and negative crystals, brown mica, tremolite-actinolite rods and needles, and undetermined colourless crystals.

Norway

A useful note and map of the Norwegian emerald deposit can be found in the excellent series Mineral-Fundstellen (Band 4, 1976); Websky describes the deposit in his paper *Ueber Beryll von Eidsvold in Norwegen, Tschermak's mineralogische und petrographische Mitteilungen* 6.117-8, 1876 is an earlier paper. The deposit is at Akerhus on the western shore and at the southern end of Lake Mjosen close to the Byrud farm. The location appears to be in a pegmatite and is compared to the Crabtree mountain site in North Carolina, USA (Kunz, 'Precious stones', chapter in US Geological Survey, *Mineral Resources of the United States for 1901*, pp. 729–71, 1902).

North Carolina, USA

Emeralds locally and picturesquely called 'green bolts' have been recovered sporadically near the village of Hiddenite. The history of the deposit can be found in the first two volumes of *Gemstones of North America* (Sinkankas, 1959, 1976). The second volume illustrates an emerald crystal group found near Hiddenite and considered to be the finest ever found in North America. In 1969 old emerald workings were reactivated and new deposits found immediately east-north-east of Hiddenite. Comparatively few North Carolina emeralds are of gem quality. The occurrence is in a dark mica associated with quartz crystals (some quartz inclusions have been found in the emeralds). Emeralds are sometimes cut with white albite.

Mozambique

Emeralds have been found both in the mica schists of the Morrua district, or the Maria 111 mine associated with the Alto-Lingonha pegmatite region, 260 km west-south-west of Maputo. Mozambique emerald has an SG of 2.73, mean refractive indices of 1.592 for the ordinary ray and 1.585 for the extraordinary ray, and a DR of 0.007. It is included by brownish mica, and has *c*-axis-oriented two-phase negative crystals and growth tubes that often have basal fractures. A fine green cabochon with pyrite inclusions and a crystal similarly included have been described by MO'D.

Afghanistan

Emeralds from the Panjshir valley are described by Bowersox and Chamberlain (1995; ISBN 0945005199). The crystals, investigated by Bowersox and Chamberlain, occur in carbonate skarn. Crystals contain multiphase inclusions.

Absorption Spectra of Emerald in General

The chromium absorption spectrum shows fine lines in the red, a weak diffuse central absorption with absorption of the violet, and weak lines in the blue. In chromium-dominant emerald there are variations of the absorption spectrum depending on the direction of observation. In the ordinary ray are two narrow lines in the red, equally strong. These two lines consist of the main doublet (683 and 680 nm) and a clear-cut line at 637 nm; the central weak absorption band covers from about 625 to 580 nm, and there is a narrow line in the blue at 477.5 nm, which may only be seen in particularly chrome-rich stones, when another line at 472.5 nm may be present. The absorption of the violet begins at about 460 nm. In the spectrum of the extraordinary ray, the doublet is rather stronger, particularly the 683 nm line of the pair; the 637 nm line is missing and in its place are two rather diffuse lines at 662 and 646 nm, these being bordered on the short-wave side by characteristic transparency patches. The broad absorption region is now nearer to the red and is much weaker, and there are no lines in the blue.

Green beryl coloured by vanadium rather than chromium may look like emerald but unless the Cr absorption spectrum is present the stones cannot be called emerald but the trade is not always exacting on this! Chromium does, however, efficiently mask absorptions due to both vanadium and iron as the approximate positions of the green colour producing absorptions in the violet and orange-red of the emerald absorption spectrum are nearly identical for chromium (ordinary ray 430 and 593 nm, extraordinary ray 416 and 628/644 nm) and vanadium (ordinary ray 434 and 618 nm, extraordinary ray 426 and 630 nm). In addition, vanadium does not produce the 680, 683 nm doublet and 660 nm absorption in the extraordinary ray, 637 and 476 nm absorptions due to chromium. The 371 and 426 nm absorptions, in the respective ordinary and extraordinary rays, due to iron are also effectively masked by the general absorption of chromium below 460 nm. A good deal of information on the use of spectroscopy in gem quality and other minerals can be found in the highly recommended

Spectroscopic Methods in Mineralogy and Geology (Mineralogical Society of America, *Reviews in Mineralogy*, vol. 18, 1988; ISBN 0939950227).

Fluorescence

Chromium-rich, iron-poor emerald transmits deep red as well as green and shows red under the Chelsea filter. Emeralds from India and South Africa (Cr-poor, Fe-rich) usually show green. Soudé stones and some natural emerald-green minerals may show red or reddish; in all ambiguous cases examination of other properties is necessary. In my experience even Cr-rich emerald does not show a very strong red response to UV – a strong response certainly suggests a synthetic emerald.

Cutting

Emerald is usually cut, preformed, on a copper lap charged with diamond dust and the polishing is carried out on a similar lap with rotten-stone as the agent. The best-quality emeralds are almost universally fashioned in the trap-cut style with the corners truncated, giving an elongated octagonal outline, a style which has, owing to the common use for emerald, become known as the emerald cut.

This style of cutting shows the deep green of emerald to best advantage. Although the mixed cut, with a brilliant-cut top and a step-cut pavilion, has been used for emerald it tends to give a glassy look. Poor-quality and badly flawed emeralds are cut as cabochons or beads, and Indian jewellery is set with poor-quality emeralds of this type.

Native-cut stones may be boiled in green-coloured fat with a view to colour improvement. Specimens may develop surface spotting so it is best to soak the stones for some time in warm alcohol when some of the fat will dissolve. Emeralds may be oiled to diminish light scattering from inclusions; the reader is referred to O'Donoghue and Joyner (2003, *op. cit.*) for details and identification.

Synthesis and simulation (see Chapter 24).

Other Colours of Beryl

Aquamarine

Brazil

Crystals of beryl other than emerald occur most commonly in pegmatites whose cavities give the crystal the opportunity to grow to large size and allow clear areas to develop. This is why fashioned beryls can be among the larger gemstones. The finest aquamarine is found in Brazil, where the pegmatite deposits of Minas Gerais produce the finest blues, blue-greens and yellow-green as well as pale green and golden beryls. Sinkankas (1981) describes the producing areas in some detail, giving in addition notes on some exceptional and named stones. Some of this detail is taken from the very useful series of surveys by J.G. Oakenfull, *Brazil in 1909 (-22)*; each dated issue deals with the year preceding and notes on gemstones, while appearing in all the editions, are most plentiful in the issue for 1922 (covering 1921). There are four chapters on mineralogy and an additional chapter on diamonds.

Beryls are also described in *Principais depositos minerals do Brasil*, vol. V, Parte A, published in 1991 by the Companhia de Pesquisa de Recursos Minerais of the Depto Nacional da Producão Mineral. This is by far the most complete survey of Brazilian gem minerals, details and maps of the major mines accompanying statistics of production and virtually anything else the student might need. The text is in Portuguese – I have been unable to find an ISBN in my copy (MO'D).

We have noted that the state of Minas Gerais produces the finest non-emerald beryl: Oakenfull cites several large crystals. One more recently discovered crystal was named Marta Rocha after a Miss Brazil of the middle 1950s, the name persisting as a description of an especially fine dark blue. Large golden beryl crystals also occur in the state as well as pink beryl (morganite) whose colour does not quite reach the excellence of those from Madagascar. Fortaleza aquamarines come from the north of the state. The Maxixe blue beryl, a perpetual fascination for gemmologists, comes from mines in the Piauí valley of Minas Gerais; the deep blue colour is subject to fading, though not quite so dramatically as popular legend suggests. The finer blue is shown by the ordinary ray – in other blue beryls the colour is in the extraordinary ray.

The hexagonal crystals of aquamarine are often of large size and, owing to oscillation between the first- and second-order prisms, may be striated parallel to the prism edge. At times this striation may be sufficiently pronounced to obscure the hexagonal outline, the crystal assuming a ribbed cylindrical form. Further, due to erosion, some aquamarine crystals may be tapered. However, Hintze, in *Handbuch der Mineralogie*, Bd 2, *Silicate und Titanate* (1897), does not emphasize this observation.

Pale aquamarine may be heated to improve the colour but disclosure is not necessary nor is any test needed at least in the ornamental context. SG will lie between 2.66 and 2.80, specimens from some localities showing the higher values. RI will be in the range 1.572–1.590 for the extraordinary and ordinary ray respectively, uniaxial negative with DR 0.007–0.008. Dispersion at 0.014 is insignificant. Dichroic colours are a darker and a paler blue with the better colour seen in the extraordinary ray (compare Maxixe beryl). Two absorption bands, not notably strong or sharp, are at 456 and 427 nm. In Maxixe beryl there are strong bands in the red at 697 and 657 nm, with a weaker band in the orange at 628 nm. Greenish blue aquamarine which has escaped heating may show an absorption band in the green at 537 nm. This aquamarine is a subtle colour and may well be preferred to the perhaps more metallic blue of some heated stones. Like other beryls aquamarine is brittle.

Apart from Brazil aquamarine is found in Namibia and crystals of weak colour in Zimbabwe. The Mourne Mountains in Northern Ireland have produced attractive aquamarine – a fine crystal is in the Natural History Museum in London.

The only natural stone closely resembling aquamarine is blue topaz, which gives refractometer indices ranging from 1.610 and 1.620 appreciably higher than the values for aquamarine.

Pink Beryl (Morganite)

The colour is due to manganese. Caesium and rubidium often occur as impurities and may raise the SG and RI. Most pink beryls have an SG from 2.80 to 2.90 although some may be as low as 2.71. The RI is in the range between 1.578 and 1.600 for the ordinary ray and between 1.572 and 1.592 for the extraordinary ray, the DR varying between 0.008 and 0.009. Dichroism is distinct with a pale pink and a deeper bluish pink, the stronger colour in the extraordinary ray. There is no characteristic absorption spectrum and the luminescence under ultra-violet light is a weak lilac, but under X-rays there is an intense, but not bright, crimson glow.

The pegmatites of Pala, southern California, have produced very fine morganites which can be polished into large gemstones. Crystals from the pegmatites of Madagascar have also produced gemstones. While place names for gem deposits are often ambiguous or misspelt in translation, in the case of morganite and the non-emerald beryls the student has no excuse today for pleading misinformation as *ExtraLapis in English* no. 1 (2001; ISBN 0971537100) is devoted to the minerals and gemstones of Madagascar and *ExtraLapis* no 23 (2002; ISBN 39216566130) to non-emerald beryls. Those able to consult Lacroix, *Minéralogie de Madagascar* (1922), will find good descriptions of pegmatite deposits containing other gem minerals. Californian deposits are described by Sinkankas in the editions of *Gemstones of North America* (*op. cit.*). In the second volume (1976) he states that the finest morganite to be found in California had been recovered from the White Queen claim on Hiriart Hill, Pala, San Diego county (a location famous for fine tourmaline); work did not begin on the pegmatite until 1959. Crystals are tabular and peach-coloured.

Golden beryl (heliodor) and yellow beryl are also pegmatite-hosted; crystals of this and other beryl varieties are charmingly reproduced in colour in Reinhard Brauns, *Das Mineralreich* (1903; Sinkankas #908): the keen interest shown in crystals (in particular Russian ones) by Nikolai Ivanovich Koksharov is reflected in his *Materialen zur Mineralogie Russlands* (1853; Sinkankas #3499) in which beryl (including emerald) is included, accounts for the profusion of crystal drawings in his book.

Red beryl, transparent and of gem quality is coloured by manganese; small tabular crystals were first found in a rhyolite at a site in the Thomas Range of mountains in Utah, USA. Larger crystals were later found in the Wah Wah mountains in the same state. The Wah Wah crystals, occurring in a white volcanic rhyolite, were longer, elongated rather than tabular and more transparent. Crystals are prismatic showing the first-order prism and a base. The colour has been described as raspberry-red. In 1968 Nassau and Wood described the red beryl in *American Mineralogist*, 53, 801–806, manganese being cited as the cause of the colour. No water was found to be present and small clear areas occurred in some crystals. Faceted stones are very beautiful; in *Standard Catalog of Gem Values*, second edition, 1994, Miller and Sinkankas suggested that up to $100 a carat might be asked for crystals of the finest quality up to 4 ct. For faceted material up to 2 ct, up to $5000 a ct might be asked.

Yellow beryls are obtained from practically all the localities mentioned for aquamarine. Particular localities are Madagascar, Brazil and Namibia. In Namibia, heliodor is found in association with aquamarine and a yellowish green variety at Klein Spitzkopie near Rossing on the Otavi railway east of Swakopmund, and also between Aiais and Gaibes on the Fish river. Some yellow beryls have been reported to show radioactivity due to their containing a trace of uranium oxide. A deep yellowish red variety of beryl, called 'berilo bocade fogo' (fire-mouth beryl), comes from Santa Maria do Suassui in Minas Gerais. Many localities in the USA produce yellow beryl but the only one of importance is the Merryall or Roebling mine at New Milford, Connecticut.

The new beryl-group mineral pezzottaïte is described in a major paper by Hawthorne *et al.* in *The Mineralogical Record*, 35, 369–378, 2004. The mineral is found in the Sakavalana pegmatite, Fianarantsola province, Madagascar. Rhombohedral crystals occur in three different habits; flat masses filling cavities between 'cleavelandite', quartz and tourmaline; subhedral to euhedral hexagonal tabular crystals up to 10 cm in diameter; and as small, flat to equant to elongated crystals attached to the faces of large tourmaline crystals. The colour is moderate red to pink with dichroic colours pink to orange and purplish pink to pinkish purple for the ordinary and extraordinary rays respectively. The RI for the same two rays is 1.612–1.620 and 1.601–1.611. No reaction to UV radiation has been reported. Crystals are brittle and show an imperfect cleavage in one direction. Pezzottaïte differs from other minerals of the beryl group by its essential Cs content and a superstructure arising from the ordering of Be and Li in tetrahedral coordination. Some chatoyant specimens and faceted stones are depicted in the Record paper.

Dark Brown Beryl

An unusual dark brown beryl with asterism and a bronzy schiller (moonstone-like effect) was reported from the Governador Valadares area of Minas Gerais about 1950. The weak asterism is said to be due to oriented ilmenite and these and other coarser agglomerates cause the dark brown colour, for clear patches show that the body colour is that of pale green aquamarine. The schiller appears to be due to the structure of thin layers parallel to the basal plane, and these seem to act as mirrors. These star beryls neither show fluorescence nor exhibit any typical absorption spectrum. Black star beryl is reported from Alto Ligonha, Mozambique.

Inclusions

Various but largely non-specific inclusions can be seen in non-emerald beryls. The most common examples are long straight tubes parallel to the vertical axis of the crystal. The tubes may be hollow or liquid-filled to give an indistinct yellowish brownish colour. The name chrysanthemum inclusion has been used to describe a snowflake-like arrangement.

Some beryls show feathers of negative crystal cavities, some of which may contain a bubble of gas. An intriguing object sometimes seen in aquamarines is a flat inclusion, which under low-power magnification looks like a snowflake and when viewed by oblique illumination looks like a metallic disc. These inclusions have been aptly described as 'chrysanthemum' inclusions.

Cutting of Beryl

Aquamarine, morganite and yellow beryl need to be cut as stones of some size if the colour is to be strong enough for the stones to be attractive. In modern practice it is usual to cut beryls into the trap-cut style (the emerald cut) and such stones are mounted as important centre stones in rings, brooches, pendants and earrings.

Sinkankas, *Emerald and Other Beryls* (1981) as well as the items cited in this section, should be consulted for further information.

7

Chrysoberyl

Chrysoberyl is beryllium aluminium oxide $BeAl_2O_4$ and forms highly characteristic transparent to translucent pseudo-hexagonal, multiple contact and penetration twinned or prominently striated short prismatic crystals in the orthorhombic system. The hardness is 8.5 and there is a distinct cleavage in one direction. While the lustre is vitreous crystals may be opaque and chatoyant with oriented inclusions. Alexandrite is most commonly dark green with a colour change to raspberry red when viewed successively in daylight or its artificial near-equivalent and incandescent lighting. Chatoyant chrysoberyl shows an eye when translucent and cut with a dome (cabochon) which allows incident light to reflect with a sharp eye from parallel included crystals or channels to the upper surface of the finished stone. Transparent chrysoberyl apart from alexandrite is often very bright and pleasing golden to greenish yellow, sometimes dark orange or brown, rarely colourless.

The RI for the alpha, beta and gamma rays is 1.746, 1.748 and 1.756 respectively, biaxial positive with birefringence 0.008–0.010. The SG is in the range 3.71–3.72.

Pleochroism is strong, and in alexandrite, shows red, yellow-orange and emerald-green colours. The absorption spectrum of yellow-green or dark brown chrysoberyl is notable for the broad absorption band centred at 444 nm in the violet. In dark specimens the observer may also see bands at 505 and 485 nm and in very brown stones the general absorption of the blue and violet conceals the 444 band. Before examining the absorption spectrum of alexandrite we should first remember to consult Anderson and Payne's text (passim) on absorption spectra and then the stone's strong pleochroism where the highest RI direction is green, the lowest red or purple and the intermediate ray orange. The observer will need to note that the absorption spectrum shows some differences depending upon the direction examined. In the green ray the narrow doublet at 680.5 and 678.5 nm is particularly prominent with the 680.5 nm band the stronger. Weaker lines can be seen at 665, 655, 649 and 645 nm. The broad central absorption extends from about 640 to 555 nm and appears to show two deeper areas of concentration. Blue and violet are completely absorbed from about 470 nm.

In the red or purple ray the doublet is weaker and the 678.5 nm band is now the stronger. Two other lines, at 655 and 645 nm are seen in the red. The broad absorption is now 605–540 nm, well away from the red and accounting for the change to red with appropriate lighting. A clear line can be seen at 472 nm, sometimes accompanied by a weaker line at 468 nm. The absorption of the blue is now seen at 460 nm. Anderson and Payne (1998) carried out the investigations so far quoted on Siberian alexandrite in which the colour change is stronger (more chromium) than that shown by Sri Lankan alexandrite. In Sri Lankan alexandrite the broad absorption zones are less intense than in Siberian (and the recently discovered Brazilian) alexandrite.

Between crossed filters most alexandrite will show a distinct red (more than might be expected in Sri Lankan specimens) and any such response from a chrysoberyl entitles it to the name alexandrite, providing there is some colour change. Compared with ruby the emission doublet in the red is far less strong but can be seen with careful observation and appropriate lighting. The strong pleochroic colours red, yellow-orange and green are not associated with the colour change.

Cat's-Eye

Cat's-eye chrysoberyl is the finest of its kind; while a number of other species may show eyes when appropriately cut, their eyes are far coarser and the flat back of many chatoyant specimens enables an RI test to be made. Chrysoberyl can be tested with the spectroscope. The name 'cat's-eye' is used on its own for chrysoberyl only, all other chatoyant stones needing the species to be used adjectivally. Occasionally a chrysoberyl will show a four-rayed star. The finest cat's-eyes should show a very sharp blue-white eye against a dark honey-coloured background.

While chrysoberyl is a characteristic mineral of granite pegmatites associated with mica schists or reaction zones in ultramafic rocks, gem crystals are found most often in placers. Brazil, long a producer of the best and brightest yellow-green chrysoberyl, has in recent years become the producer of magnificent alexandrite from the Lavra de Hematita field in Minas Gerais. The best material from this deposit equals or even surpasses alexandrite from the Takovaya emerald mines, north-east of Yekaterinburg [Sverdlovsk], Russia. The colour change is pronounced in the best alexandrite from both sources. Alexandrite was named in 1982 for Tsar Alexander II. Crystals are most commonly found as star-shaped twins (trillings) reaching 2 cm across in some examples. The host rock is a mica schist.

Fine yellow chrysoberyl is found in the Sanarka district of the southern Urals though the best examples come from Jacuda, Brazil. Interesting small, included alexandrites but with a marked colour change have been mined in Zimbabwe in the Smabula Forest area and from Lake Manyara, Tanzania. A colourless chrysoberyl has been found in Myanmar and a faceted example forms part of the ACD Pain collection in the Natural History Museum, London. A brown specimen from a pegmatite in the Harts Range of central Australia gave an RI of 1.765, 1.772 and 1.777 and showed an intense absorption band in the violet centred near 447 nm with another at 508 nm. Chrysoberyl has also been found at Anakie, Queensland.

The finest cat's-eyes are found in Sri Lanka though many commercial specimens are exported from Brazilian sources.

Inclusions

Minute rutile crystals forming groups running parallel to one of the crystal axes are characteristic of many chrysoberyls from India and may occur in material from elsewhere. Liquid-filled cavities with gas bubbles are also characteristic. Stepped planes, not always easy to see, are characteristic of some yellow crystals.

Synthesis and Simulation

Alexandrite has been synthesized by a number of different processes, including Czochralski, floating-zone and flux. Inamori in Japan has also produced an alexandrite cat's-eye. Pulled crystals sometimes advertized as alexandrite in rockhound-type journals are traps for the unwary amateur as they contain insufficient chromium to produce an alexandrite colour.

Identifying features of these synthetic chrysoberyls include undigested flux from crystals grown in precious metal crucibles, some gas bubbles, wavy fibres, weak yellow fluorescence from the crystal surface and sup-plemented by a red-orange response below the surface; this effect is seen under SWUV. The reader may consult O'Donoghue and Joyner (2003 passim) for further information.

Crystals grown in Russia may show flux-filled or partly filled negative crystals, some showing surface crazing and appearing white or pale yellow. Some crystals have shown 'Venetian blind' or shuttering effect. No natural chrysoberyl shows any such inclusions.

Imitations

Alexandrite is well worth imitating and prospective purchasers should bear in mind that even the finest natural specimens do not show a very marked colour change; the green is dark and the 'candle-light' so often recommended for viewing alexandrite might make the stone as a whole hard to see at all. The flux-grown synthetics show too strong a colour change but the colours are fairly accurate. This cannot be said of the flame-fusion-grown vanadium-doped alexandrite simulant which looks far more like amethyst. Even today judicial proceedings feature this well-known (and quite attractive) product and there is always some expert witness to demonstrate how far gemmological education still has to go. Sometimes the specimen in question is not even forthcoming. A dark green Verneuil-grown spinel said to be made to imitate green tourmaline did show a dull red colour change and this could be passed off as alexandrite. Neither birefringence nor dichroism would be possible.

Composites with a dyed central gelatine filter have also been used as alexandrite simulants.

While a number of natural yellow stones may be chatoyant enough to provide a reasonable cat's-eye imitation, in no examples is the eye really sharp. In the ingenious 'Cathay stone' the eye is sharp though its host-glass is soft. Nonetheless this is an attractive and convincing product, at least through the shop window! Two types of glass are used, one in the form of hexagonal fibre-optic bundles. 'Victoria cats-eye', another glass product, is subject to devitrification and may on occasion show chatoyancy but it would be vague at the best. Transparent green and yellow chrysoberyl is neither imitated nor synthesized.

8

Spinel

Spinel is $MgAl_2O_4$ and commonly forms octahedra in the cubic crystal system though some specimens may show dodecahedron or cube faces. Twinning on {111} is frequent and spinel-law, penetration- or contact-group types have been recorded. This direction sometimes shows parting but there is no cleavage. The hardness is 7.5–8, and the SG is 3.5–4.1 increasing with the iron or zinc content. Spinel may be transparent to nearly opaque and colours include brown, red, black, orange, yellow, green, blue, indigo and violet. Colourless spinel, at least of any size, will almost certainly be artificial. The single RI is 1.718. Spinel is formed at high temperatures as an accessory in igneous rocks, in regionally and contact metamorphosed limestones and in other contexts. Spinel forms three series, with magnesiochromite, with hercynite and with gahnite, the latter itself providing the occasional dark blue zinc-rich gemstone.

Very occasionally spinel may show an optical phenomenon, including a four- or six-rayed star, but the effect is adventitious. Red spinel is by far the most valuable, and as the finest red crystals occur with ruby in Myanmar, confusion can easily arise. With the spectroscope red spinel shows the chromium absorption spectrum but the bands in the blue, familiar in the ruby spectrum, are absent. In addition the emission doublet in the red spinel is replaced by several (up to five) emission lines, none of them forming a doublet. The strongest, at 686 nm and separated from the next strongest at 675 nm by a dark interval, is best seen with strong illumination and through a blue filter. Some Verneuil flame-fusion red spinels have been found to show a single emission line at 686 nm.

The only other feature of the red spinel absorption spectrum is a wide band with its centre at 540 nm (the broad absorption band in ruby is centred at 550 nm). Blue spinel can cause trouble for students as the absorption spectrum appears unexpectedly complex (this may be a purely subjective impression). Almost all blue spinels are a rather dark blue yet not really reminiscent of blue sapphire – some iolite may appear a similar colour but that stone of course has its startling pleochroism to distinguish it. The ferrous iron spectrum shows bands in the orange to green, at 635, 585, 555 and 508 nm; there is a narrow band in the blue at 478 nm and a stronger one at 458 nm; there are other lines at 443 and 433 nm.

Red and pink spinels will show a crimson fluorescence under LWUV and a weaker response of similar colour under SW. Under X-rays there is a moderate glow of similar colour (but other less expensive testing methods should serve to identify spinel). A reddish response to LWUV has been noted in some purple to mauve specimens which show little reaction to SW but glow plum-colour to lilac under X-rays. Other speci-mens have been found to respond with a orange to red glow under LW, little response to SW and green under X-rays. Some pale mauve spinels may glow greenish under LW, a response also shown by some pale blue or violet blue stones.

Common inclusions in spinel are octahedra of other spinel group mem-bers, for example magnetite. Apatite crystals have been identified in red spinel both from Myanmar and Sri Lanka. In spinels from a zircon-rich, silica-poor context crystals of the zirconium oxide baddeleyite occur: Koivula and Gübelin (1992) illustrate examples and also show inclusions of dolomite, calcite and phlogopite (mica group) in Myanmar stones. Chains of three-phase inclusions are characteristic of spinel from Sri Lanka. Also characteristic of spinel are parallel rows of minute octahedra with either a white or a black filling. Quartz crystals have been found in Myanmar spinel. Koivula and Gübelin show a photograph of titanate crystals in a blue spinel from Sri Lanka; they also show an unhealed (dry) fissure in a spinel from the Hunza valley of Pakistan.

Spinel is a versatile species and may display 6-rayed stars and sometimes a colour change (dark blue in daylight to purplish blue in incandescent light). One such specimen showed an absorption spectrum with bands at 640, 580 and 540 nm in addition to the usual ones associated with ferrous iron. These bands are more often seen in the cobaltian synthetic spinel. Until a few years ago any blue spinel showing red through the Chelsea filter would be considered a flame-fusion synthetic but some fine blue natu-ral cobaltian blue spinels have been found in Sri Lanka. They show a very bright red (for a natural stone) through the filter and the blue is more attrac-tive than that of the blue synthetic version.

The first report on this material was made by Shigley and Stockton in 1984; they concluded that the presence of an absorption band at 460 nm was an indication that a specimen was natural whether or not cobalt was found to be present.

A report by GIA in 1986 further investigated cobalt-coloured blue spinel, finding that in the two specimens examined the RI was 1.720 and absorption bands showed at 622, 595, 575, 559, 489, 480 and 434 nm. Bands at 480, 460 and 434 nm were not found in synthetic material. GIA have also reported a natural twin blue star spinel of 2.12 ct; the star was doubled through twinning.

The finest red spinel comes from Myanmar and is found in association with ruby; attractive red and more rarely orange crystals are found in the gem gravels of Sri Lanka where the only reported cobaltian spinels occur. From Sri Lanka, too, comes the spinel at one time described as gahno-spinel, with a composition intermediate between spinel and gahnite, the zinc spinel. This name is unnecessary; though interesting work was carried out by Anderson and Payne on the absorption spectra of the blue spinels (Anderson and Payne, 1998), they concluded that the visible absorption spectrum did not indicate the presence or absence of zinc. It is interesting to note that when ruby and spinel crystals are present together in gem gravels the spinel crystals frequently appear to have kept their shape better than the rubies; the reverse seems to be the case with blue sapphire and blue spinel.

Red spinel of gem quality has been found in Pakistan, the main locations in the Hunza Valley producing crystals of fine colour and generally larger than the Myanmar examples (Kazmi and O'Donoghue, 1990). The spinel is found with corundum in marble forming massive intercalations within garnetiferous mica schists and biotite-plagioclase gneisses. Characteristic of Pakistan spinels is a plum red-to-brown colour which I have not encountered from other locations. Characteristic inclusions in the Pakistan spinel are long prismatic crystals of a green amphibole and fine crystals of rutile.

Nigeria, whose gem species are long overdue for a general description, has produced a fine blue spinel-type mineral from pegmatites in the Jemaa area: this has been identified as gahnite. The RI is in the range 1.793–1.794 and the SG in the range 4.4–4.9. This material has been found to glow red under incandescent lighting and shows two sharply defined areas with no absorption between them. Within the areas absorption is mixed in intensity. The defined areas are at 580 and 555 nm. Absorptions are also found in the red-yellow area, in similar locations to those found in other blue spinels and gahnites. Specimens heated in an oxidizing environment to 1000 °C for one hour showed a change from blue to blue-green; other stones heated to 1400 °C changed from blue to dark olive-green. A columbite group mineral has been found in the Nigerian spinels, with beryl and hematite.

Deposits of red spinel from Badakhshan, Afghanistan, are of particular importance since one of them provided, among other celebrated stones, the Timur Ruby, now in the private collection of her Majesty Queen Elizabeth II. It forms the central stone of a necklace and weighs 352.54 ct. Bowersox and Chamberlain, *Gemstones of Afghanistan* 1995; ISBN 0945005199, deals at some length with red spinel from these deposits and illustrates some of them. In addition, sale catalogues of Indian jewellery often give useful illustrations and captions – one example is Christie's Important Indian jewellery sale held in London on 8 October 1997, in which the Carew spinel is illustrated and described. The spinel is inscribed following a Timurid fashion (the Timurid dynasty was founded by Tamerlane (Timur) and lasted from the late fourteenth to the late fifteenth centuries, ruling over Afghanistan, parts of Iran and central Asia).

The Kuh-i-Lal spinel deposit is 47 km south of Khorog, Tadjikistan, across the Amu Darya (Oxus river) from Shighnan, Afghanistan, on the edge of the Pyandzeh river valley (Bowersox and Chamberlain, 1995). These spinel mines are said to have been worked since about AD 950. In 1947 Fersman claimed that the white marble mines at the mouth of the Kuga-Lal river in the Pamirs had been producing red stones for 1000 years – bright rubies and pink to red spinels to which the name lal was given (Jewels of the Russian Diamond Fund, *Gems & Gemology*, 5 (11) 1947).

Other sites for the ancient mines have been suggested and it is possible that ruby and red spinel were confused with one another, as well as the names of the sites where they were found. Bowersox and Chamberlain made their own expedition to the area in 1994. They interviewed some of the inhabitants of the area and concluded that the site of the Kuh-i-Lal mine was across the Amu Darya from Shighnan; the authors give the coordinates. At the time of their visit military operations prevented a closer approach. Bowersox and Chamberlain quote from several authorities who give accounts of their own findings or speculations. It is possible that spinels may be found on both sides of the Ab-i-Pandja (Amu Darya), the sides today lying in both Tadjikistan and Afghanistan. The topography of the area has probably been affected by earth movements and landslides.

Spinel is not a particularly well-known species despite the fine red examples found in Mughal jewellery and (in England) the Black Prince's 'ruby' found in the Crown Jewels. Spinel has often been referred to as 'balas ruby' and although the name should not be used in the scientific context it is often found in books on jewellery.

In *Crown Jewels of Iran* (1968) (ISBN 802015190) Meen and Tushingham state that the largest red gem spinel known is a 500 ct specimen in the collection. Like most red gem spinels from Afghanistan the stone is not faceted but roughly polished, faceting not being found in general before the end of the eighteenth century. It is accompanied in the Iranian crown jewels by other red spinels similarly fashioned. Meen and Tushingham illustrate these stones in colour.

9

Topaz

The resemblance between topaz and citrine has naturally led to confusions both in determination and more often in nomenclature which need not be investigated yet again. Those who feel a need to trace the derivation and usage of the name must enquire elsewhere; there are many sources. Those who wish to look at topaz in considerable depth should consult Don Hoover's *Topaz* (1992; ISBN 0750610875; Butterworth-Heinemann Gem Books).

'Topaz' was used as a name for aluminium fluosilicate in a description of the deposits and products of the Schneckenstein area of Saxony, Germany, by Henckel in 1737 (some crystals are still found here and some gemstones have been fashioned from them).

Chemical composition: An early survey by Becke can be found in Doelter's *Handbuch der Mineralchemie*, vol. 2 pt 1, 1914, in which a yellow topaz from the Thomas Range, Utah, USA, has a density of 3.565, a fluorine content of 20.37 and a hydroxyl content of 0.19, and a specimen from Brazil with a density of 3.532 has a fluorine content of 15.48 and a hydroxyl content of 2.45. Becke also describes the effect on the refractive index of this variation in composition, the Thomas Range specimen giving an RI of 1.6072, 1.6104 and 1.6176 for the alpha, beta and gamma rays respectively, with a birefringence of 0.0104, compared with the Brazilian (Minas Gerais) specimen's 1.6294, 1.6308 and 1.6375, DR 0.0081.

Summary: Topaz has the composition $Al_2(F, OH)_2SiO_4$ with some of the hydroxyl substituting, though not completely, for the fluorine, the physical and optical properties varying with the proportions of F and OH.

The crystallization of topaz is described in Hintze's *Handbuch der Mineralogie*, 2 Ed., pp. 102–127, 1897. This work, the all-important survey of mineral species and the different forms shown by specimens of differing composition and location, shows many forms taken by crystals of topaz.

Topaz is also covered by vol. 1A of the second edition of Deer, Howie and Zussman's *Rock-Forming Minerals* [DHZ] 1982, not cited in Hoover's bibliography. DHZ assign topaz to the orthorhombic crystal system. Hoover (1992) acknowledges this but suggests that while topaz crystals may belong to space group Pbnm [indicating the presence of a centre of symmetry with three axes of twofold symmetry (about each of the crystal axes) and three planes of symmetry]. If a crystal was doubly terminated it should not show piezoelectricity or pyroelectricity. Doubly terminated and apparently non-symmetric topaz crystals showing both of these properties have been identified. Some authorities have suggested that the apparent anomaly is liked with OH substitution and ordering on the F sites. This could mean that some topaz crystals, at least, may belong to a lower order of symmetry than the rhombic-dipyramidal class.

Hoover (1992) discusses this, citing several authorities but concludes that more work is needed. This area should not affect the gemmologist, prospector or mineral collector for whom topaz crystals are easy to identify. Crystals with a lozenge (or diamond-shaped) cross section show prism faces which are often striated parallel to their length and with varying terminations. Brazilian crystals show low-angled pyramidal terminations with crystals from Utah showing a steeper angle. Some crystals from Africa (perhaps Nigeria) show a chisel-like termination due to enlargement of dome faces at the expense of the pyramids. In other crystals the pinacoid gives a flat-topped effect. Crystals cleave parallel to the basal pinacoid which is often present. Doubly terminated crystals are mentioned above. A square cross-section is due to a dominating second-order prism.

Topaz occurs most commonly in such igneous rocks as granites, granite pegmatites and rhyolites, occupying veins or cavities. Resulting most often from late-stage pneumatolytic action, topaz is commonly found in greisen. Associated minerals may include quartz, fluorite, tourmaline and beryl. Crystals may reach large sizes though some are found in gravels, their size much diminished.

The hardness is 8 and the specific gravity in the range 3.49–3.57; most gem topaz is about 3.53. The cleavage has already been noted but wearers of topaz jewellery should remember that this cleavage is very easy to start and that the resulting pearly lustre of the cleavage faces is little compensation for the accident.

The familiar colour of most topaz, ranging from colourless through to dark orange and pale yellow, results from the activity of colour centres (Nassau, 2000). Some natural brown topaz may fade on exposure to strong light (some specimens from Utah may show this). Colourless topaz crystals when fashioned make attractive and often large gemstones though their dispersion is less than that of diamond. Pink topaz contains chromium as a foreign element.

Blue topaz in Nature is attractive but generally pale; the dark blue stones now familiar in jewellery have been artificially irradiated. Colourless material when irradiated gives unstable yellow or brown colour centres/stable brown or blue colour centres. These cannot be distinguished from natural topaz by gemmological testing. Nassau (2000) tells us that most colourless topaz when irradiated will turn to a combination of brown and blue – the brown colour predominates. When heated the brown disappears, leaving the blue which is unlike any natural blue so far reported.

Sources of topaz are not rare though the finest examples are usually Brazilian. Looking first at pink topaz, quite exceptional crystals are found in Pakistan, where examples of yellow to sherry-coloured topaz are found in the Shingus pegmatite and in similar conditions at Bulechi in Gilgit district. Other topaz-bearing pegmatites occur at Niyit Bruk and Gone, near Dusso in Skardu district (Kazmi and O'Donoghue, *Gemstones of Pakistan*, 1990).

However, Pakistan provides a uniquely coloured topaz not occurring in a pegmatite. This is the hydrothermal vein-hosted 'pink' topaz which so far has been found only in Mardan district of northern Pakistan. A hillock called Ghundao is surrounded by a village of the same name. The site is about 4 km north of the small town of Katlang, whose name has become associated with the exceptional topaz found in stockworks of quartz and calcite veins. Hydrothermal or pneumatolytic origin is suggested by the absence of fluorite. Tectonic movements may have fractured the crystals which were then incorporated into later-formed vein calcite.

The gem-bearing rocks at Ghundao hill, visited and examined by the author (MO'D) in the late 1980s, consist of the following lithologic units: light grey thick-bedded largely autoclastic limestone; dark grey medium-bedded, medium- to fine-grained crystalline limestone with the lower part largely algal; thin-bedded limestone and calc shale. Further discussion of the lithology can be found in Kazmi and O'Donoghue (1990).

The topaz mineralization in this deposit is also described by the same two authors. The mineralization is structurally controlled and forms characteristic saddle-reef structures. The whole sequence has undergone north–south compression leading to tight folding, intense fracturing and faulting. In a number of places the limestone has been invaded by stockworks of calcite and quartz which contain fine, perfectly euhedral crystals of quartz and topaz. The calcite veins have been largely emplaced along a series of parallel normal and reverse faults along the crests of tightly folded anticlines. The topaz-containing calcite veins show a pinch-and-swell structure though some are only a few centimetres wide. Others are larger and extensive. Topaz is not confined to these veins but is found in calcite which forms small lenticular tension gashes in limestone at the crests of folds. The mineralized fault zones commonly contain limonite or a thick coating of clay. They contain vugs and cavities which often contain euhedral topaz crystals with quartz and calcite. The colour of the finest topaz crystals is unlike any pink topaz yet reported, being closer to an orange-red rather than the sometimes pale pink shown by Brazilian material. In a table listing chemical composition (Kazmi and O'Donoghue) it can be seen that the colouring agents are chromium and manganese though in some samples Mn is absent.

Topaz from Brazil, the chief source, is described in many publications; pink topaz from Brazil is often heated to deepen the colour. The publication *Topas*, forming *ExtraLapis* no. 13 (1997; ISBN 3921656427), describes and illustrates topaz crystals from Ouro Preto in Minas Gerais, as well as specimens from Schneckenstein, Saxony, Germany. To take one Brazilian occurrence described by Bancroft in *Gem and Crystal Treasures* (1984, no ISBN) the topaz mines to the west of Ouro Preto contain the Don Bosco, Rodrigo Silva and Saramenha occurrences providing fine topaz [featuring in the museum at Ouro Preto]. Ouro Preto topaz is orange to pinkish purple. Fine golden topaz is also found in the Ouro Preto area, the names Lavro do Moraes and Antonio Pereira being familiar to connoisseurs of topaz. This topaz does not occur in pegmatite but is found in kaolinite clays penetrated by quartz and also in sedimentary iron formations consisting mainly of hematite and silica. The Antonio Pereira mine has produced rich sherry-coloured topaz.

An official government account of the production of Brazilian precious stones is given extensively, from the economic standpoint but with considerable detail in what must be the all-time most important description of the gem minerals of a single major producing country. This is *Principais depositos minerais do Brasil* vol. 4, part A (1991). There are nearly 500 pages with mine and area plans and maps – everything the prospector could want. This is not an easy book to find; the publishers are the Departamento Nacional da Producão Mineral, a dependent department of the Ministerio da Infra-Estrutura.

The Murzinka area of Russia is another of the classic topaz producers. It forms part of a region which is a watershed of the Neva river of the western Urals. It includes the Alabashka field of granite pegmatites which lies in the northern part of the Murzinka sub-zone of the Ural Gemstone Belt and the Murzinka pegmatite field which is in the south of the area, in the Neva river valley. The best account in English forms vol. 5 of *Mineralogical Almanac* (2002; ISBN 5900395324).

This study includes an account of the morphology and structure of the topaz-beryl granite pegmatites of the Alabashka field. Topaz is a major mineral of the Mokrusha area and appears to form four main types of crystal which the survey illustrates. Colours include blue and pink as well as near colourless but apparently not the characteristic orange-yellow-brown colours.

While on the gemstone deposits of Russia it might be useful to mention the now very rare survey by A.E. Fersman, published in two parts, in 1922 and 1925. Only 1000 copies were printed of the 1925 edition of this title, *Dragotsennye I tsvetnye kamni* SSR [Sinkankas #2042]. Maps show and the text describes Russian gem deposits in a way that has not been repeated since, though the *Mineralogical Almanac* series is beginning to improve this situation.

Though the effect is not vital in everyday gem testing it is interesting to note that in general in the higher SG range, SG figures appear to indicate a higher F content. In such specimens the RI may be lower in the range and the reverse seems to be the case with (OH).

Pleochroism in topaz is not strong, though more perceptible in the heated pink stones. Specimens from Pakistan show yellow, purple and deep mauve to violet. Sherry-coloured topaz shows brownish yellow, straw yellow and orange-yellow. Blue specimens show colourless, pale pink and blue. Through the Chelsea colour filter, the blue stones may show a greenish blue, much less pronounced than the strong blue shown by some aquamarine.

Absorption Spectra

Only the pink topaz provides the investigator with a distinctive absorption spectrum. Chromium in topaz is indicated by an absorption/emission line at 682 nm. This is rarely pronounced. Sherry-brown topaz from Ouro Preto shows the effect and no doubt it can be seen in some of the darker pink crystals found at Katlang – I have not seen a note so far.

Luminescence

Topaz shows a variable luminescence which is not of particular use in determining an unknown topaz-like specimen. At best the glow when seen is weak yellow or greenish yellow; some blue crystals appear to show an oriented glow. Topaz can be electrified when heated or rubbed and we have seen that pink topaz is often heated to deepen the colour. Some topaz fades when subjected to strong sunlight. The whole area of heating, irradiation, luminescence and fading (or discolouration) is well covered by Nassau (2000; *op. cit.*).

Hoover in *Gems* 5 (1992) tells us that in 1981 measurably radioactive dark blue topaz was reported. The process used involved neutron bombardment in a nuclear reactor. Subsequent research has shown that the main emitters are the radionuclides tantalum-182, scandium-46 and manganese-54, which are produced during the treatment process.

Nuclear research reactors are now regularly used to produce blue colouration in topaz. Material so treated generally comes out blue, and therefore does not require a subsequent heating step. The most typical reactor-produced colour is a medium to dark greyish blue that is often described as 'inky' in appearance. A trade name often heard for this material is London Blue. Heating may be used, however, to remove the inky cast, resulting in a lighter but more saturated colour.

Because treatment in a nuclear reactor renders topaz radioactive, it must be stored for a period of time until this induced radioactivity decays to a safe level, normally one to two years. Many countries have regulations to control the sale and distribution of reactor-treated gems. In the USA, where there is a large market for blue topaz, the Nuclear Regulatory Commission (NRC) has licensed a number of facilities, including the Gemological Institute of America, to test topaz and other gems for residual radioactivity prior to their distribution to the public.

Most recently, combined treatments have been used to produce some darker blue colours in topaz that do not suffer from the inkiness of the London Blue material. Such a combination of treatments usually begins with irradiation in a nuclear reactor, this being followed by linac treatment, and then heating to remove unwanted secondary hues. Among the trade names used for such material are Super Blue, Swiss Blue and American Blue.

The radiation treatments used to produce attractive blue colours in topaz have greatly increased the availability of this material, which has become one of the most popular of all coloured gemstones. It should be noted that irradiation in nuclear reactors has also been used to produce green colours, such material being marketed as Ocean Green Topaz. Additionally, a predominantly blue colouration has been produced in topaz by coating the surfaces with a thin film of gold. These stones, marketed as Aqua Aura topaz, display both the blue to greenish blue transmission colour of the gold as well as superficial thin-film iridescence.

Inclusions

Topaz does not show a wide variety of inclusions. Koivula and Gübelin (1992) note that topaz forms a link between the pegmatitic and hydrothermal phases allowing it to grow from a solution which could accommodate a variety of components. Solid inclusions are those characteristic of pegmatites – apatite, albite, goethite, for example – as well as species from the late pegmatitic period such as fluorite and monazite.

Topaz is richer in liquid inclusions; the easy cleavage allowing multiple cleaving to take place during growth, the partial healing showing the investigator two- and multiphase inclusions. Many of these contain two immiscible liquids, saline aqueous solution and fluid carbon dioxide.

Much topaz is inclusion-free. Internal features that may be seen in the colourless, blue and brown fluorine-rich types of topaz are usually tiny cavities containing two or even three immiscible liquids with distinct dividing lines. These cavities are often drop-shaped and the bubbles, one of which could possibly be carbon dioxide, may be inside one another. Such inclusions may also contain a solid phase. Occasionally these cavities may be flattened out into thin films of liquid. Because of its easy cleavage, flat breaks may be noted in fashioned gems.

Crystalline inclusions typical of topazes of pegmatitic origin include monazite, muscovite mica, spessartine garnet, albite feldspar, brookite and quartz. A rare type of inclusion which has been observed in topaz from the tin workings of Nigeria are cubic crystals either singly or as groups. These as well as octahedrons, rhombic dodecahedrons and their combinations have been identified as fluorite crystals. The hydroxyl-rich sherry, imperial and pink stones from Ouro Preto usually show long tube-like cavities running the length of the c-axis of the crystal.

In the mid-1980s, faceted Brazilian topazes containing eye-visible brownish yellow acicular inclusions appeared on the gem market. These stones – mostly colourless but also some blue – were being marketed under the name 'rutilated topaz' because of their superficial resemblance to rutilated quartz. Investigations carried out at the GIA revealed the essentially parallel inclusions to be not rutile needles but surface-reaching ribbon-thin etched dislocation channels that were partially to totally filled with epigenetic iron staining. Subsequent experimentation showed that heat treatment could alter the limonite staining to hematite, thereby changing the colour to a dark red-brown and making them more prominent.

Occurrences

The earliest known source of topaz was probably that of Schneckenstein, near Auerbach in Saxon Voigtland (eastern Germany), where the crystals, of pale wine-yellow colour or completely devoid of colour, are found in drusy cavities in association with quartz and tourmaline. Traditionally, the most sought-after topaz for jewellery purposes has been the sherry-brown material from Brazil which makes such attractive pink stones when heat-treated. The locality for these crystals and including the colour now called imperial topaz is just west of Ouro Preto in the state of Minas Gerais. The topaz is found as groups of detached crystals embedded in clay or scaly kaolin in cavities all along a range of hills that form a belt about 20 km long and 6 km wide running east–west from the Saramenha mine, roughly 4 km west of Ouro Preto, to the Olaria mine, about 25 km to the west of the city.

Natural-colour pale blue as well as colourless topaz, in the form of crystals and rolled pebbles, are abundant in many places in Brazil. Colourless material, used to produce blue stones through irradiation, is found throughout the Teófilo Otoni-Marambaia and Araçuaí-Salinas pegmatite districts. Very large quantities of colourless material are also mined in Rondonia in western Brazil. Yellow and colourless topaz is found at La Paz, San Luis Potosi, and Durango in Mexico. Colourless, pale blue, reddish and wine-yellow crystals are found in the Pikes Peak region of Colorado, in southern California, Texas, New Hampshire and in a rhyolite in the Thomas Range of Utah.

Blue, colourless and pale brown topazes are found in Tasmania and, with tin, in the Gilbert Ranges of northern Queensland. Here the crystals are commonly highly included, but the rolled pebbles are flawless. New South Wales and northern Queensland also produce topaz. Well-formed crystals occur in pegmatite veins and as pebbles in the river gravels near Takayama, Naegi, and Hosolcute in Mino province, Tanokamiyama in Omi province, and Ishigure in Ise province of Japan.

The gem gravels of Sri Lanka produce a fair amount of good-quality topaz; colourless materials such as that mined at Polwatta near Matale in central Sri Lanka is especially prized for irradiating to blue colours. The ordinary alluvial deposits of the ruby mines of Mogok in Myanmar produce similar material. A considerable occurrence of fine yellow, blue and (mostly) colourless topaz of magnificent gem quality and large size has been recovered from a large pegmatite dyke at Sakangyi some 32 km west of Mogok. We have already referred to the Pakistan topaz deposits at Katlang.

In India, colourless topaz is mined at Patnagarh in Orissa and has been found in Quaternary sediments in an area of the Thuong Xuan region of Thanh Hoa province of central Vietnam.

Topaz in many colours is found in Russia. Blue and green crystals are found on the eastern slope of the Urals at Alabashka near Sverdlovsk, the gem centre once again known as Ekaterinburg, and beautiful magenta-coloured crystals are found near the Sanarka river in the Ilmen mountains. Pale brown crystals, which like those from Japan and Utah fade on exposure to sunlight, are found near the Urulga river north of the Borshchovochnoi mountains in the Nerchinsk district of eastern Sakha (formerly Siberia) in the Russian Federation. In the Adun-Chalon mountains of the same district topaz is found in good crystals. Kamchatka, in the extreme east of Sakha, also produces topaz of yellow, blue and greenish colour. Colourless topaz as well as blue, pink, red and yellow crystals come from the pegmatites of

Volynya, Ukraine. In the People's Republic of China, fine-quality pale blue-green and richly coloured yellow to red topaz crystals up to 240 mm long have been found in the Xilingeleimeng region of the Inner Mongolia Autonomous Region. The pegmatite region of Altay in the Xinjiang Uygur Autonomous Region as well as Guangdong, Guangxi, Hunan and Yunnan provinces have also produced topaz.

In the African continent topaz is abundant. Except for the blue-coloured material and the colourless material used to produce irradiated blue colours, it has little gem importance. Well-formed colourless and blue crystals are found near Klein Spitkopje in Namibia; the crystals are obtained from pegmatite dykes, and topaz is also obtained from the gem gravels of the Miami district in Zimbabwe. Blue and colourless topazes, as crystals and as rolled pebbles, are found around the tin workings in western Plateau state of central Nigeria. The crystals are here found in veins and fissures in the contact metamorphosed aureoles caused by the younger of the granitic intrusions of the plateau. The water-worn rolled pebbles are found in streams and alluvial gravels, and in the past local inhabitants placed colourless pebbles of topaz in the indigo dye pots which are such a feature of Nigeria. This superficially coloured the pebbles a blue shade and they were then sold to unwary tourists who thought they had a fine blue topaz. Today, however, this material is generally sold to be colour-enhanced through irradiation. Topaz crystals have also been found near Tamanrasset (Fort Laperrine) in the Hoggar district of south-eastern Sahara. The pegmatites at Mahabe and Soarano and the alluvial deposits of the Saka river of Madagascar supply topaz.

Of interest, but scarcely of gem importance, is the topaz found in the Mourne Mountains of Northern Ireland, the Cairngorms, Scotland, and St Michael's Mount and Lundy Island off the Cornish and Devon coasts.

Cutting of Topaz

The mixed cut is a style of cutting often used for topaz, and, owing to the long prismatic shape of the crystals, the stones are cut as rather longish oval or as pear-shaped (drop-shaped) stones. In order to minimize the risk of breakage, it is often recommended that the table be placed at 12–15° off the basal plane of the crystal.

As a result of its pleochroism, brown topaz shows a darker colour at each end of the long axis of the stone. Moderately rich-coloured topazes are cut in the trap-cut style (a cut-corner rectangular step cut commonly called the 'emerald cut'). Many pale pink topazes have been set, particularly in old jewellery, in a closed setting with the back of the stone painted with red colour, or with the settings containing a red foil. It has been recorded that topaz, despite its strong cleavage, has been cut as a cameo.

Because of its easy availability, irradiated blue topaz is one of the gem materials used in the mass production of so-called 'fantasy' cuts, that is, stones that combine aspects of faceting and carving.

Synthesis and Simulation

Topaz has been synthesized for academic interest, but not for commerical exploitation as a gem material. The process was carried out by the action of hydrofluorosilicic acid on silica and alumina, in the presence of water at a temperature of 500 °C. Except for natural and synthetic corundum and spinel with colours similar to those of topaz, the other natural stones which are most likely to be mistaken for topaz are tourmaline, danburite, spodumene, aquamarine, morganite, scapolite, apatite, beryllonite, brazilianite, euclase, transparent feldspars such as labradorite and orthoclase, and quartz. All of these materials will float in di-iodomethane (SG 3.32) in which topaz will sink.

Glass imitations of topaz are also encountered and may coincidentally have both an RI and an SG quite close to topaz. The single refraction of the glass, however, would serve quickly to separate it from the doubly refractive topaz.

10

Tourmaline

As with other major species showing a number of colours and properties multiple comparisons with specimens of similar appearance are not given: such information may be found in O'Donoghue and Joyner, *The Identification of Gemstones*, 2003; ISBN 0750655127. *The Tourmaline Group*, by R.V. Dietrich (1985; ISBN 0442218575), is recommended.

Tourmaline is perhaps the ideal gemstone; sufficiently hard, very poor cleavage though somewhat brittle and a wide range of colours including parti-colouration which may be found along the length of the crystal or at right angles to that direction. Several important and/or colourful monographs have been devoted to tourmaline; in 1990 *Der Turmalin*, by F. Benesch, published by UrachhausVerlag, Stuttgart, dazzled the gem book market with its great size and the magnificence of its colour photographs. In 2003 Paul Rustemeyer's *Faszination Turmalin* (ISBN 3827414245) provided fine illustrations of tourmaline crystals. Tourmaline from Maine, USA, is beautifully described by A.C. Hamlin in *The Tourmaline* (1873) and *The History of Mount Mica of Maine, USA and Its Wonderful Deposits of Matchless Tourmalines* (1895). The tourmalines from the pegmatites of southern California are also minutely described, by Jahns and Wright, in *Gem- and lithium-bearing pegmatites of the Pala district, San Diego County, California* (California Division of Mines *Special report 7-A*, 1951). Finally there is *ExtraLapis in English* no. 3, Tourmaline, 2002 (ISBN 0971537127). The author is William Simmons.

The name tourmaline is conveniently used to describe a mineral group which, at the time of the publication of Fleischer's *Glossary of Mineral Species* (the 8th edition, 1999; the most recent edition, 2004) did not set out the groups with their constituent minerals although group membership was stated in the appropriate individual entries. The tourmaline group comprises 12 species, rossmanite having been added between 1999 and 2004. The tourmaline group members with important ornamental application are elbaite, dravite, uvite and liddicoatite. Schorl, a black group member is sometimes fashioned or the crystals collected for their shiny black appearance. All tourmaline group members are borosilicates, crystallizing in the trigonal system as prismatic to acicular crystals striated lengthways and with characteristic rounded cross-sections. Hemimorphism is common. Tourmaline is pyroelectric and piezoelectric; the hardness is usually 7. Inclusions may be so placed that a cat's-eye effect is possible though the eye is less sharp than its counterpart in chrysoberyl. The chemical complexity of minerals of the tourmaline group has so far ruled out serious attempts at synthesis. Though various names have been thought up over the years for the more strikingly coloured tourmaline varieties, none of them has stuck and they are best found in Bayliss (*Glossary of Obsolete Mineral Names*, 2000).

The varying compositions of the tourmaline group minerals cause their properties to vary but in few if any cases should there be any confusion with other minerals. The gem species of the tourmaline group, already cited, have compositions well explained by William Simmons in *ExtraLapis in English* (see above). Working from a general structural formula for the group X Y3 Z6 [T6O18] [BO3]3 V3 W, where the letters represent specific crystallographic sites or positions within the tourmaline structure and showing which ions can occupy each site in the structural formula, Simmons, drawing on material from Hawthorne and Henry (*European Journal of Mineralogy* 11, 201–15, 1999), names the alkali species (X site Na) as elbaite, schorl, dravite, olenite, buergerite, chromdravite, povondraite and vanadiumdravite. Calcic tourmaline group species (X site Ca) are liddicoatite, uvite, feruvite. In rossmanite, foitite and magnesiofoitite the X site is vacant. For the sake of completeness I have included non-gem tourmaline group minerals above. Some of the species described are not diferentiated in *Glossary of Mineral Species* or in Anthony *et al.*

Following the same order, the Y3 site in elbaite contains Li 1.5 and Al 1.5; in schorl Fe^{2+} or Fe^{3+}; in dravite Mg_3; in olenite Al_3; in buergerite Fe^{2+} or Fe^{3+}; in chromdravite Mg_3; in povondraite Fe^{2+} or Fe^{3+}; in vanadiumdravite Mg_3. In the calcic tourmalines, liddicoatite $[Li_2, Al]$; in uvite Mg_3; in feruvite Fe^{2+} or Fe^{3+}; in rossmanite $[Li, Al_2]$; in foitite $[Fe^{2+}$ or Fe^{3+} Al]; in magnesiofoitite $[Mg_2, Al]$.

In the Z6 site elbaite, schorl, dravite, olenite, buergerite have Al_6; chromdravite has Cr^{3+} or Cr^{6+}; povondrite $[Fe^{3+}$ or $^{4+}$, $Mg_2]$; vanadiumdravite

$[V^{3+}$ or $^{6+}]$; liddicoatite Al_6; uvite $[Al_5, Mg]$; feruvite $[Al_5, Mg]$; rossmanite, foitite and magnesiofoitite Al_6.

All species contain $(BO_3)_3$ and $(OH)_3$. All alkali tourmalines contain (OH) except for buergerite and povondraite (of little or no ornamental significance) which contain (OH) and (O) respectively. Of the calcic tourmalines liddicoatite and uvite contain F, feruvite contains (OH). All the X-site vacant tourmalines contain (OH). It is possible that new species may be discovered.

ExtraLapis English 3 gives a good account of tourmaline's crystal structure and of the X-ray powder patterns with the seven highest intensity d-spacings with their peak intensities.

The tourmaline species with ornamental application can be separately described. Though elbaite, liddicoatite, uvite and dravite provide the majority of gemstones, brief notes on the other species can be given. Buergerite is found as transparent to translucent crystals, dark brown to yellow and would be sought as a mineral specimen. Good examples have been found in the Mexquitic area, San Luis Potosi, Mexico. Chromdravite occurs as small acicular dark green crystals and has been found in Kareliya, Russia. Feruvite occurs only as grains. Foitite forms bluish black crystals which may reach 20 mm; one location is a pegmatite at the white Queen mine, Pala, California, USA. Magnesiofoitite crystals found in Yamanashi Prefecture, Japan, have not yet been described.

Olenite forms on the outer rim of acicular crystals with elbaite cores and has been found in the Kola Peninsula, Russia. Povondraite (originally defined as ferridravite) forms very small black crystals at a deposit in Bolivia. Rossmanite has been found in granite pegmatites as colourless to pale pink crystals; locations have been identified in Canada. The black tourmaline group mineral schorl occurs as long to short prismatic crystals and occurs in granite pegmatites and elsewhere. Vanadiumdravite forms very small dark green to yellow crystals at locations in the lake Baikal area of Russia.

From the point of view of the gemmologist it is not necessary to differentiate between the different tourmaline species – the existence of solid solution series between some tourmaline group members (between elbaite and dravite, for example) complicates differentiation.

A review of the ornamental species relating specific gravity to colour can be found in O'Donoghue, *Gemstones*, 1988 (ISBN 041227390X). Red to pink specimens are usually in the range 3.01–3.06; pale greenstones 3.05, brown stones 3.06, dark green 3.08–3.11, blue 3.05–3.11, yellow–orange 3.10. Black specimens are usually 3.11–3.12. Refractive index can be placed in the range 1.610–1.675 with an average birefringence of 0.018. Some faceted dark green tourmalines have been found to show up to six shadow-edges on the refractometer. The effect (the Kerez effect) can be removed by polishing (but it seems a pity!).

Pleochroism is strong and most commonly light and dark of the same colour.

The colours of elbaite are due to the presence of trace elements: Anthony *et al.* (1995) record 0.11% content of both FeO and MnO and 0.04 TiO_2. Fe^{2+} gives a green colour and Fe^{3+} blue; Mn^{2+} gives pink and Mn^{3+} red. Cu^{2+} gives a fine blue in the Paraíba tourmalines mentioned below. Ti^{4+} and Fe^{2+} give yellow. Deeper reds are due to radiation-induced defects. A yellow elbaite reported from Madagascar was found to contain up to 8.9 wt% MnO. The location at Tsilaisina has given a largely unadopted name to tourmaline of this colour.

Despite the range of colours no really useful absorption spectrum is found in tourmaline, though some red stones may show two narrow bands at 458 and 450 nm in the blue. In green tourmaline most of the red is absorbed down to around 640 nm with a narrow iron band showing at 497 nm in the blue. This band can also be found in green tourmaline.

Colour zoning giving the familiar watermelon appearance with the red core and green rind is a feature of elbaite, the cause arising from more than one event, including a change of composition during growth. In the ordinary ray the pleochroic colours range from pink to pale green, pale to deep blue; in the extraordinary ray the crystals may show colourless to yellow, olive blue or purplish.

Liddicoatite is the calcium analogue of elbaite and is named after the late Richard T. Liddicoat, sometime President of the Gemological Institute of America. It occurs as prismatic crystals with pedions and can achieve considerable lengths. Striations are prominent and parallel to the vertical axis. The type locality where very fine crystals have been found in Anjanabonoina, Madagascar; it also occurs in the Sahatany valley in the same country. Examples have been found elsewhere, including Nigeria and Tanzania. Liddicoatite's colour may be pink, brown or colourless and multi-coloured crystals are common. The colours may be zoned along the length of the crystal with sections cut at right angles to the length showing pronounced colour zoning and triangular markings with some three-rayed stars. The pleochroism ranges from dark brown to pink, pale to deep blue in the ordinary ray; in the extraordinary ray light brown, colourless to yellow and olive green colours may be seen.

Dravite occurs as radiating crystals which may be short or long prismatic with or without striations. Equant forms do not usually show striations. Crystals may show a triangular cross-section with curved sides. The colour may be brown to black, red, yellow, blue or green, colourless to white. The lustre is greasy and the strong pleochroism shows brown to yellow in the ordinary ray and yellow to colourless in the extraordinary ray.

Uvite crystals may be prismatic to equant, tabular or pyramidal and range up to 15 cm. The colour may be brown, dark red, dark to light green, brown to black. Most crystals are translucent. The strong pleochroism is dark green and dark brown in the ordinary ray and olive green and light brown in the extraordinary ray. Uvite occurs most commonly in marbles and is found in Sri Lanka and at Franklin, New Jersey, USA.

The Usambara effect

In the past few years an unusual colour effect has been noted in some tourmaline crystals; the first report was by Halvorsen and Jensen in *The Journal of Gemmology* 25(5) 325–30, 1997. Specimens of gem-quality tourmaline crystals from the Umba Valley, Tanzania, each green by transmitted light, when placed on top of one another, gave transmitted light first of a yellow colour, succeeded by orange then red as the thickness of the pile increased. This is a change of colour with thickness.

The crystals were determined as a uvite containing some sodium and titanium, as well as chromium and vanadium which were the cause of the colour. Anomalous trichroism of two shades of red and a green were also noted (tourmaline is uniaxial). In *The Journal of Gemmology* 26(6), 1999, a single crystal was described which appeared green but when viewed along its length showed red in all types of illumination. When two green pieces were stacked the colour was red in all illuminations with a red component. The red colour was not visible in light containing little or no red wavelengths. Nassau (2000) explains the effect and suggests that it may in time be found in other species. The uvite specimen showed the alexandrite effect and pleochroism (instead of dichroism). Some green tourmaline from the John Saul mine in Kenya and some garnets from the Taita Hills, Kenya, have been found to show both the Usambara and alexandrite effects (*Journal of Gemmology* 27(1) 12–29, 2000).

A faceted stone of this materials would be likely to show a colour effect which Nassau cites as defying precise comprehension. Nassau suggests that some plastics may show a blue to red effect and a green to purple change could also occur.

The occurrence of tourmaline is covered by O'Donoghue (1988, *op. cit.*) and a summary is given below. By no means all occurrences are covered. Most commercial gem tourmaline still comes from Brazil where only some of the many deposits are listed. A survey by Cassedanne, Cleac'h and Lebrun, *Tourmalines* (1996), covers a number of localities in a monographic issue of *Revue de gemmologie* [hors serie no. 3, ISSN 0335 6566] which contains excellent colour photographs and a summary of the tourmaline group species.

Inclusions

The inclusions seen in tourmalines are in general irregular thread-like cavities called trichites that occur singly or in loose, mesh-like patterns. Under high magnification these are resolved into tubes filled with liquid and often containing a gas bubble (two-phase inclusions). Hollow tubes running parallel to the length of the crystal are also common and, if profuse, are the cause of the chatoyant effect seen in tourmaline cat's-eyes; sometimes these tubes contain mineral inclusions. Eppler, however, took the view that the inclusions causing the chatoyancy in tourmaline are crystal fibres.

Flat films are another common inclusion, most typically in red tourmalines, and if these happen to be viewed at such an angle that the incident light is totally reflected from them they appear as black patches. Crystalline inclusions identified in tourmaline include mica, hornblende, apatite, quartz, microlite, tourmaline and zircon. The tourmalines from Paraíba, Brazil, sometimes contain yellow metallic-appearing inclusions, possibly a sulphide mineral.

Occurences (by the late Robert Kammerling; revised by MO'D)

A principal locality for gem tourmaline is near Mursinka in the Ural Mountains of Russia where good-quality crystals, usually blue, red or violet-red in colour, are found in the yellow clay disintegration product of the granite in which they had formed. The village of Shaitanka, some 70 km north of Sverdlovsk, produces red crystals; these are found in druses in a coarse-grained granite. Another Russian locality is Nerchinsk in Transbaikalia where tourmaline is found near the Urulga river.

Sri Lanka, probably the original source of tourmaline as a gemstone, supplies stones of yellow and brown colour from the alluvial deposits in the south-east of the island; these have been identified as uvites. Brown to green tourmalines have also been found in the Elahera gem field in central Sri Lanka. About 32 km south-east of Mogok in Myanmar there is an extensive alluvial deposit of decomposed gneiss and granite carrying fine red tourmaline. The deposit lies on the banks of the Nampai river near the town of Mainglon and has been worked spasmodically, often in the past by Chinese miners, for in olden times this red tourmaline was much esteemed in China where it was used for making the distinctive button for a certain grade of mandarin. Mogok also produces some pink elbaites and brown uvites. Tourmaline, including pink, light blue and light green colours, is one of the gem minerals reported from the granitic pegmatite dykes in the Altay Mountains of the northern Xinjiang Uygur Autonomous Region of China. Tourmaline is also found in Inner Mongolia and Yunnan province. The pegmatites of Kolum district of the

Nuristan region in eastern Afghanistan provide some exceptional gem elbaites, primarily bright blues and greens as well as pinks and parti-coloured. Several deposits of tourmaline occur in the Shingus and Dusso areas of Gilgit division, northern Pakistan, while attractive stones have come from the Hyakule and Phakuwa pegmatite mines in the Sarikhuwa Subha district, Kosi zone of eastern Nepal.

Brazil produces tourmalines of many colours, particularly green, blue and red crystals, and many which show zonally arranged colours (parti-colouration). The crystals are found primarily in the north-eastern state of Minas Gerais, in regions encompassing what are known as the Araçuai-Itinga, Araçuai-Salinas and Governador Valadares pegmatite districts, and in areas surrounding the cites of Capelinha, Itambacuri, Minas Novas, Potê and Turmalina. In the 1980s, a new gem deposit was discovered in the state of Paraiba, also in north-eastem Brazil, near the village of São José da Batalha. Here in a decomposed granitic pegmatite were found colours not known previously in tourmaline, including deeply coloured 'electric' bluish green to greenish blue, dark violetish blue and purple. Chemical analysis has shown that copper, not previously known as a colouring agent in tourmaline, was responsible for the blue colour component, while managanese- or iron-titanium charge-transfer mechanisms shift the colour towards green. Manganese produces pink and thereby contributes to violetish blue to purple colours.

Tourmaline is found, in a wide range of colours including the dark green which lightens to an emerald-green shade on heating, in the pegmatites at Klein Spitzkopje near Rössing on the Otavi railway east of Swakopmund in Namibia. The gem gravels of the Somabula Forest in Zimbabwe and Alto Ligonha in Mozambique are other sources. Malawi produces green and pink tourmalines. In Tanzania there are found green tourmalines which contain chromium and/or vanadium. These show, unlike normal green tourmalines, a strong red residual colour when viewed through the colour filter. Tanzania has also produced yellow to orange and colour change tourmalines, the latter appearing dark green in daylight and red in incandescent light. In the early 1970s a deposit of small euhedral tourmaline crystals, transparent but very dark red in colour, was discovered by J.M. Saul under a meadow of Osarara, Narok district, Kenya. The crystals had a normal SG of 3.07 but the birefringence was found to be unusually high. Typical values for the refractive indices were 1.623–1.655. Analysis indicated that the crystals could best be described as iron-rich dravite tourmalines. Attempts to lighten the deep red colour of the stones by heat treatment or by radiation have not been successful. Kenya has produced attractive orange-yellow and brown stones from the Voi-Taveta area. At Chipata in Zambia, dark red crystals similar to those from Kenya have also been found with a similarly high birefringence (0.030) and mean SG of 3.05. Zambia has also produced some rare, bright yellow, yellowish brown and greenish yellow tourmalines (birefringence 0.023–0.028 and SG 3.13), with up to 9.2% MnO. In Nigeria, the pegmatites southeast of Keffi in western Plateau state have produced green, pink, red, violet, yellow and watermelon elbaite tourmalines.

Madagascar has supplied much gem-quality tourmaline. The crystals are obtained from pegmatites in the central part of the island, one of which is the famous Anjanaboina deposit, and include all colours, the red being the most prized; the colourless crystals are the rarest. Blue, brownish purple, rose-pink, yellow, brown and green material are found in the Madagascan deposits.

The Pala region of San Diego County, California, produced much tourmaline from the pegmatites of the district. One famous mine, the Himalaya in Mesa Grande district, continues to be worked commercially. A gem pocket found in 1989 produced about 500 kg of tourmaline, the majority of which was suitable for mineral specimens. The colours found at Pala vary from jet-black schorl, through the deep blue, to green and pink to pale red. Colourless material has also been found at Pala. It can be said that nearly all the colours assumed by tourmaline are found in the San Diego deposits, with only brown-coloured material not well represented. Two or more colours are commonly present in a single crystal, and some are characterized by concentric or layer-like zoning parallel to the c-axis of the prismatic crystals. Some of these mixed-coloured tourmalines are of deep blue to black cores surrounded by single or multiple layers that may be colourless, blue, green or pink, or any combination of these colours, such as a pink central portion surrounded by a rim of green-coloured material, a type called 'watermelon' tourmaline. Another type of parti-colouration is where the colours vary in layers parallel to the basal plane of the crystals (that is, perpendicular to the c-axis), which may be green at one end and pink at the other, with perhaps a zone of colourless material in the centre or other variations of this theme. The colours in some of the zoned (parti-coloured) crystals are sharply bounded from one another, whereas in other crystals the colours appear to merge gradually and intergrade with distances of about 2 mm or less. It should be noted that parti-coloured stones tend to be somewhat more fragile at the colour boundaries.

During the early years of the twentieth century, Pala tourmaline was principally marketed in China. This was particularly so with the pink and red varieties which were highly prized by the Chinese who carved and polished them into many different forms. With the collapse of the Manchu dynasty in 1912 this Asian market was eliminated, and much of the mining for gem tourmaline ceased in southern California.

Much gemmy tourmaline has been found in Maine, USA, mainly at Hebron and Paris (Mount Mica) where tourmaline of a bluish green colour is abundant and pink crystals are rare. At Auburn in the same state is found an attractive lilac-coloured tourmaline, as is also the deep blue, green and parti-coloured crystals. This area and its products are beautifully illustrated in Hamlin's two books. Maine gem material is described in the two volumes of *Mineralogy of Maine*, by King and Foord (1994) Maine Geological Survey, Dept of Conservation.

Haddam in Connecticut is another American locality for elbaite, while San Luis Potosi in Mexico has produced buergerite. Pegmatites in Western Australia and on Kangaroo Island have also produced gem-quality material.

Cutting of Tourmaline

The mixed cut – a step-cut pavilion with a brilliant-cut crown – is commonly used for tourmaline. With the deeply coloured blue and green crystals as well as with longitudinally zoned parti-coloured material, the trap cut is used to a great extent. This has the advantage of providing maximum yield from the rough and with the dark iron-colour stones, minimizing the negative effects of the strong pleochroism. The watermelon type of concentrically zoned material is often fashioned as thin, polished slabs for use in drop earrings and cuff-links. Much flawed tourmaline (and the crystals may be of considerable size) is fashioned into beads or carved into small figurines. Tourmaline is also a popular gem material for 'fantasy' cutting. Crystals with appropriate inclusions for showing the cat's-eye effect are cut as cabochons. The position of the optic axis, parallel to the principal crystal axis, which in tourmaline is the direction of maximum absorption of light, forces the lapidary when faced with a dark crystal, to place the table anywhere but at right angles to this direction.

Synthesis

Tourmaline has not been grown in ornamental-quality sizes.

Simulation

Details of possible confusion between tourmaline group species can be found in O'Donoghue and Joyner (2003, passim).

Some tourmaline from Sri Lanka exhibits pleochroism similar to that seen in andalusite with refractive indices in the range of tourmaline. The andalusite's lower birefringence (0.008–0.013) should serve to distinguish it. Lazulite, a deep blue mineral, also has refractive indices in the range of tourmaline with a higher birefringence, about 0.031. Bright greenish blue to bluish green apatite from Madagascar looks much like some of the distinctive tourmaline from Paraíba and has even been sold under the misleading name 'Paraíba apatite'; similarly coloured apatite has recently been reported from Minas Gerais, Brazil. Irrradiated topaz that has not been annealed after irradiation has also been used to imitate Paraíba tourmaline.

Occasionally, assembled stones employing tourmaline are encountered. One type exhibits chatoyancy, being made by cementing a transparent crown to a fibrous pavilion. Another relatively recent type is a faceted stone consisting of a tourmaline crown and glass pavilion. A related type consists of tourmaline with concentric colour zones cemented to a lighter-coloured piece of glass. Beryl triplets employing bright blue cement have also imitated Paraíba material.

11

The Garnets

Brian Jackson

The word 'garnet' is known from antiquity and is derived from the Latin word *granatum* – pomegranate, which in turn derives from *granum* – grain; meaning multi-seeded in the context of the pomegranate, the seeds of which turn a deep wine-red when ripe. The open fruit shows similar coloured seeds and surrounding pith, thus the allusion to the resemblance with the mineral. The name garnet is applied to a complex group of minerals that have common crystal structure and some similarity in chemical composition.

Garnet: Group, Series and Species: Crystal Chemistry Considerations

Garnets present a complex classification problem, since their chemical composition is a continuous admixture of end-member components, with few (if any) natural breaks that can be used to distinguish among gemmologically significant varieties. Varieties have been defined (Stockton and Manson) as subdivisions of a species that may be differentiated either by distinctive physical characteristics such as colour and phenomena or by consistent minor chemical disparity; however, some subjectivity is still prevalent and this gives rise to uncertainty. They have similar crystal lattice structure (isostructural) and can, with the exception of the hydrogarnets (hibschite and katoite), be expressed by the general formula $A_3B_2(SiO_4)_3$. Within the garnet group, the mineral species approved by the International

Mineralogical Association's Commission on New Minerals and Mineral Names (CNMMN) are:

Almandine	$Fe_3^{2+}Al_2(SiO_4)_3$
Andradite	$Ca_3Fe_2^{3+}(SiO_4)_3$
Calderite	$(Mn^{2+},Ca)_3(Fe^{3+},Al)_2(SiO_4)_3$
Goldmanite	$Ca_3(V,Al,Fe^{3+})_2(SiO_4)_3$
Grossular	$Ca_3Al_2(SiO_4)_3$
Hibschite	$Ca_3Al_2(SiO_4)_{3-x}(OH)_{4x}$ (where x is between 0.2 and 1.5)
Henritermierite	$Ca_3(Mn,Al)_2(SiO_4)_2(OH)_4$
Katoite	$Ca_3Al_2(SiO_4)_{3-x}(OH)_{4x}$ (where x is greater than 1.5)
Kimzeyite	$Ca_3(Zr,Ti)_2(Si,Al,Fe^{3+})_3O_{12}$
Knorringite	$Mg_3Cr_2(SiO_4)_3$
Majorite	$Mg_3(Fe,Al,Si)_2(SiO_4)_3$
Morimotoite	$Ca_3TiFe^{2+}Si_3O_{12}$
Pyrope	$Mg_3Al_2(SiO_4)_3$
Schorlomite	$Ca_3Ti_2^{4+}(Fe_2^{3+}Si)O_{12}$
Spessartine	$Mn_3^{2+}Al_2(SiO_4)_3$
Uvarovite	$Ca_3Cr_2(SiO_4)_3$

The similarity in structure permits elements with comparable ionic size to enter specific like-sized sites within the crystal lattice. Thus the A site can be filled by either Ca, Fe^{2+}, Mg or Mn^{2+} and the B site by either Al, Cr^{3+}, Fe^{3+}, Mn^{3+}, Si, Ti, V^{3+} or Zr without affecting the crystal structure. As a result, elements can be interchangeably substituted within each of the sites giving rise to variable chemical composition. This is termed 'isomorphous replacement'. Consequently isostructural pyrope $Mg_3Al_2(SiO_4)_3$, almandine $Fe_3^{2+}Al_2(SiO_4)_3$ and spessartine $Mn_3^{2+}Al_2(SiO_4)_3$ differ only in the element filling the A site. The percentage of the element filling the site can range from 0 to 100%. This complete substitution range constitutes an isomorphous series (*syn-solid solution series*). There are two such series within the garnet group: the pyralspite series and the ugrandite series. In the pyralspite series the A site can exhibit complete isomorphous replacement between Mg, Fe and Mn. Whereas in the ugrandite series the B site exhibits complete isomorphous replacement between uvarovite and andradite but only limited solid solution between these and grossular. This arises because the ionic radii of Cr^{3+} and Fe^{3+} are similar and are significantly larger than that of Al^{3+}.

Size discrepancy between the A and B site would seem to prohibit substitution between members of the two series. This, however, contrasts with analyses of garnets where limited substitution between the two series can occur. Factors that contribute to substitution between the two series include distortion of the crystal lattice arising from rotation of the [SiO$_4$] tetrahedral group and temperature and pressure of formation. Coexistence of Ca^{2+} with Mg^{2+} correlates, albeit weakly, with anomalous double refraction (ADR) but overall this weak and aberrant birefringence is attributed to strain within the crystal lattice. Thus ugrandite series garnets often show marked anamolous birefringence.

The names of the series are derived from the first few letters of the species names.

	Pyralspite series (A site replacement element)		Ugrandite series (B site replacement element)	
Species	Pyrope	Mg	Cr	Uvarovite
	Almandine	Fe$_3$$^{2+}$	Al	Grossular
	Spessartine	Mn$_3$$^{2+}$	Fe$_2$$^{3+}$	Andradite

When a species in an isomorphous series shows no substitution in the replacement site it is termed an 'end-member'. As substitution is the norm, classifying a garnet can be problematic. The mineralogical solution is to name the garnet according to which of the end-members is present in the greatest quantity. Thus a pyralspite garnet that contains Mg, Fe and Mn would be represented as percentages of the end-member, e.g. Al$_{60}$Sp$_{30}$ Py$_{10}$ and, in this case, as the almandine component is the greater it would be called almandine. This solution does not sit comfortably with gemmological needs and the International Mineralogical Association's (IMA) Commission on New Minerals and Mineral Names do not approve variety names.

Garnet Classification Systems

Following on from the previous section, it may be helpful to the reader to consider the various attempts to produce a classification for gem-quality garnets. It is generally held that existing gemmological methods of classifying garnets have proved inadequate when dealing with some new types of garnets discovered recently. There are too many ill-defined terms and fuzzy boundaries. Given their access to sophisticated analytical equipment, mineralogists are comfortable with their classification system. Any new gemmological system has to take account of standard gemmological testing (there is no point in devising a system that requires electron-probe microanalysis when this is not available to the vast majority of gemmologists).

The foundations for classifying garnets in gemmological terms were laid by Anderson in 1942 and developed by others such that by 1980 it was generally accepted that, by measuring the RI and SG, noting the colour and observing the absorption spectrum, garnets could be classified. Whilst this, by and large, held true, the stumbling block was the assortment of stones between almandine and pyrope having intermediate compositions and therefore different properties. Easing coloured varieties into a scheme that had overlapping properties ensured its demise. The work of Stockton and Manson in the 1980s addressed these problems. They proposed a new classification system based on RI, colour and absorption spectra. To achieve this, they created, in contradiction to mineralogical nomenclature, new intermediate species, *pyrope-almandine*, *almandine-spessartine* and *pyrope-spessartine*. *International Gemological Symposium Proceedings* (1982): *Gems & Gemology* 20(3) 1984: *Gems & Gemology* 191–204, 1981.

Hanneman, in response to the classification system proposed by Stockton and Manson, has developed since 1983, his own scheme based on RI, colour and spectral considerations. SG was deemed to be unimportant. The significant difference between the Hanneman scheme and that of Stockton and Manson is the RI ranges proposed for the different types of garnet, which Hannemen says eliminated contradictions in the Stockton and Manson system.

It will be useful if the garnet types and associated RIs proposed by the above schemes are set out as their application is still largely extant.

	Stockton and Manson (1985)	GIA (1998)	Anderson and others (pre-1980)
Almandine	1.785–1.830	1.76–>limit of refractometer	1.780–1.810
Pyrope-almandine	1.742–<1.785		1.750–1.780
Rhodolite		1.740–1.770	1.750–1.780
Pyrope	1.714–<1.742	1.720–1.756	1.730–1.750
Pyrope-spessartine	1.742–<1.780		
Malaia		1.742–1.799	
Almandine-spessartine	1.810–1.820		
Spessartine	1.810–1.820	1.790–1.814	1.790–1.810
Grossular	1.730–1.760	1.730–1.760	1.742–1.748
Grossular-andradite			
Andradite	1.880–1.895	>limit of refractometer	1.888–1.889

When giving RI ranges, Hanneman (1999) chooses to define garnet types with reference to series:

Series	Name	Refractive index
Pyrope-almandine	Pyrope	1.714–1.749
	Pyrope-almandine	1.749–1.795
	Almandine	1.795–1.830
Pyrope-spessartine	Pyrope	1.714–1.740
	Pyrope-spessartine	1.740–1.774
	Spessartine	1.774–1.800
Almandine-spessartine	Spessartine	1.800–1.809
	Almandine-spessartine	1.809–1.821
	Almandine	1.821–1.830
Grossular-almandine	Grossular	1.734–1.763
	Grossular-almandine	1.763–1.801
	Almandine	1.801–1.830
Grossular-spessartine	Grossular	1.734–1.754
	Grossular-spessartine	1.754–1.780
	Spessartine	1.780–1.800
Pyrope-grossular	Pyrope	1.714–1.720
	Pyrope-grossular	1.720–1.728
	Grossular	1.728–1.734
Grossular-andradite	Grossular	1.734–1.770
	Grossular-andradite	1.770–1.841
	Andradite	1.841–1.887

Hanneman proposes a solution based on end-member:intermediate:end-member ratio (a 30:40:30 proportional division) that is a more aesthetic presentation than that proposed by Stockton and Manson which has less regard for chemical composition.

Dealing with Garnet Nomenclature

To paraphrase Humpty Dumpty:

'When we use a word,' said the dealer, 'it means just what we choose it to mean – neither more nor less.'

'The question is,' said the gemmologist, 'whether you can cause simple words to have such complex meanings when every mineralogist follows the scientific definition.'

Terminology in gemmology is often driven by market forces, and the branding of coloured gems is seen by the trade to be one path to success in a crowded marketplace: it is easier to sell a tsavorite garnet than a vanadium-grossular or $Gr_{86}Go_{13}Sp_1$. Gemmologists need to take account of garnet nomenclature as applied by the trade at the same time as balancing the scientific approach of the mineralogists.

The RI and SG of garnet species are directly linked to their chemistry. Whilst mineralogists can more readily determine the chemistry of a garnet, the equipment necessary for this is not readily available to the vast majority of gemmologists, and hence gemmologists rely primarily on RI, SG and absorption spectra to define a species. To apply these criteria successfully, it is necessary to define mutually exclusive limits for each different species. Because of isomorphous replacement this is not possible in all cases. Some of the difficulties in determining the species were overcome by means of characteristic visible (750–400 nm) spectral properties. There are, however, inherent difficulties as not all garnets exhibit characteristic absorption spectra, and that some spectra, particularly that of almandine, is dominant even though the end-members' presence is considerably less than 50%.

Colour is a consequence of absorption in the visible spectrum. The body colour of a garnet reflects the bulk chemistry of the stone; however, trace elements, particularly chromium, can substantially affect the outcome. Once the link between body colour, bulk composition and trace elements was established, the gem trade relied heavily on colour to assign species status – a highly subjective method. To further assist, variety names were created to describe differently coloured garnet of the same species. When applied to garnets of intermediate composition this was substantially less effective. Nevertheless, based on colour, variety names for garnets were adopted through common usage. Although not approved by the mineralogical community, these variety names have been introduced into the literature. Historically, whilst initially helpful, this too ultimately proved less effective as new complex garnets, similarly coloured to the older established varieties, were discovered. Variety names dependant upon subjective colour descriptions based on flowers, fruits and other variable objects inevitably lead to inexactness.

Complex isomorphous replacement therefore produces highly variable RI, SG properties and colour variation. For the mineralogists this is not a problem as they relate these properties to define end-members. The gem trade has additionally adopted variety names and by and large treats these on an equal footing as species names with the emphasis on a saleable name. Gemmologists have to answer to both parties. There is a need for scientific accuracy and a need to establish envelopes defining the properties of the different varieties. This requires to be accomplished within the scope of normal gemmological testing. The fact that the envelopes can overlap and change with new discoveries creates a problem. When applying mineralogical species criteria there is the potential for a variety to span more than one species. In gemmology the matter of nomenclature is never far from the centre of misunderstanding. Litigation awareness has served to highlight this particularly where the name clearly misrepresents the nature of the material. Market-driven nomenclature, where there is no accountability, is largely responsible. CIBJO, the international jewellery confederation of national trade organizations attempts to harmonize acceptable nomenclature use to protect consumer confidence in the industry. They do, however, accept variety names where they believe misrepresentation does not occur.

Without chemical analysis it is quite difficult to get information about the proportion of the end-member components of a garnet sample. Plotting RI against SG shows a correlation between increasing RI and increasing SG. This arises because of the retardation in the speed of light caused by the higher electron density of the heavier atoms. The two solid solution series pyralspite and ugrandite show a different correlation between the two variables RI and SG. In the pyralspite group the RI values increase more slowly with increasing SG values. This is due to substitution of the lighter by heavier elements in the A site in the pyralspite series compared with B site substitution in the ugrandite series (*Plate 1*).

By drawing lines linking ideal end-members on an X–Y plot of RI against SG, the chemical composition of an unknown garnet can be estimated by plotting its RI–SG coordinates. The robustness of estimation decreases with increasing numbers of end-member components particularly when end-member proportions are not extreme.

Garnet Crystals

The three-dimensional order of atoms in the internal geometric structure of garnets produces a repeatable lattice structure that results in isotropic crystals. Thus garnets belong to the cubic system. It should be noted, however, that distortion of the crystal lattice in ugrandite garnets results in anomalous double refraction. Predominant morphologies are the trapezohedron (icosotetrahedron) and rhombic dodecahedron: crystals with 24 trapezoidal faces or 12 rhombic faces respectively, or combinations of these and some other forms. Factors controlling the morphology have variously been attributed to the A^{2+}/B^{3+} ionic radius ratio, unit cell dimension, temperature, pressure and admixture. This has resulted in the generalization that ugrandite garnets normally form as rhombic dodecahedrons whilst pyralspite garnets normally form as trapezohedrons. However, field evidence for almandine, linking morphology to geological environment, indicates that the rhombic dodecahedral form is predominant in schists whilst trapezohedrons are predominant in pegmatites and granites.

Rhombic dodecahedron and trapezohedron combination forms are also common. Hexoctahedrons are occasionally seen whilst other forms are rare.

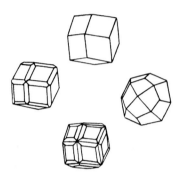

Figure 11.1   Crystal diagrams for garnet

Garnet Glossary

Species are listed alphabetically. Where a colour variety exists entirely within a species, it is subsumed within the description of that species, e.g. demantoid is described under andradite. Where a colour variety is representative of an intermediate composition, it is described under the largely dominant species, e.g. rhodolite is described under pyrope.

Almandine $Fe_3^{2+}Al_2(SiO_4)_3$

Name approved by the CNMMN of the IMA. Synonyms are alamandine, almandite, oriental garnet, alabandine ruby; precious garnet refers to deep red transparent almandine.

Almandine was named after the now ruined ancient city called Alabanda, 10 km west of present-day Cine, Aydin, south-west Turkey. From the writings of Theophrastus (315BC) and almost 400 years later of Pliny, this opulent city was known to be a cutting centre for quartz gems and garnet.

Occurrence

Almandine is the commonest species in the garnet group, typically occurring in schists and gneisses, resulting from regional metamorphism of largely clay rich sediments and rocks. It also occurs in granites and in granite pegmatites and some volcanic rocks, as well as in sedimentary deposits as detrital grains.

Whilst occurrences are worldwide (total world industrial garnet production in 1999 was estimated to be 335 000t), the gem material is less abundant. India, for example, produced 677 kg of gem garnets in 2001–2002. Notable localities for gem-quality almandine include: South Africa, Zimbabwe, Tanzania, Madagascar, Kenya, Sri Lanka, Myanmar, Brazil, Argentina, Australia, Bohemia, Russia, Pakistan, India, Canada, Mexico and the US. In 1998, a new garnet deposit was found in Brazil; it is amethystine garnet and has been identified as almandine. It is located in Fazenda Balisto in Tocantins.

Properties

Almandine generally contains significant amounts of both the pyrope and spessartine molecules forming a series with these, and may also have noteworthy amounts of the grossular molecule. Consequently the physical and chemical properties and appearance can vary appreciably.

Absorption Spectra

Pure almandine is $Fe_3^{2+}Al_2(SiO_4)_3$. Isomorphic replacement gives rise to variable amounts of magnesium, manganese, calcium and ferric iron (Fe^{3+}) replacement. The ferrous iron (Fe^{2+}) produces a conspicuous absorption spectrum that was first described by Church in 1866, and provides a ready test for identifying the presence of the almandine molecule. The spectrum consists of three main absorption bands. A 30 nm broad band in the yellow centred at 576 nm, and two bands in the green at 526 and 505 nm that, in strongly coloured stones, may not be resolved and appear as a single broad band. There are several other bands in the complete spectrum, a weak broad band in the orange at 617 nm and another in the blue at 462 nm. Other weak bands have been measured at 476, 438, 404 and 393 nm and may only, in many cases, be recorded by spectrophotometers.

Burgundy-coloured garnets from alluvial deposits in Tocantins are members of the almandine-spessartine series and comprise mainly almandine with absorptions at 692, 574, 503, 460, 422, 398 and 365 nm.

In addition to the amount of almandine molecule present, the intensity of the bands depends on the size and thickness of the specimen and the intensity of the light source. Thus the length of the light path through the stone will have an effect on the intensity and resolution of the spectrum. The presence of an almandine molecule spectrum does not necessarily indicate that the stone is an almandine, but merely that the almandine molecule is present.

Refractive Index

The theoretical value of the end-member has been variously reported as 1.827 and 1.830. Most authorities quote 1.830 as the maximum value. Lower values are more problematic given ionic substitution between the other pyralspite series members, limited ugrandite components and the mineralogical dominant molecule compositional criteria. Bearing this in mind, assigning the name almandine to Fe/Al garnets with RI values lower than 1.773 should be avoided unless supported by chemical analysis. A rule of thumb is that almandine garnets have RI values greater than 1.78. Some authors, however, make reference to a lower limit for almandine as 1.75. This approximates to the lower limit of the 'intermediate category'. The RI values for almandine garnets are not exclusive with some grossular-andradite garnets approaching 1.78.

The limitations of practical gemmological gem testing and the unclear nature of the lower limit for almandine has prompted the acceptance of an intermediate category. This has been variously called 'pyralmandine' (proposed by Fermor), 'rhodomacon' (Campbell), 'umbalite' or 'pyrandine' (Anderson) and 'pyrope-almandine' (Stockton and Manson). None of these names has generated universal acceptance.

Density

Pure end-member (calculated) 4.313. As with RI, the values for density vary with the amount of ionic substitution. Extracting values from the literature, where chemical information is provided, indicates that for pyralspite garnets containing ≥50% almandine molecule the lower limit of density approximates to 3.95. It should be borne in mind that inclusions can affect measured density and, if in abundance, give rise to altered values.

Hardness

Generally held to be 7.5 on Mohs' scale but has been reported to range from 6.81 to 7.48.

Dispersion

0.024–0.027. This modest amount of dispersion is masked by the body colour.

Cleavage/Fracture

Fracture is conchoidal to even sometimes exhibiting flat surfaces due to parting parallel to the dodecahedral face {110}.

Appearance

Colour

Colour is a significant consideration. The colour of almandine has been described as various shades of brownish red, orange-red, red and violet-red. Colour is also related to size and large stones are darker. In order to lighten darker stones they have been cut as hollowed out cabochons called carbuncles: a practice introduced in Roman times.

Colour Varieties: Almandine (allegedly) of a rich red grape colour has been reported from Tanzania, Madagascar and Brazil (see pyrope–grape garnet pp. 224–226).

Inclusions

Almandine is usually included with a variety of minerals. Commonest amongst these are needle-like crystals of rutile aligned parallel to the dodecahedral face. Planar intersections are at angles of 70° and 110°. This gives rise to two sets of long needles accompanied by a set of apparently shorter needles as the latter are inclined to the plane of the other two. Other common inclusions are euhedral crystals of zircon with accompanying tension halos and apatite. Less common inclusions include quartz, monazite, ilmenite, spinel, biotite, titanite (sphene), phlogopite, muscovite, plagioclase, chalcopyrite and sphalerite.

Optical Effects: Chatoyancy, Asterism, Colour Change, ADR

Chatoyancy and asterism: When in abundance, rutile inclusions can give rise to asterism. Four-ray stars occur when the needles are parallel to the edges of the dodecahedral faces and six-ray stars when parallel the edges of the octahedral faces. Complete spheres show multiple stars. India, Emerald Creek area of Idaho, USA, and Ambatondrazaka, near Lake Alaotra, Madagascar, are reported as sources of star almandine garnet. The star garnets from Ambatondrazaka, Madagascar, are intermediate members of the almandine-spessartine series with an appreciable pyrope component. Almandine from the latter is also reported to exhibit reddish brown cat's eyes arising from orientated sillimanite lamellae. Star garnet has also been reported from Sri Lanka but the species has not been defined.

There is no dichroism, although there may be anomalous double refraction.

Andradite $Ca_3Fe_2^{3+}(SiO_4)_3$

Name approved by the CNMMN of the IMA. Synonyms are allochroite, aplome, calcium-ferrigranat, polyadelphine and polyadelphite.

Andradite is named after the Brazilian statesman and geologist Jose Bonifacio d' Andrada e Silva (1763–1838), who first described the mineral.

Occurrence

Frequently found in metamorphosed limestones and metamorphosed impure carbonate-bearing rocks; particularly in metosomatic skarns. The light-coloured andradite varieties topazolite and demantoid occur mainly in chlorite schists and serpentinites respectively. The darker andradite variety melanite and the related schorlomite occur mainly in alkaline igneous rocks such as nepheline syenites and phonolites. Andradite also occurs in metamorphosed igneous rocks, and addition-ally as detrital grains in placer deposits.

Notable localities for gem-quality andradite include: Russia (Ural Mts), USA (Arizona), Namibia (Erongo Mts) for green (vanadium rich–chromium poor) andradite, some of which may be considered demantoid; Iran (Kerman), Afghanistan (Khost, Kunnar) and Pakistan (Kashmir) for demantoid; USA (California), Italy (Piedmont) and Switzerland (Zermatt) for topazolite; Greenland, Norway, Sweden, Germany, Uganda, Madagascar, Sri Lanka, Russia and Zaire for andradite of non-specific colour variety or melanite.

Semi-transparent light apple-green demantoid with significant grossu-lar/hydrogrossular content has been reported from Kamchatka and Chukotka, Soviet Union.

Properties

Andradite generally contains significant amounts of the grossular mole-cule, and a continuous isomorphous series exists between these two components. This has been termed the *grandite* series. A second series exists with schorlomite that includes *melanite* (titanium andradite). It is sometimes in solid solution with spessartine and uvarovite. Consequently the physical and chemical properties and appearance can vary appreciably.

Absorption Spectra

Pure andradite is $Ca_3Fe_2^{3+}(SiO_4)_3$; however, ferric iron (Fe^{3+}) may be replaced by aluminium and chromium, whilst calcium can be replaced by ferrous iron (Fe^{2+}), magnesium and manganese (Mn^{2+}). This will have an effect on the body colour of stones. In andradite, where substitution is minor, isolated ferric iron (Fe^{3+}) causes a yellow-green to brown colour. This produces an absorption spectrum with bands in the blue and violet. The strongest of these is a narrow intense band at 443 nm; however, general absorption of the violet may obscure this and in darker stones mask it altogether. This diagnostic band is more readily seen in the yellow-coloured variety *topazolite*. The other fainter and less diagnostic bands occur at 485 and 464 nm.

The spectrum of the vivid green coloured variety *demantoid* characteristically shows absorption bands due to both ferric iron (Fe^{3+}) and chromium (Cr^{3+}). The characteristic 443 nm (Fe^{3+}) band is present but as Cr^{3+} produces strong general absorption of the violet, this band appears as an abrupt cut-off to the spectrum. In the finest greenstones, Cr^{3+} absorption lines have been reported at 701 (unresolved doublet), 693, 640 and 622 nm. These may not be seen in lighter- and yellowish greenstones. Chromium absorption lines have been recorded at 622, 640 and 685 nm in demantoid from San Benito, California, USA. In addition there is broad absorption from the violet through to 520 nm. Namibian demantoid shows a variable spectrum: a sharp band at 425 nm, a band at 445 nm, general absorption below 450 nm. Diffuse bands at 580 and 630 nm are not visible in all stones and this may reflect more than one occurrence.

Refractive Index

The calculated value of the end-member is 1.889. Actual values will vary due to isomorphous replacement and literature values for 50–100% andradite range from 1.82 to 1.89 respectively. Melanite, with its high titanium content, gives elevated RI values up to 1.97 where $Fe^{3+} \geq Ti$. Demantoid values are reported as ranging from 1.880 to 1.889; the higher value relating to dark-green material. Topazolite RIs ranging from 1.84 to 1.89 have been recorded for Alpine material.

Density

Pure end-member (calculated) 3.859. Extracting values from the literature, where chemical information is provided, indicates that for ugrandite garnets containing $\geq 50\%$ andradite molecule the lower limit of density approximates to 3.77. Measured values for SG normally range from 3.82 to 3.85. It should be borne in mind that inclusions could affect measured density and, if in abundance, give rise to skewed values.

Hardness

Hardness on Mohs' scale ranges from 6.5 to 7.5 with demantoid at the lower end of this range.

Dispersion

Very high at 0.057; greater than that of diamond (0.044) or titanite (0.051). The effect is partially masked by body colour especially in intensely coloured demantoid.

Cleavage/Fracture

Conchoidal to uneven. Parting parallel to the dodecahedral face {110} is sometimes distinct.

Appearance

Colour

Occurs in various hues: red, brown, yellow (topazolite) and green (demantoid) and shades in between these; also as black (melanite). Also reported as colourless and grey. Tsavorite-like coloured andradite has been reported from Madagascar.

Colour Varieties: Demantoid colour arises from the presence of small amounts of chromium. Whilst the colour of demantoid never approaches that of the finest emerald, favourable comparison can be made especially in vividly coloured stones that are not overly dark or contain yellow hues. Chemical analysis of demantoid from Kashmir, Pakistan, showed relatively pure andradites with 0.25 wt% of Cr_2O_3. The demantoid from Panshir Valley, Kunnar, Afghanistan, is transparent but very dark green. The green andradite from Farm Tubussis, Erongo Mountains, Namibia, has only minor chromium but a significant vanadium content and is being traded as demantoid.

Topazolite

Topazolite was named after its similarity to yellow topaz but is now used to include also more orangy material. Some andradite from the Urals is a strong yellow component and is closer to topazolite than demantoid.

Melanite

Melanite is from the Greek word *melanos* meaning black, referring to its colour. This opaque material was historically used for mourning jewellery, or as inlay.

Chelsea Colour Filter

Demantoid looks red or orange under this filter with various tints of these hues; variously described as reddish or orangy red of varying tones.

Lustre

Adamantine to resinous dull.

Inclusions

Significant, and diagnostic when seen, are wispy, horse-tail like, fibrous inclusions of chrysotile (serpentine asbestos) inclusions that historically had been referred to as *byssolite*, a now obsolete term, or fibrous asbestiform actinolite. Seen in demantoid from Russia, Eritrea and Italy but not Namibia. Opaque chromite crystals may form the nucleation point for the chrysotile. Tiny fluid inclusions are seen in Namibian demantoid, also negative crystals with two phase inclusions and pronounced angular or straight transparent growth zoning which rarely may be iridescent ('rainbow graining'). Acicular diopside crystals have been reported as inclusions in demantoid from the Urals. Yellowish green to emerald-green demantoid from Kashmir, Pakistan, had acicular, fibrous and felt-like mineral inclusions that were identified as chrysotile.

Partially healed fractures and liquid inclusions plus small unidentified needle-like inclusions and transparent crystals have been reported from andradite from the Yemen. Healed fractures, growth banding (best seen in polarized light) and two phase inclusions have been reported in Arizona andradite. Yellowish brown Namibian andradite had yellow needles.

Optical Effects: Chatoyancy, Asterism, Colour Change, ADR

Iridescent andradite from Sonora, Mexico, is sometimes referred to as 'rainbow garnet'. The iridescence is caused by cylindrically stratified growth lamellae in the {211} and {110} sectors. The stratification is between $An_{79}\%$ and $An_{94}\%$.

Green andradite from Farm Tubussis, Erongo Mountains, can often have an unappealing colour shift towards brown in tungsten light.

Some demantoid show a colour change from yellowish green to a bright sparkling pastell green (daylight). Rarely, some show a colour change from green to orange.

Treatment

Colour enhancement of demantoid has been achieved by heat treatment under reducing conditions. Brown and yellow hues are removed and green enhanced by converting some Fe^{3+} to Fe^{2+} (Alferova M.S. *pers comm.*).

Calderite $(Mn^{2+},Ca)_3(Fe^{3+},Al)_2(SiO_4)_3$

Name approved by the CNMMN of the IMA. Named after James Calder, an early writer on the geology of India; name first applied to a rock, later transferred to its predominant mineral.

Occurrence

Found in regionally metamorphosed siliceous iron and manganese-rich formations. Type locality is at Otjosondu, Namibia, and at Usakos, Namibia. In Katkamsandi, Hazaribagh district, India, this rare garnet occurs as brownish vitreous masses scattered in matrix. Additionally it occurs at Bihar, and at Netra, Balaghat district, Madhya Pradesh, India. Also reported as dark yellow to reddish yellow.

Properties

Whilst compositions approaching ideal calderite are unknown, a complete solid solution exists between calderite, andradite and spessartine. Thus the calderite molecule is more likely to be found in calderitic andradite and calderitic spessartine, where its presence will affect the properties of these species.

Refractive Index – 1.872–1.890 (natural), and 1.97 (calculated for pure calderite). SG – 4.45 (pure calderite).

Not reported in any gem material to date.

Goldmanite $Ca_3(V,Al,Fe^{3+})_2(SiO_4)_3$

Name approved by the CNMMN of the IMA. Named after Marcus Isaac Goldman (1881–1965), sedimentary petrologist, US Geological Survey.

Occurrence

Found in vanadium-bearing sedimentary rocks or contact-metamorphosed carbonaceous and calcareous shales and sandstones. Also in magnetite-bearing skarns. Type locality is Laguna District, Cibola Co., New Mexico.

It occurs at Otjosondu, Namibia, and also at Usakos, Namibia. In Katkamsandi, Hazaribagh district, India, this rare garnet occurs as brownish vitreous masses scattered in matrix. Additionally it occurs at Bihar, and at Netra, Balaghat district, Madhya Pradesh, India. Also reported as dark yellow to reddish yellow.

Properties

Solid solutions exist between grossular, andradite and goldmanite. Manganese can replace calcium and a manganoan goldmanite exists tending towards the hypothetical end-member 'yamatoite' $[Mn_3V_2(SiO_4)_3]$. Vanadium can also replace the aluminium in spessartine and pyrope, producing colour change garnets. Idiochromatic vanadium gives rise to dark green to brownish green or grass-green colours.

Refractive Index – 1.792–1.834 (natural) and 1.834 (synthetic); for manganoan material RI is 1.855. SG – 3.765 (calculated).

Reported as a minor component in some garnets, e.g. vanadium-bearing tsavorite. Not reported as gem material to date.

Grossular $Ca_3Al_2(SiO_4)_3$

Name approved by the CNMMN of the IMA. Synonyms are grossularite, hessonite, essonite, cinnamon stone, rosolite, tsavorite, tsavolite, wiluite and mali. Colour varieties are hessonite (orange-brown), tsavorite (green) and mali (yellowish green to greenish yellow), leuco-garnet (colourless), imperial garnet (light pink), raspberry garnet (raspberry-red), gooseberry garnet (gooseberry-green), tangerine garnet (bright orange), merelani (mint-green). 'Grandite' is the term that has been applied to grossular garnets with significant, although normally less than 50%, andradite molecule.

Grossular is so named in allusion to the resemblance of pale green specimens to the gooseberry (*Ribes grossularia*).

Occurrence

Grossular is especially characteristic of both contact (skarns) and regionally metamorphosed impure calcareous rocks or rocks that have undergone calcium metasomatism. It also occurs in association with serpentinites and rodingites. Additionally as detrital grains in placers deposits.

Notable localities for gem-quality grossular include Sri Lanka (hessonite), India, Tanzania (tsavorite, hessonite), Kenya (tsavorite), Mali, Namibia, Madagascar (hessonite), Pakistan, Russia (chrome-grossular), Italy, USA (California), Canada, Mexico and Brazil. Massive material similar in appearance to hydrogrossular occurs at: Yukon, Canada; California and other states in the USA; Myanmar; and South Africa.

Properties

Grossular forms a series with andradite and uvarovite, and also the hydrogarnets. Moreover, solid solution to varying extent occurs with the pyralspite series. Consequently the physical and chemical properties and appearance can vary appreciably.

Absorption Spectra

Pure grossular does not have a characteristic absorption spectrum; however, the extensive solid solutions that exist can give rise to weak spectra characteristic of other garnets such as almandine, andradite and spessartine. Brownish greenish yellow grossular-andradite garnets from Mali display a cut-off at 415 nm, a band between 440 and 450 nm and a faint band at 500 nm. These vary in intensity depending upon the dominant tint. Mali pale greenish yellow grossular showed an absorption band at 435–445 nm, the intensity of which correlated with increasing RI and probably represents increasing andradite component (characteristic Fe^{3+} andradite band at 443 nm). Faint absorptions attributed to Fe^{3+} have been recorded at 434 nm in yellow to orange stones and even fainter bands at 418, 489, 503 and 529 nm in a light orange stone.

Emerald-green grossular garnets near Bekily in Southern Madagascar exhibit two broad bands in red and violet. Tsavorite, intense green V^{3+}-bearing grossular from Kenya and Tanzania exhibit noticeable absorption at 610 nm and also at 430 nm. Pakistan (Swat Valley) material shows general absorption of the far blue and violet beginning at about 460 nm. In addition to this they show very faint diffuse absorption of the red/orange and yellow. Chromium-bearing tsavorite can show a doublet at 697 nm with weaker lines at 660 and 630 nm and diffuse bands near 605 and 505 nm: these are often difficult to observe.

Pale yellow-green grossular from East Africa, in addition to weak V^{3+} absorption at 610 nm, shows weak but sharp Mn^{2+} absorption at 409, 422 and 431 nm superimposed on a poorly resolved 425 nm vanadium band.

Yellow-orange hessonite shows no spectrum in lighter tones; however, with increasing colour saturation, absorption towards the blue end of the spectrum becomes more intense, albeit with no distinct bands. Colour-saturated gems have bands at 407 and/or 430 nm.

Refractive Index

The RI of ugrandite garnets is strongly dependent on the calcium content, but does not vary significantly with changes in the $Fe^{3+}:Cr^{3+}$ ratio.

The calculated value of the end-member is 1.732; however, actual values will vary due to isomorphous replacement and measured literature values for 50% ≈ 100% grossular range from 1.796 to 1.737 respectively based on predominant andradite substitution.

Mali: 1.770 (greenish yellow); 1.768 (brownish yellow)
Mali pale greenish yellow: 1.739–1.750
Colourless Mali grossular: 1.742–1.751
Emerald-green grossular (Bekily, Madagascar): 1.741–1.746
Gogogogo, Madagascar: 1.742
Manson and Stockton study of 105 grossular garnets gave range from 1.731 to 1.754.
Refractive index values for African tsavorite range between 1.739 and 1.744 whilst that Swat material is 1.743.
Hessonite from Paskema and Okkampitiya, Sri Lanka, ranges from 1.731 to 1.735.

Dispersion

Tsavorite: 0.28

Specific Gravity

Mali: 3.66 (greenish yellow); 3.64 (brownish yellow)
Mali pale greenish yellow: 3.61–3.63
Emerald-green grossular (Bekily, Madagascar): 3.58–3.62
Gogogogo, Madagascar: 3.62

Manson and Stockton study of 105 grossular garnets gave range from 3.57 to 3.67.

African tsavorite values range from 3.57 to 3.65 and that of Swat material is 3.64.

Hessonite from Paskema and Okkampitiya, Sri Lanka, ranges from 3.598 to 3.622.

Hardness

Tsavorite $\approx$7.25

Colour

Grossular garnets range from colourless (rare) through hues, shades and various tints of brown, yellow, orange and green. Mali, West Africa, for example, has provided gemmy rough in a range of colours including dark brown, orangy brown, light yellowish brown, yellowish green, light peridot-green, olive-green and intense bluish green. A direct correlation exists between increasing green tone and vanadium concentration; however, chromium, even in very low concentrations, can have significant effect on the green colouration. With increasing Fe^{3+}, colour changes from yellow through orange to orange-red.

Colour varieties:

Tsavorite

Originally discovered (1967) and described from the Tsavo National Game Park in Kenya from which it took its name, this green vanadium-bearing (goldmanite component) grossular garnet, with minor chromium, is more prevalent in Tanzania where production from Tunduru, Ruangwa, Umba, Merelani Hills and Komolo has outstripped that from Kenya. Also found at Gogogogo, Madagascar.

Colours range from bluish green to yellowish green, the former, sometimes called forest green, being the most highly regarded. Tsavorite is uncommon in sizes above 3 ct.

Vanadian grossular has also been reported from Swat, Pakistan.

Meralani Mint Green

This is the trade name for bright green garnet discovered in the same mines that produce Tanzanite, about 25 miles south-east of Arusha, Tanzania. It is essentially a light coloured tsavorite.

Hessonite

Hessonite is the name given to the yellow orange to brownish orange colour variety. The colour is attributed to traces of iron. Major localities are Sri Lanka and India. Lighter oranges, yellows and peachy-coloured stones occur in deposits in Alberta and Quebec, Canada. Other localities include Mexico, Brazil, Tanzania and Madagascar.

Leuco Garnet (colourless)

Leuco garnet is the name that is normally applied to the colourless, transparent variety of grossular that usually contains <2% iron. Main occurrences are Sri Lanka and Quebec, Canada.

Tangerine Grossular

This is the trade name for bright orange garnet, sometimes confusingly termed 'mandarin' grossular that has been found as water-worn alluvial pebbles from the Ruvuma River, Tanzania.

Orange grossular from Tanga has been confusingly termed 'malaya'.

Mali

In September 1994 a new deposit of garnet was discovered in the republic of Mali, in Western Africa. At first identified as chrysoberyl on the basis of colour, it was later found to be a combination of andradite and grossular ($Gr_{80}An_{20}$). The bright colours range from chartreuse to a yellow-green to an almost honey green. A rare chrome green colour has also been reported.

Inclusions

Apatite inclusions are common and calcite has been reported from a number of localities.

Hessonites usually have a characteristic grainy and/or swirly internal appearance known as 'treacle': latest examinations show that this is the result of a mosaic structure caused by intergrown hessonite grains. Scapolite inclusions have been reported in massive hessonite from Maligawila, Sri Lanka, and also from Lelatema, Tanzania. Lamellar growth structures in Mexican hessonite produce iridescent colours similar to that reported for Mexican andradite.

In tsavorite, fingerprint and geometric patterns arising from tiny droplets, negative crystallites and solid particles are reported.

Additionally, graphite scales and asbestiform actinolite fibres have been observed.

Optical Effects: Chatoyancy, Asterism, Colour Change, ADR

Ultraviolet Fluorescence: some green garnets (grossular and tsavorite) may show a weak orange in LWUV and weak yellow in SWUV.

A colour shift has been reported from Mali garnets: under incandescent light they have a more yellow tint, and in daylight they are more green.

Treatment

Some colourless to pale tone grossular from various localties in East Africa will turn light yellow-green when irradiated (gamma rays) but will fade to their original colour within hours to days in daylight and within two months in the dark. These are characterized by a transmission window near 560 nm. Some material has been dyed green to imitate jadeite; this shows an absorption band from 630 to 670 nm and red under the Chelsea Colour Filter.

Hydrogarnets

Hibschite $Ca_3Al_2(SiO_4)_{3-x}(OH)_{4x}$ (where x is between 0.2 and 1.5)

Katoite $Ca_3Al_2(SiO_4)_{3-x}(OH)_{4x}$ (where x is greater than 1.5)

Henritermierite $Ca_3(Mn,Al)_2(SiO_4)_2(OH)_4$

The hydrogarnet names that are approved by the CNMMN of the IMA are hibschite and katoite. These are named after Joseph Emanuel Hibsch (Czech, who found the original specimen) and Akira Kato (Japanese mineralogist) respectively. Henritermierite (CNMMN approved; named after the French mineralogist Henri Termier) is a tetragonal mineral that is structurally related to garnet.

Hydrogarnet synonyms: hydrogrossular and plazolite (impure hibschite). Also grossularoid and garnetoid.

Hydrogarnets

Involve the substitution $4H^+ <=> Si^{4+}$, which occurs to a variable extent and may give rise to lower symmetry. This has commonly been assumed to represent the mechanism of entry of hydroxyl (OH) into their structure and $(OH)_4^{-4}$ substituting for $(SiO_4)^{-4}$ is normally referred to as hydrogarnet-type substitution. The hydrogarnets encompass the series hibschite, katoite and grossular. Water content ranges from about 13% in katoite, 1.5% in hibschite to anhydrous grossular. Gemmologically katoite and hibschite are brigaded together and referred to as hydrogrossular. Hydroandradite is also known. The rare clove brown to apricot brown manganoan hydrogarnet, henritermierite, is only found as small grains. The amount of hydrogarnet component will decrease in the sequence grossular, spessartine, almandine and finally pyrope.

Hydrogrossular-Vesuvianite Mixtures

Garnet-vesuvianite ranges from opaque to translucent, depending on quality and thickness when cut. Mostly cut cabochon-style, the gem could be confused with jade because of its mottled appearance and colours. The green portion of the gems have been identified as vesuvianite and the pink to red portions as hydrogrossular garnet. The gems contain areas where the two species are mixed to such an extent they are almost impossible to separate gemmologically. The red and pink colours are attributed to manganese and the green to iron, not chromium as was previously suggested. California jade or californite, or vesuvian jade, is massive green grossular mixed with vesuvianite: RI 1.71–1.72, SG 3.25–3.32.

Pseudomorphous substitution of wadalite (protoachtarandite) by hydrogarnet gives rise to white, predominantly hibschite pseudomorphs.

Localities include Pakistan and California, USA.

Occurrence

Occurs principally in rodingites, contact-metamorphosed marls and skarn deposits. Also as layers coating andradite and grossular. 'Transvaal jade' is reported to be formed by combined contact-metamorphism and metasomatism.

Major localities include Rustenburg, South Africa;

Refractive Index

Hibschite: 1.67–1.75
Katoite: 1.632
Hydrogrossular: 1.69–1.73

Specific Gravity

Hibschite: 3.24–3.28
Katoite: 2.76
Hydrogrossular: 3.3–3.48 and up to 3.523

Hardness

Hibschite: 6.5–7
Katoite: 5–6
Hydrogrossular: 6.5–7.5

Colour

Hibschite and katoite can be colourless, white, pale grey, light yellow, brown and pink. Transvaal jade can be green (chromium bearing) and pink (manganese bearing) to light red pink: also red brown, yellow blue to purple and sometimes brown.

'Transvaal Jade'

Massive translucent to opaque grossular often with a hydrous component is known as 'Transvaal jade'. The original material was reported to be found at Buffelsfontein and the adjacent farm of Turffontein, some 65 km west of Pretoria, South Africa. It is a compact homogeneous rock included with black specks of chromite and magnetite. The grey material is reported to contain up to 25% zoisite. RI range from 1.70 to 1.73; the lower value correlating with pink colour and increased hydrous values. Similarly SG correlates with colour and hydrous values; pink material with elevated hydrous values varying from 3.36 to 3.41 and green from 3.42 to 3.55. Hardness varies between 7 (pink) and 8 (green). A weak chromium absorption spectrum, with strongest lines at 630 and 461 nm, has been reported from green specimens. Massive hydrogrossular fluoresces a strong orange-yellow under X-rays: a feature not seen in jadeite, nephrite and jade simulants. Bi-coloured green and pink material has been marketed as watermelon garnet.

Kimzeyite $Ca_3(Zr,Ti)_2(Si,Al,Fe^{3+})_3O_{12}$

Name approved by the CNMMN of the IMA (1961). Named after the Kimzey family, mineral collectors of Magnet Cove, Arkansas, USA, where the mineral was found by Joe Kimzey.

Occurrence

Found in carbonatite at Magnet Cove, Arkansas, USA (dark brown), in shoshonite basalt at Stromboli, Italy (brown), and in lamprophyre dykes, McKellar Harbour, Ontario, Canada (pale reddish orange). Crystals up to 5 mm are found at Magnet Cove.

Properties

Kimzeyite has zirconium as a major component; however, zirconium has been noted as a minor constituent of titanian andradites. Isomorphic replacement exists such that the kimzeyite molecule can occur in andradite and garnets in the schorlomite–kimzeyite series where its presence will affect the properties of these species.

Hardness: 7

Refractive Index: 1.94

Specific Gravity: 4

Not reported as gem material to date.

Knorringite $Mg_3Cr_2(SiO_4)_3$

Name approved by the CNMMN of the IMA (1968). Named after Oleg von Knorring, Professor of Mineralogy, Leeds University, Leeds, England.

Occurrence

Knorringite is exceptionally rare and is only known from kimberlite pipes. Kao kimberlite, Lesotho; as an inclusion in a type IIa diamond from a kimberlite in the Mirny area, Yakutia, Russia.

Properties

Knorringite, which is found as tiny blue-green to lilac grains, forms a series with pyrope. There also exists a knorringite–uvarovite solid solution series. The original material comprising $Kn_{34}Py_{30}Uv_{19}Al_{13}An_4$, or alternatively as 52.7% knorringite if none of the chromium is cast as uvarovite; and the highest knorringite component so far observed is in an inclusion in diamond (66.4 mol.%). The importance of knorringite is largely as a component in chromium rich pyrope where it affects the colour and physical properties.

Hardness: 7

Refractive Index: 1.790–1.803 (1.875 for calc. end-member)

Specific Gravity: 3.86 (3.835 calc. end-member)

Not reported as gem material to date.

Majorite $Mg_3(Fe,Al,Si)_2(SiO_4)_3$

Name approved by the CNMMN of the IMA (1970). Named after Alan Major, scientist participating in a programme of garnet syntheses.

Occurrence

The first majoritic garnets were discovered in ultramafic xenoliths of South Africa kimberlites. Majorite is also known from peridotites in Western Norway and majoritic-like garnets can be found in various lithologies in the Bohemia Massif. It is also known from meteorites. Majorite is exceptionally rare.

Properties

Majorite is found as minute purple, pale yellowish brown or colourless grains. The tiny blue-green to lilac grains forms a series with pyrope.

Hardness: 7–7.5

Refractive Index: 1.80

Specific Gravity ≈ 4 (4 calc. end-member)

Not reported as gem material to date.

Morimotoite $Ca_3TiFe^{2+}Si_3O_{12}$

Name approved by the CNMMN of the IMA (1995). Named after Nobuo Morimoto, Mineralogy professor, Osaka University, Japan.

Occurrence

Occurs in skarns at Fuka, Okayama Pref., Japan, as black grains upto 15 mm. Also as tiny grains in volcanic rocks of the Mata Da Corda formation, Brazil: an area long known as a source of alluvial diamonds. Morimotoite is exceptionally rare.

Properties

Morimotoite is considered to be derived from andradite by the substitution $Ti + Fe^{2+} = 2Fe^{3+}$, giving the solid solution andradite, through melanite to morimotoite.

Hardness: 7.5

Refractive Index: 1.955

Specific Gravity: 3.75–3.8

Appearance

Colour – black
Lustre – adamantine

Not reported as gem material to date.

Pyrope $Mg_3Al_2(SiO_4)_3$

Name approved by the CNMMN of the IMA. Synonyms include names incorporating ruby prefixed by locality, for example Arizona ruby, Bohemian ruby, Cape ruby and Elie ruby.

Named in 1803 by Werner from the Greek word for *fire* and to *appear* in allusion to its characteristic red colour.

The composition of the rose-red to violet-coloured variety, known as rhodolite, approximately equates two parts pyrope and one part almandine. The original material from Macon Co., North Carolina, USA, has a distinctive purplish red ('rhododendron red') colour.

The dark brownish red pyrope–spessartine–grossular combination is referred to as malaya (or malaia).

Occurrence

Unlike other garnets the most common origin of pyrope is igneous rather than metamorphic. It occurs in peridotites, kimberlites, eclogites, serpentinites and other ultramafic igneous rocks and sediments derived from these. Metamorphic pyrope arises from metamorphism of the aforementioned rocks or from magnesium-rich rocks subjected to high-grade metamorphism.

Notable localities include Bohemia, Czech Republic and the Dora Maira Massif, Western Alps, Italy. In the USA, the most significant deposits are in Arizona near San Carlos (in the San Carlos Indian Reservation), Gila and Graham counties and near Fort Defiance (Buell

Park and Garnet Ridge), Apache Co. Notable deposits exist in the Four Corners area, where the four states of Colorado, Arizona, Utah and New Mexico meet. In Africa, pyrope is found in many of the South African diamond mines, along the Umba River and in the Pare Mountains, Tanzania, at Mandera, Hargeisa, Somaliland; the Lundaze area, Zambia and Madagascar. In Australia at localities in New South Wales and Queensland, and in Argentina at Quines. Small amounts of pyrope also occur in Sri Lanka and Minas Gerais, Brazil. In Canada, pyrope is found in Joli Township, Quebec, and in Russia at Urals, and Yakutia. Other European localities include Vetarella, Vico, Lazio, Italy; Gorund, Switzerland; Arguenos, France; Saxe, Germany; and Elie Ness, Fife, Scotland ('Elie Ruby').

The classic locality for rhodolite is Cowee Creek, Macon Co., North Carolina, USA. Rhodolite is also reported from Tanzania, India (Orissa), Sri Lanka, Malawi, Brazil and Madagascar. In Mozambique, red-violet rhodolite is found from Cuamba (previously Novo Freixo). Rhodolite has been reported from the Lokirima area, Kenya.

Malaya has been recorded from various localities straddling the Kenya–Tanzania border, Mozambique, Nigeria and also Madagascar.

Properties

There is complete solid solution amongst the garnets of the pyralspite series and pyrope commonly contains significant amounts of the almandine molecule. It can often contain small amounts of grossular. Chrome pyropes tend to have a greater percentage of pyrope molecule but all have $\geq 4\%$ of the uvarovite (or knorringite) molecules. Consequently the physical and chemical properties and appearance can vary appreciably.

Although less favourable in comparison with grossular, hydroxyl (OH) component has been recorded in pyrope. The purplish pink pyrope from the Dora Maira Massif, Italy, uniquely shows four sharp OH vibration bands in the infrared spectrum between 3660 and 3600 cm^{-1}. Pyrope from the Monastery Mine, South Africa (containing 56 ppm H_2O), exhibits broad OH absorption between 3500 and 3600 cm^{-1}. The quantities are negligible and have no discernable effect on the properties.

Absorption Spectra

Pure pyrope is colourless and all known red pyropes have varying but significant amounts of the almandine molecule. Consequently the typical absorption spectrum of almandine with the three main bands at 575, 527 and 505 nm is seen in practically all stones to varying degrees of intensity. Significant spessartine replacement can also occur and manganese-related absorption in the violet can be intense. These can both exist in the same spectrum.

Large near end-member pyrope crystals from the Dora Maira Massif, Western Alps, Italy, have absorption spectra that suggest the colour is related to Fe^{2+} (bands around 495, 501–503, 521–524 and 575–579 nm) and Mn^{2+} (363–365, 398, 419–422 and 458–462 nm).

Malaya garnets from Bekily, Madagascar, show absorption features due to Fe^{2+} (503, 610, 687 nm) and Mn^{2+} (483 nm). Other non-attributable bands at 430, 459, 525 and 569 nm have been observed.

The rich red chromium-bearing pyropes contain significant amounts of chromium and effect the absorption spectrum. In these, the ubiquitous almandine lines are largely masked by the dominant chromium feature, a broad 100 nm wide band centred at about 570 nm. The only remaining strong almandine band at 505 nm can usually be seen albeit usually as a weak feature. Beyond 440 nm there is general absorption of the violet. Other chromium lines occur as a weak doublet in the red at 687 and 685 nm with similar weak lines at 671 and 650 nm in some stones. These are often too faint to be noticeable.

Refractive Index

The RI of pyralspite garnets is strongly dependent on the magnesium content and is less influenced by the iron–manganese ratio. The RI of synthesized pyrope is 1.714 and the theoretical value reported as 1.705. The measured values of most natural pyrope vary due to isomorphous replacement, and literature values range from 1.73. The upper limit is more problematic due to variable cross-series isormorphous replacement and the value of 1.76 has been quoted for a pyrope of composition $Py_{51}Al_{37}Sp_5Gr_7$. The purplish pink pyrope from the Dora Maira Massif, Italy, is unusually pure, up to $Py_{97}Al_2Gr_1$, giving an RI range from 1.717 on the palest specimens up to 1.730 with increasing almandine molecule. The largest supposed Bohemian pyrope is in a Czech museum and weighs 13.21 ct; it has an almandine spectrum and RI of 1.79, which suggests almandine rather than pyrope.

The RI range for rhodolite has evolved such that 1.740–1.770 is acceptable to some authorities. For example, the brownish red pyrope-almandine ($Py_{52}Al_{39}Gr_7An_2$) with a purple component from Kalalani, Tanzania, with RIs ranging from 1.763 to 1.77 has been termed 'rhodolite'.

Malaya garnets from Bekily, Madagascar, have an RI range for Py_{24-68} from 1.782 to 1.740. Malaya from Mozambique, which is an apricot/ orangy red coloured material, has an RI range from 1.751 to 1.755.

Colour change pyrope-spessartine is obtained from Sri Lanka with RI 1.77.

Grape garnet describes an intense purple-red variety without delimiting RI boundaries although to date the RI range indicated by Hanneman is 1.765–1.805. Other authors cite RIs between 1.758 and 1.780 for pyrope-almandine from the Orissa state of North-west India, but are equivocal about linking the specific locality, colour and percentage of pyrope molecule, which ranged from 80 to 56%.

Density

Pyrope has the lowest density of all the garnets. The calculated SG of the pure end-member is 3.563. The measured values of natural pyrope vary due to isomorphous replacement that can also be cross-series and normally range from 3.65 to 3.84: the average being 3.74. The unusually pure purplish pink pyrope from the Dora Maira Massif, Italy, give values from 3.58 to 3.67 with increasing almandine molecule.

Malaya garnets from Bekily, Madagascar, have an SG range for Py_{24-68} from 4.04 to 3.78. Malaya from Mozambique has an SG range from 3.844 to 3.847.

Colour change pyrope-spessartine is obtained from Sri Lanka with SG 3.93.

Specific gravity quoted for grape garnet from Orissa, India, ranged from 3.82 to 3.94 and expanded to 3.77 for pyrope-almandine with unrecorded colour.

Hardness

Hardness on Mohs' scale ranges from 7 to 7.5.

Dispersion

Moderate at 0.022; however, the effect is masked by body colour especially in intensely coloured stones.

Cleavage/Fracture

Conchoidal to uneven with a tendency towards brittleness.

Appearance

Colour

Pure pyrope is colourless; however, it is typically pinkish red, ranging from orange to brownish red, through an almost crimson colour, to a purplish red with increasing amounts of the almandine molecule. Small amounts of manganese can eliminate the purple component resulting in an intense red. Normal chromium pyropes are blood red with a purple tint, and with increasing chromium content the hue is lilac or purple passing ultimately to a greenish hue (knorringite). Pyrope with strong tonal qualities is blackish red.

Fe^{2+} in pyrope garnet produces the near-red colour but often with an orange tint, whilst Cr^{3+} is responsible for producing a richer red colour similar to that seen in ruby or spinel, but the effect is muted by the ever-present iron.

Colour Varieties

Rhodolite

This variety derives its name from the mountain rhododendron (*Rhododendron catawbiense*), a magenta-coloured bloom which grows in the mountains of North Carolina, where rhodolite was first discovered in 1882. The original colour is described as 'pale rose-red inclining to purple like that of certain roses and rhododendrons': a rich rhododendron purple colour without any brown tint. The colour sets the range of RIs for this variety. However, almandine-bearing pyrope garnets whose constants fell within the range but whose colour was adulterated by, in particular, brown tints and variable tonal qualities have been traded as rhodolite such that the name no longer defines the colour. As a consequence almandine-bearing pyrope garnets whose colour matched that of the adulterated rhodolites but whose RIs were out of the range of rhodolite (*sensu stricto*) were also traded as rhodolite. The outcome being that over time the RIs range for 'rhodolite' has increased. Additionally, adjectival colour prefixes have been introduced such that the integrity of the name rhodolite is no longer robust. The most desirable colour for rhodolite is purplish red of a medium tone. Large gems tend to be darker and more cranberry red with beautiful violet flashes; however, larger stones >5 cts, with poor transparency tend to be over-dark and unattractive.

A light, almost purple-pink coloured fine 'silk-free' rhodolite occurs in the Tanga region (towards the Kenyan border), Tanzania.

Some colour modifiers have been added to the term 'rhodolite': these trade names, apart from grape garnet, retain the financial advantage of rhodolite soubriquet.

Raspberry Rhodolite

This material is called 'raspberry' because its fine purplish pink colour resembles that of the fruit. Discovered in early 1987, it is a member of the pyrope-almandite series found in the Kangala area of Tanzania, and has an RI of 1.76.

Cranberry Rhodolite

In 1998 in the Slocan Valley, located in south-eastern British Columbia, Canada, garnet-bearing feldspar-rich pegmatite sills and dikes were discovered that produced cranberry-red pyrope-almandine crystals, frequently >10 cm in diameter, that are compositionally similar to rhodolite from Tanzania.

Cranberry to pinkish red rhodolite found in Nigeria is mostly high-quality cabochon grade as most stones tend to be very slightly to moderately included.

Cherry Rhodolite

Cherry rhodolite is found in the Umba River Valley region of Tanzania and displays a bright cherry-red colour.

Grape Garnet

The name was introduced to address the problem associated with colour variation and need to retain a robust definition of rhodolite. By branding the colour, dealers were able to develop quality standards for a gem that had the colour of grape juice but had previously been brigaded with rhodolite (*sensu lato*).

One confusing aspect is that the original purplish red to violet garnet mined in the Orissa state of North-west India and known as grape garnet has been variously reported as an intermediary between both spessartine and almandine and pyrope and almandine. However, published chemical data states that these contain >50% pyrope molecule.

Similarly coloured garnet discovered later in Fazanda Balisto in Tocantins, Brazil (1997), has been described as pyrope-almandine. Garnet production has been described as 40–50% rhodolite-pyrope colour and about 50% grape colour. However, analysis of 'burgundy' coloured garnets from alluvial deposits in Tocantins indicated that the 'burgundy' garnets are members of the almandine-spessartine series and comprised mainly of almandine.

Similarly almandine of a rich grape colour has been reported from Tanzania and Madagascar.

The issue of colour varieties is further muddled by modifying colour descriptions such as rich cranberry. Material can be traded indiscriminately as facet-rhodolite garnet-grape, grape garnet (rhodolite garnet) or rasp-berry/grape based solely on colour.

Malaya (Malaia) Garnet

The name 'malaya', a Swahili word meaning 'out of the family', came to be used for garnets that did not fit the colour requirement for rhodolite or because of the properties, did fit into traditional garnet categories. The names pyralspite and umbalite have been used synonymously but have largely fallen into disuse in preference to the more marketable name Malaya. Malaya is an intermediate between spessartine and pyrope, originally from the Umba River valley in Tanzania. The colour of malaya ranges from yellowish brown and brownish pink through a cinnamon to a crisp honey brown and red-dish brown to a brick or brown orange: also orange, red-orange, peach and pink. Lighter tones are considered more desirable, and true pinks are among the rarest of all garnets, regardless of hue. Honey peach and pinkish orange tints are the most desirable. Stones of a browner orange colour resemble hessonite or spessartine and are more common. Brown is the least desirable component. Umbalite, however, is normally described as having a purple component approaching that of light 'raspberry' rhodolite or a light reddish pink with a hint of purple. However, more confusingly, they have occasionally been erroneously described as grossular by the trade. 'Imperial malaia' and 'champagne garnet' are different marketing terms for the same brownish pinkish orange material from Madagascar.

Malaya garnets from Bekily, Madagascar, are intermediate mem-bers of the pyrope-spessartine series with variable contents of alman-dine and subordinate grossular. Compositions fall within the range $Py_{24-68}Sp_{13-59}Al_{4-25}Gr_{3-5}Go_{0.1-0.5}Uv_{0.1-0.5}$ with the pyrope molecule only rarely <50%. The colours of these garnets are related to their iron and manganese contents and within the Py_{24-68} range are orange to light pink. With increasing iron, the pink component is intensified and with increasing manganese the orange colour becomes stronger.

Arizona Chrome Pyrope (Anthill) Garnet

In colour saturation, there is only modest comparison to normal pyrope. The Anthill garnet has a rich red hue described as ruby red with purple to orange highlights. Because of the strong colour tone, cut stones over 1.5 cts are dark.

Crimson Garnet

A deep red-coloured pyrope-spessartine-almandine garnet, with flashes of orange, discovered near an area called Tiriri in North-east Tanzania is being marketed as crimson garnet.

Blue Garnet

In daylight, colour change pyrope-spessartine garnets from Bekily, Madagascar, have been described variously as blue, grey-blue and greenish blue.

Chelsea Colour Filter

Although pyrope can contain chromium there is usually sufficient iron content to quench any colour modification.

Lustre

Vitreous.

Inclusions

Pyrope is normally free of inclusions. If any, inclusions are usually small rounded irregular crystals with low relief. Bohemian garnets are reported to contain quartz or possibly augite crystals arranged in a circular pattern. Some Arizona pyrope has in addition to needle-like and octahedral shaped crystal inclusions, bright green diopside. Irregular rod-like inclusions, mainly comprising nickeliferous pyrrhotite, have occasionaly been seen in pyrope from Elie, Scotland. With increasing almandine component more inclusions are recorded. Inclusions reported from the purplish pink pyrope from the Dora Maira Massif, Italy, include rutile, apatite, mica, zircon and the rare pinkish purple mineral ellenbergerite. Inclusions reported in Sri Lankan colour change pyrope-spessartine include long rutile needles orientated in three directions, large transparent angular to rounded crystals of a carbonate, small transparent colourless crystals often surrounded by stress halos which have been determined as spinel, stacks of hexagonal black platelets of hematite.

Rhodolite inclusions are generally too fine to be readily seen by the naked eye. These can impart a velvety appearance. Inclusions of zircon, apatite and rutile have been reported.

Inclusions in malaya garnets from Bekily, Madagascar, are rutile needles and platelets, graphite, quartz, apatite, zircon and sillimanite. Negative crystals were also recorded.

Optical Effects: Chatoyancy, Asterism, Colour Change

Some star rhodolite is known from Tanzania and India.

Colour change garnets from Madagascar belonging to the pyrope-spessartine solid solution series with >50% pyrope molecule change their colour moderately to distinctly from bluish to purplish (daylight/tungsten light): their colour saturation ranging from medium to strong. Pinkish orange malaya stones appear more colour saturated, often reddish, in tungsten light.

Colour-saturated grape garnet from Brazil shows weak colour change, from deep purple (daylight) to red (tungsten light).

The colour change (bluish green in daylight, reddish purple in tungsten light) in Sri Lankan pyrope-spessartine garnet is attributed to vanadium and manganese; no chromium was detected.

Colour change pyrope has also been reported from Linhorka in Bohemia and various islands in western Norway.

Treatment

Although rhodolite from Brazil has had its colour altered to hessonite-like orange brown (700 °C) and to red (900 °C), treatments are, on the whole, ineffective and not attempted.

Schorlomite $Ca_3Ti_2^{4+}(Fe_2^{3+}Si)O_{12}$

Name approved by the CNMMN of the IMA. So named in 1846 because of its resemblance to schorl.

Occurrence

Occurs principally in alkaline igneous rocks such as nepheline-syenite and related types; also in skarns. Originally found in carbonatite at Magnet Cove, Arkansas, USA; also in the USA in the mafic alkaline complex at Rainy Creek, Montana. Crystals more than 8 cm diameter from the Ice River alkaline complex, BC, Canada. It also occurs at: Iivaara (Ijola), Lääni, Finland, in an alkaline complex; in a spurrite skarn at Fuka, Japan. Complexly zoned Ti-rich melanite-shorlomite garnets are obtained from Ambadungar carbonatite-alkali complex, Deccan Igneous Province, Gujarat State, India.

Properties

Forms a series with andradite: $Ca_3Ti_2^{4+}$ $(Fe_2^{3+}Si)O_{12}$–$Ca_3Fe_2^{3+}(SiO_4)_3$ and is closely related to melanite (titanian andradite).

Refractive Index: 1.94–1.98

Specific Gravity: 3.77–3.93

Hardness: 7–7.5

Cleavage/Fracture

Conchoidal – Fractures developed in brittle materials characterized by smoothly curving surfaces.

Appearance

Colour – Dark brown, reddish brown to black.

Lustre – Vitreous to metallic

Rarely utilized as a gemstone although it has been used as black cabochon material.

Spessartine $Mn_3^{2+}Al_2(SiO_4)_3$

Name approved by the CNMMN of the IMA. So named in 1832 after its discovery locality, in the Spessart region, north-west Bavaria, Germany.
Synonyms: Spessartite, Malaya (malaia) Garnet, Fireball Garnet.

Occurrence

In igneous rocks spessartine is known from mainly granite pegmatites and aplites but also rhyolites. In granite pegmatites, the garnets are solid solutions between spessartine and almandine, with spessartine as the dominant component: they also generally contain a few percent of the pyrope component. Formed in some skarns and metasomatic manganese-rich rocks adjacent to igneous intrusions or in regionally metasomatized areas sometimes embedded in mica and mica slate. Also found in sedimentary deposits.

So far Spessartine had been found only in Sri Lanka, Myanmar, Madagascar, Brazil and Australia as well as in Kenya, Nigeria and Tanzania, and in Kunene River, Namibia (mandarin garnet). It is mined in central India, Madagascar and a number of localities in the USA, including

Silver Cliff, Colorado; in rhyolite from Garnet Hill, Nevada; Ramona, California; Amelia, Virginia (in 1991, a single piece, dubbed the Rutherford Lady, was found, which weighed more than 2800 carats). Other occurrences include: the LeChang Mine, GuangDong Province, China; various localities in the North-west Frontier Province, Pakistan, such as the Shigar Valley, Skardu; Nepal and Nuristan, Afghanistan.

Properties

Within the pyralspite series, spessartine forms a continuum with almandine; however, substitution by pyrope is usually less extensive largely due to incongruent geochemistry. Commonly, spessartine can also contain appreciable amounts of grossular and more rarely andradite molecules. Consequently the physical and chemical properties and appearance can vary appreciably. An example of this is an intermediate ($Sp_{49}Gr_{41}Al_5Py_5$) yellowish orange 'Mandarin' garnet from Madagascar, with properties that do not sit well with either spessartine or grossular but fall within the ranges for pyrope-spessartine and pyrope-almandine. Orange malaya garnets from Bekily, Madagascar, with Sp_{59} can have up to Py_{24}, whereas in pink stones this ratio is reversed.

Absorption Spectra

Strong bands in the blue-violet denote the presence of manganese. In orange spessartine these occur at 485 and 461 nm with a very strong band at 430 nm.

Some authors give slight variations on these stating that the strongest of these manganese absorptions occur at 432 and 412 nm. Other absorptions at 495 (weak), 489 (very weak), 462 (medium) and 424 nm (weak) have been observed. Two faint bands at 406 and 394 nm have also been reported.

Substitution of spessartine by almandine is common, and consequently the typical absorption spectrum of almandine, with the three main bands at 575, 527 and 505 nm, is seen in most red stones to varying degrees of intensity. Absorption spectra of spessartine from a new occurrence in Nigeria indicate that the yellow and golden-yellow specimens (low almandine content ≈ 1%) are very similar to those of spessartine from Kunene (Namibia), while those of brown-orange colour (higher almandine content ≈ 6%) are similar to those from Ramona (USA).

Refractive Index

Largely within the range 1.79–1.81 but rare intermediates may show values out of this, e.g. yellowish orange ($Sp_{49}Gr_{41}Al_5Py_5$) from Madagascar with RI 1.77.

Some locality values are: Nigeria RI 1.801–1.803; Virginia 1.802; Taita Hills, Kenya 1.795–1.809; Minas Gerais, Brazil 1.803–1.805; Ramona, California >1.81; Kunene, Namibia 1.801–1.803; Madagascar 1.800–1.808.

Density

4.179 (calculated)

Some locality values are: Nigeria 4.15–4.22; Minas Gerais, Brazil 4.15; Ramona, California >4.17; Kunene, Namibia 4.15–4.22.

Hardness

6.5–7.5

Dispersion

0.027

Cleavage/Fracture

None; parting sometimes distinct in six directions/Conchoidal to uneven.

Appearance

Colour

Due to its idiochromatic manganese content, spessartine is always some shade of orange. This colour is frequently tempered by isomorphous replacement, particularly by the almandine molecule, which increases the brown and red component.

Spessartine can therefore be found in a range of colours: red, reddish orange, orange, yellowish brown, reddish brown, yellow and pink. Nigerian stones tend to be darker being variously described as a very deep pure red of dark tone and strong saturation, rich, deep golden orange with a touch of fiery red, intense reddish orange and deep reddish orange or *burnt* orange.

Colour Varieties

Malaya (Malaia) Garnet

Malaya is an intermediate between spessartine and pyrope. The colour ranges from orange, red-orange, peach and pink. Only the rarer orange-coloured stones have the spessartine molecule dominant.

Mandarin, Hollandite or Kunene Spessartine

From the original mine in Kunene, Namibia, spessartine with rich orange colour was marketed first as hollandite but the name mandarin gained ascendancy. This material is a near end-member of the species, which accounts for its intense orange colour. Most of the material is clean, although larger sizes have veils which can give some stones a foggy look. The most common inclusions are small black spots of manganese. The RI is 1.80, the hardness is 7.25 and the SG is 4.04–4.15.

Kashmirine

Kashmirine is the trade name for orange-red spessartine mined in the Neelum valley in Kashmir, Pakistan.

Grape Garnet (see pyrope)

Blue Garnet (see pyrope)

Lustre – Vitreous to resinous

Inclusions

Most mandarin garnets from Namibia have a sleepy appearance. This is due to numerous small crystal and needle inclusions. Wispy veils can be visible to the naked eye. Even in relatively clean stones, graining is often strong. Black manganese oxide specks have been reported.

Inclusions recorded from Ramona spressartines include wavy two-phase partially healed fractures, negative crystals, needles and/or tubes and growth patterns.

The main inclusions in the Nigerian spessartines are healed cracks representing thin liquid-filled cavities or fingerprint-type feathers.

Optical Effects: Chatoyancy, Asterism, Colour Change, ADR

Pyrope-spessartine mixed garnets containing vanadium have been found that display a dramatic green (daylight) to red (incandescent light) change of colour. Such stones come from East Africa, Madagascar and Sri Lanka.

Anomalous double refraction. *Gems & Gemology* Summer 1982.

Treatment

None reported.

Uvarovite Ca$_3$Cr$_2$(SiO$_4$)$_3$

Name approved by the CNMMN of the IMA. *Synonyms*: Chrome-garnet, uwarowite.

Named in 1832 by Hess for Count Sergie Semeonovich Uvarov (1786–1855).

Occurrence

Uvarovite is found in association with chromite, serpentinites, ultra-mafic rocks and other chromium-bearing rocks; also in skarns. Notable occurrences are in the Ural mountains, particularly at the original locality of Sarany near Bisersk and in the vicinity of Kyshtymsk, north-east of Zlatoust. The largest crystals occur at Outukumpu, Finland. Uvarovite also occurs in Val Malenco, Lombardy, Italy; and in the Kop Krom mine, Erzerum, Turkey. In Canada, it was found in Quebec at Magog, Stanstead Co., Thetford, Megantic Co., and Wakefield. In the US, it occurs in California near Livermore, Alameda Co., and in Jackson, Amador Co.

Properties

Within the ugrandite series whilst there is extensive solid solution between uvarovite and grossular only limited solid solution exists between uvarovite and andradite. Limited solid solution also exists between uvarovite, pyrope and knorringite and the hydrogarnets. Consequently the physical and chemical properties and appearance can vary appreciably. Because of the solid solution between uvarovite and grossular many green garnets are mistakenly termed 'uvarovites' when they are really chromian grossular, particularly when they are from established uvarovite localities such as Sarany.

Absorption Spectra

Uvarovite is characterized by two intense absorption regions, one stretching from the violet to the blue, and the other from the red to the yellow.

Refractive Index

The calculated value for the end-member is 1.865 but with iso-morphous replacement, particularly by grossular, this can fall to 1.798 for $\approx$Uv$_{50}$.

Density

The calculated value for the end-member is 3.848 but with iso-morphous replacement, particularly by grossular, this can fall to 3.712 for $\approx Uv_{50}$.

Colour

Uvarovite has an intense green color and fine pieces feature a sparkling, bright-green surface of small crystals that cover the rock matrix thickly, evenly and smoothly. Large crystals are significantly darker.

Use

Uvarovite is seldom found in gem crystals of cuttable size and conse-quently gems are both rare and tiny. Because the overall appearance of matrix specimens is so striking, thin-shaped pieces for setting in jew-ellery are produced where much of the supporting matrix is cut away leaving an uninterrupted thick-pile, medium-grain, bright-green vitreous drusy crystal covering from edge to edge.

Colour Change Garnets

Colour change arises because stones are able to transmit two different ranges of wavelengths of light between areas of absorption. Depending on the light source, which may be enhanced in certain wavelengths, one of the ranges will be dominant. Thus when viewed in daylight, which is enhanced in the short wavelength part of the spectrum, the stone will be either violet, blue or green or some combination of these colours, and in incandescent light, which is enriched in long wavelengths, either red or orange or a combination of these. The ele-ments largely responsible for this 'alexandrite effect' are chromium and vanadium generally in conjunction with manganese. Colour change behaviour has been correlated with vanadium, chromium, manganese, magnesium and iron components (see also optical effects under indi-vidual species name).

Occurrence

Colour change garnets have been reported from Madagascar (Ilakaka and Bekily), Tanzania (Umba Valley and Tunduru) and Sri Lanka (Athiliwewa and Embilipitiya). Also recorded from East Africa and Russia.

Properties

Colour change garnets fall into two distinct groups. The most common are those of the pyrope-spessartine series where extensive substitution of magnesium by manganese occurs. These also contain traces of chromium and/or vanadium up to 2 wt%. Less common are the chromium-rich pyrope with $Cr^{3+} > 3$ wt%. Colour change has also been reported for pyrope-almandine, spessartine-grossular-almandine, spessartine-grossular-pyrope and andradite. Thus physical and chemical properties can vary appreciably.

Absorption Spectra

Colour change garnets fall within the general criteria for colour changing materials which produce an absorption spectrum with the maximum absorption between 580 and 560 nm with variable flanking minima that are generally between 665 and 625 nm and between 510 and 470 nm.

Low iron content gives weak Fe^{2+} absorption whilst the spessartine component produces strong Mn^{2+} absorption; however, the dominant absorption feature is a broad V^{3+}/Cr^{3+} absorption maximum centered at 571 nm.

East African material in the compositional range $Sp_{59}Gr_{21}Al_{12}Py_7Uv_1$ to $Sp_{43}Py_{30}Gr_{16}Al_6Uv_2An_1$ exhibited absorption spectra with absorption maxima in two zones the first ranging from 450 nm to end of violet and the second from 563 to 573.5 nm. The presence of manganese has been cited as responsible for the colour change (strawberry pink to green) in garnets found near Voi, Wandanyi Mountains, Kenya.

The absorption maxima in colour change pyrope-spessartine material from Madagascar within the compositional range $Sp_{80}Al_7Gr_7Py_4Go_1Uv_1$ to $Py_{55}Sp_{30}Gr_{10}Al_2Go_1Uv_1$ was recorded as approximately 450 nm to the end of the violet and between 568 and 583 nm.

Differences in colour change were noted between transmitted and reflected light.

Refractive Index/Density

Studies of the East African and Madagascar material above gave RIs between 1.773 and 1.763 and between >1.81 and 1.748 respectively with decreasing spessartine component; likewise SGs varied between 3.98 and 3.89 and between 4.1 and 3.784.

Appearance

Colour

Manson and Stockton recorded that stones from Madagascar exhibited, with increasing iron content, colour changes from greenish grey (daylight) to pink or red (incandescent) and greenish blue (daylight) to reddish violet (incandescent).

Colour change is highly variable in the breadth of hues, tones and tints. These depend on the degree of molecule admixture and trace element impurities.

12

Feldspar Group

Brian Jackson

Although feldspar is the most abundant mineral group on Earth accounting for around 50–60% of the Earth's crust, gem-quality feldspars are rather rare.

Mineralogically, the group has been comprehensively studied. Consequently, numerous scientific terms have been coined to describe concepts, phases and features shown by the group. Feldspars show more types of phenomena than any other gem and naming may be difficult as some feldspars are not homogeneous. Many of the descriptive scientific terms used have little or no gemmological significance but in preparing full gemmological descriptions some may be appropriate and are applied here for clarity.

The feldspar group is also inundated with variety of names e.g. peristerite, moonstone, sunstone, amazonite, adventurine, perthite and spectrolite. Whilst the International Mineralogical Association Commission on New Minerals and Mineral Names (IMA CNMMN) does not recognize or define varieties, within the gem trade variety names are frequently applied to feldspars that exhibit characteristic properties and/or colour. This is helpful where a particular optical property such as the 'moonstone effect' (schiller) is exhibited by heterogeneous feldspars that are not readily classified. Consequently, sunstone and moonstone are dealt with separately as effects and not specifically under species or intermediates. Some varieties have been characterized in the gemmological literature but often the gem trade assigns a variety name on the basis of its selling appeal, e.g. 'rainbow moonstone'. In order to avoid confusion, it is helpful to link commercial names with correct mineralogical terminology.

The group comprises two closely related families that are allied in habit; however, the crystal symmetry is controlled by both the internal structural framework arising from aluminium–silicon ordering (distribution of atoms within the structure) and distortion arising from the effect of the variation in cation (sodium, calcium, potassium and barium) size. Temperature is an essential factor in determining the resultant species. The resultant properties are further affected by intimate lamination of differing species that arise during cooling. Zoning is common due to the changing chemical composition of the melt during the growth history. Thus chemistry, temperature and ordering/structure affect the properties that are measured ahead of naming a feldspar.

Chemical Composition and Nomenclature

Feldspar is the mineral name given to a group of aluminium silicate minerals whose general chemical formula is $XAl(Al,Si)Si_2O_8$, where X is potassium, sodium, calcium or barium. The majority of feldspars are classified chemically into alkali feldspars and plagioclase feldspars.

The **alkali feldspars** range in composition from sodium bearing albite to potassium feldspar species and are identified on the basis of their composition, temperature of formation range, ensuing structure and properties.

- *Albite* $NaAlSi_3O_8$
- *Anorthoclase* $(Na,K) AlSi_3O_8$ Potassium feldspars
- *Sanidine* $KAlSi_3O_8$ or
- *Orthoclase* $KAlSi_3O_8$ K-feldspars
- *Microcline* $KAlSi_3O_8$

The **Plagioclase Series** range in composition from the end members species albite to anorthite with arbitrarily defined intermediates with fixed compositional ranges.

- *Albite* $NaAlSi_3O_8$
- Oligoclase $Ab_{90}-An_{10}$ to $Ab_{70}-An_{30}$
- Andesine $Ab_{70}-An_{30}$ to $Ab_{50}-An_{50}$
- Labradorite $Ab_{50}-An_{50}$ to $Ab_{30}-An_{70}$ named intermediates
- Bytownite $Ab_{30}-An_{70}$ to $Ab_{10}-An_{90}$
- *Anorthite* $CaAl_2Si_2O_8$

Application of International Mineralogical Association (IMA) nomenclature rules means that intermediates within the plagioclase series do not have species status. Intermediates were named for convenience.

Temperature Effects and Nomenclature

The temperature of crystallization and subsequent thermal history influences the structural state of feldspar. This yields species with similar chemistry but differing properties. Even within individual species, properties are variable depending upon temperature of crystallization. These are known as high (high temperature) and low (low temperature) feldspars. Thus, crystal symmetry and physical properties vary.

Whereas the plagioclase group varies in composition, the K-spars all have the same formula, $KAlSi_3O_8$, but they vary in crystal structure depending on the temperature of formation. Microcline is the stable form below about 400 °C. Orthoclase and sanidine are stable above 500 °C and 900 °C, respectively, but they endure as long as they need to at the surface as metastable species (see *Figure 12.1*).

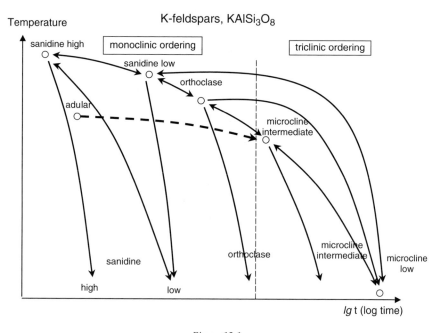

Figure 12.1

In assigning an identification based on RI, it should be noted (see *Figure 12.2*) that there is a variation in RI with composition and structure such that for the series sanidine-anorthoclase-high albite (dotted lines) the values of α and γ are lower than those for the orthoclase-low albite series (solid lines). However, the β values are essentially the same in both series.

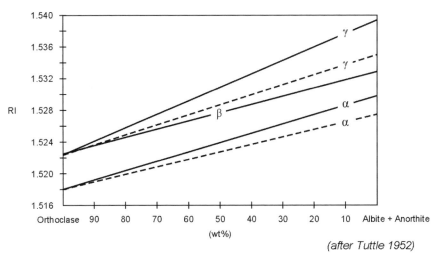

Figure 12.2

(after Tuttle 1952)

Ordering Effects

The internal structural framework arising from aluminium–silicon ordering influences the crystal symmetry (see *Table 12.1*).

Figure 12.1 shows the relationship of species and structure with respect to temperature and time. It is complementary to *Table 12.1*.

Table 12.1

Alkali feldspar	Crystal symmetry	Crystallization temperature	Cooling rate	Ordering Al/Si distribution	Exsolution lamellae – visible sign
Sanidine	Monoclinic	High	Extremely rapid	Perfectly disordered	Require electron microprobe – *moonstone*
Anorthoclase	Monoclinic	High	Extremely rapid	Perfectly disordered	Require electron microprobe – *moonstone*
Orthoclase	Monoclinic	Intermediate	Slow cooling	Partially ordered	Microscopically visible – *moonstone*
Microcline	Triclinic	Low	Extremely slow cooling	Perfectly ordered	Macroscopically visible – *amazonite* 'streaky' **structure**

The lack of symmetry gives rise to the thought that the triclinic system is the closest system to amorphism, having no form or order. It is actually the other way around. Triclinic minerals have a more sophisticated ordering of the atoms in their structure than other minerals. Ordering is a way of saying that the atoms are precisely placed into specific sites and these sites, in the case of triclinic minerals, are not symmetrically arranged.

As an analogy think of an alphabetic filing system with lots of random, unsorted papers scattered among the files. This might produce a more or less even (symmetrical) distribution of papers among the files. In other words, the A file has just as much chance of having, say 10 papers, in it as the B file or C file, etc. But if the papers are sorted alphabetically, the distribution of papers is now irregular or asymmetrical, because files such as A might contain many papers and files such as Q, hardly any.

This kind of ordering occurs in mineral structures where various positions for atoms are assigned to specific elements. The more specific the site and the more specific the element that occupies those sites then the less symmetrical, but more ordered, the structure. The degree of ordering is a function of temperature. This is important gemmologically for while in high temperature, fully or highly disordered alkali or plagioclase feldspars exist as homogeneous single-phase crystals, in the low-temperature alkali feldspar or plagioclase feldspars separation occurs on a macro-, micro- or sub-microscopic scale. When the intergrown lamellae are of appropriate thickness, interference effects can result in iridescence, variously described as 'moonstone' or 'labradorescence'.

Isomorphous Replacement, Solid Solution and Exsolution in Feldspars

In feldspars with similar structure but different chemistry, there exists a seamless series of compositions between two extreme compositions called end-members, as one element gradually replaces another. This concept is called **isomorphous replacement** and the series produced are known as **solid solution series**. Thus in feldspars with similar structure but different chemistry, there is a correlation between physical properties, especially those of RI and density, and the chemical composition (see *Figure 12.2*).

This is the case for plagioclase feldspars. However, there are three regions in the plagioclase solid solution series where a lamellar structure created by phase separation (**exsolution**) of fully ordered albite and anorthite is more stable than the solid solution structure, which consists of randomly distributed clusters of both phases. These regions are

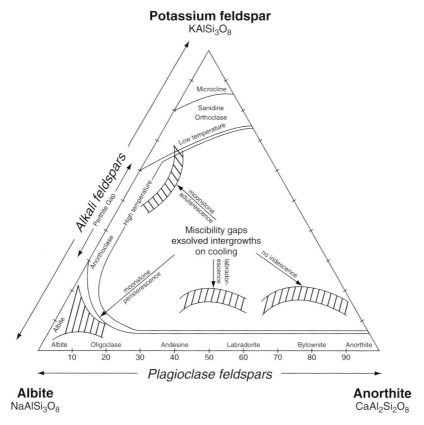

Figure 12.3

known as miscibility gaps. Two of these are important gemmologically as they produce the effects seen in moonstone and labradorite (see *Figure 12.3*).

Figure 12.3 shows *Peristerescence* in plagioclase feldspars restricted to the albite–oligoclase range (An$_1$–An$_{17}$), appearing as subtle interference colours on a white- or buff-coloured surface. Alternating thick and thin lamellae consist of exsolved oligoclase in albite; the albite lamellae being almost twice as thick as the oligoclase lamellae. The compositional difference between lamellae is $\approx 20\%$ with an interval of 50–300 nm for the albite–oligoclase paired unit (periodicity). This produces a moonstone effect. Peristerite is the iridescent variety named from the Greek word *peristera* (pigeon) in allusion to its play of colours somewhat resembling those on a pigeons's neck. These arise due to exsolution, but it has also been reported that where exsolved areas are irregular and small compared to the wavelength of light, white to bluish colours may be produced by incoherent light scattering such as the Tyndall effect (see also moonstone effect). Peristerite RI around 1.531–1.539; DR 0.008; SG around 2.63. The colour is generally brown-pink, white or cream with blue reflections, greenish or green-yellowish, yellow.

Labradorescence occurs in the andesine–labradorite range (An_{45}–An_{60}), appearing as intense interference colours that vary with increasing anorthite content; (blue $An_{48.5-52}$, green–yellow $An_{52-55.5}$, orange–red $An_{55-55.8}$). The structure responsible for the iridescence is related to immiscibility which itself is directly related to ordering. Thereby producing low energy structurally favourable configurations based on two phase lamellar intergrowths of Na-rich and Ca-rich blocks of alternating thick and thin lamellae that produce iridescence. The compositional difference between lamellae being 12–15% with an interval of 50–280 nm for paired andesine–labradorite units.

Huttenlocher intergrowth, the last of the three plagioclase miscibility gaps occurs in the labradorite–bytownite range (An_{67}–An_{90}). The alternating lamellae are much coarser and predominantly produce no interference colours although labradorescence has been noted in compositions close to the labradorite–bytownite boundary.

In the alkali feldspar series (Albite to Orthoclase/Microcline/Sanidine), complete solid solution (substitution of sodium and potassium) exists at high temperatures, but not at lower temperatures, producing a miscibility gap. Thus, during cooling and crystallization the feldspar solution, rather than crystallizing as a single feldspar of intermediate composition, separates into two feldspars with different chemistry producing alternating lamellae that can range in size from visible to the naked eye to those that are too small to see even with a high-powdered optical microscope. Thus, exsolution in the alkali feldspars gives rise to intergrowths; the submicroscopic variety is visible as coloured sheen (adularescence) producing the moonstone effect.

When albite separates from a microcline host and appears as small grains or streaks, the resulting mixtures are termed 'perthites': amazonite with a green microcline host and exsolved steaks of white albite is an example of this. When orthoclase separates from a sodium-rich plagioclase host and appears as small grains or streaks, the resulting mixtures are termed antiperthites. These textures may be coarse and easily visible as colour variations in the crystal or may be microscopic and practically invisible (microperthite). The coarseness of the texture depends mostly on the rate at which the feldspar was cooled; the slower the cooling, the coarser the texture. If the feldspar solution cools so rapidly that it does not separate and remains a single phase, it is called anorthoclase.

Feldspar Optics and Density

The optical properties of feldspar vary in response to chemical substitution, the degree of atomic ordering in the structure and intergrowths arising from exsolution.

In the **plagioclase series**, the optical properties are directly related to their chemical composition with RIs increasing steadily and nearly linearly with increasing calcium content. Thus, RI can be used to determine the composition. RI measurement to accuracy of ± 0.001 equates to a compositional accuracy of $\pm 2\%$ anorthite content (see *Figure 12.4* and *Table 12.1*).

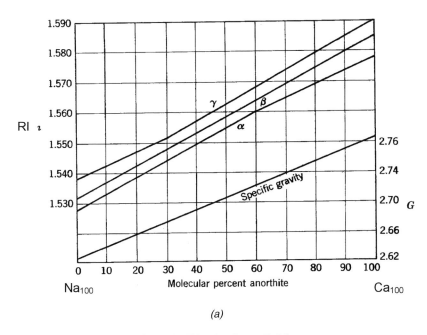

(a)

Optic Sign in Plagioclase Feldspars

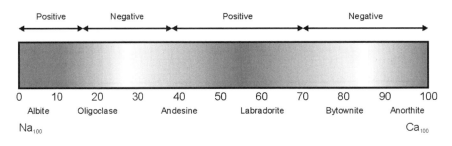

(b)

Figure 12.4

245

Table 12.2
Plagioclase series optical data

OPTICS	Albite	Oligoclase	Andesine	Labradorite	Bytownite	Anorthite
α	1.528–1.533	1.533–1.543	1.543–1.554	1.554–1.563	1.563–1.572	1.572–1.576
β	1.532–1.537	1.537–1.547	1.547–1.559	1.559–1.568	1.568–1.578	1.578–1.583
γ	1.538–1.542	1.542–1.552	1.552–1.562	1.562–1.573	1.573–1.583	1.583–1.588
DR	0.009–0.01	0.009	0.008–0.009	0.008–0.01	0.01–0.011	0.011–0.012

Optical data reproduced from http://webmineral.com/

High temperature albite has the values α 1.528; β 1.532; γ 1.534: DR 0.007 and it is optically $-$ve (Table 12.2).

Figure 12.4 The lightest areas represent $-ve_{max}$ and the darkest grey represents $+ve_{max}$. Thus it can be seen how the optic sign for the plagioclase series changes, being ($+$ve) for albite, changing to ($-$ve) in the more calcic oligoclase range, becoming ($+$ve) again for most andesine. The group comprises two closely related families and labradorite, and reverting to ($-$ve) in bytownite and anorthite.

The RIs within the **alkali feldspars** show little variation from each other and are lower than those in the plagioclase series (see *Table 12.3*). However, substitution of potassium and sodium by barium will raise the RI. When the barium component is between 2 and 80% the feldspar is called *hyalophane*. When barium is present above 80% it is called *celsian*. Other substitutions such as iron and strontium will also raise the RI and Si–Al substitutions required to maintain electrostatic neutrality will influence the optics. The optical properties of alkali feldspars are also affected both by the degree of ordering and by the presence of exsolved cryptoperthitic intergrowths (submicroscopic separating of plagioclase from a potassium rich-host). These numerous factors mean that published data for optical properties can, when taken together, give rise to an

Table 12.3
Alkali feldspars

OPTICS	Microcline	Orthoclase	Sanidine	Anorthoclase
α	1.518	1.518	1.518–1.525	1.519–1.529
β	1.522	1.522	1.523–1.53	1.524–1.534
γ	1.525	1.523–1.524	1.525–1.531	1.527–1.536
DR	0.007	0.005–0.006	0.006–0.007	0.007–0.008

Optical data reproduced from http://webmineral.com/

expanded range of values that make identification problematic. Ranges are given in the text on individual species.

Density

Density in the plagioclase series is governed by percentage cation (sodium and calcium) components with a continuous increase in RI as Ca substitutes for Na (Table 12.4). In the alkali feldspars, the major variation arises from K–Na substitution but superimposed on this is a minor variation due to monoclinic–triclinic structural change that mirrors the change for RIs (Table 12.5).

Table 12.4
Plagioclase series

SG	Albite	Oligoclase	Andesine	Labradorite	Bytownite	Anorthite
	2.61–2.63	2.64–2.66	2.66–2.68	2.68–2.71	2.7–2.72	2.72–2.75
Average	2.62	2.65	2.67	2.69	2.71	2.73

Table 12.5
Alkali feldspars

SG	Microcline	Orthoclase	Sanidine	Anorthoclase
	2.56	2.56	2.52	2.57–2.6,
Average				2.58

Feldspar Twinning

Subtle structural changes frequently result in twinning which is a common feature of the feldspar group and a useful identification aid.

Alkali feldspars exhibit normal simple twinning based on various twin laws and is common in monoclinic feldspars.

Lamellar twins in albite are often visible under the microscope and are a distinguishing feature of triclinic plagioclase. The twin lamellae show typical extinction pattern. Albite law twinning produces stacks of twin layers that are typically only fractions of millimetres to several millimetres thick. These twinned layers can be seen as striation like grooves on the surface of the crystal and unlike true striations these also appear on the cleavage

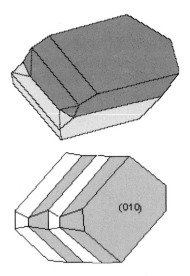

Figure 12.5

surfaces (Figure 12.5). Albite twinning is so common in plagioclase that its presence is a diagnostic property for identification of plagioclase. This can be recognized even in transparent cut stones by using the polariscope or polarizing microscope.

Alkali Feldspars

The four feldspars, orthoclase, sanidine, microcline and anorthoclase comprise a group called the potassium or alkali feldspars. They all have the same chemical composition, potassium aluminium silicate ($KAlSi_3O_8$), but different crystal structure and are therefore polymorphs. The difference in crystal structure is reflected in crystal symmetry. The differences among these feldspars result from the different ways in which aluminium is distributed, or ordered, in the aluminosilicate atomic structures of each. Sanidine is stable at the highest temperatures. It can accommodate sodium in all proportions and a complete compositional series (with a slight structural change) runs from pure potassium sanidine to pure sodium albite. Members of this series that are more than two-thirds sodium are called anorthoclase and have triclinic symmetry.

Orthoclase

Name approved by the CNMMN of the IMA. *Orthoclase* takes its name from the Greek word *orthos* (upright) and *klasis* (fracture) in allusion to its two prominent cleavages at right angles.

Synonyms: Potash feldspar; adularia; moonstone; noble orthoclase; valentianite (a 'moonstone' from Russia that is named after the last Czar's personal cutter Valentin who specialized in cutting this material).

Occurrence

Orthoclase is a component of many rocks and is typically found in acidic and alkaline igneous rocks particularly in granite pegmatites and syenites. It also occurs in Alpine-type metamorphic rocks and hydrothermal veins – additionally as detrital grains in placers deposits. In Madagascar, the gemmy 'orthoclase' found in pegmatite veins associated with medium to coarse grained marble lenses has since been found by X-RD to be high to low sanidine.

Orthoclase is one of the most common minerals and is found worldwide; however, it is very rare to find orthoclase as transparent gem-quality crystals. Only a few notable localities are mentioned: The gem gravels of Myanmar and Sri Lanka Similar produce similar material. Much adularia is found in the Alps in Switzerland, especially at Adular, the locality this variety was named after. 'Carlsbad Twins' are famous from Carlsbad, Czechoslovakia and fine 'Baveno' twins have come from Baveno, Italy. Large, well-formed crystals were found in Disentis, Switzerland. Light-blue crystals occur in the Lake Baikal area in Russia, and in Udacha, Russia, yellow-brown transparent orthoclase up to 8–10 cm is known. Large masses occur on Mt. Kilimanjaro, Tanzania. In the USA, enormous crystals were found on Sandia Mountain, Bernalillo Co, New Mexico. Famous collector specimens have come from Robinson, Colorado and Goodsprings, Clark Co., Nevada. Orthoclase has also come from the French Creek mine, St. Peters, Chester Co., Pennsylvania. Much adularia moonstone has come from New Mexico. Brazil pegmatites, some K-feldspar (potassium-feldspar) crystals, roughly 3 m long have been reported. Brownish transparent crystals around 5 cm have been reported from Greenland.

Appearance
Opaque to translucent, rarely transparent

- *Lustre*: Vitreous but may be pearly on cleavage surfaces.
- *Colour*: Commonly colourless and champagne to yellow but also pink, orange, light blue, light green, brown and grey. Massive orthoclase is generally white or pale pink in colour.

    Predominantly, opaque green orthoclase (monoclinic) with minor ruby inclusions from the Luc Yen area, Vietnam, has been reported. The colour is attributed to its high lead content (0.5wt% PbO). Some translucent and transparent pieces have been polished. White plagio-clase with ruby inclusions has also been reported from Myanmar.

- *Varieties: (see also moonstone) Adularia* – A moderate to low temperature, more completely ordered orthoclase. It is weakly triclinic (formerly regarded as apparently monoclinic) and typically occurs in well developed, usually transparent and colourless to milky white (and often opalescent (see also moonstone)) pseudo-orthorhombic crystals in fissures in crystalline schists, especially in the region of the Swiss Alps (from Adula Mts, St Gotthard massif, Ticino (Tessin), Switzerland; the type locality). Adularia displays pearly internal reflections. It typically has a relatively high content of barium.

  'Rainbow lattice sunstone' is the trade name given to orthoclase that exhibits iridescence from hexagonal platelets of hematite on intersecting cleavage planes producing a criss-cross network of colour in addition to adularescence. The material, which is exceptionally rare, comes from the Harts Range about 60 to 70 miles NE of Alice Springs in the Northern Territory, Australia.

Properties

- *Crystal system*: Monoclinic
- *Refractive index*: α 1.518–1.529; β 1.522–1.533; γ 1.522–1.539. Biaxial −*ve*
- *Birefringence*: Weak; 0.005–0.010
- *Pleochroism*: Weak
- *Density*: 2.56 but has been reported in the range 2.55–2.63
- *Hardness*: Hardness on Mohs' scale ranges from 6 to 6.5
- *Dispersion*: Low: 0.012
- *Cleavage/fracture*: Perfect prismatic cleavage in 2 directions at 90° to each other. Fracture is uneven, somewhat brittle
- *Inclusions*
- *Optical effects: chatoyancy, asterism, colour change, ADR*: Yellow and colourless cat's eye gems are known from Myanmar and Sri Lanka. Some Sri Lankan stones exhibit asterism. An undifferentiated dark brown K-feldspar (RI 1.52; SG 2.6) was reported to exhibit both asterism and chatoyancy. Yellow cat's eye without adularescence has been reported. (see also moonstone)
- *Absorption spectra (400–700 nm)*: Absorption spectra described for yellow orthoclase from Madagascar but the material is now known to be high to low sanidine (see sanidine)
- *Fluorescence*: SWUV – commonly red or pink but white, whiteish green, pale green, greyish green, blue and orange have all been reported. LWUV – weak blue or cream. Occasionally weak-to-strong reddish orange under LWUV

- *Simulant*: An undifferentiated feldspar has been reported to be dyed blue in an attempt to simulate lapis lazuli. The dye could be easily seen in the cleavages and fractures.

Microcline

Name approved by the CNMMN of the IMA. The name microcline comes from two Greek words *mikro* meaning small and *kleinen* meaning inclined, referring to the slight deviation from 90° of intersecting cleavages.

Synonyms: Potash feldspar, alkali frldspar, K-feldspar. Amazonite is sometimes called amazonstone.

Occurrence

Microcline commonly occurs in plutonic igneous rocks such as granites and syenites that cooled slowly and at considerable depth. It is the common potassium feldspar in pegmatites where it takes the place of orthoclase. It can form the largest known crystals of any mineral (in a pegmatite in Kareliya, Russia, a microcline mass weighing over 2000 tons showed the form of a single crystal!). Microcline is also common in metamorphic rocks in gneisses and in sedimentary rocks in arkoses and conglomerates.

Notable deposits are found in Italy, Norway, Madagascar, Namibia, Zimbabwe, India, Russia, Brazil, Australia, Canada and the USA.

Amazonite localities include USA (Colorado, California, Montana, Pennsylvania, Virgina–Amelia County, deep, blue-green, translucent), Canada (Ontario–Renfrew and Parry Sound districts and Quebec–Kipawa), Brazil, India (Kashmir district and elsewhere), Kenya, Tanzania, Madagascar (Anjanabonoina, Andina, Immody and Mahabe), Namibia and South Africa. Russia the biggest gem amazonite deposits in the world are located in an area 1 km^2 at Ploskaya Gora and Parus mountain, 80 km east from Lovozero, Kola Peninsula (there are more then ten pegmatite veins from 10 to 300-m long and from 0.5 to 30 m thick), also in Russia relatively large quantities of amazonite used as gemrock has come from pegmatite masses near Miass in the Ilmen range of the southern Ural mountains, Afghanistan at Madan Shar, Kabul Province in small deep blue-green, semi-translucent crystals and Australia (Broken Hill, New South Wales – also in semi-translucent masses). Green microcline is reported from various localities in Brazil.

Appearance

Opaque to translucent, rarely transparent. Simple and polysynthetic twinning are ubiquitous and perthitic intergrowth with albite, extremely common. Reflections from incipient cleavages give polished surfaces a shimmering effect.

- *Lustre*: Vitreous
- *Colour*: Commonly white to pale yellow or salmon, also may be blue to green in the microcline variety *amazonite*. It should be noted that although nearly all green potassium feldspar is microcline, much microcline, as well as most orthoclase, is commonly white, flesh- or salmon-pink.

Varieties

Perthite

Named after the town of Perth, Ontario, Canada, where it was first found, Perthite is the general variety name for alkali feldspar consisting of parallel or subparallel intergrowths in which the potassium-rich phase (commonly microcline but may be orthoclase) appears to be the host from which the sodium-rich phase (commonly albite but may be oligoclase) exsolved; such exsolved areas may be visible to the naked eye, typically forming strings, lamellae, blebs, films or irregular veinlets (the term 'flame perthite' has been used in this context). Perthite classification is based on the size of the exsolved phase. Thus, where the texture is invisible to the naked eye but can be resolved with a microscope, it is called *microperthite*. Amazonite (*macro-perthite*) is an example of perthitic structure that can be seen by the naked eye and moonstone (*crypto-perthite*) is an example where the perthitic structure cannot be resolved with the microscope. If there is more plagioclase than potassium feldspar, it is called 'antiperthite'. Moonstone from Moonstone Hill, Queensland, has been classified as an anorthoclase-crypto-perthite antiperthite. Perthite is often pink, red or honey coloured and materials exhibiting a rich golden labradorescence against a flesh-red to reddish brown or white background have been used as ornamental stones. Perthites often show silvery reflections. A glassy tan microcline Perthite from a pegmatite in Custer, South Dakota, USA, shows blue-white fluorescence. Fine perthites occur at Dungannon, Ontario, and other localities in Canada. Perthite is used for carving and also cabochoned. Parting may be seen in some perthites.

Perthite Refractive Indices

$\alpha = 1.522$–1.527
$\beta = 1.526$–1.532
$\gamma = 1.530$–1.538
$DR = 0.005$–0.013

Indices between those of microcline and albite. The orthoclase (K-feldspar) will have the lower RI and the albite will have a higher RI.

Graphic granite

Quartz and microcline often crystallize simultaneously in pegmatites in an interpenetrating pattern known as *graphic granite* as the appearance resembles cuneiform script. This rock typically consists of nearly parallel skeletal rod-like masses of quartz (colourless, white or slightly smoky) surrounded by cream or salmon-pink microcline or amazonite. The lustre is pearly to dull vitreous and the material opaque to translucent.

Graphic granite as a decorative stone has been used since antiquity and, more recently, is reported as being used by cooperatives and amateur craftsmen as a decorative material in the Chelyabinsk region of the southern Urals, Russia – also used to make carvings and cabochons.

Synonyms: Hebraic pegmatite or Hebrew stone, corduroy rock, graphic pegmatite and runite.

Amazonite

The name amazonite appears to have evolved from the term 'Amazon stone' which, perhaps through confusion, was applied originally to another green mineral from that region as amazonite is not found anywhere in the Amazon basin.

Synonyms: Amazon stone; amazonstone; mother of emerald; green feldspar; microcline, Colorado Jade, and Amazon Jade.

Amazonite is an opaque to translucent, blue to green, variety of microcline characterized by streaks of white exsolved albite. The colour has been attributed to three variables: lead, structural water and natural radiation. Hoffmeister and Rossman (1985) proposed that while lead does indeed play a role in colouring amazonite, both natural radiation and structural water are necessary to produce chromophoric monovalent or trivalent lead. The natural decay of potassium^{40} reduces Pb^{2+} to Pb^{1+}, produces hydroxyl radicals that oxidize nearby oxygens to form hole centres and oxidizes Pb^{2+} to Pb^{3+}. Comparison of color intensity with Pb concentration indicated no correlation. It has been suggested that if there is a higher degree of Al/Si disorder then green is the dominant colour. Heating will remove the blue–green colour. Use mainly as a gem material for beads, cabochons, the perfect cleavages making it somewhat unsuitable for carving. Reflections from incipient cleavages give polished surfaces a shimmering appearance.

Properties

- *Crystal system*: Triclinic
- *Refractive index*: α 1.514–1.529; β 1.518–1.533; γ 1.521–1.539 (Amazonite 1.522–1.530). Biaxial $-$ve
- *Birefringence*: 0.008–0.010 (Amazonite 0.008)

- *Pleochroism*: Weak
- *Density*: 2.55–2.63 (Amazonite 2.56–2.58)
- *Hardness*: 6–6.5 (Amazonite 6.5)
- *Dispersion*: Weak
- *Cleavage/fracture*: Characteristic cleavages are prismatic perfect, inter-secting at approximately 90°. Conchoidal to uneven fracture
- *Optical effects*: *chatoyancy, asterism, colour change, ADR*
  See moonstone
- *Absorption Spectra (400–700 nm)*: Amazonite usually shows blanket absorption throughout the visible spectrum with a minimum absorp-tion at 550 nm increasing towards both the red and violet
- *Fluorescence*: SWUV – microcline may fluoresce with a weak red or moderate blue colour, bluish white or pale greyish green. Microcline from Franklin, New Jersey, is reported as having a medium blue–grey fluorescence under SWUV. Some amazonite will fluoresce medium or weak red under SWUV. A yellowish green fluorescence has been reported under LWUV. Amazonite from Konso, Southern Nations Regional State, Ethiopia, fluoresces a deep maroon colour under SWUV whilst amazonite from Broken Hill, New South Wales, Australia, has been reported with unusual white/greenish fluorescence under SWUV. Heat lessens fluorescence
- *Simulant*: Occasionally, in the past, initated by special types of glass
- *Treatments*: Amazonite – colour is lost on heating to over 300 °C and can be restored by irradiation if heating was insufficient (<500 °C) to cause water loss.

Sanidine – KAlSi$_3$O$_8$

Name approved by the CNMMN of the IMA. Sanidine comes from the Greek word *sanida*, a board alluding to its typical tabular habit.

Synonyms: Sanidine is an end member of a series of the alkali feldspars whose series ranges from pure NaAlSi$_3$O$_8$ to pure KAlSi$_3$O$_8$. This series only exists at high temperatures. Optical properties and X-ray techniques are the only good ways to distinguish sanidine from orthoclase, micro-cline and anorthoclase. Sanidine has a flattened crystal habit and simple twinning is not infrequent.

Occurrence

Sanidine is a common constituent in acid and intermediate volcanic rocks such as rhyolites, dacites and trachytes; where the rock cooled quickly at above approximately 900 °C, sanidine is the stable structure.

Also in lamproites, a leucite-nepheline dolerite and in a basaltic tuff. Notable occurrences include Germany, USA – Idaho, Colorado and New Mexico – Russia, Italy and Madagascar (transparent colourless to yellow gemmy crystals have come from pegmatite derived feldspar veins cutting coarse marbles at Ampandrandrava, Toliara (Tuléar) Province and Itrongahy Fianarantsoa Province. The yellow gemmy crystals, originally described as orthoclase and now identified as high to low sanidine, are actually found over a rather large area, from the classic Itrongahy region to the south-east for at least 40 km. Rich yellow facetable pieces up to 2,500 cts have been reported.

Appearance
Transparent to translucent

- *Lustre*: Vitreous
- *Colour*: Colourless, yellow, light brown, light grey, smoky; the tone being defined by the size of the stone.

Properties

- *Crystal system*: Monoclinic – biaxial −ve
- *Refractive index*: α 1.518–1.527; β 1.523–1.532; γ 1.524–1.534. Yellow sanidine (formally identified as orthoclase) from Madagascar with iron up to $\approx$3% will show increases in α and γ of $\approx$0.003/wt% of Fe_2O_3. Iron rich material having values in the upper part of the range. The RI range for brown Rhineland, Germany, material have been reported as α 1.516–1.520; β 1.521–1.525; γ 1.522–1.526 and DR –0.005 to 0.007. Colourless crystals up to 1 cm; RI α 1.516–1.519; β 1.520–1.522; γ 1.521–1.523; DR 0.003–0.005 have been reported from volcanic tuffs at Ashton, Idaho, USA
- *Birefringence*: 0.006–0.008
- *Pleochroism*: Weak
- *Density*: 2.56–2.62 (SG Rhineland material 2.57–2.58)
- *Hardness*: 6
- *Dispersion*: Weak
- *Cleavage/fracture* is prismatic, excellent to good in two directions forming nearly right angled prisms. Fracture is conchoidal or uneven – may be somewhat brittle
- *Inclusions*: Not diagnostic though some melt inclusions may be visible
- *Optical effects*: *chatoyancy, asterism, colour change, ADR* See moonstone
- *Absorption Spectra (400–700 nm)*: Ferric iron (Fe^{3+}) may substitute in the tetrahedral Si/Al site producing a yellow hue. Absorption spectra of yellow sanidine (previously identified as orthoclase) from Madagascar show a weak line or band at 448 nm and a slightly stronger one at

420 nm in the blue and violet respectively. A stronger band at 375 nm can be observed using a spectrophotometer. The smoky colour in some sanidine is the result of ionizing radiation interacting with Al which produces strong absorption increasing from 500 to 400 nm in smoky coloured sanidine from Eifel, Rhineland, Germany. Not diagnostic

- *Fluorescence*: Weak reddish orange under both LWUV and SWUV (Madagascar)
- *Treatment*: Some sanidine subjected to iridation turns amber or smoky coloured.

Anorthoclase – $(Na,K)AlSi_3O_8$

Name approved by the CNMMN of the IMA. *Anorthoclase* takes its name from the Greek words *orthos* (right) and *klan* (to break) and the negative prefix *an* alluding to the fact that the two prominent cleavages are not at right angles.

Synonyms: K-analbite, K-monoalbite, K-albite (also see larvikite). Anorthoclase is the name given to a member of the alkali feldspar series that contains more than two-thirds sodium and has triclinic symmetry. Anorthoclase sometimes shows twinning, but generally not the multiple twinning seen in plagioclase.

Occurrence

As phenocrysts in high-temperature volcanic rocks such as trachytes, phonolites and andesites and in tuffs.

Rather abundant worldwide. Some localities for well-characterized material include: Pantelleria and Ustica Islands, Italy; Larvik, Norway; Berkum, North Rhine-Westphalia, Germany; Kilju, Hamgyongbukto, Democratic People's Republic of Korea; Mt. Erebus, Ross Island, Antartica; Camargo, Chihuahua, Mexico; Stettin, Marathon Co., Wisconsin, USA; Somali Republic; Australia and New Zealand.

Appearance
Opaque to translucent , rarely transparent

- *Lustre*: Vitreous, may be pearly on cleavages
- *Colour*: Colorless, also white, pale creamy yellow, occasionally greenish or pink, red.

Properties

- *Crystal system*: Monoclinic
- *Refractive index*: α 1.518–1.527; β 1.522–1.532; γ 1.522–1.534 Biaxial $-$ve

- *Birefringence*: 0.006–0.008
- *Pleochroism*: Weak
- *Density*: 2.56–2.62
- *Hardness*: 6–6.5
- *Dispersion*: Weak
- *Cleavage/fracture*: Perfect to good, prismatic partings. Uneven to conchoidal fracture
- *Optical effects*: *chatoyancy, asterism, colour change, ADR* See moonstone – Australia.

Barium Feldspars

Despite low levels of barium being present in most K-feldspars the barium-rich feldspars, hyalophane and celsian, are rather rare and gems even rarer. The optical and physical properties vary considerably; generally increasing linearly with barium content.

Hyalophane – $(K,Ba)Al(Si,Al)_3O_8$

Name approved by the CNMMN of the IMA. Hyalophane gets it name from the Greek word *hyalos*, a glass, plus *phanes*, appearance, alluding to its transparent crystals. It is intermediate in the series orthoclase-celsian. Although gems are remarkably rare, some orthoclase may actually be hyalophane and the Ba content may be required to identify this species. It can also be distinguished from celsian by its optic sign.

Occurrence

Hyalophane commonly forms in magmatic environments or by contact metamorphism. Most commonly associated with manganese deposits, sometimes in phonolites. Also in hydrothermal Alpine type veins and some pegmatites. It has been reported in veins in pyroxene-amphibole gneisses, in metasedimentary barite deposits (Spain). Notable localities include Switzerland, Australia, Bosnia and Hercegovina (specimens from Alpine type veins at Zagradski Potok near Busovaca are numerous, huge (exceptionally well-developed crystals reach dimensions of over 20 cm and crystals in the 5–10 cm range are not unusual), lustrous and complexly twinned, commonly gemmy), Germany, Japan, Brazil and Slyudyanka, Transbaikal, Siberia, Russia. An exceptional 5-cm transparent crystal, from the ancient mine Trou des Romains, in Val Sapin, France, has been reported. At Franklin, New Jersey, USA, and in Sweden the hyalophane is red granular.

Appearance
Transparent to translucent. Massive or as prismatic crystals which may show various types of feldspar twinning (Carlsbad, Baveno, etc.)

- *Lustre*: Vitreous
- *Colour*: Colorless, yellow, white and red. A green transparent Ba-orthoclase has been reported from south of Felicio dos Santos, Minas Gerais, Brazil
- *Other*: Hyalophane probably exhibits the most complex twinning of any species.

Properties

- *Crystal system*: Monoclinic
- *Refractive index*: α 1.518–1.580; β 1.521–1.583; γ 1.524–1.586 Biaxial ($-$ve)
- *Birefringence*: 0.006–0.10
- *Density*: 2.55–3.1 $Celsian_{20}$ – $Celsian_{80}$
- *Hardness*: 6–6.5
- *Cleavage/fracture*: Prismatic, perfect to good in two directions at 90°
- *Inclusions*: Zagradski Potok material may have epigenetic ferruginous staining
- *Fluorescence*: Some specimens have a weak red fluorescence in SWUV
- *Note*: May contain ammonium ions that can be detected by IR spectroscopy.

Celsian – $BaAl_2Si_2O_8$

Name approved by the CNMMN of the IMA. Celsian is named after Anders Celsius (1701–1744), the Swedish chemist and astronomer. Gems are exceedingly rare.

Occurrence

Typically found in medium-grade contact or regionally metamorphosed rocks with significant barium particularly associated with manganese deposits. In barite-carbonate veins. Rare in igneous rocks. Localities include Broken Hill, N.S.W., Australia. Alaska range, Alaska, U.S.A. Otjosondu, Namibia. Benallt manganese mine, Rhiw, LLeyn Peninsula, Caernarvonshire, U.K. Jakobsberg, Sweden. Kaso mine, Totiki Pref. Japan and Zamora, Spain.

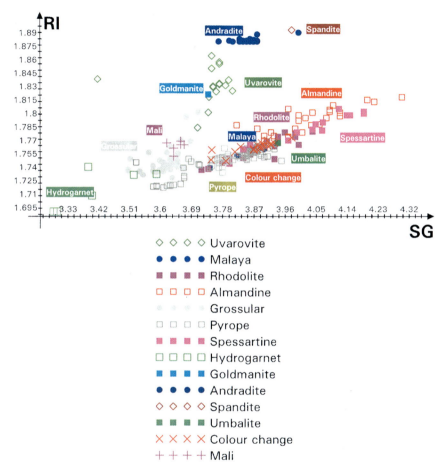

Plate 1
Circa 300 garnets of known composition and known RI and SG were plotted and best fit lines drawn to indicate trends within series (see Chapter 11)

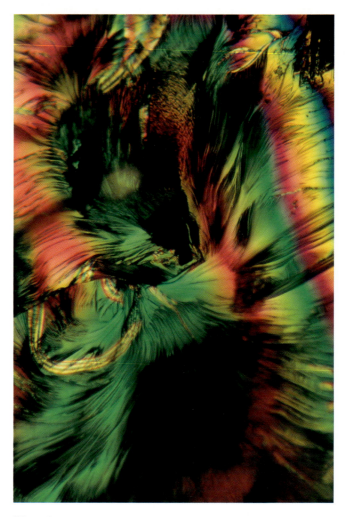

Plate 2
A green fluorite from Colombia, South America, enriched
along one octahedral plane by an iridescent feather-
textured *cleavage*. Oblique illumination. 85x (Reproduced
with kind permission from *Photoatlas of Inclusions in
Gemstones*, 1992, E. J. Gübelin and J. I. Koivula,
Gemological Institute of America).

Plate 3
Actinolite crystals in quartz

Plate 4
Zoning in synthetic amethyst

Plate 5
An unusual iolite necklace

Plate 6
Radiating *byssolite fibres*, a hair-fine fibrous asbestiform actinolite (amphibole), traverse the entire crystalline show-case of this demantoid garnet from the Ural Mountains, making further identification redundant. Darkfield illumination. 20x (Reproduced with kind permission from *Photoatlas of Inclusions in Gemstones*, 1992, E. J. Gübelin and J. I. Koivula, Gemological Institute of America).

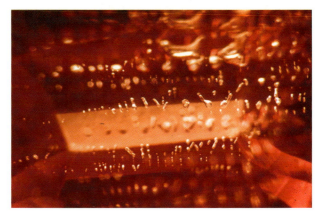

Plate 7
Flux in early Kashan synthetic ruby

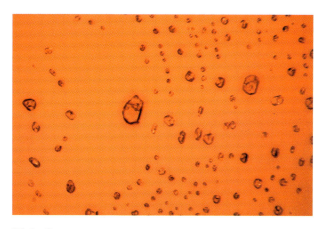

Plate 8
A plane of tiny *three-phase inclusions* in a brownish-orange Sri Lanka spinel. Fluid inclusions in spinels are very rare. Transmitted light. 100x (Reproduced with kind permission from *Photoatlas of Inclusions in Gemstones*, 1992, E. J. Gübelin and J. I. Koivula, Gemological Institute of America).

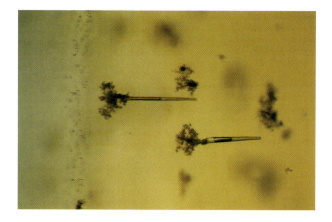

Plate 9
So-called *"bread-crumb" inclusions* – remnants of the seed plate – in this *synthetic*, hydrothermally grown *citrine quartz*, have blocked the crystal's growth away from the seed. The results are the *two-phase spicules* we see here. Such spicules and "bread-crumbs" can occur in any synthetic quartz regardless of colour. Transmitted light 70x (Reproduced with kind permission from *Photoatlas of Inclusions in Gemstones*, 1992, E. J. Gübelin and J. I. Koivula, Gemological Institute of America).

Plate 10
Synthetic moissanite

Plate 11
Synthetic moissanite

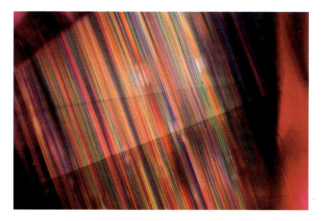

Plate 12
Layers of very fine straight and parallel *growth bands* are the only inclusions visible in this *Ramaura* flux grown synthetic ruby. As a result of the apparent distortion of these growth bands by the neighbouring facets, they appear slightly bent, as in "Verneuil ruby". In darkfield illumination it displays a bright array of vivid interference colours by its *growth layers* 40x (Reproduced with kind permission from *Photoatlas of Inclusions in Gemstones*, 1992, E. J. Gübelin and J. I. Koivula, Gemological Institute of America).

Plate 13
Stat quartz-star shown by transmitted light

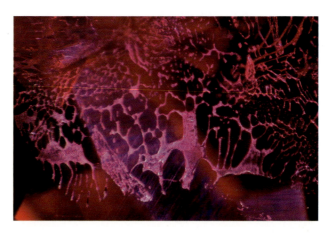

Plate 14
This photomicrograph is an attempt to advertise the exemplary *net-like patterns* of the *synthetic "Chatham rubies"* manufactured by the flux fusion process. It bears a misleading similarity to the characteristic healing halos and fissures in Thai ruby, yet differs somewhat from the flux remnants in other synthetic rubies. Darkfield 50x (Reproduced with kind permission from *Photoatlas of Inclusions in Gemstones*, 1992, E. J. Gübelin and J. I. Koivula, Gemological Institute of America).

Plate 15
Two natural blue sapphires

Plate 16
Russian synthetic emerald showing 2-phase
inclusions and platelets

Plate 17
Double refraction in zircon

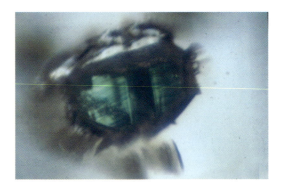

Plate 18
Chrome diopside in diamond

Plate 19
Heat-treated purple Montana
sapphire

Plate 20
Heat-treated yellow Montana
sapphire

Plate 21
'Horsetails' in demantoid garnet

Plate 22
Brazilianite

Plate 23
Benitoite

Plate 24
Andalusite

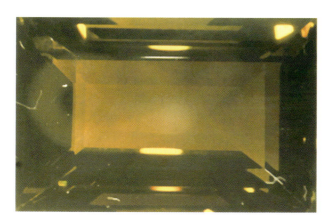

Plate 25
Enstatite

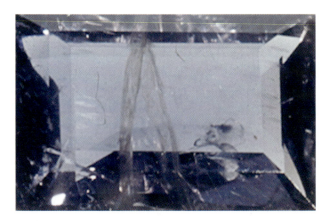

Plate 26
Iolite

Plate 27
Tremolite in emerald

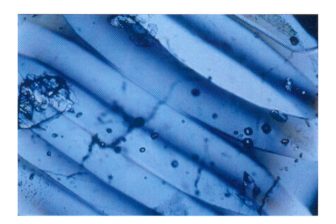

Plate 28
Twin planes in natural blue sapphire

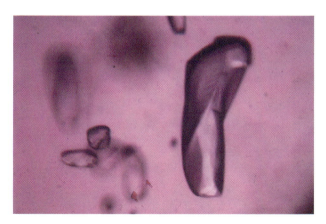

Plate 29
Crystals in Myanmar ruby

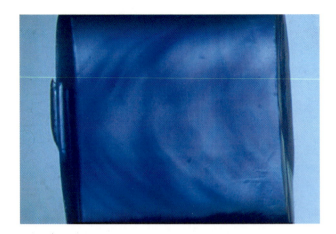

Plate 30
Kyocera sapphire

Plate 31
Stress fractures in moonstone

Appearance
Transparent to translucent. Forms short prismatic crystals; also massive. Twinning is very common.

- *Lustre*: Vitreous
- *Colour*: Colourless, white, pale yellow and yellow.

Properties

- *Crystal system*: Monoclinic
- *Refractive index*: α 1.580–1.589; β 1.583–1.593; γ 1.586–1.599 Biaxial +ve
- *Birefringence*: 0.009–0.011
- *Density*: 3.10–3.39
- *Hardness*: 6–6.5
- *Cleavage/fracture*: In two prismatic directions ranging from perfect to good. Fracture is uneven, very brittle producing uneven fragments
- *Fluorescence*: Non-fluorescent
- *Simulant*: The colour of some transparent iridescent yellowish Czech glass beads has been called 'celsian'.

Plagioclase Feldspars

The various plagioclase feldspars are identified from each other by gradations in RI and density in the absence of chemical analysis and/or optical features.

Albite – $NaAlSi_3O_8$

Name approved by the CNMMN of the IMA. Named from the Latin word *albus*, white, alluding to its usual colour.

Occurrence

Found widely in greenschist facies metamorphic rocks. In igneous rocks, it occurs principally in pegmatites and in alkali-rich rocks such as nepheline syenites and other feldspathoidal types. Common in spilites and keratophyres. Albite occurs throughout the world, with major locations in the Alps, Urals, Harz Mountains, France, Norway, Madagascar and the USA (Maine, Virginia, Colorado). Gems are normally <3 cts.

Appearance
Opaque to translucent, rarely transparent. Simple and polysynthetic twinning are ubiquitous, the latter producing fine striations on surfaces.

- *Lustre*: Vitreous to pearly
- *Colour*: Usually white, rarely colourless or pale shades of blue, green, pink–orange, reddish or brown. The colour of some bluish green albite-oligoclase material is attributed to Pb and OH^- involving radiation-induced transitions
- *Varieties*: *(see also moonstone)* Maw-sit-sit is a rock mainly composed of albite with lesser amounts of Cr-pyroxenes and Cr-amphiboles: RI $\approx$ 1.54, SG $\approx$ 2.7. It is named after the locality of origin in Myanmar. **Peristerite** is the name given to iridescent albite-oligoclase lamellar intergrowths (see also peristerescence and moonstone). The most beautiful peristerite is said to be found at Monteagle, Hastings County, Ontario, and this material has been fashioned into beads and cabochons. An ivory-white peristerite is found in some quantity in the Villeneuve mine, Labell County, and at Buckingham, Quebec. Also found at various other localities in Ontario. Platy albite is called cleavlandite.

Properties

- *Crystal system*: Triclinic
- *Refractive index*: α 1.528–1.533; β 1.532–1.537; γ 1.538–1.542 (Madagascar α 1.530–1.531; β 1.532–1.533; γ 1.539–1.540)
- *Birefringence*: Weak; 0.009–0.010
- *Density*: 2.61–2.63 (Madagascar 2.62)
- *Hardness*: 6
- *Dispersion*: Weak
- *Cleavage/fracture*: Perfect to excellent prismatic cleavages in two directions/brittle with a conchoidal to uneven fracture
- *Fluorescence*: SWUV – white, green, LWUV – green (peristerite – very weak brown fluorescence under both LWUV and SWUV)
- *Phosphoresence*: Golden brown, green
- *Inclusions*: (see moonstone)
- *Optical effects*: *chatoyancy, asterism, colour change, ADR* (see moonstone) Cat's eye gems around 50 cts are known
- *Treatment*: Irradiation may turn colourless material blue if lead and water impurities are present (see amazonite).

Oligoclase – Na(90–70%) Ca(10–30%) (Al,Si)AlSi$_2$O$_8$

Name not approved by the CNMMN of the IMA. The name is derived from the Greek words *oligos* meaning small and *klan* to break, because it was thought to have less perfect cleavage than albite. *Synonyms*: Peristerite, moonstone.

Occurrence

As a major part of many rocks, especially igneous rocks such as granites, syenites and pegmatites. Also in intermediate lavas, schists and some amphibolites. Some major localities are: Sri Lanka; USA – Oregon sunstone (localities in New York, Maine, New Mexico, North Carolina – pale faceted stones from Hawk mica mine, Bakersville {circa 1887} RI 1.537–1.547; SG 2.651, Pennsylvania and Virginia are relatively unimportant); Brazil; Kenya (colourless with a blue or yellow tinge – RI range α 1.535–1.540, β 1.539–1.544, γ 1.544–1.550, DR 0.010–0.011; SG 2.63–2.64, Russia (plagioclases of Dlinnolambinskoye pegmatitic field, Kareliya); Sweden, Norway and Canada (Baffin Island gems up to 5 cts). Gems, other than moonstone and sunstone (*sensu stricto*), are uncommon but can be large.

Appearance

- *Lustre*: Vitreous to dull
- *Colour*: Is usually off-white or grey or pale shades of green, blue, yellow or brown
- *Varieties*: *Sunstone* is the name given to oligoclase with reflective hematite (or goethite) inclusions (aventurine-feldspar) (see sunstone section), *peristerite* is the name given to iridescent albite-oligoclase lamellar intergrowths (see also peristerescence and moonstone), the term 'moonstone' has been applied to both peristerite and alkali feldspars with perthitic intergrowths that exhibit sheen (see moonstone)
- *Other*: Lamellar twinning is ubiquitous and may cause a grooved effect on crystal and cleavage surfaces that appear as striations.

Properties

- *Crystal system*: Triclinic
- *Refractive index*: α 1.533–1.543; β 1.537–1.559; γ 1.542–1.552 Biaxial −ve
- *Birefringence*: 0.009
- *Density*: 2.64–2.66
- *Hardness*: 6–6.5
- *Dispersion*: Weak
- *Cleavage/fracture*: Perfect in one and good in another direction forming nearly right-angled prisms/conchoidal fracture
- *Absorption spectra*
- *Fluorescence*: LWUV – pale brownish white, pale brown. Kenyan material fluoresces a faint white under LWUV and is inert under SWUV
- *Inclusions*: Hematite, goethite (see sunstone section)
- *Optical effects*: *chatoyancy, asterism, colour change, ADR etc.* Sunstone effect (see sunstone section)

- *Treatment*: Irradiation may turn colourless material blue if lead and water impurities are present (see amazonite). Oligoclase has also been reported to turn orange and green upon irradiation.

Andesine – Na(70–50%) Ca(30–50%) (Al,Si)AlSi$_2$O$_8$

Name not approved by the CNMMN of the IMA. Andesine is named after its discovery locality, in lavas from Marmato, Andes Mts., Bolivia.

Synonyms: Congo sunstone.

Occurrence

Andesine is only a minor constituent in most granites and syenites but is the dominant feldspar in a wide range of igneous rocks of basic to intermediate and certain igneous rocks called appropriately andesites. It is also found in some metamorphic rocks as a minor constituent.

Localities include Greenland, Andes Mountains (hence the name Andesine) and Norway. Large crystals are found in the San Gabriel Mountains of California and pale yellow to colourless andesine (5 cm+) is found at deposits in north-eastern Idaho, China (red and greenstones – occasionally referred to as 'sunstone') and Democratic Republic of Congo (red stones up to 30 cts), Japan, India, South Africa, Argentina, France, Italy and Germany.

The properties of weakly pleochroic (orange-red to red), andesine from the Democratic Republic of Congo lie close to the andesine–labradorite boundary. Red feldspars with labradorite composition and properties have been described from this locality (see labradorite and sunstone). The inference is that this material has a wide range of properties and chemical composition that span the andesine–labradorite boundary.

Appearance
Crystals are translucent to transparent, polysynthetic twinning is ubiquitous.

- *Colour*: Mostly white or grey; can be red, orange-red, greenish or yellowish
- *Lustre*: Vitreous to dull
- *Varieties*: Lavanite (an andesine feldspar similar to sunstone, but without the schiller inclusions).

Properties

- *Crystal System*: Triclinic
- *Refractive index*: α 1.543–1.554; β 1.547–1.559; γ 1.552–1.562; α 1.551–γ 1.560 Democratic Republic of Congo Biaxial (+ve/−ve)

- *Density*: 2.66–2.68: 2.67 (Congo)
- *Hardness*: 6–6.5
- *Dispersion*: Weak
- *Cleavage/fracture*: Is perfect in one and good in another direction forming nearly right-angled prisms/conchoidal to uneven fracture – brittle
- *Colour varieties*: Red andesine, up to 33.42 cts (very rarely green or yellow) from Democratic Republic of Congo. Irregular colour distribution has been noted
- *Inclusions*: Lath-like groups of minute reflective inclusions (presumed to be copper by analogy to Oregon sunstone) producing schiller observed in red andesine from the Democratic Republic of Congo. Twin planes are also reported
- *Fluorescence*: SWUV – blue, pink, yellow, yellowish brown. Weak red emission with an even weaker blue surface related luminescence (SWUV) has been reported for red andesine from the Democratic Republic of Congo which under LWUV fluoresced a weak to medium orange
- *Absorption spectra*: Red andesine from the Democratic Republic of Congo showed increasing absorption across the visible spectrum from red to violet on which was superimposed a moderately broad band with a max at ≈565 nm attributed to minute, non light scattering, copper particles.

Optical effects: Chatoyancy, Asterism, Colour Change, ADR

Red and green andesine from the Democratic Republic of Congo has been reported as showing an 'Alexandrite effect', showing alternating colours depending on the light source. Green material from China ($\approx An_{44}Ab_{54}Or_2$ on the basis of analyses on a single specimen) shows a similar effect, turning red and losing transparency under incandescent incident light. The loss of transparency appears to be associated with light scattering and the scattering may also be responsible for the colour change. No copper was detected in the specimens analysed by SEM-EDAX. There has been speculation as to where the red/green material actually comes from; one or other or both localities.

Labradorite – Na(30–50%) Ca(70–50%) (Al,Si)AlSi$_2$O$_8$

Name not approved by the CNMMN of the IMA. Labradorite is recorded as having been discovered by a Moravian missionary in 1770 on the Isle of Paul, Nain, Labrador, Canada. And is named for the locality.

Synonyms: Sunstone, heliolite, or plushstone; labrador moonstone – synonym of black moonstone.

Occurrence

A common constituent of anorthosites, norites, basalts and gabbros as well as other igneous rocks. In the metamorphic environment it occurs in gneisses derived from basic rocks. It is found in Canada (Tabor Island, Nain area of Labrador, Newfoundland Province – extensive rock-like masses), Madagascar (striped labradorite from Bekily, the bands, due to twinning showed exsolution labradorescence). Blue 'flashing' labradorite from India is reported to show a similar effect as the light source, relative to the stone, is moved. Blue labradorescent twinning bands become dark and vice versa. Found in Tanzania; Mexico, Russia, Brazil and widely distributed throughout the USA (in addition to the commercial Oregon deposits of faceting-grade labradorite (see sunstone). Gem-quality material has been recovered from localities in Arizona, California – Mode County (sunstone), New Mexico, Nevada and Utah – Sunstone Knoll, Millard County – yellow transparent; *RI* 1.565–1.572; *SG* 2.68 {similar material comes from Mexico and Australia, Hogarth Range, New South Wales, and near Springsure, Queensland – pale yellow transparent *RI* 1.556–1.564; *SG* 2.695}). Anorthosite rocks host iridescent labradorite at the very large Golovinskoe deposit in the Volyn district which part of the western Ukraine shield and at Dzhugdzhurskoe in the east of the Aldan shield, Russia.

The normal rock-like labradorite is also found at other localities in Newfoundland, along the shore of Lake Huron, at Cape Mahul, at Abercrombie and at Morin in Quebec Canada. In Russia it occurs in the Ukraine especially at Gorodishch in the Zhitomir district, and in the Ural Mountains. In the USA small quantities occur in Arkansas, New Mexico and Vermont.

Appearance

Transparent to translucent. Simple and polysynthetic twinning are ubiquitous. The latter may cause a grooved effect on crystal and cleavage surfaces that appear as striations. Usually massive.

- *Lustre*: Vitreous
- *Colour*: Usually colourless, white, greyish, pale yellow, bluish grey or greenish, with greyish massive material often showing a distinctive play of colours ('labradorescence') normally blue and green but can be yellow, golden, red and purple.

Varieties

Sunstone
See sunstone section.

'Sunspar'
(*RI* 1.560–1.568) with a chemical composition between andesine and bytownite exhibits hues ranging from colourless to light yellow on the smaller stones to champagne or straw yellow on the largest.

The variety known as 'rainbow moonstone' exhibits multi-coloured sheen. Semi-transparent blue and multi-coloured sheen moonstone from Patna, Bihar, India (RI 1.56; SG 2.69) marketed as 'Blue Rainbow' moonstone and 'Rainbow' moonstone may be either labradorite or bytownite. The finest stones have a reddish orange or sometimes a lavender sheen with areas of green and blue. The colour effects shown by rainbow moonstone from India arise due to diffraction from intergrowths of repeated twin lamellae and not from exsolution. They appear white to almost colourless because they lack the ilmenite inclusions that give most labradorite a dark greyish body.

Black moonstone
Name given to colourless labradorite (anorthosite) which exhibits only sporadic bluish labradorescence and is darkened by needle-like inclusions, that give some degree of chatoyancy when appropriately cut. However, the name has also been applied to material from Myanmar (formerly known as Burma) that shows a similar effect to moonstone but with more 'play of colour' (SG 2.69).

Bull's eye
Name sometimes applied to relatively dark labradorite (anorthosite).

Lynx eye
Labradorite with a predominantly green iridescence.

Opaline feldspar
Name sometimes given labradorite from anorthosites.

Ox-eye
Name sometimes applied to labradorite (from anorthosites) that exhibits dark reddish hues.

Spectrolite
Name originally applied to labradorite from Ylämaa in the south-east of Finland. A feature of Finnish spectrolite, compared with labradorite from other sources, is, subjectively, its greater range of colours, richness of colours and the number of colours seen at any one time. As a result all further sources of comparable material, regardless of locality, have also

been termed 'spectrolite'. Its composition is $\approx An_{55}$, RI 1.56–1.57; DR 0.008; SG 2.69–2.70; hardness 6–6.5. Some labradorite of spectrolite quality has also been found in Madagascar.

Properties

- *Crystal system*: Triclinic
- *Refractive index*: α 1.554–1.563; β 1.559–1.568; γ 1.562–1.573 Biaxial +ve/−ve
- *Birefringence*: 0.008–0.01
- *Pleochroism*: From the Congo, red stones show weak pleochroism whilst greenstones have distinct greenish yellow to bluish green pleochroism
- *Density*: 2.68–2.71
- *Hardness*: 6–6.5
- *Dispersion*: Low (0.012), but the combination of a high polish and minute inclusions can simulate the effect of moderate dispersion
- *Cleavage/fracture*: Excellent {001}, very good {010} and good {110} cleavages/brittle with an uneven fracture
- *Colour varieties*: Red and green transparent labradorite from the Democratic Republic of Congo (see also andesine) have chemistry (An_{52}^{red}–An_{55}^{green}) and properties close to the labradorite–andesine boundary (An_{50}). RI α 1.553–1.555; RI γ 1.562–1.563; SG 2.68–2.70. Congo labradorites have outstanding clarity, are of a deep reddish orange colour, and more deeply colour saturated than normally seen in the sunstone from Oregon which they strongly resemble. The cause of colour, largely due to copper, is complex and the intensity of colour is attributed to relatively high copper levels compared with that of Oregon material: Congo labradorites contain about twenty times more copper (see Oregon sunstone). The differences in colour are attributed to tiny copper colloids of different size (see Oregon sunstone)
- *Inclusions*: Exsolved black needle-like crystals of ilmenite are common, particularly in material that exhibits labradorescence where they, along with platy magnetite (which sometimes produce iridescent colours), impart a grey colour. When seen in association with rutile, are locality-typical for Malagassy labradorite moonstone. Transparent, blue flash, labradorite from Madagascar containing needle-like crystals similar to that in 'black moonstone' has been reported. Zircon inclusions have also been reported in material from Madagascar. Platy hematite crystals, often exhibiting interference colours, are not uncommon. Tiny disc-like epigenetic exsolution crystals of native copper are common in sunstone from Oregon.

From the Congo, red stones are relatively clean with minor hollow channels parallel to the twin lamellae and step-like healing fissures along cleavages. Tiny inclusions impart a milky turbidity. Greenstones, under high magnification, were seen to contain parallel fluid channels and trails. Distinct lamellar zoning has been observed. Stones are described as rather clean with only some twin lamellae, hollow channels and fine undifferentiated inclusions

- *Optical effects (chatoyancy, asterism, colour change, ADR)*: Labradorescence – play of colour produced by light interference from lamellar intergrowths. Labradorite is a grey feldspar showing signature labradorescence (a type of iridescent schiller). Most common colours of labradorescence are blue and green. When many spectral colours are seen simultaneously the term 'spectrolite' has been coined.

  In white incident light a milky turbidity with a reddish hue has been reported for greenstones from China. Colour change from green (daylight) to red in unspecified African material has been reported (rare): this should be considered alongside the 'colour change' reported for Chinese andesine and the milky turbidity with a reddish hue that has been reported for Congo material. No aventuresence has been reported in Congo material

- *Absorption spectra (400–700 nm)*: From the Congo, visible spectra of red stones are characterized by a general increase in absorption towards the violet (attributed to copper) with distinct bands at 565 (attributed to Cu^0 particles) and 380 nm (due to Fe^{3+}). Visible spectra of greenstones vary with pleochroic directions: the yellowish green direction showing a weak band at 430 nm (due to Fe^{3+}) followed by strong absorption towards the violet (attributed to copper) and broad absorption towards the red with a minimum at 620 nm. The bluish green direction shows no distinct minimum at 620 nm and a broad absorption centred at 680 nm

- *Fluorescence*: Rainbow moonstone shows an even, chalky, moderately blue fluorescence under LWUV. Under SWUV the intensity is diminished and fluorescence is a weak chalky pinkish orange. Under LWUV red stones (Congo) fluoresced weak to distinct orange and greenstones (Congo) appeared distinctly orange.

Bytownite – $Ca(70–90\%)$ $Na(30–10\%)$ $(Al,Si)AlSi_2O_8$

Name not approved by the CNMMN of the IMA. Named after its discovery locality, in Bytown (named after Lt. Col. John By, who founded the village of Bytown in 1826), the original name for Ottawa, Ontario, Canada. *Synonyms*: Golden labradorite; sodium anorthite.

Occurrence

Bytownite (70–90% anorthite) is the least common plagioclase. It is usually found as compact masses and grains in gabbros, norites, allivalites and anorthosites. It also forms from sodic metasomatism of metamorphic rocks. Often it occurs as part of compositionally zoned feldspars. Notable localities include Canada (Ottawa); USA (small to medium-sized pieces of light brown bytownite from Arizona and New Mexico can be faceted into 0.5–2 ct, eye-clean stones; RI $\approx$ 1.57; SG $\approx$ 2.73) (Oregon); Mexico (large stones (yellow) 50 + cts; rough 100 + cts) (Casas Grande, Chihuahua; Santiago de Papasquiaro, Durango); India (Trichy) and South Africa. It is important to note that compositional variation may occur within any given calcium rich plagioclase provance such that intermediates may range from andesine through labradorite to bytownite. Some material previously reported as bytownite has been shown to be labradorite and vice versa. (Pueblo Park, New Mexico bytownite was reputedly analysed by EPMA and found to be on the labradorite side of the labradorite–bytownite boundary.)

Appearance

Crystals are translucent to opaque and only sometimes transparent. Polysynthetic and simple twinning ubiquitous. Lamellar twinning may cause a grooved effect on crystal and cleavage surfaces that appear as striations.

- *Colour*: Usually white, grey or colourless but can be pale shades of other colours or light pastel yellow
- *Lustre*: Vitreous to dull.

Properties

- *Crystal system*: Triclinic
- *Refractive index*: α 1.563–1.572; β 1.568–1.578; γ 1.573–1.583 Biaxial (+ve/−ve)
- *Birefringence*: 0.01–0.011
- *Density*: 2.7–2.72
- *Hardness*: 6–6.5
- *Dispersion*: Low (0.012)
- *Cleavage/fracture*: Prismatic; perfect in one and good to excellent in another direction forming nearly right-angled prisms/brittle with uneven to conchoidal fracture
- *Inclusions*: Schiller-like silk.

Anorthite – $CaAl_2Si_2O_8$

Name approved by the CNMMN of the IMA. The name is derived from the Greek word *an* meaning not (without), and *orthos* right; alluding to its oblique crystals.

Occurrence

Anorthite is usually found in contact metamorphic limestones and as a constituent in some basic calc-alkali plutonic rocks and lavas. Notable localities include USA (Alaska; California; Franklin, New Jersey; Nevada); Italy (Trentino, the lavas of Vesuvius and Monte Somma); Finland; Sweden (Södermanland); India (Tamil Nadu); and Japan (Miyake – very rare, red transparent crystals have been facetted as gems). Gems are rare and normally only cut for collectors. The Smithsonian has a colourless 1.58 ct anorthite from Nevada.

Appearance

Crystals are translucent to opaque and only sometimes transparent. Lamellar twinning, which is very common, may cause a grooved effect on crystal and cleavage surfaces that appear as striations.

- *Colour*: Usually white, grey or colourless but can be pale shades of other colours. Occassionally reddish
- *Lustre*: Vitreous to dull
- *Varieties*: Anorthite-pargasite-ruby rock; colourless anorthite matrix 93–96%An – locality uncertain – Myanmar or Longido, Africa. Cut as cabochons. An atypical anorthosite, also used as a gemrock, comes from a locality recorded only as the Philippines (Johnson and Koivula, 1996). This rock is described as granular, overall white and containing sporadic green crystals up to 2.0 mm across with its granular material consisting largely of nearly pure anorthite plus minor interstitial oligoclase and crystals of zoned garnet, 'primarily uvarovite with varying amounts of andradite in solution'. It is said to roughly resemble jade.

Properties

- *Crystal system*: Triclinic
- *Refractive index*: α 1.572–1.576; β 1.578–1.583; γ 1.583–1.588 Biaxial −ve
- *Birefringence*: 0.011–0.012
- *Density*: 2.72–2.75
- *Hardness*: 6–6.5
- *Dispersion*: Weak
- *Cleavage/fracture*: Perfect in one and good in another direction forming nearly right-angled prisms/brittle – conchoidal to uneven fracture
- *Fluorescence*: LWUV – cream, pale brown.

Moonstone

The variety name moonstone is used to describe an optical effect and unlike most variety names it is not confined to a single species. Moonstone gets its name from its appearance which is likened to the reflected shine produced by moonlight. It is normally characterized by a milky white to bluish shimmering sheen effect that seems to float across the convex surface of the stone as it is turned and moved: the more intense the colour and the larger and more transparent the stone, the more valuable the gem. Top quality fine blue moonstones show an incredible 'three dimensional' depth of colour. This effect has been variously termed 'adularescence' or 'peristerescence' depending upon the dominant species. The term 'schiller' has been used to describe both of these. The term 'adularescence' is derived from the variety of orthoclase feldspar, known as adularia. Peristerescence is derived from peristerite, a variety of plagioclase feldspar of approximate composition $An_1–An_{17}$ (see explanation of Figure 3 – *Peristerescence*). Thus moonstones, which are usually cut in a smooth-domed cabochon shape to maximize the effect, can be either K-feldspar moonstones (normally adularia) or plagioclase moonstones (normally peristerite): each having RI and SG commensurate with the dominant species, normally orthoclase or albite/oligoclase. The K-feldspar moonstones have lower physical constants with RI around 1.518–1.526 and SG between 2.56 and 2.59 whilst the physical constants of peristerite moonstones have been reported around 1.531–1.539 for RI and 2.63 for SG with DR 0.008.

The Moonstone Effect

This effect is generally attributed to the interference of light reflecting from submicroscopic lamellar structures (comprising two types of feldspar)

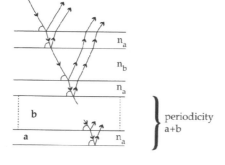

Figure 12.6 The layered feldspars are labelled 'a' and 'b'. Their respective RIs are n_a and n_b. The inteference phenomena occurs because the lamellar intergrowths are on the same cryptoperthitic (Submicroscopic) scale as wavelengths of visible light. The interference colour seen is dependant on the wavelength of the periodicity and is a function of the viewing angle.

producing iridescence (see also Isomorphous replacement, Solid solution and exsolution in feldspars).

Interference and Iridescence

When a ray of white light falls upon a transparent medium, some of the light is reflected, whilst the remainder is refracted within the optically denser medium. When the refracted ray meets another surface, such as exsolved lamellae, some is reflected back from the surface and some is refracted, penetrating deeper into the medium where, at the next interface, the process is repeated.

The speed of light is retarded in the optically denser medium. Each constituent colour of the white light is dispersed such that each colour follows a different path and is retarded to a different degree. These will be reflected from the various interfaces. As they exit the medium similar colours will try and recombine but because they followed different paths and experienced different retardation they may exhibit phase variance. Where there is no phase variance the rays reinforce each other (constructive interference). Where there is phase variance the rays will have a cancelling effect on each other (destructive interference). Thus the emerging combined rays may not exit as white light but exhibit colour. The colour effects caused by this phenomenon are termed 'iridescence'.

Other Factors

Light scattered from exceptionally low concentrations of very fine colloidal particles in a transparent medium produces a milky white to bluish body colour that is termed *opalescence* (not to be confused with the play of colour in opal) such as seen in unhomogenized milk or in some moonstone imitations made by heat-treating synthetic spinel (presumably to precipitate a second phase). This term has also been applied to moonstone to describe the milky or pearly appearance. It has been suggested therefore that the white to bluish moonstone effect is caused by light scattering from very fine particles. In support of this it has been reported that in feldspars, where exsolved areas are irregular and small compared to the wavelength of light, white to bluish colours may be produced by incoherent light scattering such as the Tyndall effect. Further research has shown that very fine exsolution lamellae can form along planes other than that normally associated with exsolution in peristerite. One such plane has a periodicity that has been shown to correlate with iridescence.

It is therefore possible that more than one phenomenon contributes to the moonstone effect.

Occurrence

The most prominent source is in the gravels and pegmatites of the Dumbara district, Sri Lanka. Many gem moonstones come from India, Brazil and Madagascar. Other localities include Switzerland, Tanzania, Pennsylvania, New Mexico Colorado, Indiana, New York, North Carolina, Pennsylvania, Wisconsin and Virginia in US, and Myanmar, Canada, Australia and Russia.

Colour

Moonstones come in a variety of colours. The body colour can range from colourless to grey, brown, yellow, pale orange or peach, green, plum-blue, reddish or pink. Inclusions can effect the overall body colour; goethite inclusions, for example, imparting red colouration. The clarity ranges from transparent to translucent. The best moonstone has a blue sheen, perfect clarity and a colourless body colour. This material has become increasingly rare and the term 'blue-sheen' has been stretched to include gems of lesser quality. Consequently the term 'Royal Blue Moonstone' has been introduced in an attempt to preserve the quality differential.

The K-feldspar moonstone has no characteristic absorption spectrum but may show weak bluish fluorescence (LWUV) and weak orange (SWUV).

Properties

Inclusions

Commonly, especially in Sri Lankan material; abundant short parallel stress-related elliptical incipient cleavages at right angles to a 'spine'. Stress-related lath-like cracks, often in pairs, with numerous branching cracks that extend for a short distance and taper off obliquely such that they resemble a multi-legged insect and consequently often termed 'centipede inclusions' or occasionally 'Chinese aeroplanes' Negative rectangular inclusions that are thought to possibly be stress-related have been reported. Some Myanmar stones may show orientated needle-like inclusions which when abundant may well produce a cat's eye effect. Rare bamboo-like inclusions in K-feldspar moonstone from Sri Lanka arise from sub-parallel tube-like inclusions (stalks) with intersecting cleavage cracks (nodes).

Varieties

- Rainbow moonstone (see labradorite)
  An iridescent labradorite from Madagascar is being marketed as rainbow moonstone. The material has a slightly grey body colour and dark inclusions but is otherwise similar to the rainbow moonstone from southern India
- Norwegian moonstone (see labradorite)
- Black moonstone (see labradorite).

Optical Effects: Chatoyancy, Asterism, Colour Change, ADR

Sometimes moonstone will have a cat's eye as well as a sheen and occasionally a four-rayed star may be present. Brownish grey moonstone from Sri Lanka with RI 1.53 and SG 2.6, SWUV red fluorescence, slightly translucent with silvery schiller shows a combination of both Cat's eye and six-rayed star, the latter attributed to micro-inclusions of hematite or ilmenite. Chatoyant moonstone with a typical darker orange body colour is reported from a deposit near Chennai, India.

Simulants

'Celestial moonstone' simulates natural cat's eye moonstone. It is constructed by applying a backing to natural moonstone. This treatment produces moderate colour in the stone as well as a coloured cat's eye effect. A range of colours has been produced. Untreated material only exhibits a white band regardless of body colour. Moonstone can be confused with opalescent glass imitations, heat-treated flame fusion synthetic spinel (excess aluminium gives an RI of 1.727), and both milky and bluish white chalcedony. California moonstone is a name given to chalcedony simulating moonstone. Quartz, both blue quartz, milky quartz and heat-treated amethyst, have also been so used. Some faceted thaumasite with a slightly satiny appearance and a colourless to slight white sheen can imitate moonstone. Some pink-violet scapolite can have a milky appearance and when cut cabochon can imitate moonstone. Also pink and violet scapolite produces a beautiful cat's eye effect which has sometimes been mislabelled as 'pink moonstone'.

The best simulant is white flame fusion synthetic spinel.

Additional Locality-specific Information

Moonstone – Sri Lanka

Sri Lanka is an important source of moonstones which occur in feldspar-rich dykes in the Central and Southern Provinces and also in abundance in the gems gravels of the latter. It has been reported that the principal area is the Dumbara district of the Central Province where it occurs as pebbles and irregular masses in the gem gravels and clay deposits, and is also obtained by quarrying an adularia leptite.

In Sri Lanka colours range from misty white to soft grey, pale orange, with a silver-white or blue sheen. Stones with a fine blue sheen are now rare. Meetiyagoda and Tissamaharama are notable localities. Smoky* moonstone, from a locality near Imbulpe, east of Ratnapura, is the name given to moonstones with grey body colour which is attributed to Fe^{2+}–Fe^{3+} intervalency charge transfer. The dark body colour is said to enhance the blue sheen.

Metiyagoda in southern Sri Lanka was an important twentieth-century locality producing fine material. The pegmatite is potassium-dominant (approx K/Na ratio 2:1); however, the correlation of colour and composition revealed that white moonstone is K-rich whilst the more desirable blue is sodium-dominant i.e. more albite than orthoclase. Similar moonstone deposits have been reported from the areas around Balangoda and Kundasale.

Most Sri Lankan moonstones have SG ≈ 2.56–2.58 (generally nearer 2.56); RI usually 1.520–1.525; DR 0.005 indicative of K-feldspar.

Moonstone – Myanmar

Noteworthy among the many minerals of the Mogok Stone Tract are large alkali feldspar moonstone crystals (to >20 cm), typical of volcanic environments, but here hosted in coarse marble. The best material exhibits a strong blue sheen over a colourless body.

Moonstone – USA

New Mexico

At various localities in Catron County and Grant County, New Mexico. High-temperature pegmatites in porphyritic rhyolite injected into Tertiary rhyolite tuffs at Faywood, Catron County, contain sanidine that exhibits

---

* As most Sri Lankan moonstones are colourless or white this new name was introduced (1994).

a moonstone effect. The material has a high degree of transparency and displays 'cool' blue to silvery white colours. Single moonstone crystals at Rabb Canyon, Grant County, have been reported as large as (33 cm × 33 cm × 51 cm); these too are found in high-temperature pegmatites. Smoky moonstone ($Or_{49}Ab_{48}An_3$) has been reported from Black Range, New Mexico, USA.

The quality of the adularia moonstone from Catron County is comparable to the best material from India and Sri Lanka. The best of the New Mexico material is water clear, nearly colourless, with only a hint of a soft-tan or brown tone, and has a wonderful blue or silver sheen. Unfortunately, only about 10% of the moonstone has the more desirable blue adularescence. The deposit has produced eye-clean, blue sheen, faceted stones as large as 5 carats, silver sheen faceted stones of almost 10 carats, and good-grade cabochons of 15–20 carats.

Virginia

Orthoclase-dominant moonstone from Allen's Mica Mine at Amelia Courthouse, Virginia, is comparable to that of Sri Lankan material: RI 1.581–1.524; DR 0.006.

Moonstone – India

Characteristically reddish brown, but can be white or plum-blue, moonstone occurs in the Coimbatore district of Tamil Nadu. Material from the Kangayam district in southern India includes both cat's eye and star varieties. SG ≈ 2.58−2.59. Moonstone from the Karur–Kangayam belt shows various hues including pearly white, grey, bamboo-green, light brownish pink and black. The gem varieties are characterized by a milky white sheen that moves across the crystal when rotated against the light, and is identical to the chatoyancy observed in cat's eye chrysoberyls from southern Kerala. Green Indian moonstone has been reported. 'Royal blue' sheen high calcium moonstone (probably labradorite *cf* 'rainbow moonstone') is reported from Patna, Bihar. The best material is 1 ct or less.

Moonstone – Russia

Largely translucent or semi-transparent peristerite-oligoclase with an attractive light blue or yellowish light blue iridescence (known as 'belomorite') is found in pegmatites of the Dlinnolambinskoye field north Kareliya for example at Chupa on the western White Sea coast. It can occur as large blocks and fine material is transparent with a strong gentle blue schiller.

The Kyrgyz Republic (Kyrgyzstan) (Central Asia)

Iridescent feldspar is known from four deposits; total reserves and forecasted resources are estimated at 1,780.2 tons. At Ottuk – fine iridescent light blue to bright-blue material (crystals from $0.5 \times 0.5 \times 1$ cm to $10 \times 20 \times 25$ cm) occurs in a syenite-diorite dyke.

Moonstone – Australia

Moonstone has been found in pegmatite in the Olary Province and at Gumeracha, Lobethal, Kangaroo Island, Mount Painter and south of Williamstown, South Australia.

Adularescent moonstone from Moonstone Hill, north of Hughenden in North Queensland, is an anorthoclase cryptoperthite antiperthite that displays very fine albite-pericline 'tartan' twinning when examined between crossed polars; RI is reported as α 1.526; β 1.531; γ 1.534; DR 0.008. Inclusions included cleavages that intersected at 89°, iron oxide staining on surface-reaching cleavages, 'centipede' type patterns.

Moonstone – Canada

The pegmatites at Quadeville, Ontario, Canada, have produced superb colour play material with either a reddish or white background and a fine blue or golden colour play. This material is fairly easily collected, and while much of it is too fractured to cut, solid material produces striking cabs.

Moonstone – Africa

Tanzania

Found in pegmatites in the Kondoa District, Mkoyo, Zoissa, Mahenge and Morogoro. Fine material is translucent to sub-transparent with a brilliant blue sheen. Also as pale grey with lesser sheen and translucency.

Some Tanzanian moonstones (alkali feldspar) show weak asterism: the stones exhibit cat's eyes with a weaker ray perpendicular to the eye producing a four-ray star in some specimens. It has been called 'sunstar'.

High-quality albitic (RI 1.530–1.540; SG 2.62–2.64) moonstone exhibiting weak red fluorescence (SWUV) in pieces up to 300 ct and good transparency has been reported from the Morogoro region, Tanzania. Needle-like inclusions and twin planes can be seen particulary in cabochon material. It has, unconventionally, been used as a doublet top, altering the appearance of the pavilion colour and adding schiller. Previously the region only produced low-grade material.

Madagascar

A labradorite moonstone with blue sheen has been reported. RI α 1.550–1.553; γ 1.560–1.561; DR 0.008–0.010; SG 2.70.

Sunstone

Nomenclature

The term was originally coined for near colourless oligoclase feldspar in which aligned red or orange platy hematite (and/or goethite) inclusion (platelets may be so numerous that they also influence the material's body colour) that produced a specular reflectance or spangle effect that was named aventurescence (a sort of schiller but more scintillating). Light interference on hematite or goethite plates can also result in green or blue colours.

Over time, aventurescent plagioclase feldspars with compositions other than oligoclase and/or inclusions other than hematite were discovered. In each instance the name sunstone was also applied.

Thus by common usage the name sunstone has come to mean any aventurescent feldspar. Attempts to preserve the original definition of sunstone and adopt the term 'heliolite' for other aventurescent feldspars have, by *force majeure*, failed to gain acceptance.

The red copper-bearing, mainly labradorite (but bytownite, andesine and oligoclase have all been reported), plagioclase from Oregon that exhibits aventurescence/schiller was so dealt with; the name labradorite all but being ousted except when applied to the dark grey massive material. The sunstone name was further extended to incorporate material without aventurescence/schiller and even subsumed non-red material: presumably arising from the success of product branding. Similar stones from other localities worldwide have likewise been named sunstone. Given this scope for variation a range of properties, chemistry and features for sunstone exists.

Sunstone – Norway

The best-known locality is Tvedestrand, near Arendal, in south Norway, where masses of fine quality aventurescent sunstone occur embedded in a vein of quartz running through gneiss. The material is translucent oligoclase with an orange/brown background colour containing small exsolved aligned hematite and/or goethite crystals. Also at Bjordammen, Telemark, in a biotite-quartz-feldspar-pegmatite which was mined in the 1970–1990s for the 'Aventurine-Feldspar' and several other Norwegian localities including fine material from Hiterö.

Sunstone – Russia

Chatoyant grey-brown perthitic feldspar from Vishnevye Mts, Urals, Russia – RI 1.530–1.537, SG 2.62, with red fluorescence under SWUV – shows a 'sunstone' effect from three sets of crystallographically aligned hexagonal crystal platelets of hematite. The chatoyancy is attributed to 'fibrous' feldspar lamellar intergrowths. Sunstone, aventurescent oligoclase, is also found at Verkhne Udinsk on the Selenga River near Lake Baikal, in Siberia. Prior to the discovery of the vein source in 1831 sunstone was found in only one location, on Sattel Island in the White Sea.

Sunstone – Oregon, USA

The 1987 Oregon Legislature designated Oregon sunstone as the official state gemstone. Uncommon in its composition, clarity and colours, it is a large, brightly coloured transparent plagioclase feldspar gem that occurs in Lake and Harney Counties at the Plush area Dust Devil mine and Ponderosa mines respectively. The largest pieces are recovered from the soil and the underlying, primarily weathered, basaltic lava flows.

Compositionally, the Oregon sunstones are mixed crystals in the plagioclase solid solution series with a range of Na:Ca ratios. The most abundant type is labradorite (An^{70}) but bytownite, andesine and oligoclase have all been reported.

Oregon sunstone contains native copper inclusions that produce a variety of colours. Additionally these millions of tiny colloidal copper particles, when sufficiently large (>22 nm), reflect light with varying intensities resulting in a golden red play of colour known as schiller (a shining surface); akin to aventurescence (a spangling or glittering caused by light reflecting of small or minute particles respectively).

Commonly it is straw-yellow coloured, but it can also be pink, peach, red, salmon red-orange, red-green and blue-green. It also can be bicoloured and tricoloured in combinations of yellow, green and red arranged concentrically in the rough. A small percentage is di- and tri-chroic.

Copper content is variable and affects the colour: pale yellow stones have a copper content as low as 20 ppm, greenstones about 100 ppm and red stones up to 200 ppm. Colloidal size is also a factor affecting colour and intervalency charge transfer (Cu^0–Cu^+) and Cu^0 pairs may also play a part. The reported colour change may be due to the effect large copper colloids has when the stones are viewed in incident light.

Ponderosa sunstone can be found as incredibly saturated reds, with a lot of copper schiller. Dust Devil sunstones appear more in peach and green colours, with less schiller, and in a wider range of colours than Ponderosa material. Both localities produce gems with strong red-green pleochroism.

Plush material $RI^{average}$: α 1.564; β 1.569; γ 1.576. Biaxial +ve/−ve. SG: 2.725 average over a range of 2.712–2.736. Ponderosa mine $RI^{average}$ α 1.564; β 1.568; γ 1.573. SG 2.67–2.72. Biaxial +ve. SG 2.70. Oregon oligoclase $Ab_{86}An_{13}Or_1$ – RI α 1.535–1.536, β 1.538–1.539, γ 1.545–1.547.

Sunstone ('Golden' Sunstone) – Mexico

At the time of writing, the renowned gemmologist, Joel Arem, kindly provided information on gem-quality labradorite occurring as phenocrysts (up to 150 grams) in a black basaltic lava flow, in northern Mexico that is to be launched as 'Golden' sunstone with larger 20 ct+ stones being called 'Royal sunstone'. The names have been formally trademarked. The lava flow is remarkable in that perhaps as much as 25–35% by volume is feldspar. Surprisingly the labradorite is exceptionally consistent: the medium-dark straw yellow hue (attributed to traces of iron) exhibiting no colour variations whatsoever save that accorded as a consequence of the size of the stone. Likewise physical properties (not disclosed) have shown no variation. There are no associated phenomena such as schiller and all materials are totally transparent save for some occasional fine silk-like fibres or tubes. Only extremely rarely will stones show iridescent reflections from incipient cleavages. No fluorescence has been detected. Hardness testing put the material closer to 7 than 6. Another notable feature is the lustre which is exceptional for feldspar. This unusual reflectivity, combined with the relatively low refraction typical of feldspar, creates a phenomenon for which the name 'pseudo-dispersion' has been coined. This phenomenon is attributed to the 'capture' and reflection, of colours in the surrounding environment by a gemstone's facets. These colours are transmitted to the observer's eye and *appear* to be generated *by and within* the cut stone, thus resembling dispersion. The *average* size of crystals from this locality would yield cut gems between 3 and 10 ct, with stones over 20 ct available on a regular basis. The largest stone produced, an astonishing 258 cts, has been donated to the Smithsonian Institute, National Museum of Natural History. It is the world's largest plagioclase feldspar gem, by a factor of nearly five times.

Sunstone – Canada

Sunstone is found in several places near the contact of the Bancroft sodalite and nepheline syenite in Hastings County, Ontario, and there are other Canadian localities in Lanark, Renfrew and Haliburton Counties of Ontario. Brownish pink coloured sunstone is found in a pegmatite dyke east of the French river on the north-east side of Lake Huron.

Sunstone – Congo

Plagioclase feldspar from the Democratic Republic of Congo was first reported in the gemmological literature as andesine and in a later study as labradorite thereby inferring a compositional range. The gem trade has variously referred to it as sunstone, Congo-sunstone or andesine-sunstone (see andesine and labradorite).

Sunstone – China

Red and green andesine from China, because of its similarity to Oregon schillerless labradorite 'sunstone' has also been called 'sunstone' (see andesine). The material, however, bears a stronger resemblance to the red and green labradorite reported from the Congo.

Sunstone – Tanzania

Light greyish green oligoclase sunstone (RI 1.537–1.547; SG 2.64) from feldspar lenses in biotite schists at Engare Naibor, Arusha, Tanzania, contains eye-visible iridescent orange hematite platelets in variable distribution density that produce a 'confetti' sunstone effect. Lamellar twin planes are observed. Near-colourless acicular anthophyllite inclusions are also reported. The names 'Maasai Sunstone', 'Illusion Sunstone' and 'Tanzasun' have also been applied to the material. Finished stones up to 120 cts have been reported.

Honey to reddish brown K-feldspar sunstones show multiple phenomena, asterism, aventurescence and iridescence; eye-visible platy inclusions of hematite are responsible for the iridescence and aventurescence. The bright golden yellow-brown four-rayed star, comprising one strong but fuzzy cat's eye intersecting a less prominent ray at slightly less than 90°, was also produced by the flaky to acicular (depending upon orientation) inclusions of hematite aligned along cleavage directions. Close to the girdle, light blue adularescence was observed aligned along the strongest ray. RI 1.522–1.528; SG 2.6–2.59.

Opaque gold-brown to black cabochon material without aventurescence that exhibits either stars or cat's-eyes is being sold as Tanzanian sunstone.

Sunstone – India

Weak four-rayed asterism has been reported in 'sunstone' reputedly from southern India. Translucent to semi-transparent Indian sunstone, with rich reddish orange colour and golden orange iridescence, has been cut into spheres or produced as tumbled pebbles at Kangayam, southern India. Coarse cordierite crystals at Lachmanapatty and Kiranur are associated with feldspar of sunstone variety. The feldspars contain numerous oriented and thin, elongate inclusions of brown biotite which impart a bright reddish brown glitter in reflected light. Composite grains of iolite and sunstone from these localities are used to make single stones with coupled deep blue and orange-brown hues.

Sunstone – Australia

Orthoclase feldspar from the Mud Tank Zircon Field, in the Harts Range, Northern Territory, contains crystallographically orientated exsolutions of ilmenite and hematite that form a criss-cross lattice pattern which produces spectral colour aventurescence in reflected light. The material has been named 'Rainbow Lattice Sunstone' RI α 1.520; β 1.525; γ 1.527; DR 0.007; SG 2.57; Hardness 6–6.5.

Sunstone has also been reported from Mexico (oligoclase sunstone from Madera, Chihuahua) and Namibia.

Sunstone Simulant

Goldstone is a glass containing randomly dispersed fine hexagonal and/or triangular platy copper crystals which produce aventurescence. The metallic copper is formed when cuprous oxide, introduced during manufacture, is reduced during annealing. Bubbles and swirl structures are often visible in the glass. It is usually a golden brown, but may be stained blue to create 'bluestone'. Aventurescence can also be seen in other gem materials such as scapolite and iolite.

Feldspar Gem-Rocks

Unakite

The feldspar-epidote mixture known as 'unakite' or epidotized granite is often used for carvings, cabochons and beads. Spheres are sometimes cut. It consists largely of a salmon-pink feldspar (microcline or orthoclase); nearly colourless milky or smoky quartz, with or without white albite-oligoclase plagioclase and yellow-green epidote that is typically dispersed irregularly through the rock.

Larvikite

Larvikite is the name applied widely to a relatively coarse-grained anorthoclase-rich igneous rock from the vicinity of Larvik, Norway. Anorthoclase, which constitutes 90 or more percent of this rock, is in essence an intimate, submicroscopic intermixture of a plagioclase and an alkali feldspar. In petrographic circles this rock has been equated with both alkalic syenite and monzonite. Larvikite has the following properties:

- *Color* – bluish grey with two more-or-less distinct colour varieties, one light bluish grey, the other dark bluish grey with golden brown overtones
- *Hardness* $6-6^{1}/_{2}$
- *Specific Gravity* ~2.59
- *Light transmission* – subtransparent to subtranslucent
- *Lustre* – overall pearly
- *Miscellany* – on some surfaces, the large anorthoclase grains exhibit a chatoyant-like appearance frequently referred to as a pearl-grey iridescence, opalescence or even labradorescence (in allusion to the appearance of labradorite – see anorthosite).

Other names

- Birds eye granite – a misnomer used in the marketplace
- Blue granite – another misnomer used in the marketplace
- Blue Norwegian moonstone – trade name used to direct attention to its country of origin and some people's idea of its appearance
- Norwegian Pearl granite – trade name sometimes applied to the lighter coloured variety of larvikite. The 'granite' portion of this designation is incorrect so far as accepted petrographic nomenclature
- Royal blue pearl granite – trade name sometimes applied to the darker coloured variety of larvikite. The 'granite' portion of this designation is incorrect so far as accepted petrographic nomenclature.

Uses

Larvikite is a popular facing stone that is used throughout the world. It also is used, albeit in much smaller quantities, as the rough from which ornaments, such as paper weights, bookends, *etc.* are formed.

Nebula Stone

Nebula stone comprises mainly quartz and anorthoclase, with lesser amounts of riebeckite and aegirine. It is a dark green rock that contains golden green spheroidal masses (the 'nebula').

Anorthosite

An atypical anorthosite, also used as a gemrock, comes from a locality recorded only as the Philippines. It is described as granular, overall white and containing sporadic green crystals up to 2.0 mm across with its granular material consisting largely of nearly pure anorthite plus minor interstitial oligoclase and crystals of zoned garnet, 'primarily uvarovite with varying amounts of andradite in solution'. It is said to roughly resemble jade.

The information in the text has been complied from numerous journals, principally the *Journal of Gemmology* and *Gems and Gemology*, and numerous reference works and scientific papers. Copious web sites have been visited to see what material is currently avilable on the market and robust reference information assimilated. As far as possible dubious data and inaccuracies and false information perpetuated in all types of references have been rejected.

13

Zircon

While zircon in fashioned form can be found in a wide range of colours the crystals are usually reddish brown and heat treatment is needed to produce the colourless, golden yellow and bright blue colours most seen in jewellery. Other colours, often very pleasing, might be called fancy zircons as the range is considerable. We should understand from the first that some zircons are much older than others and, like older people, behave in different ways. Over time some zircons have suffered a breakdown of their crystal lattice to the extent that in extreme cases they are to all intents and purposes amorphous. They show little or no birefringence and their physical properties are different from those of other zircons. They are 'low' zircons and come virtually exclusively from the gem gravels of Sri Lanka. Zircons from elsewhere, with their crystal lattice intact, are 'high' zircons. The name metamict (first used by Brögger in 1893) is applied to low zircons, which approach a glassy amorphous state, are almost always green and can be radioactive though gem specimens do not normally show this property to any significant extent.

The progressive disintegration of the crystal lattice in metamict zircon arises from the presence of decaying nuclei of uranium and thorium, the process emitting alpha particles. With each emission the nucleus recoils, causing the displacement of lattice atoms. This process is described in Deer, Howie and Zussman, *Rock-Forming Minerals*, second edition, vol. 1A, *Orthosilicates*, 1982.

Zircon is zirconium silicate $ZrSiO_4$ and forms prismatic crystals of the tetragonal system with square cross-sections and is sometimes attractively terminated by pyramids; large reddish brown crystals from Nigeria can be especially attractive. Geniculate twins are also found. Crystals show a markedly oily lustre. Most gem zircon occurs as pebbles in gem gravels. Zircon is often brittle and faceted stones have traditionally been kept separate from one another so that their facets do not become 'paper worn'. Thai dealers for years used 'zircon twists' in which several stones were kept in screws of black paper.

The hardness is 7.5 and the SG 4.5–4.7. The RI for the ordinary and extraordinary rays is 1.925–1.961 and 1.980–2.015 respectively with a birefringence of 0.059, uniaxial positive. Metamict zircon may have SG around 4 and RI as low as 1.78 with scarcely perceptible birefringence. The dispersion, seen at its best in the heat-treated colourless material when faceted, is about 0.039. Properties to some extent identify the degree of metamictization and gem zircons have been classed as high, low and intermediate, the absorption spectrum often providing a useful indication of a specimen's position on this arbitrary scale. Zircon shows very weak pleochroism in general though blue heat-treated stones will give a blue and near-colourless response.

In high zircon and particularly those from Myanmar there is a beautiful and near-diagnostic uranium absorption spectrum with fine though strong lines and narrow bands extending across the entire width. The strongest band is at 635.5 nm. This phenomenon was first described by Church, with the spectrum of almandine, in *The Student's Guide and Intellectual Observer*, 1866. Myanmar greenish brown specimens may contain at least 40 lines and bands. Details of the absorption spectra of gem zircons were collected by B.W. Anderson and C.J. Payne over many years between and after the Second World War while working at the London Gem Testing Laboratory. Their work was collected by Anderson in MS notes, copies of which are held by GemA in their library in London. Between 1954 and 1957 the journal *The Gemmologist* published some of these notes and in 1998 R.K. Mitchell edited them for a monograph *The Spectroscope and Gemmology*, GemStone Press, Woodstock, Vermont and Robert Hale, London; ISBN 071980261X. In the section on zircon Anderson & Payne list bands shown by different zircons, making the point that some are best seen in the ordinary or extraordinary rays. Some showed up best on a photograph taken with a small grating spectrograph using plates sensitive to the deep red. Not only do zircons show a variety of absorption bands and lines corresponding to their low or high state; when heated, some specimens show an anomalous spectrum. To an extent the absorption spectrum may give some clue to origin as specimens from Uralla, New South Wales, show only a few bands, and red specimens from sites in the Auvergne, France, have shown no absorption in the visible. When zircons from the Indo-China region are heated to give commercial colourless, golden yellow and blue their absorption spectrum is very weak and may show only the persistent band at 653.5 nm.

The main bands in the spectrum of zircon are at 691, 683, 662.5, 660.5, 653.5 (the strongest and most persistent), 621, 615, 589.5, 562.5, 537.5, 516, 484, 460 and 432.7 nm. Metamict zircons show a woolly band at 653.5 nm.

Heating these zircons may cause this band to sharpen and other lines to appear. There are some rare variations of the low-type absorption spectrum, one of which shows three broad strong bands in the red at 691, 669 and 653.5 nm, the centre band being the strongest. Another type, which curiously has been found only in zircons with an RI of 1.82 and an SG of 3.98, shows a vague band at 655 nm and another at 520 nm.

Fluorescence

Some zircons show cathodoluminescence and thermoluminescence but the responses are not usually valuable in gemmological testing. Intermediate types may give an orange incandescence in a Bunsen flame – thorium has been suggested as the cause. Neutron irradiation of yellow crystals causes their colour to change to bright green (the colour of metamict zircon). Subjecting zircon to any form of irradiation may cause them to discolour. This is especially true of the colourless and blue specimens whose colour has been arrived at by heating. Though gemmologists speak of colour fading this is, like so many other pronouncements, incorrect; dark patches appear in the stone and spread as the treatment continues. Heating usually restores their original appearance though not, it can be imagined until some heart-searching has taken place.

Heating

Both the low and intermediate zircons when heated to about 1450 °C may achieve an SG of 4.7 with high-type RIs and sharper absorption lines. Heating the reddish brown Indo-Chinese crystals gives blue, colourless or yellow specimens according to whether or not an oxidizing atmosphere is employed. Colourless and sky-blue zircons need reducing conditions while golden yellow colours are achieved in oxidizing conditions. Gem zircon is usually the high type.

The best account of treatment can be found in Nassau, *Gemstone Enhancement*, second edition, Butterworth Heinemann, Oxford, 1994; ISBN 0750617977. Nassau reminds readers that many references to the methods used in heating zircon can be found in Eppler, *Das Geheimnis des Zircons*, in *Goldschmiede Zeitung 51*, 531 (1936) and by Buckingham in the *Journal of Gemmology* 2, 177 (1950).

Inclusions

Metamict zircon shows the most characteristic inclusions. Most prominent are tension fissures meeting at an angle of 57°5′ and probably due to the degeneration process. They echo the original crystal's prism and dipyramidal faces. Also due to isotropization are disc-shaped tension fissures. Unequal isotropization may give rise to parallel stripes which also suggest tetragonal form (*Figure 7.3*). Ilmenite may be found in fractures and some healing fissures are reported.

Occurrence

Zircon is unusually common and widely distributed though most gem crystals occur in the gem gravels of Sri Lanka and some are found in the Mogok area of Myanmar and in Nigeria. The red zircon of Expailly, Auvergne, France, has already been mentioned. Gem crystals in a variety of colours and sizes have been found in the Mud Tank area of the Harts Range, Northern Territory, Australia. These crystals appear to have undergone less structural damage than the Sri Lankan material. Fine crystals are reported from Slyudanka, Irkutsk Oblast, Russia, and from Seiland Is., Alta Fjord, northern Norway; this area has produced beautiful orange crystals up to 2.5 inches in length.

The gem zircon locations in South-east Asia include the Pailin district of Kampuchea and the Chiang Mai district, Thailand. The zircon-producing sites overlap political boundaries in many cases so that names can be hard to locate. In all cases material is sent to Bangkok for fashioning.

Reddish brown zircon is reported from decomposed alkali basalts near Jemaal Kaduna, Nigeria. This is a source of fine crystals.

Fashioning

In order to obtain the best optical effect with zircon, the stones are sometimes cut into a modified brilliant cut which has a second set of pavilion facets. The octagonal and four-sided trap cuts are now extensively used for specimen zircons of the blue and golden colours, for some of the greens and yellows, and also the browns of natural colour. The natural coloured stones are, however, more usually cut in the mixed-cut style. The high birefringence can affect the appearance of the finished stone if care is not taken by the lapidary.

A basaltic rock with groups of flower-like radiating crystals of zircon and xenotime-(Y), (YPO$_4$), is found in the hill Maru-Yama, just north of Mount Funabuse, Gifu prefecture, Japan. The rock is locally called kiku-ishi (chrysanthemum stone). A similar rock is found, usually as boulders, in Vancouver Island of British Columbia, Canada.

Small crystals of zircon have been grown hydrothermally and are described in the chapter on synthetics (Chapter 24).

14

Peridot

The minerals forsterite and fayalite in the olivine mineral group (there is no individual mineral olivine) form the end-members to the series Mg_2SiO_4 (forsterite) and Fe_2SiO_4 (fayalite). An iron content of 12–15% gives a good, bright green and traces of chromium or nickel may enhance it. Peridot is a member of the orthorhombic system in which it forms typically thick crystals with wedge-shaped terminations; twinning is common and some faces are striated. Hardness is 7 and there is an imperfect cleavage; the SG for peridot is usually near to 3.34 and the RI for the alpha, beta and gamma rays is 1.654, 1.671 and 1.689 respectively, biaxial positive with birefringence usually 0.036. This can easily be detected by the $10\times$ lens. The oily lustre can be characteristic. Pleochroic colours are weak with one ray inclining more to yellow than the others. The occasional cat's-eye is found. The iron content inhibits luminescence.

The absorption spectrum shows three bands due to iron in the blue area; the bands are distinct and well separated at 493, 473 and 453 nm. In the 493 nm band a narrow core can be seen at 497 nm. Anderson & Payne (q.v., passim) report vague, weak bands in the orange at 635 nm and in the green at 529 nm. Users of the spectroscope should observe in all directions to ensure a correct absorption picture is given.

Inclusions

Peridot hosts a number of inclusions, some of them characteristic for particular areas. Rectangular 'biotite' [name no longer used – refer to dark monoclinic trioctahedral Li-free micas] crystals characterize stones from Mount Kyaukpon, Myanmar. Blebs of natural glass have been found in Hawaiian stones together with grains of chromite; these latter, however, occur in peridot from other places, including Arizona and the Island of St John (Zeberget). The characteristic discs known as lily-pads are decrepitation haloes and often contain a chromite crystal at the centre. Curly hair-like ludwigite and near-black ludwigite rods may be characteristic of peridot from Pakistan.

Chromian spinel seems to characterize peridot from San Carlos, Arizona. Fluid inclusions are found in some stones but do not indicate particular deposits. Hercynite is reported from peridot found at Kilbourne Hole, Arizona. Chalcopyrite has been found in a peridot from Arizona. Curly hair-like ludwigite, and near-black ludwigite rods are characteristic of peridot from Pakistan.

Occurrences

The Zeberged occurrence (the name is a translation and can be variously spelt) is about 100 km south-east of the Râs Banâs peninsula in the Red Sea. Peridot crystals are found in outcropping peridotites with the larger ones occurring on the eastern slopes of Peridot Hill. Much of the finest Arizona peridot comes from the area of Peridot Mesa on the San Carlos Apache Reservation in Gila County, where the crystals occur in vesicular basalt. The location is described in *Mineral resources of the San Carlos Indian Reservation, Arizona, US Geological Survey Bulletin* 1027-N, 1956.

Bright green well-formed crystals are found north of Mogok, Myanmar, on the northern slope of Kyaukpon in a weathered serpentine in which the crystals are loose. Volume 2 part 2 of Hume's *Geology of Egypt* (1935) gives some details of the occurrence but takes details from W. F Moon's *Preliminary Geological Report on Saint John's Island (Red Sea)*, 1923 [Sinkankas #4563]. This is the first mineralogical and geological study of this area and shows mining in operation at the peridotite cavities containing crystals. The report is published by the Geological Survey of Egypt.

Hawaiian peridot is found on beaches. Stones from Ameklovdalen, Sondmore, Norway, are Fe-poor and a bright yellowish green. Gem-quality peridot is found in China in the Zhangjikou-Xuanhua area of Hebei province. It is reported from Antarctica and from alluvial deposits near the Usambara Mountains in the Umba district of Tanzania.

Cutting of Peridot

Peridot is best cut in the trap-cut (step) style, although oval, round and pendeloque-shaped mixed-cut stones are common. In some peridot the large table facet is cut not flat but with a slight doming. It may be remarked that peridot may lose its polish after contact with hydrochloric or sulphuric acids. Over the past decade fine quality peridot from a new source, Pakistan, has been on the market. Well formed crystals are a feature of this deposit.

Simulation

The characteristic oily lustre and strong DR of peridot serve to identify the stone from its counterfeits, although there are convincing imitations made in glass and in the composite stone made with a garnet top on a suitably coloured glass. A peridot-coloured synthetic corundum has been produced and also a peridot imitation based on the soudé emerald type of composite stone, the top and base being synthetic colourless spinel and the join across the girdle containing a suitable green colouring matter. In the Summer 2004 issue of *Gems & Gemology* a large, apparent peridot measuring 38 mm × 23.95 mm × 16.30 mm was found to be a glass whose RI was over the limits set by the standard refractometer and contact liquid. The usual RI for peridot falls between 1.65 and 1.69. EDXRF spectroscopy showed that the material consisted mainly of zirconium with some titanium and silicon with no correspondence with any known mineral. IR spectroscopy showed two broad peaks associated with glass. X-ray diffraction analysis showed that the material had no crystal structure and was therefore amorphous.

15

Spodumene

Spodumene, a member of the pyroxene group, is lithium aluminium silicate $LiAlSi_2O_6$ and forms highly characteristic flattened and striated prismatic transparent crystals of the monoclinic system with good cleavages in two directions intersecting nearly at right angles and a notably brittle tendency. The hardness is 6.5–7 and the SG 3.18. The RI for the alpha, beta and gamma rays is 1.648–1.663, 1.655–1.669 and 1.662–1.679 respectively with biaxial positive DR 0.014–0.027. Crystals are strongly pleochroic with lilac, green or yellow in one direction and colourless in the other. The varieties include a yellow to pale green spodumene with no varietal name, the lilac-pink kunzite and a very rare chrome-green hiddenite.

Kunzite shows no useful absorption in the visible but the chrome-green hiddenite shows a characteristic chromium spectrum with the customary strong doublet in the deep red at 690.5 and 686 nm and weaker lines at 669 and 646 nm. A broad absorption band has its centre near 620 nm but no Cr lines are visible in the blue. The iron band is strong and well-defined at 437.5 nm and is accompanied by a weaker band at 433 nm. A narrow line at 505 nm is reported from a yellow spodumene from Myanmar. The band at 437.5 nm is seen in a number of pyroxenes, including jadeite and is due to iron.

Luminescence

Under LWUV kunzite glows golden-pink or orange with a weaker effect under SW; it shows a very strong response to X-rays with a very strong orange with persistent phosphorescence. After this irradiation the specimen may have changed from lilac-pink to bluish green. Kept away from strong light sources this induced colour will be maintained. If such a specimen is exposed to strong sunlight for some hours or if it is heated to around 200 °C the colour will discharge with emission of orange luminescence. Yellow-green spodumene may give a weak orange-yellow under LWUV; a much weaker though similar effect is seen under SW.

Yellow-green spodumene gives a fairly strong glow under X-rays but with no phosphorescence or change of colour.

A brownish orange spodumene was found to have been irradiated, primarily by the isotope scandium-46. A Geiger counter should be used routinely on spodumene of citrine-like colour. Gamma ray spectroscopy also detected radio-active isotopes. Wearing is not recommended.

Occurrence

Spodumene is a common constituent of lithium-rich granite pegmatites. Fine kunzite is found in southern California, in particular at the White Queen mine, Hiriart hill, San Diego County, and elsewhere in the area. It is also found in many Brazilian pegmatites, particularly at Urupuca, at Lavra do e Mario, Corrego do Urucum, Galileia and as thick, pale pink and pale green crystals at Resplendor. The pegmatites of Kunar, Lagham and Nuristan, Afghanistan, provide good-quality crystals. Yellow-green spodumene is found in the Brazilian pegmatites and also at locations in Afghanistan. In the early days of its appearance yellow spodumene was thought to be yellow chrysoberyl. Various colours of spodumene have been found in the pegmatites of Madagascar, in particular Anjanabonoina, Mt Bity and Tilapa. Hiddenite spodumene has been found only at Stony Point, Alexander County, North Carolina. The variety was identified in the late nineteenth century. Spodumene has been reported from the Mogok area of Myanmar. A blue spodumene weighing 35 carats has been reported. The stone had RI in the spodumene range.

Note on the Cutting of Kunzite

Owing to the strong cleavage, brittleness and heat sensitivity, spodumene is a difficult stone to facet. Further, owing to the thin nature of some of the crystals it is not possible to cut well-coloured stones, since to get stones of good colour advantage should be taken of the strongest pleochroic colour. The deepest colour is seen approximately parallel to the length of the crystal, so for the best result the stones should be cut with the table facet at right angles to this direction. This is, of course, more important in the case of kunzite, which is usually cut very deep in order to enhance the colour.

Some kunzites have a tendency to fade on exposure to strong sunlight. Some yellowish brown spodumenes turn purple after heat treatment.

Simulation

Kunzite is simulated by synthetic pink spinel and by a suitably coloured glass, but distinction is easy as both of these stones are singly refractive, whereas in kunzite the DR is strong enough for the doubling of the opposing facet edges to be seen with a hand lens in appropriate directions. Further, glass and spinel are not dichroic. What may be a more convincing simulator of kunzite is a bluish pink amethyst, but even here, apart from the differences in the RI and SG, the lustre of the amethyst is not nearly that of kunzite.

16

Quartz

The quartz gems comprise the fully crystalline quartz varieties, amethyst, rock crystal and so on, and the crypto-crystalline quartz or chalcedonies. To settle, at least temporarily, whether or not quartz should be sought in reference books under oxides or silicates, Anthony *et al.* (1995) place it with the silicates. As I write in August 2004, vol. 4B of the second edition of Deer, Howie and Zussman, *Rock-Forming Minerals,* has just been published by the Geological Society of London and I can recommend this to readers – ISBN 1862391440. The volume also covers feldspathoids and the zeolites. For amethyst I can recommend the study *Amethyst: Geschichte, Eigenschaften, Fundorte* by Werner Lieber, 1994; ISBN 3921656338.

At temperatures up to 570 °C quartz is the stable modification of crystalline silica, SiO_2. At approximately 573 °C the structure alters from the stable (alpha quartz or low quartz) to beta quartz or high quartz. This alters to the modification tridymite at 870 °C, with an increase in volume; at 1470 °C a change to cristobalite takes place. Quartz (trigonal), tridymite, which may be monoclinic and pseudo-hexagonal (or triclinic), and the tetragonal cristobalite are all polymorphs of SiO_2; other polymorphs exist. At temperatures over 1728 °C cristobalite melts.

Quartz is a member of the trigonal system and can be morphologically complex. It occurs as enantiomorphic prismatic crystals; Anthony *et al.* (1995) state that over 500 forms have been noted. Prism faces are laterally striated and are terminated most commonly with the near-universal forms of the prism with positive and negative rhombohedra. Where the two types of rhombohedron are present the positive usually predominates; the names major and minor are sometimes used. Small modifying forms are also seen; these include the trigonal pyramid and the positive trigonal trapezohedron, with other less common trigonal trapezohedra with positive right and positive left faces. Overall the identifying of forms can be difficult in quartz, particularly in striated or rounded zones. Some crystals have equant habit in which both positive and negative rhombohedra are equally developed with no prism form present or, when present, poorly developed. The name quartzoid has been used for this type of crystal which resembles a hexagonal pyramid. Prism faces are nearly horizontally always striated and this feature helps in correct identification of the morphology; the notably brighter lustre and larger size of the positive rhombohedron compared with its negative counterpart is also useful. The presence and situation of vicinal faces and growth hillocks are closely studied in the context of quartz crystal synthesis. Collectors and lapidaries know that colour in quartz crystals concentrates within the rhombohedra and this is seen at the crystal tips.

Twinning in quartz is almost universal. The various types include Dauphiné and Brazil twinning with parallel axes: in the former type the twinned portions are related geometrically and are of the same hand. They are related by a rotation of 180° around the c-axis. In general, both portions in Dauphiné twinning are of similar size. Its presence can only be detected by chemical etching or by X-ray diffraction. Brazil twinning, also including penetration, involves the relationship of two portions, of opposite hand. Brazil and Dauphiné twinning are present simultaneously in most crystals. Brazil-law contact twins are known with right- and left-hand individuals joined with large re-entrant angles. Brazil-law twins can be recognized in polarized light.

Twinned crystals with inclined axes include the Japan-law type and its various subtypes. All twinned crystals with inclined axes are contact twins with the c-axes inclined at 84° 33'.

Quartz belongs to the trigonal trapezohedral class of the trigonal system. Crystals in this class show the property of enantiomorphism – a relationship between two identical but non-superposable objects or mirror-images. This relationship can give crystals a right-handed or left-handed form. In quartz it appears as a helical arrangement of the SiO_2 tetrahedra along threefold axes. The analogy of a corkscrew may be useful; its helix is right-handed when it is thrust into the cork away from the observer and turned clockwise; seen from the side with the axis of the helix held vertically, the thread of the corkscrew slopes upwards to the right (Frondel, 1962). The identification of hand can usually be done by examining the position of the trigonal pyramid in the upper corner of the prism face below the positive rhombohedron. If the pyramid is in the upper right corner the crystal is morphologically right-handed; if it is in the upper left corner the crystal is left-handed. The two types are found in almost equal numbers in nature.

If a quartz crystal is mechanically stressed along certain directions it will develop a surface electrical charge (piezoelectricity). Applying an electrical field will produce mechanical strain in the crystal. This phenomenon is the basis of many electronic devices. Quartz will also develop a surface electrical charge with change of temperature (pyroelectricity).

The hardness of quartz is 7 varying slightly with direction. There is no distinct cleavage and the fracture of single crystals is conchoidal. The SG is close to 2.651 at room temperatures, varying only slightly with the presence of impurities. The RI is 1.544 for the ordinary ray and 1.553 for the extraordinary ray, with a birefringence of 0.009. The dispersion is 0.013. Quartz shows no fluorescence. Quartz rotates the plane of polarization of light travelling parallel to the c-axis either to the right or to the left. The normal uniaxial interference figure is virtually unique for quartz in that the arms of the black cross do not meet in the centre. This is a most useful test for rock crystal spherical beads which can quickly be passed between crossed polars when the characteristic figure can be seen. In recent years the aluminium phosphate berlinite (with which quartz is isostructural) has been found to exhibit a similar figure but the likelihood of confusion with quartz is remote, at least in the ornamental context.

Occurrence

Quartz occurs in epithermal veins; it is a characteristic mineral of granites and granite pegmatites and is found in sandstones and quartzites. It occurs in hydrothermal metal deposits and in carbonate rocks where it is common. Details of a selection of significant locations of ornamental-quality material are given with the type of quartz being discussed.

Single-Crystal Quartz Varieties

Rock Crystal

Colourless water-clear quartz is known as rock crystal. Crystals of colourless quartz are abundant in many places on the Earth and show considerable variation in size, some weighing more than 500 kg. The exhibition *Cristeaux géants* at the Musée Nationale d'Histoire Naturelle, Paris, shows some fine examples. Rock crystal is often attractively included with golden needles of rutile or more substantial green tourmaline crystals. Various names have been used for these materials. We deal with inclusions below.

Rock crystal may be faceted, cut as cabochons to display asterism or carved. Crystal balls are ground and polished, particularly in Japan. Quartz, except for one position along the optic axis will, owing to DR, show as doubled a mark or line over which the ball is placed. This cannot occur with glass.

Inclusions

Rock crystal may contain two-phase inclusions (often gas or liquid CO_2 with varying mineral neighbours) but several minerals give rock crystal an unexpectedly exotic appearance. Reddish or golden rutile crystals in hair-like form produce rutilated quartz or Venus hair stone and other long slender invaders may include dark colours of tourmaline or actinolite. In some cases the actinolite fibres are sufficiently dense to colour the host green. Reticulated (net-like) needles give sagenitic quartz. Other rock crystal inclusions are blue anatase (particularly in Norwegian material), ilmenite, the sulpho-salts boulangerite, giessenite and heyrovskyite (these particularly from Alpine cleft material), brownish yellow bundles of goethite crystals, red plates of lepidocrocite, reddish hematite, yellowish green chlorite group minerals, golden yellow pyrochlore and orange spessartine. The copper silicate hydroxide chrysocolla may be brown or blue; also giving blue inclusions is dumortierite. Thin cracks may give interference colours in white light (rainbow quartz). Sometimes the cracks may be filled with dyes (firestones).

Occurrences

Rock crystal is found widespread in the world. Especially fine material is found in clefts in the European Alps.

White Quartz

White quartz (often called milky quartz) has profuse minute cavities containing water or liquid carbon dioxide. Much vein quartz is white quartz and this type is often gold-bearing. Such material with included gold grains has been cut as plates or cabochons, perhaps as souvenirs of the mining camps.

In examining the colouration of transparent single-crystal quartz below I have given some leads into some of the (fairly) recent literature. Inevitably the following paragraphs can be a simplification at best.

Brown and Smoky Quartz

Quartz crystals of a clear brown colour and varying from a light brown to so dark a brown as to appear black (morion) are found in the debris of weathered granite in the Cairngorm mountains of the Scottish Highlands. Faceted stones of this brown quartz are known as cairngorm and were traditionally used to embellish Highland dress; much of the cairngorm used for Scottish jewellery today is amethyst from Brazil which has had the colour altered to brownish yellow by heat treatment. Some brown quartz has a smoky tinge and is then known as smoky quartz. Distinctions between 'original' yellow and heated amethyst are never made in the commercial context.

The usual inclusions seen in brown quartz are the two-phase negative crystal cavities in which the liquid phase is usually carbon dioxide, as shown by the disappearance of the bubble when the quartz is heated to between 30 and 31 °C, which is the critical temperature for carbon dioxide and above which it cannot exist in the liquid state. Brown quartz shows quite a distinctive dichroism, one ray being brown and the other pinkish brown. The stones show no absorption spectrum of diagnostic value, nor do they luminesce under UV rays or X-rays.

The colour of smoky quartz arises from the operation of a colour centre involving an aluminium impurity. Natural irradiation ejects and traps one of a pair of electrons, leaving a hole colour centre with absorption of light giving the brown colour. Heating restores the trapped electron and removes the colour. Details of the mechanisms involved are covered by the second edition of Nassau, *The Physics and Chemistry of Color*, 2001 (ISBN 0471391069).

Natural smoky quartz or irradiated synthetic quartz turns greenish yellow on heating before it loses its colour entirely. The greenish yellow colour is stable to light. In this case the cause of the colour centre is not known and no Fe is present. Re-irradiation returns it to the smoky colour. Brown quartz is found in the Alps; fine crystals of smoky quartz are found at Pike's Peak, Colorado, USA. It is common elsewhere.

Citrine

The yellow quartz called citrine varies in colour from a light golden yellow to a reddish yellow and has been thought to owe its colour to a trace of ferric iron (Fe^{3+}). The best material comes from Brazil. Citrine may show perceptible dichroism. Most of the yellow and brownish yellow quartz which in the past was often sold under the misnomer 'topaz' is heat-treated amethyst.

Heaney *et al.* (*Silica*, 1994, passim) gives more detail on the colouration of citrine. Citing Schmetzer (*Journal of Gemmology* 21, 368–91, 1989), five types of citrine are given: orange-brown natural quartz containing iron; yellow to orange-brown colours produced by the heating of amethyst; yellow to orange-brown colours in synthetic quartz grown in the presence of iron; light yellow radiation-induced colour found naturally in quartz or produced by heating smoky quartz; greenish yellow colour produced by irradiation (with or without subsequent heat treatment) either of natural colourless quartz or synthetic colourless quartz. Since the publication of the paper the precise nature of some of the specimens has become unclear.

Naturally occurring citrine with orange-brown colour and iron content is found in Brazilian deposits, and, as ametrine (part citrine, part amethyst) from the Anahi mine, Bolivia. Early on the cause of the yellow colour was believed to be substitutional Fe^{3+}. This does occur in citrine but has no significant role in producing the colour, though electron paramagnetic resonance (EPR) has proved its presence. The optical spectrum of undisputedly natural Fe-containing citrine shows strong absorption in the UV region with the tail of the absorption band extending to the short wavelength region of the visible spectrum. On the tail an absorption band at 476 nm is superimposed (Heaney *et al.*, *Silica. Reviews in Mineralogy* 29, 1994).

Lehmann and Moore (in *Journal of Chemical Physics* 44, 1741–45, 1966) attribute the colour of citrine to sub-microscopic particles of a presumed iron oxide dispersed in the quartz. This observation is based on the similarity of the EPR and optical absorption patterns with heated amethyst.

Neumann and Schmetzer (*N. Jb. Min. Mh.* 6, 272–82 and *Z. Deut. Gemmol. Ges.* 33, 35–42) describe the heating of amethyst over the range 330–500 °C during which it loses its amethyst colour and may become yellow, yellowish brown, brown, green or colourless. On initial heating the violet colour begins to bleach; as heating proceeds a specimen may become completely colourless or yellow or other colours may begin to develop. At higher temperatures some specimens develop yellow colour from colourless or light green regions, or brown colour from yellow regions. These changes are accompanied by the loss of the amethyst absorption band at 545 nm and increases in strength of the green quartz band at 741 nm and the yellow quartz band in the UV region. However, the part played by water and (OH) in the colour changes in amethyst-citrine were not investigated in this study.

The absorption spectrum of heated amethyst consists of intense, isotropic absorption in the UV which tails into the visible. The colour of the sample is determined by the rate of increase in the absorption over the SW region of the spectrum. This is determined by the amount of iron contributing to the absorption. Rossman, in Heaney *et al.* (1994), lists, with figures, some of the colour changes induced in amethyst from different localities by heating. Readers are referred to the Rossman paper in Heaney *et al.*, in general, for coverage of the whole topic of single-crystal quartz colouration.

Amethyst

The violet and purple varieties of quartz are the most prized. The name amethyst with the sense of preventing or alleviating drunkenness has persisted from Classical times; Pliny stated that the gem was so called from the colour being near to, but not quite reaching, that of wine (*Historia Naturalis* 37, 121–24 for those who really want to check the (supposed) origin of observations made in book after book on gemstones).

Amethyst has been held in high esteem in the church and many fine specimens are set in episcopal rings. Two fine amethysts are in the English regalia: a large faceted orb set below the cross and surmounting the Star of Africa in the Sceptre with the Cross, and another fine stone set below the cross which surmounts the orb.

Amethyst is found as a lining to the inside of hollow cavities; such crystals characteristically showing only the rhombohedral terminations. In crystals from South America the three alternate rhombohedral faces may be the only ones developed, giving cube-like shapes. Sometimes the faces of the rhombohedra are arranged in parallel position. It is also characteristic of amethyst that quite often only the tips of the crystals are deep coloured, the remainder grading into milky quartz or rock crystal. While amethyst is always violet the range of colour is wide and may vary from nearly colourless with a faint mauve tint to a fine deep purple.

Colour

While the yellow colour of citrine may be linked with an iron compound the colour of amethyst arises from the natural (or artificial) irradiation of Fe-bearing quartz. Either divalent or trivalent Fe may be involved, and provided that growth occurs in certain directions rather than others (e.g. in the positive or negative rhombohedron) a successful purple colouration will be achieved. The mechanism is a colour centre.

Much amethyst on heating lightens in colour: if Fe^{3+}, in whatever state, was present before irradiation, a yellow citrine colour will develop. If Fe^{2+} was present, a green colour develops ('prasiolite'). Heating to 350–400 °C may produce a bicolour amethyst-citrine ('ametrine') occasionally. Some amethyst when heated will give the prasiolite colour, perhaps via a colourless intermediate stage. The change to green occurs at 400–500 °C.

Rossman in Heaney *et al.* (1994) describes the colouration of amethyst. In describing amethyst as the violet variety of alpha-quartz, the point is made that the amethyst colour can only occur when iron is present in the crystal although the presence of iron does not by itself guarantee the characteristic colour. To some extent this colour is unstable to light, the process being observable most notably in the change of colour in large ornamental crystal groups subjected to bright lighting. While re-irradiation can restore the amethyst colour it will be accompanied by some of the smoky quartz colour which does not make for so pleasing an appearance and in any case the attempt to restore colour is hardly likely to be made very often, certainly not in the ornamental context. In the laboratory temperatures greater than 400 °C have been found necessary for the bleaching of amethyst.

Nassau (2000), citing Rossman (passim) in Heaney *et al.* (1994) shows that if quartz is grown in the laboratory with a small amount of aluminium present the colourless crystal will have Al^{3+} ions substituting for about one out of every 10,000 of the Si^{4+} ions. If the crystal is now irradiated (X-rays or gamma rays) a dark brown to black smoky quartz colour will appear (it can be removed by heating and restored by irradiation). Most (natural) rocks crystal contains sufficient Al for the smoky colour to be induced.

When, instead of Al^{3+} ions, Fe^{3+} ions are present in the quartz, the crystal will be a pale yellow colour. On irradiation an $[FeO_4]^{4-}$ colour centre absorbs light to produce the amethyst colour. Nassau (2000) makes the point that although a transition element is involved (in the formation of the amethyst colour centre) the effect is not similar to that which would obtain where a ligand field is involved (as in transition element colouration). The strong colour of amethyst arises even when the iron concentrations are low, as the transitions are allowed ones in this case.

The colours of the crypto-crystalline varieties of quartz are fairly easy to understand. Jasper owes its range of colour from red through yellow to brown to admixed minerals; hematite absorption bands can be seen in the optical spectrum and it is possible that goethite causes yellow and brown colours, White or grey colours may come from clay minerals. In general, iron oxides are the main cause of colour in these quartz varieties.

Effects of Light

Amethyst shows distinct dichroism, bluish violet and reddish violet for the ordinary and extraordinary rays respectively. Brown heat-treated amethyst shows no dichroism so treated stones may be distinguished from those of natural brown or yellow by their lack of dichroism. Under the Chelsea colour filter amethyst appears a reddish colour, the stronger the more pronounced the body colour of the stone. There is a wide absorption of varying intensity in the yellow-green region of the spectrum from about 550 to 520 nm. Amethyst is practically inert to UV though some blue response to SW has been reported.

Inclusions

The most common inclusions are feather-like structures formed of negative cavities. A quite common imperfection in amethyst is a mark like a thumb-print or tiger stripes which arise from rhombohedral twinning or to the partial healing of separations occurring along the planes of the major rhombohedral faces. Very few amethysts are free from parti-colouration with angular zones of colour, and even in heat-treated brown specimens this zonal colour marking may be seen. The iron aluminium phosphate cacoxenite may occur as yellow sheaf-like crystal groups in amethyst.

Occurrences of Amethyst

Amethyst is found in the Mursinka and other areas in Russia; some of it is reported to appear more crimson when viewed by tungsten light. Most deposits were and are in the Brazilian states of Minas Gerais, Rio Grande do Sul, Goias and Bahia. Fine amethyst is also found in the Artigas area of Uruguay. Details or rather lists of amethyst occurrences can be found in the literature and an uncritical or detailed list cannot be accommodated in the present text.

New locations for amethyst are constantly being discovered and readers are advised to keep up with the literature – journals such as *Lapis* are especially recommended.

Simulation of Amethyst

Amethyst is imitated by glass which usually gives a low RI and pronounced swirl marks. Synthetic corundum is made in a colour to imitate amethyst, and a soudé-type composite stone is made which can be very deceptive, for the top, if not the base as well, is made of quartz and will thus give a refractometer reading for amethyst. Pale amethyst is often mounted in a closed setting with paint or foil below in order to enhance the colour.

Interestingly, the flame-fusion corundum so often offered as 'alexandrite' is much more like amethyst in colour.

Rose Quartz

Pink quartz (rose quartz) may occur either as masses or as crystals which often reach quite large sizes. The cause of colour, once ascribed simply to manganese impurities, is more likely (Nassau, 2000) to result either from Ti^{3+}–Ti^{4+} charge transfer, inclusions of pink dumortierite or a radiation-induced colour centre which rapidly fades in light involving Al^{3+}–P^{5+} charge transfer. Specimens of rose quartz may fade.

Rose quartz is found either as large crystals or masses. Some specimens show asterism either by transmitted or reflected light; despite the older gem textbooks many rose quartz star stones show the star by either form of light as simple experiments show. Rose quartz may show a strong pleochroism in different shades of pink. Specimens deserving a particular mention are found in the pegmatites of Madagascar and those in Brazil, in particular in the state of Rio Grande do Norte.

Blue Quartz

Though a good deal of the transparent blue quartz on the market is a synthetic product coloured by trivalent cobalt and thus showing the Co absorption spectrum and a strong red through the Chelsea filter, natural blue transparent quartz gains its colour by the scattering of light from minute inclusions. Viewing the same slice by transmitted and reflected light will show orange-brown and blue respectively. The synthetic Co^{3+}-doped product is obtained either by incorporating Co^{3+} in an interstitial site to form yellow quartz which is then heated to 500 °C to obtain the blue colour by the formation of Co^{2+} at a different site. This material has no natural counterpart. Some synthetic quartz containing iron may show a pale blue colour. The original colour would be pale green from Fe^{2+} in an interstitial position. As less yellow is present than in the green quartz more blue is transmitted. Irradiation of synthetic quartz may also give a blue colour (Nassau and Prescott, *Mineralogical Magazine* 41, 301–12, 1977).

Green Quartz

Green transparent quartz has been called prasiolite and is obtained naturally or artificially by heating amethyst. The process is described by Neumann and Schmetzer (1984, *op. cit.*). Natural transparent obviously green rather than yellow-green quartz is uncommon in nature. From an examination of specimens found in Poland it was proved that iron occupies the same sites in both natural and synthetic green quartz. Green can be obtained by the irradiation and subsequent gentle heating of natural or synthetic quartz – the colour tends to a yellowish green. A synthetic green quartz has been grown in steel autoclaves (hydrothermal method) (Tsinober and Chentsova, *Sov. Phys. Cristalog.* 7, 113–14, 1959). The best green obtainable from single-crystal quartz does not compare with that of emerald or peridot but nonetheless I find the colour attractive. In the Summer 2004 issue of *Gems & Gemology* a transparent light greyish yellowish green faceted stone of 9.70 ct is described, the specimen showing notable Brazil twinning (particularly characteristic of natural amethyst). This effect, where present, can be found when the specimen is examined in the optic axis direction between crossed polars (immersion of the specimen in water helps). The specimen was determined as natural quartz.

Chatoyant Quartz and Quartz Pseudomorphs

Quartz cat's eyes, in general, show a less sharp eye than those of chrysoberyl, nor is the background colour very similar. Nonetheless there will be times when one species needs to be distinguished from the other; when the specimen has a flat back the RI can be found and quartz will not show the characteristic absorption band at 444 nm. Most specimens come from the Sri Lanka gem gravels and they have also been reported from the Bavarian Fichtelgebirge.

Tiger's-Eye

The popular and attractive gold, yellow and brown banded tiger's-eye and its blue counterpart hawk's-eye are formed by the silicification of previously existing softer asbestiform mineral matter whose characteristic structure is retained. The original mineral is currently believed to be a blue variety of riebeckite which occurs in iron formations. The varietal name crocidolite is still used for the asbestiform material. In this instance quartz is said to be pseudomorphous after riebeckite. Sources of tiger's-eye are usually centred on Griqualand West, South Africa. Appropriately fashioned tiger's-eye may show chatoyancy but most of the material is cut into flat plates, beads and cabochons or carved as cameos.

Quartzites

The rock quartzite consists of a granular interlocking mass of quartz crystals with irregular boundaries. Often the quartzite rock contains small crystals of a mica or iron mineral. The name aventurine is used for such a rock when transparent to translucent and exhibiting attractive colours: one variety of aventurine quartz contains included platy crystals of the green chrome mica group mineral formerly known as fuchsite, and this green aventurine quartz is often used for beads and other small articles of jewellery. Aventurine quartz varieties may be coloured by iron to give a rich white to reddish brown. Locations are numerous. Most green aventurine comes from India and varieties of patterning and shades of colour are constantly reported. In Europe reddish brown varieties occur in the south of Spain and a bluish white quartzite with red to brown patches and pyrite inclusions has been found in the Idar-Oberstein area, Rheinland Pfalz, Germany.

In general, quartzite will show a single RI near 1.55 on the refractometer and a variable SG between 2.64 and 2.69. Chromium may cause some green aventurine to show a red glow through the Chelsea filter and/or a vestigial absorption spectrum with vague lines in the red and a weaker than usual absorption of some of the yellow-green. Some of this material may show a greenish glow under UV, an effect not shown by other varieties of quartzite.

The Crypto-Crystalline Quartz Varieties

Chalcedony

Generally speaking, the name chalcedony is applied to any fine-grained quartz with a fibrous microstructure. It almost always displays a more or less distinct banding, though a microscope may be needed to detect some examples. The name agate is used for the important subvariety of chalcedony which displays a distinct banding with successive layers differing in colour and degree of translucency. All gradations exist between agate and ordinary chalcedony.

The individual fibres of chalcedony and its subvarieties are not physcially separable; they occur as parallel or subparallel aggregates, as laths or as spherules. The fibre direction is perpendicular to the layering and to the free surface while the banding is parallel to the free surface or parallel to the walls of the cavity in which the chalcedony occurs.

The hardness of chalcedony is somewhat variable and lower than that of coarsely crystallized quartz, at 6.5. The SG, which also varies owing to porosity and water content as well as to the presence of other substances, is usually 2.57–2.64.

Chalcedony is about 90–99% SiO_2 with the higher content in the lighter coloured material. Brown and reddish subvarieties are relatively high in Fe_2O_3. Water in chalcedony is not essential but is held by capillary forces; (OH) is also present and may be held as a substitution for O in (SiO_4) tetrahedra in a strained region between adjacent interlocking fibres. The hypothesis that opal is contained in chalcedony as an interstitital material between the fibres of quartz is not borne out by investigation.

The RI of chalcedony lies in the range 1.530–1.539, most commonly near 1.534, for the ordinary ray. It is hard to state a precise value for the extraordinary ray but values range upward from about 1.538. In practice, a small birefringence can sometimes be seen but this amounts only to 0.004 at the highest. This and other anomalous optical properties arise from the disoriented aggregates of fibres.

Chalcedony fractures readily across the banding to give a splintery surface with a waxy lustre. Most chalcedony shows pale colours, often with a greyish tinge. Tyndall scattering is responsible for a bluish colour seen in some specimens by reflected light. As banding becomes more distinct the boundary between chalcedony and the agate subvariety is crossed.

When chalcedony or agates display spectral colours (iris effect) the structures contain elements giving a periodic effect and acting as a diffraction grating. Twisting of the fibres is known to occur and if adjacent fibres are in twist phase a periodic planar variation in RI is produced perpendicular to the fibre elongation direction; this may produce the diffraction effects. The name iris agate has been used for specimens with this property.

The fibrous nature of chalcedony enables it to take up dyes and such material, often with brighter colours than untreated chalcedony, is widely available commercially.

The luminescent response of chalcedony to UV varies considerably. Mostly the glow seen is bluish white, but in some varieties, particularly those from Wyoming in North America, which contain traces of uranyl, (UO_2^{++}), ions, the glow is bright yellowish green. Fluorescence in chalcedony is best seen with the SW source, for many agates do not glow at all under LW. The banded agates will often show the fluorescent light in bands or in patches.

Chalcedony is deposited in a variety of environments and deposition is commonly at low temperatures. Light-coloured material with no marked banding is often found as a late hydrothermal deposit or an alteration product in acidic to basic igneous rocks, tuffs and breccias. It is vary common as crusts, as vein and cavity fillings.

Chrysoprase

The finest translucent bright apple-green chrysoprase is highly valued as far as quartz gemstones go. It occurs in quartz veins in nickel-bearing rocks or in veins in laterites overlying Ni-bearing serpentinites (Queensland) and associated with jadeite in veins in serpentinized gabbro and ultramafic inclusions in Lower Silesia. These and later summaries are taken from Heaney, Prewitt and Gibbs, *Silica; its behaviour, geochemistry and materials applications* (*Reviews in Mineralogy no. 26*, Mineralogical Society of America, 1994). The colour arises from admixed fine-grain nickel compounds in the silica matrix rather than from substitutional nickel in the silica itself. Heaney *et al*. list some of the papers discussing models for the origin of colour. The absorption spectrum shows features of octahedral Ni^{2+} with bands centred at about 1105 and 652 nm. Chrysoprase, popular in the nineteenth century, was often cut as low cabochons with a narrow rim of facets around the edge.

It was mined in the Urals, though probably the most important European deposit was and to some extent still is in Silesia, Poland. Good quality chrysoprase was found in Tulare County, California, and the deposit worked until the early years of the twentieth century when popularity declined. At the time the colour was reported to have been near emerald-green. Probably the main source of chrysoprase today is the Marlborough Creek district, about 145 km from Rockhampton, Queensland, Australia, where it occurs in veins up to 200 mm in thickness. Nickel content is high and colour consequently good.

Chrome Chalcedony (Mtorolite)

Green chalcedony coloured by chromium was found in 1955 in the area of the Great Dike, Zimbabwe, north of Mtoroshanga. It shows red under the colour filter and a sharp one-band absorption in the red. The material has locally been called mtorolite.

Simulation of Green Chalcedony

The dearth of fine chrysoprase has been compensated for by staining agate green, either by nickel or by chromium. The latter is more common, and may be detected by the brownish red residual colour shown when the stone is viewed through the Chelsea colour filter (chrysoprase shows green) and by three vague chromium lines in the red. Nickel staining is more difficult to determine, but whatever type of staining is used, it is sometimes possible to see mosaic-like veining. This effect is due to the sheaf-like nature of the crystal fibres which lie at right angles to the layers of the chalcedony. In order to get the best penetration of the dye solutions in staining agate, it is necessary to slab the mineral parallel to the layers and then the outlines of these aggregates of crystals tend to show up. Chrysoprase is imitated by suitably coloured glass with its customary bubbles and swirls.

Other Chalcedonies

Plasma is microgranular or microfibrous chalcedony coloured various shades of green, including apple-green and leek-green. Particles of various silicate minerals (chiefly of the chlorite group) disseminated through the material are responsible for the colour, which may be varied by the presence of white or yellowish patches. India, Madagascar, Egypt and the north-western states of the USA have commercial deposits of plasma.

More translucent than plasma, and coloured leek-green, is prase, the colour coming from fibrous ferrohornblende aggregates or from disseminated chlorite group minerals. Vermont and Pennsylvania have deposits of prase, and the material is found worldwide. The name buddstone has been given to a green chlorite-rich chalcedony with white veining, found in southern Africa.

The names heliotrope, now hardly used, and bloodstone, still in common use, are used for a type of chalcedony or plasma with spots of iron oxide or red jasper resembling blood spots against a dark green background. Good-quality material comes from the Deccan trap-rocks, India, from Brazil and many other countries.

Uniformly coloured light brown to dark brown translucent chalcedony is known as sard. It may have reddish or orange tints. Good-quality material is blood-red in transmitted light and blackish red in reflected light. Uniformly coloured red to reddish brown or flesh-coloured chalcedony is known as cornelian or carnelian; the boundary between sard and cornelian is a matter of individual judgement. The colour of cornelian is caused by colloidally dispersed hematite. India has produced most of the finest sard and cornelian since the fourth century BC but there are many other sources worldwide. Much commercial material comes from various deposits in Brazil and Uruguay. Some beaches on the east coast of England produce cornelian. Other reddish chalcedonies may have been coloured by heating since if they contain disseminated iron compounds the heating will oxidize them.

Amethystine chalcedony is reported from a single locality in central Arizona. It has been marketed under different names.

The red mercury sulphide cinnabar is an occasional inclusion in chalcedony. Such a material can be a translucent white chalcedony with streaks and clouds of bright red or pink cinnabar, or it can be so impregnated with cinnabar as to be completely red. Blue and green chrysocolla is another mineral which impregnates chalcedony to give a stone of intense sky-blue colour which has been called chrysocolla quartz. Most of these colourful impregnated chalcedonies come from the USA.

Some translucent chalcedony when cut in the cabochon form has a schiller which resembles to some extent the true moonstone. A translucent chalcedony of fine blue colour is found about 130 km north-east of Okahandja, Namibia, and an attractive blue and white banded chalcedony also comes from Namibia. This is sometimes called lace agate. A blue chalcedony has also been found in Cornwall, England. Chalcedonic pebbles abound along the Vaal and Orange rivers in South Africa.

Fire agate is a botryoidal growth consisting of platy crystals of an iron mineral over layers of chalcedony. With careful polishing away of the brown outside layers, the rainbow colours of the iridescent layers appear. This material has been set in jewellery.

Agate

Agate is a variety of chalcedony in which successive layers in the banding are different in colour and translucency. It occurs in filling cavities, and individual bands are concentric to the external surface of the mass or nodule. Layers show sharp angles on occasion, echoing the plan of a fort, and the name fortification agate may be used; landscape agate, ruin agate and star agate are also familar names.

Sizes of agate vary a good deal and may reach hundreds of kilograms. Colours are generally pale though varied; natural colours are green, yellow, red, reddish brown, white and bluish white, among others. Heating may change brown colours to red, and oil immersion may heighten translucency.

Some types of agate have their own names. Onyx is used for agate with milky-white or white bands alternating with bands of black or dark brown; white with brown bands may also be called sardonyx. Cameos are made from onyx with the figure carved in white in relief against the black background. The now dated name jasp-agate refers to the alternation of translucent chalcedony with opaque red jasper layers.

Much agate is dyed to give strong and commercially acceptable colours. The material is sawn and shaped before dyeing. Immersion in a hot sugar solution followed by immersion in concentrated sulphuric acid and heating gives black; a blue colour, once called Swiss lapis, is produced by immersion in potassium ferrocyanide and subsequent warming in a solution of ferrous sulphate to give a precipitate of Berlin blue. Chromium alum or potassium dichromate may be used to give green or bluish green, and nickel compounds give a brighter, apple green. Impregnation with iron compounds and heating may give reddish brown and red colours. Yellow to greenish yellow is obtained by heating dry agate that has been treated with concentrated hydrochloric acid.

A pagoda-like shape may be obtained from some Myanmar agate when the piece is slit across the whitish bands. The name pagoda stone has been used for these attractive curiosities.

Agate geodes are found in basic lavas and other igneous flow rocks and have probably been formed by silica deposition in cavities created by gases. The smaller cavities are near-spherical; larger ones, known as amygdules, are elongated and flattened. Characteristic chalcedony nodules, locally known as thunder eggs are recovered from rhyolitic lavas and tuffs in the north-west of the US. Some show a central five-pointed star or a cavity in the middle. Agate banding involves a rhythmic deposition of silica on cavity walls according to one of several theories. Alternatively a silica gel process is possible. Here a cavity may contain liquid colloidal silica with metallic salts; percolating acid water may later enter the cavity and alter the colloidal solution to a gel. Diffusion within the gel leads to the formation of Liesegang bands giving bands of colour. This process can be repeated experimentally. Iris agate shows rainbow-coloured bands when sliced across the layering.

Occurrences

Agate nodules are found worldwide and many countries will claim the finest specimens so any attempt at a complete list or evaluation of deposits would not significantly advance knowledge. Material from deposits in Egypt features in antiquities over 3000 years old and it is likely that the name agate is derived from the Drillo (formerly Achates) river in Sicily. Agate from the hills surrounding Idar in the Nahe region of Germany gave the impetus to the development of the twin towns Idar-Oberstein. Most agate on the market will have come from Brazil (the state of Rio Grande do Sul) or from Uruguay. Deposits in Rio Grande do Sul are attractively described by Reinhard Balzer in *Rio Grande do Sul, Brasilien*, 2003; ISBN 3882931361.

There are deposits in some of the western states of the USA. The collector Matthew Heddle is invariably linked in Scotland with a fine collection of agates some of which are now in the Royal Museum of Scotland, Edinburgh. 'Heddle localities' and their specimens are still keenly sought, not always with success though this makes the search more interesting!

In south-west England agate is found at the Dulcote site between Shepton Mallet and Wells where mining for limestone has made accessible a thick bed of geode-bearing clay. The geodes may have variously coloured fillings.

Thin-walled chalcedony geodes containing a large air space may also contain water whose movement can be heard or seen if the shell is sufficiently thin. Evaporation of the liquid is responsible for the air space. The best examples are from the agatized areas of Brazil and Uruguay; they occur also at Monte Tondo, Vicenza, Italy and elsewhere. The name enhydros is still used.

Moss agate is a pale bluish, grey or yellow translucent to subtransparent chalcedony containing dendritic inclusions of various oxides. Their colour varies; most are black but red and green dendrites are also seen. The name Mocha stone is used synonymously with moss agate and comes from Mocha in the Yemen. The best-known locality for green moss agate is the Deccan traprocks of India. Dendritic agates are found in a number of localities in North America, particularly in Montana, Oregon, and Idaho and in the Sweetwater agate beds near Granite Mountain in Wyoming, where the local agate contains traces of a uranium mineral and in consequence fluoresces a strong green colour under UV. Moss agate is found in many places and some of the Scotch pebbles from the Ochils, Scotland, show red and green moss. Moss agate is imitated by placing some chemical, like manganese dioxide, with thin gelatine on one surface of a glass plate. When a chemical 'tree' has grown on the gel the excess water is driven off by gentle heat and a second plate of glass to act as a cover is cemented on. The whole may then be ground and polished in cabochon form. The readily seen join between the two pieces of glass exposes the fake.

Chalcedony Pseudomorphs

Chalcedony often replaces other materials and takes their form. Agatized wood is thus a chalcedony pseudomorph after wood and is cut and polished. Bone has also been fossilized with agate, and at Tampa Bay, Florida, there is much chalcedony pseudomorphous after coral. Cabochons are cut and frequently dyed blue or pink. Algae (seaweed) and small mollusc shells may be preserved by silica and used for ornamentation. The process of replacement by silica, as in the case of the petrified wood, is that the original substance is replaced particle by particle with silica so slowly that the original form remains. In the case of fossil wood this is done so that even the tree rings are still visible.

Jasper

Jasper, which is the archetypal collectable beach pebble, consists of massive, fine-grained quartz, fairly dense, containing significant amounts of other materials, particularly iron oxides. Most commonly, jasper is a dark brownish red but it may be yellow or black. Variegated types are found. Banding is often present, the bands being planar rather than concentric as in agate. Finely divided hematite is responsible for the red colour and goethite for the brown and yellow colours. The presence of clay may give white, yellowish or greyish material with a porcellaneous appearance.

When the agent causing colour is present only in small quantities, jasper grades into a translucent fine-grained material for which the names chert, hornstone and novaculite have been used. This material may range from near-colourless through yellow to reddish or brown.

Jasper, unlike agate and chalcedonies, occurs as extensive beds of sedimentary or metamorphic origin. Jasper grades into or is combined with other types of quartz which may traverse it or be combined with it as a breccia. Jaspilite is a metamorphic rock containing alternating layers of jasper with black or reddish hematite. Jasponyx is thin-banded jasper with alternating dark or light bands; the name band jasper or ribbon jasper is also used. Lyclian stone or touchstone (basanite), a black fine-grained jasper, is used for testing the streak of gold alloys.

Jasper occurs as a cavity filling or as nodules or veins in iron ores. It occurs in altered igneous rocks and in detrital deposits. Large-bedded deposits provide a good deal of ornamental jasper. It may occur in variegated red to brown colours as a petrifying agent of wood; orbicular varieties and jaspers displaying spherules of different colours are used as ornaments.

Chemical and Physical Properties

The variable composition of jasper accounts for its wide range of SG, between 2.58 and 2.91. Identification should present few problems.

Specimens of the red and green ribbon jasper from Russia have an SG between 2.7 and 2.8, and one recorded specimen of heavily pigmented brown material from Siberia gave the high value of just over 2.9. The hardness of jasper is very little below that of quartz, and as a rule the material is tough although it may break easily along veins or laminations. The RI is approximately 1.54.

Petrified dinosaur bone from Colorado, Utah and Wyoming in the USA is used for ornamental purposes. Fossil 'dinny bone' is used in West Germany for such carvings as small tortoises where the markings of the fossil bone give some realism to the carving. Turritella agate, often used for ornament, consists of agatized shells of the turritella. The picture jasper at Biggs Junction in Oregon is probably a jasperized rhyolite, and when suitably cut shows scenic designs in light and dark-brown colours, and sometimes with black 'trees' formed by dendritic inclusions. A similar material is found in the Namib desert.

Occurrences

Jasper occurs worldwide. Ribbon jasper is reported from Okhotsk in eastern Siberia and other varieties are found in the vicinity of Troitsk and Verkhne Uralsk in the sourthern Urals. India and Venezuela supply a red jasper but the most varied examples are found in the USA. Orbicular jasper is found at Morgan Hill, Santa Clara County, California; the Arizona petrified forest is the prime location for jasper replacing wood.

17

Opal

The most up-to-date account of opal for the mineralogist is vol. 4B of the second edition of Deer, Howie and Zussman, *Rock-Forming Minerals*, published (2004) by the Geological Society of London; ISBN 1862391440.

Opal's play of colour seen against a dark or light background (black opal, white opal) can raise it to the level of a major gem species as prices asked in the saleroom will attest. While rumours persist that opal is unstable and will inevitably break, crack or craze, specimens should survive normal use without difficulty, although extremely hot and dry conditions of storage should be avoided as should violent temperature/humidity swings. British textbooks have in the past wearisomely trotted out legendary and forgettable properties for opal but mishandling is the most likely cause of damage.

Fire and water opals are also very beautiful at their best. Fire opal relies on its red-orange-yellow body colour, with or without play of colour; in water opal the play of colour appears to be suspended in a colourless transparent body. Opal with a play of colour was for years known as precious opal as distinct from common opal but the adjective is superfluous in a text dealing only with gemstones. A green translucent, rarely transparent opal, coloured by nickel, can be very attractive.

Opal is silica SiO_2 with a variable amount of water, usually from 6 to 10%. Some opals are amorphous and some have a disordered crystalline structure. Gem opals are the least crystalline type of opal (Gaines *et al.*, *Dana's New Mineralogy*, 1997). Opals are formed of structural units of amorphous silica and crystalline cristobalite and/or tridymite [polymorphs, with quartz, of SiO_2]: precious opals are mixtures of amorphous and crystalline silica (low tridymite with the amorphous silica being dominant) – this is especially characteristic of opal from sedimentary deposits.

Opal has a hardness of 5.5–6.5 and SG in the range 1.98–2.20. The 1.44–1.46 but in practice the contact liquid might very well har porous specimen.

Cause of the Play of Colour

Opal's play of colour (this is not 'opalescence', a term that should be reserved for the milky effect seen in many translucent substances) was not thoroughly investigated satisfactorily until the 1970s when Darragh and Sanders published their findings in *Scientific American* 234(4), 84–5, 1976. Darragh and Sanders found that in opal with a play of colour the structure consists of a randomly faulted close-packed ordered array of minute transparent silica spheres which also contain a small amount of water. Additional amorphous silica cements the spheres together; in this silica there is a slightly different water content so that there is a small difference between the RIs of the spheres and of the cementing material. Optical diffraction takes place from the array; Nassau, in *The Physics and Chemistry of Colour*, second edition, 2000, reminds those readers with rather older textbooks that the theory of thin-film interference does not, after all, explain the play of colour.

Those familiar with opal will know that the colour shown by the different patches will change as the viewing angle alters. Only UV rays will be diffracted by very small spheres and as the size of the spheres increases, longer (visible) wavelengths will be seen until red appears. Nassau (2000) gives the relevant equations.

It is easy to recognize that a sheet of small spheres of similar size will give predominantly blue to green colours and that a mixture of sphere sizes gives the best chance for a beautiful array of colours – a stone with *value* red predominating will always cost more than another in which blue takes up most of the viewing area. For this reason opal with a play of colour is hardly ever faceted, although I have seen some superb faceted water opals. Under both LW and SWUV opal may luminesce white to bluish, brownish or greenish with often persistent pale phosphorescence. The black opals are, however, generally inert, and the fire opals usually show a greenish brown response. Much common opal shows a green fluorescence, and often phosphorescence, probably from uranium.

Opal may be found pseudomorphous after wood or fossils. I have seen a most attractive opal belemnite mollusc. Opal with a dark background and a notably fine play of colour is found at Virgin Valley, Humboldt County, Nevada, USA; this opal has often said to be rather more prone to cracking than some others. However, the Roebling Opal, a 2402 ct black opal from the Rainbow Ridge mines and on display at the Smithsonian

Institution, has shown very little if any degradation in more than 75 years (Gaines *et al.*, *Dana's New Mineralogy*, 1997).

Opal may also be found as pseudomorphs of other minerals, including gypsum and calcite. Opal pseudomorphous after aggregates of the sodium calcium sulphate glauberite has been called pineapple opal. Hyalite (Muller's glass) is a colourless transparent opal variety; with a play of colour it becomes water opal. In the same state a green-fluorescing yellow, black-streaked opalized wood occurs in Washoe County.

The name hydrophane has been used for a light-coloured opaque opal that becomes transparent and exhibits a play of colour after immersion in water. Cacholong is an opal with either a porcellanous or a mother-of-pearl-like lustre and is so porous that it will adhere to the tongue. Other types of opal include an opaque greyish to brownish concretionary type; tabasheer is an opal-like silica found within the joints of bamboo. Jasper opal is red, reddish brown or yellow-brown and closely resembles ordinary jasper. A violet-coloured opal is found in Mexico, and a blue-green chrysocolla opal has been reported. Marcus McCallum showed MO'D a necklace of spherical beads of a fine, almost chrysoprase colour but stronger, with more blue. The source was reported to be Peru. There was no play of colour but the colour was fine and even.

Occurrence

Much of the finest opal occurs in thin seams and has to be recovered together with its underlying rock (matrix). The significance for fashioning is discussed below. Alternatively some fine opal is found as nodules ('nobbies' in Australia). Australian opal is found chiefly in sedimentary host rocks though much precious opal, including that from Mexico, occurs with volcanic rocks.

Locations

Australia

Australia and especially the states of New South Wales, Queensland and South Australia produce, unquestionably, the finest black and white opal. The recent books of Len Cram, illustrated with superb colour photographs, are intended, on completion, to serve as a history of opal throughout the world. Cram has illustrated and discussed the opal mines of Lightning Ridge, New South Wales, and locations in Queensland and elsewhere in *Beautiful Opals, Australia's National Gem* (2000; ISBN 0958541426) and particularly in Lightning Ridge *in Beautiful Lightning Ridge* (2000; ISBN 0958541534). Lightning Ridge opal pseudomorphs after fossils are discussed in an attractive book by Elizabeth Smith, *Black Opal Fossils of Lightning Ridge*, 1999 (ISBN 086417909X).

In December 2004 I received a copy of what must be the finest illustrated account of *Lightning Ridge*; this is by Len Cram, *A History of Lightning Ridge*, forming the second part of the promised trilogy, *A Journey with Colour, Covering Opal in Australia*. The book is published by its author at PO Box 2, *Lightning Ridge*, NSW 2834, Australia. The ISBN is 0958541493. The illustrations are Cram's best and all that the reader will want to know about this classic locality can be found here.

From the jewellery point of view, black opal from Australia was not fully accepted by the European gemstone trade on its first appearance in London; dealers were accustomed to the 'Hungarian' opal with its white background to the play of colour. The book *Opal, the Gem of the Never-Never* by Tullie Wollaston (London, 1924) tells this story. Wollaston himself helped a great deal to bring black opal into favour. This rare book was reprinted by Flinders Press, Adelaide, in 1995 in a limited edition which is now itself rare.

A 1977 study by Eckert *The World of Opals* (ISBN 0471133973) gives a fair overview of the world's major opal locations. A good account of the history of Australian opal in the early days is *They Struck Opal*, by E.F. Murphy, published at Sydney in 1948. This, together with the book by Wollaston, is rare today but still very well worth obtaining.

Lightning Ridge is now the centre of a large opal field and sites such as the Coocoran and New Coocoran, the Lunatic Hill field on top of the Three Mile and Allah's Rush have been very productive over the last few years.

Other major fields in Australia include Coober Pedy, South Australia, which is 750 km north by north-west of Adelaide. Here the opals most commonly show a light background to the play of colour. A variety, contra luz, which shows its play of colour by transmitted light is also found. The first opal was found there in 1915 and the area is still producing fine specimens. It is the largest opal field in the world. The Andamooka field is also in South Australia; it lies nearly 600 m north of Adelaide and is particularly celebrated for crystal opal, a name given to a variety in which the play of colour appears to be set inside a transparent or translucent medium.

In New South Wales the discovery of the White Cliffs opal field led to the development of the first Australian opal mining town. A good deal of the opal found is pseudomorphous after wood, plants and animals.

Queensland has for years been known as the source of boulder opal in which the play of colour is combined with brown streaks of iron-bearing sandstone (ironstone). Cram (1999) illustrates some magnificent specimens of this material. The discovery of opal on Listowel Downs in the west of the state in 1869 began the development of mining. Until today such sites as the Hayricks on the southern side of Mount Canaway are producing some of the world's finest opal. The Yowah nuts are also found in Queensland; these are opal-filled nodules; the Black Gate mine also produces similar material.

Mexican Opal

In 1901 George Frederick Kunz presented a paper on Gems and precious stones of Mexico to the American Institute of Mining Engineers later printed in their Transactions vol. 32, 55–93; as far as I know this paper has not been reprinted since its appearance in *Comptes rendus Internat. Geol. Congress*, Mexico City, 1907. Kunz begins by stating that the opal is the only gemstone to be systematically mined in Mexico. The names lechosos and zeasite were given by the Mexicans to those varieties of opal which showed deep green flashes of colour; the name harlequin meant an opal with small, angular, variously coloured patches. Opal was mined in the states of Querataro, Hidalgo, Guerrero, Michoacan, Jalisco and San Luis Potosi, the most extensive deposits being in Queretaro and near La Esperanza. Deposits at Zimapan in Hidalgo have been known for a long time. In Guerrero the chief localities are at Huitzuco and San Nicolas del Oro, where the opals are described as transparent, streaked with red, green and blue.

Kunz goes on to state that Del Rio mentions in 1802 that in Zimapan, near the sanctuary of Guadelupe Hidalgo, hyacinth-red fire opal was found in abundance in a red trachytic porphyry. Queretaro was the chief opal-producing state at the time of the Kunz paper; he was able to visit the mines of Esperanza, ten leagues north-west of San Juan del Rio. Kunz quotes a report by Mariano de la Barcena in 1873 (*La Naturaleza* 2, 297–302) in which he said that he had discovered ten veins.

Kunz also reports on a violet-blue colour reflection by an opal from the Rosario mine and continues his description of Queretaro opal by recounting the cutting (in the city of Esperanza) and the prices paid for the rough and fashioned material. The common opinion that Mexican opals eventually lost their colour was current at the time the paper was published in 1901. The avoidance of dry atmospheres for opal storage was enjoined.

The all-time most important study of minerals as far as locations (and most other data) are concerned is still the multi-volume *Handbuch der Mineralogie* (Hintze): opal (as well as quartz and corundum) are covered by the first volume of *Abteilung* 2, published at Leipzig in 1915. Complete sets of Hintze would be unobtainable today (it is not, even now, complete) and few libraries hold sets. Nonetheless the information contained is often to be found nowhere else.

Opal from Mexico is best described in the editions of John Sinkankas's *Gemstones of North America* (1959, 1976); by 1969 opal production was at a very high level with Japanese buyers especially prominent. New fields were opened, especially in the state of Jalisco where, at the largest mine, La Unica, Japanese buyers were buying about 75% of the product. In the state of San Luis Potosi some opal with the colour of smoky quartz was discovered in the 1960s. This material was found as large botryoidal masses with some play of colour. In this instance the colour arose from diffraction from growth bands in the same way as in iris agate. The RI, at 1.4625, and SG, 2.257, are high for opal.

In the state of Jalisco the opal districts around the town of Magdalena contain large igneous intrusions, generally reddish rhyolites in whose gas cavities some fine opal can be found. The rock, at one time at least, was sold directly to field buyers. In Eckert (1997, *op. cit.*) a list of Mexican locations is given and in the third volume of *Gemstones of North America* (Sinkankas, 1997) the author cites Zeitner ('*The opal of Queretaro*', *Lapidary Journal* 33(4), 868–80, 1979), who claims that crazing is much less common than is often claimed. Further brief notes on Mexican opal can be found in Panczner, *Minerals of Mexico* (1987; ISBN 0442272855).

Honduras

In the third volume of *Gemstones of North America* (1997) Sinkankas states that opal from Honduras was known in the gem trade well before the Australian production. Opal from Las Colinas is reminiscent of Hungarian opal. Many mines exist around the town of Erandique: host rocks include basalt, andesite, rhyolite and altered rhyolite, and dark and light opals of good quality are found. Sinkankas quotes a price of $50–$120 per carat for the black matrix opal from the Tablon mine near Erandique.

Czech Republic and Slovakia

European opal sites have never been plentiful and accounts vary a good deal. The name Hungarian opal is used little today and in any case the political changes in this part of Europe have altered boundaries over the years. Opal mines in the former Czechoslovakia still produce some specimens but they do not turn up as a result of serious mining. Opals from this area date back to Classical times; today the opal sites are in the Czech Republic and Slovakia. Work on these materials has been carried out by Kourimsky and others; and details can be found in his books

Die Edelsteine der Tschechoslowakie (Mainz, 1981) and *Katalog sbirky drahych kamenu Narodniho muszea v praze* (Praha, 1968). The largest recorded opal from the Czech Republic/Slovakian mines weighs 3000 ct and is in the Vienna Museum.

Work by Rudolf Duda; of the Vychodoslovenske Museum at Kosice, Michal Hamet; of the Czech Geological Survey; and Kourimsky has shown that nine different types of opal can be found there; precious and common opal, bog opal, moss opal, wax opal, hyalite, chloropal, hydrophane and what is named as ungvarite (said to be a green opal coloured by inclusions of the smectite group mineral nontronite). Few of these opal varieties can have much ornamental value.

The opals occur in vugs, veinlets and seams and cavity fillings (Eckert, 1997), and while the common opals are usually opaque the better ornamental specimens may be translucent to transparent. The background is usually white. The play of colour may be 'broadflash' or 'pinfire' (self-explanatory terms). Eckert (1997) lists several of the precious opal deposits in the Czech Republic and Slovakia. Among them are Cervenica and Dubnik.

Kazakhstan

Kazakhstan has produced some very beautiful gem-quality fire opal but I have seen only specimen material when in that country. Turkey has produced some very fine fire opal from a site in the central part of north-west Anatolia; also at Polatli in Ankara province, near Karaman in Karaman province, at Simav in Kutahya province and at Bayramic in Qanakkale province.

Indonesia

Opal with a good play of colour on a brown to black background occasionally comes on to the market. The occurrence varies from basaltic rocks and rhyolites to seams and pockets. Workings at Labak in West Java were reported by Eckert (1997) to be the only site in production.

Brazil

Opals of good gem quality from unspecified localities in the state of Piauí in Brazil appeared on the London market in the early 1970s. During 1973 at the request of the state government the geology of the district was investigated and published in 1980. The chief primary source for these opals was in the Pedro II area in the state of Piauí and occurs where flat-lying Mid Devonian sandstones are intersected by intrusions of quartz-dolerite. The opals were found in and above the altered zone and in veins in the sandstone, while alluvial opals of poorer quality derived from the original sources were found up to 10 km distant, the area being drained by the Rio de Matos and its tributaries.

The best Brazilian stones are similar to some Australian stones of medium quality and have similar properties, some figures quoted being SG 2.09 and R1 1.458. A recent DVD, *The Precious Opals of Pedro Segundo*, depicts attractive specimens from NE Brazil.

Large flawless opals comprising the contra luz, hydrophane and crystal opal are reported from Opal Butte, north-eastern Oregon, USA. Fire, blue and dendritic varieties are also found. Opal occurs in rhyolite geodes embedded in decomposed perlite. The material varies in stability. A dendritic opal, white and multi-coloured, occurs in many places in the US, Zimbabwe and South Africa. The non-iridescent green prase opal coloured by nickel is found at Kosemutz and Zabkowice in Silesia, Poland.

Fashioning

Thin seams of opal on a dark- or light-coloured matrix are often cut together so that the opal is protected by the matrix base (true doublet). This can in turn be capped by any transparent material which acts as a magnifying lens. Examined from the side the join may be apparent.

Likewise, a thin piece of good opal on potch (opal with no play of colour) may be cut so that the potch acts as a backing. The term 'matrix' is used for a natural substance supporting precious opal – that is, the rock upon which the opal occurs. Such opals might well be mistaken for opal doublets but careful examination of the edge of the stone by a lens will show the slight irregularity of the matrix and opal and the absence of a distinctive join as seen in opal doublets. Such stones are rightly termed 'true opal'. Opal when in small areas in the other rock may be cut complete with the matrix and such stones are known as opal matrix. Sometimes opal forms in ironstone and the opal matrix cut from this shows patches of coloured opal in a chocolate-brown matrix, and a similarly appearing type of opal occurs where the ironstone and opal have impregnated roots of the gidgee tree (an acacia). The stones cut from this material are called gidgee opals. Such iron-impregnated stones have a higher density than true opal and may range in SG from 2.65 to 3.00.

In order to make use of thin films of opal, which often show a fine play of colour, they have been made into doublets by backing them with pieces of potch, black onyx or a black glass called 'opalite', which was at one time obtained from Belgium. The cement used to join the pieces has a high melting point so that it will not be affected by such heat as boiling water. An ingenious triplet is made where an ordinary opal doublet is completed with a cover-glass of rock crystal which fits over the top of the opal. Water opals always show their play of colour better when they are on a dark background, so it is common to mount such opals in a closed setting with a piece of black, or preferably purple, silk below. An imitation opal made of small fragments of real opal set in a black resin has been encountered.

18

Turquoise

Turquoise, whose blue has been ineffectively compared to many substances and phenomena whose appearance must mean different things to different people, is a beautiful material at its best and cloying when it is not. The further the colour is from what I have called barclaysbankite (this must be dear to some) the more attractive it becomes. Turquoise, with amber, is an ornamental material which you might give (in profusion) to a rival gem testing laboratory. The porosity easily allowing dyeing and/or impregnation, it has been synthesized (or at least imitated by) an artificial material of similar composition. There has been no adequate study in recent years (at least in English) and probably the best way of finding out more about it is to consult mineralogies or gemstone studies (where they exist) of turquoise-producing countries.

Turquoise is $CaAl_6(PO_4)_4(OH)_8 \cdot 4H_2O$ and though it forms triclinic crystals the massive gem material occurs as nodular or globular crusts or veinlets. The rare crystals are steep, pinacoidal, but though small they are sought by collectors at the classic site at Lynch Station, Virginia, USA. The hardness is 5–6 with an SG in the range of 2.60–2.91. The RI for the massive material is usually near 1.62: values for the alpha, beta and gamma rays shown by the crystals are 1.61, 1.62 and 1.65 respectively. The presence of ferric iron may alter the values to a small extent.

As turquoise is found at several locations in the US it is not surprising that several books have been published there. The one-time all-important book, however, is Joseph Pogue's *The Turquoise* [Sinkankas #5176] published in 1915 as *Memoirs of the National Academy of Sciences* vol. 12, parts 2 nd 3. To quote Sinkankas '[this is] the most comprehensive, carefully researched and endless source of information on turquoise ever published and with an unparalleled bibliography leading to far more information'. Fortunately it has been reprinted as a facsimile (1972).

A book in Russian, *Biryuza*, was published in 1981 by Nedra, Moscow. Despite the possible language difficulties it is surprising how items can be tracked down and there is a very extensive bibliography. The text is authoritative.

With gem turquoise RI determinations are not usually possible as the liquid used may enter the pores of the stone. The absorption spectrum of copper may be seen most effectively in deeper coloured specimens: two similar bands in the violet, one narrow and strong at 432 nm, the other fainter at 460 nm. The bands are best seen by reflected light but finding them (or, usually, the stronger of the two) should be practised when possible. Some turquoise may show greenish yellow to a bright blue (this last, presumably, when little or no iron is present: Anthony *et al.* (2000) give a possible iron content of 0.21%, this figure taken from *Dana's System of Mineralogy*, 7th edition). In general, turquoise simulants may give a suspiciously strong bright blue fluorescence under LW though a Gilson synthetic/imitation gives a dull blue glow which is indeterminate. Under SW no response has been recorded.

Turquoise can be polished quite effectively though the presence of matrix, usually a dark material long known as limonite, a name used for massive brown powdery aggregates of undifferentiated ferric iron oxides, may hinder the work; in some cases veining can be attractive. Though soft, turquoise will sustain the carving (often of characters) on the surface.

Turquoise in the US is popular and accounts for a notable proportion of US gem mineral production. In the case of the Americas the relationship between turquoise and the early and present inhabitants is strong; Pogue covers this aspect well and it is regrettable that a similar treatment was not accorded long ago to the use and recovery of turquoise in the Near and Middle East.

Probably the most significant turquoise-producing state of the Union is New Mexico from which a large amount of material has been recovered over the years; the states of Nevada, Colorado and, in particular, Arizona rival it today.

New Mexico has been fortunate in its authors; Northrop's *Minerals of New Mexico* (revised edition, 1959) gives useful accounts of turquoise deposits.

The Cerillos Hills in Santa Fe county are one of the major deposits where native Indian workings of considerable complexity can be found in Mount Chalchihuitl. A good account of this area and the presumed method of working can be found in Sinkankas, *Gemstones of North America* (first edition, 1959). Turquoise occurs along seams and cracks in a country rock composed largely of monzonite porphyry. The colour ranges from green to greenish blue and sometimes a fine sky blue, equal to the Iranian turquoise (Sinkankas, *op. cit.*). In volume 3 of the same title

(1997; ISBN 0945005229) Sinkankas highlights the 'faddish peak' of silver and turquoise jewellery and its fall in popularity in the 1980s. However, the selling term of 'South-Western jewellery' and the establishment of small workshops in many areas in south-western turquoise-producing states have kept the demand high.

In Arizona the main turquoise-producing site is the copper mines in the Globe-Miami district, Gila county, where turquoise occurs in the orebodies; turquoise from the Sleeping Beauty mine was reported to be treated with paraffin to deepen the colour (see below). In Anthony *et al.*, *Mineralogy of Arizona* (1995; ISBN 0816515557), the beauty of this turquoise is emphasized and some other locations given; Arizona is probably the leading turquoise-producing state today.

Colorado turquoise is described in Eckel, *Minerals of Colorado*, 1997; ISBN 1555913652. At least one of the leading turquoise deposits is at Cripple Creek in Teller county.

The blue colour of turquoise arises from idiochromatic copper and many specimens are porous. Stones with a high porosity may scatter white light from the pore structure to such an extent that the blue colour cannot be seen. If the pores are filled with a substance with an RI higher than that of air the blue becomes visible. Though soaking in water will produce the colour the process has no practical application; oil, wax or plastic may be used. Such treated turquoise is called 'stabilized' and in general it does not have to be disclosed. A change from blue to green may arise from contact with acids and even from the natural oils of the skin. The ligand field about the copper in the turquoise surface is changed in these circumstances; the nature of the ligand field is described in Nassau, *The Physics and Chemistry of Colour*, second edition, 2000; ISBN 0471391069 and by R.G. Burns in *Mineralogical Applications of Crystal Field Theory*, second edition, 1993.

In *Historia Naturalis* book 37,110, Pliny says "Comitatur eam similitudine proprior quam auctoritate callaina, e viride pallens" (with this stone [peridot has just been described] is associated a stone similar in appearance but of less value, of the pale green callaina). Authorities have suggested that turquoise is meant. When Pliny continues, giving some idea of the location of the mineral, he states that areas above India were meant. He later refers to another stone called callaica (described as a turbid callaina), and that turquoise is the mineral in question. Pliny quotes from Democritus that the stone is without transparency, agreeable and uniform in colour, satisfying the vision without penetrating it.

There is little available literature on the turquoise of Iran; there are many deposits including one at Ma'dan, 45 km north-west of Neyshabur (Nishapur). Some of the finest material from this area and others in the world is described by various authors in *extraLapis* 16 (1999;

ISBN 3921656486). In a table comparing the contents of alluvial and primary turquoise, the copper content of the primary blue material is 8.77 and 7.34 for a greenish blue turquoise from alluvial deposits. A survey by Bauer (1896, translated 1904) is taken by Scalisi and Cook (1983) and is quoted in the next page.

Cutting of turquoise is reported to be centred on Mashhad and sales on Tehran. In Meen and Tushingham, see below, *Crown Jewels of Iran* (1968; ISBN 0802015190), the illustrated gold cups and bowls studded with turquoise are presumably making use of Iranian material.

In *Standard Catalog of Gems*, second edition (1994; ISBN 0945005164), Miller and Sinkankas allow that Iranian turquoise may sometimes be equalled by North American material. The best qualities, often described by the adjective 'Persian' irrespective of origin, should be used to describe pure medium blue-coloured specimens, with the colour uniformly distributed and with no spots, flaws or inclusions visible. The gems should be translucent on thin edges; the name Persian spiderweb is used to describe stones in which a network of fline black lines divides the surface into evenly sized patches. Persian matrix describes a lower grade in which the dividing lines are coarser. Greenish tinges rapidly lower the value and lack of translucency is also deprecated.

Very porous turquoise is likely to be treated as described above and it is reported that a 'tongue test' will identify porous specimens, siliva rapidly drying – this test cannot be applied to stones that have already been treated. Yasukazu Suwa, in *Gemstones, Quality and Value*, second edition vol. 1 (1999; ISBN 4418999027), claims that both Arizona and Chinese material tend towards fragility and that much Chinese turquoise is green.

In *Gem and Crystal Treasures* (1984) Peter Bancroft describes the Abdurezzagi mine at Midan, Iran. The mine is in the province of Khorasan about 200 km south-west of the common border of Iran with Afghanistan and the former USSR. The deposit was originally named the Abu Ishagi deposit and mining takes place within tunnels and stopes. Here as doubtless at many other turquoise mines, turquoise mined from deeper levels may lose colour when exposed to light, so dealers keep the stones in moist conditions until the moment of sale. Dyeing is routinely practised.

Some interesting reports on turquoise occurrences can be found in Eugenii Ya Kievlenko, *Geology of Gems*, 2003, no ISBN but published by Ocean Pictures Ltd of Littleton, Colorado 80123-1942, USA. As this is a translation of a Russian text, readers can find descriptions of numerous locations of many gem materials not covered by standard gem books.

As far as synthetic turquoise is concerned there is an interesting account of work by Hofman who published recommendations for turquoise synthesis in 1927 (not included in the otherwise extensive bibliography); the recommendation is that a mixture of aluminium hydroxide with carbonic copper and phosphoric acid be heated and pressed. This may be a method similar to that adopted by Gilson. Balitsky and Lisitsyna announced in 1981 that they had synthesized turquoise which was virtually indistinguishable from natural turquoise. (*Synthetic analogs and imitations of natural precious stones* (in Russian) [a monograph].)

Kievlenko appears to concur with some other mineralogists when he suggests that turquoise is a supergene mineral characteristic of weathered crusts (zones of oxidation) of rocks containing copper sulphides (chalcopyrite) and phosphates (apatite and phosphorites). This appears to be confirmed by the sinter forms of turquoise and the poor crystallization of turquoise aggregates; also by the paragenetic relationship with opal, limonite, jarosite, halloysite and other supergene minerals. Kievlenko considers that a postulated hydrothermal origin for turquoise is less sustainable.

Kievlenko introduces the reader to the turquoise of Uzbekistan where fine deep blue gem-quality turquoise occurs at Ungurlikan. The turquoise often shows a green tint.

In a chart illustrating a geological genetic classification of turquoise deposits, Kievlenko places some of the best-known sources; for example, Nishapur turquoise is exogenic (as are other deposits), weathered crust of phosphorus-bearing rocks with copper sulphide mineralization, found in magmatic hydrothermally altered rocks, with hosting rocks including rhyolites, trachytes, quartz porphyries and andesites with apatite. The structural-morphological types of orebodies are extended or less often equidimensional stockwork zones with the turquoise occurring as lenses and nodules with jarosite, goethite, halloysite and quartz as accompanying minerals. Kievlenko places other turquoise deposits in this classification.

Scalisi and Cook, *Classic Mineral Localities of the World: Asia and Australia* (1983; ISBN 0442286856), the turquoise section drawing largely on Max Bauer's *Precious Stones*, 1904, this being the English translation of *Edelsteinkunde* (1896), state that Iranian turquoise is divided into three classes, Augushtari, Barkhaneh and Arabi, the Augushtari material being the finest and most valuable. Barkhaneh turquoise shows fine patterning of matrix; Arabi turquoise shows spotting with matrix and has a generally pale colour. All deposits are government-owned. The Nishapur district is 400 miles east of Tehran and 15 miles west of Meshed in Khorasan province. The neighbouring mountains, the Kuh-i-Binalud, form a chain tending E–W between Kotsham and Nishapur and consist of limestone, sandstone and clay-slates interbedded with large masses of gypsum and halite. These masses are intruded by younger tertiary volcanics (trachyte and felsic porphyries).

Many mines have been established over centuries on the southern slopes of Ali-Mersai, a mountain of 6655 feet, the turquoise occurring as veins 2–20 mm thick, between layers of brecciated porphyries cemented by limonite; it is also found pseudomorphous after feldspar. Weathered masses are mined directly from matrix and are also recovered from sediments and from talus (scree) at the base of the mountain. Further accounts by Gübelin may be found in *Gems & Gemology*, Spring 1966, 3–13.

The village of Maaden is the centre of mining activity and is situated on Ali-Mersai mountain. There were more than 200 active mines in 1876.

Egypt

Reports on turquoise from the Sinai peninsula of Egypt have been available for many years, useful examples being Barron, *The Topography and Geology of the Peninsula of Sinai* (western portion), published by the Egypt Survey Dept, Cairo, 1907, and Davey, *Notes on the occurrence and origin of turquoise in the Sinai Peninsula*, in *Transactions of the Royal Geological Society of Cornwall* 16(2) 1929. Kievlenko (*op. cit.* 2003) gives a useful and up-to-date account.

The Sinai turquoise mines are situated on the coast of the Gulf of Suez, south-west Sinai Peninsula. Cretaceous-early paleogene red Nubian sandstones and limestones, locally covered by young basalts, are exposed in the northern termination of the Precambrian Nubian-Arabian shield. Canyon-like faults are widely distributed. Turquoise deposits have been operated during historical times and a turquoise-bearing horizon has been located in the upper portion of the Nubian sandstone sequence. The mineral is found under a friable coarse-grained quartz sandstone. Turquoise is indicated when iron-enriched rocks of this kind are located. It occurs as lenses 1–1.5 cm thick especially where these 'baluta' sandstones contain intercalations of clay-enriched shale. The turquoise is usually greenish blue, often showing bright blue spotting.

Embrey and Symes, *Minerals of Cornwall and Devon* (1987; ISBN 0565010468), mention turquoise from the West Phoenix mine, Linkinhorne, Cornwall. This forms attractive blue-green spherules but the speciems are of interest chiefly to mineral collectors. No Cornish turquoise is of ornamental significance despite occasional claims.

Many other sources for turquoise are known but to detail them all would lend most of them undue significance.

A number of treatments to deepen the colour of pale turquoise can be found, with details of imitations and attempts at manufacturing a synthetic turquoise, in O'Donoghue and Joyner (2003) and in O'Donoghue (1997) as well as in Nassau, *Gemstone Enhancement* (1994).

19

Lapis Lazuli

The ornamental material lapis lazuli can be classed as a rock rather than as a single mineral species. It consists of lazurite, sodalite, nosean and haüyne, the four members of the sodalite mineral group. These species occur mainly in association with leucite or nepheline in nepheline syenites, phonolites and related undersaturated igneous rocks (Deer, Howie and Zussman [DHZ] *Rock-Forming Minerals*, second edition, vol. 4B (2004; ISBN 1862391440)). *Framework silicates. Silica minerals, feldspathoids and the zeolites.* Lazurite and sometimes sodalite are occasionally found in contact metamorphosed limestones. The name lapis lazuli is usually given to the typical lazurite-rich rock which usually includes brassy yellow pyrite crystals, calcite and a colourless pyroxene. Nassau (2000) describes how it was discovered that the colour of lapis lazuli was derived from three polysulphide units of three sulphur atoms having a single negative charge. The S_3^- ion in sulphur has a total of 19 electrons in molecular orbitals and a transition among these orbitals produces a strong absorption band in the yellow at 600 nm, giving a blue colour with yellow overtones. The blue colour is increased by increasing the sulphur and calcium contents. A green colour results from insufficient sulphur. The process can be described as an anion–anion charge transfer.

Lapis lazuli has been mined in the Sar-e-Sang district in Badakhshan, north-east Afghanistan, for many centuries. The lazurite occurs in a rock with diopside, calcite and pyrite. Lapis lazuli is localized in two horizons of magnesian marbles near the centre of the Hindu Kush granitic massif (DHZ, *op. cit.*). Beds and lenses of lapis lazuli at Sar-e-Sang were described as up to 4 m thick and to extend for 400 m within a skarn formed under relatively high conditions of pressure and temperature

(Wyart *et al.*, Lapis lazuli from Sar-e-Sang, Badakhshan, Afghanistan, *Gems & Gemology* 17, 184–90, 1981).

Lapis lazuli in the Baffin Island deposit at Lake Harbour shows a primary coating of deep blue to a paler blue and a teal green, not usually recorded for lapis lazuli. In this deposit Hogarth and Griffin (*Lapis lazuli from Baffin Island-a Precambrian meta-evaporite, Lithos* 11, 37-060, 1976) and Cade *et al.*, *Abstracts of the Eighteenth General Meeting of the International Mineralogical Association, Edinburgh, 2002*. The deposit at Lake Harbour is in a dolomitic marble.

Lapis lazuli is found in the North Italian Mountains of Colorado, USA, where the occurrence is in impure marble layers near their contact with Tertiary quartz monzonite and quartz diorite of the Italian Mountain stocks (Hogarth and Griffin, *Contact-metamorphic lapis lazuli: the Italian Mountain deposits, Colorado, Canadian Mineralogist* 18, 59–70, 1980).

Lapis lazuli is being mined from a limestone-granite contact (elevation 3500 m) in the headwaters of the Cazadero and Vias rivers, Ovalle, Coquimbo, Chile, close to the border with Argentina. The lapis here is paler in colour and associated with wollastonite rather than diopside as in the Baikal deposits, Russia. The Chilean lapis contains phlogopite, sodalite, calcite and pyrite and is described by Borelli *et al.*, *Caratterizzazione del lapis-lazuli* (Borelli *et al.*, *La Gemmologia* 11(4), 24–27, 1986 and Coenrads and Canut de Bon, *Lapis lazuli from the Coquimbo region, Chile, Gems & Gemology* 36, 28–41, 2000.

The properties of lapis lazuli are helpful in identification. The hardness is usually between 5 and 6, the RI near 1.50 and the SG 2.7–2.9; if a significant amount of pyrite is included the figure will be higher. Under LWUV some specimens may show a streaky orange fluorescence (Chilean specimens in particular). Under SW lapis may show a pinkish response. If touched with HCl a characteristic rotten egg smell will result.

Using a ceramic technique Pierre Gilson has manufactured a product known for convenience as synthetic lapis. Specimens may or may not contain added pyrite. The material has been found to be porous with a lower SG than natural lapis, near 2.46. The RI is close to 1.50 but the shadow-edge is vague. The Gilson material is discussed in a paper in the *Journal of Gemmology* 19(7), 1985: the material should be regarded as an imitation though it has been offered as synthetic. Ultramarine and hydrous zinc phosphates are the main components. An imitation of lapis by Verneuil-grown spinel has higher constants (SG near 3.6, RI near 1.73) – interestingly gold flecks have sometimes been added to simulate pyrite!

Dyed howlite has also been used as an imitation though it is easily detected by an orange fluorescence after the dye has been removed. A dyed jasper with the quartz SG near 2.65 was at one time known as Swiss lapis.

A cobaltian polycrystalline material has a granular structure and shows red through the Chelsea filter. A dyed blue quartzite has also been used as a lapis imitation. In many cases of dyed specimens acetone will be found to remove the dye which has been found to penetrate about 1.5 mm from the surface. One interesting imitation was found to be dyed dolomite with RI between 1.50 and 1.68 and SG 2.85.

The best contemporary account of lapis from the classic deposits of Afghanistan is given by Bowersox and Chamberlin, *Gemstones of Afghanistan*, 1995 (ISBN 0945005199). The Badakhshan deposits are in the north-east of the country in a mountainous area and occur in mainly plutonic and metamorphic rocks which are cut by deep valleys. The lapis deposits are situated along the valley of the Kokcha river, on the Blue Mountain on which folding and faulting occurred in the Cretaceous period when intrusion by a diorite porphyry took place. The steep valley slopes of the upper Kokcha river in the eastern Hindu Kush are gneisses with thick intercalations of greyish white dolomitic marble. Lenses of light to dark blue lapis frequently occur. In *Lapidary Journal* 38(11) Emmett explains that the lapis occurs in grey lenses of calcite-dolomite skarn formed by contact metamorphism of impure limestone and high-grade metamorphic sequence of Precambrian rocks resulting from the intrusion of masses of molten granite, causing the formation of marble. The skarn lenses are more than 1–4 m thick and are underlain by gneiss. One of the classic mines is set in strata of black and white limestone. The crystalline series in which the lapis is found consists of gneiss, leptinites and cipolin marble in a heavy layer, amphibolites, pyroxenites and peridotites in which the lapis is found in disseminated veins.

Efinov and Suderkin, in *Vestnik Akademii Nauk, Kazakhshoi SSR* no 8, pp. 64–66, 1967, published a paper The *Sary-Sang lapis lazuli* deposit of northern Afghanistan in which they identify nine producing zones.

In 1841 John Wood described a visit to the lapis mines (*A personal narrative of a journey to the source of the river Oxus*, Murray, London.) Some of his observations are recorded in Bowersox and Chamberlain (*op. cit.*), together with useful notes on mining, lore and legend in the area.

Lapis deposits have been found at Sludianka in the Lake Baikal area of Russia and were worked in the late nineteenth century. The paragenesis is described by Voskoboinikova in *Mineralogy of the Sludianka lazurite deposit, Zap. Vses. Min. Obshch ser.* 2, 67, 601–622, 1938 (in Russian with English abstract). A later discussion can be found in Ivanov *et al.*, *The mineralogy of lazurite deposits of the USSR*, in *Gem Minerals* (*Proc. XI Gen. Meeting IMA Novosibirsk*, published in *Acad. Sci USSR* 97–104, 1980 (in Russian with English abstract). A useful general guide is C. da Cunha, *Le Lapis Lazuli* (1984: ISBN 2268008444).

20

JADE – Geology and Mineralogy

Andrew Middleton

Introduction

Jade (*Figure 20.1*), the toughest of all the natural materials to have been worked by man, has been highly prized for millennia by societies around the world. It has been fashioned into tools and weapons, carved as

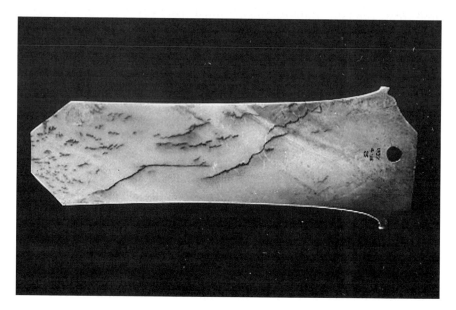

Figure 20.1   Votive 'jade' axe, probably jadeite, H = 29 cm; Olmec (British Museum Reg. No. Ethno. St 536). Copyright, British Museum. [MM033502]

figures and ornaments, and made into items for personal adornment of both the living and the dead. Jade and the artefacts made from it have frequently enjoyed a special significance, over and above their simple utilitarian value and intrinsic worth. Beck (2002: 9) in his discussion of the use of jade by Maori in New Zealand commented that jade has fulfilled a role played by no other stone, meeting the need for tools, weapons, religion and art. Jade was held in high esteem by the indigenous civilizations of Mesoamerica, but nowhere was this esteem for jade more highly developed nor maintained so consistently for so long as in China (*Figure 20.2*). Here, jade has been prized for more than seven thousand years, from the earliest Neolithic through until the present day. It has been regarded as the most precious of all materials, above gold and other gems; it has been held to possess a symbolic and spiritual, almost magical, significance; from the middle of the Zhou period (1050–256 BC) its physical properties have been used as a metaphor to describe the human soul (Michaelson and Sax, 2003: 3). The aim of this chapter is to explore the nature of jade, its geological occurrence, the ways in which it has been worked and used, and how it may be identified and distinguished from the numerous materials that have been used as imitations, almost since jade was first worked.

Terminology

'Jade' is a well-known term which is used widely to refer to a variety of coloured stones, mainly white to green, and typically having a tough and compact texture. However, the nomenclature of jade and jade-like materials is not altogether straightforward, and the term 'jade' has been applied by different specialist groups in different ways. A commonly adopted approach is to follow mineralogical terminology in which jade encompasses two quite distinct minerals, nephrite (tremolite-actinolite) and jadeite (Clark, 1993). Unfortunately, this approach is not without difficulty. If strictly applied, it excludes many objects traditionally regarded as being carved from jade: for instance, the Chinese word *yu* (jade) was used in ancient China to refer to nephrite jade but also to a range of other, more or less jade-like, materials. Thus, *yu* may be a mixture of nephrite with a variety of other minerals (i.e. a metamorphic rock), or it may be comprised of minerals and rocks containing neither nephrite nor jadeite.

In an attempt to resolve these problems and provide a consistent nomenclature, Wen and Jing (1992: 257–8) proposed the terms *true jade, hemi-jade* and *pseudojade*. True jade refers to nephrite and jadeite jade. Hemi-jade was used to describe rocks composed of mixtures of nephrite and other minerals (there seems no reason why this term

Figure 20.2   Ritual sceptre of nephrite jade; Zhou dynasty (British Museum, Reg. No. OA1937.4–16.25); L = 23 cm. Copyright, British Museum. [CM908]

should not be applied also to jadeite rocks). The terms *monomineralic* and *polymineralic* have also been used to refer to jade composed solely of jadeite and rocks composed of jadeite along with other minerals, respectively (see for example, Win Htein and Aye Myo Naing, 1994:

274); this is perhaps more precise mineralogically but it remains to be seen whether it proves to be generally acceptable. Minerals that may resemble jade and were 'mistaken for jade by ancient people' were referred to by Wen and Jing as pseudojade.

Although the use of jade in Europe for objects such as ceremonial axes can be traced back to the Neolithic, the word 'jade' is of considerably less antiquity. Apparently it was not used in print until 1727, when it appeared in *Chambers Encyclopedia*. Before then it had been referred to widely as jasper, derived from the Latin word *iaspis*, described by Theophrastus, writing in the late fourth century BC, and by Pliny several centuries later, in the first century AD. Pliny may have been using *iaspis* to refer to jade from the East (see Caley and Richards, 1956: 108). In the sixteenth century, after Europeans became aware of the use of jade by the indigenous peoples of Central America, the stone seems rapidly to have acquired a reputation as a treatment for kidney disorders and became known in Spanish as *piedra de hijada* (loin stone) or *piedra de los riñones* (kidney stone). Later, in the early eighteenth century, Sir Hans Sloane, founder of the British Museum, still referring to it as 'green jasper', wrote about *piedra de ijada*. This was translated into French as *pierre de l'ejade* and at some point *l'ejade* was corrupted to *le jade*, from which the English word derived. The term *nephrite* also alludes to the supposed medicinal properties of jade and was derived from the Latin *lapis nephriticus* (kidney stone); it was first applied in 1789 by the mineralogist A.G. Werner to early Chinese jades. It was not until 1863 that Alexis Damour, a French chemist and mineralogist, recognized that the term 'jade' encompassed two distinct materials of different chemical composition. For one of these two types he retained the term *nephrite*, for the other he introduced the additional term *jadeite* (Desautels, 1986: 2–3).

Mineralogy and Geology

Nephrite

The term 'nephrite' includes two silicate minerals, tremolite and actinolite, both members of the amphibole group of minerals. They are hydrous, calcium-magnesium-iron silicates and differ essentially in their ratio of magnesium to iron, with compositions lying in the range between $Ca_2Mg_5Si_8O_{22}(OH)_2$ (tremolite) and $Ca_2Fe_5Si_8O_{22}(OH)_2$ (ferroactinolite). Nephritic jades are typically magnesium-rich, with a magnesium to iron ratio that typically is greater than 4:1. Most would be classified as tremolite but some have more than 10% of the iron-rich end-member: these should be classified as actinolite.

Nephrite is formed geologically by metamorphism of a pre-existing, magnesium-rich rock; during this process the mineralogy and texture of the original rock are modified by high temperatures and pressures. Magnesium-rich rocks are not common and none is chemically identical to nephrite, so that some selective changes to the composition of the host rock are required during metamorphism. Such changes may take place by the passage of very hot fluids through the rock, a process referred to as metasomatism. During metasomatism, there must be a nearby source of the additional chemical elements required for the production of tremolite/actinolite, so that nephrite is commonly found close to the geological boundary or contact between two or more distinctive rock types. It is now recognized that there are two principal types of nephrite body. One type is associated with serpentinite, a magnesium-rich rock composed mainly of serpentine minerals, and well known as a jade simulant; these are known as ortho-nephrite. The second type is associated with dolomitic (magnesian) marbles; these are known as para-nephrite (Nichol, 2000). The recognition of these two types offers some possibilities for tracing nephrite jade sources, for nephrite derived from dolomitic marbles typically has a lower ratio of iron to magnesium; moreover, nephrites associated with serpentinites usually have higher levels of the trace elements Cr, Ni and Co (see for example, Wen and Jing, 1992: 268; Nichol, 2000).

The particular geological conditions necessary to form tremolite-actinolite rocks are unusual but not rare. However, there are two factors that make the occurrence of nephrite of gem quality relatively uncommon. First, the tremolite-actinolite is typically mixed with substantial quantities of other minerals, so that the valued jade properties of colour and translucency are not developed. Secondly, the formation of gem-quality nephrite jade depends not only upon the formation of relatively pure deposits of tremolite-actinolite but also upon the texture of the rock. Gem-quality nephrite is characterized by a dense intergrowth of fine, randomly oriented bundles of fibres (*Figure 20.3*). Frequently, the amphibole mineral grows in other forms, such as more prismatic shapes often displaying a marked preferred directional orientation of the fibres. The term *semi-nephrite* was introduced by F.J. Turner for materials having a texture intermediate between true jade (random orientation of fibres) and a schist rock (preferred orientation) (Clark, 1993: 630). The factors determining the texture of nephrite are still a matter of discussion and are not fully understood. Observations by Wen and Jing (1992: 263), made using a scanning electron microscope, have shown that the quality of nephrite jade can be related to its microstructure, with the finer jades having felted fibres of smaller average size.

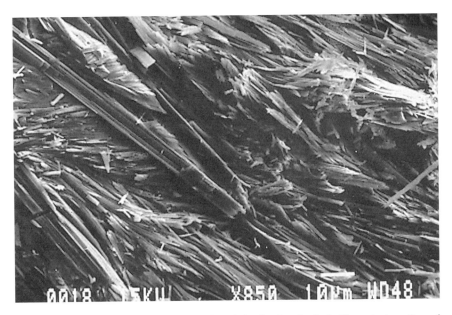

Figure 20.3   Optical photomicrograph of nephrite jade, showing the finely fibrous texture. Crossed polars, width of field c. 1 mm. Copyright, British Museum. [CU337]

Jadeite

Jadeite, a sodium aluminium silicate, is a member of the pyroxene group of minerals. The formula of jadeite is $NaAlSi_2O_6$; other components that may be incorporated include $CaMgSi_2O_6$ and $NaFeSi_2O_6$. Like nephrite, jadeite is formed by the metamorphism of pre-existing rocks, in this case under unusual conditions of moderate temperature, coupled with very high pressure. In addition, the formation of jadeite, like nephrite, requires metasomatic chemical changes to the composition of the parent rock. The formation of jadeite rock typically involves an interaction between a rock which is particularly rich in a source of sodium and one which is low in silica, such as where a rock rich in the feldspar mineral, albite ($NaAlSi_3O_8$), is enclosed by or adjacent to serpentinite, which is low in silica.

Jadeite does not occur as finely fibrous crystals like nephrite, instead it has a texture in which the individual crystals are more granular but closely interlocked (*Figure 20.4*). The term *semijade* is sometimes applied to less compact examples of jadeite. As for nephrite, not all rocks that contain jadeite are of gem quality, and high-quality jadeite is truly rare.

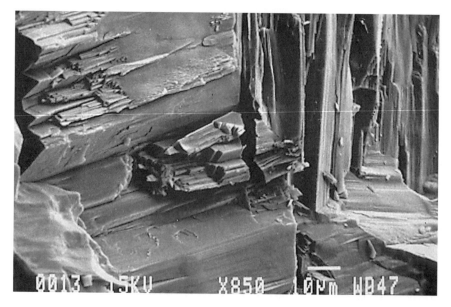

Figure 20.4   Optical photomicrograph of jadeite jade. Note the interlocking granular texture. Crossed polars, width of field c. 3 mm. Copyright, British Museum. [CU336]

Sources

Because of the geological conditions required for their formation, deposits of gem-quality jade are often quite limited in extent; they may be in the form of thin veins, sheets or blocks only metres in diameter, within or at the margins of other rocks. Much jade has been recovered from secondary alluvial deposits. Consequently, the identification of ancient sources is difficult and some deposits exploited in the past have yet to be located; some may have been completely worked out in antiquity. It is only in the last two decades that the deposits of nephrite likely to have been the sources for many early Chinese jades have begun to be been identified (Wen and Jing, 1992, 1996, 1997; Jing and Wen, 1996). The location of a possible source of the distinctive but elusive 'Olmec Blue' jade (jadeite) is equally recent (Seitz *et al.*, 2001).

Nephrite

Because nephrite is so tough, pebbles and even boulders may be carried by rivers for many miles from their original source outcrops, to be deposited in river gravels. The passage downstream provides a most rigorous natural 'quality control process', which only the toughest survive, so that it is perhaps not surprising that in China nephrite from river gravels was valued more highly than the quarried material (see Hansford, 1950: 38). Again, in New Zealand, most of the jade used by the Maori was (and continues to be) found as boulders.

China   The jade used in China was almost exclusively nephrite until the eighteenth century, when jadeite from Burma began to be imported. This is often called *feicui* (kingfisher), although this term was used originally to refer to bright green nephrite. It is often suggested that the ancient sources of nephrite lay in the Lake Baikal region of Siberia and in the Khotan area in the western province of Gansu. However, whether these were the earliest sources to have been exploited has been discussed recently by Wen and Jing (e.g. 1997). They have shown that most of the jade used in the Neolithic period was para-nephrite (i.e. derived from dolomitic deposits). Their analyses caused them to question whether conditions of regional metamorphism are necessary to the formation of nephrite, leading to the location of previously unrecognized deposits of nephrite in a zone of contact dynamothermal metamorphism in Jiangsu Province, in the east of China. This is close to the main centres of jade working in the Neolithic period. There were also supposedly sources in north eastern China, the area of the earliest jade working. Nevertheless, the Black Jade River and the White Jade River that run through the area of Khotan (Xinjiang Province), in the far west of China, have been important sources of alluvial nephrite jade for millennia (*Figure 20.5*). The primary sources (para-nephrite) in the Kunlun Mountains have also been exploited (see Hansford, 1950: 36ff. for an historical account of these sources).

New Zealand jade (*pounami*), also known as 'New Zealand greenstone', has been fashioned by Maori artisans into a variety of implements, weapons and personal ornaments. The Maori term *pounami* is applied mainly to nephrite, with different varieties being recognized by their colour and texture, but also includes *tangiwai* which is in fact bowenite (i.e. a variety of serpentinite, well known as a jade simulant; see below). Nephrite is found as boulders and cobbles in rivers (and on beaches) in several areas of South Island. Some of the sources yield very fine, translucent, pale-coloured varieties (*inanga*) sometimes with chatoyant bands. The primary geological sources are associated with ultramafic rocks, including serpentinite, so that these deposits are classified as are ortho-nephrites. For a recent and comprehensive account see Beck (2002).

New Caledonia   Several alluvial and *in situ* sources of ortho-nephrite have been reported on the east side of the island and also inland. Nephrite and semi-nephrite were used by the local population for implements and ornaments.

Figure 20.5   Pebbles of jade collected from Khotan, China. All are of nephrite except for the large black triangular fragment (lower left), which is serpentinite. [CU898]

Australia   A pale-coloured nephrite has been reported from Tamworth in New South Wales, associated with serpentinite (i.e. ortho-nephrite); dark green and black para-nephrite are mined commercially at Cowell, South Australia.

Zimbabwe   Good-quality dark green nephrite has been reported from Mashaba district.

Brazil   Sources of nephrite have been reported from the Amargosa district of Bahia State but these have not been confirmed, and Beck suggests these are probably artefact finds rather than geological sources. There is no confirmed source for nephrite in Mesoamerica or South America.

USA   Wyoming is said to be the source of the finest jade in the western hemisphere (Desautels, 1986: 18), though only a small proportion is of a fine green colour; some are chatoyant. It occurs as alluvial boulders near Lander and also as *in situ* deposits, associated with serpentinite (i.e. ortho-nephrite); these include 'black jade' found at Crooks Mountain.

Nephrite was formed in association with serpentinite in the coastal ranges of California and several *in situ* (e.g. Bagby, in Mariposa County) and alluvial (e.g. Placer County) sources have been reported. The coastal source at Jade Cove, Monterey County, is well known to collectors.

Nephrite (said to be associated with jadeite) is known to be obtained from Mount Vernon in Washington State.

There are extensive deposits of ortho-nephrite in the area of Jade Mountain (and more widely in the Jade Hills), in Alaska, to the west of Kobuk; *in situ* deposits are known but much has been recovered from river gravels; some show chatoyancy when cut as beads or cabochons (Webster and Read, 1994: 271). This source was first exploited by the Inuit.

Canada   There are numerous deposits of nephrite, both alluvial and *in situ*, associated with ultramafic rocks lying between the Coast Mountains and the Rocky Mountains of British Columbia. Boulders from the Fraser River system have been used by the indigenous population for thousands of years. These deposits are said to have been re-discovered in 1890 by Chinese placer miners, but modern exploitation did not begin until the 1960s. Canada is now the main world source of nephrite; high-quality material from Dease Lake is known as 'Polar Jade'.

Europe   Because there were no known sources of jade in Europe, it was initially thought that prehistoric artefacts of jade recovered from archaeological sites were made from material imported from China. However, subsequent discoveries have revealed several sources; the first to be located was at Jordansmuhl (*Poland*) in 1889. This and other ortho-nephrite sources in Silesia form the largest and most significant deposits in Europe. Other European sources include alluvial nephrite in Bavaria and *in situ* deposits in the Harz mountains of *Germany*. Worked nephrite has been found in Swiss lake dwellings, and a source located at Poschiavo in eastern *Switzerland*. *Italian* sources have been reported in the Appenine Mountains.

Russia   There are huge deposits of nephrite in Siberia that are mined commercially. To the west of Lake Baikal, in the Sayan Mountains, ortho-nephrite occurs both as *in situ* and alluvial deposits. These were re-discovered in the mid-nineteenth century and are characterized by dark flecks (chromite), which gave rise to the name 'spinach jade'. Some of the nephrite from Sayan are chatoyant. To the east of Lake Baikal, in the area of the Vitim River, are deposits of para-nephrite, again both alluvial and *in situ*, and including some exceptionally white material.

Mongolia   An ancient alluvial source of nephrite has been reported from the Prikhubsugul area.

Japan   Nephrite occurs as boulders and pebbles in the Matsukawa River, Nagano Prefecture, and as beach pebbles at Itoigawa, Niigata Prefecture. It was used along with jadeite by the ancient Jomon culture.

Korea   An almost pure tremolite, para-nephrite jade from Chuncheon (South Korea) has been mined since 1976.

Taiwan   Ortho-nephrite, some of it chatoyant, has been used for more than 5000 years.

Jadeite

There are far fewer sources of jadeite than nephrite, because of the very particular geological conditions required for its formation. Like nephrite, much jadeite has been recovered from river gravels and alluvial deposits, rather than from primary, *in situ* deposits.

Mesoamerica   Jadeite rocks (jadeitites) were valued in Mesoamerica from the time of the Olmecs (second millennium BC), who had access to supplies of fine blue-green jade known as 'Olmec Blue'. The Motagua Valley of Guatemala was identified as the probable source area for jadeite some years ago but, as noted already, a possible source for Olmec Blue jade has been identified only very recently. This work has revealed previously unknown resources of jade in the Sierra de las Minas to the south of the Motagua River (see Seitz *et al.*, 2001).

Burma   Jadeite occurs *in situ*, associated with serpentinites, to the north of Tawmaw, but most comes from boulders in the Uru Conglomerate, in a sense a 'fossil' alluvial deposit. There are also several secondary sources to the west of the Uru River. Burma is the world's main source for jadeite and the material is renowned for its quality and colour. Translucent to near translucent, bright emerald to blue green varieties are prized in China and are known as Imperial jade. A variety known as 'maw-sit-sit' was first reported by Gübelin (1965) from Namshamaw in upper Burma: it is characterized by dark green-black spots of chromite, often rimmed by dark green kosmochlor (= ureyite, the Cr analogue of jadeite; see *Table 20.1*, page 349), and associated with several other minerals (Webster and Read, 1994: 274; Colombo *et al.*, 2000). A particularly green variety with a 'looser' granular texture (which facilitates enhancement by impregnation with epoxy resin) has been referred to as 'hte long sein' (Ou Yang and Li Jian, 2001). The name 'inky jadeite jade' has been applied to a very dark variety, rich in omphacite (a pyroxene mineral; see Ou Yang *et al.*, 2003). Hansford (1950: 43–7) gives a brief historical account of the discovery and exploitation of the Tawmaw jadeite deposits; geological information is provided by Chibber (1934) and Bender (1983).

Japan   Jadeite occurs in the bed of the Kotai River as huge alluvial boulders, mostly white, but with some emerald green and mauve material. This is believed to have been the source for jade used by the Jomon culture; the deposits are now protected from mining.

USA   Jadeite has been reported from several localities in California.

Russia Beck (2002) comments that jadeite occurs in the northern Ural Mountains and also in the Sayan Mountains of Siberia.

Appearance and Physical Properties

Colour

Jade is prized for the range and quality of its colours and textures. The development of these colours depends upon two main factors: the chemical composition of the jade mineral itself and the presence of impurities. Neither nephrite nor jadeite has a unique chemical composition: each is a member of a complex chemical series, and their colour may be modified by the substitution of different metal ions (see further discussion below). Because the genesis of jade involves the alteration of pre-existing rock, the overall chemical composition of the resulting rock may not be precisely that of nephrite or jadeite, and the 'balance' of the chemical elements will have crystallized as other minerals. For example, albite, calcite, talc and antigorite (a serpentine mineral) have all been reported as impurities in Chinese nephrite (Wen and Jing, 1992). Not all these impurities may be detrimental to the appearance of the jade. The mineral kosmochlor imparts a desirable bright emerald-green colour to jadeite, and the black specks of chromite in so-called 'spinach jade' from the Lake Baikal area are not considered to spoil the appearance of their nephrite host.

Nephrite may range from near white ('muttonfat') through many shades of green to almost black (see *Figure 20.5*). Typically, nephrite that is low in iron and composed almost entirely of tremolite is white or pale coloured, whilst iron-rich nephrite containing more of the actinolite component is darker green. This relationship is illustrated by a series of analyses of New Zealand nephrites made by Finlayson early in the twentieth century and quoted by Woolley (1983), e.g. deep green 5.61% FeO; pale green 1.35%; greenish white 0.35%. Where nephrite shows a yellow or brownish colour, this may be due to the presence of iron in its oxidized, rather than reduced, state. Small amounts of chromium may also contribute to the green colour of some nephrites.

Jadeite occurs in a wider range of colours: like nephrite, it may take many shades of green, but pink, mauve, blue, grey, white, red, yellow, black and mottled variations are also found. The bright emerald green colour, so sought after in jadeite is due to the presence of a small amount of chromium. Small amounts of iron and nickel may modify the green colour caused by chromium; the oxidation state of any iron may also be significant, with white jadeite containing iron mainly in its ferric form; manganese may contribute towards the purple colour of some jadeites (Harder, 1995).

343

Refractive Index

The maximum and minimum refractive indices of individual crystals of *nephrite* vary between 1.600–1.627 and 1.614–1.641, depending upon composition, but when measurements are made on the surface of a polished nephrite jade only a single, rather diffuse reading in the region of 1.62 will be obtained. Similarly, for *jadeite*, although the minimum and maximum refractive indices for individual crystals vary between 1.640 and 1.692, measurement of a polished surface will typically yield only a single diffuse value of c.1.66 (Woolley, 1983: 258, 264).

Translucency and Lustre

The highest quality jade is valued for its translucency; some jadeites are almost transparent. Translucency varies with density of colour and the presence of flaws and impurities; impure hemi-jades will generally be less translucent. Jadeite takes a high polish, although it frequently presents a 'dimpled' surface because of variations in hardness across its surface; nephrite typically has a 'softer' lustre.

Surface Alteration

As a result of oxidation of iron during natural weathering, alluvial nephrite and jadeite often have a brown skin or rind that may penetrate more deeply along fractures and flaws in the stone. This is often incorporated to advantage by the jade carver. Alteration may occur also during burial and a distinctive yellowish or greenish brown colour taken on by some Chinese nephrite artefacts has given rise to the term 'burial (or buried) jade' (see Woolley, 1983: 257; also Desautels, 1986: 7). A whitening of the surface of some nephrite jades, often referrred to misleadingly as 'calcification', appears to be due largely to an increase in surface porosity in response to natural weathering processes (Wen and Jing, 1997: 119). Analysis and replication some years ago suggested that surface whitening might arise also from the action of highly alkaline body fluids on nephrite (Gaines and Handy, 1975).

Whitening of the surface of nephrite may be caused by heating. Such heating may have occurred incidentally when jade artefacts were included as part of ritual practices involving cremation. However, the heating may have been more deliberate. It has been suggested that heating nephrite would have made it easier to work, but Douglas (2001) concluded that this was not the case. Again, it has been suggested that nephrite was deliberately heated to produce white, so-called 'chicken-bone jade' (Tsien and Ping, 1997: 126).

Hardness, Toughness and Specific Gravity

Nephrite (H on Mohs' scale = 6.5) and jadeite (H = 7) are both relatively hard and dense materials (SG of nephrite = 2.9–3.1; jadeite = 3.24–3.43) but, apart from their range and quality of colour, perhaps their most striking physical property is toughness. Jadeite is tough, and nephrite exceptionally so, giving to it the permanence and enduring quality that has been valued so highly for millennia. This extreme toughness arises from its finely felted texture of fibrous crystals (*Figure 20.6*). Jadeite has a finely granular texture in which the individual crystals are closely interlocked (*Figure 20.7*).

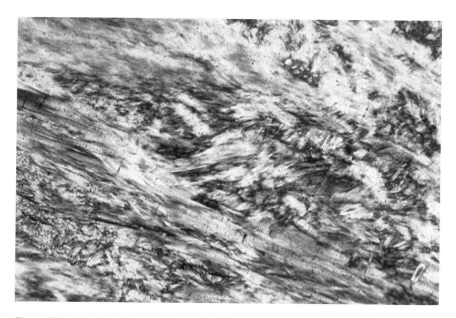

Figure 20.6   Scanning electron micrograph of fractured surface of nephrite jade, width of field c. 0.15 mm. Copyright, British Museum. [CU343]

Simulants

Many materials have been used at various times as substitutes for jade; they have been termed *simulants, false jade* or *pseudojade*. Many are minerals or rocks which are quite distinct from both nephrite and jadeite. Some resemble jade only in a relatively superficial way, others can be difficult to distinguish visually; most simulants will have carving qualities distinctively different from those of true jade. Some of the better-known simulants are listed in *Table 20.1*. The use of special treatments and modern

Figure 20.7   Scanning electron micrograph of fractured surface of jadeite jade, width of field c. 0.15 mm. Copyright, British Museum. [CU340]

synthetic materials, which are perhaps better described as outright forgeries or deceptions, will be mentioned briefly in the next section.

Several metamorphic rocks have been used as jade simulants. One of these, bowenite (a serpentine rock), is probably the most widespread and most satisfactory substitute for jade. It has been obtained from a number of different sources. Bowenite from China, Kashmir and Afghanistan is often referred to as 'new jade', those from Korea as 'Korea jade'; 'tangiwaite' from Milford Sound, New Zealand, used extensively by the Maoris, has been mentioned already. Williamsite, a similar though somewhat softer variety of serpentinite with distinctive octahedral inclusions, has also been used. Other metamorphic rocks that have found use include smaragdite, a metamorphic rock rich in amphibole (actinolite), which can closely resemble jade; and saussurite, a feldspar-epidote rock, described originally as jade. Some simulants that are less jade-like include 'verd-antique' or green marble, Connemara Marble and Iona Stone (both owe their green colour to the presence of serpentine minerals), and agalmatolite or 'figure stone', a name which has been widely applied to various rather soft stones including steatite (soapstone) and pyrophyllite.

A high-quality mineral simulant of jade is vesuvianite, also known as idocrase. In its compact form (californite) vesuvianite can be quite difficult to distinguish visually from true jade. The garnet minerals, grossular and hydrogrossular, may occur in association with vesuvianite in rocks known as rodingites. These may be either green (when vesuvianite is present) or pink in colour and have been used under the name 'Transvaal jade'. Several varieties of quartz have been used in imitation of jade. These include the green varieties of chalcedony, chrysoprase and plasma and also green aventurine quartz (otherwise known as 'Indian jade'), which is coloured by inclusions of the chromian muscovite mica, fuchsite. An impure fuchsite rock known as 'verdite' has also been used for carving, as has green microcline (amazonite, often known as 'Amazon jade'); smithonsite (also known as 'bonamite'); it is said that prehnite can be a most convincing jade imitation. Even fluorite is reported to have been used: as Desautels (1986: 10) comments, '... almost every known greenish mineral that lends itself to carving has been used and confused at one time or another with jade'.

Glass has been used as a substitute for jade at least since it was first made in China during the Warring States period (fifth to third centuries BC). However, it appears to have been regarded as a very inferior material, and 'jades' made from it were clearly recognized as imitations. In a clear reference to this practice, An Jiyao (1991) quotes the Northern Song poet, Su Dongpo: 'It is deceitful to make jade by boiling lead with white stone.' Early Chinese glass is commonly based upon a distinctive combination of barium, lead and silicon oxides, which is easily distinguished from jade by relatively simple tests. Nevertheless, this glass is often turbid, resulting in a convincing jade-like appearance, which has led to its being confused with jade (and occasionally *vice versa*; see Brill *et al.*, 1991).

Deception and Enhancement

Coloured plastics have been used to make objects more or less resembling jade in appearance; these should usually be easily detectable by simple tests of their physical properties. Jade doublets have been made to deceive. Similarly, coloured and opacified leaded-glass is used in imitation of jade, a practice that originated in antiquity (see above); such deceptions will usually be detectable once pieces are suspected and tested.

More subtle are various surface treatments applied to jade, particularly jadeite because it is more amenable to dyeing and because of its high value in the marketplace (see Nassau, 1984: 139–41, who reviews methods for enhancing both jadeite and nephrite). The use of a light wax polish to enhance the appearance of the polished jade is often regarded as being acceptable. Similarly, bleaching using a mild traditional acid such as plum juice (Sun, 2001), followed by light waxing (so-called 'A-jade') is sometimes acceptable. Other more far-reaching and less acceptable treatments may involve bleaching with strong acids before impregnation of the surface and any cracks or fissures with colourless polymer (B-jade); jadeite may be dyed without being bleached (C-jade) or after bleaching (B + C-jade). Detection of some treatments may be possible using a UV lamp, or inspection with a hand lens or simple optical microscope but, for more definitive determinations, analysis by techniques such as infra-red or Raman spectroscopy may be required. A more comprehensive account of the enhancement of jadeite jade, approaches to its detection, and the effects of such deceptions on the jade market was published recently by Sun (2001).

Some jades may have been heated to whiten them, thus enhancing their value by giving them the appearance of naturally weathered, ancient jades (see above). Detection of such treatment may be possible by analysis, because heating above c.800–900 °C causes a breakdown of tremolite-actinolite, producing minerals such as diopside, which can be identified by X-ray diffraction or infra-red spectroscopy.

The dyeing of precious and semi-precious stones has been practised for millennia, and jade has not escaped. Paler-coloured varieties have been dyed green or purple; conversely coloured jades have been bleached. Organic dyes may fade in time and once suspected they can often be detected by careful examination of the distribution of the colour.

Identification

Visual assessment alone may not be a sufficient basis for identification, and because true jade is not a single, well-defined mineralogical entity, there is not really a single simple test that can be applied to establish the true identity of a 'jade' artefact. However, there are several non-destructive or only minimally destructive tests that can be applied to assess the physical properties of the object. Often it will be necessary to use the results from several tests to achieve a positive identification.

The relative *hardness* of nephrite ($H = 6.5$) and jadeite ($H = 7$), whilst not diagnostic, will help to distinguish jade from many of its simulants (see *Table 20.1*). These tests can be carried out under a binocular microscope and, if carefully executed, will inflict a minimum of damage to the artefact.

Specific gravity can be extremely useful for distinguishing between jade and many of its simulants; it will permit also the distinction of nephrite, which typically has an SG of c.3.0, from jadeite, which is more dense with an SG of c.3.3.

Table 20.1
Physical properties of nephrite, jadeite and selected simulants
(from Middleton and Freestone, 1995)

Mineral/Rock	Synonyms	Refractive index	Hardness	Specific gravity
Nephrite	New Zealand jade	1.62	6.5	2.9–3.1
Jadeite	Chinese jade, Yün-nan jade Astridite (impure jadeite)	1.66	7	3.30–3.36
Amazonite	see Microcline			
Chlorite	see Pseudophite			
Chloromelanite (var. of jadeite)				3.38–3.5
Cosmochlore	see Kosmochlor			
Epidote	Tawmawite	1.7	6.5	3.6
Fluorite		1.43	4	3.18
Fuchsite (impure) (var. of muscovite)	Verdite	1.55–1.70	3.0	2.75–3.25
Hydrogrossular	Transvaal jade British Columbia jade	1.73	6.5	3.5
Idocrase	see Vesuvianite			
Kosmochlor (impure) (= cosmochlore, ureyite)	Maw-sit-sit Szechenyite	1.53	6	2.77
Marble (impure)	Verd-antique	1.6	3	2.7
Connemara marble	see Serpentine			
Iona stone	see Serpentine			
Microcline (var. Amazonite)	Amazon jade	1.53	6	2.56
Pectolite		1.60	5	2.86
Pinite (impure muscovite; see also pyrophyllite, steatite)	Agalmatolite, pagodite, figure stone	1.6	2.5	2.8
Prehnite		1.63	6	2.9
Pseudophite (compact chlorite, esp. clinochlore, pennine)	Styrian jade	1.57	2.5	2.7
Pyrophyllite (see also pinite, steatite)	Agalmatolite, pagodite, figure stone	1.6	1–1.5	2.84
Quartz				
var. Aventurine	Indian jade	1.55	7	2.6
var. Jasper		1.54	7	2.6–2.9
var. Chalcedony (chrysoprase, prase)	Swiss jade	1.54	7	2.6
var. Chalcedony (Moss agate)		1.54	7	2.6
Saussurite (epidote-plagioclase rock)		1.55–1.70	6.5	2.7–3.2

Table 20.1 (*cont.*)

Mineral/Rock	Synonyms	Refractive index	Hardness	Specific gravity
Serpentinite (Serpentine rock)				
var. Bowenite	New jade, Korea jade, Soochaw jade, Manchurian jade, Mukden jade, tangiwaite	1.55	5	2.6
var. Williamsite		1.57	4	2.6
Smaragdite (actinolite rock)				3.25
Smithsonite	Bonamite	1.73	5	4.4
Steatite (massive talc, soapstone; see also pinite, pyrophyllite)	Agalmatolite, pagodite, figure stone	1.55	1	2.82
Ureyite	see Kosmochlor	1.7	6	3.3
Vesuvianite (= idocrase)	Californite, Californian jade			

The determination of *refractive index* of jade using a refractometer is not straightforward. Apart from problems that may arise because of the size or geometry of the object, or weathering of the surface of ancient jades, the finely granular or fibrous texture of both jadeite and nephrite jade leads to rather diffuse readings (nephrite c.1.62, jadeite c.1.66; see p. 344). However, even an approximate reading can often make a useful contribution to distinguishing nephrite and jadeite from each other and from many simulants.

Observation of the *absorption spectrum* can be particularly informative for the examination of green jadeite jades, which owe their colour to the presence of chromium. This gives rise to characteristic absorption bands in the red region of the spectrum. There is some possibility of confusion with emerald but this has much sharper absorption bands. Examination with the spectroscope may also allow the recognition of staining using organic dyes, which give a broad absorption band in the red region. For more details the reader is referred to texts such as O'Donoghue and Joyner (2003).

Perhaps the most definitive technique for the identification of 'jade' is *X-ray powder diffraction*. Crystalline materials, including true jade and most of its simulants (glass is a notable exception) give unique diffraction patterns. By reference to suitable standards, these can be used in the manner of a 'fingerprint' to identify the material of the object. The technique is minimally destructive, requiring only a very small sample (considerably less than pin-head size). However, it is of course necessary that the sample removed for analysis be representative, and it is also the case that problems can arise in the interpretation of patterns from the complex mixtures of minerals encountered in some simulants.

Raman spectroscopy offers many of the advantages of X-ray diffraction, providing a definitive 'fingerprint' of jadeite, nephrite (tremolite/actinolite) and most of the simulant materials (including glass). Coupled to a good optical microscope, Raman spectroscopy offers the considerable advantage that the analysis may be carried out directly on the surface of the object, without the need to remove a sample. Furthermore, the technique is rapid, usually providing definitive spectra in minutes or even seconds. Problems arise sometimes from surface treatments, which may give rise to fluorescence, making it difficult to obtain a clear spectrum.

Infra-red spectroscopy can also be used to provide diagnostic spectra, in the same way as X-ray diffraction and Raman spectroscopy, but is probably not as convenient to use in a non-destructive manner as Raman.

Jade Carving

Because jade cannot be worked effectively by the techniques of chipping or flaking that can be applied to more brittle stones such as flint and obsidian, it has almost always been worked by abrasion. Jade is usually carved with electrically driven, rotary carborundum or diamond tools, sometimes mounted on flexible drives; even using these modern methods, finishing may still be time-consuming, and Beck has noted that the completion of a sculptural nephrite carving may take hundreds of hours. Chinese records from the eighteenth century provide details of the resources used to produce carved jade pieces, and it is apparent that even a small item might have involved 50–60 workers and taken several months to complete (Till and Swart, 1986: 38).

Nevertheless, traditional techniques of carving are still valued (see Sinkankas, 1984: 34; Beck, 2002), and there has long been an interest in how this exceptionally tough material was worked in antiquity, before the advent of metal tools. Hansford (1950: especially 57–90) published an extensive account of the historical evidence and recorded his own observations of traditional jade carving workshops in Beijing in 1939. He noted the use of quartz, garnet, emery, carborundum and diamond abrasives, applied to a range of tools, either hand-held or mounted on treadle-driven lathes.

Based on observations made in the mid-nineteenth century and published by Chapman in 1892, Hansford commented also on the traditional methods used by the Maori in New Zealand. Saws made from thin laminae of quartzose sandstones and siltstones were used to cut nephrite into slabs, which were then shaped and finished with stone tools; perforations were made using flint tips mounted in string-driven rotary drills. There is no direct knowledge of this sort for the methods used in Mesoamerica but Hansford refers to evidence for the use of hollow drills made from bone or bamboo. Recent work on the carving of prehistoric ancient Chinese jades (e.g. Lu and Hang, 2002; Sax *et al.*, 2004) is revealing evidence for the use of rather similar stone tools in the form of saws, files and drills, sometimes used with loose abrasives.

Drilling is a particularly efficient way of removing material, and the introduction of hollow 'core' drills of bamboo or bone, and later of metals such as bronze or iron, would have speeded the process and minimized wastage of valuable jade, and must have represented a significant advance in technique. Although there is evidence that a form of rotary grinding disc was in use during the late Neolithic period in China, the date when the jeweller's engraving wheel, another major technical innovation, was introduced is a matter of considerable discussion.

Acknowledgements

It is a pleasure to acknowledge assistance from several colleagues: Anthony Milton, Antony Simpson and Trevor Springett for their help with illustrations; Carol Michaelson and Margaret Sax for numerous discussions on the carving of jade; Janet Ambers for the Raman analyses of the pebbles shown in *Figure 4* and many other 'jades'. In particular, I would like to thank Ian Freestone who read the manuscript and on whose contributions to a previous collaborative paper some parts of this article draw heavily.

References

Beck, R.J., 2002, *Mana Pounamu New Zealand Jade*. Auckland: Reed Books.
Bender, F., 1983, *Geology of Burma*. Berlin: Gebrüder Borntrager.
Brill, R.H., Tong, S.C.C. and Dohrenwend, D., 1991, Chemical analyses of some early Chinese glasses. In R.H. Brill and J.H. Martin (eds), *Scientific Research in Early Chinese Glass*. New York: Corning Museum of Glass, pp. 31–58.
Caley, E.R. and Richards, J.F.C., 1956, *Theophrastus on Stones*. Columbus, Ohio: Ohio State University.
Chapman, F.R., 1892, On the working of greenstone or nephrite by the Maoris. *Transactions of the New Zealand Institute*, **XXIV**, 479–539.

Chibber, H.L., 1934, *The Mineral Reserves of Burma*. London: Macmillan.

Clark, A.M., 1993, *Hey's Mineral Index*. Third edition, London: Chapman and Hall.

Colombo, F., Rinaudo, C. and Trossarelli, C., 2000, The mineralogical composition of *maw-sit-sit* from Myanmar. *Journal of Gemmology*, **27**, 87–92.

Desautels, P.E., 1986, *The Jade Kingdom*. New York: Van Nostrand Rheinhold.

Douglas, J.G., 2001, The effect of heat on nephrite and detection of heated Chinese jades by X-ray diffraction (XRD) and Fourier-transform infrared spectroscopy (FTIR). In *Proceedings of Conference on Archaic Jades Across The Taiwan Strait*, Taipei, pp. 543–54.

Gaines, A.M. and Handy, J.L., 1975, Mineralogical alteration of Chinese tomb jades. *Nature*, **253**, 433–4.

Gübelin, E., 1965, Maw-sit-sit: A new decorative gemstone from Burma. *Gems and Gemology*, **11**, 227–38.

Hansford, S.H., 1950, *Chinese Jade Carving*. London: Lund Humphries.

Harder, H., 1995, Trace elements as colouring agents in jadeites. *Journal of Gemmology*, **24**, 508–11.

Win Htein and Aye Myo Naing, 1994, Mineral and chemical compositions of jadeite jade of Myanmar. *Journal of Gemmology*, **24**, 269–76.

Jing, Z. and Wen, G., 1996, Mineralogical inquiries into Chinese Neolithic jade. In S. Bernstein (ed.), *The Journal of Chinese Jade*. San Francisco: S. Bernstein, pp. 135–51.

Jianfang Lu and Tao Hang, 2002, Prehistoric jade working based on remains at the site of Dingshadi (trans. E. Childs-Johnson). In E. Childs-Johnson (ed.), *Enduring Art of Jade Age China, vol. II Chinese Jades of Late Neolithic through Han Periods*. New York: Throckmorton Fine Art, pp. 31–42.

Jiyao, An, 1991, The early glass of China. In R.H. Brill and J.H. Martin (eds), *Scientific Research in Early Chinese Glass*. New York: Corning Museum of Glass, pp. 5–20.

Michaelson, C. and Sax, M., 2003, 7000 years of Chinese jade from the collection of Sir Joseph Hotung. *Apollo*, **CLVIII**, 3–8.

Middleton, A. and Freestone, I. 1995, The mineralogy and occurrence of jade. In J. Rawson (ed.), *Chinese Jade*. London: British Museum Press, pp. 413–23.

Nassau, K., 1984, *Gemstone Enhancement*. London: Butterworths.

Nichol, D., 2000, Two contrasting nephrite jade types. *Journal of Gemmology*, **27**, 193–200.

O'Donoghue, M. and Joyner, L., 2003, *The Identification of Gemstones*. Oxford: Elsevier.

Ou Yang, C.M. and Li Jian Qi, 2001, Hte long sein – A new variety of chrome jadeite jade. *Journal of Gemmology*, **27**, 321–7.

Ou Yang, C.M., Li Jian Qi, Li Hansheng and Bonita Kwok, 2003, Recent studies on inky black omphacite jade, a new variety of pyroxene jade. *Journal of Gemmology*, **28**, 337–44.

Sax, M., Meeks, N.D., Michaelson, C. and Middleton, A.P., 2004, The identification of carving techniques on Chinese jade. *Journal of Archaeological Science*, **31** (10), 1413–28.

Seitz, R., Harlow, G.E., Sisson, V.B. and Taube, K.E., 2001, 'Olmec Blue' and Formative jade sources: New discoveries in Guatemala. *Antiquity*, **75**, 687–8.

Sinkankas, J., 1984, *Gem Cutting*. Third edition, New York: Van Nostrand Reinhold.

Sun, T.T., 2001, Disclosure – gemstones and synthetics. Does the jewellery industry care? *Australian Gemmologist*, **21**, 67–75.

Till, B. and Swart, P., 1986, *Chinese Jade. Stone for the Emperors*. Victoria, British Columbia: Art Gallery of Greater Victoria.

Tsien, H.H. and Ping, T.L., 1997, Alteration of *Yu* Artefacts. In R.E. Scott (ed.), *Chinese Jades*, Colloquies on Art and Archaeology in Asia No. 18, University of London, Percival David Foundation, pp. 123–31.

Wen, G. and Jing, Z., 1992, Chinese Neolithic jade: A preliminary geoarchaeological study. *Geoarchaeology*, **7**, 251–75.

Wen, G. and Jing, Z., 1996, Mineralogical studies of Chinese archaic jade. *Acta Geologica Taiwanica*, **32**, 55–83.

Wen, G. and Jing, Z., 1997, A geoarchaeological study of Chinese archaic jade. In R.E. Scott (ed.), *Chinese Jades*, Colloquies on Art and Archaeology in Asia No. 18, University of London, Percival David Foundation, pp. 105–22.

Woolley, A.R., 1983, Jade axes and other artefacts. In D.R.C. Kempe and A.P. Harvey (eds), *The Petrology of Archaeological Artefacts*. Oxford: Clarendon Press.

Further Notes for the Gemmologist

Western students are inevitably faced with language difficulties when studying the jades. It is tantalisingly clear that much vital information has not been translated from the Chinese (or other relevant languages). Readers unfamiliar with the invaluable journal *Bulletin of the Friends of Jade* may well fall upon a set with cries of joy as it contains a number of translations from Chinese academic sources as well as up-to-date reports on any activity involving jade (the level is well above 'rockhound'). The Bulletin appears irregularly, the editorship passing from one individual to another so that it is not easy to find a set – the best way to track it down would be to ask Oriental departments in major museums. The ISSN is 0261–7080. Stop press: try fred@fredwardgems.com.

Ms C.M. Ou Yang continues to write useful books on jade and I can add to the bibliography her *Jadeite Jade* (2003; ISBN 9628698710). The same author has published the two-volume *Jadeite Jade* (2000; ISBN 9629930811; text in Chinese). Chinese-language publications deal with the mining of and trading in jadeite; an example (in Chinese) is Jackie W. Hwang and Dinna C. Chang, *Natural Bleach Jadeite Identification* (1995, ISBN 9579983119).

In the first volume of Yasukazu Suwa's *Gemstones Quality and Value* (1993, ISBN 44188999019), the author discusses jadeite qualities and possible pricing, illustrating the acknowledged most desirable properties and ways in which some specimens, though acceptable, fall short of the highest rank. Suwa's photographs, viewed as a single page, give a good idea of jadeite grading. Similarly, Miller and Sinkankas, in the *Standard Catalog of Gem Values*, second edition (*passim*) give criteria for the finest jadeite: the colour is an intense pure green, evenly distributed and without mottling, veining or other conspicuous defects and the cabochon or carving is semi-transparent. Value decreases as translucency decreases even if the colour is excellent. Value also decreases when individual veins become apparent or when veinings or areas of paler green or colourless jadeite appear. Value decreases even further when the colour becomes excessively dark or takes on greyness or tinges of olive-green. Other colours are less prized though pale pure greens are highly valued, also rich red, mauve, yellow or orange colours. Dark olive-green, grey-green, indistinctly coloured or black jade specimens are the least esteemed.

Michael O'Donoghue

21

The Natural Glasses

Natural glasses occur in a number of forms and compositions while consisting mainly of silica. Even today the precise origin of some of the natural glasses is uncertain. Many are used ornamentally and some large, clear pieces have been faceted. There is considerable variation in the chemical composition, but obsidian may contain up to 77% silica with 10–18% alumina. Had a rock formed instead of obsidian it would have consisted of quartz, feldspar and mica.

Obsidian

Obsidian is formed by the rapid cooling of volcanic lava, which, had it been allowed to cool slowly, would have developed a crystalline structure. A network of curving cracks can be characteristic of some varieties of obsidian. The resulting texture is known as perlitic, and disintegration along the cracks gives bead- or ball-shaped pieces. The name marekanite has been used for these pieces, the name being taken from a Siberian locality. The name has also been used for similar material from Mexico which produces spherulitic textured obsidian with radiating feldspar fibres, known as peanut obsidian. This is found in the neighbourhood of Sonora.

Obsidian is normally an opaque black or grey but may be yellow, red or greenish brown. It may have an iridescent sheen caused by reflections from minute inclusions as seen in the silver and golden obsidians which are often polished into beads for necklaces. In the USA a variety of obsidian having spherulitic inclusions of a white mineral on the black groundmass is cut and polished. This material is called 'flowering' obsidian. A variety banded black and red has been called 'mountain mahogany' and is sometimes used for ornamental objects and for carvings. Transparent leaf-green obsidian has been mentioned, but green is a very rare colour and most transparent green obsidians are usually found to be moldavites or merely green glass. Red and blue obsidian have been reported.

Moldavite consists of approximately 75% silica with 10% alumina; it may also contain oxides of iron, potassium, calcium, sodium, magnesium, titanium and manganese. The hardness is near 5.5, the SG ranges from 2.30 to 2.50 and the RI is 1.48–1.52. The absorption spectrum with vague bands in the orange and blue is of little help in identification. A faint yellowish green under X-rays is the only fluorescence. Profuse rounded or torpedo-shaped bubbles with swirls and striae are very characteristic of moldavite; the crystallites seen in obsidian are absent here. The name marekanite has also been applied to material from Mexico. Classy pebble-like solid cores of unaltered glass, 25 mm or more across, from the decomposed obsidian of the American south-west, are known as 'Apache tears'. The transparent ones when cut produce stones of a grey or light grey colour and may show fine silky striations which give to the stone a cat's-eye effect. A brown-black obsidian from Hungary has been called 'Tokay lux-sapphire'.

Chemical and Physical Properties

Obsidian has a hardness of near to 5. The fracture of obsidian is notably conchoidal, and is due to the facility with which the material can be broken into sharp-edged flakes that obsidian was so valued by Stone Age people who lived in areas where obsidian was common, the easily controlled flaking allowing the production of keen-edged spear points, knives and tools.

The lustre of obsidian is vitreous; the RI lies between 1.48 and 1.51, but is commonly near 1.49. Thin fragments viewed between crossed polars show the groundmass of the material to be isotropic but the field is speckled with bright points of light due to very small crystalline particles, which are usually present in volcanic glass. The SG of obsidian varies between 2.33 and 2.42, at least for the types used in jewellery.

When a thin sliver of obsidian is examined under the microscope the texture is seen to consist of a clear glassy ground containing many small crystallites. These may be round bodies (globulites) or may be rod-shaped (belonites) or may be coiled or twisted hair-like bodies (trichites). They may be fused into chains (margarites), probably in fanciful analogy to a string of pearls. It may take a fairly high magnification to resolve these bodies. Round or torpedo-shaped bubbles are present in some obsidian. These, which can generally be seen clearly with low-power magnification, are often in parallel arrangement, and it is this paralleliza-tion of the bubbles and the crystallites which give to some obsidian the prized sheen.

Occurrences

Obsidian is widespread in occurrence, but most of the material used in jewellery is obtained from North America. Important localities include the Glass Buttes, Lake County, where iridescent material is found, and at Hampton, Deschutes County, Oregon. At Obsidian Cliff, Yellowstone National Park in Wyoming, obsidian has been quarried since the days of the North American Indians, who used it for arrowhead material. Arizona, Colorado and Nevada have deposits of obsidian, while California has many sites where the material has been quarried from ancient times. The best known of these Californian sources are Glass Mountain, Modoc County; Little Lake, Inyo County; and Obsidian Butte on the south-west side of Salton Sea in Imperial County. Apache tears are usually found in New Mexico.

Obsidian was used by the Aztecs and their predecessors for the sharp points of their weapons, for mirrors and masks, and for ornament. They called the material iztli and surnamed it teotetl (divine stone) because of its many uses. The ancient Aztec obsidian pits were rediscovered by Humboldt during the opening years of the nineteenth century at Sierra de la Navagas in the state of Hidalgo, Mexico. Obsidian is found at sev-eral other places in Hidalgo, and in the other Mexican states of Jalisco, Queretaro, Michoacan and Vera Cruz. The Maya obtained their obsidian from ancient quarries at La Joya, some 30 km east of Guatemala City, and the Indians of Equador had a quarry at Guamani.

Basalt Glass

Basalt glass, which is found mainly on the chilled margins of basaltic intrusions, or lining or filling vesicles or cavities in basalt, is semi-translucent to opaque and usually black or brown in colour. It contains about 50% of silica. Basalt glass, which has a hardness of about 6, does not afford large flakes or show extensive conchoidal fracture when struck with a hammer, but breaks up into small irregular fragments and splin-ters. These glasses have an RI between 1.58 and 1.65 and a range of SG between 2.70 and 3.00. Under low-power magnification the texture is seen to be uniform and the structure cannot be resolved except under high power. Basalt glasses are not normally cut as gem material but some experimental cabochons have been cut from the light grey, navy-blue, bluish green and brown pieces found near the headwaters of the Flinders river in northern Queensland. The pieces are often attractively mottled and spotted.

Meteorites

Meteorites are hardly expected to enter the field of gems, but, as will be explained, one such type of suspected meteorite has been used for ornamentation.

A meteorite is understood to be part of a disintegrated stellar body which, travelling through space, has entered the gravitational field of the earth and crashed on to its surface; thus it may be said to have cosmic origin. There are two general types of meteorites: those whose constitution is mainly iron and nickel, known as siderites; and those which consist of olivine, pyroxene and a little feldspar, and thus approach the basaltic rocks in composition. These are termed 'aerolites'.

Tektites

It is not the metallic or stony objects from the skies which interest those who study gems, but a type having a glass-like constitution rich in silica, and which can be likened to the volcanic glass obsidian. These pieces, which are only suspected as being of meteoritic origin, have been given the name tektites. The first noted occurrence of these glassy pieces was in 1787 in western Moravia and around České Budějovice near the Bohemian river Moldau, a river of which the Czech name is Vltava. Derived from the former name of the river the pieces were called moldavites. These lumps of glass, often peculiarly fissured on the surface – either reminiscent of alpine topography with craggy mountain tops, or with rounded crests having a botryoidal surface – are transparent green, greenish brown or brown in colour. The material has been cut as gemstones producing bottle-greenstones. Cut moldavite has been sold under a variety of fancy and completely misleading names, such as 'bottle stone', 'obsidian', 'water chrysolite' and 'pseudo-chrysolite'. The stone has also been known under the name bouteillenstein.

Czech moldavites fall into three categories. Oval or spherical specimens from the Radomilice area of southern Bohemia are pale or bottle-green with few included bubbles and less obviously flattened. These are silica-rich and alumina-poor with little lechatelierite (i.e. near the silica glass composition). Moldavites from most other southern Bohemian finds are bottle-green, more flattened and with profuse bubbles. Their lechatelierite content is higher, with Si, Al and Fe content intermediate between the first type and a type found in Moravia, coloured brown and with more Al and Fe than other types. Moravian specimens have fewer bubbles and are spherical. Four different occurrences are reported. Some are found in Upper Miocene sediments and others in Pliocene and Pleistocene sediments. Slope lavas, possibly Pleistocene, and Holocene or Pleistocene alluvia house the remainder. Moldavite from Chlum nad Maisi in southern Bohemia has been used in jewellery in recent years.

Button-shaped glass pieces (australites) have been found in South Australia and Tasmania. The variety billitonite was found on Billiton Island between Sumatra and Borneo, and specimens tend to be dark brown.

Colombia produces colourless tektites but whatever the origin they tend to be button-shaped. The Jukes-Darwin field near Queenstown, Tasmania, produces silica-rich and alumina-poor colourless to olive-green to black specimens with SG 1.85–2.30, with the varietal names darwinite or queenstownite.

Crater Glass

Other types of natural glasses exist which, although not of meteoritic origin, have had their genesis in meteoritic action. One type is obtained from meteoritic craters, the glass being formed by the intense heat generated by the impact of a meteor on to the earth's surface where the surface cover is siliceous, such as in a sandy desert.

Such glass, white, greenish yellow or black in colour, is, in general, simply a fused quartz, about 90% silica with some impurities, of which the most common is iron. The material is slaggy and contains numerous vesicles. The SG of these glasses varies from 2.10 to 2.31, a range which is lower than that for the previously mentioned moldavites and tektites. The value is in many cases lower than that for pure fused quartz (silica glass, 2.203), probably owing to the included vesicles. The RI varies from 1.46 to 1.54, the high values probably being due to a richness in iron. This crater glass has been found in many widely different places, mostly in localities known for their sandy deserts, such as Wabra in Saudi Arabia, Henbury in central Australia, and at Meteor Crater, Arizona, in the USA.

Fused Sand Glass

Fulgurites or lightning tubes are thin tubes of fused sand caused by the intense heat set up when a lightning flash strikes and enters the sand of the desert. Scarcely a natural glass, but certainly a crater glass, is the fused sand formed by the experimental atomic bomb which was exploded in New Mexico during 1945. This material, which has been named trinitite, is vesicular and greenish in colour, and the constants for the material are similar to those for ordinary crater glass. Even after many years much of this fused sand is still radioactive. It may appear unlikely that such material would be used for ornament, but it is on record that some has been fashioned into gems – mainly as publicity for film stars. In view of the radioactivity present it would certainly be a risk to wear such objects for too long at any one period.

Silica Glass

A nearly pure silica glass of uncertain origin found in 1932 in the wastes of the Libyan desert is known as desert glass. The material is found as sand-blasted lumps up to 7 kg in mass. It is of a greenish yellow colour and is usually cloudy rather than clear owing to the inclusion of vast numbers of irregularly shaped vesicles.

Silica glass consists of nearly pure silica (98% SiO_2), and has an RI of 1.46, an SG of 2.21, a hardness of 6 on Mohs' scale and a vitreous lustre. The stones cut from this material lack brilliancy and have a dispersion of 0.010. Occasionally there are found included in such glass, either as strings or as single individuals, spherulites of the cubic modification of quartz known as cristobalite. A little of the material has been cut and polished for collectors, but its insipid tint and poor brilliancy does not allow the stones cut from silica glass to be at all attractive.

Lumps of glassy material, usually striped with greyish blue and dark brown or black, are occasionally encountered. The material looks like an obsidian but it is actually a furnace slag and has an SG near 2.82. The properties of a green glass reportedly made from ash from the Mount Saint Helens eruption in 1980 were compared with those shown by a sample of melted black ash known to have been from that source. The green glass gave an RI of 1.508 while the black sample showed a variation over the range 1.500–1.526; SGs were 2.448 and 2.485 respectively. It was concluded that the green glass contained from 5 to 10% of Mount Saint Helens ash at most, chemical analysis giving a conclusive separation from known material. A transparent brownish natural glass from

San Luis Potosi, Mexico, showed layered and botryoidal textures giving rise to an iridescent effect. Similar material has been described, misleadingly, as iris opal. IR spectroscopy showed that SiOH groups were present but that water was not.

The origin of the Libyan desert glass has been discussed in a book by J. McCall (*Tektites in the geological record*, 2001: ISBN 1862390851) who suggests that it may be impact-related and that the glass originated in two craters both just over 100 km from the strewn field.

22

Marble*

Robin Sanderson

Marble, which derives from the Greek word *marmaros* (Latin *marmor*, shining stone), may be defined as a crystalline aggregate of calcite and/or dolomite which often has a granular or saccharoidal texture. Classical writers appear to have used the name for many stones used for decorative purposes, regardless of composition, even including the porphyries ('marmor lacedaemonium viride', the green porphyry from Greece) and granites ('marmor claudianum', a fine, pale granite from Egypt). This imprecise usage is still current among antiquarians, and even in modern commerce to a lesser extent.

The 'marbles' of commerce come from all over the world, being found in areas where detritus is poor and mainly marine sedimentation has occurred at some time. Taken in isolation the crystalline marbles may appear to be lithologically distinct, but they merely represent a product of metamorphism, the end result of a post-depositional and gradational mineralogical reorganization of a limestone, and it is often a matter of subjective choice to separate them. The calcareous rocks are relatively easy to work, even with stone tools, and all cultures have used these stones for decoration to some extent. Their present variety and popularity, however, may be said to derive from the architectural fashions of Imperial

I am indebted to Robin Sanderson for permission to reproduce his chapter first published in the 5th edition of Gems (1994). He adds, in a letter addressed to me on 7 January 2003, that the best method of distinguishing one marble from another is by pattern recognition and colour combinations. Work described in specialist journals will be in general too detailed for the gemmologist who in any case will not often be asked to pronounce on hardstones [MO'D].

Rome. Many of the stones to be found in Western object d'art and furnishings, both antique and modern, still originate from within the bounds of the Empire, or its sphere of influence, and Italy remains the major marble processing country.

The specimens described below are grouped as the true calcareous marbles, the compact limestones and the travertine/onyx marbles.

Nomenclature

Names given to marbles are innumerable and confusing. Synonyms abound, and are coined at will to satisfy commercial pressures. Names commonly derive from the source, e.g. 'breccia di Seravezza', or the appearance, e.g. the Belgian 'rouge-et-gris', or are entirely fanciful, e.g. 'pavonazzo' (peacock) and 'Napoleon' marbles. They should be treated with caution as, for example, the 'Griotte d'Italia' is from France, and 'Molina rosa', named after a Spanish locality, is of Italian origin.

There is no universal or consistently employed system of nomenclature for marbles, and ancient names are 'pirated' for use by suppliers of stone from new sources which may have only a superficial resemblance to the original. The 'verde antico' (verd-antique) marble may be taken as an example. The original is a beautiful dark green serpentinite breccia containing fragments of white calcite marble from Thessaly, Greece. This was so prized in classical times that the name has become familiar and has been applied to most other green breccias, even some from the USA.

Nevertheless, there are a small number of terms which have relatively consistent meanings, although they are of little use alone in specific identification. Among these may be cited 'arabescato', patterns of swirling lines reminiscent of Arabic script; 'breccia', stone composed of angular fragments; 'cipollino', finely parallel banded stone, likened in structure to a section of an onion (Italian cipolla); 'Griotte', dark red marbles of the colour and often the appearance of a mass of Griotte cherries; 'lumachella', stones composed of oyster-like shell fragments; 'perlato', pale stones containing scattered rounded algal or shell fragments; and 'verde' or 'vert', mostly applied to serpentinite breccias. Many of the names are of Italian derivation, reflecting the pre-eminence of that country in marble working since ancient times.

Such is the variation in colour and textural combinations that it is impossible to adequately, and unequivocally, describe specific varieties. Identification therefore requires experience and access to good, well-documented reference collections.

Although marbles from around the world are common in developed countries, most are used for large-scale architectural purposes, and will be unfamiliar even to experts. The marbles of most decorative artefacts of fine quality which will be encountered will still be from established sources; these will include India as well as European countries.

True Marbles

True marbles are the product of recrystallization of existing calcareous rocks by the agency of pressure and possibly heat produced by regional metamorphism. When the country rock which suffers this metamorphic change is a very pure limestone, the resulting rock is a pure white, crystalline and translucent marble such as the famous Carrara statuary marble from the Apuan Alps of Tuscany, Italy.

Pure marble is white; other colours derive from impurities in the original limestone. Iron is the most abundant source of strong colour, giving red, brown, yellow, grey and green tints, depending on its quantity and oxidation state. Blue is a rare colour in marble. Most marbles which are traditionally described as blue, such as 'bleu turquin', a French synonym for the Bardiglio marbles from Carrara, and 'bleu belge', from Belgium, are really soft greys. A recent introduction is the 'azul Macauba' marble from Brazil, a white crystalline stone streaked and veined with brilliant blue, which is probably derived from sodalite impregnation. The less brilliant 'African blue' comes from Kenya.

Limestones often contain impurities rich in heavy elements, either disseminated through the stone or concentrated in original sedimentary layers. Such impure rocks may produce attractively patterned and coloured marbles. Because of their origin in mobile parts of the earth's crust, marbles have often been shattered and fragmented, resulting in brecciated or breccia marbles. The fissures so formed, subsequently filled with consolidating, often coloured mineralization, produce the beautiful variegated or veined marbles. Examples such as the spectacular purple, grey and white 'breccia di Seravezza' from north-eastern Italy, and the pale pink and white 'breche rose' of Norway, may be cited.

Many limestones contain some magnesia in the form of dolomite $(CaMg(CO_3)_2)$, and when quartz (silica) is also present the two may combine with water to form the mineral serpentine. Serpentinous crystalline marbles are known as ophicalcites and commonly show swirling, irregular banding in shades of green and white. Possibly the most familiar variety is the 'Connemara marble' from the Republic of Ireland, but other similar varieties occur in Sweden ('Ringborg' or 'Swedish green') and the

USA, and the Scottish isle of Iona. The name ophicalcite has also been applied erroneously to the serpentinite breccias (the 'verde' marbles) because of their serpentine-carbonate mineralogy. The latter stones are hydrothermally altered ultrabasic igneous rocks of quite different origin from the true ophicalcites.

Some rare marbles contain scattered euhedral crystals of non-carbonate minerals, and can be extremely decorative, although only of use for small objects. Among these are the ruby marbles of the Mogok Stone Tract and Sagyin, Myanmar. Large crystals of pink grossular garnet occur in a creamy marble at Xalostoc, Mexico, the stone being known as 'xalostocite', 'landerite' or 'roselite'. In Britain, the Tiree marble from the Western Isles of Scotland contains small crystals of diopside in a pink to white marble groundmass.

Compact Limestones

Origins

Those compact limestones which are capable of taking a polish are known to the French as pierres marbrier, for which there is no English equivalent term, and so they are included with the marble for practical purposes. They are limestones, sedimentary rocks, which have become compacted through diagenetic processes which are less destructive to the original structures than true metamorphism. Apart from their ability to take a polish, they depend upon the original sedimentary structures and colourful variation in composition for their decorative effect, rather than the crystalline translucency of true marbles. Many primary limestones owe their origin to organic processes, either biomechanical or biochemical, although the influx of weathered detritus is also important in giving variety of colour and structure. Secondary reworking and fragmentation of these limestones produce more heterogeneous sedimentary breccias and conglomerates. The stones dealt with in this section are mostly of shallow-water, submarine, mechanical origin. Biochemical rocks will be treated below with the travertines.

Sedimentary Breccias and Conglomerates

These are coarsely fragmental rocks produced by vigorous and rapid erosion or by slumping or collapse of existing strata, resulting in a chaotic mixture of fragments. In practical terms, breccias are composed of angular fragments, whereas in conglomerates the fragments are rounded. Because of their coarse structure, these stones show their best in large objects such as monolithic vases, columns and fire surrounds, and have been extensively employed as wall cladding in banks and other public buildings.

In nature they tend to form around unstable continental margins, and to be associated with mountain ranges such as the Appalachian Mountains of the south-eastern USA, and the Pyreneean-Alpine ranges of southern Europe. From the latter region come the pale pink and grey 'arabescato' from Bergamo, Italy, the coarse, more strongly coloured 'machiavecchia' from the Italian Alps, and 'libeccio' from Sicily. The strong colouration of the last two is due to the admixture of dark red and brown terra rossa sediment which contrasts with the usually grey limestone fragments.

In the shallow waters of continental shelves, coral reefs may form, and the fragmented coral and coralline algal detritus from these, mixed with other sediment, forms important decorative marbles which are commonly highly coloured. Beligum has long been a major producer of such marbles from restricted reef knolls of Devonian age in the Namur region. Their predominant colours are dull red and grey, and they are known collectively as 'Belgian rouge-et-gris', with more specific varietal names such as 'rouge de rance' or 'rouge byzantin'.

Conglomeratic marbles are uncommon and represent near-shore deposits of the more stable continental shelf seas. They were more used in antiquity than at present, but can be found reused in later objects. Confusingly they are often designated as breccia in literature, as with the pale-coloured 'breccia polychroma' and 'breccia d'Aleppo' from northern Italy, and the dark green 'breche universelle' from Egypt. Pietra dura inlay work from India often includes a brown stone crowded with small yellow and green pebbles, from Sabalghar in Gwalior. This stone was popular with the seventeenth- and eighteenth-century Mughal craftsmen and continues to be used.

Fossil Marbles

Here are included a large variety of stones which are composed predominantly of accumulations of obvious fossil skeletal remains of a restricted variety of invertebrate animals and plants. The fossils visible may be shells (of the phyla Mollusca and Brachiopoda), coralline (corals, bryozoa and algae) or encrinital (the remains of crinoids, also known as sea-lilies). Representatives of other phyla are rarely sufficiently abundant to form rock-like masses.

Shelly Marbles

Species of the phylum Mollusca often live in communities of innumerable individuals, and probably form the most abundant group of seashells still to be found. The important rock-forming types lived in shallow waters affected by tidal and current action which concentrated the skeletons of dead individuals into masses. Oyster and clam shell accumulations lithify to form the type of stone known as lumachella. In these stones the arcuate white, grey or black shell may contrast with a residual matrix of buff or brown sediment. They have been exploited for use in small-scale items in Italy, in Spain and along the Mediterranean coast of north-west Africa. In England the Jurassic Forest marble has been employed for ecclesiastical and domestic decoration. From Tortosa in Spain comes a type, brightly coloured in red and yellow, known as 'broccatello', and golden varieties called 'castracane' have been found in Asia and India.

Lumachella is also a name given to a variety of marble from Bleiberg, Carinthia, in Austria, and from Astrakhan, in Russia. The marble consists mainly of fossil shells with iridescent surfaces so prominent in reds and greens that it has also been called 'fire marble'. Similar effects are shown by fossil ammonites from southern Alberta, Canada. The ammonites (Placenficeras meeki) occur in Upper Cretaceous dark grey shales, and the iridescent colours come from the nacreous layer of the shell. Again the common colours are red and green, but the better-quality pieces may show a complete range of colours. The shell consists chiefly of calcium carbonate in the form of overlapping platelets of aragonite. Trade names for this iridescent material include 'korite', 'ammolite' and 'calcentine'.

The gastropods, water snails, may also occur in rock-forming quantities, but such stones are very rare and of unusual patterning. The 'Purbeck' and 'Sussex' marbles of southern England, composed of masses of shells of the genus Viviparus with a grey, green or reddish matrix, may be mentioned for their importance in architectural decoration of the Early English period.

Coralline and Algal Marbles

Fossil coral marbles produce intriguing marbles with diverse large-scale patterns. The more fossil-rich varieties of the Belgian 'rouge-et-gris' marbles, such as the 'rouge royal', may be the best known, with irregular pale grey, often black-rimmed, fragments of coral or coralline algae set in a dull red matrix. Other coral limestones, also of Devonian age from south Devon, England, were exploited during the nineteenth century.

Perhaps more common than true coral rocks are limestones resulting from the activities of lime-concentrating marine algae, with their muted grey and brown colouration and smaller, but commonly still irregular, patterning. The Lower Carboniferous strata of western Europe are especially rich in these rocks; the 'Napoleon' and 'lunel' varieties from northern France are well known. Some of the English stones have similar appearances, but the rare brown 'rosewood' marble from Ashford in Derbyshire shows a regular fine parallel lamination reminiscent of wood graining.

Like the algae, lime-concentrating bacteria also inhabit very shallow waters, and develop spectacular mushroom-shaped masses known as 'stromatolites'. These masses are uncommon and of only limited extent, but have given rise to the 'Cotham' or 'landscape marble' from the Rhaetic of Gloucestershire, England. The term 'landscape' derives from the appearance of the stone in vertical section, where dark grey dendritic masses of the stromatolites rise from parallel laminated paler mudstone, giving the impression of views across ploughed fields with distant hedgerows. A contrasting pictorial effect is developed in the 'Florentine' or 'ruin marble' or 'paesina'. This is a fine even-grained yellow to brown calcareous mudstone where orthogonal fractures have allowed box-stone-like patterns of iron staining to occur, an effect, however, of chemical and not organic origin. Appropriate sections through the stone give images reminiscent of a ruined town. From the neighbourhood of Sao Paulo, Brazil, comes another pictorial limestone, with diffuse pink to purple areas in a grey ground. The effect here is of a sunset. All these stones are rare oddities, and may be found as furnishing inlay.

Encrinital or Crinoidal Limestones

The encrinital limestones are usually fawn to dark grey in colour, prettily marked in contrasting tints by fragments of the cylindrical stems of crinoids (sea-lilies). Such marbles are common in the Lower Carboniferous strata of Derbyshire, England – 'bird's-eye', 'Hopton Wood' and 'Derbyshire fossil' marbles – and Belgium. In the latter area the stones are known as 'petit granit', referring to their granularity rather than to any supposed resemblance to granite. Northern Italy is the source of white speckled reddish brown limestones called 'porfirico' from their remarkable superficial similarity to the purple imperial porphyry of Egypt.

Stones showing small-scale sedimentary structures supply attractive types. The well-known 'rosso Verona' and 'porfido ramello' types from north-eastern Italy, brownish red intraformational conglomerates of rounded but irregular limestone fragments set in a contrasting marly matrix, are good examples. The French Pyrenees provide the 'campan' types, which have suffered a degree of metamorphism and shearing. In these, the pebbles are often flattened to almond-like shapes, giving their alternative and perhaps rather misleading name of 'amygdaloidal marbles'. They usually show grey-green colours but may be dark red, or of mixed colours as in the dark red and green banded 'campan mélange'. Of more sombre but equally striking appearance is the 'portor' or 'black and gold', a dolomitic dark grey limestone with sinuous chrome-yellow veining.

Many varieties of marble are of uniform appearance, being used for statuary, and as a foil to more spectacular material as background to inlaid table-tops and pietra dura work, boxes and plinths for sculptures. They are usually fine-grained stones with little or indistinct patterning, and soft, pale and essentially monochrome colouration. From Italy may be mentioned the 'botticino' and 'filetto' marbles, cream to yellowish stones marked by widely spaced stylolitic streaks of contrasting colour, 'Roman stone', despite its name, comes from Istria, and includes the greyish-buff 'auresina', 'nabresina' and 'repen zola' varieties. These are shell detritus stones similar in appearance to the English 'Hopton Wood' marble, but the shell fragments are rarely identifiable.

Fine-Grained Limestones

The so-called 'lava cameos'[1] which are occasionally met as brooches or bracelets, and are either opaque grey or brownish yellow, appear to be nearly always a fine-grained limestone. Sometimes they are imitated by glass.

Oolitic limestones (formed of ovoid or spherical concretionary carbonate grains) of jurassic age may also be treated as marbles. A number of these stones are found in the upper reaches of the Loire valley of France. 'Comblanchien' and 'Nuits St George' are two varieties. The latter is appropriately the colour of a pale red wine. In England, Portland stone, more familiar as a constructional stone, is sometimes polished. Monochrome black marbles are often found in late nineteenth-century clock cases and as tombstones of all dates. In Europe the majority of these stones have come from Belgium, the famous 'Tournai marble' being only one variety. 'Noir Belge' (Belgian black) was found along the Sambre and Meuse rivers, and much black marble is also found in the departments of Nord and Pas-de-Calais in northern France. That from Nord is known as 'grand antique' and that from Pas-de-Calais as 'Noir Francais'. The Republic of Ireland provides the 'Kilkenny black', England the 'Derbyshire black' and also formerly the 'Poolvash' marble from the Isle of Man. These stones are all relatively unaltered compact limestones, but the Carrara area of Italy can supply a black crystalline true marble.

Travertines and Onyx Marble (Stalagmitic Calcites)

Origins

Unlike the marbles of the previous section, the travertines and onyx marbles are predominantly of non-marine physiochemical or biochemical origin. They have formed subaerially or in shallow lakes on the land surface from lime-saturated waters, and the commercial varieties occur in limestone areas where flowing groundwater can dissolve the stone to deposit the lime elsewhere. Calcite is virtually insoluble in pure water, but all groundwater contains dissolved carbon dioxide, which makes it slightly acid and able to dissolve limestone. The resulting solution of calcium bicarbonate may become oversaturated when the waters reach the surface and carbon dioxide is lost to the atmosphere, precipitating the lime as a fine powder.

This process may be accelerated and the lime concentrated by algal or bacterial activity in lakes when sunlight is available, the resulting rocks being known as 'travertine'. The stones are fine-grained, irregularly laminated and often cavernous. Great thicknesses of travertine occur in Italy near Rome, where the stone is usually of a greyish cream colour, and Siena. In the latter locality a wider range of colouration occurs, red, brown and yellow varieties being present. There is also a beautiful variety, 'silver travertine', of a clear silvery grey.

Travertines are usually opaque, but can grade into crystalline translucent types known as 'stalagmitic calcite' or 'onyx marble' where either the water current or lack of sunlight prevents the growth of algae. They are commonly found as cave deposits on rock surfaces, and show a fine concentrically laminated structure, strikingly coloured by periodic variation in the composition of the waters from which they are derived. The colours of the laminations come mainly from iron and manganese salts, and sudden alternations from colourless to white to browns, reds and black create striking effects. As in all marbles, the translucency is a factor of crystallinity: the coarser the calcite crystals, the more translucent the stone. Some finer varieties were even used for window 'glazing' by the Romans, much as white true marbles were used in the Middle East and India.

The nomenclature of the stalagmitic calcite stones is unsatisfactory. 'Stalagmitic calcite' is more a mineral than a rock name. 'Onyx marble' harks back to the ancient method of naming stones from their outward appearance, and not their composition. It is therefore ambiguous. Onyx, as now understood, is a variety of chalcedony, a banded silica hardstone, the only connection with the marble being in the structure. One of the early names is 'alabaster' or 'alabastron', derived from the ancient district of that name in central Egypt, where it was exploited at an early date. This usage is still current in the Mediterranean area. The meaning was extended to include the vases and jars made from the stone, and was later translated by northern Europeans to mean the gypsum rock of similar patterning. Although these marbles are travertines in the broadest sense, better definition is needed.

The earliest, and still active, large-scale exploitation of 'onyx marbles' was in Egypt, whence it has been traded as 'alabaster', 'Egyptian alabaster' or 'oriental alabaster', the last name becoming generic for all varieties, wherever they came from. The expansion of international trade during the Roman Empire gave access to supplies in Turkey, and in Algeria and Morocco. From the latter areas come dark brown, almost red types. Considerable quantities come from the provinces of Oran and Constantine. The major quarries are situated close to Bon-Hanifa, a village about 15 km north-east of the town of Mascara. Bon-Hanifa lies near the ancient Roman town of Aquae Sirensis, which was noted for its hot springs, and it may well be that the deposits originated from these springs. The Constantine deposits lie about 15 km from Constantine, the capital of the department of the same name, and much of the marble from this locality is bright red in colour. On the Moroccan side of the border with Algeria lies another deposit of stalagmitic marble, which may be considered an offshoot of the Algerian deposits. Like the Oran marble, the colour of Moroccan marble is normally white with coloured veins, but material which is blue, grey, red or pink mixed with green is also found. Smaller deposits occur in Italy, the 'stalattite' of Fruili-Venezia Gitilia; in France, the 'stalagmite de Bedat' of Ariege; and in the valley of Vintschgau near Laasee, Austria, where it is known as 'Tyrolese onyx'. 'Gibraltar stone' is from small deposits found within the caves of the rock, and has been used for tourist souvenirs.

The modern age has found deposits further afield in the New World. These stones are often of a pretty white to pale green, marked with brown and yellow, such as the 'Pedrara onyx' of northern Mexico, and other 'mexican onyxes' from Puebla and Oaxaco. In America these stones are sometimes called 'Tecali marble', a name originating from the Aztecs who deemed the stone too sacred for it to be used by the common people. The marble was almost solely devoted to the ornamentation of religious buildings or for the making of sacrificial vessels. Indeed 'Tecali' is but a corruption of the Aztec name Teocali given by the Mexican Indians to their temples. It is from the Mexican localities that most of the onyx marbles now used in England for ornamental pieces are derived. The province of San Luis in Argentina supplies the so-called 'Brazilian onyx', and limited occurrences are known in the USA, including the lemon-yellow 'Utah onyx' from near Lehi, some 30 km from Salt Lake City. Marble similar to that from Mexico has also come from Yavapai County, Arizona and is called 'Yava onyx'.

'Java onyx' is a stalagmitic marble usually dull white in colour or variegated with amber-coloured wavy banding. The rock, which is found around the town of Kediri (Wadjak) some 100 km south-east of Surabaja (Surabaya) in the Toeloeng Agoeng district of Java, does not have the translucency of many other 'onyxes'. A beautiful translucent banded material of bright yellow colour comes from the Karibib district of Namibia. It is locally called 'aragonite' but is actually a stalagmitic calcite.

Staining and Coating of Marbles

The true marble is a granular material and will be susceptible to staining; this is artificially carried out, usually by aniline dyes. Marble may also be coated by essentially colourless substances to mask small scratches or other irregularities and improve the surface appearance. Among the most commonly used materials are waxes, paraffin and plastics. Under magnification the effect of a needle probe in depressing a surface may reveal how soft it is, while the proximity of a heated needle or tiny hot filament may cause a coating to soften and flow.

Calcite

Chemical and Physical Properties

Pure calcite is colourless and the mineral, which is calcium carbonate ($CaCO_3$), crystallizes in the trigonal or rhombohedral system, commonly with rhombohedral or prismatic habit. The hardness of the pure calcite is 3, and is the standard 3 on Mohs' scale, but from the admixture of impurities some marbles may be slightly harder. The purest true marbles, such as the white statuary marbles from Carrara, break with a saccharoidal or granular surface, while the fracture in limestones is usually much smoother. The RIs of calcite are 1.658 for the ordinary ray and 1.486 for the extraordinary ray, but in the compact and granular marbles the most pronounced edge seen in the refractometer is that of the extraordinary ray at 1.48, which can sometimes be seen to move upwards on rotation of the specimen. The ordinary ray is seen with difficulty, much depending on the granular size of the calcite crystals. The limestones generally show an indistinct edge at about 1.50.

The SG of pure calcite is 2.71, but in marbles and limestones there may be, and usually is, a lowering of the SG owing to the granular nature of the crystal aggregates, or the SG may be raised owing to the inclusion of other minerals. The range of SG for most marbles and limestones lies between 2.58 and 2.75.

As calcite is a carbonate, a spot of dilute hydrochloric acid placed on marble, especially a damaged or broken surface or powdered fragment, will produce an effervescence. This is a sure test for calcite marble against some other, but not all, marble-like rocks.

Effects of Light

The absorption spectra and the fluorescence of marbles under UV are too vague to be of value in identification. The luminescent glow exhibited under SWUV or LWUV, if any fluorescence is shown at all, is patchy, some of the veined material showing up the veins as a whitish glow. Some of the travertines, including Gibraltar stone, show a bright greenish yellow glow and a strong phosphorescence under both wavelenghts of UV. Under X-ray stimulation most true marbles and many limestones show the typical orange glow shown by calcite.

Species analogous to calcite include sphaerocobaltite $CoCO_3$ which has sometimes been called caobaltocalcite. Rose-red specimens have been used ornamentally. Sphaerocobaltite occurs most commonly as small spherical masses or crusts with a concentric radiating structure.

Varieties of Calcite

Concretions of calcium carbonate with pearly lustre, which have been formed by the agency of water in limestone caves, are called cave pearls. Despite its low hardness and easy cleavage, golden yellow and colourless calcites have been fashioned.

One kind of crystal aggregate consisting of fibrous crystals of calcite is known as satin-spar. It is found in veins and the fibrous crystals stretch across the vein from side to side. When polished with a flat surface the material shows a silky lustre. Calcite satin-spar – which occurs at Alston Moor in veins from 20 to 100 mm in thickness, at Glen Tilt, Perthshire, and in several places in North America. It is now scarce.

Aragonite

Aragonite, the orthorhombic dimorphous form of calcium carbonate, has also been fashioned. Crystals are often twinned to form pseudohexagonal prisms; they are usually white or colourless but may show tinges of brown, grey, green or violet. The hardness is 3.5–4; SG 2.93–2.95, RI 1.530, 1.680 and 1.685 for the alpha, beta and gamma rays respectively, biaxial negative with DR 0.155.

The yellow stalagmitic marble found at Karibib, in Namibia, sometimes sold as aragonite is stalagmitic calcite. The groups of pseudohexagonal crystals from the sulphur mines in Sicily show a rose-pink fluorescence under LWUV and a green phosphorescence. Transparent crystals of aragonite are found in many places, particularly in Bohemia, Spain and the north of England. Aragonite is a major constituent of pearls in concentric layers of tiny overlapping crystal platelets.

Werner Lieber's book on calcite, *Baustein des Lebens* (1990; no ISBN but published by Munchner Mineralientage Fachmesse GmbH) gives a useful overview with fine photographs. Calcite also features in one of the excellent *ExtraLapis series* (in English, no. 4, 2003, ISBN 071537135).

Note

1. Recent research now categorizes 'lava cameos' as made of a 'carbonate cemented tuff'.

23

Less Common Species

In this section I have included virtually any natural material that has been fashioned and whose descriptions have reached significant literature. I have had to be strict over localities since the literature cites so many and names of countries have changed in some cases. For older mineral localities, there is no substitute for the volumes of Hintze's *Handbuch der Mineralogie* 1889–1971. At any time, a new mineral species or variety may be reported and large transparent varieties of known but hitherto small species may alert the gem community. In any case, collectors of gemstones take every opportunity to obtain whatever is new, sometimes regardless of appearance.

Species are described in alphabetical order. Crystal systems and group membership are taken, in the main, from Fleischer's *Glossary of Mineral Species*, 2004 and Anthony *et al.* (both *op. cit.*). There may be occasional 'hexagonal/trigonal' inconsistencies but they are unlikely to confuse gemmologists. I am indebted to Ulrich Henn for useful notes.

Actinolite

Actinolite forms a series with the minerals tremolite and ferro-actinolite within the amphibole mineral group. This group has been revised in recent years and readers need to consult up-to-date mineralogical literature to keep up with name changes. Actinolite gemstones, with the exception of nephrite (see Chapter 20), are usually small and dark green and are fashioned from bladed monoclinic crystals. The latest version of the composition is given by the 2004 edition of Fleischer's *Glossary of Mineral Species* as $Ca_2 (Mg,Fe_5^{2+}) Si_8O_{22}OH_2$. The hardness is near 5.5, the SG around 3.05

and the RIs in the range 1.620–1.642 with DR 0.022. Specimens may show yellow to dark green pleochroism and an absorption line at 503 nm. Actinolite gemstones found in Tanzania showed an RI for the alpha, beta and gamma rays respectively of 1.619–1.622, 1.632–1.634 and 1.642–1.644. Chatoyant material can be quite attractive; one cat's-eye gave an SG of 3.00 and RI near 1.63. A brighter green than the commoner iron-coloured material is shown by a chromium-bearing variety to which the confusing name smaragdite was at one time given.

Adamite

Adamite is $Zn_2(AsO_4)(OH)$ and forms transparent to translucent pale yellow through green to bluish green elongated orthorhombic crystals and radial aggregates with a hardness of 3.5 and SG 4.32–4.48. The RI for the alpha, beta and gamma rays is 1.707–1.722, 1.734–1.744 and 1.758–1.773, biaxial positive or negative and birefringence 0.031–0.050. Copper and cobalt content may alter properties and some cobaltian adamite from Tsumeb, Namibia, may be purple. Adamite may show an intense yellow or green fluorescence with lemon-yellow phosphorescence under SWUV. Adamite occurs as a secondary mineral in the oxidized zone of zinc- and arsenic-bearing hydrothermal mineral deposits. Reddish to pink and violet transparent crystals are found at the Ojuela Mine, Mapimi, Durango, Mexico.

Aegirine

Aegirine is $NaFe^{3+}Si_2O_6$ and forms translucent to opaque dark to bright green, brown to black prismatic monoclinic crystals with blunt to steep terminations and lengthwise striations. The hardness is 6 and the SG 3.50–3.60. The RI for the alpha, beta and gamma rays is 1.772–1.776, 1.780–1.820 and 1.795–1.836, biaxial negative with birefringence 0.045.

Aegirine occurs in alkalic igneous rocks and in pegmatites. Good crystals come from several Norwegian locations, including Langesundfjord and large crystals have been found at Mt Malosa, Zomba district, Malawi.

Alabaster

The name alabaster is used (incorrectly) for a variety of massive calcite and correctly for a massive variety of gypsum, q.v. (see also Chapter 22). Alabaster is the massive variety of the mineral gypsum, ($CaSO_4 \cdot 2H_2O$), which can occur as transparent monoclinic crystals often twinned in swallow-tailed forms. The crystals show an easy cleavage in one direction. The name selenite has been used for fine crystals, the name referring to the moonlight effect from cleavage surfaces.

The satin-spar variety of gypsum is generally white and occurs filling veins in rock formations with parallel fibrous crystals stretching across the vein from side to side. A similar formation of calcite is also called satin-spar. The low hardness (2) of gypsum precludes its use to any extent except as beads or slabs.

Massive alabaster has been known from many civilizations and is usually deposited by the evaporation of an enclosed sea basin or desert lakes. It may be formed in mineral veins where sulphuric acid, derived from the oxidation of pyrite and other sulphides, has acted upon limestone.

The purest form of alabaster is white and translucent but the material is often associated with a trace of ferric oxide which produces light brown and orange-coloured bands and veins, or with other impurities which colour the stone in yellows, browns and black in veins or patches. The softness of the rock, especially when freshly quarried, enables it to be carved easily. Indeed, alabaster can be scratched by a fingernail, but to some extent the mineral hardens after exposure and will then take a good polish. Due to its comparative porosity alabaster can be dyed.

The hardness of gypsum is given as 2, but most alabaster is found to be harder. The fractured surface is more finely granular than the saccharoidal fracture of marble. The translucency varies from highly translucent to practically opaque and the lustre is glistening. The RIs are 1.52–1.53 and in the case of the massive variety – alabaster – a shadow edge about this value is moderately clearly seen on the refractometer scale. The SG of alabaster lies between 2.30 and 2.33. There is no absorption spectrum of value to be seen in alabaster and the luminescence under UV, usually a brownish shade, gives little aid to distinction. Under X-rays the material is inert.

Massive gypsum is of world-wide occurrence, but only in relatively few localities are quarries worked for ornamental alabaster. One of the important localities for alabaster is the quarries situated at Castellina, in the district of Volterra in Tuscany, Italy. The alabaster from this district is found in nodular masses embedded in limestone interstratified with marls. The mineral is mainly worked by mining through underground galleries. The carving of Tuscan alabaster is carried out at Florence, Pisa and other cities of central Italy, the industry dating from Etruscan times. Pure white Tuscan material is often carved into figurines with religious significance. Some of this pure white material is, after carving, treated by immersion in cold water, which is then slowly raised to boiling point.

The stone is allowed to cool very gradually and is afterwards thoroughly dried. This treatment is said to make the alabaster scarcely distinguishable from white marble. The material may be stained.

The most important quarries for ornamental alabaster in England are in Derbyshire and Staffordshire. The Derbyshire quarries are in the neighbourhood of Chellaston, where the mineral is found in thick nodular beds or 'floors' and in small lenticular masses termed 'cakes'. These deposits are close to the surface and are worked by open-pit mining. The major Staffordshire quarries are at Fauld near Hanbury and near Weston, which lies close to the Derbyshire border. Occasional thin bands of fine granular gypsum of a pinkish colour are found along the coastal plains of South Wales. It is known as 'pink Welsh alabaster' but is not commercially important.

Algodonite

Algodonite is Cu_6As and occurs as opaque metallic granular, often reniform masses, or as incrustations of minute crystals of the hexagonal system. The hardness is 3–4 and the SG 8.38. The RI is outside the range of gemmological instruments. Algodonite associates intimately with the copper arsenides domeykite-α and domeykite-β, with arsenian copper and with silver, in hydrothermal deposits. Locations for those specimens which are polished to produce bright silvery ornaments (which may tarnish) include Painsdale, Michigan, USA (good crystals) and (in masses) from the Algodones silver mine, Coquimbo, Atacama. Chile. Domeykite forms and locations are described below.

Amblygonite

Amblygonite $(Li,Na)Al(PO_4)(F,OH)$ forms a series with montebrasite in which the hydroxyl component exceeds that of the fluorine. Almost all faceted stones are montebrasite. They are an attractive, bright yellow-green, crystallizing in the triclinic system in which they form equant to short-prismatic crystals with twinning common. There is one direction of perfect and one of good cleavage; the hardness is 5.5–6 and the SG 3.00–3.10 (values for amblygonite-montebrasite is 2.98). The RI of amblygonite ranges from 1.578 to 1.591 for the alpha ray and from 1.598 to 1.612 for the gamma ray. The corresponding values for montebrasite are 1.594–1.611 and 1.616–1.633. The optic sign is positive for montebrasite and negative for amblygonite. Some variable but unimportant luminescence may be seen.

Both minerals are characteristic of granite pegmatites, the clear Brazilian lemon-yellow montebrasite occurring in particular from Linopolis, Galileia; also from Aracuai, Governador Valadares and Itinga, Minas Gerais. Gemstones from Mogi das Cruzes, Sao Paulo are reported to be amblygonite rather than montebrasite.

Analcime

Small colourless transparent crystals of analcime have occasionally been cut. Analcime, a mineral of the zeolite group, forms icositetrahedral crystals of the cubic system and has the composition $NaAlSi_2O_6 \cdot H_2O$. The hardness is from 5 to 5.5, the SG ranges from 2.22 to 2.29 and the RI is 1.487. Though analcime is found world-wide, the most likely source for the occasional colourless-faceted stone is Mt Saint Hilaire, Quebec, Canada; some coloured material with RI 1.530–1.536 and 1.532–1.538 has been produced there. Crystals, including specimens from Croft Quarry, Leicester, England, are often well formed and collectable.

Anatase

One of the polymorphs, with rutile and brookite, of titanium oxide, TiO_2, anatase is brown or blue in colour and crystallizes in the tetragonal system, usually as bipyramids. The form often resembles the octahedron. Anatase has a hardness of 5.5–6, an SG between 3.82 and 3.95, and RI 2.493 for the extraordinary ray and 2.554 for the ordinary ray. The negative birefringence is 0.061. Anatase results from the alteration of other titanium-bearing minerals and is also found in hydrothermal veins. Localities for anatase include Isère, France; gemmy material from Binn, Valais, Switzerland, the Urals and Minas Gerais in Brazil. Other localities are Massachusetts, North Carolina and Colorado in the USA, where blue crystals are found.

Andalusite

Andalusite's colours range from a greenish brown to a rich green in colour, the stones owing much of their attractiveness to their pleochroism which shows as red, yellow and green colours. The mineral is, with kyanite and sillimanite, a polymorph of Al_2SiO_5. While the crystals, members of the orthorhombic system, have a prismatic habit with vertically striated prisms which are nearly square in section and capped with pyramids, most gem material is found as water-worn pebbles.

The hardness of andalusite is 7.5 on Mohs' scale and the SG varies from 3.15 to 3.17. The least and greatest RIs vary from 1.634 to 1.641 and 1.644 to 1.648, the biaxial negative, DR varying from 0.007 to 0.011. The pleochroic colours are yellow, green and red which may vary in intensity.

The absorption spectrum varies: a deep green variety, believed to come from Brazil, and in which Mn occupies the octahedral Al site, shows a manganese spectrum with an absorption band, graded in intensity, which ends in a knife-sharp edge at 553.5 nm and which is accompanied by fine lines at 550.5 and 547.5 nm and fainter ones at 518 and 495 nm. There is a strong absorption in the blue and violet but a band at 455 nm can just be seen. This has been noted in the normal types of gem andalusite and these stones, particularly those from Sri Lanka, show this band to be accompanied by a narrow band at 436 nm. These last two bands are probably due to iron. Andalusite exhibits no luminescence under LWUV but the brownish greenstones from Brazil often show a dark green to yellowish green glow under SW and greenish yellow under X-rays.

While andalusite of a dull green is found in the gem gravels of Sri Lanka, Brazil is a more common source of the gem material, where the stones are found in secondary deposits, either on the stream beds or on the slopes of the hills under several yards of clay and gravel. This is in an area some 15 km wide and 40 km in length near the town of Santa Tereza in the state of Espirito Santo. Andalusite is also found in the state of Minas Gerais. The gem mineral is rarely found in crystals with good form and most of the gemmy material is found as water-worn pebbles. The Brazilian stones usually show a very strong flesh-red and olive-green pleochroism, which is especially well brought out when the stones are cut with the table facet nearly at right angles to the vertical axis of the crystal. Such stones may closely resemble certain types of Brazilian tourmalines. A dark green andalusite is also found in Brazil and so too are stones of a rose-red colour, although these are rare.

The opaque variety of andalusite in which the crystals, cut across their length, display carbonaceous inclusions in the form of a cross is known as chiastolite. Sections are cut and polished for amulets and charms in several countries, and particularly in the Pyrenees. Owing to impurities chiastolite may have a lower hardness and SG than transparent crystals. Chiastolite crystals are found in the Nerchinsk district of Transbaikalia in Siberia, at Sallas de Rohan near Brieux in Brittany, France, and Mount Howden, north of Bimbowrie in South Australia. Other localities are at Hof in the Fichtelgebirge, Bavaria, Germany, and in the slates around Skiddaw in Cumberland, England, but here the needle-like prisms are too slender for fashioning into gems. Sources in the USA are Arizona, and Madera in Madera County, California. Other localities are in Maine, Connecticut, Massachusetts and New Mexico, and Fannin County, Georgia. Good examples of the chiastolite variety are found in the Chantaing U range, near Kyankse, south of Mandalay, in Myanmar and Zimbabwe.

Anglesite

Orthorhombic anglesite is $PbSO_4$ and may be colourless and transparent. Shades of yellow, green or blue may be found, the mineral has a hardness of 3 on Mohs' scale, an SG of 6.30–6.39 and RIs of α 1.877, β 1.882 and γ 1.894, biaxial positive with DR 0.017. The lustre may be adamantine inclining to resinous or vitreous; and anglesite fluoresces with a weak yellow glow under SWUV. The mineral has a world-wide occurrence and is found in large crystals at various places in the USA. Anglesite was originally found in the isle of Anglesey and fine crystals have been found in the Leadhills district of Scotland. The best cuttable material is most likely to have come from the Touissit mine, Oujda, Morocco.

Anhydrite

Anhydrite is $CaSO_4$ and forms tabular or equant crystals of the orthorhombic system with hardness 3.5 and SG 2.95–2.96. There is one direction of perfect cleavage. Transparent to translucent material may be colourless to light pink, pale blue to violet. Anhydrite is usually formed by the dehydration of gypsum. Good cleavages in pale purple have come from the Simplon tunnel, Valais, Switzerland and some material has been fashioned; similar specimens have come from the Faraday mine, Bancroft, Ontario, Canada. Large and attractive blue masses have come from Naica, Chihuahua, Mexico. Some German anhydrite may show a red fluorescence under LWUV.

Apatite

Once assigned a group of their own the apatite minerals chlorapatite, fluorapatite and hydroxylapatite stand alone as independent species. They are monoclinic or hexagonal phosphates. Gem material apatites are usually hydroxylapatite or fluorapatite; the fine transparent yellow

Mexican material is hydroxylapatite and the purple material from Maine is fluorapatite. For the gemmologist, there is no need to make mineralogical distinctions providing that an apparent apatite does not prove to be something quite different. Apatite is found in a wide range of situations; gem apatite usually occurring in pegmatites, hydrothermal veins or alluvially.

Apatite can be found in a wide range of colours with violet (especially from Maine) being notably attractive. The Mexican yellow apatite is familiar to students for its rare earth fine line absorption spectrum and a rather beautiful quiet green has been in and out of the connoisseurs' market in the past few years. A vibrant blue, resembling and rivalling the bright copper blue of the Paraíba tourmaline, has been found in Madagascar.

Apatite's softness at 5 rules out hard wear: the hexagonal crystals are often quite large. The SG falls in the range 3.17–3.23 with some blue specimens near 3.18 and yellow ones 3.22. RIs are in the range 1.63–1.64 uniaxial negative with birefringence low at 0.002–0.004. Dichroism is not usually prominent though distinct in the blue material. Both blue and yellow apatites may show the fine-line RE absorption spectrum in which elements of both neodymium and praseodymium are combined. Yellow apatite's absorption spectrum is characteristic, consisting of two groups of fine lines, one group in the yellow, the other in the green. Blue apatite shows broader elements with strong bands at 512, 491 and 464 nm.

The apatite minerals respond variously to UV with some yellow specimens showing a lilac-pink under both LW and SW and some blue stones giving bright, deep violet-blue to sky blue. There would be little need for evaluating apatite's response to X-rays.

Beautiful blue stones are found in the Mogok Stone Tract of Myanmar and in the gem gravels of Sri Lanka, some of which have a fibrous structure making a cat's-eye possible. Yellowish green crystals found in the Spanish province of Murcia, probably in the Serra de Espuna, have been called 'asparagus stones', from their characteristic colour. From Arendal, in southern Norway, comes a beautiful bluish green variety. Well-formed prismatic crystals of yellow colour come from Cerro de Mercado, Durango, Mexico. Very fine violet-coloured crystals are found at the Pulsifer Quarry, Androscoggin County, Maine, USA, one of the great mineral localities.

A deep rich green apatite is found in Quebec and Ontario, Canada, and marketed as 'trilliumite'. There are many other sources of apatite, which is a widespread mineral, but the localities given are the better-known places where gem material is found. Well-formed green crystals in an orange calcite come from Mozambique and form fine mineral specimens, the apatite often being clear enough to produce cut stones. Yellow and green cat's-eye apatite are not uncommon and an intense sapphire-blue material comes from Minas Gerais, Brazil. Yellowish green, greenish brown and reddish brown apatites and apatite cat's-eyes have been found in the Umba valley in Tanzania. Green apatites from Rössing in Namibia have high RIs of 1.649–1.655 with a birefringence of 0.006.

A massive sky-blue variety of apatite has been polished as an ornamental stone, and a variety of lapis lazuli from Siberia has been named lazurapatite.

Apophyllite

Apophyllite, a name well known to mineral collectors rather than gemmologists on account of the beautiful green crystals from India, is not now a valid mineral species under that name. The names fluorapophyllite, hydroxyapophyllite or natroapophyllite should be used according to the composition of particular specimens. These species are members of the apophyllite group. Most apophyllite likely to reach the collector or gemmologist will be fluorapophyllite which forms a series with hydroxyapophyllite. Fluorapophyllite has the composition $KCa_4Si_8O_{20}$ $(F,OH)\cdot 8H_2O$. It is a member of the tetragonal system, occurring in various habits with a perfect basal cleavage. The hardness and SG are 4.5–5 and 2.33–2.37. The RI is 1.530–1.536 for the ordinary ray and 1.532–1.538 for the extraordinary ray. The birefringence is small and may be positive or negative. Neither characteristic absorption spectrum nor luminescence has been recorded. Apophyllite occurs in basalts and granites, specimens from the Poona, Nasik and Mumbai (Bombay) areas of India easily excelling all others. Other world localities can be found in mineral textbooks but their specimens are rarely used ornamentally. Mont Saint Hilaire, Quebec, Canada, has produced variously coloured specimens.

Apophyllite has a notably pearly lustre and may be colourless, pink, yellow or green, none being a strong shade.

Aragonite

The orthorhombic polymorph of $CaCO_3$, aragonite, has been found in transparent colourless, yellowish, yellowish green and brownish yellow crystals at Horuenec near Bilina in the former Czechoslovakia. Aragonite was named after the original locality in Aragon province, Spain and has a hardness of 3.5 to 4 on Mohs' scale. The SG is 2.93–2.94 and the RIs are 1.530–1.540; 1.680–1.690 with a birefringence of 0.150; the optic sign is negative. An opaque, iridescent aragonite of ammonite shells comes from a location between Cardstone and Lethbridge in Alberta, Canada. RI's are α 1.522, β 1.672–1.673 and γ 1.676–1.679 with a birefringence of

0.154–0.155. This material, known by the trade name of ammolite or korite, will be discussed below. Aragonite frequently shows pseudo-hexagonal habit which is caused by combination and twinning. As a carbonate mineral, it is easily soluble in acids.

Arandisite

The presumably local name arandisite was given for a while to an attractive apple-green material, described only in a paper by Partridge in *Trans. Geol. Soc. South Africa* 32 (for 1929) which gave the composition as a tin silicate. The material had hardness 5, SG 4 and RI 1.70. The green material was sometimes cut with surrounding brown limonite (name used for undifferentiated oxides or hydroxides of iron). The source was the Arandis tin mine, Namibia.

Augelite

Colourless to yellowish, pale rose or pale blue crystals of augelite, $Al_2(PO_4)(OH)_3$, have been fashioned. The monoclinic crystals have a tabular habit, a hardness of 5 and an SG near to 2.7. The RIs are α 1.574, β 1.576 and γ 1.588. The mineral is biaxial and optically positive with a birefringence of 0.014. The stones have a vitreous lustre and show no luminescence under either form of UV. Augelite is found at White Mountain, California, in Bolivia and in Spain. Material from California has RIs of α 1.570–1.575 and γ 1.590–1.591, with a birefringence of 0.02. Large crystals have been found at the Champion mine, Mono County, California; some have been fashioned but the source has been reported to be exhausted.

Aurichalcite

Aurichalcite is a zinc–copper carbonate $(Zn,Cu^{2+})_5(CO_3)_2(OH)_6$. Ornamental semi-translucent to opaque white and greenish blue banded material has been reported. The structure is fibrous botryoidal and the hardness and SG 1–2 and 3.64–4.23. The RI is approximately 1.63–1.75. Aurichalcite is a member of the monoclinic system, orthorhombic by twinning. Fine examples have been reported from a number of sources, the Bisbee copper mine, Arizona, USA and the Kelly and Graphic mines, Magdalena, Socorro County, New Mexico, USA. Good material has been found at the Loktevskii mine, west Altai mountains, Russia. The name zeyringite should be reserved for aragonite coloured by oriented inclusions of aurichalcite.

385

Axinite

Mineralogists now differentiate axinite group minerals into ferro-axinite, magnesio-axinite, manganaxinite and tinzenite. However, the transparent axinite of the gemmologist, as shown below, can most conveniently be described as a calcium borosilicate of a mineral group whose members show considerable compositional variation.

Axinite is notable for its strong pleochroism in which the colours olive-green, violet and cinnamon may be seen when transparent specimens are examined in different directions. Axinite is a member of the triclinic crystal system: crystals are characteristically wedge- or axe-shaped. This is a gemstone that can be fairly easily spotted by the unaided eye. The hardness is 7 and the SG in the range 3.27–3.29. The RI for the alpha, beta and gamma rays respectively is 1.674–1.693, 1.681–1.701 and 1.684–1.701 with birefringence 0.010–0.012. The optic sign is biaxial negative. Absorption lines may be seen at 532, 512, 492, 466, 444 and 415 nm. The absorption at 492 and 466 nm is sometimes broad and the 415 nm absorption may be intense.

Axinite as a mineral may show variable luminescence though this is not common in gem-quality specimens. Some yellow crystals from Franklin, New Jersey, may show red under SWUV. A variety of magnesio-axinite has been found to show an orange-red response to LW. These phenomena are not generally shown by gem-quality material. Pale brown transparent axinite has recently been reported from Khapalu, Ghauche District, Pakistan.

While axinite has traditionally been written calcium aluminium borosilicate ($Ca_2Al_2BSi_4O_{15}(OH)$), there is considerable variation of the composition through replacement of the calcium by iron, magnesium and manganese. Ferro-axinite forms a series with manganaxinite.

Axinite is a mineral of contact metamorphism and metasomatism. Magnificent crystal groups have been found at St Cristophe, near Bourg d'Oisans, Isère, France. This material is not often fashioned as crystals are so fine. Axinite is found in the state of Bahia, Brazil. Gem-quality axinite of reddish brown colour, and in quite large crystals, was discovered at Mina la Olivia in Baja California, Mexico, in the 1960s.

Magnesio-axinite from Tanzania has RIs of 1.656, 1.660 and 1.668 with the unusually low SG of 3.18. Analysis showed the specimen to have $Al_2O_3SiO_2$ (44%), MgO (6.9), MnO (0.4), FeO (0.03) and V (0.4).

A cinnamon-brown ferro-axinite has been found in Sri Lanka. The SG is 3.178–3.314 and the RIs are α 1.656–1.675, β 1.660–1.685 and γ 1.668–1.685. The pleochroism is α reddish brown, β dark violet, γ colourless to yellowish.

Azurite

An attractive blue opaque mineral (though superb small transparent crystals are found at Tsumeb) azurite is found in the oxidized zones of copper deposits and is very often intimately associated with malachite to the extent that the name azur-malachite has been used. Azurite may alter to malachite over time. The composition is $(Cu_3(OH)_2(CO_3)_2)$.

Azurite is found as prismatic monoclinic crystals of varied habit, often in spherical radiating groups, and as such is found at Chessy, near Lyons in France. It is common in botryoidal or stalagmitic groups and massive in veins with malachite. Some crystals become pseudomorphs of malachite after azurite. The SG of azurite lies between 3.77 and 3.89 and the principal RIs for the crystals are 1.73 and 1.84. The refraction is biaxial positive and the birefringence is 0.11. The hardness is 3.5–4 on Mohs' scale. Azurite is easy to recognize. Green malachite and blue azurite in bands are cut as 'azur-malachite'.

Azurite is affected by acids and will indicate the presence of copper. A fragment moistened with hydrochloric acid and introduced into the edge of a Bunsen flame will turn the flame blue. These are not appropriate tests for ornamental materials.

Azurite is found at Chessy, France, Bannat, Romania, Bisbee and Morenci in Arizona in the USA. The mineral is also found in New South Wales and South Australia, and in Siberia. Magnificent though small translucent to transparent crystals from Tsumeb, Namibia, are the source of the very few faceted specimens.

In 1971, samples of azurite recovered from an abandoned source at the Copper World Mine, between Barstow and Las Vegas, US, were found to be an unusually tough form of the mineral, mixed in some cases with malachite and other copper minerals. This soon became a popular lapidary material under the name 'Royal Gem Azurite'.

Barite

Barite is barium sulphate $(BaSO_4)$ at one time known among miners as heavy spar. Fine and often well-formed transparent to opaque orthorhombic crystals (more than 70 forms are recorded) are most commonly white though yellow, green, red, blue or brown colours are sometimes found. The massive white material may resemble marble.

The hardness is 3 and the SG 4.3–4.6; for gemmy material the values are near 4.47; the RIs are α 1.636, β 1.637 and γ 1.648, DR 0.012 biaxial positive. The crystals show two directions of perfect cleavage. Barite sometimes fluoresces and often phosphoresces with a faint blue or light green colour under UV. Barite has a worldwide occurrence, gemmy material comes from Colorado and Sardinia. Beautiful blue crystals from the Sterling area of Colorado could be faceted but such crystals should be left alone. Fine yellow material with clear areas, from Cumbria, England, tends to be more massive and faceting could be appropriate here. A brown stalagmitic variety when sectioned shows a concentric structure. Barite is the most common barium mineral and occurs in low-temperature hydrothermal veins and in residual deposits from weathered barium-bearing limestones. Transparent orange barite is found at the Rosh Pinah mine, Namibia, and a deposit in Nevada, USA, has produced yellow specimens.

Bastnasite-(Ce)

Bastnasite-(Ce) has been found as dark orange brown to brown facetable crystals in Pakistan. The composition is $(Ce,La)(CO_3)F$ and crystals belong to the hexagonal system. The hardness is 4–4.5 and the SG 4.78–5.20. The RI for the ordinary and extraordinary rays is 1.72 and 1.82 respectively, uniaxial positive. Specimens show a characteristic RE absorption spectrum with bands at 690, 660, 610, 585–560, 530 and 525–500 nm.

Bayldonite

The green lead–copper arsenate bayldonite $Pb,Cu_3(AsO_4)_2(OH)_2$ has sometimes been used ornamentally as somewhat metallic cabochons. Though members of the monoclinic system, crystals occur most commonly as mammillary crusts or powdery fine-grained masses. The hardness is 4.5, SG is 4.35 and the three RIs are 1.95, 1.97 and 1.99, biaxial positive. Bayldonite is found at Tsumeb in Namibia and in Cornwall, occurring in the oxidized zone of polymetallic deposits.

Benitoite

The original paper describing the discovery of benitoite was the work of Louderback; *Benitoite, a new California gem mineral, Univ. Calif., Bull. Dept. Geol. 5, 149–53, 1907.* W. C. Blasdale provided a chemical analysis. The same author later published, in the same series, 5, 331–380, 1909; *Benitoite, its paragenesis and mode of occurrence.* In a paper by Laird and Albee (*American Mineralogist 57, 85–102, 1972*), it is reported that benitoite may be milky white, colourless or blue but that the composition does not vary with colour. The blue is believed by Nassau (*Physics and Chemistry of*

Colour, second edition, 1983) to arise from intervalence change transfer between iron and titanium (as with blue sapphire). It is likely that the blue of tanzanite and of iolite arises from the same process. Excellent summaries of the sources of benitoite can be found in editions of *Gemstones of North America* (Sinkankas), several accounts of the early twentieth century disagree with one another over the first benitoite location to be found. It seems fairly certain that during the autumn of 1906, a prospector searching for mercury and copper minerals in the area of the headwaters of the San Benito river of California discovered blue crystals in a vein of white natrolite. They were subsequently considered to be sapphire. The strong dichroism shown by the stones, however, led to further investigation in the course of which specimens were sent to Louderback, of the University of California, who identified them as a new mineral to which the name benitoite was given, after the locality in which they were discovered. Kunz, on the other hand, states [in *Gems and Precious Stones of North America* 1890] that the crystals were found in 1907 by Hawkins and Edwin Sanders who were prospecting in the southern part of the Mount Diablo range near the San Benito-Fresno border, and that these crystals were brought to the attention of Dr Louderback by Shreve and Company, a San Francisco firm who had purchased one of the cut stones from a lapidary and who were later offered some of the rough material as sapphire.

Benitoite is a fine bright near sapphire-blue and crystallizes uniquely in a class of the hexagonal system which has a trigonal axis of symmetry and a plane of symmetry at right angles to it, a form that was only postulated in nature but not found until the discovery of benitoite. Colourless crystals of benitoite are sometimes cut, perhaps for their occasionally orange fluorescence.

Benitoite is barium titanium silicate with the formula $BaTiSi_3O_9$. The SG lies between 3.65 and 3.68, but the latter value is more common. The hardness is about 6.5 on Mohs' scale. The RIs are 1.757 for the ordinary ray and 1.804 for the extraordinary ray, uniaxial positive. The birefringence is strong at 0.047. The pronounced dichroism shows twin colours of blue for the extraordinary ray and colourless for the ordinary ray, so that to obtain the best colour the table facet should be cut parallel to the principal crystal axis, and as most of the crystals have a tabular habit this precludes the cutting of large stones. Benitoite has a high colour dispersion, 0.039 for the ordinary ray and 0.046 for the extraordinary ray (compare the 0.044 of diamond).

The lower density allows distinction from sapphire, but if the stone is set determination of this constant would not be possible and a refractometer reading would be necessary. Care is needed, however, in observing the shadow edges as the maximum value of the extraordinary ray is beyond the range of the refractometer and may be missed unless the stone is rotated fully. The pronounced dichroism may also distinguish it.

Benitoite shows a bright blue fluorescence under SWUV but is inert to LW. Some colourless specimens have been found to give a reddish orange glow under LW.

At the time of writing, gem benitoite still comes only from mines in San Benito County, California, USA, where it occurs with joaquinite and neptunite in natrolite veins cutting glaucophane schist.

Beryllonite

Beryllonite is sodium beryllium phosphate, $NaBePO_4$ and is a mineral of granite and alkalic pegmatites. It forms pseudo-orthorhombic crystals in the monoclinic system with hardness 5.5–6, SG in the range 2.77–2.84, RI 1.552, 1.558 and 1.561 for the alpha, beta and gamma rays; the birefringence is 0.009. Crystals are biaxial negative. The dispersion is low at 0.010. Faceted beryllonite is almost always transparent, colourless. The classic location for beryllonite is Stoneham, Maine, USA, where it occurs at Sugarloaf Mountain, Brazil.

Beryllonite is inert under UV, but under X-rays there is a dark sky-blue fluorescence with slight phosphorescence. Other sources are Finland, Zimbabwe and Minas Gerais, Brazil. Some chatoyant specimens have been reported from Afghanistan.

Bismutotantalite

Bismutotantalite is $Bi(Ta,Nb)O_4$ and forms stout prismatic crystals or masses of the orthorhombic system, usually opaque and coloured light brown to smoke-grey with adamantine or submetallic lustre, hardness 5, SG 8.15–8.89.

Facetable material is rare but the RI for the alpha, beta and gamma rays is 2.388–2.395, 2.403–2.408 and 2.426–2.428, biaxial positive with birefringence 0.040. Bismutotantalite occurs in granite pegmatites and as water-rounded pebbles. Sources of ornamental material are the Muiane mines, Mozambique and Acari, Brazil.

Boléite

Boléite forms deep Prussian blue to indigo cubic crystals which are often surficially overlaid with or overgrown by pseudoboleite. The hardness is 3–3.5, the SG 5.05 and there is one direction of perfect cleavage. The RI is 2.05. Boléite occurs as a secondary mineral formed by the reaction of chloride with primary sulphides in the oxidized zone of lead–copper deposits. Facetable material will almost certainly have come from the Amelia mine, Boleo, Santa Rosalia, Baja California, Mexico. The composition is $KPb_{26}Ag_9Cu_{24}Cl_{62}(OH)_{48}$.

Boracite

Boracite is magnesium chloroborate $Mg_3B_7O_{13}Cl$ and may produce pale green ornamental specimens. While some crystals, though members of the orthorhombic system, are pseudocubic others can be referred to pseudotetragonal morphology. Boracite is very slowly soluble in water, has a hardness of 7–7.5 and an SG of 2.91–3.10. The RI is 1.662–1.658, 1.662–1.667 and 1.668–1.673 for the alpha, beta and gamma rays respectively with birefringence 0.011, biaxial positive. The lustre is vitreous to adamantine and a very weak greenish fluorescence has been reported. Facetable boracite is found in the Stassfurt and Hannover areas of Germany.

Bornite

Bornite is Cu_5FeS_4 and forms opaque compact copper red to bronze masses with a metallic lustre and pseudocubic crystals of the orthorhombic system with hardness 3–3.25 and SG 5.08. The fresh red to brown surface colour does not persist and an iridescent purple soon develops. The RI is out of range of gemmological testing. Bornite is occasionally fashioned into cabochons; the mineral occurs in mafic igneous rocks, in pegmatites and in contact metamorphic (skarn) deposits. Fine examples have come from the Redruth area of Cornwall, England, and from Butte, Silver Bow County, Montana, USA.

Bowenite

See Serpentine.

Brazilianite

Brazilianite, fine transparent yellow at its best and still a comparative rarity was first reported in 1944. It occurs in phosphate-rich zones of granite pegmatites, the only gem-quality material coming from locations in Brazil. The original find was in the Conselheira Pena area of

Minas Gerais. Brazilianite is a hydrous sodium aluminium phosphate $NaAl_3 (PO_4)_2 (OH)_4$. It crystallizes in the monoclinic system forming characteristically equant or spear-head shaped crystals. There is a good cleavage and the hardness is 5.5. The SG is usually 2.98 and the RI for the alpha, beta and gamma rays is 1.602, 1.609 and 1.621–1.623, biaxial positive with a birefringence of 0.019–0.021. The dispersion is low at 0.014.

Brazilianite is brittle and shows a conchoidal fracture. There is no recorded response to UV and there is no useful absorption spectrum in the visible. The dichroism is insignificant. Crystals are reported from Corrego Frio, north of São Tomé, Minas Gerais. In 1947, another source of brazilianite was discovered at the Palermo mine, North Groton, Grafton County, New Hampshire, USA.

Breithauptite

Breithauptite is NiSb and forms thin tabular crystals of the hexagonal system though the massive variety provides the only material of ornamental interest. The colour is copper-red often with a violet tint, the hardness 5.5 and the SG 7.59–8.23. The RI is not obtainable by normal gemmological testing. Breithauptite occurs in hydrothermal calcite veins associated with cobalt–nickel–silver ores; most ornamental material coming from the Cobalt district, near Red Lake, Ontario, Canada – there are other locations in the Harz Mountains, Germany.

Bronzite

Bronzite is not used as a mineral name today, having been dropped from the latest revision of the composition of the pyroxene group of minerals (1988). The name was used to denote a member of a solid solution series with end-members enstatite, $Mg_2Si_2O_6$ and ferrosilite $(Fe^{2+},Mg)_2Si_2O_6$. The iron content is between 5 and 13% (this applies to the 'bronzite' of old).

Crystals are orthorhombic, the hardness is 5–6 and there is a distinct cleavage. Yellowish brown bronzite has been reported to come from Sri Lanka with SG 3.25 and RI α 1.663, β 1.669 and γ 1.677 with a birefringence of 0.014, the optic sign positive.

Brown six-rayed star bronzite has been found in Sri Lanka. The SG is 3.41 and the RIs are α 1.680, β 1.689 and γ 1.693 with a birefringence of 0.013. The star effect is caused by scattering of light at oriented hollow tubes.

Brookite

One of the polymorphs of titanium dioxide (TiO_2), with rutile and anatase, brookite, which crystallizes in the orthorhombic system with varied habit, has been used ornamentally. It is yellowish brown to reddish brown with hardness of 5.5–6. The SG varies from 3.87 to 4.08 and the RIs are near to 2.583 and 2.705 for the principal rays; the biaxial and optically positive DR is about 0.122, but it can reach 0.157 in certain cases. The lustre of brookite is metallic adamantine. Examples are found in France, Switzerland, Russia (Urals) and the USA.

Brucite

Brucite is $Mg(OH)_2$ and forms transparent tabular crystals, platy or foliated or granular masses of the trigonal system with one direction of perfect cleavage, hardness 2.5 and SG 2.39. The RI for the ordinary and extraordinary rays is 1.56–1.59 and 1.58–1.60, uniaxial positive (anomalously biaxial) with birefringence 0.010–0.020. The colour ranges from white through pale green or blue to brownish red and brown. Brucite occurs as a common alteration of periclase in marble and in low-temperature hydrothermal veins in metamorphic limestones and chlorite schists. Good examples are found at Asbestos, Quebec, Canada.

Burbankite

Burbankite from the intrusive alkalic gabbro-syenitic complex at Mt Saint Hilaire, Quebec, Canada, has the composition $(Na,Ca)_3$ $(Sr,Ba,Ce)_3(CO_3)_5$. Ornamental specimens have been found to show a colour change from greenish yellow in fluorescent to yellowish orange in incandescent light. Burbankite is a member of the hexagonal system, forming long dihexagonal prismatic crystals with shallow pyramidal terminations. The hardness is 3–3.5 and the SG near 3.53. The RI is 1.630 and 1.632, DR 0.012, uniaxial negative. There may be a weak pink fluorescence under LW and SWUV. Sharp absorption lines may be seen at 578, 522, 510, 488 and 444 nm. Remondite-(Ce), an associated mineral, has the composition Na_3 $(Ca,Ce Na,La)_3(CO_3)_5$ and is found at Mt Saint Hilaire with burbankite. Remondite-(Ce) has similar properties to burbankite. Rare earth elements may be present in either species.

Bustamite

Usually massive, transparent pink bustamite $(Ca,Mn^{2+})Si_3O_9$ occurs at Broken Hill, NSW, Australia. The hardness is 5.5–6.5, the SG about 3.3–3.4 and the RI α 1.683, β 1.698 and γ 1.701 with a birefringence of 0.018. Gem-quality bustamite is also found at Daghazeta in Tanzania. Bustamite can be confused with rhodonite from its colour and common occurrence, but the RIs are lower.

Canasite

Canasite is $Na_3K_3Ca_5Si_{12}O_{30}(OH,F)_4$ and occurs as greenish yellow granular blocks of the monoclinic system with hardness 5–6, perfect cleavage in two directions and SG 2.70. The grains are usually twinned. The RI for the alpha, beta and gamma rays is 1.534, 1.538 and 1.543, biaxial negative with birefringence 0.009. Canasite occurs in pegmatites in a differentiated alkalic massif at Khibiny, Russia. A purplish material is sometimes referred to as 'canasite' or 'stichtite' (q.v.). Work on differentiation continues.

Cancrinite

A semi-opaque yellow-coloured fibrous variety of cancrinite, $Na_6Ca_2Al_6Si_6O_{24}(CO_3)_2$, has been cut into beads and cabochons. Cancrinite is occasionally found as crystals of the hexagonal system; the RI approximate to 1.50, which is all that would be expected to be seen from the massive gemmy material. The crystals of cancrinite have RI for the ordinary ray of 1.515–1.524 and for the extraordinary ray of 1.491–1.502, the DR of 0.022, uniaxial negative. The SG varies from 2.42 to 2.50 (a specimen of Canadian gem material gave a value of 2.435). The hardness is 5–6.

Yellow gem cancrinite does not luminesce under UV but an orange-yellow response has been observed under X-rays. Cancrinite effervesces with acids. While cancrinite is found in several places, gem-quality material usually originates from Bancroft and French River, Ontario, Canada. Cancrinite is found in pegmatites in nepheline syenites. Other colours, including light greyish purple, yellow, violet and orange have come from Canadian sources and a blue variety from Greenland, this material being reported to show dark violet under LWUV and dark red to purple under SW.

Carletonite

Pale blue transparent carletonite has been found at Mt Saint Hilaire, Quebec, Canada. The composition is $KNa_4Ca_4Si_8O_{18}(CO_3)_4 (OH,F){\cdot}H_2O$ and it crystallizes in the tetragonal system. The hardness is 4.5 on Mohs' scale with perfect cleavage. The SG is 2.44 and the RIs are 1.517 and 1.521 with a birefringence of 0.004. Carletonite is found in nepheline syenite. Some Mt Saint Hilaire material has been found in opaque to translucent pink, white and grey with colour zoning common. Some of the blue specimens may show blue to pinkish brown pleochroism.

Cassiterite

Cassiterite, SnO_2, is the principal ore of tin. When transparent material is faceted its high dispersion (0.071) and uniaxial positive RI 2.003 and 2.101 for the ordinary and extraordinary rays respectively, combine to produce an attractive and brilliant stone; most cassiterite crystals are more or less heavily included or opaque black. Cassiterite forms prismatic crystals of the tetragonal system, characterized by square cross-section and steep pyramidal forms; geniculate twins are common.

Cassiterite has a hardness of about 6.5 and SG of 6.95. The lustre is adamantine. Cassiterite shows no luminescence and no distinctive absorption spectrum. Colourless specimens rarely occur, most gemquality specimens inclining to brown.

Cassiterite can be distinguished from brown diamond, brown zircon and sphene by its greater density; from zircon by the absence of any typical absorption lines; from sphene by the absence of (or a much weaker) dichroism and from diamond by the DR of 0.098.

Cassiterite occurs in medium- to high-temperature hydrothermal veins. In Cornwall, England, a source has been known since Roman times. Facetable cassiterite is usually from Bolivia though some dark-coloured Cornish crystals have been cut. Orange to purple cassiterite has been found in Sri Lanka and a botryoidal laminated variety in the Bolivian tin deposits.

Catapleiite

Catapleiite is found at Mt Saint Hilaire, Quebec, Canada. The hexagonal crystals show a perfect cleavage with hardness of 6 and SG of 2.72. The RIs are α 1.590, β 1.609 and γ 1.629 with a birefringence of 0.039 and positive optic sign. The composition is $Na_2ZrSi_3O_9{\cdot}2H_2O$. Only small colourless faceted stones are known.

Cavansite

Cavansite, resembling turquoise and with the composition $Ca(V^{4+}O)Si_4O_{10} \cdot 4H_2O$ is occasionally fashioned, material coming from the Poona area of India. It occurs as attractive radially fibrous crystal clusters of the orthorhombic system and can be cut into blue to green cabochons. The hardness is 3.5, the SG 2.24–2.27. The RI reading of 1.54 is all that can be seen on the refractometer. Resin impregnation has been resorted to for stabilization, the porosity of the material making the process simple.

Celestine

Celestine is strontium sulphate, $SrSO_4$, and crystallizes in the orthorhombic system with tabular habit. The transparent crystals have two directions of cleavage and stones are faceted from colourless, yellowish or bluish green crystals. The hardness is 3.5. The SG is 3.97–4.00 and the RIs are α 1.619, β 1.623 and γ 1.631 biaxial positive with birefringence of 0.012. Gem-quality material may have slightly higher values. A scraping taken from the girdle of a cut celestine may colour the Bunsen flame red. A whitish or bluish white glow is shown by celestine under UV.

The origin of celestine is chiefly sedimentary. It occurs as cavity or fissure fillings, particularly in limestones and in hydrothermal veins. Ornamental-quality material is found at Yate, Bristol, England; most faceted celestine comes from Madagascar (blue) or Canada (orange). Facetable crystals from Tsumeb, Namibia.

Ceruleite

Opaque turquoise-blue ceruleite $(Cu_2Al_7(AsO_4)_4(OH)_{13} \cdot 12H_2O)$ is found in the gold mine Emma Luisa, Guanaco District, Antofagasta, Chile and at an unspecified locality in Bolivia. The triclinic material occurs as earthy masses which nonetheless take a reasonable polish. The hardness is 5–6, SG 2.70–2.80 and mean RI 1.60. Specimens have been resin-stabilized. Ceruleite occurs in the oxidized zone of arsenic-rich polymetallic mineral deposits.

Cerussite

Cerussite is lead carbonate, $PbCO_3$. It is usually white or grey in colour, but black, blue and green colours are known, these two latter owing their colour to copper; faceted cerussite is almost invariably transparent and colourless. The hardness is 3.5; it is notoriously brittle and heat-sensitive. The SG is in the range 6.46–6.57. The lustre is adamantine and the RIs are α 1.804, β 2.076 and γ 2.078. The birefringence is 0.274, biaxial negative. The luminescence is variable but under SWUV. It usually shows a pale blue or green glow. Some cerussite from Utah is said to show a bright orange under LW.

Most facet-quality material comes from Tsumeb, Namibia. Despite difficulties in cutting, the dispersion of 0.055 makes the occasionally encountered gemstone quite spectacular. Collectors value the orthorhombic crystals which are frequently twinned to give pseudo-hexagonal shapes.

Chalcopyrite

Chalcopyrite is $CuFeS_2$ and forms equant tetrahedral-shaped crystals often with modification by scalenohedral faces; it also forms compact masses. Chalcopyrite has a metallic lustre though brass-yellow often shows tarnish and iridescence. The hardness is 3.5–4 and the SG 4.1–4.3. It is sometimes used for cabochons though specimens are more often sought in their original state. Chalcopyrite occurs as a primary mineral in hydrothermal veins and in other environments. The occurrence is worldwide.

Chalcosite

The name chalcosite has been given to rocks found in Ambatofinandrahana, Madagascar, which have been used ornamentally. They are composed of reddish brown potassium feldspar, green plagioclase, grey quartz and black mica. The name is unnecessary and incorrectly used in this connection (if it ever is) since it is the validated name for copper sulphide, Cu_2S.

Chambersite

Chambersite, $Mn_3B_7O_{13}Cl$, is unlikely to rank high in the gem world since material of sufficient size for cutting has (so far) been found only at one site, the Barber's Hill salt dome, Mont Belvieu, Chambers County, Texas, USA. The very small transparent stones recovered have been pale brown to deep purple with a hardness of 7, SG 3.49, RI 1.732, 1.737, 1.744 for the alpha, beta and gamma rays respectively. The birefringence is 0.012, biaxial positive. The orthorhombic crystals are pseudocubic. Recovery which entails descent into brine–diving equipment is necessary.

Charoite

A mainly purple opaque mineral, forming a rock with orange tinaksite, dark aegirine-augite and green microcline was reported from the Murun massif, between the Chara and Olekma rivers, south-west of Olekminsk, Yakutia, Russia, in 1978. Charoite is $K(Ca,Na)_2Si_4O_{10}(OH, F) \cdot H_2O$, probably monoclinic but found massive. The colour can be described as shades of deep violet to lilac; the hardness is 5–6 and the SG near 2.68 at its highest. The RI is in the range 1.550–1.559. Charoite occurs in potassic feldspar metasomatites at the contact of nepheline and aegririne syenites with limestones. The combination of charoite with some of its associated minerals makes attractive ornaments.

Chiolite

Chiolite is $Na_5Al_3F_{14}$ and forms very small dipyramidal crystals of the tetragonal system and, more commonly, transparent to translucent near-colourless to snow-white masses with a perfect cleavage in one direction, hardness 3.5–4 and SG 2.90. The RI for the ordinary and extraordinary rays is 1.349 and 1.342, uniaxial negative with birefringence 0.007. Chiolite occurs in some granite pegmatites; any fashioned specimens are most likely to have come from the Ivigtut cryolite deposit, south-west Greenland.

Chlorastrolite

The name chlorastrolite has been used for an opaque ornamental variety of the monoclinic mineral pumpellyite. The ornamental variety forms fibrous aggregates with other minerals and shows green and white patterns. The most desirable pattern resembles tortoiseshell.

Some of the green is intense and could be mistaken for one of the jade minerals. However, the RI differs from those of nephrite and jadeite at a mean value of 1.7. The hardness is 5–6 and the SG in the range 3.1–3.5. The crystal system is orthorhombic. Pumpellyite occurs in a number of different environments; most of the chlorastrolite variety occurs in the Lake Superior area of the USA.

Chondrodite

Facet-quality chondrodite of a transparent deep garnet-red is found in the Tilly Foster mine, Brewster, New York, USA and is one of the rarest and most beautiful of gemstones. Chondrodite, a member (with clinohumite, another rare gem mineral (q,v.) of the humite group) has the composition $(Mg,Fe^{2+})_5(SiO_4)_2(F,OH)_2$ and belongs to the monoclinic system in which it crystallizes as characteristically flattened crystals. The hardness is 6–6.5 and the SG 3.16–3.26. The RI for the alpha, beta and gamma rays is 1.592–1.643, 1.602–1.655, 1.621–1.636 with a birefringence of 0.028–0.034, biaxial positive. Chondrodite may give a weak orange fluorescence under LWUV. Transparent to translucent orange-brown specimens have recently been reported from Sri Lanka and are said to show a distinct cleavage and orange-brown to yellow-brown pleochroism.

Chromite

Chromite is $Fe^{2+}Cr_2O_4$ and provides translucent to opaque material in a lustrous black which has sometimes been fashioned as cabochons. The crystal system is cubic and the masses finely granular with hardness 5.5 and SG 4.5–4.8. The lustre is submetallic to metallic and the RI 2.08–2.16. Chromite occurs in olivine-rich igneous rocks. Distribution is widespread and there is no obvious place of origin for the odd cabochon.

Chrysocolla

Chrysocolla has the composition $(Cu,Al)_2H_2Si_2O_5(OH)_4 \cdot nH_2O$ and can be found as translucent to opaque blue to blue-green acicular crystals of the orthorhombic system though ornamental varieties are crypto-crystalline. The texture can resemble enamel. The hardness is 2–4 and the SG 1.93–2.4. Specimens are notably brittle. Sources differ on RI values but for practical purposes an RI of around 1.5 will be obtained on the refractometer. Specimens can show considerable colour variation with oxides of copper, iron and manganese included with the green. Chrysocolla is often found in a matrix of quartz, in which case the constants may be those of quartz.

Chrysocolla is a mineral of secondary origin occurring in the oxidized zones of copper deposits where it may be mixed with malachite or turquoise. Occurrences are widespread but most ornamental chrysocolla comes from the south-western US or from Israel where 'Eilat stone' is found near Eilat in the Gulf of Aqaba. Much of the chrysocolla of jewellery consists of very attractive cabochons of green or blue chrysocolla impregnating either rock crystal or opal.

Cinnabar

Cinnabar is HgS and forms rare cochineal-red translucent rhombohedral crystals of the trigonal system with an adamantine to metallic lustre; masses and encrustations are more common. There is one direction of perfect cleavage, the hardness is 2–2.5 and the SG 8.20. The RI for the ordinary and extraordinary rays is 2.905 and 3.25, uniaxial positive with birefringence 0.351. Cinnabar occurs in low-temperature ore deposits and is also found in veins and around hot springs. Cinnabar crystals, some of which may be faceted, usually come from a notable deposit at the Tsar Tien mine, Hunan Province, China.

Clinohumite

Transparent gem-quality bright yellow to deep orange clinohumite, a monoclinic member of the humite group of silicates with the composition $((Mg,Fe^{2+})_9(SiO_4)_4(F,OH))_2$ was found in recent years in the Pamirs of Tadzhikistan. The hardness is 6, the SG about 3.18 and the RI α 1.623–1.702, β 1.636–1.709 and γ 1.651–1.728 with a birefringence of 0.024–0.036; biaxial positive. A yellow fluorescence may be seen under SWUV. The occurrence is in contact zones in limestone or dolomite. A dark brown variety has been found in the Taymyr region, Siberia. The RI is in the range 1.640–1.671 with birefringence 0.039 and SG 3.35.

Clinozoizite

Clinozoisite is a monoclinic epidote group mineral $Ca_2Al_3(SiO_4)(Si_2O_7)$ O·OH and usually occurs in pale shades of green, pink or yellow and is sometimes colourless. It occurs in typically elongated transparent prismatic crystals with one direction of perfect cleavage. The SG is in the range 3.21–3.38 and the RI for the alpha, beta and gamma rays is 1.670–1.715, 1.675–1.725 and 1.690–1.734, biaxial positive. The birefringence is 0.005–0.015. The absorption spectrum is similar to that of epidote. Clinozoisite comes from Austria, Italy, Switzerland, Bohemia (Czech Republic) and Baja California, Mexico. A grey–green clinozoisite is found in Kenya.

Cobaltite

Cobaltite is CoAsS and forms opaque silver to greyish black pseudocubic or pseudopyritohedral crystals of the orthorhombic system with a perfect cleavage in one direction, hardness 5.5, and SG 6.3. The RI will be out of the range of gemmological instruments. Cobaltite occurs in high-temperature hydrothermal deposits and in veins in contact metamorphosed rocks. Specimens, which may show a metallic redness, are found at Cobalt, Ontario, Canada, and fine crystals are found at Tunaberg, Sodermanland, Sweden.

Colemanite

Colemanite is monoclinic hydrous calcium borate, $Ca_2B_6O_{11} \cdot 5H_2O$ and is commonly colourless to milky white. The hardness is 4.5, the SG 2.42. The RI is α 1.586, β 1.592 and γ 1.614, biaxial positive. Colemanite usually fluoresces whitish or green under UV. Equant to short-prismatic crystals are brittle and fashioning of transparent specimens is difficult. California produces most facetable specimens, especially from the Death Valley and Boron areas. Colemanite is a common constituent of borate deposits formed in sodium- and carbonate-poor lake environments.

Cordierite

See Iolite.

Covellite

Covellite is CuS and forms notably iridescent opaque masses in the hexagonal system with a highly perfect cleavage in one direction, hardness 1.5–2 and SG 4.6–4.7. The RI for the ordinary and extraordinary rays is 1.45 and 2.62, uniaxial positive with high birefringence (not of importance to gemmologists). Covellite occurs in the secondary enrichment zones of copper deposits. Ornamental material, sometimes large, is found at a number of copper mines in the US. Fine crystals have been found in the primary ore at Butte, Silver Bow County, Montana; also at Kennecott, Alaska. Large crystals have also been found at the Calabona mine, Alghero, Sardinia, Italy.

Creedite

Creedite is $Ca_3Al_2 (SO_4)(F,OH)_{10} \cdot 2H_2O$ and forms monoclinic crystals of prismatic habit or radiating masses. Some light purple transparent stones have been cut. The hardness is 3.5 and there is a perfect cleavage. The RIs are α 1.461, β 1.478, γ 1.485, biaxial negative and the SG is 2.71. Facetable material has been found at a fluorite/barite mine at Santa Eulalia, Chihuahua, Mexico. Creedite is a rare mineral, occurring in fluorite-rich hydrothermal mineral deposits.

Crocoite

The bright orange near transparent prismatic to monoclinic acicular crystals of chromite, $PbCrO_4$, are occasionally cut but crystal groups are too highly prized by mineral collectors for many specimens to be faceted. The hardness is 2.5–3 and the SG in the range 5.9–6.1. The RI is 2.29, 2.36 and 2.66 for the alpha, beta and gamma rays respectively with a birefringence of 0.370, biaxial positive. A reddish to dark-brown fluorescence may sometimes be seen and thin sections may show an absorption band at 550 nm. Crocoite of facetable dimensions and quality occurs only at Dundas, Tasmania, where it is found in the oxidized zone of lead deposits in association with chromium-bearing rocks.

Cryolite

Colourless transparent monoclinic crystals of cryolite, Na_3AlF_6, have occasionally been cut despite their softness and hazy appearance. The hardness is 2.5, the SG 2.97. The RI is 1.338, 1.338 and 1.339 for the alpha, beta and gamma rays, with a birefringence of 0.001, biaxial positive. Monoclinic crystals are short prismatic or pseudocubic. Cryolite occurs in some granite pegmatites, in topaz-bearing, fluorine-rich rhyolites and elsewhere. Fashioned material is most likely to have originated at Ivigtut, Greenland.

Cuprite

Though fine transparent red crystals of the red copper oxide (Cu_2O) occur naturally at the Onganja mine, Namibia, and have been faceted, synthetic cuprite has been grown and some faceted examples are magnificent. Both natural and synthetic cuprites are liable to undergo some surface alteration to malachite after long exposure to light. Cuprite crystals occur as cubes, octahedra or combinations of forms of the cubic system. The hardness is 3.5–4 and the SG 6.14. The RI is 2.84. Cuprite occurs in the oxidized zones of copper deposits.

Danburite

Danburite, most often colourless or pale yellow (orange specimens may have been treated), is calcium borosilicate $CaB_2 (SiO_4)_2$. Crystals of the orthorhombic system are prismatic with rhombohedral to square cross-section. The hardness is 7–7.25 and the SG 3.00. RI for the alpha, beta and gamma rays is 1.627–1.633, 1.630–1.636, 1.633–1.639 with birefringence 0.006–0.008. As the angle between the optic axes is nearly 90°, danburite is optically positive for blue to violet light and negative for red to green light. Some danburite may show the fine line absorption spectrum of RE. Under LWUV danburite will glow a bright sky-blue to blue-green. Red thermoluminescence has been reported.

Danburite occurs in granites and in metamorphosed carbonate rocks associated with hydrothermal activity. Specimens suitable for fashioning come from Sri Lanka, Myanmar (in particular, yellow material), Mexico, where a light pink variety has been reported; Madagascar, where yellow crystals are found at Mt Bity and from Russia. Colourless crystals have been found at Obira, Bungo, Oita Prefecture, Japan; some have been faceted. Danburite will float in di-iodomethane while topaz will sink. In bromoform (tri-bromomethane), whose use is subject to restrictions, similarly coloured citrine will float.

Datolite

Datolite is $CaBSiO_4(OH)$ and forms characteristic short prismatic monoclinic crystals. Transparent colourless crystals have been faceted and other pale colours turn up from time to time. The hardness is 5–5.5 the SG 2.96–3.00 and the RI for the alpha, beta and gamma rays is 1.622–1.626, 1.649–1.654, 1.666–1.670 biaxial negative with birefringence 0.044–0.047. Some specimens may fluoresce blue under SWUV, perhaps from europium impurities. Datolite occurs as a secondary mineral in mafic igneous rocks, in geodes in tuffs or skarns in limestones, in serpentinites and hornblende schists and in some ore veins. Gem-quality datolite has come usually from the Lane quarry, Westfield, Mass., from other workings in the eastern US (pale green specimens from Paterson, New Jersey) or from Charcas,

San Luis Potosi, Mexico. Some datolite from Michigan, USA, contains included copper. Datolite is also found in the Habachtal, Austria.

Diaspore

Some examples of diaspore, AlO(OH), have been found to show a colour change from greenish in daylight to pinkish brown under tungsten light, though much paler than the similar effect seen in alexandrite. Stronger colours are visible when crystals are cut into thin sections. Diaspore forms platy crystals in the orthorhombic system with a perfect cleavage in one direction and a distinct cleavage in another. This makes fashioning quite difficult as the hardness is only 6.5–7. The SG is 3.2–3.5 and the RI 1.682–1.706, 1.705–1.725 and 1.730–1.752 for the alpha, beta and gamma rays. The birefringence is 0.048, biaxial positive. Absorption bands noted in some specimens were at 471, 463 and 454 nm in the blue.

Diaspore may be formed by the hydrothermal alteration of aluminous minerals and may be a hydrothermal mineral in some alkalic pegmatites. Large gem-quality crystals have been found from the area of Mamaris, Yagatan, Mugla Province, Turkey, and this seems to be the source of all faceted stones to date.

Dickinsonite

Dickinsonite is $KNa_4Ca(Mn^{2+},Fe^{2+})_{14}Al(PO_4)_{12}(OH)_2$ and forms very small monoclinic commonly rhombohedral olive green tabular transparent to translucent crystals with a perfect cleavage in one direction with striations giving a triangular pattern. The hardness is 3.5–4 and the SG 3.34–3.41. The RI for the alpha, beta and gamma rays is 1.648–1.658, 1.655–1.662, 1.662–1.671, biaxial positive with birefringence 0.013. Dickinsonite, of which some faceted stones turn up occasionally, occurs as a high-temperature mineral in granite pegmatites. Crystals are from Branchville, Connecticut, USA.

Diopside

Diopside is a member of the pyroxene group of silicates and provides transparent mostly green to brown monoclinic prismatic crystals with nearly square cross-section and twinning common. There is a distinct cleavage in one direction and the hardness is 5.5–6.5. The SG is in the range 3.22–3.38, increasing with increasing iron content; the composition is $CaMgSi_2O_6$; diopside forms two series, one with hedenbergite and one with johanssenite. The RI for the alpha, beta and gamma rays is 1.664, 1.672 and 1.694 with a mean birefringence of 0.026, biaxial positive. Pleochroism for the important chrome green material is yellow and green.

While gem diopside is most commonly green, this colour may result from either chromium (chrome diopside) or iron impurities. The absorption spectrum of chrome green diopside shows a doublet at 690 nm with other absorptions at 670, 655, 635, 508, 505 and 490 nm; paler green specimens show absorptions at 505, 493 and 446 nm.

A bluish, sometimes a fine violet variety of diopside has been called violane and comes from St Marcel, Val d'Aosta, Italy. Star stones with four notably sharp rays come from Myanmar; the rays are alternately strong and weak. The background may be near-black. Four-rayed star diopside has been reported from southern India, some specimens responding to a magnet; magnetite inclusions may be responsible. Chatoyant diopside is also found. Some diopside may luminesce but the iron content prevents significant response in most cases. Under LWUV some green specimens show green, do not respond to SW and show a dull mustard-yellow glow under X-rays. Fine emerald-green chrome diopside is found in association with diamond from the Kimberley mines and in the Mogok Stone Tract of Myanmar. Fine green diopside is also found at Outukumpu, Finland and at De Kalb, New York, USA. Diopside of a more sober green is found in the Sri Lankan gem gravels and in the state of Minas Gerais, Brazil. Smoky yellow or brownish diopside is found at locations in Ontario and Quebec, Canada. A fine green mineral at one time reported from the Northern Areas of Pakistan as diopside is pargasite (q.v.).

Dioptase

Dioptase is copper silicate $CuSiO_2 (OH)_2$ and forms prismatic or rhombohedral crystals of the trigonal system. Due to the low hardness of 5 and the perfect cleavage, faceted stones are comparatively rare; crystal groups on the other hand may command high prices. The green is very strong and bright even when compared to the perhaps more subtle green of emerald. Many specimens are translucent at best.

The SG ranges from 3.28 to 3.35 and the RIs are 1.644–1.658 for the ordinary ray and 1.697–1.709 for the extraordinary ray, uniaxial positive. The DR is about 0.053. The absorption spectrum shows only a broad band in the yellow and green centred at about 550 nm and there is a strong absorption in the blue and violet.

Dioptase is found in the oxidized zones of some copper deposits. Some of the best crystallized specimens of fine green dioptase are found near

Altyn-Tyube, east of Karaganda, Khirghiz Steppe, Kazakhstan, and in the areas of Renéville, Mindouli and Pimbi, Congo Republic. Fine crystals are also found at Tsumeb, Namibia.

Dolomite

Dolomite is $CaMg(CO_3)_2$ and forms transparent colourless to light yellow rhombohedral crystals of the trigonal system. The hardness is 3.5–4 and there is one direction of perfect cleavage. Faces may sometimes be curved. The SG is 2.8–2.9, and the RI 1.502 and 1.681 with birefringence of 0.179, uniaxial negative. Dolomite occurs in a number of conditions: in sedimentary rocks, in altered Mg-rich igneous rocks or in geodes.

Perhaps the finest transparent dolomite crystals are those found at Eugui, Navarra Province, Spain. Massive dolomite or 'dolomitic marble' is often banded and in colours of red, brown, yellow, green and white. Such material comes from the US and Brazil.

Domeykite-alpha

Domeykite-alpha is Cu_3As and forms reniform or botryoidal masses of the cubic system with hardness 3–3.5 and SG 7.2–7.9. The ornamental value is due to the surface which shows a metallic lustre and is tin-white to steel-grey, slowly tarnishing to yellowish, then to orange-brown, finally developing an attractive iridescence. Above 90° the material inverts to domeykite-beta, a hexagonal form. Domeykite is usually of hydrothermal origin and occurs with algodonite (q.v.) with which it is often intergrown. A major location is the Mohawk mine, Keweenaw County, Michigan, USA.

Domeykite-alpha is the cubic form of Cu_6As and forms reniform or botryoidal masses of the cubic system with an opaque steel-white colour and metallic lustre.

Dumortierite

Dumortierite is $Al_7(BO_3)(SiO_4)_3O_3$ and forms fibrous or columnar crystals in the orthorhombic system with, in general, poor development of faces. Ornamental dumortierite occurs most commonly in blue or violet masses, faceted examples being quite rare due to the scarcity of individual crystals. The hardness of the massive material is 7; some crystals have hardness over 8. The SG is 3.21–3.41. The RI would show as a blurred shadow edge on the refractometer at about 1.68; transparent specimens are small but might show birefringence which may be in the range 0.15–0.37.

A red-brown faceted Sri Lankan specimen of dumortierite had SG of 3.41 and RI 1.686, 1.722, 1.723, biaxial negative with birefringence 0.037, pleochroic colours black, deep red-brown and brown. The hardness was 8. Dumortierite occurs in pegmatite veins or in aluminium-rich regionally metamorphosed rocks. Ornamental specimens are mainly from Sri Lanka or Madagascar. Quartz impregnating dumortierite has been found in Arizona, USA.

Durangite

Durangite is $NaAl(AsO_4)F$ and occurs characteristically as transparent to opaque euhedral monoclinic crystals which may show an attractive orange, orange-brown or pale yellow. The hardness is 5–5.5 and the SG about 3.9. Crystals are brittle. The RI is 1.597–1.634, 1.636–1.673, 1.647–1.685, biaxial negative. Birefringence is near 0.050. Pleochroic colours are shades of orange-yellow with one direction near colourless. Durangite occurs in pegmatite dikes in granite or in tin placers derived from alkalic rhyolite. The main source of ornamental durangite is the Barranca tin mine, Durango, Mexico.

Ekanite

In 1961, a green translucent water-worn stone showing some asterism, found some years earlier in a Sri Lanka gravel pit, was published as a new mineral with the composition $Ca_2ThSi_8O_{20}$. The colour was dark, dull green and the specimen was found to be metamict and radioactive, emitting alpha, beta and gamma radiation.

Ekanite occurs in poorly formed prismatic crystals of the tetragonal system. The hardness is 4.5–5 and the SG in the range 2.95–3.36. The RI (uniaxial but may be biaxial) is 1.595–1.597, 1.580–1.568, negative. A crystalline (i.e. non-metamict) variety of ekanite has been recovered from a glacial syenitic boulder in the Tombstone Mountains in Yukon Territory, Canada. Justification for the name 'ekanite' being used for this variety of the mineral was based on identity of its chemical composition with that of type specimens from Sri Lanka and the exact correspondence of its X-ray powder pattern with that given by samples of the metamict material to which crystallinity had been restored by heat treatment. The crystalline mineral is straw-yellow in colour when pure, though sometimes dark red owing to inclusions. The largest crystals are only some 2 or 3 mm in length and are tetragonal in symmetry. RIs were measured as 1.568 and 1.580 for the extraordinary and ordinary rays respectively. The measured SG of 3.08 was much lower than the calculated figure of 3.36, owing to impurities and inclusions.

The name given to the mineral was derived from that of its original discoverer in Sri Lanka, and the data given above for the crystallized form of the mineral were taken from a paper by Szymanski *et al.* in *Canadian Mineralogist* (1982).

Enstatite

Enstatite is $Mg_2Si_2O_6$, dimorphous with clinoenstatite and forms a series with ferrosilite in the pyroxene mineral group. Use of the names hypersthene and bronzite has been discontinued. Crystals of the orthorhombic system are prismatic with two directions of good cleavage, hardness 5–6, SG 3.20–3.30. Both lamellar and simple twinning is common. The colour ranges from colourless through yellow, green to brown and there is a characteristic pink to green pleochroism. A strong absorption line at 506 nm is also characteristic. The RI for the alpha, beta and gamma rays is 1.649–1.667, 1.653–1.671 and 1.657–1.680, biaxial positive with birefringence 0.010. Star stones with 4-rays are fairly common.

Enstatite occurs in pyroxenites, peridotites and dunites and in ultramafic inclusions in alkalic olivine basalts and kimberlite and in several other contexts. A rare transparent emerald-green variety occurs with diamond at Kimberley, South Africa and different colours occur in Sri Lankan alluvial deposits. A yellowish green variety has been reported from Mairimba Hill, Kenya.

Eosphorite

Forming a series with childrenite which has Fe > Mn, eosphorite is $Mn^{2+}Al(PO_4)(OH)_2H_2O$, in which manganese >Fe. Attractive transparent pinkish orange crystals have been faceted. Crystals are monoclinic, pseudo-orthorhombic, hardness 4.5, SG near 3.1. RI is 1.63, 1.65 and 1.66 for the alpha, beta and gamma rays respectively. A specimen of yellowish brown colour from Itinga, Minas Gerais, Brazil, was found to have an SG of 3.08 and RI of α 1.645, γ 1.680, giving a birefringence of 0.035. A brownish pink stone examined by a laboratory had an SG of 3.06 and RI of 1.640 and 1.668. This stone showed an absorption spectrum consisting of a very strong absorption line at 410 nm and a moderate one at 490 nm. Eosphorite occurs in granite pegmatites in association with Mn phosphates.

Epidote

Epidote, $CaAl_2$ $(Fe^{3+},Al)Si_3O_{12}(OH)$ forms a series with clinozoisite and shares group membership with that mineral and with zoisite both of which are dealt with under their respective headings. Prismatic monoclinic crystals of epidote show one direction of perfect cleavage and have a hardness of 6–7. Gem epidote is transparent and a characteristic green known as pistachio. The SG is 3.38–3.49, the RI for the alpha, beta and gamma rays is 1.715–1.751, 1.725–1.784 and 1.734–1.797 with a birefringence 0.015–0.049, negative. Epidote shows pleochroic colours colourless, pale yellow or yellow-green, and greenish yellow. A band at 455 nm is characteristic and may be accompanied by another band at 475 nm. Specimens need to be examined in different directions for the bands to become visible.

A chrome epidote ('tawmawite') is a deep greenstone showing strong pleochroism (emerald-green and bright yellow) which is found at Tawmaw in the Kachin Hills of Myanmar; gem-quality crystals are said to come from the Fungwe area of Zimbabwe. Specimens come also from Outukumpu, Finland. The name unakite has been used to denote rocks consisting of pink feldspar and green epidote; most specimens have originated in Zimbabwe.

Epidote group minerals have a wide range of modes of occurrence. Most originate in low- to medium-grade metamorphic rocks.

Epidote is a fairly common mineral and is found at many places in the alpine system of Austria (Knappenwand), Switzerland and France. It is also found in Norway and there are a number of localities in the North American continent.

Ettringite

Ettringite occurs in striated prismatic colourless to pale yellow transparent to opaque crystals of the hexagonal system with a perfect cleavage in one direction. The hardness is 2–2.5 and the SG 1.77. The RI for the ordinary and extraordinary rays is 1.491 and 1.470, uniaxial negative, altering to positive on dehydration; birefringence 0.021. Ettringite occurs in metamorphosed limestones near igneous contacts or in xenoliths. While specimens from the Ettringer-Bellerberg volcano, Mayen, and from Schellkopf, Brenk, Eifel, Germany, are too small to be ornamentally useful, larger crystals have been found at the N'Chwaning mine, Kuruman district, Cape

Province, South Africa. Ettringite cores coated with light yellow sturmanite (an ettringite group species) are reported from the Jwaneng mine, South Africa. The composition can be written $Ca_6Al_2(SO_4)_3(OH)_{12}26H_2O$.

Euclase

Euclase is $BeAlSiO_4$ (OH) and forms flattened transparent prismatic monoclinic crystals with some morphological complexity and one direction of perfect cleavage. The colour is most commonly a pale aquamarine or green but crystals of a very fine dark blue (from iron) have been found at the Miami mine in Zimbabwe and a number of faceted specimens exist, despite the pronounced cleavage. The hardness is 7.5 and the SG usually near 3.10. The RI for the alpha, beta and gamma rays is 1.651–1.653, 1.655–1.657 and 1.669–1.675 with a birefringence of 0.020, biaxial positive.

In the deeper-coloured stones, two vague absorption bands at 468 and 455 nm may be seen as well as a strong doublet at 705 nm in the red, due to chromium which is sometimes present. Euclase is found in pegmatites and is formed from the decomposition of beryl. Apart from the Zimbabwe deposit gem euclase usually comes from Minas Gerais, Brazil.

Eudialyte

Transparent to translucent red to reddish brown eudialyte (very fine though small bright red crystals come from Mt Saint Hilaire, Quebec, Canada) has sometimes been faceted. Eudialyte is $Na_4(Ca,Ce)_2(Fe^{2+},Mn^{2+})$ $ZrSi_6O_{17}(OH,Cl)_2$ and forms short rhombohedral to prismatic crystals of the trigonal system. The hardness is 5–6 and the SG 2.74–3.10. The RI is 1.588–1.636 and 1.588–1.638 with a birefringence of 0.004, uniaxial positive or negative. Absorption lines may be seen at 595, 527 and 523 nm. Eudialyte is soluble in acids. It occurs in nepheline syenites, alkalic granites and associated pegmatites. Apart from the Mt St Hilaire deposit, facet-grade material has also been found in Quebec at Kipawa Complex, Sheffield Lake, Temiscamingue County.

Euxenite-(Y)

Euxenite–(Y) is $(Y,Ca,Ce,U,Th)(Nb,Ta,Ti)_2O_6$ and occasionally provides opaque greenish black, brownish black or black cabochon material. Euxenite is a member of the orthorhombic system and is typically metamict with hardness 5.5–6.5 and SG 5.16. Material may be radioactive. The RI range 2.06–2.24 is beyond the refractometer limit. Euxenite-(Y) occurs in granite pegmatites or as detrital grains. Locations are widespread; large crystals have been reported from Vohimasina and other sites in Madagascar.

Faustite

Faustite (($Zn,Cu)Al_6(PO_4)_4(OH)_8 \cdot 4H_2O$) is a zinc-rich member of the turquoise group and crystallizes in the triclinic system, occurring as masses or nodules in argillized shales associated with copper mineralization. Gem materials of apple-green and yellow-green colour have been described. Light bluish green massive samples of turquoise-faustite aggregates occur at the turquoise deposit of Neyschabour, Iran. The hardness is about 4.5. The SG is near 2.9 and the RI about 1.612. The colour is caused by the copper of the turquoise content.

Fluorite

Fluorite, CaF_2, for which fluorspar is an old name, forms fine transparent crystals in a range of colours which include colourless, yellow, brown, green, blue, violet and pink. The green may resemble that of emerald in specimens found near Otjiwarongo, Namibia.

Blue John is the traditional name for a highly characteristic massive-banded translucent white/blue/violet/purple variety of fluorite found in the Castleton area, Derbyshire, England. Fluorite occurs as cubic crystals with an octahedral perfect and easy cleavage which often gives rise to the mistaken impression that the mineral forms octahedra in preference to cubes. Interpenetrant twinning is common and dodecahedra and combinations can be found as well as a number of other forms.

Virtually all fluorite octahedra found on sale have been cleaved by dealers. Fluorite has a hardness of 4 and SG usually near 3.18; the RI is 1.432–1.434.

Under LWUV blue, violet, red, green and yellow responses have been reported. It may show thermoluminescence or triboluminescence and some specimens phosphoresce. Blue John shows no response to energies above those of visible light. Fluorite is named for its low melting point.

Fluorite occurs in low-to-high temperature hydrothermal veins, granites and granite pegmatites. Details of the fluorite occurrences in Derbyshire, England, can be found in different works by Trevor D. Ford including *Derbyshire Blue John* (2000: 1873775199) and the fluorite deposits of the Northern Pennine Orefield, England, are covered by successive editions of

Geology of the Northern Pennine Orefield, the latest being volumes 1 (1990) and 2 (1985) by K.C. Dunham in the series *Economic Memoirs* of the British Geological Survey. The Friends of Killhope's *Fluorspar in the North Pennines* (2001) is a useful guide to the fluorite of that area as are *British Mining* 47 and 57, dealing with Alston Moor (1993) and the Weardale Mines (1996), the publishers of the *British Mining* series are the Northern Mine Research Society.

Most fine crystals of fluorite are not faceted (though faceted stones can be very beautiful) as fine groups sell for high prices. Examples of deep pink to red fluorite from the Chamonix area of the European Alps were exhibited at the Natural History Museum in London some years ago and attracted great interest. Fluorite showing smoky-brown in daylight and mauve in incandescent light has been reported from India.

Gem-quality fluorite from deposits in Hardin and Pope counties, Illinois, USA, is discussed in detail by Bastin in *Bulletin* no. 58 (1931) of the State Geological Survey of Illinois; the deposits extend into Crittenden county, Kentucky. The Illinois mines are centred in the Rosiclare district and also at Cave-in-Rock. Crystals show a wide variety of colours with characteristic zoning parallel to crystal faces.

The cause of the violet colour in some fluorite has been ascribed to Mie scattering from microcrystals of calcium metal; this is cited by Nassau in *The Physics and Chemistry of Colour* (2000). The blue colour arises from the presence of an F-centre. Colourless fluorite can be irradiated with energetic radiation to give F-centres, named from the German Farbe (colour). The specimen will turn purple. Nassau (2000) gives an explanation of the mechanisms involved.

The blue luminescence seen in fluorite, on the other hand, is believed to arise from the presence of Eu^{2+} or an associated colour centre. Bill, in *Physics and Chemistry of Minerals* 3, 117–31 (1978) cited by Nassau (*op. cit.*), states that a colour centre combination of Y^{3+} with a fluorine vacancy can also give a blue colour and that a combination of this with Ce gives a yellowish green. O_3^--substituting for F^--gives yellow and $Y^{3+} + O_2^{3-}$ a pink.

In green fluorite, weak bands have been reported at 634, 610, 582 and 445 nm, and there is also a strong broad band at 427 nm which may be seen in sizeable pieces.

A part of the *extraLapis* series (no. 4, 1993, ISBN 3921656273) includes valuable papers and beautiful photographs.

Friedelite

Friedelite is $Mn_8^{2+}Si_6O_{15}(OH,Cl)_{10}$ and fashioned specimens are transparent to translucent rose-red to orange with hardness of 4–5, SG 3.04–3.06. RI for the ordinary and extraordinary rays is 1.654–1.656 and 1.620–1.625

with a birefringence of 0.030, uniaxial negative. The crystal system is monoclinic, crystals occurring most commonly as pseudohexagonal tablets. Absorption bands may be seen centred at 556 nm and near 456 nm and some specimens have shown reddish fluorescence; other effects have been reported, including yellow (LWUV) and green (SW). Most ornamental specimens of friedelite have come from Franklin, New Jersey, USA, where they occur in metamorphosed manganese–iron deposits.

Gadolinite-(Y)

Gadolinite-(Y) is $Y_2Fe^{2+}Be_2Si_2O_{10}$ and usually occurs in very dark green to black metamict (originally monoclinic) masses from which the occasional large cabochon has been fashioned. The hardness is 6.5–7 and the SG 4.41. The RI for the alpha and gamma rays is 1.77–1.78 and 1.78–1.82 (no data for the beta ray), biaxial positive though isotropic when metamict. Any birefringence will be in the range 0.01–0.04. Gadolinite-(Y) occurs in granite and alkalic granite pegmatites. Specimens occur at several locations in Norway and in Texas, USA.

Gahnite

Gahnite, a cubic member of the spinel group, forms two series, one with that mineral and the other with the iron aluminate, hercynite. Gahnite is $ZnAl_2O_4$ and occurs characteristically as octahedra which may be transparent dark blue or bluish green. Some green crystals have been reported to resemble tourmaline and most cut blue specimens have come from Nigeria where fine examples have been found in the Jemaa district. These stones have a hardness of 7.5–8, SG 4.40–4.59, RI 1.805. Artificial material with similar composition is known (O'Donoghue and Joyner, 2003). Natural gahnite occurs as an accessory mineral in granites and granite pegmatites.

Gaspéite

Gaspéite is a member of the calcite group with the composition $(Ni,Mg)CO_3$. It occurs as translucent bright grass-green to olive- or yellowish green nodular concretions or as aggregates of rhombic crystals (too small for use) of the trigonal system. The hardness is 4.5–5 and the SG near 3.7. The RI for the ordinary and extraordinary rays is 1.83 and 1.61, uniaxial negative with a birefringence of 0.220. Gaspéite is uncommon, occurring as a secondary mineral in a Ni-sulphide vein in metamorphosed siliceous dolostone at the Gaspé Peninsula, Quebec, Canada.

Gaylussite

Colourless transparent gaylussite, $Na_2Ca(CO_3)_2 \cdot 5H_2O$ occurs as flattened wedge-shaped crystals of the monoclinic system with a hardness of 2.5–3 and one direction of perfect cleavage. The SG is near 1.99 and the RI for the alpha, beta and gamma rays is 1.44, 1.516 and 1.523, biaxial negative, birefringence 0.077. Gaylussite occurs in evaporates or shales from alkali lakes, notably Searles Lake, San Bernardino County, California, USA. The mineral does not make an ideal gemstone since surfaces may well develop a white appearance as the crystals dry out after recovery. Dehydration rules out ornamental use as specimens need to be kept in sealed containers.

Goodletite

The name goodletite is a commercial name for attractive ornamental boulder material showing red, blue and green from corundum varieties and green tourmaline. These minerals are embedded in a fine-grained mica and tourmaline matrix which occurs in the alluvial gold deposits of Westland district, South Island, New Zealand. Cabochon materials are usually found to be composites. Some specimens have a goodletite centre with top and bottom of colourless quartz. Various unhelpful names have been used.

Grandidierite

Grandidierite is $(Mg,Fe^{2+})Al_3(BO_3)(SiO_4)O_2$ and has provided rather characteristic translucent bluish green or greenish blue elongated anhedral crystals of the orthorhombic system with hardness 7.5 and SG 2.98–3.00. The RI for the alpha, beta and gamma rays is 1.590–1.602, 1.618–1.636, 1.623–1.639 biaxial negative with birefringence 0.039. Pleochroic colours are dark blue-green, colourless and dark green. Grandidierite occurs in aluminous boron-rich rocks and in pegmatites. Gem-quality material comes from Andrahomana, southern Madagascar.

Gypsum

See Alabaster.

Hambergite

Hambergite is $Be_2BO_3(OH,F)$ and occurs as transparent colourless orthorhombic prismatic crystals which are usually well crystallized, sometimes twinned. There is one direction of perfect cleavage and the hardness is 7.5. The SG is 2.35–2.37 and the RI for the alpha, beta and gamma rays is 1.54–1.560, 1.587–1.591, 1.628–1.631, biaxial positive with a high birefringence of 0.072. Hambergite occurs in granite pegmatites, ornamental-quality material coming usually from Anjanabanoana, Madagascar.

Haüyne

Haüyne is $(Na,Ca)_{4-8}Al_6Si_6(O,S)_{24}(SO_4,Cl)_{1-2}$ and forms very bright and attractive though uncommon transparent blue dodecahedral or pseudo-octahedral crystals of the cubic system. It is also an important constituent of lapis-lazuli. The hardness is 5.5–6 and the SG 2.40–2.50. The RI is usually near 1.502 for the material from the Eifel. Some of this material has been reported to show an orange-red fluorescence under LWUV. Haüyne occurs in alkaline nepheline or leucite-rich igneous rocks; fine bright blue crystals from Mendig and Mayen in the Eifel region of western Germany.

Hematite

See Metallic species.

Hemimorphite

Hemimorphite is $Zn_4Si_2O_7(OH)_2H_2O$ and is found either as orthorhombic thin tabular crystals with double terminations, showing hemimorphism, or more commonly in masses. Attractive translucent light blue cabochons have been fashioned from Mexican specimens and some transparent faceted stones are known. The hardness is 4.5–5 and there is one direction of perfect cleavage. The SG is 3.4–3.5 and the RI for the alpha, beta and gamma rays is 1.614, 1.617 and 1.636, biaxial positive with a birefringence of 0.021. Hemimorphite occurs as a secondary mineral in the oxidized zone of zinc-bearing mineral deposits. Apart from the facetable crystals the masses, often sheaf- or fan-like, have their own attraction. The main source is Santa Eulalia, Chihuahua, Mexico.

Herderite

Herderite is recorded by Anthony *et al.* in *Handbook of Mineralogy* (2000) as CaBe(PO$_4$)F but there is some doubt about its species status as reported by Mandarino and Back, Fleischer's *Glossary of Mineral Species* (2004). Here the same composition is given but queried. There have been no reports of data with F > OH. Anthony *et al.* (2000) report an electron microprobe of a gemstone 'from Brazil' which showed F 7% (thus with F > 5.86%, the series [with hydroxylherderite] midpoint). Herderite forms pseudo-orthorhombic crystals of the orthorhombic system by analogy to hydroxylherderite with a hardness 5–5.5. Gem material is transparent and pale in its range of colours which include pink, violet and green, some crystals are colourless. The SG 2.95–3.02 and the RI for the alpha, beta and gamma rays is 1.556–1.59, 1.578–1.61 and 1.589–1.62 respectively, biaxial negative, birefringence 0.029–0.030. A green gemstone from Brazil is noted by Arem, *Color Encyclopedia of Gemstones*, 2nd edition, 1987; the RI is given as 1.581, 1.601, 1.610 for the three rays with a birefringence of 0.029–0.030. The Fe content was 7%. Perhaps this was the specimen described by Anthony *et al.* (*op. cit.*).

Some herderite shows a green or violet fluorescence in both LW and SWUV; some specimens are reported to have shown orange fluorescence and persistent phosphorescence. Herderite is believed to occur in complex granite pegmatites. A transparent green specimen (perhaps the stone already described) weighing 5.90 is in the National Museum of Natural History, Smithsonian Institution, Washington DC, USA. Most gem herderite has come from locations in Minas Gerais, Brazil. Specimens from Maine, USA, described by Bank and Becker (1977) were violet or colourless with RI 1.587–1.590, 1.609–1.610 and 1.619–1.621 with SG fairly constant at 3.00. Brown material from the Virgem do Lapa mine in the same Brazilian state has the high RI 1.610, 1.630 and 1.642. F content was reported to be 0.2%.

Hodgkinsonite

Pink transparent hodgkinsonite, $Mn^{2+}Zn_2SiO_4(OH)_2$ occurs as stout prismatic crystals of the monoclinic system with steep pyramids or with dominating pyramids. There is one direction of perfect cleavage; the hardness is <5 and the SG 4.06–4.08. The RI for the alpha, beta and gamma rays is 1.720–1.724, 1.741–1.742, 1.746, biaxial negative with birefringence 0.026. Pleochroic colours are lavender and pale purple. Specimens may show a dull red fluorescence under LWUV. Hodgkinsonite occurs as seams in massive willemite-franklinite granular ore in a metamorphosed stratiform zinc deposit. The only significant occurrence is at Franklin, New Jersey, USA.

Holtite

Holtite forms semitransparent light buff to cream-buff or deep olive-buff blocky orthorhombic crystals with a resinous to vitreous lustre with hardness 8.5 and SG 3.90. The RI for the alpha, beta and gamma rays is 1.705–1.746, 1.728–1.759 and 1.730–1.761, biaxial negative with birefringence 0.015. Holtite may fluoresce bright yellow in LW and dull orange in SWUV. Holtite has been found replacing a tantalite in a pegmatite at Greenbushes, Western Australia.

Hornblende

New names resulting from amphibole group nomenclature revision, details of which can be found in *American Mineralogist* 63, 1023–1052, 1978, have replaced 'hornblende'. The species actinolite, (q.v.) ferrohornblende, magnesiohornblende, pargasite, tremolite may include ornamental quality varieties.

Howlite

Howlite is $Ca_2B_5SiO_9(OH)_5$ and while found as white monoclinic tabular crystals, most ornamental material is nodular with H 3.5 and SG 2.53–2.59. Howlite is opaque white and may be porcellanous. The hardness is 3.5 and the SG 2.53–2.59. The RI for the alpha, beta and gamma rays is 1.583–1.586, 1.596–1.598, 1.605 biaxial negative with a birefringence 0.022 but in practice a value of about 1.59 may be obtained. For ornamental purposes howlite is dyed to give an imitation of turquoise. Luminescent effects for the undyed material include a deep orange under LWUV for some Californian specimens and a brownish yellow under SW for some other material. Howlite is found in borate deposits, most ornamental pieces coming from various Californian locations.

Hübnerite

Hübnerite is $Mn^{2+}WO_4$ and is found as transparent to translucent prismatic monoclinic yellowish brown to reddish brown crystals with striations in the direction of elongation, with one direction of perfect cleavage and a hardness of 4–4.5, varying with direction. The SG is 7.12–7.18 and the RI for the alpha, beta and gamma rays is 2.17–2.20, 2.22 and 2.30–2.32, biaxial positive with birefringence 0.13. Deep blood-red internal reflections are attractive, combined with the metallic-adamantine-resinous lustre.

Hureaulite

Hureaulite is $Mn_5^{2+}(PO_4)_2(PO_3OH)_2 \cdot 4H_2O$ and forms transparent to translucent light rose to pink to rose-violet, orange or yellowish brown long to short prismatic elongated monoclinic crystals with a somewhat greasy lustre, hardness 3.5 and SG 3.19. The RI for the alpha, beta and gamma rays is 1.647–1.654, 1.654–1.662 and 1.660–1.667 with birefringence 0.012. Hureaulite occurs as a late-stage secondary mineral formed by the alteration of primary phosphates in complex granite pegmatites. Possible ornamental material from Branchville, Fairfield County, Connecticut and from the Fletcher and Palermo #1mines, North Groton, Grafton County, New Hampshire, USA.

Hurlbutite

Hurlbutite is $CaBe_2(PO_4)_2$ and forms stout monoclinic transparent to translucent colourless to greenish white crystals with hardness 6 and SG 2.90. The RI for the alpha, beta and gamma rays is 1.595, 1.601 and 1.604, biaxial negative with birefringence 0.009. Hurlbutite occurs as a late-stage mineral in complex granite pegmatites. Some material has been fashioned from the G.E. Smith mine, Newport, Sullivan County, New Hampshire, USA.

Idocrase

See Vesuvianite.

Inderite

Inderite scarcely has ornamental potential; it is colourless, soft (H2.5) and the surface is likely to suffer cloudiness or to develop a whitish appearance after fashioning. Inderite is $MgB_3O_3(OH)_5 \cdot 5H_2O$ and forms long prismatic crystals of the monoclinic system with one direction of perfect cleavage which leaves behind a notably pearly surface. The SG is near 1.79 and the RI for the alpha, beta and gamma rays is 1.488, 1.491, 1.505, biaxial positive with a birefringence of 0.017. Inderite is dimorphous with kurnakovite (q.v.) and occurs in lacustrine borate deposits. The main source of both minerals is Inder Lake, Kazakhstan.

Iolite

Though the name cordierite is used by mineralogists, the name iolite has become established among gemmologists so it is used (perhaps inconsistently!) here. Iolite is $(Mg,Fe^{2+})_2Al_4Si_5O_{18}$ and forms short prismatic transparent crystals of the orthorhombic system and may be pseudo-hexagonal. The cross-section is rectangular. Some crystals show twinning. Most gem iolite occurs as water-worn pebbles. Iolite has a hardness of 7–7.5 and SG 2.53–2.66, values increasing with higher iron content. The RI for the alpha, beta and gamma rays is 1.527–1.560, 1.532–1.574, 1.537–1.578, biaxial positive or negative.

Iolite is fashioned so that the blue colour is uppermost on a faceted stone or in the front of a statuette. The pleochroism is strong with violet-blue, pale blue and pale yellow to green colours. The iron absorption spectrum also varies with direction. Bands may be seen at 645, 593, 585, 535, 492, 456, 436 and 426 nm. The absorption spectrum varies with the direction, so that in the direction of the violet-blue colour the 645 and 426 nm bands are masked by the general absorption and appear as cutoffs which shorten the spectrum.

The variety bloodshot iolite found in Sri Lanka shows a red colour from oriented hematite platelets. The occasional four-rayed star stone turns up.

Apart from alluvial, iolite occurs in a variety of environments, including altered aluminous and igneous rocks. Sri Lanka is the main source with some good crystals coming from Thompson, Manitoba, Canada. The gem gravels of Myanmar also produce iolite.

Jeremejevite

Jeremejevite is $Al_6(BO_3)_5(F,OH)_3$ and some pale blue-green, cornflower-blue, brown or colourless transparent stones are found. The crystals of the hexagonal system occur as hexagonal prisms, sometimes tapered with vicinal forms and with pyramidal terminations; sector zoning is characteristic. The hardness is 6.5–7.5 and the SG 3.28–3.29. The RI for the ordinary and extraordinary rays is 1.644–1.647 and 1.637 respectively. Jeremejevite, however, may be anomalously biaxial with the alpha, beta and gamma rays giving the RI 1.637, 1.644–1.647 and 1.644–1.652. In either case the optic sign is negative. The birefringence is 0.007–0.013. Jeremejevite is a rare late hydrothermal mineral formed in granite pegmatites: most facetable crystals come from Cape Cross, Swakopmund, Namibia. Large crystals have come from Mt Soktui, Nerschinsk district, Adun-Chilon Mountains, Siberia, Russia.

Johachidolite

Johachidolite is $CaAlB_3O_7$ and forms transparent pale yellow equant subhedral crystals of the orthorhombic system. The hardness is 7.5 and the SG 3.45. The RI for the alpha, beta and gamma rays is 1.717, 1.720, 1.724, biaxial positive with birefringence 0.007. A strong whitish blue fluorescence is seen under LWUV; under SW a weak blue has been reported. Johachidolite occurs in a nepheline dike cutting limestone in the Johachido district, North Korea. A faceted stone is reported from the Mogok region of Myanmar and the constants and properties described above relate to that specimen.

Kakortokite

Kakortokite is the name given to an ornamental rock found in the Julianehaab area of southern Greenland. The varied colours arise from the different minerals in its composition: white nepheline, black arfvedsonite and red from crystals of eudialyte (q.v.) which occurs in the nepheline. The SG of the rock is 2.7–2.8.

Kammererite

The name kammererite has been abandoned by the IMA, CNMMN, in favour of clinochlore, variety chromian clinochlore. Clinochlore is $(Mg,Al)_6Al(Si,Al)_4O_{10}(OH)$ and forms thin to thick pseudohexagonal crystals of the monoclinic system. Crystals of the rose-red chromian clinochlore are usually thin, micaceous with a perfect cleavage and hardness 2–2.25. The SG is 2.60–2.64 and the RI for the alpha, beta and gamma rays is 1.571–1.588, 1.571–1.588, 1.576–1.597, biaxial positive or negative with a birefringence of 0.003. Pleochroism is notable, violet to hyacinth-red. Fashioning is very difficult and specimens are rare. The only crystals likely to attract the attention of the lapidary come from chromium ore veins at the Kop Krom (chrome mine) area in the Kop Mountains near Askale, Turkey.

Kogarkoite

Kogarkoite is $Na_3(SO_4)F$ and forms transparent to translucent pale sky blue to colourless pseudorhombohedral crystals of the monoclinic system with frequent twinning; the hardness is 3.5 and SG 2.68. The mineral is water soluble. The RI for the alpha, beta and gamma rays is 1.439, 1.439 and 1.442: as kogarkoite though biaxial positive is nearly uniaxial positive, the RI for the ordinary and extraordinary rays is given, at 1.435 and 1.439 with birefringence 0.003. Under LWUV there may be a white fluorescence and under SW a pale blue fluorescence. Kogarkoite occurs in sodalite-syenite xenoliths associated with an intrusive alkalic gabbro-syenite complex at Mt Saint Hilaire, Quebec, Canada.

Kornerupine

Kornerupine is $(Mg,Fe)(Al,Mg,Fe)_9(Si,Al,B)_5(O,OH,F)_{22}$ and occurs as transparent prismatic crystals of the orthorhombic system with hardness 6–7 and a good cleavage in one direction. Colours include a characteristic dark green inclining to brown and a rare and attractive bright vanadium green. The SG is 3.28–3.35 and the RI for the alpha, beta and gamma rays is 1.660–1.671, 1.673–1.683, 1.674–1.684, biaxial negative (may be pseudouniaxial) with birefringence 0.013–0.017. Pleochroic colours are brown to greenish brown. A weak absorption band may be seen at 503 nm. Chatoyant stones show a bright eye against a dark green background. Vanadium-coloured stones are a very attractive bright apple-green and were rare at the time of writing.

Kornerupine occurs in boron-rich volcanic and sedimentary rocks subjected to metamorphism. Most gem material is found in the Sri Lanka gem gravels in the Matale and Ratnapura districts and gem crystals (some a pale blue) also occur at Itrongahy, Betroka, Madagascar. The Myanmar gem gravels have produced greenish brown specimens. The vanadium-green specimens appear to come only (so far) from the Kwale area of Kenya; some may show a yellow fluorescence.

Kurnakovite

Kurnakovite is $MgB_3O_3(OH)_5 \cdot 5H_2O$ and forms colourless transparent roughly prismatic crystals of the triclinic system with hardness 2.5–3 and SG near 1.83. The RI for the alpha, beta and gamma rays is 1.488–1.491, 1.508–1.510, 1.515–1.525, biaxial negative with birefringence 0.030. Kurnakovite occurs in lake-bed borate deposits and the occasional faceted specimen will probably have come from Boron, Kern County, California, USA. Kurnakovite is dimorphous with inderite (q.v.), found at Inder Lake borate deposit, Kazakhstan.

Kyanite

Kyanite with andalusite and sillimanite (both q.v.) is a trimorph of Al_2SiO_5. It occurs as characteristic bladed and tabular triclinic crystals which are often bent or twisted. Lamellar twining is common. The hardness is markedly directional, 5.5 parallel to the long direction and 7 at right angles to it. There is one direction of perfect cleavage which leaves a pearly lustre. The SG is 3.53–3.65 and the RI for the alpha, beta and gamma rays is 1.710–1.718, 1.719–1.724, 1.724–1.734, biaxial negative, birefringence 0.017. The colour is green to blue-green, frequently zoned, with pleochroism near-colourless, violet blue and cobalt blue or different shades of green. Some bluish green kyanite may show an absorption line in the deep red at 701 nm and these and other specimens may also show two lines in the blue and a cut-off in the violet. Some dim red fluorescence may be seen under LWUV.

Kyanite occurs in gneisses and schists and in granite pegmatites. The main sources of gem kyanite include Minas Gerais, Brazil, Machakos District, Kenya (some colourless specimens are reported as well as a fine blue; some striped crystals come from this source). Some dark blue specimens from Mozambique contain chromium and titanium. Facetable crystals have come from Yancy County, North Carolina, USA. A very fine deep sapphire-blue oval of 6.72 ct was shown to MO'D by Marcus McCallum in December 2004.

Langbeinite

Langbeinite is $K_2Mg_2(SO_4)_3$ and usually forms as bedded masses or nodules from which transparent colourless white or grey, pale pink or pale greenstones have been faceted. The hardness is 3.5–4 and the SG 2.83. The RI for the cubic material is 1.532. Langbeinite is a principal ore of potash and is mined from marine salt deposits. Commercial quantities are mined at the Carlsbad potash district, Eddy County, New Mexico, USA.

Lawsonite

Lawsonite is $CaAl_2Si_2O_7(OH)_2·H_2O$ and forms colourless or pale blue translucent prismatic or tabular crystals of the orthorhombic system.

The hardness is 7–8 and there is one direction of perfect cleavage. The SG is 3.08–3.09 and the RI for the alpha, beta and gamma rays is 1.663–1.665, 1.672–1.675, 1.682–1.686, biaxial positive, birefringence 0.019. Lawsonite occurs in low-temperature metamorphic rocks, metamorphic and glaucophane schists. The source of any fashioned lawsonite is likely to be one of the large crystals occurring at Covelo, Mendocino county, California, USA.

Lazulite

In recent years, fine transparent blue crystals of lazulite have been found and fashioned. Lazulite is $MgAl_2(PO_4)_2(OH)_2$ and occurs as stubby to acute dipyramidal crystals of the monoclinic system with hardness 5.5–6 and SG 3.122–3.240. The RI of the alpha, beta and gamma rays is 1.604–1.626, 1.626–1.654 and 1.637–1.663, biaxial negative with birefringence 0.031–0.036. It is strongly pleochroic in dark blue, light blue and colourless directions. Lazulite occurs in granite pegmatites and in quartzites. Fine blue transparent lazulite may resemble Paraíba tourmaline (q.v.), this material originating probably from west of Dattas, Minas Gerais, Brazil. Fine green material is also found in the Chilas area of the northern areas of Pakistan and Angola is another possible source. Lazulite forms a series to scorzalite (q.v.).

Lazurite

See Lapis Lazuli (Chapter 19).

Legrandite

Legrandite is $Zn_2(AsO_4)(OH)\cdot H_2O$ and forms transparent bright yellow elongated prismatic crystals of the monoclinic system. The hardness is 4.5 and the SG 4.01. The RI for the alpha, beta and gamma rays is 1.702, 1.709, 1.740, biaxial positive with birefringence 0.060. Pleochroism shows as yellow to colourless. Legrandite occurs in facetable quality only from the Ojuela mine, Mapimi, Durango, Mexico.

Leifite

Leifite is $Na_2(Si,Al,Be)_7(O,OH,F)_{14}$ and forms transparent to translucent light yellow, light violet or colourless acicular, deeply striated hexagonal prismatic crystals of the trigonal system with hardness 6 and SG 2.57. The RI for the ordinary and extraordinary rays is 1.515 and 1.519, uniaxial positive, birefringence 0.004. A green fluorescence under SWUV has been reported. Leifite occurs in association with an intrusive alkalic gabbro-syenite complex at Mt Saint Hilaire, Quebec, Canada.

Lepidolite

Recent revision of minerals of the mica group has renamed lepidolite as 'lepidolite' (a series name). Mandarino & Back, Fleischer's *Glossary of Mineral Species*, 2004 states 'used for monoclinic trioctahedral micas on or close to the trilithionite-polylithionite join: light micas with substantial lithium . . . Various polytypes are known.' For the purposes of this description the former name will be retained in order to avoid circumlocutions.

Ornamental lepidolite is massive, fine-grained and a pale rose colour. It has been polished into various shapes despite the hardness of 3.5 and the easy and perfect cleavage. The SG is 2.5–4 and the RI (transparent to translucent material is sometimes faceted) for the alpha, beta and gamma rays is 1.525–1.548, 1.551–1.585, 1.554–1.587, biaxial negative with birefringence 0.018–0.038. The occurrence is chiefly, if not exclusively, in granite pegmatites, ornamental specimens coming from various sites in California and Maine, USA. Some of the transparent lepidolite comes from Brazil.

Leucite

Leucite is $KAlSi_2O_6$ and occurs commonly as transparent to translucent colourless or white euhedral pseudocubic crystals of the tetragonal system. The hardness is 5.5–6 and the SG 2.45–2.50. The RI for the ordinary and extraordinary rays is 1.508 and 1.509, uniaxial positive (may be biaxial) with birefringence very low to measure with gemmological instruments. Some but not all specimens may show a medium to bright orange under LWUV and a bluish colour under X-rays. Leucite occurs with potassium-rich mafic and ultramafic lavas, most specimens coming from the Alban Hills, Rome, Italy; this site provides virtually all leucite that can be fashioned.

Leucophanite

Leucophanite is $(Ca,Ce)CaNa_2Be_2Si_4O_{12}(F,OH)_2$ and forms semitransparent light yellow to light green short prismatic to tabular pseudotetragonal crystals of the orthorhombic or triclinic systems with one direction of perfect cleavage, hardness 3.5–4 and SG 2.96–4.07. The RI for the alpha, beta and gamma rays is 1.565–1.573, 1.590–1.595 and 1.593–1.598, biaxial negative, birefringence 0.028. A pink fluorescence under LWUV and a green response to SW have been reported. Leucophanite occurs at Mt Saint Hilaire, Quebec, Canada.

Linarite

Linarite is $PbCu(SO_4)(OH)_2$ and forms elongated and characteristically tabular transparent to translucent bright to dark blue monoclinic crystals, which may be deep blue in transmitted light. There is one direction of perfect cleavage and the hardness is 2.5. The SG is 5.35 and the RI for the alpha, beta and gamma rays is 1.809, 1.838 and 1.859, biaxial negative with birefringence 0.050. Linarite is not often found in faceting quality; it occurs as an uncommon secondary mineral in the oxidized zone of lead–copper deposits. Probably the finest crystals are those found at the Mammoth mine, Tiger, Arizona, USA.

Lithiophilite

Lithiophilite is $Li(Mn^{2+},Fe^{2+})PO_4$, forming a series with triphylite (q.v.), and forms transparent to translucent clove brown to greenish brown to yellow sharply terminated crystals of the orthorhombic system with one direction of perfect cleavage, hardness 4–5, SG 3.4–3.5. The RI for the alpha, beta and gamma rays is 1.663–1.696, 1.667–1.700 and 1.674–1.702, biaxial positive with birefringence 0.005. Lithiophilite occurs as a late-stage mineral in some complex granite pegmatites. Gem-quality material has come from Brazil.

Ludlamite

Ludlamite is $(Fe^{2+},Mg,Mn^{2+})_3(PO_4)_2 \cdot 4H_2O$ and forms apple- to bright-green prismatic crystals in the monoclinic system with one direction of perfect cleavage and a hardness of 3.5. The SG is 3.12–3.19. The RI for the alpha, beta and gamma rays is 1.650–1.653, 1.669–1.675, and 1.688–1.697, biaxial positive with birefringence 0.038–0.044. Ludlamite occurs in granite pegmatites, resulting from the alteration of primary phosphates. Specimen material in fine crystals comes from the Blackbird mine, Lemhi County, Idaho, USA.

Magnesite

Magnesite is $MgCO_3$; crystals are rare, the common occurrence being as translucent, sometimes transparent colourless, white, yellowish to brown or grey masses with an earthy or chalky appearance. The crystal system is trigonal; there is one direction of perfect cleavage and the hardness is 3.5–4.5 with SG 3.0–3.1. The RI for the ordinary and extraordinary rays is 1.710 and 1.509, uniaxial negative with birefringence 0.202, very easy to see in the occasional faceted specimen. Under SWUV blue green or white fluorescence may be seen with greenish phosphorescence. Magnesite occurs as a primary mineral in igneous or sedimentary rocks or may be formed by the metamorphism or alteration of serpentine and peridotite. Fine crystals come from Brumado, Bahia, Brazil.

Malachite

Malachite is $Cu_2(CO_3)(OH)_2$ and occurs as acicular to prismatic or thick tabular monoclinic crystals but these are too small to be used as ornaments. Botryoidal masses or stalactitic malachite is plentiful. The masses are a deep saturated green and show characteristic banding; the hardness is 3.5–4.5 and there is one direction of perfect cleavage. The SG is usually just over 4 and the RI of the massive material is 1.85. Malachite occurs as a common secondary mineral in the oxidation zone of copper ore deposits and can be an ore of copper. From the ornamental point of view the most productive deposits include large pure masses formerly (Anthony *et al.*, 2003) from mines around Nizhni Tagil and Ekaterinburg, Ural Mountains, Russia and from various deposits in Africa, including especially mines in Katanga Province, Congo. Crystals from Tsumeb. Namibia.

Malachite occurs frequently with azurite (q.v.) and the name azurmalachite has been used for intimate combinations. Pseudomalachite is $Cu^{2+}_3(PO_4)_2(OH)_4$ and is found in association with malachite; it may have been used ornamentally as the hardness is higher than that of malachite at 4.5. Malachite is also found in association with the copper silicate chrysocolla and they may be fashioned together.

Manganotantalite

Manganotantalite is $(Mn^{2+},Fe^{2+})(Ta,Nb)_2O_6$ and forms short prismatic crystals of the orthorhombic system, some transparent and scarlet red. There is one direction of distinct cleavage and the hardness is 5.5–6.5.

The SG is 7.7–7.9 and the RI for the alpha, beta and gamma rays is 2.14, 2.15 and 2.22, biaxial positive with birefringence 0.150. Manganotantalite occurs in granite pegmatites. Most gem material has come from Salinas, Minas Gerais, Brazil and from Morrua, Mozambique. Large crystals have been found at Amelia, Virginia, USA, a major deposit for the manganese garnet, spessartine.

Manganotychite

Manganotychite is $Na_6Mn^{2+}{}_2(SO_4)(CO_3)_4$ and occurs in pink or white semitransparent massive aggregates or as irregular grains of the cubic system with hardness 4 and SG 2.75–2.79. Manganotychite is weakly magnetic and slightly water soluble. The RI is 1.544–1.552. A sharp absorption line can be seen at 414 nm. Manganotychite is found at Mt Saint Hilaire, Quebec, Canada.

Marcasite

Marcasite is the orthorhombic form of FeS_2 (pyrite, q.v. is the cubic dimorph). Marcasite crystals are typically tabular but may be pyramidal or prismatic and may show curved faces. Fine-grained masses may be globular or resemble cockscombs. The hardness is 6–6.5 and the SG 4.88–4.92. The colour is tin-white on fresh surfaces then pale bronze-yellow darkening on exposure with an iridescent tarnish. The lustre is metallic.

The jeweller's 'marcasite' is typically pyrite. Marcasite occurs in sedimentary environments formed characteristically under low temperature, highly acidic conditions. It is abundant throughout the world.

Meliphanite

Meliphanite is $(Ca,Na)_2Be(Si,Al)_2(O,F)_7$ and forms colourless or lemon-yellow transparent thin tabular crystals of the tetragonal system. The hardness is 5.5 and the SG 3.01–3.03. The RI for the ordinary and extraordinary rays is 1.593 and 1.613, uniaxial positive with birefringence 0.019. Meliphanite is found in augite syenite, largely from sites in Norway, including the Langesundsfjord and also from the Kola Peninsula, Russia.

Mellite

Mellite is organic with the composition $Al_2C_6(COO)_6 \cdot 16H_2O$ and forms transparent honey yellow or reddish prismatic or dipyramidal crystals or masses of the tetragonal system. The hardness is 2–2.5 and the SG 1.58–1.64. The RI for the ordinary and extraordinary rays is 1.539–1.541 and 1.509–1.511, uniaxial negative with birefringence 0.028–0.032. Weak fluorescent effects may be seen. Mellite is found in brown coal and/or lignite deposits, material coming from Artern, Thuringia and Bitterfeld, Saxony-Anhalt, Germany.

Mesolite

Mesolite is $Na_2Ca_2(Al_6Si_9O_{30}) \cdot 8H_2O$ and a orthorhombic member of the zeolite mineral group. It forms translucent to transparent elongated white or colourless invariably twinned prismatic crystals, may be stalactitic or occur in porcellanous masses with hardness 5 and two directions of perfect cleavage. The SG is 2.27 and the RI for the alpha, beta and gamma rays is 1.5048, 1.5050 and 1.5053, biaxial positive with birefringence 0.001.

Mesolite occurs with natrolite (q.v.), scolecite, other zolites and calcite in cavities of volcanic rocks, especially basalt. Fashioned crystals with a characteristic silky lustre may have come from the outstanding crystals found in the Pashan Hills in the Poona area of India.

Metallic Species

I have described algodonite, bornite, cobaltite, columbite, domeykite, euxenite, hematite and samarskite here together as well as individually: they have scant use in ornament but some occur in association with gem minerals and some have occasionally been fashioned.

Hematite is the alpha-form of Fe_2O_3 and forms rhombohedral crystals in the trigonal system; they may sometimes appear pseudocubic or prismatic. The name kidney ore describes a common habit. The cherry-red to brown streak is a useful test when carefully used; the hardness is 5–6 and the SG 5.26. When appropriately lit crystals appear a lustrous black, sometimes with iridescent tarnish. Cumbria, England, has produced notable specimens and exceptional crystals have been described from Val Tavetsch, Grisons, Switzerland, and from locations in Minas Gerais, Brazil; those at Mesa Redonda and Congonhas do Campo are notable. Hematite is not normally attracted by an iron magnet though artificial forms of similar composition do respond. The material hemetine is one example. A material reported to be from Zimbabwe and to carry the name hematite garnet has an SG of 4.16 and gives a notably intense almandine spectrum; it shows a deep red on passage of a strong ray of light.

Cobaltite is CoAs and forms cubic or pyritohedral crystals of the cubic system. Ornamentally the silver-white colour with a tinge of pink may be found attractive enough to fashion. The lustre is metallic, the hardness 5.5 and the SG in the range 6.0–6.3. Notable crystals are reported from Skutterud, Norway, and Tunaberg, Sweden.

Other minerals with a metallic appearance though of more interest to mineral collectors, include domeykite, algodonite, bornite, columbite, samarskite and euxenite, breithauptite and pentlandite. Few, if any, gemmological tests will distinguish them and, if their identity is suspected, a museum should be consulted for a determination. Details of their composition and properties can be found in mineralogical literature.

Pyrite is included in the main alphabetical sequence of this chapter as it is far more likely to be encountered. The 'peacock ore' variety of bornite, with an attractive surface tarnish, has been used ornamentally. Bornite is Cu_5FeS_4.

Microlite

Microlite is $(Ca,Na)_2Ta_2O_6(O,OH,F)$ and forms a series with pyrochlore. It occurs as euhedral to subhedral generally opaque to translucent pale yellow to brown, reddish or green octahedral crystals of the cubic system. The hardness is 5–5.5 and the SG normally about 5.5. The RI is 1.993–2.02; a degree of metamictization may lower the value to 1.93–1.94. Microlite occurs as a primary mineral in lithium-bearing granite pegmatites. Gem-quality crystals have been found at the Rutherford mine, Amelia, Virginia, USA; a transparent garnet-red specimen found in 1885 was faceted into a stone resembling red zircon (Arem, *Color Encyclopedia of Gemstones*, second edition, 1987).

Milarite

Milarite is $KCa_2AlBe_2Si_{12}O_{30}·0·5H_2O$ and a member of the osumilite group in which it forms transparent to translucent colourless to pale yellow or green prismatic crystals, typically well formed, of the hexagonal system. The hardness is 5.5–6 and the SG 2.46–2.61. The RI for the ordinary and extraordinary rays is 1.532–1.551 and 1.529–1.548, uniaxial negative with birefringence 0.003. Milarite is found in vugs in granites and syenites or in hydrothermal veins. Large crystals are found in the Jose Pinto pegmatite, Jaguarcu, Coronel Fabriciano, Minas Gerais, Brazil and smaller examples at several Swiss localities though not in the Val Milar after which it was named! A sapphire-blue scandium-bearing variety with RI 1.549 and 1.560, birefringence 0.011 with pleochroism strong blue to green has been reported from Tamil Nadu, India. The Sc content was 1.9%. *The Mineralogical Record* 35(5) carries a useful paper showing Alpine milarite crystals up to 1 cm.

Millerite

Millerite is NiS and occurs as slender elongated opaque brassy yellow crystals of the trigonal system with two directions of perfect cleavage and a metallic lustre which may tarnish to iridescence. The hardness is 3–3.5 and the SG 5.3–5.6. The RI is not available from the standard literature. Though there has been occasional report of faceted stones, it is more likely that any fashioned millerite will be slabs or cabochons. Millerite occurs as a low-temperature mineral in cavities in limestones and carbonate veins. Any fashioned examples are most likely to have come from large cleavages at the Marbridge mine, Malartic, La Motte Township, Quebec, Canada, and similar occurrences at Kotalhti, Finland.

Mimetite

Mimetite is $Pb_5(AsO_4)_3Cl$ and forms transparent and very attractive prismatic crystals of the hexagonal system in yellow to orange-yellow to brown, sometimes colourless specimens with a characteristic resinous to subadamantine lustre. The hardness is 3.5–4 and the SG 7.24–7.26. The RI for the ordinary and extraordinary rays is 2.147 and 2.128, uniaxial (commonly anomalously biaxial) negative. The birefringence is 0.019. Mimetite occurs as a secondary mineral in the oxidized zone of arsenic-bearing lead deposits. Especially fine examples are found at Tsumeb, Namibia (one specimen reported to show an orange-red fluorescence under LWUV), and globular masses at Mapimi, Durango, Mexico.

Monazite

The latest authorities (Anthony *et al.*, 2000) postulate three different species in place of the single undifferentiated monazite. The three minerals are monazite-(Ce), monazite-(La) and monazite-(Nd). All form crystals of the monoclinic system, the Ce and La form usually tabular and the Nd forms prismatic. Hardness is about 5.5 and SG 5.4; the Ce and La minerals may be radioactive from thorium content. Colours range from rose-red to a more common reddish brown. Only monazite-(Ce) is transparent, the other two translucent; RI for the alpha, beta and gamma rays is 1.774–1.800, 1.777–1.801, 1.828–1.849, biaxial positive with birefringence 0.049–0.055. Very complex absorption spectra provide the best means of separating the three minerals from one another and spectral data handbooks need to be consulted. For convenience, a portmanteau formula can be written as $(Ce,La,Nd,Th)PO_4$. The minerals are found in igneous rocks including granite pegmatites. Most examples reaching the gemmologist will have come from Madagascar or Sri Lanka, which produces orange pebbles. Colour change (reddish orange in daylight to pinkish brown in tungsten light) is reported but is common enough when RE are involved in the composition. Some specimens have shown a greenish fluorescence under LWUV.

Montebrasite

See Amblygonite.

Mordenite

Mordenite is $(Na_2\,CaK_2)_4[Al_8Si_{40}O_{96}]\cdot28H_2O$ and an orthorhombic member of the zeolite group of minerals in which it forms compact fibrous aggregates with a perfect cleavage in one direction. Specimens may be colourless or stained yellowish or pinkish with hardness 3–4 and SG 2.12. The RI for the alpha, beta and gamma rays is 1.472–1.483, 1.475–1.485 and 1.477–1.487, biaxial positive or negative with birefringence 0.005. Mordenite occurs in veins and amygdules in various igneous rocks and may be found as an alteration of volcanic glasses. Mordenite is found at, among other places, Morden and along the Bay of Fundy, Nova Scotia, Canada.

Nambulite

Nambulite is $(Li,Na)Mn^{2+}_4Si_5O_{14}(OH)$ and occurs as triclinic prismatic to tabular transparent reddish brown crystals with an orange tint and a perfect cleavage in one direction. The hardness is 6.5 and SG 3.51–3.55. The RI for the alpha, beta and gamma rays is 1.707, 1.710 and 1.730, biaxial positive with birefringence 0.023. Nambulite occurs in veinlets cutting manganese oxide ores. Rare fashioned specimens may have come from the Funakozawa and Ohtaniyama mines, Ohno, Iwate Prefecture, Japan.

Narsarsukite

Narsarsukite is cited in Fleischer's *Glossary of Mineral Species* (2004) as $Na_2(Ti,Fe^{3+})Si_4(O,F)_{11}$ and is described by Gaines *et al.* in *Dana's New Mineralogy* [*System of Mineralogy* 8th edition] 1997 as $Na_2(Ti,Fe,Zr)$ $[Si_4O_{10}](OH,F)$. Olive-green tetragonal crystals of small size with a hardness of 6–7 and SG 2.78 show a perfect cleavage in one direction. The RI for the ordinary and extraordinary rays is 1.610–1.614 and 1.644–1.647, uniaxial positive with birefringence 0.033. Narsarsukite occurs at Mt Saint Hilaire, Quebec, Canada.

Natrolite

Natrolite is $Na_2[Al_2Si_3O_{10}]\cdot 2H_2O$ and a member of the zeolite group in which it occurs as orthorhombic pseudotetragonal short to long striated prismatic crystals, transparent to translucent, usually colourless but sometimes pale shades of grey or yellow to red. There is a perfect cleavage in two directions; hardness is 5 and SG 2.2–2.26. The RI for the alpha, beta and gamma rays is 1.473–1.483, 1.476–1.486, 1.485–1.496, biaxial positive with birefringence 0.012. Natrolite occurs in cavities in amygdaloidal basalts and related igneous rocks in association with other zeolites. Very large crystals are found at the Johnson asbestos mine, Thetford, Quebec, Canada and also from Mt Saint Hilaire in the same province. Many other locations for fine crystals exist. Milky white transparent natrolite is reported from Pakistan.

Nepheline

Nepheline is $(Na,K)AlSiO_4$ and forms transparent to nearly opaque prismatic crystals of the hexagonal system, usually colourless but coloured variously from impurities and sometimes notably large. The hardness is 5.5–6 and the SG 2.55–2.66. The RI for the ordinary and extraordinary rays is 1.529–1.546 and 1.526–1.542, uniaxial negative with birefringence 0.004. Some German specimens may show a light blue and some from Ontario an orange fluorescence in LWUV. The variety elaeolite may be red, green, brown or grey; when minute crystals are present they may produce chatoyancy. Nepheline occurs in alkalic rocks, including nepheline syenites and pegmatites associated with them, also in gneisses. Large crystals occur particularly from Davis Hill, Bancroft, Ontario, Canada; Katzenbuckel, Saxony, Germany.

Neptunite

Neptunite is $KNa_2Li(Fe^{2+},Mn^{2+})_2Ti_2Si_8O_{24}$ and forms opaque black or very dark red monoclinic prismatic crystals with square cross-sections and one direction of perfect cleavage; crystals show piezoelectricity. The hardness is 5–6 and the SG 3.19–3.23. The RI for the alpha, beta and gamma rays is 1.692, 1.702 and 1.734, biaxial positive with birefringence 0.029–0.045. Neptunite occurs in natrolite veins cutting a glaucophane schist in a serpentinite body at the Gem mine and Mina Numero Uno, San Benito County, California, USA.

Nickeline

Nickeline is NiAs and the occasional opaque cabochon is fashioned from granular aggregates of the hexagonal system. The colour ranges from a pale copper-red, tarnishing grey to near-black. The hardness is 5–5.5 and the SG 7.78–7.83. Nickeline occurs in vein deposits in igneous rocks, with silver minerals, bismuth and bismuthinite. Large masses occur in the Cobalt district, Ontario, Canada, and elsewhere.

Odontolite

Odontolite is not a distinct mineral species *sensu stricto* but a phosphate arising from fossil bones and teeth whose colour is provided by the iron phosphate vivianite (q.v.). It may resemble turquoise and has been called bone turquoise in the past. The hardness has been reported as near 5 and the SG near 3.00. The RI is 1.57–1.63; comparable figures for turquoise are close but the turquoise copper absorption spectrum is absent from odontolite. Included calcite may cause local effervescence with acids. Apatite may also be present.

Odontolite has been found at Simmore, near Auch, Gers, France. Imitation odontolite has been made by calcining contemporary ivory and staining blue by soaking in a solution of copper sulphate. The SG is 1.8.

Painite

Painite is $CaZrBAl_9O_{18}$ and occurs as elongated, pseudo-orthorhombic transparent deep garnet-red crystals of the hexagonal system with hardness near 8 and SG 4.01–4.03. The RI for the ordinary and extraordinary rays is 1.8159 and 1.7875 uniaxial negative with birefringence 0.029. Pleochroism is ruby-red and pale brownish orange or pale red-orange. So far painite has been found near Ohngaing, Myanmar and nowhere else. Less than 10 specimens are known so far to exist though some may be masquerading happily as garnet! Accounts of the discovery of painite can be found in the mineralogical literature, e.g. Claringbull, Hey and Payne, *Mineralogical Magazine* 31, 420–425, 1957. As I write a deposit of slender prismatic reddish pink brown crystals is being examined at GemA, London. They originate from Myanmar.

Palygorskite

Palygorskite is $(Mg,Al)_2Si_4O_{10}(OH)\cdot4H_2O$ and usually occurs as fibrous mats often called 'mountain leather'. The crystal system is monoclinic with a hardness of 2–2.5 and SG 2.0–2.6. The colour may be white, grey, shades of pink or yellowish and the RI is near 1.55. This, when observed, may arise from quartz which is frequently used to impregnate the soft palygorskite. Palygorskite is an alteration product of magnesium silicates in soils and sediments, and occurs in hydrothermal veins. Locations include Attapulgus, Georgia and Sapillo, New Mexico, as well as Palygorskaya, Perm, Russia.

Papagoite

Papagoite is $CaCuAlSi_2O_6(OH)_3$ and forms crystalline coatings and aggregates of the monoclinic system; crystals are too small to be fashioned. The colour is cerulean blue. The hardness is 5–5.5 and the SG 3.25. The RI for the alpha, beta and gamma rays is 1.607, 1.641 and 1.672, biaxial negative with birefringence 0.065. Papagoite is often mixed with quartz and occurs in narrow veinlets in altered granodiorite porphyry at the New Cornelia mine, Ajo, Pima County, Arizona, USA.

Pargasite

Pargasite is $NaCa_2(Mg_4Al)Si_6Al_2O_{22}(OH)_2$ and forms stout prismatic crystals of the monoclinic system with twinning frequent and a perfect cleavage in one direction. The hardness is 5–6 and the SG 3.04–3.17. Crystals are transparent to translucent bluish green or chrome (emerald)-green to light brown. The RI for the alpha, beta and gamma rays is 1.613–1.65, 1.618–1.66 and 1.635–1.67, biaxial positive with birefringence 0.022. Pleochroic colours include colourless, light brown, bluish green and greenish yellow. Pargasite, a member of the amphibole group, occurs in igneous and metamorphic rocks and fine emerald-green crystals have been found in the northern areas of Pakistan; some have been faceted.

Parisite

Parisite may be parisite-(Ce) or parisite-(Nd). It is convenient to write the composition as $Ca(Ce,Nd,La)_2(CO_3)_3F_2$. Monoclinic, pseudohexagonal crystals are transparent to translucent brownish yellow, orange to brown with hardness 4.5 and SG 4.36. The RI for the ordinary and extraordinary rays is 1.671–1.676 and 1.771–1.757, uniaxial positive with birefringence 0.081. Parisite occurs in carbonaceous shales including those hosting emeralds in the Muzo district, Boyaca, Colombia, and as inclusions in the emeralds. The occasional star stone has been reported. An RE absorption spectrum could be expected when the Nd content is significant.

Pectolite

Pectolite is $NaCa_2Si_3O_8(OH)$ and a member of the wollastonite mineral group in which it forms a series with sérandite (q.v.) whose Mn is borrowed to provide shades of pink in an otherwise colourless, white or grey mineral. Pectolite occurs most commonly as translucent globular masses, the rarer crystals showing tabular habit of the monoclinic or triclinic system. There is a perfect cleavage in two directions.

The hardness is 4.5–5 and the SG 2.84–2.90. The RI for the alpha, beta and gamma rays is 1.592–1.610, 1.603–1.615 and 1.630–1.645, biaxial positive with a birefringence of 0.036. Usually a spot reading of 1.60 is obtained. Pectolite may show orange-pink fluorescence under LWUV (material from Bergen Hill New York, USA) and a yellow fluorescence with green phosphorescence (Paterson, New Jersey, USA). Massive pectolite coloured a strong blue may be offered as a turquoise substitute under the trade name Larimar. This material comes from the Dominican Republic. Pectolite occurs in nepheline syenites and fine crystals (colourless, light pink and yellowish brown) are found at Mt Saint Hilaire, Quebec, Canada, and from the Jeffrey mine, Asbestos, in the same province. This location produces fine pale green prismatic crystals. Pectolite from Alaska has been offered as a jade simulant.

Pentlandite

Pentlandite is $(Fe,Ni)_9S_8$ and forms masses in the cubic system with a light bronze colour, hardness 3.5–4 and SG 4.6–5. No data is given for the RI. Cabochons have been fashioned from deposits at Sudbury, Ontario, Canada, where it occurs as a major ore mineral.

Periclase

Periclase is MgO and occurs as masses or small octahedra and other forms of the cubic system with a perfect cleavage in one direction, hardness 5.5 and SG 3.56–3.68. The RI is 1.735–1.745. Periclase is usually colourless or white to pale yellow, green or brown but gem-quality adularescent orange-yellow material is known from Mt Saint Hilaire, Quebec, Canada.

Petalite

Petalite is $LiAlSi_4O_{10}$ and forms transparent colourless tabular monoclinic crystals or masses. The hardness is 6.5 and there is one direction of perfect cleavage. The SG is 2.3–2.5 and the RI for the alpha, beta and gamma rays is 1.504–1.507, 1.510–1.513, 1.516–1.523, biaxial positive with birefringence 0.013. Some material may show an orange fluorescence. Petalite is a mineral of granite pegmatites where it may be found with other Li-minerals including spodumene and tourmaline group species. Faceted material is conspicuously limpid. Large crystals of facet quality have been produced at San Miguel de Paracicaba and elsewhere in Brazil. There have been reports of reddish specimens from Russia.

Phenakite

Phenakite is Be_2SiO_4 and occurs as transparent to translucent colourless rhombohedral, tabular to prismatic crystals of the trigonal system with hardness 7.5–8 and SG 2.93–3.00. The RI for the ordinary and extraordinary rays is 1.650–1.656 and 1.667–1.670, uniaxial positive with birefringence 0.016. Coloured specimens result from surface staining. Various fluorescence effects have been reported under UV and under X-rays specimens show a distinct blue. The occasional chatoyant specimen is reported; four-rayed star stones with a brownish tint are reported from Sri Lanka.

Phenakite occurs in granite pegmatites and fine crystals have been found at the São Miguel de Piracicaba east of Belo Horizonte, Minas Gerais, Brazil, and at Amelia, Virginia, US. Fine crystals also occur along the Takovaya River near Ekaterinburg, Russia. Phenakite crystals occurring as inclusions in emerald usually indicate man-made origin.

Phosgenite

Phosgenite is $Pb_2(CO_3)Cl_2$ and forms transparent to translucent colourless short prismatic to prismatic crystals of the tetragonal system with adamantine lustre, hardness 2–3 and SG 6.12. The RI for the ordinary and extraordinary rays is 2.118 and 2.145 m, uniaxial positive, anomalously biaxial, birefringence 0.028. A pronounced yellow fluorescence can be seen under UV and X-rays.

Phosgenite occurs from the alteration of galena in the oxidized zone of lead ore deposits. Fine crystals are found in the Monteponi and Montevecchio areas. Fine crystals are also found at Tsumeb, Namibia, and there are occurrences in Namibia, Morocco and Tasmania.

Phosphophyllite

Phosphophyllite is $Zn_2(Fe^{2+},Mn^{2+})(PO_4)_2\cdot4H_2O$ and forms bluish- or sea-green monoclinic thick tabular crystals with numerous other forms, fish-tail and penetration twinning, a perfect cleavage in one direction and hardness 3–3.5, the SG is 3.08–3.13. The RI for the alpha, beta and gamma rays is 1.595–1.599, 1.614–1.617 and 1.616–1.620, biaxial negative, birefringence 0.021–0.033. Some specimens may fluoresce violet in SWUV. Phosphophyllite occurs as a secondary mineral in zoned complex granite pegmatites or as an alteration product of sphalerite. Fine specimens come from massive sulphide deposits in the Unificada mine, Cerro Rico, from the Siglo XX mine, Llalagua, Potosi and in the Morococala mine, Oruro, all Bolivia.

Pollucite

Pollucite is $(Cs,Na)(AlSi_2)O_6\cdot H_2O$ and forms transparent colourless cubes, dodecahedra or trapezohedra of the cubic system, commonly striate and it may also occur as fine-grained masses. The hardness is 6.5–7 and the SG 2.85–2.94. Orange to pink fluorescence may be seen under UV and X-rays. The RI is 1.518–1.525. Pollucite occurs in lithium-rich granite pegmatites.

The area of Newry, Maine, USA, provides facet-grade material and fine crystals are found round San Piero, Campo, Elba, Italy.

Porcelanite

Porcelanite is a name given to an ornamental rock found at Bucnik, Czech Republic. The material is an altered marly clay with characteristic sedimentary textures.

Porphyries

Porphyries are types of igneous rocks which show comparatively large and well-formed crystals embedded in a groundmass of much finer texture. The porphyritic structure is considered to be due to a two-stage crystallization, or solidification, of the igneous magma at the time of its intrusion into the surrounding rock or extrusion on to the earth's surface, when the magma contained already formed crystals enclosed in the molten liquid which subsequently solidified as a fine-grained groundmass. Porphyritic structure is common in rocks and it might be expected that a wide range of such a type of rock could well be used as an ornamental stone. This does not seem to be the case and the two porphyries first considered have more historical interest than as ornamental stones of modern application.

The first of these is the green porphyry found in the province of Laconia in Greece, a rock with an olive-green groundmass with light green feldspar crystals sprinkled abundantly through it. The green colour is due to included epidote and chlorite throughout the rock, which was known in classical times as 'marmor lacedaemonium viride'. It was later known as 'perfidio serpentino', but the rock is not a serpentine. The quarries yielding this rock lie between the towns of Sparta and Marathonisi.

The other and perhaps more important porphyry is the famous red porphyry of Egypt, 'porfido rosso antico', which was known in classical times as 'porphyrites leptosephos'. This rock has a dark red groundmass, the colour being due to included piedmontite, a manganese mineral. In this groundmass is an abundance of small white and light pink feldspar crystals. The rock is quarried from a dyke some 25 m thick on the Jebel Dhokan mountain which lies some 40 km inland from the junction of the Red Sea with the Gulf of Suez and about 80 km eastwards from the Nile. It is doubtful whether the rock was known to the Egyptians but probably it was discovered in the reign of Claudius by the Romans who took it to Rome, where it was called 'lapis porphyrites' and later 'The Stone of Rome'. During the Roman occupation of Egypt, thousands of workmen were employed in the quarries and the stone was transported to the Nile en route for the imperial city of Rome. A carved head of the Emperor Hadrian in Egyptian porphyry may be seen in the British Museum at Bloomsbury, London. The head is said to have been made in Egypt in about AD 130 and the carving was carried out by using copper tools fed with sand and emery.

Llanoite or llanite is a reddish porphyry which contains mainly not only reddish microcline crystals in a dark brown matrix, but also small patches of quartz which show bluish gleams. This rock is found in Llano County, Texas, in the US.

Although not strictly a porphyry, the rock called corsite, or sometimes napoleonite, might conveniently be mentioned here. It is a greyish rock containing lighter-coloured oval rings showing some radial structure. The material is an orbicular diorite or hornblende gabbro which occurs in the island of Corsica. There are two fine examples of this material in the form of vases in the Chateau de Malmaison outside Paris.

Poudrettite

Poudrettite is $KNa_2B_3Si_{12}O_{30}$ and forms transparent violet to pink roughly equant barrel-shaped prismatic crystals of the hexagonal system with hardness about 5 and SG 2.51–2.53. The RI for the ordinary and extraordinary rays is 1.511 and 1.532, uniaxial positive with birefringence 0.021. Poudrettite is found at Mt Saint Hilaire, Quebec, Canada, in marble xenoliths within nepheline syenite breccias in an intrusive alkalic gabbro-syenite complex. Violet crystals with strong purple-pink to colourless to light brown pleochroism are reported from the Mogok area of Myanmar.

Powellite

Powellite is $CaMoO_4$ and is isostructural with scheelite with which it forms a series. Powellite occurs as transparent to translucent straw yellow, greenish yellow, greenish blue or blue dipyramidal crystals of the tetragonal system; the lustre is characteristically greasy or subadamantine and the hardness 3.5–4. The SG is near 4.23 and the RI of the ordinary and extraordinary rays is 1.974 and 1.984, uniaxial positive. Under both forms of UV, specimens may fluoresce yellowish white to golden yellow. Powellite occurs as an uncommon secondary mineral in the oxidation zone of molybdenum-bearing hydrothermal mineral deposits. Facetable material is rare as crystals are usually small but some specimens from the Isle Royal, Calumet and Hecla mines, Houghton

County, Michigan, USA, have been fashioned. Notable crystals are found at Pandulena Hill, Nasik, Maharashtra, India.

Prehnite

Prehnite is $Ca_2Al_2Si_3O_{10}(OH)_2$ and is most commonly found in translucent light to dark green or yellow reniform or globular masses of the orthorhombic system with a characteristic oily appearance. The tabular or prismatic, steeply pyramidal crystals are uncommon. The hardness is 6–6.5 and the SG 2.802–2.95. The RI for the alpha, beta and gamma rays is 1.611–1.632, 1.615–1.642, 1.632–1.665, biaxial positive with birefringence 0.021–0.033. Prehnite occurs as a secondary or hydrothermal mineral in veins or cavities in mafic volcanic rocks or in granite gneisses or syenites. The most attractive green specimens come from the Fairfax quarry, Centreville, Virgnia, USA; colourless crystals from the Jeffrey mine, Asbestos, Quebec, Canada.

Prosopite

Prosopite is $CaAl_2(F,OH)_8$ and forms transparent small tabular colourless crystals of the monoclinic system. Crystals may become translucent on exposure and have one direction of perfect cleavage. The hardness is 4.5 and the SG 2.88–2.89. The RI for the alpha, beta and gamma rays is 1.500–1.501, 1.503–1.509, 1.509–1.510, biaxial positive with birefringence 0.009. Prosopite occurs in granite pegmatites, sometimes in base metal deposits or fluorine-rich granites and greisens. Prosopite from the Santa Rosa mine, Mazapil, Zacatecas, Mexico, occurs with azurite and can resemble turquoise though its higher RI adequately separates the two materials.

Proustite

Proustite is Ag_3AsS_3 and forms beautiful prismatic transparent adamantine dark red rhombohedral crystals of the trigonal system. Crystals are translucent, often darkening with exposure to light. The hardness is 2–2.5 and the SG 5.6–5.7. The RI for the ordinary and extraordinary rays is 3.08 and 2.79, uniaxial negative with birefringence 0.296. Proustite occurs as a late-forming mineral in hydrothermal deposits in the oxidized and enriched zone, with other silver minerals and sulphides. Exceptional crystals are found at the Dolores mine, Chanarcillo, Atacama, Chile. Fine faceted stones may turn out to be (admittedly rare) synthetic products.

Pyrargyrite

Pyrargyrite is Ag_3SbS_3 and forms translucent deep red prismatic crystals of the trigonal system, showing hemimorphism and adamantine lustre. They may darken on exposure to light. The hardness is 2.5 and the SG 5.85. The RI for the ordinary and extraordinary rays is 3.08 and 2.88, uniaxial negative, birefringence 0.200. Pyrargyrite occurs in hydrothermal veins as a primary late-stage low-temperature mineral. Fine crystals are found at Chanarcillo, Atacama, Chile and from Colquechaca, Potosi, Bolivia.

Pyrite

Pyrite is the cubic form of FeS_2 and dimorphous with marcasite (q.v.). It forms numerous forms of which the cubic pyritohedral, octahedral and combinations are frequent. Striations conform to pyritohedral symmetry; penetration and contact twins and granular and stalactitic masses are also common. The hardness is 6–6.5 and the SG 5.01–5.03. Pyrite is paramagnetic and a semiconductor. The colour is pale brass-yellow, tarnishing to a darker colour with iridescence and the lustre metallic and splendent. Pyrite forms under a number of different conditions and in many different rock types. Material appropriate for ornament may be found in almost any deposit. The 'marcasite' of the jeweller is pyrite.

Pyrophyllite

Pyrophyllite is $Al_2Si_4O_{10}(OH)_2$ and forms compact translucent or opaque spherulitic aggregates of acicular radiating crystals or foliated laminae in the monoclinic or triclinic systems. There is one direction of perfect cleavage and the lustre is greasy or pearly. Colours are white or pale shades of blue, yellow or green. The hardness is 1–2 and the SG 2.65–2.90. The RI for the alpha, beta and gamma rays is 1.534–1.556, 1.586–1.589, 1.596–1.601, biaxial negative with birefringence 0.050. Pyrophyllite occurs in hydrothermal veins and in deposits of schistose rocks.

The ornamental interest in pyrophyllite arises from its similarity to soapstone (q.v.) from which it may be distinguished by chemical tests for aluminium whose presence indicates pyrophyllite. Large crystals have been found in the Ibitiara area of Bahia, Brazil. China produces materials suitable for carvings.

Pyroxmangite

Pyroxmangite is $Mn^{2+}SiO_3$ and forms frequently twinned translucent red tabular crystals of the triclinic system with a perfect cleavage in one direction, hardness 5.5–6 and SG 3.61–3.80. The RI for the alpha, beta and gamma rays is 1.728–1.748, 1.730–1.742, 1.746–1.758, biaxial positive with birefringence 0.016–0.020. Pyroxmangite occurs in regionally metamorphosed manganese ore deposits and in manganiferous rocks. Fine crystals come from the Taguchi and other mines, Shidara, Aichi prefecture, Japan, and large crystals from Broken Hill, New South Wales, Australia.

Pyrrhotite

Pyrrhotite is $Fe_{1-x}S(x = 0$–$0.2)$ and provides bronze-yellow to pinchbeck-brown masses whose surface quickly tarnishes, sometimes to give an attractive iridescence. Pyrrhotite is monoclinic, pseudohexagonal with hardness 3.5–4.5 and SG 4.58–4.65. RI details not published. Specimens may be attracted by a magnet. Pyrrhotite occurs mainly in mafic igneous rocks and in pegmatites and high-temperature hydrothermal and replacement veins. Material is widespread around the world, localities including Sudbury, Ontario, Canada, and Herja (Kisbanya), Romania, where crystals have been found.

Realgar

Realgar is AsS and forms striated red to orange-yellow prismatic crystals of the monoclinic system. The crystals when fresh are transparent with a greasy lustre but disintegrate to a powder after long exposure to light. The hardness is 1.5–2 and the SG 3.56. The RI for the alpha, beta and gamma rays is 2.538, 2.684 and 2.704, biaxial negative with birefringence 0.166. Gem-quality crystals have been found at the Getchell mine, Humboldt County, Nevada, and at locations in Washington state, especially at the Green River Gorge, Franklin, King county, USA.

Rhodizite

Rhodizite is $(K,Cs)Al_4Be_4(B,Be)_{12}O_{28}$ and is found as well-formed transparent to translucent colourless to white dodecahedra and tetrahedra of the cubic system with hardness 8–8.5 and SG 3.44. The RI is 1.694 but rhodizite frequently displays anomalous birefringence. Some specimens give a strong greenish or yellowish response to X-rays with some phosphorescence and a weaker yellowish response under SWUV. Rhodizite occurs as a late-stage accessory mineral in alkali-rich granite pegmatites. Good crystals from Antandrokomby near Mt Bity and other locations in Madagascar and from near Sarapulka and Shaitanka, Ural Mts, Russia. Some Russian material is reported to show a rose-red colour.

Rhodochrosite

Rhodochrosite is $Mn^{2+}CO_3$ and forms pink, rose- or cherry-red transparent or translucent rhombohedral or scalenohedral crystals of the trigonal system with one direction of perfect cleavage. The hardness is 3.5–4 and the SG 3.70. The RI for the ordinary and extraordinary rays is 1.810 and 1.597, uniaxial negative with birefringence 0.201–0.220. Absorption bands can be seen in the visible at 565, 535 nm (both faint lines) and at 551 and 410 nm. Under both forms of UV there is a pink response from the mineral as a whole though individual specimens may not show much. Rhodochrosite occurs as a primary mineral in low- to moderate-temperature hydrothermal veins, also in metamorphic deposits and in carbonatites. Exceptional transparent scalenohedra of magnificent orange-pink colour have been found comparatively recently from the Hotazel and N'Chwaning mines in the Kuruman area of Cape Province, South Africa; fine rose-red crystals are found at the Sweet Home mine, Alma, Park County, Colorado, USA, and from the Huallapon mine, Pasto Bueno, Ancash province, Peru. The Inakuraishi and Yakomo mines, Hokkaido, Japan, also produce cabochon-quality material. Specimens from the Capillitas mine, San Luis, Catamarca province, Argentina, show characteristic banding from whitish calcite inclusions. Chatoyant rhodochrosite is reported from Kazakhstan; the eye is caused by an agate-like structure. Star rhodochrosite is reported from Russia and Argentina. A major paper on the Sweet Home mine can be found in *The Mineralogical Record* 29(4), 1998.

Rhodonite

Rhodonite is $CaMn_4Si_5O_{15}$ and forms transparent to translucent rose pink to brownish red triclinic crystals. Commonly tabular or cleavable masses. The hardness is 5.5–6.5 and there are two directions of perfect cleavage. The SG is 3.67 for crystals and 3.57–3.76 for masses. The RI for the alpha, beta and gamma rays is 1,711–1.734, 1.716–1.739, 1.724–1.748 biaxial positive with birefringence 0.013. Absorption band at 548 nm with strong line at 503 nm and weak band at 455 nm. Masses often show black veining from undifferentiated manganese oxides. Facetable crystals are rare but show a beautiful red. Rhodonite occurs in manganese deposits formed by hydrothermal contact and regional metamorphism. Gem crystals usually from Broken Hill, New South Wales, Australia, or from Morro da Mina, Conselheiro Lafaiete, Minas Gerais, Brazil. Some facetable material also from Honshu, Japan, and good colour masses from Daghazeta, Tanzania. Rhodonite from Ekaterinburg, Urals, Russia, features in Fabergé and other artefacts.

Richterite

Richterite is $Na(NaCa)Mg_5Si_8O_{22}(OH)_2$ and a member of the amphibole group. It forms flattened prismatic monoclinic crystals with one direction of perfect cleavage, hardness 5–6 and SG 2.97–3.45. Opaque blue material has been fashioned. The RI for the alpha, beta and gamma rays is 1.605–1.624, 1.618–1.64 and 1.627–1.641, biaxial negative. Richterite occurs in contact-metamorphosed limestones and in alkalic igneous rocks and carbonatites. The blue material has been found at the Kalahari manganese orefield, South Africa.

Rinkite

Rinkite is $(Ca,Ce)_4Na(Na,Ca)_2Ti(Si_2O_7)_2F_2(O,F)_2$ and forms opaque to translucent reddish brown masses in the monoclinic system with hardness 4–5 and SG 3.18–3.44. The RI for the alpha, beta and gamma rays is 1.643–1.662, 1.645–1.667 and 1.651–1.681, biaxial positive with birefringence 0.009. Absorption lines have been recorded at 878, 806, 749, 743, 683, 585, 527 and 513 nm. Rinkite occurs in nepheline syenites and related pegmatites. Gemstones have been fashioned from specimens collected at sites at the Khibiny massif on the Kola Peninsula, Russia.

Rutile

Rutile with anatase and brookite is a trimorph of TiO_2 in the rutile group and forms characteristic slender prismatic striated lengthways, variably terminated and often geniculate twinned crystals of the tetragonal system with a metallic to adamantine lustre and most frequently pale brown to yellow. The hardness is 6–6.5 and the SG 4.23. The RI for the ordinary and extraordinary rays is 2.605–2.613 and 2.899–2.901, uniaxial positive with birefringence 0.287 and dispersion, very notable in synthetic transparent faceted stones, 0.280. Rutile occurs as a common high-temperature, high-pressure accessory mineral in igneous rocks and granite pegmatites and in regionally metamorphosed rocks including crystalline limestones. Rutile needles are common inclusions in a number of gem minerals but fine unenclosed crystals are found at a number of Swiss locations. Larger crystals are found at Conquista and Ibitiara, Bahia, Brazil.

Samarskite-(Y)

Samarskite-(Y) is found either as characteristically rough crystals or as opaque granular, velvet-black masses which may be externally yellowish brown. Samarskite is orthorhombic and metamict; the hardness is 5–6 and the SG 5.0–5.69; the RI is 2.1–2.2, the mineral behaving as if isotropic. Samarskite occurs as an accessory mineral in rare earth-rich granite pegmatites. Caochons may come from any of a number of localities, including Miass, Ilmen Mountains, Russia. Samarskite's composition is spelt out in *Fleischer's Glossary of Mineral Species* 2004 (passim).

Sapphirine

Sapphirine is $(Al,Mg)_8(Al,Si)_6O_{20}$ (several polytypes are recognized) and forms indistinct tabular transparent to translucent light to dark blue crystals of the monoclinic or triclinic system. The hardness is 7.5 and the SG 3.4–3.5. The RI for the alpha, beta and gamma rays is 1.701–1.726, 1.703–1.728, 1.705–1.734, biaxial positive or negative, birefringence 0.006. Pleochroism is strong with a variety of shades of blue, also pale reddish or yellowish green. Sapphirine occurs in high-temperature rocks or xenoliths with abundant aluminium and magnesium and low silicon. Sakena, Vorokafotra, Bekily, Anjamiary and Betroka, Madagascar have provided crystals. GIA have reported a purplish pink Sri Lankan faceted stone of 1.54 ct with mica inclusions. An orange-red variety has been reported from Tanzania.

Sarcolite

Sarcolite is $NaCa_6Al_4Si_6O_{24}F(?)$ and forms transparent flesh-pink to colourless equant pseudocubic hemihedral crystals of the tetragonal system with hardness 6 and SG 2.82–2.90. The RI for the ordinary and extraordinary rays is 1.604 and 1.615, uniaxial positive with birefringence 0.011. Sarcolite occurs in contact metamorphosed limestone-bearing volcanic ejecta; the only locations are Monte Somma, and Vesuvius, Campania and Anguillara, Lazio, Italy.

Scapolite

For convenience, the gem-quality varieties of the tetragonal isomorphous series marialite-meionite are conflated here. Marialite is $3NaAlSi_3O_8 \cdot NaCl$ and meionite $3CaAl_2Si_2O_8 \cdot CaCO_3$. Crystals are prismatic, usually with flat pyramidal terminations; masses are also known. The hardness is 6 and the SG 2.50–2.78 varying with composition. The total RI range for the ordinary and extraordinary rays is 1.546–1.600; crystals are uniaxial negative. Quartz varieties with overlapping values and resembling scapolite in colour are optically positive. Pleochroism for pink and violet crystals is dark blue or lavender to colourless or violet. Pale yellow stones show colourless to pale yellow. Some pink and violet specimens show absorption bands at 663 and 652 nm, from Cr and the yellow is often strongly absorbed.

Scapolite varieties may exhibit fluorescence, yellow specimens from Myanmar giving yellow to orange and some East African yellow stones a very bright yellow-green under LWUV. There are variable responses to SWUV and to X-rays with some phosphorescence reported. Marialite-meionite specimens occur in regionally metamorphosed rocks, particularly in marbles. The Myanmar stone tract at Mogok has produced fine pink chatoyant specimens some of which may incline to violet; fine golden yellow to orange stones have come from Mpwapwa, Morogoro region, Dodoma, Tanzania. Massive yellow scapolite from Quebec is particularly attractive and may show a bright yellow fluorescence.

The most topical account of the scapolite group minerals can be found in the recently published (2004) vol. 4B of Deer, Howie and Zussman, *Rock-Forming Minerals*, published by the Geological Society of London; ISBN 1862391440.

Scheelite

Scheelite is $CaWO_4$ and forms characteristic colourless to yellow transparent to translucent pseudooctahedral crystals in the tetragonal system with twinning common, hardness 4.5–5, SG 5.9–6.03. The lustre may approach adamantine. The RI for the extraordinary and ordinary rays is 1.935–1.938 and 1.918–1.921, uniaxial positive with birefringence 0.016 and dispersion 0.038. Scheelite shows a conspicuous blue-white fluorescence under SWUV and faint RE lines may be seen in the yellow and green sections of the absorption spectrum.

Scheelite occurs in contact metamorphic deposits, hydrothermal veins or pegmatites. Fine-quality transparent crystals are found at Mitopilas, Sonora, Mexico (orange specimens), large crystals from the Bisperg iron mine near Sater, Dalarne, Sweden. There are locations in California and Arizona, USA.

Large crystals from the Tae Wha mine, Chungju, North Chungchong Province, South Korea, may not be facetable but are collectable. Both chatoyant and four-rayed star scheelite are known. Scheelite has been made synthetically.

Scolecite

Scolecite is $Ca[Al_2Si_3O_{10}]\cdot3H_2O$ and is a member of the zeolite group, forming pyroelectric and piezoelectric colourless to white slender pseudotetragonal crystals of the monoclinic system with two directions of perfect cleavage, hardness 5–5.5, SG 2.25. The RI for the alpha, beta and gamma rays is 1.507–1.513, 1.516–1.520, 1.517–1.521, biaxial negative, birefringence 0.007. There may be a yellow to brown fluorescence under LW and SWUV. Scolecite occurs with other zeolites primarily as a cavity mineral in basalts. Transparent crystals of high quality are found from the Teigarhorn, Berufjord, Iceland and exceptional specimens from the classic localities in the Poona, Nasik and Mumbai areas of India.

Scorodite

Scorodite is $Fe^{3+}AsO_4\cdot2H_2O$ and forms tabular, prismatic or pyramidal pale leek green, blue to brown semitransparent crystals of the orthorhombic system. The hardness is 3.5–4 and the SG 3.28–3.29. The RI for the alpha, beta and gamma rays is 1.741–1.784, 1.744–1.805, 1.768–1.820, biaxial positive with birefringence 0.027–0.030. Pleochroism is strong, purple to blue and a strong absorption line can be seen at 450 nm with broad absorption in the green. It is soluble in HCl. Scorodite is a secondary mineral formed by the oxidation of As-bearing sulphides. Some large pleochroic blue crystals are found at Tsumeb, Namibia and scorodite also occurs at the Ojuela mine, Mapimi, Durango, Mexico.

Scorzalite

Scorzalite is $(Fe^{2+},Mg)Al_2(PO_4)_2(OH)_2$ forming a series to lazulite (q.v.) and occurs as semitransparent dark blue masses in the monoclinic system with hardness 6 and SG 3.32. The RI for the alpha, beta and gamma rays is

1.626–1.645, 1.654–1.674, 1.663–1.680, biaxial negative with birefringence near 0.040. Scorzalite occurs as a secondary mineral in complex-zoned granite pegmatites. Gem specimens most likely to be from Brazil, perhaps from the Corrego Frio pegmatite, Divino das Laranjeiras, Linopolis, Minas Gerais.

Sellaite

Sellaite is MgF_2 and occurs as short prismatic colourless or white transparent crystals of the tetragonal system with perfect cleavage in two directions, hardness 5–5.5 and SG 3.15. The RI for the ordinary and extraordinary ray is 1.378 and 1.390, uniaxial positive, birefringence 0.012. Facetable crystals occur in a metamorphic magnesite deposit at the Brumado mine, Bahia, Brazil.

Senarmontite

Senarmontite is Sb_2O_3 and commonly forms transparent to translucent colourless or greyish white octahedra in the cubic system. The lustre is characteristically resinous or subadamantine. The hardness is 2–2.5 and the SG 5.3. The RI is 2.087. Facetable specimens are not common as crystals are brittle and appear colourless only by transmitted light. Senarmontite occurs by the oxidation of antimony, stibnite and other antimony minerals. Large crystals are reported from the Haminate mine, Sensa, Ain-el-Bebbouch, Qacentina, Algeria, though I have seen no specific reports of gem material being fashioned from the crystals.

Sepiolite

Sepiolite is $Mg_4Si_6O_{15}(OH)_2 \cdot 6H_2O$, the compact opaque white variety meerschaum occasionally being used ornamentally though pipe bowls appear to have been its commonest use. It has a hardness of 2 and SG of 2 and it may float on water due to its porosity when dry. It is found as a sedimentary clay mineral, the meerschaum variety coming from Eskishehir, Turkey.

Sérandite

Sérandite is $Na(Mn^{2+},Ca)_2Si_3O_8(OH)$ forming a series with pectolite and exceptionally beautiful faceted specimens have been obtained from transparent rose-red, orange or salmon-pink elongated prismatic crystals of the triclinic system. There are two directions of perfect cleavage; the hardness and SG are 5–5.5 and 3.42. The RI for the alpha, beta and gamma rays is 1.668, 1.671 and 1.703 biaxial positive with birefringence 0.028. Sérandite occurs in pegmatites cutting syenites and nepheline syenites in an intrusive alkalic gabbro-syenite complex at Mt Saint Hilaire, Quebec, Canada.

Serendibite

Serendibite is $Ca_2(Mg,Al)_6(Si,Al,B)_6O_{20}$ and forms small greyish blue-green to deep blue transparent tabular crystals of the triclinic system with hardness 6.5–7 and SG 3.43–3.44. The RI for the alpha, beta and gamma rays is 1.700–1.738, 1.703–1.741 and 1.706–1.743, biaxial negative with birefringence 0.005. Strong pleochroic colours are yellowish green, bluish green and violet-blue. Serendibite from Gangapitiya, Ambakotte, Sri Lanka, has presumably been found as alluvial material though the mineral, a member of the aenigmatite group, occurs generally in skarns affected by boron metasomatism.

Serpentine

Group Status and Nomenclature

Minerals once classed as members of the serpentine group now find themselves in the kaolinite–serpentine group which comprises hexagonal, trigonal, orthorhombic, monoclinic or triclinic silicates with the general composition $M_{2-3} Z_2O_5(OH)_4 \cdot nH_2O$ where M = Al, Fe^{3+}, Fe^{2+}, Mg, Mn^{2+}, Ni, Zn; Z = Al, Fe^{2+}, Si. Bowenite is antigorite as are the minerals for which the name 'serpentine' has long been in general non-specific use.

Kaolinite–serpentine group minerals are polished primarily for their attractive patterning though some may be intended to resemble the jade minerals nephrite and jadeite. Distinction is not too difficult by gemmological testing as both nephrite and jadeite have SG (3.0 and 3.3 respectively) higher than kaolinite–serpentine minerals. Using the validated names of today antigorite, clinochrysotile, and lizardite the SG and RI range for antigorite is 2.61, 1.560–1.571 respectively with birefringence 0.014. For clinochrysotile RI is 2.53–2.55 and 1.545–1.569, birefringence 0.001. For lizardite, RI is 1.546–1.560, birefringence 0.008.

Nomenclature problems from which 'bowenite' may suffer also affect 'williamsite', also a variety of antigorite, occurring in notably translucent apple-green.

Kaolinite–serpentine group minerals occur from the alteration of ultra-mafic rocks. Bowenite in translucent green colours is found in New Zealand; attractive green williamsite is found at Rock Springs, Maryland, USA. Banded material is found in many places.

Shattuckite (Planchéite)

Shattuckite is $Cu_5(SiO_3)_4(OH)_2$ and most frequently forms translucent orthorhombic deep to medium aggregates of spherulitic blue masses with two directions of perfect cleavage, hardness 3.5 and SG 3.8–4.12. The RI for the alpha, beta and gamma rays is 1.753, 1.782 and 1.815, biaxial positive. Pleochroism is strong ranging from very pale to deep blue. Shattuckite occurs as a secondary mineral in oxidized copper deposits. Probably the best examples come from the Shattuck mine, Bisbee, Cochise County, Colorado, USA.

Shomiokite-(Y)

Shomiokite-(Y) is $Na_3(Y,Dy)(CO_3)_3 \cdot 3H_2O$ and forms short transparent rose-coloured prismatic pseudohexagonal crystals or granular masses of the orthorhombic system with one direction of perfect cleavage, hardness 2–3 and SG 2.59–2.64. The RI for the alpha, beta and gamma rays is 1.528, 1.529 and 1.531, biaxial positive with birefringence 0.009. Characteristic RE fine absorption lines can be seen and specimens may show a rose-coloured fluorescence under UV. Shomiokite-(Y) occurs with K-feldspar in pegmatites at Mt Alluaiv on the Kola Peninsula, Russia.

Shortite

Shortite is $Na_2Ca_2(CO_3)_3$ and forms characteristic transparent yellow wedge-shaped crystals of the orthorhombic system with hardness 3 and SG 2.60. The RI for the alpha, beta and gamma rays is 1.531–1.532, 1.555–1.556 and 1.570, biaxial negative with birefringence 0.039. Shortite slowly decomposes in water and occurs in association with an intrusive alkalic gabbrio-syenitic complex at Mt Saint Hilaire, Quebec, Canada.

Siderite

Siderite is $Fe^{2+}CO_3$ and forms translucent yellow-brown rhombohedral to steep scalenohedral crystals in the trigonal system with hardness 3.5–4.5 and one direction of perfect cleavage. The SG is 3.83–3.96 and the RI for the ordinary and extraordinary rays is 1.875 and 1.633, uniaxial negative with birefringence 0.240. Siderite occurs in sedimentary deposits and hydrothermal ore veins. Large brown rhombs are found at Mt Saint Hilaire, Quebec, Canada.

Orange and pinkish orange specimens from this locality show weak absorptions at 520–510 nm, 485 and 450 nm and brown transparent crystals have been found at the Morro Velho gold mine, Nova Lima, Minas Gerais, Brazil. Some facet-quality specimens are reported from Panasqueira, Portugal.

Sillimanite

Sillimanite (name preferred to fibrolite) with andalusite and kyanite is a trimorph of Al_2SiO_5. Transparent, rather slate-blue or blue-green slender poorly terminated prismatic crystals of the orthorhombic system show one direction of perfect cleavage and some material is fibrous. The hardness is about 7.5 for the crystals and from 6 to 7 for the fibrous material. The SG is 3.23–3.24 and the RI for the alpha, beta and gamma rays is 1.653–1.661, 1.657–1.662 and 1.672–1.683, biaxial positive with a birefringence of 0.020. Pleochroic colours, often distinct, are pale yellow to green; brown or greenish; dark brown or blue, sometimes violet. Absorption bands may be seen at 462, 441 and 410 nm. Sillimanite occasionally shows a red fluorescence which is apparently recorded for Myanmar specimens only.

Sillimanite is found in high-grade metamorphic schists and gneisses, sometimes in pegmatites. Probably most gem sillimanite is found in gem gravels. Chief occurrences are in the Mogok Stone Tract, Myanmar, where violet–blue stones have been reported and in the Sri Lanka gem gravels, where a greyish green chatoyant material is obtained.

Small-sized material similar to the Myanmar specimens is reported from Kenya. Chatoyant stones turn up from time to time. Sillimanite is one of the most difficult transparent gem materials to fashion.

Simpsonite

Simpsonite is $Al_4Ta_3O_{13}(OH)$ and forms euhedral to subhedral yellow to orange to brown tabular to short prismatic transparent crystals of the

trigonal system with an adamantine lustre and hardness 7.5. The SG is
5.92–6.84. The RI for the ordinary and extraordinary rays is 2.035 and
1.977, uniaxial negative with birefringence 0.058. Crystals from western
Australia show a bright blue-white under SWUV. Simpsonite occurs as an
accessory mineral in some Ta-rich granite pegmatites. The most attractive
simpsonite, of strong bright orange, is found at Tabba Tabba, western
Australia; other material from the Alto do Giz pegmatite near Parelhas and
the Orca mine in the same general district of Rio Grande do Norte, Brazil.

Sinhalite

Sinhalite is $MgAlBO_4$ and is most commonly found as transparent honey
yellow to brown grains or pebbles of the orthombic crystal system. The
hardness is 6.5–7 and the SG 3.475–3.50. The RI for the alpha, beta and
gamma rays is 1.667–1.676, 1.697–1.704 and 1.705–1.712, biaxial negative
with birefringence 0.035–0.037. The absorption spectrum shows bands at
493, 475, 463, 452 and 435 nm with general absorption of the violet: the
band at 463 nm is not present in the peridot variety of olivine (q.v.) with
which there might otherwise be confusion. Sinhalite occurs in boron-rich
skarns at the contact of limestones with granite or gneiss; most gem
sinhalite is found as alluvial pebbles. One crystal from Myanmar was
described in 1958. A transparent golden brown chatoyant variety with a
silver eye is reported from the main locality for sinhalite, Sri Lanka.

Smithsonite

Smithsonite is $ZnCO_3$ and in the trigonal system forms typically reniform or botryoidal translucent masses; fine transparent to translucent
crystals may be rhombohedral or scalenohedral with one direction of
nearly perfect cleavage. The colour is pale to deep yellow, sometimes
pink, more commonly blue-green. The hardness is 4–4.5 and the SG 4.43.
The RI for the ordinary and extraordinary rays is 1.842–1.850 and
1.619–1.623, uniaxial negative with birefringence 0.227. Smithsonite
occurs as a secondary mineral in the oxidized zone of zinc-bearing
deposits. Notably fine, near-transparent crystals are found at the Kelly
mine, Magdalena, Socorro County, new Mexico, USA; pink smithsonite
is found in some Mexican deposits and transparent yellow crystals occur
at Tsumeb, Namibia. The classical slags at Lavrion, Greece, provide
attractive translucent material. A beautiful translucent green smithsonite,
which could be mistaken for one of the jade minerals, occasionally
appears on the specialist collectors' market.

Soapstone

Soapstone, $Mg_3Si_4O_{10}(OH)_2$, is the popular name for ornamental varieties of talc, which forms triclinic or monoclinic crystals and fine-grained compact masses with a characteristic soapy feel, one direction of perfect cleavage and hardness 1 when more or less pure; the SG is 2.20–2.83. The colour ranges from light to dark green to brown, white or colourless. The RI for the alpha, beta and gamma rays is 1.539–1.550, 1.589–1.594, 1.589–1.600, biaxial negative with birefringence 0.039. In practice the massive material will show a single reading near 1.54 on the refractometer.

Soapstone occurs in talc-rich schists from the hydrothermal alteration of mafic rocks. Material used for carvings may have come from a number of sources including Egypt (ancient deposits) and the European Alps.

Sodalite

Sodalite is $Na_8Al_6Si_6O_{24}Cl_2$ and forms characteristic light to dark blue masses in the cubic system; the typically dodecahedral crystals are uncommon. The hardness is 5.5–6 and the SG 2.27–2.33. The RI is 1.483–1.487. The transparent colourless variety hackmanite from Dungannon Township, Ontario, Canada, shows bright orange under LWUV and bright pale pink under SW; after exposure to SW the colour may reach raspberry-reed but fades rapidly in sunlight. Further irradiation of the sample under SWUV repeats the cycle. Near-transparent dark blue crystals from Ohopoho, northern Namibia, have been faceted. Sodalite is found in the Ilimaussaq intrusion, southern Greenland and in the Khibiny and Lovozero massifs, Kola Peninsula, Russia. Sodalite is a major ingredient of the opaque ornamental rock lapis lazuli (q.v.).

Sogdianite

Sogdianite is $KNa(Zr,Fe^{3+},Ti,Al)_2$ Li_3 Si_{12} O_{30} and forms transparent to translucent violet crystals of the hexagonal system with a perfect cleavage in one direction, hardness 5–6 and SG 2.76–2.90. The RI for the ordinary and extraordinary rays is 1.606 and 1.608, uniaxial negative with birefringence 0.002. Sogdianite fluoresces dark violet under LWUV and very dark red under SW. The absorption spectrum shows an absorption area between 645 and 630 and between 493–488 nm, with additional lines at 437, 419 and 411 nm. Sogdianite occurs in a pegmatite vein with a composition near that of an alkalic granite. Locations include the Dara-i-Pioz massif, Tadzhikistan, and the Wessels mine, Kuruman, Cape Province, South Africa.

Sphalerite

Sphalerite is (Zn,Fe)S and forms tetrahedral, dodecahedral, often complex transparent to translucent colourless to yellow, red, green and brown crystals of the cubic system with resinous to adamantine lustre, perfect dodecahedral cleavage, hardness 3.5–4 and SG 3.9–4.1. The RI is 2.37–2.43 and the dispersion notably high at 0.156. Some specimens fluoresce a bright orange-red to red under both forms of UV. Sphalerite occurs in a variety of hydrothermal conditions and is the chief ore of zinc. Very fine material comes from Picos de Europa, Santander, Satander Province, Spain, and near-colourless to pale green crystals from Franklin, New Jersey, USA. Some crystals show absorption bands in the red; the strongest and clearest is at 665 nm and a much weaker one is at 651 nm. A fairly strong broad band is centred near 692 nm. A red filter serves to diminish the glare from the yellow part of the spectrum, assisting observation of this band. Cadmium may be the cause.

Sphene

Sphene is $CaTiSiO_5$. The name titanite is universally used by mineralogists today. It forms characteristically wedge-shaped crystals of the monoclinic system; cleavage is good rather than perfect but crystals may show parting due to twinning. Colours may be pale yellow to brown, green or, rarely, emerald green. The hardness is 5–5.5 and the SG 3.45–3.55. Sphene has a near adamantine lustre and a notably high dispersion, 0.051. The RI for the alpha, beta and gamma rays is 1.843–1.950, 1.870–2.034, 1.943–2.110, biaxial positive with birefringence 0.100–0.192. Some degree of metamictization may be present. Some specimens may show faint RE absorption lines. Sphene occurs as a common accessory mineral in igneous rocks, also in schists and gneisses. Fine crystals are found in Baja California, Mexico, at El Rodeo, La Huerta and in particular at Pino Solo where emerald-green crystals coloured by Cr have been found. Sphene also provides fine crystals at several locations in Switzerland, including Tavetsch, Graubunden and sites in Valais and Ticino cantons. Transparent orange sphene has been found in Myanmar. Emerald-green specimens have been found in the Urals, Russia.

Spurrite

Spurrite is $Ca_5(SiO_4)_2(CO_3)$; the monoclinic, pseudo-orthorhombic crystals are anhedral but massive opaque to translucent lavender-grey to purple masses have been used ornamentally. They show one direction of perfect cleavage; the hardness is 5 and the SG 3.02. The RI for the alpha, beta and gamma rays is 1.637–1.641, 1.672–1.676, 1.676–1.681, biaxial negative with birefringence 0.039. Spurrite occurs as a product of high-temperature thermal metamorphism along the contact between carbonate rock and mafic magma. Most ornamental spurrite has come from Mexico, probably from the Santa Juana mine, Velardena, Durango or from the Encantada district, Coahuila.

Staurolite

Staurolite is $(Fe^{2+},Mg)_2Al_9(Si,Al)_4O_{20}(O,OH)_4$ and forms highly characteristic monoclinic, pseudo-orthorhombic prismatic, typically rough reddish brown crystals, often forming 60° twins. A few crystals are just about transparent but the majority are translucent with hardness 7–7.5 and SG 3.68–3.83. The RI for the alpha, beta and gamma rays is 1.736–1.747, 1.742–1.753, 1.748–1.761, biaxial positive with birefringence 0.011–0.015. Distinct pleochroism gives colourless to yellow and reddish brown. Absorption bands may be seen at 632 and 610 nm (these in zinc-rich material) and sometimes at 531.5 nm.

Staurolite occurs most commonly in schists and gneisses; cross-like twins from Pizzo Forno and Alpe Piona, Ticino, Switzerland, and large crystals from the Finistere and Morbihan areas of France. Some rare facetable crystals from Brazil, perhaps from Rubellita and Ardenella, Minas Gerais. Reddish brown to brownish black staurolite is reported from Myanmar.

Stibiotantalite

Stibiotantalite is $Sb(Ta,Nb)O_4$ and forms semi-transparent yellowish to brownish equant or flattened prismatic crystals of the orthorhombic system with hardness 5.5 and SG 7.3. The RI for the alpha, beta and gamma rays is 2.374, 2.404, 2.457, biaxial positive with birefringence 0.90. Some specimens may show a faint RE absorption spectrum. Stibiotantalite occurs in complex granite pegmatites. Uncommon facetable material is most likely to have come from the Muiane and other mines, Alto Ligonha, Mozambique, or from the Himalaya mine, Mesa Grande district, SanDiego County, California.

Stichtite

Stichtite is $Mg_6Cr_2(CO_3)(OH)_{16}\cdot4H_2O$ and is dimorphous with bar-bertonite. It forms commonly matted lilac to rose-pink aggregates of fibres or plates in the trigonal system with a perfect cleavage in one direction and feeling greasy to the touch; hardness 1.5–2, SG 2.16–2.22. The RI for the ordinary and extraordinary rays is 1.545 and 1.518, uniaxial negative (may be anomalously biaxial) with birefringence 0.02. The absorption spectrum is characteristic of Cr with lines in the red. Stichtite occurs as an alteration product of chromite in serpentinite; fine material from the Adelaide Ag-Pb mine, Dundas, Tasmania, Australia, and from asbestos mines at Stolzburg, Barberton district, Transvaal, South Africa.

Stolzite

Stolzite is $PbWO_4$ and is dimorphous with raspite. It forms translucent reddish brown to yellow, sometimes bright orange dipyramidal crystals of the tetragonal system with hardness 3 and SG 8.3–8.4. The RI for the ordinary and extraordinary rays is 2.27 and 2.18–2.19, uniaxial negative with birefringence 0.080. Stolzite occurs as an uncommon mineral in the oxidized zone of hydrothermal tungsten-bearing lead deposits. Localities include the Kola peninsula, Russia and Broken Hill, New South Wales.

Strontianite

Strontianite is $SrCO_3$ and a member of the aragonite group, forming typically short to long prismatic crystals of the orthorhombic system with a nearly perfect cleavage in one direction. The colour may be a pale grey or pale green, sometimes pale red or colourless. Strontianite is an uncommon low-temperature hydrothermal mineral formed in veins in limestone. Fine crystals are found at Aschberg, North Rhine-Westphalia, Germany; the occasional small stone may be faceted.

Sugilite

Sugilite is $KNa_2(Fe^{3+},Mn^{3+},Al)_2Li_3Si_{12}O_{30}$ and most commonly occurs as interlocking aggregates of subhedral grains forming compact masses in the hexagonal system with a light brownish yellow to a bright magenta colour, hardness 5.5–6.5, SG near 2.74. The RI for the ordinary and extraordinary rays is 1.595–1.611, 1.590–1.607, uniaxial negative, birefringence 0.003–0.005. The absorption spectrum shows a weak diffused band centred at 570 nm, with a strong band at 419 and a weak band at 411 nm. Sugilite occurs in an aegirine-bearing syenite stock in biotite granite in bedded manganese deposits; ornamental translucent facetable sugilite comes from the Wessels and N'Chwaning mines, Kuruman, Cape Province, South Africa. A further source has been found in Tadjikistan and at Mt Saint-Hilaire, Quebec, Canada.

Sulphur

Sulphur, S, forms transparent to translucent yellow crystals of the orthorhombic system which are in general too heat-sensitive to be fashioned for ornament. The crystals are dipyramidal, thick tabular and often well formed. The hardness is 1.5–2.5 and the SG 2.07. The RI for the alpha, beta and gamma rays is 1.958, 2.038 and 2.245, biaxial positive with birefringence 0.291. Sulphur is a sublimation product of volcanic fumaroles. The finest crystals are found at Cianciana, Agrigento and Racalmuto, Sicily, Italy.

Taaffeite

Recent authorities, including Fleischer's *Glossary of Mineral Species* 2004 now use the name magnesio-taaffeite but for the purpose of this listing I have retained the original name. Taaffeite is $Mg_3Al_8BeO_{16}$ and forms mauve or lilac, red to brown transparent to translucent hexagonal crystals of the hexagonal system with hardness 8–8.5 and SG 3.605–3.613. Unusually the type species is a faceted stone, originally believed to be spinel. The RI for the ordinary and extraordinary rays is 1.721–1.726 and 1.717–1.727, uniaxial negative (may be anomalously biaxial). The birefringence is 0.004–0.009. Taaffeite occurs in gem gravels or in skarns at the contact between dolostones and limestones with beryllium-bearing granite. The productive gem gravels are located in Sri Lanka; an *in situ* occurrence in China is located on Hsianghualing Ridge, China. Taaffeite has also been found 6 km west of Mt Painter, Southern Australia. Recently, a purplish brown chatoyant taaffeite has been found in Sri Lanka, the eye arising from parallel reflective planar inclusions with striations.

Tanzanite

Probably the most important in the gem context of the epidote group minerals is zoisite $Ca_2Al_3(SiO_4)(Si_2O_7)O(OH)$ as this includes the magnificent transparent blue material known universally by the varietal name tanzanite. Tanzanite is especially notable for its pleochroism which in one specimen I (MO'D) have examined shows the now familiar deep blue in one direction but also ruby red and emerald green in the other two directions. The pleochroic green is replaced by rich gold in many specimens. As well as tanzanite whose occurrence is till apparently limited to Tanzania, a transparent green zoisite has been found in the northern areas of Pakistan. Tanzanite is very brittle and should be handled with care; it should never be placed in ultrasonic cleaners. The hardness is 6–7, SG is 3.35, the RI 1.692, 1.693 and 1.700 for the alpha, beta and gamma rays respectively, biaxial positive with DR about 0.009. Tanzanite occurs in calcareous rocks, including metamorphosed dolomite and calcareous shales subjected to regional metamorphism. Occasional reports have claimed other localities for tanzanite but so far finds have been confined to the Lalatema area of Tanzania. A recent study by Valerio Zancanelli, *Tanzanite*, 2004; ISBN 8890150904, includes some useful and interesting information.

Tephroite

Tephroite is $Mn^{2+}SiO_4$ and forms characteristically short prismatic or anhedral transparent to translucent olive green to reddish brown crystals or compact masses of the orthorhombic system with hardness 6 and SG 4.15. The RI for the alpha, beta and gamma rays is 1.770–1.788, 1.807–1.810 and 1.817–1.825, biaxial negative with birefringence 0.037–0.047. Tephroite occurs in Fe–Mn ore deposits and related skarns, also in metamorphosed Mn-rich sediments. Ornamental-quality material is found at Franklin and Sterling Hill, Ogdensburg, Sussex County, New Jersey, USA.

Thaumasite

Thaumasite is $Ca_3Si(CO_3)(SO_4)(OH)_6 \cdot 12H_2O$ and forms transparent to translucent white, yellowish, pink or brown acicular prismatic crystals of the hexagonal system or compact masses. The hardness is 3.5 and the SG about 1.90. The RI for the ordinary and extraordinary rays is 1.498–1.507 and 1.458–1.470, uniaxial negative with birefringence 0.036. Thaumasite occurs as a very late-stage mineral in some sulphide ore deposits. As one of the zeolite mineral group, it associates with other group members.

Yellow material from the Wessels and N'Chwaning mines, Kuruman, Cape Province, South Africa, has been faceted.

Thomsonite-(Ca)

Thomsonite, $NaCa_2Al_5Si_5O_{20}6H_2O$, is a common member of the zeolite group of minerals and forms transparent to translucent orthorhombic pseudotetragonal thin or bladed prismatic crystals with one direction of perfect cleavage and a range of pale colours, notably yellow, pink, greenish or grey. Fibrous masses are more common and occur as radiating spherical or columnar aggregates. The hardness is 5–5.5 and the SG 2.25–2.40. The RI for the alpha, beta and gamma rays is 1.497–1.530, 1.513–1.533 and 1.518–1.544, biaxial positive with birefringence 0.021. Thomsonite occurs in amygdules and fractures in mafic igneous rocks, most commonly basalts. The eye-like material usually comes from Isle Royale, Michigan, USA.

Thorite

Thorite is $(Th,U)SiO_4$ forms nearly opaque yellow-orange or brown square prisms or pseudo-octahedral crystals of the tetragonal system though the mineral is commonly metamict and measurably radioactive. The hardness is 4.5–5 and the SG 4.0–6.7. The RI for the ordinary and extraordinary rays is 1.78–1.82, uniaxial positive. Thorite occurs as an accessory mineral in felsic igneous rocks and their associated pegmatites. Some gem material is reported from the Mogok area of Myanmar.

Tremolite

Tremolite, $Ca_2(Mg,Fe^{2+})_5Si_8O_{22}(OH)_2$, is a member of the amphibole mineral group in which it forms a series with actinolite and with ferro-actinolite. Stout elongated crystals of the monoclinic system are typical as are fibrous, granular or columnar aggregates. There is a perfect cleavage in one direction; the hardness is 5–6 and the SG 2.99–3.03. The colour is grey, green or lavender to pink and the RI or the alpha, beta and gamma rays is 1.605–1.613, 1.616–1.624 and 1.630–1.636, biaxial negative with birefringence 0.027. The transparent pink variety hexagonite shows pleochroic colours, bluish-red to deep reddish violet. The name nephrite is not now used for a valid mineral species (see Chapter 20). The best current account of the reviewed structure of the minerals of the amphibole group can be found in Deer, Howie and Zussman, *Rock-Forming Minerals* vol. 2B, Double-chain silicates, Second edition, The Geological Society, London, 1997.

Tremolite is formed by contact metamorphism of Ca–Mg siliceous sediments and in metamorphics derived from ultramafic or magnesium carbonate rocks. Hexagonite is found at Fowler, St Lawrence County, New York, USA; a green facetable material has been found in the Lalatema area of Tanzania.

Triphylite

Triphylite is $Li(Fe^{2+},Mn^{2+})PO_4$ and forms transparent to translucent bluish grey to greenish grey to brown or black compact masses of the orthorhombic system with a perfect cleavage in one direction, hardness 4–5, SG 3.50–3.58. The RI for the alpha, beta and gamma rays is 1.675–1.694, 1.684–1.695 and 1.685–1.700, biaxial positive with birefringence 0.006–0.008. Triphylite occurs in complex-zoned granite pegmatites in which it is the most abundant primary phosphate. Large crystals are found at the G.E. Smith quarry, Newport, Sullivan County and the Palermo mine, North Groton, New Hampshire, USA.

Triplite

Triplite is $(Mn^{2+},Fe^{2+},Mg,Ca)_2(PO_4)(F,OH)$ and crystals occur in a number of indistinct forms or as masses which are translucent to opaque dark red to brown, or black if altered. The crystal system is monoclinic, the hardness 5–5.5 and the SG 3.85–3.94. The RI for the alpha, beta and gamma rays is 1.643–1.684, 1.647–1.693, 1.668–1.703, biaxial positive with birefringence 0.020. Triplite occurs as a principal primary phosphate in complex-zoned granite pegmatites or in hydrothermal tin veins. Ornamental material has been reported from Brazil.

Tugtupite

Tugtpite is $Na_4AlBeSi_4O_{12}Cl$ and forms fine-grained aggregates of characteristic cyclamen colour; crystals are very small, tetragonal, pseudocubic, showing short prisms with pyramids and sphenoids. The hardness is about 4 and the SG 2.33–2.35. The RI for the ordinary and extraordinary ray is 1.492–1.496 and 1.499–1.502, uniaxial positive, birefringence 0.006–0.008.

Tugtupite shows an attractive fluorescence with the red response to UV appearing stronger in SW than in LW. I have noted crimson and salmon colours in LW and SW respectively. Tugtupite occurs in hydrothermal veins cutting sodalite syenite and syenite. The chief locality is the Ilimaussaq intrusion at Tugtup Agtakorfia, Greenland.

Ulexite

Ulexite is $NaCaB_5O_6(OH)_6 \cdot 5H_2O$ and forms masses in colourless parallel fibrous veins along which an image can be carried to the observer, thus from placing the direction of the fibres at right angles to a page of print they will carry a fibre-optic image of the print; the name television stone describes this facility. Ulexite has no other claim to fame but is a member of the triclinic system with one direction of perfect cleavage, hardness 1–2.5, SG upto 1.95. The RI for the alpha, beta and gamma rays is 1,491–1.496, 1.504–1.506, 1.519–1.520, biaxial positive with birefringence 0.023. Ulexite occurs in dry lakes and playa deposits, with other borates. Most material used as television stone comes from borate deposits in California, USA.

Unakite

Unakite is a granite containing pink feldspar and green epidote; the sources include Zimbabwe and the Blue Ridge, Unaka Range, North Carolina, USA. The SG, from reports, seems to be in the region of 2.8–3.2 and the RI 1.52–1.76.

Ussingite

Ussingite is $Na_2AlSi_3O_8(OH)$ and forms transparent to translucent pink to dark violet-red triclinic pseudocubic crystals or fine-gained compact masses with two directions of perfect cleavage, hardness 6–7, SG 2.46–2.49. The RI for the alpha, beta and gamma rays is 1.503–1.504, 1.506–1.508, 1.543–1.545, biaxial positive with birefringence 0.039. Ussingite occurs as a secondary mineral in pegmatites associated with sodalite syenite. At Mt Saint Hilaire, Quebec, Canada, ussingite is found in sodalite xenoliths in an intrusive alkalic gabbro-syenite complex.

Vanadinite

Vanadinite is $Pb_5(VO_4)_3Cl$ and forms subtransparent to opaque red-orange, deep red or brownish red well-developed hexagonal prismatic crystals of the hexagonal system with subresinous to adamantine lustre, hardness 2.5–3 and SG in the range 6.5–7.1. The RI for the ordinary and extraordinary rays is 2.416 and 2.350, uniaxial negative with birefringence 0.066. Vanadinite occurs as a secondary mineral in the oxidized zone of lead-bearing deposits. Fine crystals are found at a number of locations, including the Red Cloud mine, Silver district, Yuma County, Arizona, USA and at Mibladen, Morocco.

Variscite

Variscite is $AlPO_4 \cdot 2H_2O$, ornamental material being fashioned from fine-grained transparent to translucent pale green to blue masses of the orthorhombic system, with a waxy lustre and hardness 4.5 and SG 2.57–2.61; crystals are rare. The RI for the alpha, beta and gamma rays is 1.550–1.563, 1.565–1.588, 1.570–1.594, biaxial negative with birefringence 0.031. Variscite is typically deposited from phosphate-bearing waters in contact with aluminous rocks.

Good ornamental material is found as large nodules at the Little Green Monster mine, Clay Canyon, west of Fairfield, Utah, USA, and in the same state rich green variscite has been reported from sites in Tooele County. Variscite also occurs at the Sapucaia pegmatite mine, about 50 km east, south-east of Governador Valadares, Minas Gerais, Brazil.

Väyrynenite

Väyrynenite is $Mn^{2+}Be(PO_4)(OH)$ and has been found as occasional transparent to translucent rose-red to salmon-pink monoclinic prismatic crystals showing elongation and striations in one direction with hardness 5 and one direction of perfect cleavage. The SG is 3.22 and the RI for the alpha, beta and gamma rays is 1.638–1.640, 1.658–1.662 and 1.664–1.667, biaxial negative with birefringence 0.026. Väyrynenite occurs in complex-zoned granite pegmatites as an alteration product of beryl and triphylite. Väyrynenite was reported from the Viitaniemi pegmatite near Erajarvi, Finland, but this material was translucent at best. After Kazmi and O'Donoghue had reported on Pakistan gemstones in *Gemstones of Pakistan* (1990) a single near-transparent rose-red crystal of facet quality was noticed in the Natural History Museum London and a few years ago further reports on this material turned up in the literature but there have still not been many faceted examples reported. The location for some large crystals is Broghul Pass, Chitral Valley, Pakistan.

Verdite

The name verdite is given to a deep green opaque ornamental rock composed of finely intergrown mica minerals: mica group species have undergone extensive revision and the name fuchsite denoting chrome mica is no longer used though muscovite remains as the name for potassium mica. Verdite therefore has to be described as a mixture of chrome mica and muscovite. The hardness is 3 and the SG in the range 2.70–2.87. Some verdite from Zimbabwe may contain crystals of ruby and albite as well as bands of rutile grains. Verdite of ornamental quality comes most commonly from Barberton, Transvaal, South Africa in many shades.

Vesuvianite

Vesuvianite is the preferred mineralogical name to the gemmologists' idocrase, $Ca_{19}(Al,MgFe)_{13}OSi_{18}O_{68}(OH,F)_{10}$ which is found as short to long prismatic tetragonal transparent crystals in various colours of which green, yellow and brown are the most common. Crystals are morphologically complex and there are chemical affinities with grossular with which it can be intergrown. The hardness is 6–7 and the SG 3.32–3.47. The RI for the ordinary and extraordinary rays is 1.703–1.752 and 1.700–1.746, uniaxial positive though biaxial examples have been recorded. The birefringence is very small.

The ornamental material californite is a mixture of vesuvianite and grossular and can be quite an effective simulant of green jadeite from which it can be distinguished by the presence of an absorption band at 461 nm while the jadeite 437 nm band is absent. Some brown specimens may display fine-line RE absorptions.

Vesuvianite occurs in skarns formed during contact or regional metamorphism of limestones in serpentines and ultramafic rocks. Some brown gem-quality material comes from Italy and Switzerland. Bright green specimens from Asbestos, Quebec, Canada, may be intergrown with grossular and when coloured by Cr are a fine deep green; some pink specimens are coloured by Mn. Vesuvianite is also reported from Kenya. Classic European localities for brown, yellow and other colours of vesuvianite are in Italy and Switzerland. Californite is found in the Pulga area of California. An unusually fine oval dark green faceted stone of 3.68 ct was shown to MO'D in December 2004 by Marcus McCallum.

Villiaumite

Villiaumite is NaF and most commonly forms translucent to transparent carmine-red to lavender-pink granular masses of the cubic system; the cubic crystals are rare and show a perfect cleavage in one direction with hardness 2–2.5 and SG 2.79. The RI is 1.327–1.328. Villiaumite may show anomalous anisotropism and yellow to pink/deep carmine pleochroism. Villiaumite occurs in nepheline syenite and nepheline syenite pegmatites. Very small faceted stones have been obtained from crystals found at Rouma Isle, Los Islands, Guinea. Some gem-quality material or at least larger crystals has been found at Mt Saint Hilaire, Quebec, Canada. Villiaumite is likely to be water-soluble and may show weak red fluorescence under SWUV.

Vivianite

Vivianite is $Fe^{2+}_3(PO_4)_2 \cdot 8H_2O$ and forms flattened prismatic transparent to translucent light green oxidizing to deep blue crystals of the monoclinic system with one direction of perfect cleavage, hardness 1.5–2 and SG 2.68–2.69. The RI for the alpha, beta and gamma rays is 1.579–1.616, 1.602–1.656 and 1.629–1.675, biaxial positive with birefringence 0.040–0.059. Anomalous plechroism is strong with colours deep blue/pale yellowish green/yellowish green. Vivianite occurs as a secondary mineral in the oxidized zone of metallic-ore deposits and in complex granite pegmatites. Faceted stones are rare and are most likely to have come from Poopo, Llallagua, Avicata, Tazna, and elsewhere in Bolivia. Facetable crystals may also come from Anloua, Ngaoundere, Cameroon.

Vlasovite

Vlasovite is $Na_2ZrSi_4O_{11}$ and has been found in transparent pale brown monoclinic (triclinic) crystals with greasy lustre, hardness 6 and SG 2.92–2.97. The RI for the alpha, beta and gamma rays is 1.605, 1.623, 1.625, biaxial negative with birefringence 0.019. Vlasovite occurs as a late phase in nepheline syenites and syenitic pegmatites; faceted material from the Lovozero massif, Kola Peninsula, Russia, occurs along the contact zone of a differentiated alkalic massif.

Wardite

Wardite is $NaAl_3(PO_4)_2(OH)_4 \cdot 2H_2O$ and forms transparent to translucent colourless to white, pale green or blue-green typically striated dipyramidal pseudo-octahedral crystals of the tetragonal system with hardness 5 and SG 2.81–2.87. There is one direction of perfect cleavage. The RI for the ordinary and extraordinary rays is 1.586–1.594 and 1.595–1.604, uniaxial positive with birefringence 0.009. Wardite occurs in complex-zoned pegmatites, most faceted specimens coming from Alto Patrimonio, Piedras Lavradas, Paraíba, Brazil or from mines in Maine (the Dunton quarry), and other sites in the USA. Some wardite occurs with large green variscite nodules, these specimens usually originating from the Little Green Monster mine, Clay Canyon, west of Fairfield, Utah, USA.

Wavellite

Wavellite is $Al_3(PO_4)_2(OH,F)_3 \cdot 5H_2$ and forms very small crystals and larger characteristic radial aggregates of the orthorhombic system with a perfect cleavage in one direction, hardness 3.5–4 and SG 2.36. The masses are translucent white to greenish white, yellowish green to green to brown or black. The RI for the alpha, beta and gamma rays is 1.518–1.535, 1.524–1.543 and 1.544–1561, biaxial positive with birefringence 0.025. Wavellite occurs as a secondary mineral in low-grade metamorphic rocks and phosphate deposits. Fine examples can be found in the High Down quarry, Filleigh, South Molton, Devon, England, and at the Hot Springs area of Arkansas, USA.

Weloganite

Weloganite is $Sr_3Na_2Zr(CO_3)_6 \cdot 3H_2O$ and forms triclinic, pseudorhombohedral transparent to translucent lemon-yellow crystals with a perfect cleavage in one direction, hardness 3.5 and SG 3.20. The RI for the alpha, beta and gamma rays is 1.558, 1.646 and 1.648, biaxial negative with birefringence 0.090. Weloganite is associated with an intrusive alkalic gabbro-syenite complex at Mt Saint Hilaire, Quebec, Canada.

Whewellite

Whewellite is $CaC_2O_4 \cdot H_2O$ and forms transparent to translucent white, colourless, pale brown or pale yellow equant to short prismatic crystals

of the monoclinic system with hardness 2.5–3 and SG 2.21–2.23. The RI
for the alpha, beta and gamma rays is 1.489–1.491, 1.553–1.554 and
1.649–1.650, biaxial positive with birefringence 0.159–0.163. Whewellite
occurs as a hydrothermal mineral in ore veins and may be found in
septarian nodules. Brittleness makes faceting difficult. Large crystals
are from Freital-Burgk, near Dresden Saxony, Germany, and from the
Boldut mine, Kapnikbanya, Romania. The septarian nodules have been
found at a site about 11 km South-east of Havre, Hill County, Montana,
USA.

Willemite

Willemite is Zn_2SiO_4 and forms both stout and slender prismatic
transparent to translucent trigonal crystals with rhombohedral terminations or as coarse to fine granular masses. The colour ranges from
colourless to white though yellow- and apple-green to blue-green or
brown. The hardness is 5.5 and the SG 3.89–4.10. The RI for the ordinary
and extraordinary rays is 1.691–1.694 and 1.719–1.725, uniaxial positive
with birefringence 0.028. Willemite shows an intense green fluorescence
under SWUV – some specimens may also show green phosphorescence.
The absorption shows bands at 583, 540, 490, 442 and 420 nm, all fairly
weak, with a stronger absorption at 421 nm. Willemite occurs in zinc
deposits in limestones and is an uncommon ore of zinc. The ornamental
value of willemite is in its fluorescent effects especially when combined
with the fluorescence of red zincite and pink calcite; specimens with
this effect are particularly associated with Franklin, New Jersey, USA.
Facetable green willemite is most likely to come from Franklin though
some blue-faceted stones have come from Mt Saint Hilaire, Quebec,
Canada.

Williamsite

See Serpentine.

Witherite

Witherite is $BaCO_3$ and forms translucent to transparent grey or
white, yellow- or brown-tinged botryoidal to spherical fibrous or
granular masses of the orthorhombic system with hardness 3.5 and

SG 4.24–4.29. The RI for the alpha, beta and gamma rays is 1.529, 1.676 and 1.677, biaxial negative with birefringence 0.148. Witherite may show a green or yellow fluorescence with phosphorescence under SWUV. Witherite occurs in hydrothermal veins as a low-temperature mineral, good examples from Alston Moor, Cumbria, England; large yellow crystals from mines in the Rosiclare area, Hardin County, Illinois, USA. Crystals are pseudohexagonal dipyramidal due to repeated twinning.

Wollastonite

Wollastonite is $CaSiO_3$ and forms transparent to translucent tabular or long prismatic crystals of the monoclinic or triclinic system with hardness 4.5–5 and one direction of perfect cleavage. The colour is white, colourless or brown, red, yellow or pale green. The SG is 2.8–3.09 and the RI for the alpha, beta and gamma rays is 1.616–1.640, 1.628–1.650 and 1.631–1.653, biaxial negative with birefringence 0.148. Wollastonite occurs in thermally metamorphosed siliceous carbonates and in some alkalic igneous rocks and carbonatites. Some fibrous wollastonite may produce cat's-eyes and a few crystals from Mt Saint Hilaire, Quebec, Canada, have been faceted. Large crystals have been reported from Belafa, Madagascar. Some texts report a compact ornamental variety from Isle Royale, Lake Superior, USA.

Wulfenite

Wulfenite is $PbMoO_4$ and forms beautiful individual transparent bright orange crystals and crystal groups of the tetragonal system. Individual crystals are usually square, flat tabular with resinous to subadamantine to adamantine lustre. The hardness is 2.5–3 and the SG 6.5–7.0. The RI for the ordinary and extraordinary rays is 2.405 and 2.283, uniaxial negative (may be biaxial) with birefringence 0.122. Wulfenite occurs as a secondary mineral formed in the oxidized zone of hydrothermal lead deposits with the Mo commonly introduced externally. The finest specimens are the near-red specimens from the Red Cloud mine, Silver District, La Paz County, Arizona, USA, and similar specimens from the Glove mine, near Amado, Tyndall district, Santa Cruz County in the same state. Fine examples have also come from Bleiberg, Carinthia, Austria.

Xonotlite

Xonotlite is $Ca_6Si_6O_{17}(OH)_2$ and forms needle-like translucent chalky white to bluish grey crystals or fibrous masses in the monoclinic or triclinic system with one direction of perfect cleavage, hardness 6–6.5 and SG 2.71–2.72. The RI for the alpha, beta and gamma rays is 1.581–1.583, 1.581–1.583 and 1.591–1.593, biaxial positive. The birefringence is 0.010. Xonotlite occurs in contact metamorphic deposits within limestones and serpentines. Specimens are found at Tetela de Xonotla, Puebla, Mexico.

Yugawaralite

Yugawaralite is $CaAl_2Si_6O_{16} \cdot 4H_2O$ and a member of the zeolite group. It forms flat monoclinic tabular transparent to translucent colourless to white crystals, often nearly parallel in groups with hardness 4.5–5 and SG 2.20–2.23. The RI for the alpha, beta and gamma rays is 1.492–1.496, 1.497–1.499 and 1.502–1.504, biaxial negative or positive with birefringence 0.009. Yugawaralite occurs as crystals lining cavities, typically deposited in active geothermal areas. Faceted stones will probably have come from the Khandivali quarry near Mumbai, India.

Zektzerite

Zektzerite is $NaLiZrSi_6O_{15}$ and forms translucent colourless to pink, cream or white stout pseudohexagonal prisms of the orthorhombic system with two directions of perfect cleavage; the hardness is about 6 and the SG 2.79–2.80. The RI for the alpha, beta and gamma rays is 1.582, 1.584 and 1.584, biaxial negative (crystals nearly isotropic). There may be a light yellow fluorescence in SWUV though not in LW. Zektzerite occurs in a small number of miarolitic cavities in a riebeckite granite at Kangaroo Ridge, Washington Pass, Okanogan County, Washington, USA.

Zincite

Zincite is $(Zn,Mn^{2+})O$ and forms translucent, sometimes transparent yellow-orange to deep red hemimorphic pyramidal crystals of the hexagonal system with one direction of perfect cleavage, hardness 4 and SG 5.68. The RI for the ordinary and extraordinary rays is 2.013 and 2.029, uniaxial positive, birefringence 0.016. Zincite occurs as a primary mineral in metamorphosed stratiform zinc ore bodies, virtually the only facetable material coming from Franklin and Sterling Hill, New Jersey, USA.

Zoisite

Zoisite is $Ca_2Al_3(SiO_4)(Si_2O_7)O(OH)$ and forms transparent to translucent crystals or compact masses of the orthorhombic system. Crystals are prismatic and deeply striated; colours are white-grey, greenish brown, greenish grey, pink, blue and purple. The fine transparent blue tanzanite variety is produced by heat treatment of brownish material and so far has not been found in this colour. The hardness is 6–7 and the SG for tanzanite 3.35. The RI for the alpha, beta and gamma rays in tanzanite is 1.692, 1.693 and 1.700, biaxial positive with birefringence 0.009. Tanzanite has very pronounced pleochroism, some crystals showing ruby red, emerald green, or gold, and blue colours. Tanzanite shows an absorption band centred at 595 nm with fainter bands at 528 and 455 nm; there may be faint lines in the red.

Zoisite, a member of the epidote group, occurs in medium-grade regionally metamorphosed crystalline schists formed from igneous, sedimentary or metamorphic rocks relatively high in calcium. Tanzanite is produced from material occurring in the Merelani Hills, Lalatema Mountains, Tanzania. Some examples of the unheated crystals have been faceted to give transparent brown specimens. Transparent green crystals from Turmik, Skardu district, northern areas of Pakistan, have also been faceted. Massive dark pink to yellow zoisite is sometimes fashioned and is sometimes known as thulite. Ruby crystals in green zoisite have been named anyolite; the source is the Longido area of Tanzania.

In 2004, Valerio Zancanella published *Tanzanite; all about one of the most fascinating gemstones* (no ISBN but author's email is nfo@valeriozancanella.com). See also the entry for epidote (p. 409).

Zunyite

Zunyite is $Al_{13}Si_5O_{20}(OH,F)_{18}Cl$ and forms fine transparent colourless or grey or pale brown tetrahedral or pseudo-octahedral crystals of the cubic system with hardness 7 and SG 2.87–2.90. The RI is 1.592–1.600. Some specimens may show a strong red fluorescence under UV. Zunyite occurs in highly aluminous shales and hydrothermally altered volcanic rocks. Large crystals are found in the Dome Rock Mountains, near Quartzsite, La Paz County, Arizona, USA. There are also sites in Japan.

24

Synthetic Gemstones

In late 2004 the diamond and gemstone trade found itself in an unprecedented quandary, with serious uncertainty in gem-testing laboratories who found the issue of satisfactory certificates very difficult. Some members of the trade must have felt that only the straight identification could be relied upon and that questions of natural/artificial and treated/not-treated could not be determined cheaply, if at all. This may have had a knock-on effect among the general buying public who may well feel that the stone on whose value they have relied may not be worth as much as they thought. It will take some years for the effects of current developments to be felt.

These developments are making life hard for gemmological teaching too: specimens for teaching must be unimpeachable and, with a greater need for more expensive testing instruments, it places a strain on laboratories which teach and on gemmological teaching bodies with no laboratories. In a way the most unfavourable scenario is now in place – the one in which nobody quite knows where they are and in which no stone's identity is certain.

Nonetheless gemstones do not go away and all the familiar ones are still around. Some treatments have been applied to synthetics as well as to natural stones and we need to review both long-accepted and new materials in a comprehensive study of this kind.

Introduction

Since the time of my last major survey of artificially produced gemstones published in the fifth edition of *Gems* (1994) (but see also O'Donoghue and Joyner (2003: ISBN 0750655127)) for gemstone identification in general and O'Donoghue, *Synthetic, Treated and Imitation Gemstones*, 1997: ISBN 0750631732) for the eponymous products, I have looked closely at my account in the fifth edition of *Gems* and decided to keep some at least of the historical account unaltered. At the end of the chapter I shall produce some exotic species whose crystals will one day be sought by collectors! My recent book, *Artificial Gemstones* 2005: ISBN 071980311X, covers some products from a different standpoint.

Since the issue of *Gems,* fifth edition, major developments in synthesis have been in the field of diamond. Today synthetic diamonds of good quality are on the general market: this does not make the task of the gemmologist any easier when combined with HPHT-treated stones. In fact the work of the gemmologist will probably become, at least in the case of diamond, that of spotting stones which have to be sent elsewhere for testing. In addition, pink and green synthetic moissanites cut to resemble diamond are reported to be on the market.

Alan Collins in Chapter 4 gives a summary and some points are repeated here. The following study has been developed from the one compiled for the fifth edition of *Gems*. There have been no 'new' synthetics in the old sense of fresh products appearing to take the market by storm: the only newcomer said to have aroused the traditional 'near-panic in the trade' has been the diamond simulant synthetic moissanite, specimens of which can easily be identified. Some materials which look very beautiful when cut have not caught on, largely due to the scarcity of waste supplies from laboratories and processing plants working with them. Many such materials are soft but highly dispersive; the germanates are a good example. Proustite, quite magnificent when faceted, is likely to show surface alteration in brightly-lit display conditions. It is pleasant to review the production of crystals with optical and physical properties suitable for gemstone use if we start with the 1960s, a period during which a number of such materials were produced by Bell Laboratories (in particular) and others. This was a time during which the synthetic garnets began to appear in crystals large enough to cut and in which very large quartz crystals were grown. Interestingly, some of the crystal growth literature of the 1960s and 1970s is almost as difficult to find as some of the experimental specimens: print runs can never have been large and the second-hand market is often a desert for those seeking them. Today's crystal growth literature does not cover quite so many macroscopic crystals and access to periodicals like *Journal of Crystal Growth* depends on sets being available in the largest university and national libraries. Fewer libraries can afford the quite amazingly high subscription rates asked today. Online access would appear to be the answer. At the time of writing, *Mineralogical Abstracts*, which carries a section on Experimental mineralogy (covering synthetic materials), is on line.

Many of these rare materials are sought by collectors who in the past may have worked in research laboratories where wastage could be high. I have seen waste bins with beautiful crystals of spinel-group specimens doped to give colours not found in Nature and cupboards with large proustite crystals awaiting an unknown destination. They were all rescued.

Present thinking and commercial rules restrict the adjective 'synthetic' to materials which, while being manufactured by man, have a natural counterpart, however unornamental this might be. The adjectives 'imitation' and 'simulant' can refer to any substance, natural or man-made, which can be mistaken for another which can itself be natural or artificial. As many man-made gemstones are expensive and many are rare, they have themselves become sought by collectors and there have been instances of a natural stone being offered as a synthetic!

As a natural counterpart must exist, a synthetic stone will have the same chemical and physical properties and in fact there is only a small variation in these over the whole of the gem species. Generally speaking, synthetic materials are cheaper, larger, clearer and more spectacular than most natural stones. Gem-quality crystals may be grown by methods which echo the work of natural processes or which have no natural counterpart. Many of the more expensive gem species are grown from the melt, but some of the cheaper ones are too; others are grown from a solution, fewer by chemical precipitation, and some by growth from the vapour.

Diamond

Readers should first consult the relevant parts of Chapter 4 on Diamond, by Alan Collins, for details of the latest developments in synthetic and treated stones and how they may be identified. What follows here is a summary of what will be reasonably well known to practising gemmologists but will also be a useful introduction for the beginner as no up-to-date textbook on diamond as a whole is available at the time of writing. Amanda Barnard's survey of the manufacture of diamond is invaluable (*The Diamond Formula*, Oxford, Butterworth-Heinemann, 1999) and Robert M. Hazen's *The Diamond Makers* (1999: ISBN 0521654742) is a shorter and less technical account. ALL are drawn upon here.

Both diamond and graphite are polymorphs of carbon, but of the two it is graphite which gives the stable form at ordinary pressures. Diamond becomes stable at high pressures but, as it will not convert to graphite below about 1500 °C, it is safe enough for wear! Diamond can be described as metastable.

The very high mechanical strength and hardness of diamond are due to its bonding. All bonds are covalent and equivalent, with the only direction of weakness in the octahedral plane, which gives rise to diamond's easy cleavage in this direction. For successful diamond synthesis such a structure has to be reproduced, and very high pressures are needed. Even when this problem is solved the grower still has to be able to achieve high temperatures to enable rapid movement of carbon atoms from other carbon polymorphs to diamond. All three problems having been solved, the grower wanting large stones is then faced with further tests of ingenuity. The apparatus within which suitable diamond synthesis can be achieved needs to be very strong, and capable of withstanding extremely high pressures at which most materials will break down, so the resources of high-pressure technology are stretched.

Over the centuries from Tennant's discovery in 1797 that diamond is pure carbon, many attempts, some involving colourful personalities and eccentric techniques, have been made to grow diamond. Quite early on it was realized that pressure would be needed to convert a substance with an SG of 2.25 (graphite) to one with an SG of 3.52 (diamond). One notable experimenter was James Ballantyne Hannay (1855–1931), a Scottish chemist working in Glasgow. By sealing a mixture of lithium metal, bone oil and paraffin into a heavy wrought iron tube and heating the tube to red heat for some hours, he found at the end of three runs (in all the rest the tubes exploded) that diamonds of diameter about 0.25 mm were present. Several investigations have been undertaken into the Hannay product; the strongest probability is that the undoubted diamonds found in the tubes were natural rather than synthetic. Attempts to reproduce Hannay's results by his methods have been unsuccessful.

The experiments by Ferdinand Frédéric Henri Moissan (1852–1907) involved heating iron with sugar charcoal in a carbon crucible in a furnace; temperatures up to 4000 °C were achieved. Carbon dissolves in the iron, and Moissan solidified the molten iron by plunging it into cold water, thus solidifying it from the outside inwards. Excess carbon was subjected by the process to very high pressure. The iron was dissolved by acids and the residue sank in di-iodomethane (SG 3.32). Moissan's claim to have made diamond was both refuted and confirmed by other workers of the time, but today it is thought that his diamonds were natural specimens placed in the apparatus by a co-worker; he could not have achieved high enough pressures at the right temperatures to make diamond.

Of the many other attempts to fabricate diamond the only work of importance is that of the steam turbine inventor Charles Parsons. He tried both Hannay's and Moissan's experiments, adding some ingenious and colourful ideas of his own. One of these involved firing bullets into materials containing carbon and heating carbon rods embedded in a mixture of quartz and calcium oxide under pressure. In fact, Parsons realized that his product was colourless synthetic spinel; in 1928 he issued a disclaimer via another worker. At the close of 1954 there was no real proof that diamond had been synthesized. As early as 1941 Nickle referred to experimental research on diamond synthesis at the Schenectady Laboratories of General Electric (GE). Little more was heard of these experiments until 1955 when the company published information on the synthesis of minute diamonds at Schenectady. One-carat synthetic diamonds are reported to have been made, but of industrial quality only.

Graphite to Diamond

In 1947, Bridgman contributed valuable information on considerations of the thermodynamic equilibrium between graphite and diamond, and carried out considerable work on high pressures and temperatures. Bridgman obtained high pressures in his experiments but found difficulty in obtaining high enough temperatures simultaneously with the pressure, and for a sufficient length of time. The success of the GE team (Bundy, Hall, Strong and Wentorf) was due mainly to their overcoming this difficulty. By using a 1000 ton press and special heat and pressure resisting chambers, pressures up to 100 000 atmospheres (1 atmosphere = 101325 Pascals (Pa) or N/m^2) and temperatures up to 2760 °C were produced. An undisclosed carbonaceous compound was used to supply the carbon. Although the results of their experiments were only minute crystals, these are now made in quantity. The product is available commercially as diamond powder and as industrial abrasive grits.

The heart of Bridgman's work lies in the use of an 'anvil', in fact two anvil-shaped pieces of cobalt-bonded tungsten carbide; pressure is exerted on the sample which is contained in a catlinite container. This material squeezes out under pressure into a thin gasket adhering to the anvil; it does not thin out any further. The anvil is placed in a hydraulic press which allows pressures up to 6 million psi to be achieved. In present-day apparatus a special Bridgman hydraulic seal, the catlinite gasket, and a special technique (massive support) by which the strength of the anvils is increased greatly are retained albeit with modifications. Although Bridgman was able to achieve high pressures, the complementary high temperatures required could not be attained; he succeeded only in reaching 3 GPa (3×10^9 Pa, which approximates to 30 Kb, 30 000 times atmospheric pressure) at temperatures in the range 1000–20000 °C. A group of workers at GE began work in 1951 on the high-pressure synthesis of diamond. Hall and Wentorf (physical chemists) and Bundy and Strong (physicists) were joined by the engineers Cheney and Bovenkerk. Nearly 6 GPa were obtained with high temperatures. An innovation by Hall in 1953 gave the project the clue to its final success. This was the belt, which allowed 18 GPa to be attained at a simultaneous temperature of 5000 °C. Massive support and a non-extruding gasket were used as with the Bridgman apparatus, though catlinite was substituted by pyrophyllite as the container material. As pyrophyllite is an insulator it allows an electric current passing from the upper to the lower anvil to pass only through the graphite sample which is contained in the pyrophyllite tube.

Strongly tapered pistons give a large motion while keeping the gasket relatively thick, and the compressed pyrophyllite in turn enhances the pistons' strength. The largest diamond crystal achieved by these runs, which succeeded in producing diamond in 1954, was about 150 microns across and was octahedral. In reviewing the results it turned out that iron sulphide, the mineral troilite, had been added to the graphite; this echoed the diamond-troilite association found in the Cañon Diablo meteorite. Later, sulphur was found to be superfluous but iron necessary for success. Much later the presence of a catalyst was abandoned with no detriment to diamond growth. In 1955, the results from these experiments enabled Berman and Simon to plot a diamond-graphite equilibrium curve indicating the range of temperatures and pressures necessary to convert graphite to diamond.

These accounts so far have produced no gem-quality diamond crystals. The first announcement of this major achievement, again by GE, was not made until 1970, with the first descriptive paper published in 1971. Before that time further developments in diamond growth were assisted by processes and apparatus largely the work of Hall. From the belt he developed the tetrahedral press to give the same range of temperatures and pressures as the belt: instead of the belt and two anvils the tetrahedral press has four anvils so shaped that when they impinge upon a pyrophyllite tetrahedron about 25% larger than the final area, some of the pyrophyllite is squeezed out, thus forming a non-extruding gasket. This in turn lets the pressure rise as the pistons continue their inward motion. A graphite tube containing the sample is placed across the pyrophyllite tetrahedron and so positioned to allow passage of an electric current to heat the tube and sample. Further developments of the press and of the belt have taken place over the years. The successful growth of diamond grits by high-pressure methods has continued since the 1950s using hydraulic presses with the belt apparatus; virtually any form of carbon, including peanuts, has been pressed into service!

Crystals of diamond large enough to provide material for faceting need very slow growth under the most rigorously controlled conditions while using the belt or tetrahedral press apparatus. So far the cost would seem to outweigh the benefits of large-scale production, but crystals grown for laser windows or for high-power semiconductor heat sinks can and no doubt have been cut and polished. The best material for conducting heat is type IIa diamond, so these have a market from which there may be some 'stray' faceted stones.

The pyrophyllite tube described above contains seed crystals which may be thin pieces of natural diamond or synthetic diamond. Feed material may be grit or graphite. The ends of the tube are nickel plates; as heating gets under way, part of the nickel melts and dissolves the graphite which crystallizes as diamond. The centre of the tube is hottest, so the diamond is more soluble in the molten metal there than at the seeds placed at the ends. Transport from the centre to the ends gives a growth time of approximately one week to give a 5 mm gem-quality 1 carat diamond. The growth method and the arrangement of the pyrophyllite tube are similar to those used for the production of grit in the belt apparatus, but for the use of seeds.

Many years after the event, workers at the Swedish Allemanna Svenska Elektriska Aktiebolaget (ASEA) announced that they had produced synthetic diamonds in 1953. The report, describing apparatus built by von Platen, cited work carried out by Liander and Lundblad. The product was not of gem size or quality. Diamond crystals grown by both CE and De Beers show cube and octahedron as predominant forms, with the dodecahedron sometimes seen (*Figure 24.1*). Up to 40 faces have been reported on a crystal 5 mm in size. The first production gave colourless, canary-yellow and pale blue colours, and apparently no difference was detected by polishers between these and natural diamond crystals. By excluding nitrogen a good white resulted, and blue arose from the incorporation of boron. Dust-like inclusions with some rounded or plate-like nickel inclusions can be seen under high magnification. Some blue stones show a whitish cross. All except the yellow stones are semiconductors, a property shown only by natural blue stones. No absorption spectra have been reported and all are inert to LWUV. Under SWUV a variable response has been noted. Of eight stones tested by GIA in 1984, three cut stones and five crystals responded to SWUV in ways with some testing application. The cut stones consisted of one near-colourless, one bright yellow and one greyish blue. The largest crystal was near-colourless, two were bright yellow and one was greyish blue. The near-colourless cut stone gave a very strong yellow fluorescence with very strong persistent phosphorescence of the same colour. The yellow cut stone and crystals were inert. The greyish blue cut stone and crystal showed very strong slightly greenish yellow fluorescence with very strong persistent phosphorescence of the same colour. The near-colourless crystal gave a strong whitish yellow with long phosphorescence of the same colour.

Figure 24.1

Only blue natural diamonds will phosphoresce after exposure to both types of UV. X-ray fluorescence results were identical with those obtained under SWUV. Blue and near-colourless crystals were electrical conductors, as were the similarly coloured cut stones; yellow cut stones and crystals were not. GE stones could all be lifted by a pocket magnet, especially those with metallic inclusions. GE stones could be distinguished from natural diamonds using a superconducting magnetometer. Some useful tips for identification suggested by GIA are as follows:

1. Near-colourless stones conducting electricity.
2. Near-colourless stones fluorescing and phosphorescing strongly under SWUV but inert to LWUV.*
3. Fancy yellow stones with no absorption spectrum and inert to LWUV.
4. Near-colourless stones with no hint of blue, grey or brown and with no absorption at 415.5 nm.
5. Stones with strong yellow fluorescence under X-rays and with strong persistent phosphorescence.
6. Any yellow fluorescence under X-rays.

*In 1994 Russian-made gem-quality yellow diamonds were reported to fluoresce under LWUV.

The Japanese firm of Sumitomo manufactures yellow diamonds for non-gem purposes, but some crystals have been cut and polished. The crystals, up to 0.40 carat, grow as distorted octahedra, with modifying cube and dodecahedral faces. They are inert to LWUV but under SW they give a moderate to intense green or greenish yellow response. There is no phosphorescence, but there is a weak to moderate X-ray fluorescence coloured bluish white. No absorption bands can be seen. Stones are type Ib, as are GE yellows. They do not conduct electricity and are similar to natural stones in their heat conductivity. Vein-like colourless areas are characteristic, as are opaque black flux inclusions. Graining can be seen on the surface by reflected light, this effect sometimes surviving polishing. The interference pattern is distinctive, resembling a bow tie, with the four arms of the cross-shaped pattern either coinciding with or at 45° to the direction of the radiating internal grain lines. Stones faceted from Sumitomo crystals did not show the pattern.

Two Sumitomo gem-quality diamonds examined in 1990 showed how the manufacture of these crystals has developed over the decade. One crystal appeared to have been produced from an originally cube-shaped crystal. It was brownish yellow with visible areas of brown graining; it had a central brownish yellow area free from graining, with narrow, very light yellow to colourless to blue zones beneath the corners. Metallic inclusions were present. The second crystal examined was probably cut from an originally octahedron-shaped crystal. In reflected light dendritic patterns were observed on some faces.

Both samples are inert to LWUV, but show a weak orange-yellow response under SWUV. Natural type Ia stones either are inert or fluoresce yellow, blue or green to both kinds of UV, and some variation of luminescence exists among natural yellow type Ib diamonds. Examination of a number of natural type Ib and mixed type Ib and IaA stones showed that there is some overlap between the luminescence of the synthetic product and some natural type Ib stones. However, all synthetic diamonds so far tested were inert to LW and fluoresced under SWUV. In one of the samples the colourless centre is inert to SWUV while the yellow outer areas fluoresce weak yellow in planes paralleling octahedral faces. A stronger yellow fluorescence under the corners appears to be banded parallel to dodecahedral faces. Fluorescent areas phosphoresce a weak yellowish white lasting about 10–15 sec. This uneven distribution of fluorescent areas shows the presence of both cubic and octahedral internal growth sectors. Growth zoning appeared to be typical of this product in 1990.

Gem-quality synthetic diamonds weighing up to 11 carats have been grown by the De Beers Diamond Research Laboratory since the 1970s. Stones, not so far placed on the market, are brownish yellow, yellow and greenish yellow with metallic inclusions, uneven colour distribution and geometrical patterns of graining. Stones are inert to LWUV under SWUV but brownish yellow specimens give a moderate to strong yellow or greenish yellow, and greenish yellow stones fluoresce weak yellow. Only the latter show phosphorescence. Both this effect and the fluorescence are pronounced in the outer portions of the greenish yellow crystals. A zoned pattern seen when stones are exposed to an electron beam indicates growth sectors, and this feature of cathodoluminescence is useful rather than diagnostic when testing De Beers diamonds. It is possible that the more apparent properties of all synthetic gem diamonds so far examined (colour zoning in particular) correlate with an observed higher nitrogen content.

Cathodoluminescence (CL) tests show up a number of features in the synthetic diamond made by De Beers. Internal features show up well, and those stones which luminesce display distinct growth sectors. Greenish yellow CL of high intensity is emitted from cubic and octahedral growth sectors, the spectra giving peak wavelengths between 534 and 526. The predominating CL colour of natural yellow diamonds was blue, but their spectrum differs from that shown by blue-luminescing synthetic diamonds. Some of the De Beers stones (3 out of 10 tested) contained inclusions which showed orange-red under electron bombardment. The inclusions showed a distinct band in the near IR at 820 nm.

At this point the reader should consult the chapter on Diamonds by Alan Collins, Chapter 4.

Corundum

The first synthetic ruby was announced in 1902 but in the preceding years, at least since the 1850s, work had continued on the synthesis of corundum. The name of Verneuil is lastingly associated with ruby synthesis but others also worked towards the final successful production.

General Description of Early Synthesis Work

The factor which gives gem-quality and gem-sized corundum is the correct choice of feed powder: the alpha form of Al_2O_3 with Cr_2O_3 added to give the ruby colour and other dopants added to give sapphires of various colours. Around 2.5% of chromic oxide is needed to give a satisfactory ruby colour. One problem is that any other substance which may get into the growing crystal may discolour it; this is particularly true of iron, which gives a brownish tinge. Another problem is the mechanical stability of the grown crystal or 'boule'; this may crack if some impurities are present. Verneuil used ammonium alum to begin the preparation of his feed powder: $(NH_4)Al(SO_4)_2 \cdot 12H_2O$ is dissolved in distilled hot water with subsequent filtering to remove solid matter. On cooling, crystals of alum form from the solution and any impurities are left in it. After this process has been repeated several times the purity is high enough to allow the powder to be used in the crystal growth process. Chromium alum $(NH_4)Cr(SO_4)_2 \cdot 12H_2O$ is purified in the same way to give the dopant for ruby. The two alums are mixed and fired at a temperature of about 1000–1200 °C to give the final feed powder. Other components are vapourized at the high temperature. The powder needs to flow freely and is sieved before use. The method of preparing the feed powder is the same today.

The feed powder is melted to give a characteristic boule which is a single crystal. The smooth flow of feed powder is ensured by tapping the powder container with a hammer, and the heating is achieved by a very hot flame, produced from an oxygen stream and a stream of coal gas which was available in Verneuil's time. The powder drops down through the flame. The flame touches a ceramic pedestal upon which the powder builds up a sintered cone whose tip melts and enlarges. Originally, the pedestal was lowered to give a start to the formation of the boule, but today a small seed is attached to the pedestal which promotes epitaxial growth of the boule; the gas flow is maintained while the oxygen flow is monitored to control growth. On shutting down the flame the boules tend to crack through thermal shock; the longitudinal splitting can be done by nipping with pliers to provide boule sections for the lapidary. The boule of the early period would have yielded several cut stones up to 2 carats; most were cut in sizes of 0.5–1.5 carats.

Identification of early Verneuil rubies was by curved growth striations and by gas bubbles. Verneuil himself stated that the flame needed to be rich in hydrogen and carbon as this stopped the formation of bubbles in the melt; transparency is achieved by allowing slow growth in layers from the bottom of the boule. To prevent the formation of many crystals, and to stop cracking of the contact area of the sintered cone or seed, the boule has to be kept small.

Though the name of Verneuil is inseparable from the story of corundum growth, other workers were active in the nineteenth century, some of them with similar ideas to Verneuil. In 1817 Gay-Lussac reported that pure aluminium oxide could be obtained by heating ammonium alum. This allowed progress towards the final successful corundum growth, although a hare was started by the transient belief that silicon oxide was present in corundum. This rumour was soon laid, however, and Gaudin published his first report on ruby synthesis in 1837. By 1870, the date of his last report on the topic, gem-quality ruby had still not been grown. By using a downward-pointing flame, Gaudin obtained hexagonal platelets of corundum which were modified into ruby with the addition of a chromium salt. His product contained profuse bubbles which lowered the SG. Other workers found that the high melting point (2050 °C) made growth difficult. One idea was to dissolve Al_2O_3 with Cr_2O_3 in a high-melting solvent to give a saturated solution. This was heated, allowing the solvent to evaporate and permitting the ruby to crystallize from the solution. By using borax as a solvent, hexagonal platelets of ruby were obtained. This solvent was later termed a 'flux' and became the basis of the important flux-melt method used today for high-quality ruby and emerald. The work of Edmond Frémy is important because he was the first to produce clear red, though small, rubies. Verneuil became Frémy's laboratory assistant and worked with him for about 16 years on ruby synthesis.

Frémy's previous personal assistant, Feil, died in 1876 but he had been junior author of a paper with Frémy on ruby growth. This was published in 1877. By using large fireclay crucibles with 20 kg batches held at red heat for 20 days, thin ruby crystals were obtained. Lead oxide Pb_3O_4 reacted to form lead aluminate, which then reacted slowly with silica to form lead silicate with aluminium oxide crystallizing as sapphire plates. Potassium dichromate up to 3% gave the ruby colour.

Frémy and Verneuil worked together for 16 years after Feil's death. Frémy's book *Synthèse du rubis*, published in 1891, is the best account of their work; it is now a very rare and valuable item, containing what now seem quaintly coloured plates. Their final effort produced very clear rhombohedral crystals ranging from colourless to red, violet or blue with some giving red at one end and blue at the other. They were about 3 mm in diameter and up to one-third of a carat in weight. The process turned on the recrystallization of alumina with some potassium dichromate by using potassium hydroxide and barium fluoride. Temperatures of 1500 °C were achieved in a ceramic crucible. Nassau (in *Gems Made by Man*, 1980) considers that alumina reacts with the barium fluoride to give gaseous aluminium fluoride which then reacts with moist air to re-form alumina at a constant temperature. Vapour-phase nucleation and growth then take place in small cavities scattered in the porous mass which fills the crucible. This is possible because the reactions take place in a multiphase medium. Thus the process gives multiple nucleation with only a small amount of growth on each nucleus. It is not strictly flux growth as the phrase is understood today. Many of the crystals were used as watch bearings, though some unfaceted crystals were used in jewellery. The retirement of Frémy in 1902 signalled the end of this particular line of work until the 1950s.

The 'Geneva rubies' produced in 1885 were first thought to be made from fragments of natural rubies fused together. With properties the same as natural rubies but with profuse round gas bubbles, the stones were ruled artificial by the French Syndicate of Diamonds and Precious Stones. Many stories circulated about the stones, as in the following example from a trade paper of 1890:

> A Berlin jeweller has just been the victim of a curious hoax. He recently received a circular from a Zurich firm offering rubies at remarkably cheap rates, and therefore entered into negotiation for the purchase of some. He bought rubies for which he paid 4500 marks (£225), receiving a guarantee from the firm that the stones were genuine. Shortly after, the jeweller heard that false rubies were being manufactured so cleverly as to deceive the connoisseur, and, becoming alarmed, sent those he had purchased to Paris to be examined by the Syndicate of Dealers in Precious Stones, who are considered unimpeachable authorities. They reported that the stones were not imitation, but were real rubies, which were small and consequently of little value, fastened together so cleverly as to render detection difficult. The jeweller then wrote to Zurich requesting the firm to take back the stones. This they refused to do on the ground that their guarantee only ensured the genuineness of the gems and contained no mention of their size.

Examination by Nassau showed that the Geneva ruby is made in three distinct growth stages. This was deduced partly by the high Cr and Fe content of the 'skin' of the boule forming when the gas supply ceases and the melt solidifies. With this observation as a background it was deduced that a tiny boule was first grown on a sintered cone, broken off and grown further in a different direction. The bulk of the final boule was then grown by the use of two torches on a rotating support. The stones sold until the Verneuil process out-performed the Geneva process. The term 'reconstructed' or 'reco' was used loosely in the trade for synthetic rubies for many years. A final blow to the 'reconstruction' legend was Nassau's discovery of the notably low Fe content of the Geneva material and the highly purified aluminium oxide used. This would not be the case if natural ruby was the starting material. An attempt to fuse ground natural rubies in a Verneuil torch produced an orange-brown opaque material with high iron content. Rubies were grown in 1903–4 in Hoquiam, Washington, USA, the technical expertise being provided by one of Verneuil's assistants from Paris. They appear to have been grown by the early type of Verneuil apparatus and contain profuse gas bubbles; boules are unevenly coloured and small. Highly purified feed powder was used; the source of the gases needed is uncertain, but Nassau (1980) suggests that they may have come from the electrolysis of water into hydrogen and oxygen. Geography and cost led to the ultimate failure of the venture.

The Verneuil Furnace

In 1891 Verneuil deposited with the Paris Academy of Sciences a sealed account of a new and revolutionary process of ruby synthesis, later published in November 1902. He described the construction of an inverted oxyhydrogen blowpipe or chalumeau which is still used (*Figure 24.2*).

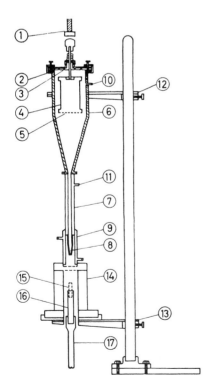

Figure 24.2   The Verneuil furnace: (1) hammer, (2) rubber gasket, (3) corrugated metal diaphragm, (4) powder hopper, (5) sieve, (6) main body of the furnace forming the expanded top of the inner tube, (7) outer annular tube carrying the hydrogen, (8) removable nozzle to inner tube, (9) water jacket, (10) oxygen inlet, (11) hydrogen inlet, (12) and (13) brackets for securing furnace to upright, (14) divided fireclay chamber, (15) window in fireclay chamber, (16) candle upon which the boule grows, (17) candle support connected to centring and raising and lowering adjustments (not shown)

The inverted furnace consists of two iron tubes (6) and (7). Tube (6) is widened into a circular chamber at the top, the lower and thinner end passing down the centre of the tube (7) into which it is tightly screwed so as to form a gas-tight joint. Both tubes are constricted to nozzles at the lower ends, that of (6) being situated close to but inside and above that of (7). Pure oxygen under pressure is admitted to the tube (6) through the pipe (10). Within the upper chamber of (6) is placed a small hopper (4) with a fine wire-mesh bottom (5) which is rigidly fixed to a metal rod which passes upwards through a block of resilient rubber (or a metal diaphragm) (3). At the top of this rod is an anvil head which is periodically struck by the hammer whose beats can be controlled.

Hydrogen under pressure mixes with the oxygen at the orifice, where it is ignited. Owing to the intense heat generated it is necessary to protect the pipes against fusion, and this is done by placing round them, just above the nozzles, a water jacket (9) with cold water continually running through it (with modern metals and construction this water jacket is omitted). Below the orifice is placed a fireclay support (16) which is carried on an iron rod (16) connected to a screw adjustment, whereby the candle may be centred below the orifice of the furnace; it may also be raised or lowered into or out of the flame as the boule grows. 'Boule' is the name given to the pear-shaped single crystal which forms on the candle: the name is derived from the French world for ball, for the small specimens first made by Verneuil were ball-shaped. To protect the growing boule from cold draughts and to maintain a regular temperature around it, a divided fireclay chamber (14) is provided.

Growth of Crystals by the Verneuil Process

In the manufacture of synthetic stones by the Verneuil process the gases and chemicals used must be of exceptional purity and have certain physical attributes. Today most factories manufacture their own gases and the chemical powders. To make corundum a fine powder of alumina as the feed powder is prepared by calcining crystals of ammonium alum in silica trays in a muffle furnace at 1100 °C for 2 hours. If ruby is required, about 8% chromic oxide is added to the ammonium alum before calcining. The charge swells to a meringue-like cake of anhydrous alumina. The cakes are tumbled until a fine powder is achieved.

The powder is placed in the hopper (4) at the top of the blowpipe and shaken through a sieve. The hydrogen is turned on and lit at the nozzle, and the oxygen is then admitted, which with the already ignited hydrogen forms the intensely hot oxyhydrogen flame. The hammer (1) is then started, at about 80 taps per minute, causing the release of a sprinkling of the alumina powder at each tap. This powder travels down the oxygen stream and fuses as it passes through the hottest part of the flame.

The molten powder is caught on the top of the ceramic candle (15), which is about 20–25 mm in diameter and is situated in the cooler part of the furnace. This fused alumina solidifies on the candle in the form of a small cone consisting of a number of small crystals of corundum. The tip of this cone is in the hotter part of the flame and remains molten. By manipulation of the speed of tapping down of the powder, and the amount of powder delivered at each tap, which can be controlled by altering the distance of the head of the peg from the hammer, and with control of the gas supply, a single central crystal grows upwards in stalagmitic fashion and opens out to mushroom shape, after which the speed of tapping is reduced in order to keep the boule diameter to about 20 mm. The rate of tapping, for ruby, is about 20 times per minute. When the boule reaches a suitable size, about 65 mm in length for ruby, and weighs from 150 to 200 carats, the gas supply is cut off and the boule is allowed to cool on the furnace. When cool it is removed from the furnace and broken away from the candle and from the fritted cone of crystal aggregates. The stem of the boule is nipped with a pair of pliers or given a slight tap with a hammer, which causes it to split into two nearly equal pieces with fairly smooth faces.

Different manufacturers have used slight differences in technique, but the essentials are the same, and there is little difference between the original blowpipe apparatus used by Verneuil and that used today.

Some of Frémy's crystals had shown a violet colour. Although a different valence state of Cr was suspected at first, a report of 1908 by Verneuil's former student Paris stated that blue sapphire crystals had been grown, coloured blue by cobalt and with magnesium oxide added to the feed powder.

Verneuil grew a number of experimental crystals with various compositions and found that Frémy's violet crystals contained iron and other elements with the chromium and that cobalt gave an uneven blue. Paris's crystals were spinel rather than corundum. When Verneuil grew corundum with added titanium and iron, blue sapphire was obtained. Intervalence charge transfer between $Fe^{2+}Ti^{4+}$ and $Fe^{3+}Ti^{3+}$ is the cause of the blue colour.

During the Second World War the US government encouraged the Linde Air Products Company to grow Verneuil ruby and sapphire in anticipation of the loss of European supplies. After the war European production was once more available; the crystals were cheaper owing to lower labour costs, and the Linde operation was no longer viable. Star corundum was made by Linde from 1947 and later the demand for laser rods helped to keep production going. Work on Verneuil corundum ceased in the USA in 1974.

Colours of Transparent Synthetic Corundum

Table 24.1 gives a fairly complete list of the colours most commonly made in transparent synthetic corundum, complete with the manufacturers' trade names. In addition, some parti-coloured synthetic corundum has been made in America with blue sapphire one end and ruby the other. I have seen similar specimens grown in Japan but not put on sale.

Star rubies and blue sapphires are made by adding 0.1–0.3% TiO_2 to the feed powder. The boule is grown in the normal way and then reheated to about 1300 °C. This causes the previously disseminated rutile to take up needle-like forms in six (sometimes twelve) rays at right angles to the longitudinal axis of the boule. Stones are cut with the cabochon top coinciding with the top of the boule. Stars in synthetic corundum are notably perfect with bright, even rays reaching the edge of the stone, and the backs are flat and smooth. Early stars were imperfect with uneven rays and areas of transparency caused by the migration of titanium to the edge of the boule. The trouble was cured by varying the oxygen flow to the flame at even intervals. This allowed each deposited layer to solidify on the growing boule before the next was deposited, thus holding the rutile in place. Diffusion of a slurry into the stone which is heated for 24 hours at 1800 °C can also give a star which is about 0.1 mm inside it. The slurry is made from two parts of alumina to one of rutile.

Table 24.1
Colours of transparent synthetic corundum

Colour	*Manufacturers' trade name*
Colourless	White Sapphire
Red	Ruby
Dark red	Garnet Colour
Deep pink	Rosaline
Pink	Pink Sapphire
Lilac-pink	Rose de France
Orange	Padparadscha
Deep yellow	Danburite
Yellow	Topaz
Yellow-brown	Madeira Topaz
Brown	Palmeira Topaz
Green	Green Sapphire
Pale green	Amaryl
Blue	Burma Sapphire
Purple	Amethyst
Purple/green	Alexandrite

Physical and Optical Properties

Inclusions

The physical and optical characteristics of synthetic corundum are those of natural stones. However, owing to their different mode of formation, inclusions are enough to identify them.

Synthetic rubies and sapphires, owing to the shape of the top of the boule and to the discontinuous feed of the powder, show curved structure lines when examined in the correct direction under a microscope. In the case of ruby these lines have been aptly likened to the grooves seen on a gramophone record. With blue sapphire the bands are wider and more diffuse and these are often better seen if the stone is placed in a highly refractive oil in a white cell and the stone viewed in different directions, as under the microscope the bands may be so wide and diffuse that they may not be seen at all clearly.

Included gas bubbles, which may be round or take characteristic 'flask' or 'tadpole' shapes, are common and give positive evidence that the stone is synthetic. Gas bubbles, whether in the Verneuil product or in natural/artificial glass, invariably show a notably bold outline and random distribution.

Clouds of very small bubbles, looking like black dots, are often seen and these sometimes tend to follow the direction of the curved bands of colour.

In natural rubies and sapphires the inclusions are largely mineral and characteristic liquid formations. Sometimes the inclusions in synthetic stones may be most deceptive, for undissolved alumina can produce natural-looking inclusions and in some cases triangular cavities containing a gas bubble. Careful observation will generally reveal that these deceptive inclusions are in conjunction with tell-tale normal gas bubbles and curved colour lines. Twinning lines in parallel planes usually indicate a natural stone, but these too have been seen, although rarely, in synthetic corundum. Plato has reported that under certain conditions some synthetic corundums show faint zonal lines with angles of 60° and 120°, but these lines are difficult to see and differ from the zonal lines observed in natural stones. To do this the stone should be immersed in di-iodomethane; find the optic axis and, using a magnification of 20 to 30 times, examine the stone in directions parallel to the optic axis. These lines are not normally seen in natural corundum.

Absorption Spectrum

The absorption bands seen in synthetic ruby are the same as those in natural ruby but, as always, the elements, compared to those of the natural material, will appear strong and well defined: this applies in particular to the lines in the blue which can sometimes be difficult to see.

In natural blue sapphire the iron lines, three bands at 450, 460 and 471 nm, are usually visible in the absorption spectrum. Sometimes only a fine but sharp line may be seen in fine blue sapphires at 450 nm, while with the greenish blue and greenstones the three bands are strong. In synthetic blue sapphire the line at 450 nm does not normally show, and although a weak smudgy line may sometimes be seen about this position, it is nothing like the sharp line of the natural stone.

The synthetic sapphire made to imitate alexandrite provides a case where the absorption spectrum is so markedly different that identification of the synthetic stone is easily accomplished by the spectroscope alone. There is an absorption line in the blue at 475 nm and there are no fine lines in the red as seen in the spectrum of the true alexandrite. This line is due to V^{3+} added to the crystal at a level of about 3%. It is rarely if ever seen in a natural corundum, and then the colour of the stone would be different. The 475 nm line has, however, been observed in a synthetic corundum of green colour. Some synthetic green sapphires show an absorption line at 688 nm which may be due to nickel.

Fluorescence

Response to UV may be helpful in identification. The glow from synthetics in a mixed parcel of natural and synthetic rubies will be perceptibly stronger. Synthetic blue sapphires under SW may give a bluish green response and, with care and turning the stone, the curved bands may be seen as bands of different intensity of light in some cases where the bands have not been observed under lens or microscope. A low-power lens is useful in these circumstances.

Synthetic corundum transmits UV much more freely than natural corundum. Natural corundum usually absorbs UV below about 290 nm, while synthetic ones generally transmit quite freely down to 220 nm. With the aid of a spectrograph the absorption edge can be measured, but an easier and more practical method for identification purposes is to use a SWUV lamp whose main emission at 253.7 nm is between the absorption edges of the natural and the synthetic corundum, in conjunction with photographic paper. The suspected stone is placed, with controls, on the bare surface of a slow photographic printing paper, which is immersed in a dish of water in a darkened room and exposed for a few seconds to the rays from a SWUV lamp placed some 38 cm above the dish. The paper is then developed. As the UV rays will be transmitted through the synthetic corundum and not through the natural, the paper will be darkened to some extent under where the synthetic stones have rested but not so under the natural stones.

As the SW lamp emits some LWUV rays the exposure time is critical and must be gauged by the use of control stones. Particular care is needed with blue sapphires in which the method is not always certain.

Pink, Yellow, Orange and Brown Corundum

These are virtually inclusion-free. When irradiated by X-rays such stones, except the brown, in common with rubies, show pronounced phosphorescence. Natural corundum does not show this afterglow as iron is usually present. Synthetic pink sapphires glow violet under SWUV while natural pinks give a crimson glow. Immersion of a difficult stone may help in identification. Di-iodomethane with an RI of 1.74 is useful. The specimen can be placed in a glass cell standing on a piece of fine-grain photographic film. From a nearly parallel light source held at some distance above (an enlarger with the lens stopped down makes an ideal source) a short exposure, about 15 s, is given. On development of the film curved lines may be visible. They may not be seen otherwise.

Green Sapphire

In the Summer 2004 issue of *Gems & Gemology* a transparent faceted light green sapphire of 8.85 ct is reported as owing its colour to a combination of trivalent cobalt and trivalent vanadium. The presence of these elements was detected by EDXRF spectroscopy.

Some ruby has been grown hydrothermally (details of the process are given in the section on quartz later in this chapter) but although this growth method parallels natural growth it has not been found economic to continue it on a commercial scale. Vessels are lined with silver or platinum to avoid a green colour from introduced iron. Most experimental crystals are small and heavily included. NaOH is added as a mineralizer and sodium bichromate gives the colour.

Ruby and some sapphire have also been grown from the vapour phase but the slow growth makes the process uneconomic. A compound such as aluminium trichloride is reacted with an oxygen-containing gas in one such process.

Both these growth methods produce stones with a strong phosphorescence under X-rays: a liquid veiling of tiny particles may show in the hydrothermal product, to some extent resembling the wisps and lacy feathers of some flux-grown rubies. There are no mineral inclusions and separation from natural stones is not difficult. Some hydrothermal stones were grown on seeds of Verneuil corundum.

Ruby and sapphire have been pulled from the melt by the Czochralski process in which the desired substance is melted in an iridium crucible.

Platinum cannot be used as it melts at 1774 °C (corundum melts at 2050 °C). A corundum seed is lowered to the melt surface and, if the temperature is right, growth on the seed will take place when the seed is vertically withdrawn (rotating at about 30 revolutions per minute). A rod-like crystal is formed. Heating of the crucible is by water-cooled RF. Pulling rates range from 6 to 25 mm per hour and quite large crystals can be grown. They are virtually inclusion-free though some may contain elongated bubbles.

Flux-Grown Synthetic Rubies

Flux-grown synthetic rubies (*Figure 24.3*) produced in a fundamentally similar manner to that employed by Frémy have been grown for scientific and commercial purposes since the 1960s (*Figure 24.4*). The procedure followed in the Bell Telephone Laboratories in employing this method has been described in detail by Nassau in *Gems Made by Man*. The flux most often used was lead fluoride, and some 20 kg of the compound were contained in a large platinum crucible, together with 3.6 kg of aluminium oxide and several grams of chromic oxide, according to the depth of colour desired in the resulting rubies. The crucible is covered with a lid and placed on a large ceramic pedestal which can be raised into the centre of a special furnace, and the contents are melted. The problem of thoroughly mixing the fused contents is solved by rotating the plinth on which the crucible rests, reversing the direction of rotation at intervals. After a long period of slow cooling during which perhaps 1.8 kg of ruby crystals may have been formed at the base of the vessel, another problem to be solved is to find a successful way of obtaining these free from the remaining flux. One drastic-sounding method is to punch a hole in the base of the crucible allowing the fluid contents to drain down a central hole in the supporting pedestal. The

Figure 24.3   Synthetic flux-grown ruby crystal

Figure 24.4   A flux-melt furnace used for the experimental production of synthetic stones (by courtesy of General Electric)

exceedingly expensive platinum crucible can later be repaired by welding, ready for use in a further experiment.

Similar methods have been used in growing ruby crystals by Carroll F. Chatham in San Francisco and by F. Truehart Brown of Ardon Associates Inc in Dallas, Texas. The latter products have been extensively marketed as Kashan rubies and owing to their high quality have achieved a considerable reputation. The price asked is far higher than that of Verneuil synthetics, but this is understandable in view of the costly nature of the process used. Uncut crystals have been occasionally available for amateur or professional lapidaries who wish to carry out their own faceting. The manufacturers of the Kashan stones at one time took the unusual step of issuing identification aids in the form of samples of their products with suggestions for recognizing their origin, and with photomicrographs of their inclusions in colour. The most typical inclusions in Kashan stones consist of flattened 'footprints' or 'paint-splashes' in small groups which contain solid flux. There are also fine pinpoint inclusions in lines sometimes described as 'rain'. Twin lamellae are frequently seen but are also common in natural stones. The fluorescence under SWUV tends to be brighter in Kashans than in Myanmar or Sri Lankan stones, and certainly more than in Siam rubies.

In a paper on the differentiation between natural and synthetic rubies of all kinds, Bosshart (*Z Dt Gemmol Ges.*, December 1981) measured the UV absorption characteristics of 94 rubies using a Pye Unicam SP 8–100

UV-VIS spectrophotometer. The specimens used included rubies chiefly from Myanmar, Sri Lanka, Kenya, Tanzania and Thailand, plus a few from Pakistan, Cambodia, Brazil and Australia, as opposed to 11 Verneuil stones, 14 Kashan, 16 Chatham and 4 Knischka synthetics. These tests were found to give a clear separation between the natural and man-made rubies. Absorption bands due to titanium and ferric iron affect the position, depth and width of the UV absorption minimum.

A report in 1985 classed Kashan inclusions into four types, all consisting of melt residues. They are feathers, fingerprints, strings of pearl-like bodies and comet or hairpin shapes.

In 1983 the Ramaura ruby was placed on the market by Overland Gems Inc. of Los Angeles. The product was grown by Judith Osmer and the growth method is a high-temperature flux process with spontaneous nucleation. Crystals show a near-equidimensional rhombohedral form and this form predominates in the majority of the other crystals. Some thin platy crystals show a predominating basal form. Growth without a seed makes for notably clear crystals as there is no need for forcing. The RI is 1.762 and 1.770 for the extraordinary and ordinary rays respectively, with a DR of 0.008. The SG is 3.96–4.00. Orange-yellow and white flux fingerprints are reported, with colour zoning, comet-tail inclusions and fracturing. A chalky-yellow and bluish white zoning can be seen under LW and SWUV with a chalky-red to orange fluorescence. Although an early statement by the grower stated that a dopant (europium oxide) had been added to give a characteristic shift in fluorescence towards the orange-yellow, this appears difficult to detect in the faceted product.

Rubies grown by the late Professor Paul Otto Knischka of the University of Steyr, Austria, are a very fine red with crystals showing many faces. Some crystals show a pseudo-cubic habit. The growth is thought to have been via a technique involving a temperature gradient with supercooling and supersaturation. Cloudy veils and healing fissures have been seen as inclusions; some early products also show metallic platelets of platinum or silver. Later specimens appear to be devoid of these crucible residues. Gas bubbles have also been noted. The SG averages 3.976. The RI is 1.760–1.761 and 1.768–1.769 for the extraordinary and ordinary rays respectively, with a DR of 0.008. There is a strong carmine-red fluorescence with no trace of chalkiness. A weak phosphorescence can be seen after subjection to X-rays. Although growth details are not released, the flux used for the first production at least is thought to be $Li_2O-Na_2O-K_2O-WO_3-PbF_2$ and/or PbO. Knischka ruby crystals over 50 mm in length have been produced, and material is being marketed as faceted stones in sizes ranging to over 11 carats, as preforms which may exceed 25 carats, and as rough macro-clusters, plates and micro-clusters. These qualities are distinguished by their relative free-dom from inclusions.

In 1985 *Gems and Gemology* reported cut Lechleitner ruby and blue sapphire. Stones showed both Verneuil-type growth features and inclusions characteristic of flux growth, suggesting that a Verneuil-grown crystal was encased by flux-grown corundum, or that a small Verneuil seed crystal was placed in the crucible used for flux growth, either to initiate growth or to influence its size and direction. The inclusions serve to distinguish these products from natural corundum.

Contact twinning on the prism faces $r(10\bar{1}0)$ or $r(11\bar{2}0)$ is frequently found in Chatham synthetic rubies. In Ramaura crystals there is penetration twinning about the *c*-axis, while Kashan stones show repetitive twinning of intercalated lamellae on $r(10\bar{1}1)$. There is also seen in Chatham rubies. Natural rubies often show repetitive twinning on $r(10\bar{1}1)$.

A star ruby manufactured by the Kyocera American Corporation and marketed under the trade name Inamori shows sharp, intense white six-rayed stars with some wavy or broken lines giving them a more natural appearance. Properties overlap those of natural star rubies and testing must rely upon observation of internal characteristics: these include smoke-like bluish white swirls running at random between the rays of the star, and round and distorted gas bubbles. The stones are probably grown from a high-temperature melt since the absence of curved striae would seem to rule out Verneuil production.

Some fluxes used by different manufacturers of synthetic ruby are as follows (Schmetzer, 1986):

Li_2O-MoO_3-PbX, where $X = F_2$ or O_2	Chatham, Gilson, Lechleitner
Na_3AlF_6	Kashan
Li_2O-WO_3-PbX	Knischka
Bi_2O_3-La_2O_3-PbX	Ramaura

Note that natural-looking fingerprint inclusions can be induced into Verneuil stones by flux healing. Also note that natural-looking needle-like inclusions identified as the edges of twinning planes may be confusing in Verneuil stones. Ruby with induced fingerprint-type inclusions and poly-synthetic twin lamellae has been reported.

Synthetic Corundums of Other Colours

Corundum of colours other than ruby has been grown by several manufacturers using flux-growth techniques. Chatham produces orange and blue sapphires as single crystals or as crystal groups, the groups being held together by a ceramic glaze which contains gas bubbles and which lowers the SG to around 3.70. Individual crystals in the groups are bladed and too thin for faceting. Orange groups show strong reddish to yellowish orange fluorescence under LWUV; most natural orange corundum is inert. Flux and platinum inclusions are characteristic of both blue and orange Chatham products as well as of similar specimens produced by Japanese growers.

Lechleitner has produced ruby and blue sapphire as well as orange, colourless and pink corundum. An alexandrite imitation is also reported. The ruby and blue sapphire both show strongly saturated colours and, though showing some haziness, are relatively inclusion-free. Flux inclusions are responsible for the haziness. In the ruby the optic axis is nearly parallel to the table while in the blue sapphire the optic axis lies about 20–30° away from the table plane. The ruby has an SG near 4.00 and an RI of 1.760–1.768 with a DR of 0.008. While the ruby shows no phosphorescence under any circumstances there is reddish fluorescence under both types of UV. Pleochroism is strong purplish red parallel to the c-axis and pale orange-pink at right angles to this direction. Wispy veils can be seen varying from transparent to opaque. Some may be colourless and some white. Some curved striae have been observed. The blue sapphire has an RI of 1.760–1.768 with a DR of 0.008 and an SG of 4.00. Similar inclusions to the ruby are reported. Neither in the ruby nor in the blue sapphire does the absorption spectrum help in identification. Growth could involve a Verneuil-grown seed.

Curved colour banding in yellow and orange synthetic sapphires may be seen if a translucent medium blue plastic filter (just over 3 mm thick) is placed over dark-field or transmitted light, the stone being positioned over the diffused illumination thus provided.

Flame-fusion-grown blue sapphire showed needle-like inclusions as well as curved growth lines. They are in fact strings of tiny gas bubbles.

Synthetic rubies with glass fillings have been reported.

Brown star corundum has been synthesized by the Linde Company in the 1960s. The stones show no absorption in the visible and are inert to both LW and SWUV. Widely spaced colour bands have been observed as well as minute gas bubbles.

Czockralski-pulled pink sapphire, recently on the market, is titanium-doped and gives a strong orange-red through the Chelsea filter. It fluoresces moderately orange under LWUV and a very strong bluish violet under SW. There is no distinctive absorption spectrum.

Blue and green sapphires have been grown by Czochralski pulling, the stones having a slightly violetish blue colour in the case of the blue sapphire. Rod fragments showed concentric colour zoning from core to rim.

Spinel

It was not until the 1930s that spinel became commercially significant. Natural spinel, except for the red variety, has no great use in jewellery and a simple simulation of the spinel would have had no especial value. The synthetic spinel is therefore made to imitate stones of important but different species, such as aquamarine, blue zircon, green tourmaline, pale green Brazilian emerald, and a greenish yellow shade which may imitate the chrysoberyl. An attractive rose colour and a royal-blue shade are also made, as is a completely colourless stone which rivals the synthetic white sapphire in brilliance. When these colourless or white spinels, as they are called, were first marketed they were the subject of a silly scare when retailed by a jeweller and antique dealer with considerable press ballyhoo as 'Jourado diamonds'. After many fruitless experiments a red synthetic spinel is now produced but this stone has properties slightly different from the usual run of the synthetic spinels of other colours. It will be mentioned later.

Physical and Optical Properties

Very rarely do the boules of synthetic corundum exhibit any crystal faces, but the spinel boules commonly show flat four-sided forms which indicate the cubic nature of the mineral, and unlike natural spinel synthetic spinel boules often show cubic cleavage cracks.

Normal natural spinel has the formula $MgO \cdot Al_2O_3$ which implies equimolecular proportions of magnesia and alumina. It was found that equimolecular amounts of the two ingredients did not produce good boules, and the synthetic spinel as now made is usually in the proportion of $1MgO$ to $2Al_2O_3$, the alumina remaining in the cubic gamma form. This extra alumina does slightly alter the physical and optical properties of the spinel. The SG varies from the value 3.60 of the normal natural spinel to 3.63 for the synthetic, and the RI is higher, from 1.72 (usually 1.718) for the natural to 1.73 (1.728) for the synthetic. The excess alumina in the synthetic spinel strains the crystal lattice and this is apparent when stones are viewed between crossed polarizing filters, for a typical anomalous DR is seen, the stone extinguishing in stripes (*Figure 24.5*). Further, many spinels show pseudo-interference figures when rotated between polarizing filters (*Figure 24.6*), and peculiar 'strain knots' may be observed (*Figure 24.7*). Also seen in synthetic spinels are worm-like gaseous tubes (*Figure 24.8*).

*Figure 24.5   Anomalous double refraction (tabby extinction)
in a synthetic spinel*

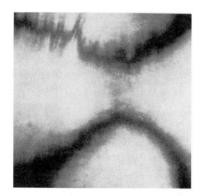

*Figure 24.6   Pseudo-interference figure shown by a
synthetic spinel between polarizing filters*

Figure 24.7   Strain knots in a synthetic spinel

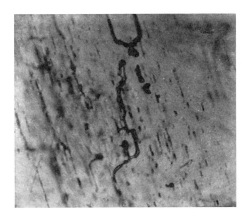

Figure 24.8   Worm-like distorted gaseous tubes in a synthetic spinel (by courtesy of E Gübelin)

Colours of Synthetic Spinels

The blue colours of synthetic spinel are due to the addition of cobalt oxide sometimes modified with other metallic oxides. The yellow and yellowish green colours are due to a trace of manganese, and the dark greenstones are coloured by chromium oxide, which in corundum gives a red colour, for at the temperature of the oxyhydrogen furnace the stable colouration of the gamma alumina and magnesia boules is green. A chromium-coloured red spinel is now marketed, but, owing to the fracturing of the boules, only in small sizes. However, red spinels of large size have been grown in the USA using the flame-fusion method. Unlike the general run of synthetic spinels, these stones are made with equimolecular amounts of magnesia and alumina as in the case of natural spinels. Their SG approximates to 3.60 and their RI from 1.722 to 1.725, the higher values, compared with those of the natural stones, being due to excess chromium.

Such red synthetic spinels vary somewhat in their internal structure, but as a rule show many gas bubbles and very pronounced curved colour lines (*Figure 24.9*) which may be visible to the unaided eye. Sometimes, too, they may show straight lines rather like twin lines seen in some red and blue corundum of natural origin. The fluorescence spectrum differs from the 'organ-pipe' pattern as seen in the natural red spinel, for one fluorescent line near 685 nm predominates markedly over the others.

The lovely pink-coloured synthetic spinels which have been marketed for some years owe their colour to a trace of iron. A type with a colour change (being made to imitate the alexandrite) is, although rarely met, a better simulation than the much more common similarly coloured synthetic corundum.

A list of the usual colours of synthetic spinels and the trade names sometimes applied to them is given in *Table 24.2*.

Figure 24.9   Curved bands (Venetian blinds) in a synthetic red spinel

Table 24.2
Colours of synthetic spinels

Colour	Manufacturers' trade name
Colourless	Synthetic White Spinel
Blue	Hope Sapphire
Blue (bright)	Azurite
Green-blue	Blue Zircon
Pale blue	Aquamarine
Yellowish green	Brazilian Emerald
Dark green	Tourmaline Green
Yellow-green	Peridot
Pink	Synthetic Pink Spinel
Red	Synthetic Red Spinel
Green/red	Alexandrite

In about 1957 a spinel was made to imitate moonstone, the schiller probably being induced by some form of heat treatment of the synthetic white spinel. The bright blue-white fluorescence shown by these stones under SWUV indicates their nature, for neither the 'heated quartz' imitation of moonstone nor the true feldspar shows this type of luminescence. A blue schillerized synthetic spinel is known.

Synthetic spinels rarely show curved colour bands or the structure lines seen in synthetic corundum. If bubbles are present they are sparse and may have a typical whorled or turned appearance with the end tending to show crystal form; they are termed 'profilated bubbles' (*Figure 24.10*). High magnification may be required to resolve this feature. A disturbing kind of inclusion sometimes seen in synthetic spinel is solid crystals of alumina (*Figure 24.11*). The synthetic spinels may be distinguished by the unusual colour (for spinel), the RI and SG, although there are certain zinc-rich spinels from Sri Lanka which have a higher RI than usual. They do not have the colour of the synthetic spinel, being mostly pale mauve to dark inky blue.

Absorption Spectrum

The blue colours of synthetic spinel show a distinctive absorption spectrum of three broad bands in the orange-red, yellow and green, the bands being centred at 635, 580 and 540 nm, the last being only about

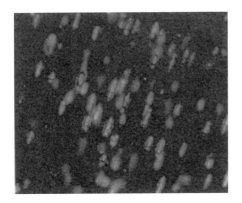

Figure 24.10   Profilated bubbles in a synthetic spinel (by courtesy of E Gübelin)

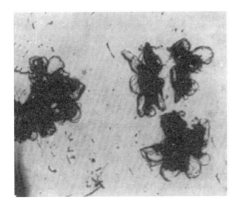

Figure 24.11   Solid crystals of alumina brought into being by exsolution. These are a disturbing kind of inclusion sometimes seen in synthetic spinel (by courtesy of E Gübelin)

half the width of the first two. A blue glass coloured by cobalt exhibits a similar spectrum but in this case the bands are slightly wider apart and, further, it is the central band which is the narrowest. Such cobalt-coloured blue stones show a red or orange residual colour when viewed through the Chelsea colour filter, unlike sapphire, aquamarine and blue zircon which show a bluish green through the filter. The synthetic spinel made to imitate the alexandrite also shows a cobalt absorption spectrum.

The yellow and greenish yellow synthetic spinels show an absorption spectrum of two bands in the blue-violet at 445 and 422 nm due to manganese. In certain shades of pale blue, the cobalt spectrum may be in association with the two bands due to manganese, and it sometimes occurs that the cobalt bands complete a spectrum which shows the bright fluorescence line of chromium.

Luminescence and Fluorescence

The luminescence of synthetic spinels has some diagnostic value. The synthetic white spinel shows only a faint misty glow under the LW lamp but under SW rays the glow is usually a bright bluish white, although only a weak but distinct blue light may be seen in some white synthetic spinels. As the synthetic white sapphire shows a dull deep blue glow under short-wave excitation, this provides a distinction between the two white synthetics. Synthetic white spinels emit an apple-green light with similar phosphorescence when excited with X-rays. A few specimens have been found to give out a blue or greenish blue light under X-rays. Synthetic white spinels quickly turn to a brown colour on irradiation with X-rays, but they may be decolourized by heating to about 250 °C.

Most blue spinels show a red glow when under the LWUV lamp. Under the SW rays the glow seen may be either red, orange or, more usually, bluish white. The yellow, yellowish blue and yellowish greenstones, those which show the manganese absorption spectrum, emit an apple-green glow with all radiations. The 'tourmaline-green' colour of synthetic spinel shows a red glow under the LW lamp, a milky greenish white glow under the SW lamp and usually a red colour under X-rays. The pink-coloured synthetic spinels are inert under any of the three radiations.

Flux-Fusion and Other Synthetic Spinels

Spinel is also grown by flux fusion. Crystals grow as octahedra or groups of octahedra, the latter having the greater gem significance as these groups have been used in the modern concept of jewellery design where uncut crystals form the central motif. Such spinel groups are grown in many colours by doping with trace elements, such as red by chromium and blue by cobalt, and there are seen groups with a yellow colour, the doping agent being nickel.

Specimens of synthetic spinel groups, all of English manufacture, included two stones of red colour having SGs of 3.592 and 3.598, both showing a fluorescence spectrum with a single strong line at 685.5 nm and traces of others. This fluorescence spectrum is similar to that shown by the Verneuil flame-fusion synthetic red spinels. Under UV of both types the glow was crimson, and under X-rays the fluorescence was more of a purplish red with a persistent phosphorescence.

The blue-coloured group had an SG of 3.60 and showed the typical cobalt absorption spectrum. Under the Chelsea filter the group showed a bright red, but the group did not show the differential response to UV as seen in the Verneuil-type blue synthetic spinels which show red under the LWUV lamp and a greenish or bluish white under the SWUV lamp. Indeed in the group examined there was very little luminescence under UV or X-rays, except for a possible dull red glow when irradiated with the SW lamp. It is always problematical whether such a glow is not reflected from the light passed by the lamp.

The yellow-coloured specimen was found to have an SG of 3.609, showed no distinctive absorption spectrum, but exhibited a green fluorescence under LWUV. It was practically inert or with a slight dull yellow glow under the SW lamp, but under X-rays the glow was a vivid green there being no phosphorescence. The pale green specimens, having SGs of 3.601 and 3.589, again showed no characteristic absorption spectrum and only slight luminescent response of indeterminate colour under UV, but a dull green glow under X-rays.

Two unusual specimens of similar habit and of a pale bluish green colour were found to have unusual characters, for the SG was about 5.7, and the absorption spectrum showed a fine line in the red and indeterminate lines in the yellow, which may be traces of the didymium series. The luminescence under UV and X-rays was a strong yellow.

The flux-grown lithium-iron spinel has an SG of 4.58 and is strongly magnetic.

Fine-quality flux-grown spinel is reported from the USSR. One red octahedron weighing 17.19 carats has been seen, and the writer's collection includes a similar crystal and also a blue octahedron coloured by cobalt and showing a bright red through the Chelsea filter.

Said to have been devised by a German dentist, a lapis lazuli-coloured synthetic spinel has been marketed since 1954. This is a coarse powder of synthetic spinel heavily charged with cobalt oxide which are sintered together in an electric furnace. Some specimens have bright specks of 'pyrites' so common in true lapis lazuli, but in the case of sintered spinel the specks are gold. These synthetic spinels show a brilliant red residual colour through the Chelsea colour filter and have RI of 1.725, but the reading is never as sharp as in transparent synthetic spinels of normal type. Sintered spinels have a hardness of 8 and an SG of 3.52. The absorption spectrum may be observed quite clearly by light reflected from the surface of the specimen. It is a typical cobalt spectrum although the distribution and strength of the bands are not quite the same to those shown by the blue spinels made by the Verneuil process. The three bands in this spinel are at 650 nm (broad and strong), 580 nm (weak) and 480 nm (strong).

A Verneuil red spinel came on to the market in 1987. It has an SG of 3.59 and an RI of 1.72, with minimal strain birefringence and no fine line absorption in the red part of the spectrum. A red fluorescing band is centred around 685 nm under LWUV. Broad curved colour banding is seen with a wide variety of bubble oriented at right angles to the bands.

A synthetic spinel crystal grown in Russia was a transparent purplish red. The RI was measured from a flat face at 1.719 and the SG was 3.58. Neither these figures nor the absorption spectrum sufficed to distinguish the synthetic material from a natural red spinel.

Under magnification the stone (represented as a hydrothermally grown product) showed what appeared to be primary flux inclusions coloured a deep orange-brown with an angular to jagged profile. Some brightly reflective and iridescent air-filled fractures had signs of strain associated with them seen under polarized light. No signs of the presence of water were found, so the material cannot have been a hydrothermal product. Should stones be faceted from the material, considerable care in examination will be needed.

Rutile

The birefringent dark red or brown mineral rutile TiO_2 is made by a modification of the Verneuil process to give a soft but highly dispersive diamond simulant. Some experiments with a flux growth of rutile were made in the nineteenth century, but the Verneuil torch has proved to be a more effective means of production. To offset the oxygen loss a third outer tube is added to give a tricone burner with a sequence oxygen–hydrogen–oxygen. Up to three volumes of hydrogen to one of oxygen may be used, this giving a strongly reducing flame in the conventional Verneuil burner, but the tricone apparatus needs only $1\frac{1}{2}$H to 1O. The oxygen deficiency results in a black crystal, this being altered by annealing at about 1000 °C to a gunmetal blue, then a paler blue and finally a pale yellow. Colourless crystals cannot be obtained owing to a strong absorption of the deep violet. Following original production by the National Lead Company in the USA, one or two other companies grew rutile for a time, production being around 150 kg annually in peak years. Today only a little is grown and specimens are becoming rare.

Rutile has a hardness of just over 6, an SG of 4.25 and an RI of 2.8 on average. The high DR of 0.287 shows very clear doubling of opposite facet edges and the dispersion of about 0.33 is so high compared with diamond's 0.044 that rutile shows an opal-like play of colour. Blue rutile is strongly electroconductive.

The addition of about 0.5% magnesium oxide to the feed powder with annealing of the finished boule gives a star effect when cabochons are cut. A thin layer of corundum deposited on the rutile surface may harden it and this has been tried, though with little commercial importance. The feed powder is made from highly purified titanium tetrachloride $TiCl_4$, and reaction with ammonium sulphate gives a basic titanyl double salt which is recrystallized and washed to remove the chloride. It decomposes at 900 °C to give TiO_2. Rutile melts at about 1860 °C.

Strontium Titanate

In nature the mineral tausonite, with no ornamental importance, Verneuil-grown strontium titanate is a highly dispersive isotropic though soft diamond simulant. Other titanates have also been grown but with less commercial success. Early work was carried out by the US National Lead Company using a three-tube burner as strontium titanate suffers from an oxygen deficiency. The grown boules are rendered colourless by annealing for 12 hours at 1000 °C. In some cases two anneals were performed, the first at 1700 °C to eliminate strain and the second at 500 °C to improve the colour. The feed powder is made by reacting highly purified titanium tetrachloride $TiCl_4$ with strontium chloride $SrCl_2$ and oxalic acid to give the basic titanyl double oxalate salt $SrTiO(C_2O_4)_2 \cdot 2H_2O$. The chloride traces are removed by washing, and on heating to over 1000 °C a free-flowing granular powder is obtained. The melting point of strontium titanate is near 2080 °C.

Strontium titanate, which had the trade name Fabulite (among others), has a hardness of 6, an SG of 5.13 and an RI of 2.41. The dispersion is 0.19, giving a very notable play of colour. When it is used as mêlée, or in the base of composites to imitate diamonds, the jeweller or gemmologist can be deceived: the surface can be marked by the point of a needle, and there is no absorption in the visible spectrum and no luminescence. It is opaque to X-rays and somewhat brittle, so ultrasonic cleaners should not be used.

Quartz

The search for a non-twinned single-crystal quartz for electronic purposes during the Second World War led to developments of a hydrothermal growth technique to which Spezia had given the greatest development impetus during 1898–1908. He used a quartz crystal as a seed which was supported by a silver wire in a steel pressure vessel with a silver lining. A solution of hot water with sodium silicate and crushed natural quartz was placed under pressure, giving up to 15 mm of growth on the seed over a period of about 200 days. Ignoring the advantages of thermal convection, Spezia placed the hotter region at the top and the cooler below; present-day runs are much quicker because these positions are reversed. Work by Nacken of flux emerald fame and by Bell Laboratories, Sawyer Research Products Inc., and others have led to the development of processes which can produce crystals several metres in length as well as amethyst, citrine, smoky and colourless gem-quality quartz.

The pressure vessel or autoclave which is sealed (and appropriately known as a bomb) is about 460 mm in diameter externally and 3.7 m long (*Figure 24.12*). Internally it is 150 mm in diameter and 3 m long; the volume is about 2 litres. The bottom feed temperature is about 390 °C, the top crystallization temperature 330 °C. An 80% fill with a 4% NaOH solution will give a pressure of about 1700 bars. A baffle is placed between the quartz source material and the seeds and has a 5% opening. The purpose of the baffle is to control the convection currents to ensure

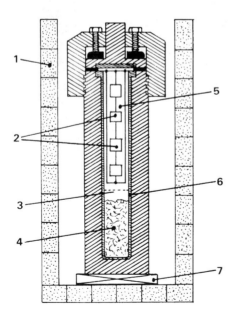

Figure 24.12   Simplified sketch of a silver lined autoclave for the production of hydrothermally grown quartz: 1, thermal insulation; 2, seed plates; 3, baffle; 4, source material; 5, silica-rich aqueous solution; 6, silver liner; 7, electrical heater

an even deposit on all the seed crystals. About 32 kg of quartz feed will be exhausted in a growth period of 20 days and will grow on to 40 seed plates which begin at 4 mm × 15 mm × 0.1 mm. The final crystals will be almost 50 mm thick and weigh about 700 g. The sodium hydroxide is used as a mineralizer and increases solubility by an order of magnitude (10×). Heat dissolves the crushed quartz at the bottom of the vessel. The saturated liquid rises by convection to the upper part of the vessel where it deposits on the seed plates. Crystals grown by this method show characteristic cobbled faces, the effect arising from small disc-like structures (*Figure 24.13*). These faces are basal ones and do not exactly correspond with faces on natural quartz crystals.

This is an efficient method of crystal growth for certain substances but it is not without problems. Apart from the possibility of a vessel failure it is necessary to achieve a reaction (melting the feed) and its reverse (recrystallization). The mineralizer which reacts with the quartz to give a more stable compound must not make it too stable or the reverse reaction will not take place. Growth of 1 mm per day would be an acceptable rate to give clear crystals.

In an earlier work the twinning-free Brazilian quartz known as lasca was used as feed material. However, a Brazilian government ban on the export of all but small quantities in 1974 forced growers to look

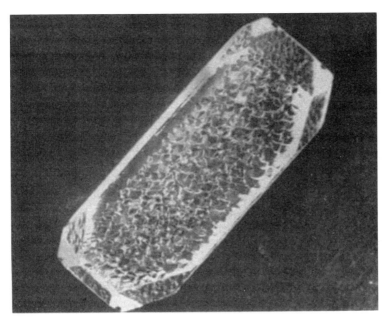

Figure 24.13 A synthetic quartz crystal showing the 'cobbled' surface of the large basal face

elsewhere. Fortunately adequate supplies of material do exist in other countries, including the USA.

The coloured varieties of quartz are also grown hydrothermally. A smoky colour is obtained when a trace of aluminium is present and the specimen is irradiated by X-rays or gamma rays. Some smoky quartz will give a greenish yellow on heating. Yellow citrine occurs when Fe^{3+} is present, and irradiation can give an amethyst colour from the citrine provided that the iron is in the appropriate site. Natural citrine has iron in a different site and cannot usually be irradiated to give amethyst, though heating will often accomplish this. A green colour may be obtained by heating some amethyst.

Synthetic smoky quartz is grown as colourless and later irradiated by gamma rays to give the smoky colour. Aluminium has to be present replacing some of the silicon. A greenish yellow quartz can be obtained by heating smoky quartz; depending on the degree of heating a range of colours, including lemon and honey, brown and greenish brown can be obtained in this way.

Adding iron oxide or hydroxide to the growth solution will give the required Fe^{3+} for citrine. Oxidizing conditions are necessary and $LiNO_3$ or $LiNO_2$ may be added to give yellow. About 0.01% Fe_2O_3 will give a medium yellow. The steel of the autoclave might contribute iron which would be undesirable when colourless quartz is to be grown, but the use of a sodium-containing mineralizer will lead to the formation of an insoluble silicate on vessel walls and stop migration of iron into the growing crystal. When iron is required the use of sodium compounds is ruled out and salts of potassium or ammonium are used instead.

When amethyst is to be produced, the crystals as grown are near-colourless and are later irradiated by gamma rays. Iron must be present in the crystals for the colour change to operate. Aluminium must be eliminated as a smoky purple colour will result. The use of lithium salts such as $LiNO_2$ will prevent the aluminium entering the quartz in large amounts since they will cause $LiAlSiO_4$ (eucryptite) to form as an insoluble compound.

A green quartz will result from the presence of iron and the absence of oxygen during growth. The green colour results from Fe^{2+}. Blue quartz can be obtained by adding Co^{2+}. If Al exists in the crystal, the more there is, the deeper the blue. Colours such as ruby red and emerald green are not possible in quartz since the structure will not accommodate the appropriate ions. Further details of the synthesis of quartz will be found exhaustively treated in Nassau's *Gems Made by Man*.

Identification of synthetic quartz is difficult since the physical and optical properties accord with natural quartz, and there are usually few if any significant inclusions. A Japanese synthetic amethyst was found to contain feather-like inclusions made up of liquid-filled and two-phase structures and sharp growth zoning parallel to one rhombohedral face. Twinning structures seen in the same stone differ from those seen in natural amethyst.

A technique involving the examination of amethyst using a horizontal microscope with the optic axis parallel to the path of light through the microscope shows Brazil twinning by its interference colours in natural specimens, while in synthetic stones a succession of broad colour bands is seen. Some Japanese amethysts have shown arrowhead or small flame-like structures. Some stones showing combined amethyst and citrine colours ('ametrine') may be synthetic and could perhaps be identified by this method.

The presence of Brazil-law twinning in an otherwise flawless amethyst virtually proves natural origin, while an acutely angled flame-like pattern seen while the specimen is immersed between crossed polars confirms artificial origin. Angular or straight zoning with juxtaposition of colourless or violet-blue zones next to purple areas indicates a natural stone, while the presence of light and dark purple zones, or the absence of zoning, indicates a synthetic amethyst.

Colourless synthetic quartz has been grown over seed plates of natural amethyst, both portions showing Brazil-law twinning. This may make previous work on distinguishing natural from synthetic quartz more difficult. In a report of 1989, GIA found such a composite to show the twinning clearly between crossed polars together with tiny fluid inclusions and breadcrumb-like particles. Earlier assertions that a flame-like appearance between crossed polars would serve to distinguish natural from synthetic quartz are disputed in the 1989 paper which holds that the flame-like appearance is a form of Brazil-law twinning.

Two crystals with 2 mm thick amethyst seed plates were used as seeds in a hydrothermal growth experiment. The seeds were Brazil-twinned and differently oriented from each other. The resulting crystals were of differing morphologies, one being tabular (with the seed oriented nearly parallel to the c-axis) and the other more characteristic of a natural quartz crystal and providing sufficient material for two faceted stones. The seed in this case was oriented nearly perpendicular to the c-axis. In both cases the seed turned yellow during the growth process; in polarized light Brazil-law twinning was visible as a phantom between the interface of seed plate and overgrowth. Further examination in polarized light with the stones immersed in benzyl benzoate showed that the colourless quartz overgrowth was Brazil-law twinned. The GIA paper shows that synthetic amethyst twinned according to the Brazil law can be grown.

Emerald

Growth Methods

The early attempts to synthesize emerald go back to the middle of the nineteenth century. In 1848 Ebelmen reported that he had obtained emerald crystals by heating powdered natural material with molten boric acid as a flux. The emerald dissolved in H_3BO_3 and recrystallized as hexagonal crystals of small size. In 1888 and 1900 Hautefeuille and Perrey found that by using Li_2O with MoO_3 or with V_2O_5 as solvents (fluxes) emerald crystals could be obtained. Their crystals were 1 mm across after a growth period of 14 days at 800 °C. At higher temperatures phenakite resulted. Chromium was added to give the emerald colour.

In 1911 the firm which was later to become IG-Farbenindustrie began research on emerald growth, and in 1924 Espig began the work which led to the growth of larger crystals. The ingredients needed for emerald growth reacted in the flux to give emerald in the growth region only, rather than allowing spontaneous nucleation to give tiny crystals all over as had happened with earlier work. A flux of lithium molybdate with extra MoO_3, and with silica glass to give the SiO_2, allowed emerald crystals to grow, but they grew on the quartz, giving rise to unacceptable inclusions and cracking.

A platinum apparatus (*Figure 24.14*) was later used with a flux SG of 2.9; which allowed the silica glass to float to the surface while the seeds for emerald growth were confined by a platinum screen so that emerald would not form on the silica. BeO and Al_2O_3, with $LiCrO_4$ to give the colour, could

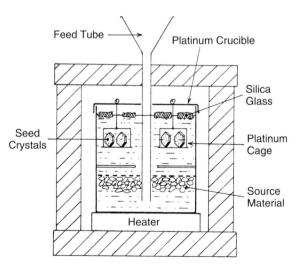

Figure 24.14 The flux-melt apparatus used to grow synthetic emerald crystals by the I.G. Farbenindustrie process

be added via a feed tube and extra SiO_2 could be added at the top. This topping-up is not possible with the sealed-vessel hydrothermal process. Growth by diffusion (SiO_2 dissolving near the top and diffusing downward, the other materials dissolving near the bottom and diffusing upward) allows emerald to form in solution in the middle where the growth materials react. Espig obtained emerald crystals up to 20 mm long in a growth period of one year. Later, commercially successful growers such as Chatham and Gilson used the same or a similar process, though the colour of their product is better than that obtained by Espig (he found that chromium did not give a satisfactory emerald colour by itself; vanadium when added improved the colour greatly). Crystal groups made by Espig and his co-workers were shown at the Paris Exhibition of 1937. Though the name Igmerald was used no crystals were ever put on the market; some were sent in attractive presentation cases to noted gemmologists of the time.

Professor Richard Nacken, a pioneer in the hydrothermal growth of quartz, also grew emerald crystals which were frequently cited as hydrothermal in gemmological literature. The absence of water (proved by an examination of the infra-red absorption spectrum) showed that this could not be so, and the crystals are in fact grown from a flux. Both natural and hydrothermal emeralds contain water. The colour of Nacken's product was better than that achieved by Espig and this is probably due to the use of a flux containing vanadium oxide (this is used in modern emerald production). Inclusions of flux resembling wreaths of smoke and nailhead-type inclusions, characteristic of later flux products, are seen clearly in Nacken crystals which are now very rare.

Modern emerald production using a flux accounts for by far the greater amount of synthetic emerald on the market, though some hydrothermal emeralds are still made. The work of Carroll F. Chatham of San Francisco (now carried on by his son Thomas) who produced the high-quality Chatham Created Emerald and of Pierre Gilson of France (his process has now been sold to Japan), lead the field in emerald production. No water is present but there are traces of Li, Mo and V, so it is assumed that a lithium molybdate-vanadate flux is used. The work of Chatham, Gilson and others probably parallels Nacken's work fairly closely, although details are unavailable owing to commercial secrecy. In turn Nacken's growth method must have been very similar to that of Espig as described above. While Chatham appears to use a flux reaction method, Pierre Gilson, after encountering trouble with spontaneous nucleation, uses a natural colourless beryl seed on which emerald forms on both sides (*Figure 24.15*). The emerald layers are then used as seeds for commercial growth. The growth rate is about 1 mm per month.

The addition of iron to eliminate the normally characteristic red fluorescence under UV shown by synthetic emerald was experimentally tried for

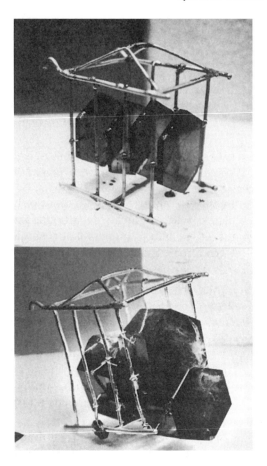

Figure 24.15 Top: Platinum cage containing Gilson synthetic seed plates after 2 months overgrowth. Bottom: Gilson synthetic emerald crystals after 10 months growth. (Laboratoire Gilson)

a few runs; an absorption line at 427 nm indicates this product. Both Chatham and Gilson have made and sold clusters of emerald crystals unlike any combination of natural crystals. Gilson has made crystals yielding faceted stones as large as 18 carats, and at its highest the quality is superb in any size. Stones are or were sold on a graded basis of quality as well as size.

Gilson appears to use or to have used a flux transport system in which the source material dissolves in one section of the vessel and grows on seeds on the other side which is cooler. The growth rate is slow and the formation of phenakite has to be avoided by careful temperature control. Also to be avoided is spontaneous nucleation of emerald. A detailed account with the history of emerald growth processes can be found in Nassau's *Gems Made by Man*.

Hydrothermal growth of emerald of commercial size and quality was not achieved until the 1960s. Johann Lechleitner of Innsbruck, Austria, grew emerald hydrothermally on to faceted natural beryl seeds. The emerald layer was thin and stones show a characteristic cracking. The lower facets were often left unpolished. The trade names Emerita and Symerald were used. It is probable that earlier work may have been hindered too greatly by the use of an alkali medium which could not produce crystals of useful size.

A beryl with vanadium only as a colouring element (though with emerald colour) was grown by A M Taylor in the 1960s, but examples are very few.

The Linde Division of Union Carbide produced hydrothermal emerald from 1965 to 1970. As always in emerald growth, any silicon in the growth material must be separated from the other elements in the feed to prevent spontaneous nucleation and allow growth to take place only on seeds. Growth is by diffusion reaction with convection. Water boils in the sealed vessel, building up pressure and filling the vessel as a single fluid. Pressure increases, dissolving feed quartz at the top and the rest of the feed at the bottom. Emerald in solution forms in the vessel centre and crystallizes on the seeds. Too much mixing will produce phenakite. Pressures are around 700–1400 bars at temperatures of 500–600 °C. Aluminium is obtained by using gibbsite $Al(OH)_3$, beryllium from $Be(OH)_2$, silicon from crushed quartz and chromium from $CrCl_3 \cdot 6H_2O$.

A variation in the original process initiates growth on a natural beryl seed. The new growth is removed to serve as a seed in its turn and can be seen as a colourless layer in the finished stone. No topping-up can take place as the vessel is sealed, so that grown crystals have to be coated up to three times in successive runs.

The Linde product was set in the company's own jewellery (Quintessa) and the process was sold to Vacuum Ventures Inc. of New Jersey who were manufacturing by the hydrothermal process during the 1990s.

Other methods of beryl growth have been tried experimentally. However, since many beryllium compounds are toxic to a dangerous degree (the toxic ones are those which are soluble in water or biological fluids) there is not much latitude in choice of growth method.

The Igmerald was copied closely for a few years by the firm of Walter Zerfass of Idar-Oberstein. One of Espig's IG-Farben colleagues in the Zerfass firm used the same flux process and a flux of Li_2O-MoO_3 with V_2O_5. Stones are notably included and are very rare. Very occasionally a small hexagonal crystal may be seen, sometimes with small crystals growing from the prism faces. Such crystals are prized collectors' items.

Characterization

Synthetic emerald shows inclusions which are characteristic in themselves, but are remarkably like inclusions in natural stones. These inclusions are

liquid-filled feathers, usually two-phase, which assume veil-like formations and may be likened to curtains waving in a draught, or they may be
in a series of parallel bands (*Figure 24.16*). Zoning is quite common and
these zonal lines may be straight or angular in conformity to the hexagonal
prism (*Figure 24.17*). Quite large crystals of phenakite may also be an
inclusion.

*Figure 24.16   Parallel arrangement of feathers in an
American synthetic emerald*

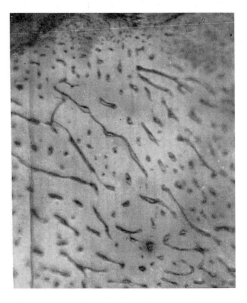

*Figure 24.17   Liquid-filled feather and zoning
in an American synthetic emerald*

The SG of synthetic emerald at 2.65 is perceptibly lower than that of natural emeralds. Thus, in a bromoform/monobromonaphthalene mixture in which quartz just floats, a synthetic emerald will either remain freely suspended or just float. Natural emeralds will sink decisively in such a liquid. The RIs also differ perceptibly from those of natural emerald. Again they are lower and are usually 1.563 for the ordinary ray and 1.560 for the extraordinary ray, the birefringence being approximately 0.003.

When viewed through the Chelsea colour filter, the synthetic emeralds show a strong red residual colour which, while indicative, must not be taken as conclusive; some natural emeralds, particularly those from the El Chivor mine in Colombia, also exhibit this strong red colour through the filter. This warning also applies to the fluorescent red glow shown by synthetic emeralds when in UV, of which the radiations from the LW fluorescent UV lamp are the more useful. Tests carried out in the London Gem Testing Laboratory in 1953 revealed that Chatham synthetic emeralds showed a far greater transparency to SWUV than natural emeralds or indeed any natural beryl. This was shown by taking photographs of the spectrum of a copper arc as transmitted by the stones tested. No natural emerald was found to transmit light of shorter wavelength than 295 nm, while Chathan stones transmitted freely to 230 nm or even beyond. A more convenient way of employing this as a test was found to be to expose any doubtful stone, resting on slow photographic film (preferably immersed in water) to light from SWUV lamp in a dark-room, using known natural and synthetic emeralds for comparison on the same film to ensure that the exposure has been of the right duration. The developed film should show a marked difference between natural and synthetic emeralds (*Figure 24.18*).

Carroll Chatham maintained that the synthetic crystals are free from impurities and strain so that they can be raised to an incipient white heat without damage, while the natural stone is destroyed far below the visible temperature range.

Figure 24.18 Differential transparency to SWUV of synthetic and natural emeralds

Emeralds produced by Pierre Gilson of France have ranges of SG and RI similar to those of Chatham's stones. The veil-like feathers are also similar (*Figures 24.19, 24.20, 24.21*). The Gilson stones do have one character different in that under the UV lamp they glow orange and not red, but recent products show red.

Linde synthetic emeralds have RI of 1.572 and 1.567, with a birefringence of 0.005, and the SG of the stones is near 2.67 or even higher. The inclusions in Linde stones are the veil-like structures so commonly seen in other synthetic emeralds, and also nail-like cavities with the head of the nail formed by a single, or a group of, phenakite crystals. However,

Figure 24.19   Feathers in a flux-grown synthetic emerald

Figure 24.20   Feathers in a Gilson synthetic emerald

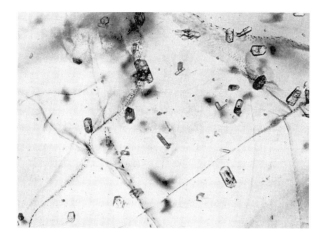

Figure 24.21   Phenakite crystals in a Gilson synthetic emerald

the most striking observations made on this type of synthetic material are the strong red light shown by the stone when it is bathed in a fairly intense beam of white light, and also the strong red fluorescence shown by the stone when viewed through a Chelsea colour filter or crossed filters or under UV.

Some Linde hydrothermal emeralds examined by workers in the USA have RI of 1.578 and 1.571, with a birefringence of 0.007. The SG was found to be 2.678. These stones showed much less fluorescence under UV, although the red appearance when the stones were in a beam of white light was still a characteristic feature, and so were the nail-like cavities capped by phenakite crystals (*Figures 24.22* and *24.23*), as well as the veil-like feathers which are so common in synthetic emeralds.

Some characteristics of Nacken, Igmerald and Symerald synthetic stones are shown in *Figures 24.24, 24.25*, and *24.26* respectively.

The Lennix synthetic emerald is a flux-grown product with low RI at 1.556–1.562 and 1.558–1.566, with DR 0.003. These figures are for the light green areas; those for darker areas are a little higher. The SG is in the range 2.65–2.66. Fluorescence is not clearly diagnostic, with stones showing bright red under LWUV and dim orange-red with transparency under SWUV. Lennix emeralds show fairly high FeO and MgO contents compared with many other synthetic emeralds, but the levels are lower than the average for most natural stones. A purple or bright violet-blue cathodoluminescence has been observed and appears to be peculiar to this product. With the microscope, a variety of inclusions has been observed: opaque tube-like structures preferentially aligned to

Figure 24.22   Feathers and needles in a Linde synthetic emerald

*Figure 24.23   Conical cavities capped with phenakite crystals
in a Linde synthetic emerald*

the c-axis; clusters of inclusions along the borders of sequential growth zones paralleling the edges of the basal pinacoid; slender crystals of phenakite and beryl; secondary flux-lined healed fractures resembling veils at low magnification; and two-phase inclusions along the edges of the basal pinacoid and sometimes parallel to the c-axis. Lennix emeralds have been found with no iron but with vanadium. Some contain three-phase inclusions similar to those seen in natural Colombian stones.

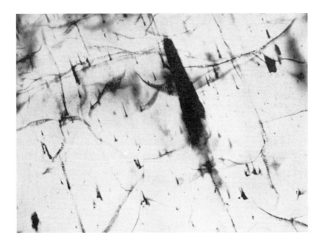

Figure 24.24   Inclusions in a Nacken synthetic emerald (circa 1926)

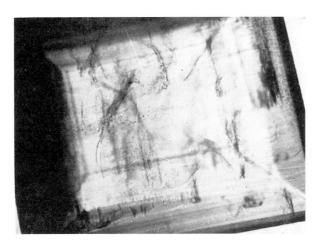

Figure 24.25   Feathers in an Igmerald (1935) (photo B W Anderson)

The Biron hydrothermal emerald contains both chromium and vanadium as well as chlorine. It was first produced in Western Australia early in the 1980s. In about 1988 the Emerald Pool Mining Company (Pty) of Perth, Australia, claimed that the source of their product was natural emerald and that a secret and unique process was used to convert low-grade material from their mine into high-grade emerald. Examination proved that Pool material was Biron-type synthetic emerald. The stones show considerable clarity but contain characteristic

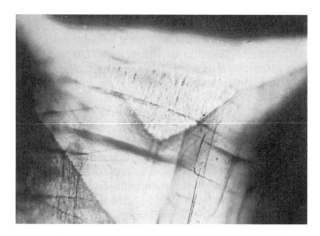

*Figure 24.26 Synthetic emerald-coated beryl (Symerald) showing
the reticulation of crack-like markings and one unpolished facet*

inclusions of gold in various forms, phenakite crystals, large two-phase
inclusions, nail-head spicules with gas and liquid phases, comet-like
white particles, parallel needle-like structures, veils of flux and dark
metallic inclusions. The SG is 2.68–2.70, the RIs are 1.570–1.571 and
1.577–1.578, and the DR is 0.007–0.008. Chlorine is present in the Biron
stones but not in natural emeralds.

Full Lechleitner hydrothermal emeralds have shown the presence of
layers and sublayers parallel to the table with colour zoning between
subsequent layers and sublayers. Schmetzer (*Journal of Gemmology*, 1990)
summarizes the Lechleitner product types as follows:

A Flux growth of emerald on seed plates of natural beryl. Only a few
   samples ever made in 1956–58.
B Pre-shaped natural beryl seeds overgrown hydrothermally. Seeds
   were colourless or slightly greenish. Made and marketed between
   1959 and 1972.
C Very few grown for study in 1961 or 1962. Flux-grown seed plates
   with hydrothermal overgrowth.
D Natural beryl seeds cut at oblique and variable angles to the *c*-axis.
   Hydrothermal emerald overgrowth. Grown in 1962 and 1963; samples sent to laboratories only.
E Fully synthetic hydrothermal emerald first made in 1964; several
   autoclave runs needed, hence the presence of layers and sublayers.
F Growth began in about 1972 and ran up to the 1980s. May be an
   improvement of type A process. Presumably flux-grown since no
   water present; basal plates of natural or synthetic beryl used as
   seeds.

Russian hydrothermally grown emerald is made in steel autoclaves without precious metal inserts using seed plates of colourless beryl. Distinct amounts of chromium, iron, nickel and copper are incorporated and characteristic microstructures are found. The RI range is from 1.580–1.586 to 1.573–1.579, with a DR of 0.006–0.007; the SG is 2.68–2.70. Step-like growth lines and colour zoning are characteristic.

Gilson has tried out an emerald containing chromium and nickel (flux-grown) which gives a yellowish green colour. The Gilson operation was acquired in the 1980s by Nakazumi Earth Crystals Corporation. The Gilson emerald had RIs of 1.563 and 1.559, with an SG of 2.65. Chromium and nickel were the dominant colour-causing elements with smaller amounts of V, Fe and Cu. Absorption bands are due to the presence of Cr^{3+}, Ni^{3+} and Ni^{2+}, differentiating this product from Russian chromium- and nickel-bearing hydrothermal emerald. The Gilson stone was yellowish green.

A quick, non-destructive separation of natural from synthetic emerald can be carried out by infra-red spectroscopy using a Fourier transform instrument to give rapid analysis of results. The spectra vary according to the orientation of the sample with respect to the incident IR beam, giving very weak to strong response with direction, owing to beryl's birefringence. It is not difficult to separate natural and hydrothermal beryl, which contains water, from flux-grown material from which water is absent. Hydrothermal emeralds show absorption features at 4375, 4052, 3490, 2995, 2830 and 2745 cm^{-1} in the far infra-red, while natural stones do not.

An epitaxial flux growth of emerald on to opaque white beryl has been sold under the trade name Emeraldolite. The material has been made in France, and so far specimens have been opaque. In contrast to the method of growth of such products as the Lennix emerald, where the Al_2O_3 and SiO_2 are dissolved from the crucible itself with a typical flux of lithium molybdate and a boron salt, Emeraldolite is made using a chemical rather than a physical reaction involving fluorine as a transport agent, giving faster growth.

The SG is estimated at 2.66 and the RI at 1.56, both characteristic of flux-grown emerald; the hardness is approximately 8 and the material is notably tough. Under the Chelsea filter surfaces appear intense brownish red. A mosaic of tiny parallel crystal faces covers the overgrowth layer, and in some specimens the overgrowth had chipped to reveal the white beryl substrate. Characteristic signs of flux growth can be seen in the overgrowth.

Emeralds grown by accelerated crucible-rotation flux techniques in Novosibirsk were reported to reach 100 mm in length and 60 mm in diameter, yielding cut stones adequately sized for jewellery. The SG and RI are in the characteristically low range for flux-grown material, but stones are more satisfactorily identified by the presence of flux inclusions which occur as secondary healed fractures and as primary void fillings.

Aquamarine of gem quality has been synthesized by hydrothermal growth in Siberia. The main constants overlap with the natural material but some specimens at least show distinct amounts of Ni and a high Fe content. Spectra in the visible and UV areas indicate the presence of Fe and Ni, and IR spectra confirm the presence of water and Li in channel sites, thus proving hydrothermal growth. Under magnification a distinct boundary can be seen between the near-colourless seed and the light blue beryl; small groups of birefringent crystals have been noted as well as opaque hexagonal platelets, perhaps of hematite. Cavities with multiphase fillings and of irregular shape, feather-like structures, virtually planar and with liquid and two-phase fillings, and residues of the growth solution trapped as twisted veils are also found.

A Biron hydrothermally grown pink beryl is stated to have Ti^{3+} as the main colouring element. This can be detected by an absorption pattern characteristic of Ti^{3+} in octahedral coordination. Orange to blue light is absorbed by a band with two apparent maxima at 550 and 494 nm. Stones show a very strong pale pink to strong purplish pink dichroism. The IR spectrum shows absorption characteristic of hydrothermally grown beryl.

The thermal conductance of synthetic emerald can be distinguished from that of natural emerald by the use of the Alpha-test™, digital readout thermal test meter, which indicates the higher thermal efficiency of synthetic emeralds by a much lower reading than that for natural emeralds. The greatest difference is seen with flux-grown emeralds but hydrothermally grown stones can also be distinguished.

Alexandrite

Synthesis of the alexandrite variety of chrysoberyl was attempted in the middle of the nineteenth century using borate fluxes. Modern production uses lithium molybdate and combinations of lead oxide, lead fluoride and boron oxide. In 1973 Creative Crystals Inc. of San Ramon, California, produced a flux-grown alexandrite with some Fe as well as Cr to give a satisfactory effect. Kyocera of Japan produces Crescent Vert Alexandrite (marketed as Inamori Created Alexandrite in the USA). Alexandrite rods have been used in lasers, but colours are too pale for the rods to be used ornamentally as only about 0.05% Cr is incorporated.

In all cases so far reported the physical and optical constants are within the range for the natural material. The inclusions, typically wisps and twisted veils of flux material, give the best guide to a man-made origin, though a startling colour change is suspicious!

Czochralski-grown alexandrite shows very strong bright red fluorescence under both LWUV and SWUV. Most natural alexandrites show only a weak to moderate response to both types. There are no apparent inclusions characteristic of Czochralski-grown stones.

A synthetic alexandrite has been marketed under the name Nicholas Created Alexandrite. Made by J O Crystals (Judith Osmer), it has had the chromium content adjusted to give a better change of colour. The centre part of the Czochralski rod has myriads of oriented gaps or negative crystals which may resemble bubbles. Faceted stones are cut from the periphery and cabochons from the centre of the rods.

A Czochralski-grown alexandrite is being marketed under the trade name Allexite. The manufacturer is The House of Diamonair, a subsidiary of Litton Airtron. From 7 to 10 days are needed to grow a large crystal. While properties accord with other synthetic and natural alexandrite, this product shows distinct curved striae. In both colours displayed (i.e. reddish purple and bluish green) there is a noticeably strong red transmission.

The first synthetic cat's-eye alexandrite was reported by *Gems and Gemology* in 1987. It is manufactured by Inamori; the growth process is by pulling, as the parent firm Kyocera states in a patent assigned to them in 1986. Stones show a dark purplish red in incandescent light and a dark greyish green in fluorescent light. Although the fluorescence under LWUV overlaps the response of some natural material, the synthetic gives a weak, opaque chalky-yellow fluorescence under SW, with the effect seeming to be confined to surface areas; there is an underlying weak red-orange fluorescence. Unevenly spaced parallel growth features with associated colour zoning and whitish particles oriented in parallel planes serve to distinguish the Inamori stones from natural alexandrite.

Infra-red spectroscopy has helped to distinguish natural from synthetic alexandrite. Fourier transform instrumentation provides a quick evaluation of results. Observation of specimens over the range 4200–2000 cm^{-1} in the far infra-red (2300–5000 nm) has shown that natural alexandrites invariably have absorption features centred approximately at 4150, 4045, 2403 and 2160 cm^{-1} and that these are not found in synthetic stones, which give features only in the range 3300–2800 cm^{-1}, with some specimens absorbing in the range 2700–2500 cm^{-1} and others with absorption between 3700 and 3300 cm^{-1}. Absorption features between 3300 and 2800 cm^{-1} are always stronger in natural stones. This may be attributed to the absence of water in the synthetics.

Cavities in an alexandrite were shown to contain a variety of substances using qualitative energy-dispersive analyses with the help of a scanning electron microscope. Fissures surrounding inclusions were found to contain tin, copper, nickel and lead (this may have come from the polishing process). The presence of potassium-rich aluminium silicates, however, proved the natural origin of the stone.

Jadeite

In 1987 GE announced the synthesis of jadeite in a variety of colours. As the main properties do not serve to distinguish it from natural jadeite, when and if the synthetic comes on to the market gemmologists will need to look for broken edges on some of the thin layers from which it is made. Colours tend to be mottled and rather more intense than natural jadeite. A belt-type apparatus is used for the synthesis.

Malachite

A synthetic malachite made in Russia exhibits three distinct types with banded, silky and bud-like structures. The malachite is synthesized from aqueous solutions and can only be distinguished from the natural mineral by differential thermal analysis, a destructive testing method.

Synthetic Garnets

These hard, highly refractive materials are not garnets in the chemical sense since they are not silicates, but they possess the same structure as natural garnets and have the general formula $A_3B_2C_3O_{12}$ or $A_3B_5O_{12}$ if the elements corresponding to B and C are the same. Substitution can also take place as in the natural garnets. The structures will accommodate a range of dopants, the rare earths being frequently used, and their high dispersion makes the colourless varieties reasonably acceptable diamond substitutes. Work on these groups began comparatively recently, in the 1960s, when $A_3B_5O_{12}$ compounds were grown by the flux method for work on magnetic and insulating properties. The crystals grown from the flux were too small to be developed for ornamental use and it would have not been worth while scaling up the process on grounds of expense.

Instead the method of crystal pulling was used. This technique, involving lowering a seed crystal to the melt surface and then slowly drawing it upwards, is also used for ruby and sapphire growth. High-quality crystals are required for laser use, and for this reason the work on crystal characterization was hastened, giving a consequent impetus to allied but less important gemstone development.

One of the earliest of the synthetic garnets was YAG (yttrium aluminium garnet) with an SG of 4.55, an RI of 1.83 and a dispersion of 0.028 (compared with diamond's RI of 2.42 and dispersion of 0.044). Some difficulty in eliminating iron was found in early work on colourless YAG but colours were easily obtained from dopants: yttrium giving green, terbium pale yellow, dysprosium yellow-green, holmium golden yellow, erbium pink, thulium pale green, ytterbium pale yellow and lutecium pale yellow. In practice few rare-earth dopants were used for gem material. Chromium has been used to give a green which in the early days of the product was mistaken for demantoid garnet; manganese gives a darkish red, cobalt blue and titanium yellow. Neodymium, often used, will give a characteristic lilac colour with distinct colour change when the type of illumination is changed; praseodymium gives a pale green. Many of these elements give an absorption spectrum with many lines and bands; no natural mineral shows a comparable spectrum. Mechanically YAG and its analogues are hard at about 8.25. Since they are isotropic, no doubling of back facet edges can be seen. Some green YAG shows red through the colour filter; some colourless stones have been reported to show yellowish under UV and mauve under X-rays. The fluorescence spectrum has been found to show a discrete band in the yellow in some cases. No really characteristic inclusions have been reported, though some specimens show angular flux particles when they have been grown from the flux rather than by pulling.

The analogous substance $Gd_3Ga_5O_{12}$ (GGG or three Gs) has also been used as a diamond simulant, though one can hardly imagine very seriously as the SG is 7.02. This material has a hardness of 7 and is isotropic; the RI is 1.97 and the dispersion 0.045. Trace elements may result in originally colourless stones turning brown; the UV component in sunlight may be enough for this, but the effect is usually reversible when the stone is removed from UV. Some GGG show a fluorescence spectrum under both types of UV and under X-rays. This is accompanied by a pale straw fluorescence under LW and a peach colour under SW. Under X-rays the stones may fluoresce lilac. GGG, like YAG, can be doped with trivalent ions such as Cr (green), Pr (yellow) and Nd (lilac with colour change). Elements with divalent ions need a coupled substitution with a four-valent ion such as silicon. Examples are blue from the replacement of

$2Al^{3+}$ with $Co^{2+} + Si^{4+}$. Manganese coupled with silicon gives a pink to red. In general, coloured GGG is rare; the growing number of collectors of synthetic gemstones will pay surprisingly high prices for them today.

Cubic Zirconia

The most successful diamond simulant so far is cubic zirconia (CZ), a hard, highly dispersive isotropic version of the ornamentally insignificant mono-clinic ZrO_2 baddeleyite. To obtain a cubic form stable at room tempera-tures, either yttrium or calcium needs to be added: these are incorporated as the oxides CaO or Y_2O_3. Though ZrO_2 has been grown by the flux method, the high melting point of 2750 °C makes that method unsuitable for the growth of large transparent crystals; most flux-grown crystals are monoclinic in any case. As no container can withstand the temperatures required, a method which has come to be called skull melting (from the shape of the apparatus) is used (*Figure 24.27*). Scientists of the former USSR devised the cold crucible (an alternative name to skull melting, and more descriptive). A block of zirconia powder incorporating some zirco-nium metal is placed inside a cup (skull) with an open top, surrounded by copper tubes (fingers) containing cooling water. Radio-frequency heating (4 MHz, up to 100 kW) melts the powder inside the block. The heating is helped by the incorporation of some zirconium metal which heats up first. Then the zirconia powder close to the heated metal also begins to conduct

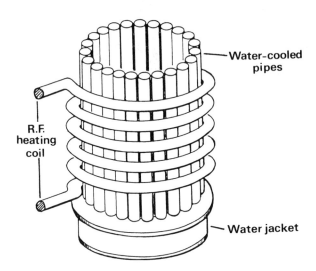

Figure 24.27   Sketch of a skull-crucible apparatus showing the vertical water-cooled pipes and the radio-frequency induction coil

electricity and to melt, the oxygen surrounding the open apparatus contributing to further zirconia formation. The powder can be topped up, but a 1 mm skin closest to the cooling water remains unmelted. The skin reduces the possibility of unwanted elements entering the melt from the apparatus materials. Crystals nucleate at the bottom of the skull, growing to form columnar crystals (*Figure 24.28*). Grown crystals are annealed in air for about 12 hours to eliminate any strain, and can then be faceted. A charge of 1 kg will give a maximum of 500 g of facetable material.

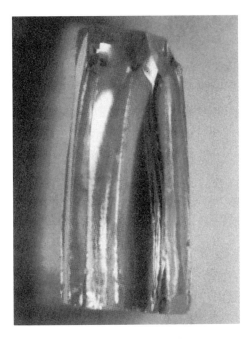

Figure 24.28   A single crystal of cubic zirconium oxide (weighing around 600 carats) as broken out of the mass of crystals in the skull crucible

Yttrium-stabilized zirconia has hardness 8.25, SG 5.95, RI 2.171 and dispersion 0.059, with some greenish yellow or reddish fluorescence. There may be parallel rows of small semi-transparent isometric crystal-like cavities extending into hazy stripes of tiny particles. Faceted stones may have rounded facet edges, small chips or percussion marks. Calcium-stabilized zirconia has hardness 8.5, SG 5.65, RI 2.177 and dispersion 0.065, with some distinct yellow fluorescence. The form is virtually inclusion-free, with faceted stones showing similar features to the yttrium-stabilized stones. Both types of CZ have a far lower thermal conductivity than diamond, and the polished surface will break an ink drop into beads, contrasting with the coherent drop seen on diamond. CZ is doped to give a variety of colours. Nassau lists the following:

CeO_2, Ce_2O_3 to give yellow, orange, red
CuO, Fe_2O_3, NiO, Pr_2O_3, TiO_2 for yellow, amber, brown
Er_2O_3, Eu_2O_3, Ho_2O_3 for pink
Cr_2O_3, Tm_2O_3, V_2O_3 for olive green
Co_2O_3, MnO_2, Nd_2O_3 for lilac, violet.

Cubic Zirconia is opaque to X-rays and is thus easily distinguishable from diamond. For everyday testing the thermal conductance probe is probably the most convenient instrument, followed by the reflectance meter. Some manufacturers have incorporated both methods of testing in the one instrument.

Emerald-green and good-quality sapphire-blue colours were achieved in CZ in the 1980s, although production of CZ in general took off early in that decade. To produce the emerald and sapphire colours, much more stabilizer is added than usual. The original Russian manufacturer called this product C-Ox.

White, pink and black non-transparent CZ has also been produced in Russia. The material is sold as cabochons or as beads. The white material is a uniform milky white and the pink is a uniform medium pink. Using strong transmitted light the cabochons show banded or striated colour distribution. Some of the black pieces appear dark brownish red with strong transmitted light. The lustre of all samples is high.

A dark yellowish green CZ appeared green through the Chelsea filter and gave absorption bands at 607, 583, 483, 472 and 450–443 nm. The product resembled good-quality tourmaline.

Opal

Once the cause of the play of colour in opal had been established in the 1950s, the way was clear for scientists to attempt the manufacture of an artificial opal-like material. Three stages are necessary for successful synthesis: the production of uniformly sized spheres (of whatever material); the settlement of the spheres in an orderly close-packed array; and the consolidation of the structure, filling the voids between the spheres with a substance that will produce a hard material that can be fashioned.

If the sphere material is silica the product can be called synthetic opal; if other substances are used, 'opal simulant' or 'imitation' is a preferable term.

Silica spheres may be prepared by dispersing an organic silicon compound in fine droplets in an alcohol–water mixture. Tetraethyl orthosilicate

$(C_2H_5O)_4Si$ has been used. When ammonia or similar mild alkali is added to the mixture, which is carefully stirred, silica spheres containing some water are formed by the reaction

$$(C_2H_5O)_4Si + 2H_2O \rightarrow SiO_2 + 4C_2H_5OH$$

(The spheres must all be the same size.) It may take one year to settle the array, which will then be sintered and subjected to a small degree of pressure before fashioning. Gilson began producing opal in various types in 1974: black, white, water and fire opals are on the market, the quality often being very high. A diagnostic lizard-skin effect (*Figure 24.29*) can be seen within the patches of colour; stones may also show a chalky-blue fluorescence with a similar but stronger surface glow, with some phosphorescence. These effects will not be seen in all Gilson opal products, but the glow where seen is stronger under LWUV. Gilson opal uses silica spheres, but the composition does vary: some yellowish brown stones contain a variety of organic compounds, and another sample has been reported to contain about 0.5% crystalline ZrO_2.

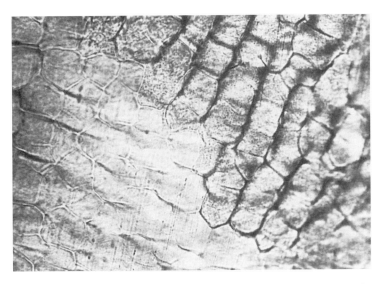

Figure 24.29   Gilson synthetic white opal, showing lizard-skin shrinkage cracks:
×118 (photo W F Eppler)

Where silica spheres are not used, a variety of substances has taken their place. Latex, for example, gives a stone with a very low SG, only just above 1.0. The stones are hardened by impregnating with another plastic with a different RI. Another product has been shown by IR spectroscopy to consist of a co-polymer of styrene and methyl methacrylate. This stone had an SG of 1.17 and an RI of 1.48.

Gilson has produced a hydrophane opal which improves in appearance after being placed in water. The play of colour in the stone examined was predominantly orange and appeared in small angular patches. Transformation on immersion took approximately 35 minutes to reach its full effect. On removal from the water the stone appeared slightly more transparent than before.

A sugar-treated opal coated with plastic has been reported, with spot RI readings on the dome of 1.45 and on the base from 1.56 to 1.57. Under LWUV the dome fluoresced a very strong yellow-green, while the base gave a very strong chalky blue-white. SWUV gave similar but weaker reactions. The surface could be indented by the point of a pin.

Examination of Kyocera products showing a play of colour (opal imitations with dark or light background) shows that they consist of amorphous silica glass with no admixture of water; thus they fall outside the compositional range of natural opal. They should be designated as opal simulants.

Less Common Synthetic Products

Materials described in this section are known to have occurred as fashioned stones and all are likely to turn up from time to time. This could be said of almost any man-made substance, but only those which are persistently though sporadically reported are included here. Some have natural counterparts, others not. All materials listed are transparent.

Bismuth germanate ($Bi_{12}GeO_{20}$ or $Bi_4Ge_3O_{12}$) These can give a fine orange with hardness 4.5, SG 7.12, RI 2.07. Isotropic.

Bismuth silicate ($Bi_{12}SiO_{20}$) Also gives brown to orange; colourless crystals may be obtained by annealing. Isotropic with RI near 2.0.

Bromellite (BeO) Very hard (hardness 9) hexagonal material with SG 3.01 and RI 1.720–1.735. Rarely cut owing to the toxicity of beryllium dust. Usually colourless.

Fluorite (CaF_2) A wide range of dopants has been used to produce coloured fluorites with the same properties as the natural mineral. Spectacular examples include a colourless crystal with a brilliant green fluorescence with exceptionally long phosphorescence under X-rays and red crystals, unknown in nature.

Germanates With similar properties to silicon, germanium has been used in a number of compounds, some of which have been transparent. As well as the bismuth germanates cited above, a lead germanate ($Pb_5Ge_3O_{11}$) of a fine yellow colour is known.

Periclase   A colourless though sometimes doped isotropic material with hardness near 5, SG 3.55–3.60 and RI 1.73. The trade name Lavernite has been used; some stones fluoresce whitish in UV.

Phenakite (Be_2SiO_4)   A vanadium-doped crystal is known with a most attractive light blue colour, and colourless crystals could easily be produced. Properties as for the natural material.

Scheelite ($CaWO_4$)   Various dopants are added to colour this material, which is a possible diamond simulant when colourless. The strong sky-blue fluorescence under SWUV is the most spectacular feature of the colourless variety.

Silicon carbide (SiC)   Very hard (hardness 9.5) anisotropic greenish material whose dispersion (twice that of diamond) may cause confusion until the DR (0.043) is spotted. SG 3.20, RI 2.648–2.691.

Yttrium compounds other than YAG   Include: (1) $YAlO_3$ with hardness 8+, SG near 5.35, RI 1.938. Isotropic; may be colourless diamond simulant or be coloured by rare-earth dopants. (2) $Y_3(Al,Ga)_5O_{12}$ with a strong red fluorescence under LWUV; has SG 5.05–5.08, RI 1.88–1.90; a possible imitation of green grossular but with observable flux inclusions. (3) Y_2O_3 with trade name, Yttralox is a ceramic product with a high dispersion, SG 4.84, RI 1.92 and hardness 7.5–8. It is isotropic.

Zincite ($(Zn,Mn^{2+})O$)   Has been doped to give a variety of colours. Hardness about 4.5, SG near 5.5 and RI 2.0; hexagonal.

Zircon   Zircon and some analogous substances including $ThGeO_4$, Mg_2SiO_4 and $ZnSiO_4$ have been grown. Hydrothermal zircon doped with vanadium gives a fine purple. As zircon is soluble in most convenient fluxes, the hydrothermal method is used for growth. Rare-earth dopants colour the analogues.

The materials described above are all capable of being grown to reasonably large sizes. Various other materials have been synthesized. Greenockite (CdS) is orange-yellow, with hardness near 3.5, RI 2.5 and SG 4.9. Powellite ($CaMoO_4$) with SG 4.34 and RI 1.92–1.98 has been produced in a light blue faceted form (see also below). Lead tungstate, highly dispersive, soft and yellow, is $PbWO_4$ (the natural mineral is stolzite) and is tetragonal. Berlinite is $AlPO_4$ and structurally isomorphous with quartz, showing left- and right-handed forms; it shares a similar interference figure with quartz, and a possible fashioned bead could be mistaken for quartz if this test were used. The hafnium analogue of zirconia, HfO_2, has been grown in a colourless form but is far more expensive to manufacture than zirconia owing to the cost of hafnium feed material. Cut stones of lithium tantalate, the monoclinic zinc tungstate sanmartinite, Nd-doped YVO_4, the lead molybdate wulfenite (highly dispersive) and the magnificent red proustite (Ag_3AsS_3) are also known.

The trigonal material lithium niobate has a dispersion in excess of that of diamond (0.130 against 0.044) but its high DR of 0.090 gives it away via the doubling of facet edges. Grown by the Verneuil process, it has a hardness of just over 5, SG 4.65 and RI 2.2–2.3. Coloured versions from doping are occasionally seen. An early trade name was Linobate.

During January 1972 there was a report of the examination of a cut powellite, a calcium molybdate-tungstate, $Ca(Mo,W)O_4$, a mineral allied to scheelite. Powellite has a hardness of 3.5 on Mohs' scale; the RIs are given as 1.974 and 1.984, and the SG is 4.34. The stone was pink in colour owing probably to doping with holmium. The fluorescence under SWUV was said to be green, and unlike scheelite, which glows blue under the SW lamp and is inert under the LW lamp, this specimen of powellite was said to glow green under the LW lamp as well as under the SW lamp. Natural powellite glows yellow to golden yellow under the SW lamp only, so this may give a clue to the synthetic nature of the stone.

Oolongolite, with RI 1.93–2.00, dispersion 0.030, SG 6.7–7.0 and hardness 7.5–8, and singly refractive, is being manufactured in Switzerland. Colours seen include dark and medium blue, medium and dark bluish green, lilac and colourless. The material is reported to have a garnet-type structure.

Sodalite has been synthesized in China. Crystals are heavily included and the blue colour arises from irradiation.

Imitation Gemstones

We have already seen that a synthetic gemstone must have a natural counterpart, but any gemstone can imitate another; to the collector of synthetic stones the natural version can imitate the synthetic as well as the other way round! The chief gemstone imitation is of course glass, but some substances described as 'synthetic' are more accurately classed as imitations; they include Gilson lapis lazuli and coral, which are described in Chapter 27.

25

Composite Gemstones

J. Rouse

A composite, or assembled, stone is no new device, for they have been known since the days of the Roman Empire; the Roman lapidaries constructed their *jaspis terebinthizusa* by cementing together three different coloured stones with Venice turpentine. Many different types of doublets were mentioned by Camillus Leonardus as early as 1502. King refers to Italian glass intaglios of the eighteenth and nineteenth centuries which could not be distinguished from the real except by the use of a file, and to baffle this mode of detection the dealers used the ingenious contrivance of backing the paste with a slice of real stone of the same colour. Presumably any test would be made on the base rather than on the incised front. King also refers to the fact that the same method is used for forging all coloured stones, a paste of proper colour being backed by a suitably faceted rock crystal.

Composite stones are generally described as *doublets* when the stone consists of two main pieces, and *triplets* when three pieces are used or when two pieces are joined using coloured cement. Internationally there are some slight differences in the nomenclature, for in North America the modern soudé-type stones are known as triplets while in Europe these stones are called doublets. Sometimes a diamond crown is placed above a much smaller faceted diamond, termed a 'piggy-back' doublet.

The fundamental reason for the manufacture of these composite stones is in many cases to obtain a larger stone from a given natural material, or to produce a stone of much better colour and appearance. Other considerations are the provision of a harder wearing and more lustrous surface to a glass imitation, which is probably the reason underlying the making of the garnet-topped doublets, although it has

531

been suggested that the use of garnet here is to overcome the testing by a file. Further, doublets may be made for the purpose of supplying a rigid support for gem material which can often be found only as thin and easily damaged slices. This is the most logical reason for the production of opal doublets.

True Doublets

The true doublet or genuine doublet consists of two parts constructed by cementing together two suitably fashioned pieces, one for the crown of the stone and the other for the pavilion, both pieces being cut from similarly coloured material of the same species (*Figure 25.1a*). Thus true doublets can be made of two pieces of diamond, or of ruby, sapphire, emerald or other species of stone. Except in the case of opal on opal doublets, true doublets are rarely encountered. However, there has been a report of a jadeite doublet in which both sides were jadeite: the top was a thin layer of green jadeite and the bottom a thicker layer of white jadeite in the form of a tablet, 20 × 6 × 2.5 mm.

Semi-Genuine Doublets

When a piece of genuine material of the stone simulated forms the crown of the composite stone and this is cemented to a pavilion cut from a less valued stone, or even glass, the stone is known as a semi-genuine doublet. The most important stone of this type is the diamond doublet (*Figure 25.1b*), a composite stone where the crown consists of a piece of real diamond and the pavilion is any suitable colourless stone, such as rock crystal, white topaz, synthetic white sapphire or spinel, or even glass.

When unset diamond doublets may easily be detected, not only by the join at the girdle which may be seen when the stone is inspected with a hand lens, but convincingly when the stone is immersed in di-iodomethane or monobromonaphthalene, when the difference in relief of the diamond top and the material of inferior refraction forming the base is readily seen. It must be pointed out that there is always a danger in immersing such doublets in these liquids as they may have a deleterious effect on the cement and cause pseudo-flaws or even cause the two parts to separate.

Most often diamond doublets are set with rubbed-over settings, which in England are often called gypsy settings, and such a setting effectively covers the join of the two pieces. Diamond doublets may be readily detected if the jewel is held so that the table facet of the stone is slightly

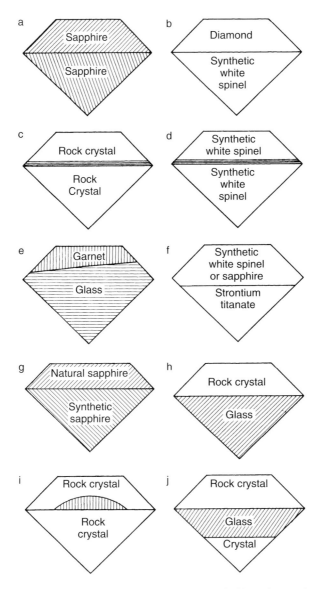

Figure 25.1 Types of composite stones: (a) True doublet with top and base of genuine stone of the species simulated; (b) Diamond doublet, real diamond top and any colourless inferior stone for the base; (c) Quartz doublet, or beryl doublet, the original soudé emerald; (d) Synthetic spinel doublet (soudé sur spinelle); (e) Garnet-topped doublet; (f) Doublet of two synthetic stones, usually made to simulate diamond; (g) Doublet with natural sapphire top and synthetic sapphire (can be synthetic ruby) base; (h) Quartz on glass doublet; (i) Liquid-filled hollow capped doublet; (j) A triplet

tilted away from the observer, when a dark border to the opposite edge of the table facet will be seen (*Figure 25.2*). This is due to reflection of the facet edge on the cement layer. There may also be seen prismatic colours due to air films which have penetrated along areas of deterioration of the cement layer. A word of warning: the so-called *lasque diamonds*, large thin parallel-sided plates, often with the top edges faceted as seen in Indian jewellery, may also show the dark shadow of the table facet edge reflected from the lower surface of the stone, which may lead to an incorrect diagnosis.

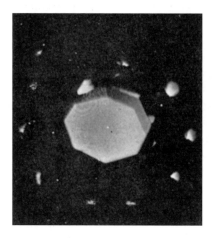

Figure 25.2 The reflection of the edge of the table facet on the cement layer, and Newton's rings where the cement has deteriorated, seen in a diamond doublet (by courtesy of the Journal of Gemmology)

Except for diamond doublets and opal doublets, semi-genuine doublets are uncommon. However, Weinstein mentions pale-coloured emeralds backed with green glass, and an American journal has reported a semi-genuine doublet made of a crown of flawed aquamarine and a base of green glass, and another in which the top was a very transparent chrysoberyl cat's-eye on a darker base, the nature of which was not recorded. Recently two new types of semi-genuine doublet have made their appearance on the market. In both cases the crown of the stone is made of natural greenish yellow sapphire and the pavilion either of synthetic blue sapphire or of synthetic ruby (*Figure 25.1g*). Again, when unset the distinction is easy, but when set and of sapphire colour the natural zoning and silk of the top portion, and the fact that the stone shows the 450 nm absorption band, tend to prove the stone genuine. The base, if one thinks that far, will show the curved bands and gas bubbles of synthetic sapphire, or the use of a SWUV lamp will show the greenish glow of the synthetic sapphire base. The same applies to the red variety, for the base will show the curved fine lines and gas bubbles, and the LWUV lamp will show the base to glow crimson and the top to be inert. A major clue to this imitation is the abnormally shallow crown of the included, often deep green, sapphire which produces the green reflections from the junction plane which are characteristic of this composite. There is also a report of a jadeite doublet whereby the quartz top is glued with a green jelly-like cement and attached to a green-dyed base.

False Doublets

'False doublets' are composite stones where the crown consists of rock crystal or other colourless stone which is cemented to a suitably coloured glass (*Figure 25.1h*). Three specimens of such doublets, one of sapphire blue colour, another a purple and the third a yellow colour, gave in all cases an RI reading for quartz from the crown, but an isotropic reading of 1.51 for the glass base. The SGs were 2.61, 2.56 and 2.55 respectively. The blue stone exhibited the absorption spectrum of cobalt; the purple stone showed a faint cobalt spectrum with a fainter band in the blue. The danger with this stone lies in the fact that if an RI measurement is taken only from the table the stone could be mistaken for a true amethyst. The yellow-coloured stone resembled the yellow synthetic spinel, and like it showed a strong yellow-green fluorescent glow under UV from the glass base. This glow, examined with a spectroscope, unlike the glow from the synthetic spinel showed a discrete spectrum, indicating that the glass was coloured by uranium. Except for the yellow stone where the strong glow masked the effect, irradiation with UV showed the cement layer as a bright line around the girdle of the stone.

Doublets have been seen where the crown is glass and the pavilion is quartz, the reason for such a type being hard to determine. A most unusual example of a doublet was one which consisted of practically all of the stone being white topaz with just the tip of the pavilion made of a piece of natural blue sapphire. Presumably the idea was that the blue base would suffuse colour through the stone rather in the same way that native lapidaries leave colour at the base of their sapphires. It did not seem to work in the case of this doublet which may have been an experimental try-out.

Hollow Doublets

Max Bauer mentions hollow doublets. They are said to consist of a crown of rock crystal or glass which is hollowed out below, the walls of the hollow being highly polished. This cavity is filled with a coloured liquid and the whole closed in with a pavilion of the same material as the crown (*Figure 25.1i*). Such composite stones, which are extremely rare,

are said to be readily spotted for the cavity of coloured liquid is visible when the stone is observed from the side.

Imitation Doublets

This term is usually applied to doublets which consists of two pieces of colourless glass cemented together with a coloured cement. Two such red-coloured specimens were found to have a glass with an RI of 1.52, and the SGs were found to be respectively 2.478 and 2.488. The only point of interest was that the red colouring matter gave an absorption spectrum reminiscent of that of almandine garnet.

Triplets

True triplets are constructed of three pieces, the crown being some real stone such as quartz, and the tip of the pavilion being of similar material, the central portion being, usually, coloured glass (*Figure 25.1j*). The idea behind this type appears to be that not only will the crown resist a file but the lower part of the pavilion also. Such triplets are not common.

Garnet-Topped Doublets

These are the more important types of doublets but are now slowly going out of circulation. They are made by fusing a thin slice of almandine garnet to a blob of coloured glass and are produced in many different colours to imitate many different stones, particularly ruby, sapphire, emerald, amethyst, peridot, topaz and even colourless stones (*Figure 25.1e*). It is the colour of the glass which controls the colour of the finished stone, the thin layer of garnet having little or no influence on the resulting colour.

When garnet-topped doublets were first made provides an intriguing problem. No specific mention was made of these doublets in the long series of editions of Herbert Smith's *Gemstones* until the 1940 edition, but Michel gives a very good account in his book published in Leipzig in 1926. A much earlier reference to the fusing of two parts of a doublet together is given in the Viennese book written by Schrauf in 1869. A little-known book *L'Industrie lapidaire* by Burdet attributes the invention of the doublet, as we know it, to Cartier, lapidary of the Comte de Mijoux, in 1845. Frank E. Goldie, who spent his early days as an apprentice in the Jura Mountains just after the First World War, tells how these stones were made:

After the day's work and our evening meal, a small pile of doublet moulds were put on the kitchen table, together with a packet or two of thin slices of garnet, and also a mound of glass squares. Everyone sat down to their respective job, even neighbours who called in to pass the time with us. Some put in the garnet, others would place a piece of glass on top. The moulds were made of baked clay, measuring about 16 in by 10 in with a number of indentations depending upon the size of doublet required. Actually it was a very pleasant way of passing an evening, refreshments were *ad lib* and the conversation always interesting and amusing.

Friday was an important day, the kiln was stacked with prepared moulds, the fire lit and heated to the correct temperature. Next morning the moulds having cooled were removed. The rough doublets were then sorted into their different colours and sizes ready to be given to the local craftsmen who cut them in their homes.

Garnet is the only stone which will easily fuse to glass, and the objects removed from the firing moulds are each composed of a thin slice of garnet which has fused to a ball-like piece of glass. When these are cut no great care is taken to get the plane of joining in the plane of the girdle. Hence, the join may be perhaps half-way up the side facets of the crown, and may not even be at a regular distance from the girdle of the stone. Examination with a hand lens will show this join and, what is just as important, will show the difference in lustre of the garnet top and the glass base (*Figure 25.3*). If the stone is loose and laid table facet down on a sheet of white paper, a red ring will be seen round the stone. This is obvious with all such doublets except those of red colour, which masks the effect.

Figure 25.3 A garnet-topped doublet showing the difference in relief between the garnet top and the glass base

Examined by a low-power microscope, garnet-topped doublets may show crystals or reticulation of needles in the garnet top, and, owing to the fusing of the glass, a layer of bubbles in the plane of joining of the two pieces (*Figure 25.4*). The garnet used for the top of these doublets has an RI of about 1.78, and the glass base usually between 1.62 and 1.69, although a glass with an RI as low as 1.51 has been met in these doublets. Thus, if such stones be immersed in oil and viewed sideways the difference in relief of the top and base will be strikingly revealed. The SG has little significance but is generally above 3.34 (di-iodomethane). SWUV usually induces a whitish or greenish glow in glass and this may show up the two parts of such a doublet as the garnet part is inert under such radiations.

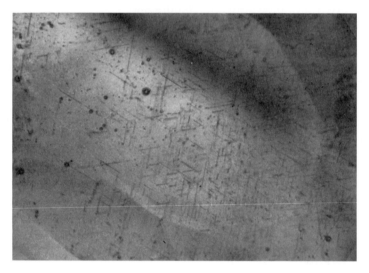

Figure 25.4 Needle inclusions in the top and a layer of bubbles in one plane seen in a garnet-topped doublet. The difference in relief of the garnet top can also be seen

Soudé-Type Stones

Early Soudé Emeralds (Quartz Doublets)

These stones consisted of two pieces of rock crystal forming the crown and pavilion of the stone cemented together by a green-coloured gelatine. These stones were first made to imitate the emerald and were called soudé emeralds from the French *émeraude soudé* (soldered emerald); under present legislation 'quartz doublets' may be the better term. Other colours are known, such as a sapphire blue and an 'alexandrite' colour resembling the colour of the alexandrite-type fancy-coloured synthetic sapphire. Both these stones showed the three-band cobalt spectrum with the 'alexandrite' one showing an extra band in the blue-violet. The green-coloured stones of the early make showed a red residual colour when viewed through the Chelsea colour filter, and a further disadvantage of the greenstones is that the colour tends to deteriorate and turn to a yellow colour, thus producing a 'citrine'. The RIs of 1.544 and 1.553 will, of course, indicate the nature of the stone, and between crossed polars the orientation of the two pieces of quartz are seen to be different. In all colours of these stones there was a significant lowering of SG, despite the thin layer of the gelatine. The value was found to be fairly constant at 2.62. Under UV the cement layer will tend to glow and is most prominent at the girdle, indicating that the stones are doublets.

Modern Quartz-Type Soudé (Modern Quartz Doublets)

Believed to have entered the market during the early 1920s these composite stones are similar in construction to the earlier type, with the exception that the unstable gelatine is replaced by a layer of coloured and sintered glass – or at least that is what it is thought to be. The colour is probably due to a metallic oxide and not an organic dyestuff. In these stones the RI is that for quartz but the SG is higher at about 2.8; this high value has been ascribed to the heavy lead glass of the coloured layer, but this has not been proved. In North America these composite stones, which show green and not red through the Chelsea colour filter, are known as triplets. The colour layer of this type does not glow under UV and it is said that colours other than green have been produced.

Synthetic Spinel Doublets

During 1951 Jos Roland of Sannois, France, produced a type of soudé stone in which the crown and pavilion were made of colourless synthetic spinel instead of quartz. These were known as *soudée sur spinelle*, but in English nomenclature are better termed 'synthetic spinel doublets'. These stones are made in all colours and the coloured layer may also be a sintered heavy lead glass, for again there is a rise in SG to 3.66–3.70, with one exception, a black-coloured stone, which had an SG of 3.63. The RI of this type of composite stone is 1.73.

When the quartz types of these doublets are immersed in water and viewed sideways the clear colourless parts of the crown and pavilion with the dark line of colour in the plane of the girdle are clearly seen. With the synthetic spinels this is not so evident if water is used, but with oils of higher refraction the effect is well seen (*Figure 25.5*). All synthetic colourless spinel fluoresces with a strong bluish white glow under

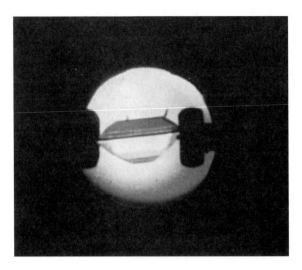

Figure 25.5   A spinel soudé emerald immersed in di-iodomethane and viewed through the side clearly shows the colourless top and bottom and the line of green colour

SWUV and this is useful for, as reported by Richard Liddicoat, the cement layer does not glow and shows up as a dark line indicating the composite nature of the stone. No characteristic absorption was seen whatever the colour of the stone, and under the Chelsea colour filter the greenstones showed a green residual colour; the blue zircon and aquamarine-coloured material showed green and *not* orange as in the same colours in synthetic spinels. The amethyst and sapphire-blue coloured stone did show an orange colour when examined through the filter.

Beryl Doublets

In 1931 Kraus and Holden illustrated a triplet consisting of a crown and pavilion of pale emerald or beryl separated by a thickish layer of green glass. It would be fair to assume that in some cases the slice of glass could be replaced by a green-coloured cement layer, and indeed such types of emerald doublets were apparently encountered in early days, but they were scarce.

During 1966 the firm of Kämmerling of Idar-Oberstein produced a doublet comprising two pieces of colourless beryl, pale aquamarine or poor-quality emerald for the crown and pavilion of the stone, the two pieces being cemented together with an emerald-coloured Duroplastic cement. These were marketed under the trade name Smaryll. When two pieces of aquamarine are used, negative crystals or canals ('rain') indicate the nature of the object, and, further, the stones show green under the Chelsea colour filter and the spectroscope shows the typical dyestuff lines in the red. When unset, examination in water will usually show the division made by the dark colour line.

More recently the crown and pavilion have been made of two parts of real emerald. These stones provide a greater difficulty in identification, especially if they are in a setting. They may show an RI for emerald and the spectroscope may show a trace of the chromium spectrum of emerald, usually and happily with the saving grace of the dyestuff spectrum. The emerald-type inclusions and the possibility of a reddish residual colour through the Chelsea colour filter help to make things difficult. UV does not seem to greatly assist in this case.

Synthetic Doublets

Apart from the sapphire/synthetic-sapphire and sapphire/synthetic-ruby doublets mentioned in the section on semi-genuine doublets which came out at about the same time, a new type of doublet appeared on the market during 1971. This new composite stone consists of a crown of synthetic colourless spinel or sapphire and a pavilion of colourless strontium titanate (Fabulite) (*Figure 25.1f*). The probable notion behind the manufacture of such stones is to give a better wearing face to the rather soft strontium titanate, and perhaps to reduce the exceptional fire of that stone. These composite stones are usually joined below the girdle so as to prevent the edges chipping. They are not difficult to identify for the RI taken from the table will be 1.72, that of synthetic spinel, while the fire of the stone is too great for spinel. Further, the SWUV lamp will reveal the synthetic spinel top by its strong whitish glow. There has also been a report of a doublet made with a synthetic sapphire crown and pavilion. The object of making such a doublet seems obscure.

Opal Doublets

Opal, which shows a fine play of colour, is often found in too thin a vein to cut any solid gemstone from it. Such pieces are then backed by other common opal (potch opal, which shows little or no play of colour), by black onyx, or by a black glass called 'opalite' which is obtained from Belgium. Such a device supports the thin piece of colourful opal and allows the stone to be set into jewellery. Opal doublets on black glass or black onyx are fairly easy to detect, but opal on opal doublets are much more difficult, and when in a setting it may be impossible to be sure. The reason for this is that many natural opals are cut with the natural backing of the original seam wall, and when this is fairly straight the stone can resemble to a great extent a doublet. Careful examination with a lens may assist in determining whether the stone is a doublet or not but in most cases it is necessary to remove the stone from its setting in order to be quite sure. When looked through from the back most opal doublets appear dark but some transmit light; and then by lens or microscope observation, through either the front or the back, flattened bubbles may be seen which indicate a doublet.

Another version is the *opal triplet*, called by the trade name Triplex Opal. This composite stone is a simple opal doublet, usually with a common opal backing, to which is cemented over the top part of the opal, a suitably fashioned cabochon of rock crystal. Similar triplets of opal made with caps of synthetic colourless spinel or synthetic colourless sapphire, have been reported. The idea seems to be to present a better-wearing face to the opal and to give the slightly magnified opal more brilliancy, as to some extent the quartz cover acts as a lens. The overall effect is a little unreal.

An imitation opal, which in fact could be termed a 'false doublet', consists of a cabochon of rock crystal or glass to the back of which is cemented a slice of iridescent MOP, either from the pearl oyster or from the colourful paua shell (*Haliotis*), or the abalone as it is called in America. Resin-topped abalone shell is also encountered.

A variation is the Schnapperskin triplet in which the colourful layer consists of fish skin with red or blue dye. They are detected by the false hue of the colours and the fact that the skin shows grey at the girdle. Another type is a hollow-backed cabochon of rock crystal which is filled with clear plastic and opal chips.

Another interesting variation is an example of assembled ironstone opal. The opal, in this case, was cemented to the brown matrix of typical ironstone opal. The cement was of the same colour as the matrix, thus effectively masking the unevenness of the opal to the matrix. The glue was checked with a thermal reaction tester and it melted when it was touched.

Jadeite Triplets

During 1958 there came on the market a jadeite triplet which has the colour of Imperial jade. These stones were made up of three pieces, the components being: a hollow cabochon of very fine translucent white jadeite about 0.5 mm thick, a cabochon of smaller size made of the same white jadeite cut to fit into the hollow cabochon, and a flattish oval piece fashioned so as to close in the back. The central cabochon is coloured with a jelly-like dyestuff of Imperial green colour. The dyed centre piece is inserted into the hollow cabochon and the bottom piece cemented on and repolished to make a perfect fit. When unset the nature of the stone is readily seen from the join of the outer pieces; it is when the stones are set with the edge concealed that they are so deceptive. They can, however, be detected by the characteristic dyestuff absorption band or bands in the red part of the spectrum.

Star Stone Doublets

A number of attempts have been made, with some success, to produce a star stone doublet. In one case a star doublet was made using a black star sapphire as a base with the cabochon top of synthetic padparadscha sapphire, and a similar doublet was constructed of a cabochon of medium blue synthetic spinel cemented to a flat base of a natural star sapphire. Another composite stone consisted of two pieces of synthetic blue spinel with a very thin foil engraved with intersecting lines at 120° cemented between the top part of the cabochon and the base part.

Some few years ago there was produced the so-called 'star rose quartz doublet', which was constructed by preparing a cabochon cut from a correctly oriented piece of pale star rose quartz, to the back of which was cemented a blue-coloured mirror, may be by a sputtering technique, and in some cases this was backed by a base plate of some other stone. Although rose quartz does sometimes show asterism by reflected light (epiasterism), the best effect is found by transmitted light (diasterism), and this is the reason for the mirror at the back. Unlike the star corundums and other natural star stones, it is found that when the star in a star rose quartz doublet is viewed under an electric lamp, an image of the lamp bulb is seen at the crossing of the rays of the star. The nearer the stone is brought to the lamp the larger the image becomes. The colour of these star doublets is not quite like that of a natural star sapphire, probably owing to the transparency of the rose quartz being greater than that of most natural star sapphires, and, of course, the lustre of rose quartz is less than that of sapphire.

Cameo and Intaglio Doublets

Cameo doublets, where the raised portion is separate and is only cemented on to the base, are only occasionally met, except perhaps in some cameos made of porcelain. Intaglio doublets are more common and date back to Roman times, if not before. Similar objects of more modern vintage are encountered today. They are usually two pieces of glass, the upper piece either engraved or, more commonly, moulded to produce the intaglio. This top piece is then cemented to a base plate, usually with a reddish brown cement in order to imitate cornelian. The pigment used is usually not very stable and it is common for the colour to deteriorate and the cement to craze. A more stable and convincing doublet has the back made of a piece of real cornelian.

Moss Agate Doublets

These doublets are produced by placing a chemical, such as manganese dioxide, with gelatine, on to a glass plate where a planar dendritic (tree-like) crystal will form on the gel. The glass plate is gently heated to drive off the excess water. Another clean glass plate is then cemented on to the top and the whole is then ground to a suitable form, either as a flat plate or as a cabochon. When the stones are unset the join between the glass plates readily reveals the fake, and even when set the greater transparency of the piece compared with true moss agate indicates that something is wrong.

Turquoise Doublets

Unimportant, but mentioned for the sake of completeness, is the turquoise doublet constructed by cementing a base of blue-stained chalcedony, or other suitable material, to a cabochon of turquoise-coloured opacified glass. Very few of these objects have been seen.

Mosaic Doublet (Triplet)

An experimental doublet was produced in 1972 by Suzuki of Australia and is unique in that between two pieces of colourless stone, usually crown and pavilion of synthetic white spinel, is sandwiched a transparent filter in which there is a mosaic of three basic colours. This mosaic is made by first producing a larger original by painting squares, rectangles or triangles of the three basic colours with the addition of areas of white in random position, presumably on a white card or even glass. This is then photographed on reversal colour film to a suitable size of pattern and this film is cemented between the crown and pavilion of the stone.

26

Gemstone Simulants

Faience

The simulation of precious objects by substances of much less value dates back to pre-history. An artificial blue frit composed of quartz, malachite, calcium carbonate and sodium carbonate had been carved into small objects as early as the 6th dynasty (3503–3335 BC). It is, however, faience, a glazed siliceous ware made in Egypt as long ago as pre-dynastic times (before 4777 BC), which is the more important as an early imitation gemstone.

Faience, which was used for beads, necklace pendants and rings, consisted of an inner core of gritty material, probably powdered quartz, and an outer coating of coloured vitreous glaze. Occasionally a thin middle layer was present between the core and the glaze, which was apparently incorporated in order to enhance the glaze. Although glaze was used in Egypt so early, glass itself was not used before the 18th dynasty (1587–1328 BC).

Glass

The glassy state arises when a melt becomes so viscous when it cools that the crystalline state cannot form and only short-range atomic order is found. During the passage of time some of the constituents of the glass may crystallize, a process known as 'devitrification'; this is positively encouraged when glass ceramics are made. The glassy or vitreous state can be described as metastable, and glass is one of the few amorphous substances known.

Silica (SiO_2) forms the basis of most glasses as it is always prone to form glassy substances on cooling. Most non-ornamental glass such as window glass is Na-Ca glass. The addition of boron or aluminium oxides gives greater resistance to heat and to chemical attack. Lead oxide increases dispersion and improves optical performance. Lead (flint) glasses are used for cut glass ('crystal') ornaments and for the glass in refractometers; a high lead content may tarnish the glass which will be very soft and brittle with a high SG. When a very high dispersion is needed, thallium may be added. Addition of a wide variety of agents such as selenium, copper or gold gives red colour, uranium compounds gives a yellow-green colour, and cobalt oxide gives blue colour. Final colour also depends upon oxidizing or reducing conditions during manufacture and upon annealing after the glass is cooled. Translucent or opaque glass may need the addition of an opacifier.

Very cheap glass gemstones will be pressed in moulds, only the more expensive varieties being faceted. Moulded glass can be identified by its rounded facet edges. The names 'paste', 'strass', 'rhinestone' and others once had specific meanings but are now used so loosely that their history is no longer relevant to modern commercial production.

Most glass used as gemstone simulants has a hardness of 5.5–6.0 and will contain tiny to large, well-shaped gas bubbles which are often elongated. A distinctive swirliness from incomplete mixing of the constituents also helps to identify a glass which, as a poor heat conductor, feels warm to the touch compared with inorganic crystalline substances. Any stone which is isotropic (especially when showing anomalous DR) and has an RI between 1.50 and 1.70 is more likely to be glass than anything else.

Some interesting glasses include beryl glass (fused emerald, scientific emerald) with the emerald composition but made by melting the constituents and then cooling the melt. This is softer than emerald and has an RI near 1.52 and an SG varying from 2.39 to 2.49, usually 2.42. Other beryl colours are obtained by adding cobalt (mass aqua) or didymium (pink). A very convincing cat's eye imitation (Cathay stone, Catseyte) is made with thin glass optical fibres in fused mosaics, the fibres being of several distinct glasses (fibre optic faceplates). Stacked in cubic or hexagonal arrays, they give a fine sharp eye as there can be about 150 000 fibres per cm^2 (over a kilometre of fibre in a small cabochon).

Fluorides and phosphates may be added to glass to give an imitation of moonstone or even pearl. The glass must contain lime so that the calcium compounds can precipitate to give the translucency desired.

Plate 32
The lack of uniformity size, shape, and colour in strands of natural pearls
(Chapter 28)

Akoya	Black lip	Brown lip	Black pen shell
Trochus	White lip	Dyeable white freshwater	Turbo/Tamagai shell

Plate 33
Shell buttons (Chapter 29)

Plate 34
Transillumination reveals the layered structure of the shell
bead that is centrally located in a bead nucleated salt water
cultured pearl

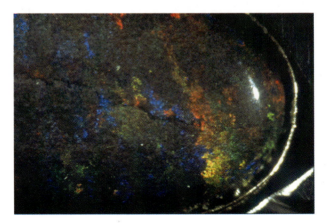

Plate 35
Treated opal matril

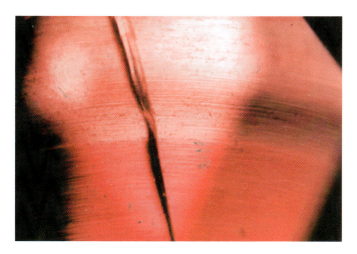

Plate 36
Curved growth lines in Verneuil ruby

Plate 37
Gas bubbles in synthetic ruby

Plate 38
Ruby with glass infilling

Plate 39
Glass infill in blue sapphire

Plate 40
Natural orange sapphire

Plate 41
Natural colourless sapphire

Plate 42
The same stone as in Plate 41, but here it has acquired a
yellow colour after irradiation

Plate 43
Rainbow-like markings show that this diamond has been
infilled

Plate 44
Flux-grown *synthetic sapphires* can display both *colour and growth zoning* with straight as well as angular features, which greatly hinder detection of their provenance. Darkfield and oblique illumination 30x (Reproduced with kind permission from *Photoatlas of Inclusions in Gemstones*, 1992, E. J. Gübelin and J. I. Koivula, Gemological Institute of America).

Plate 45
The faint green colours indicate that this diamond is infilled

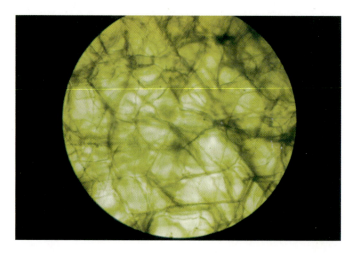

Plate 46
Flux droplets in flux-grown Zerfass emerald

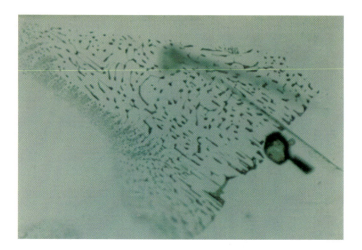

Plate 47
Two-phase inclusions at the junction of seed and
overgrowth in Chatham synthetic emerald

Plate 48
Two-phase inclusions at the junction of seed and
overgrowth in synthetic emerald

Plate 49
Mettallic fragments in Russian synthetic emerald

Plate 50
Lechleitner emerald

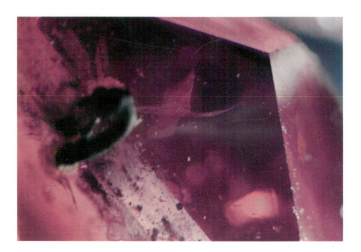

Plate 51
Two-phase inclusions and crucible fragments in synthetic
mauve beryl

Plate 52
Blue coated Aqua aura quartz crystals

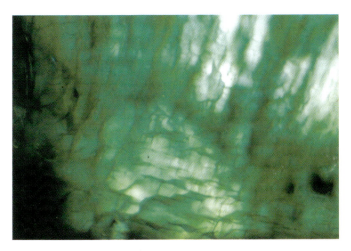

Plate 53
Crackled quartz imitating emerald-dry is introduced into cracks caused by thermal shock in this glass imitation of emerald (ruby imitations have been produced by the same method)

Plate 54
Uniform colouration shown by this group of synthetic opal beads

Plate 55
The plastic coating of this jadeite reflects the light

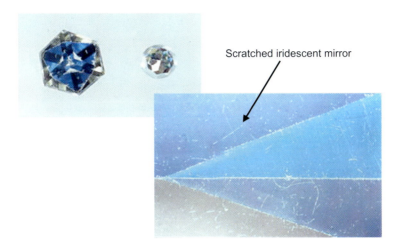

Scratched iridescent mirror

Plate 56
Mirror backed glass

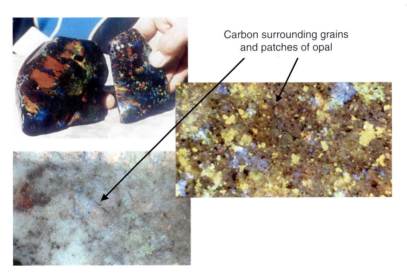

Carbon surrounding grains
and patches of opal

Plate 57
Identifying features of carbonised Anadamooka matrix opal

Plate 58
Superficial colour distribution of induced yellow colour in Be-diffused green sapphire rough. From Australia (T. Coldham, photo)

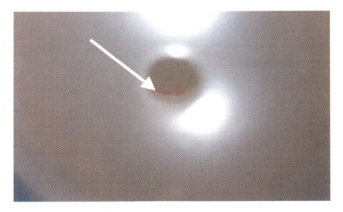

Bleached & dyed *akoya*
Bead nucleated cultured pearl

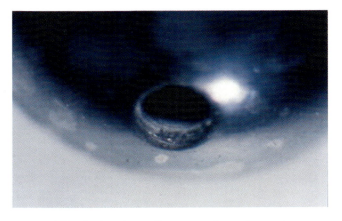

Black dyed south sea cultured pearl

Plate 59

Plate 60
Flux-healed quench-cracked Verneuil synthetic sapphire that displays induced 'fingerprints'

A star stone can be made by pressing a white opaque glass in a mould so that ridges in star ray formation are impressed into it. The cabochon is then coated with a deep blue glaze which is thinly applied, causing the star to appear below the surface. A less effective star is made by scratching star rays on the back of a glass cabochon or cementing rayed foil to the back. An opal imitation has been made by foiling with tinsel.

An interesting partly crystallized glass was made in the 1970s with the aim of simulating jade. Under the names 'Meta-jade', 'Victoria stone' or 'Kinga-stone', glass of varying degrees of devitrification and of different colours was offered. A chatoyant effect can be obtained by incorporating fibrous inclusions in parallel bands. Iimori stone, from the same source, Iimori Laboratory of Tokyo, was a clear or translucent jade imitation also with a variety of colours.

Goldstone is a glass simulant of sunstone, the glass being filled with cuprous oxide which reduces to metallic copper during later annealing. If the glass host is blue, an effect resembling lapis lazuli is obtained.

Glass with a near-metallic lustre and black body colour showed an indistinct red/green blink to the RIs of 1.59 and (a very low) 1.35. The SG was 3.21 and the hardness 4–5. It was not possible to go from the high to the low RI reading by rotating the stone. It is possible that at least some of the optical anomalies result from the addition of an opacifier to the glass and from tarnishing.

Two specimens of radioactive glass, of a fine emerald colour, gave an RI of 1.635 and SGs of 3.754 and 3.767. Chalky orange and chalky green fluorescence could be seen under LWUV and SWUV respectively. Strong absorption extended from 700 to 600 nm and from 440 to 400 nm with a set of two lines centred at 470 and 460 nm. Radioactivity was measured at 12× background levels.

Metal Imitations

Hematite is the mineral most imitated by metals. One imitation is a steel-grey titanium dioxide with a yellow-brown streak (hematite has a red streak). The hardness is about 5.5 and the SG near 4. Another imitation appears to be made from powdered lead sulphide with some added silver. The hardness is 2.5–3 and the SG 6.5–7. It is a brittle and easily fusible substance. Hemetine is the name given to a sintered material with several constituents, including lead sulphide (galena). It has a black streak and an SG of 7. Natural hematite is non-magnetic but some of the simulants will be attracted by a magnet.

A hematite imitation made from silicon produced from a refined melt by the pulling process has SG 2.33 and hardness 7.

Opal Simulants

The material 'opal essence', formerly Slocum Stone (manufactured by John S. Slocum of Rochester, Michigan), where not too gaudy, is a good imitation of opal. It is a glass made by a controlled precipitation process, with SG 2.4–2.5 and RI 1.49–1.50. The play of colour seems to come from included tinsel-like laminated material (aluminium foil) which diffracts light from the spacing between the laminations, approximately 0.3 μm. Typical gas bubbles and swirl marks indicate a glass. It is possible that some of the colour flakes arise from the disrupted remains of what were probably continuous sheets produced by a sedimentation process.

In some plastic opal imitations the lizard-skin effect cannot be seen. Another type, in which the colour domains were scattered rather than compact, transmitted wavelengths at about 430 nm, as does natural opal which it closely resembles. This opal is made from colourless poly-styrene in the main, but some structural features in both types give an absorption region at 590–565 nm. The RI is 1.485 and the SG 1.18. A strong bluish white fluorescence with no phosphorescence can be seen under LWUV.

Some synthetic opals give a blush (not necessarily of pink or red) which sweeps over the surface as the stone is turned. An early Gilson opal had colour bands rather than patches; they followed the long direc-tion of the oval cut stone.

Colour-enhanced opals are made from poor-quality yellowish to greyish Brazilian opals, plastic impregnated by n-butyl methacrylate, probably in a vacuum chamber. The RI is 1.41 to 1.46 and the SG is 1.85. Excess polymer can be scraped from the surface.

A plastic imitation opal with the trade name Opalite was reported in 1989. Sold as polished cabochons, the stones show a true play of colour with both pinfire and flash effects in various colours. Under incandescent light, the stones are milky bluish white and translucent; using transmitted light they appear pinkish orange. Stones are probably assembled, as a layer of a different plastic substance overlies part of the base. The outer layer may be an acrylic resin coated over the polystyrene imitation opal to pro-tect it. The RI is 1.50–1.51 (natural opals are in the range 1.37–1.47); the SG is approximately 1.20 (natural opal is in the range 1.99–2.25). The colour patches show characteristic lizard-skin patterning, confirming artificial ori-gin. The hardness of the Opalite was found to be 2.5, against 5.5–6.5 for nat-ural opal. Opalite is hydrophobic and is warmer to the touch than natural or synthetic opals. It will indent with the point of a pin and can be diag-nosed by IR spectroscopy.

Plastic imitation opal seen in Thailand had a vague spot RI of 1.57, with a strong chalky bluish white fluorescence under LWUV and a weaker yellowish green reaction under SW. Fine absorption lines extending from 700 to about 430 nm could be seen, with a typical plastic reaction to the hot point.

Colourless glass with fragments of synthetic opal embedded in it has been marketed under the name Gemulet. The RI of the glass is 1.47, close to that of the opal at 1.45, so giving the latter a low relief. The play of colour thus appears to be coming from the larger stone.

Chips of white opal, boulder opal and rock matrix with a transparent colourless binder have been offered in Australia as natural opalized shell. Various tests show that the binder is a plastic.

Porcelain

Porcelain has occasionally been used for the production of some semi-opaque gemstone simulants. Porcelain, or pottery, is simply a baked uniformly fine clay and the frits and faiences mentioned at the beginning of the chapter are a type of porcelain. The material is not, as a rule, cut and polished, but is moulded and then glazed with a glassy coating; thus the RI will be of little assistance in its identification. On the other hand, the SG is fairly constant at 2.3 and this will give an indication of the nature of such material, for, with the exception of a few minerals like sodalite and thomsonite, this value is not common to ornamental gem materials.

Ceramics

Finely ground inorganic powders which are heated, fired or sintered and sometimes compressed will produce a polycrystalline solid known as a 'ceramic'. A binding agent with a low melting point can be used to help the particles adhere to one another. The surface, as in porcelain, is often glazed. Natural substances like turquoise, jade and lapis lazuli have similar compositions.

A lapis lazuli simulant made by Gilson is sold either with or without pyrite inclusions. More porous than natural lapis lazuli, the Gilson product has a lower SG at near 2.46 (natural lapis lazuli is 2.81). The hardness is 4.5 compared with the 5.5 of the natural stone, and the RI is a vague edge at 1.50. Some quartz and iron has been detected by X-ray analysis. It has been considered that the product consists of ultramarine and hydrous zinc phosphates with the added pyrite.

A possible ceramic imitation of lapis lazuli seen as a rectangular bar-cut cabochon of 4 ct had RI 1.55 and SG 2.85. There was a strong chalky greenish yellow colour under SWUV but weak red colour under LW. No evidence of dye was found. EDXRF chemical analysis revealed patterns from the coloured areas consistent with phlogopite mica. The colouring agent is not so far identified.

Gilson has also manufactured a turquoise simulant which probably approaches a ceramic substance, though full details of manufacture have never been released. A medium blue is marketed as Cleopatra and a darker blue as Farah. Thought to be made by precipitation, grinding and pressing with no trace of the iron normally found in natural turquoise (suggesting the use of pure starting materials rather than ground natural pieces), Gilson turquoise comes with or without matrix; it is only slightly porous and the colour is stable. When the surface is examined with a magnification of 30–40×, a mass of angular dark blue particles against a whitish ground-mass is seen. The SG is near 2.74 and the average RI is 1.60. The absorption band at 432 nm is faint, leaving the microscope as the only simple test available.

An imitation turquoise made by Syntho Gem Company of Reseda, California, a so-called reconstituted turquoise made by Adco Products of Bueno Park, California, and Turquite made by Turquite Minerals of Deming, New Mexico, were examined together. The Adco and Syntho materials were similar to Gilson stones, while Turquite had little aluminium but a larger amount of sulphur, silicon and calcium. Though containing less iron, the Gilson product approached most closely to the composition of the natural stone. Structurally the Gilson turquoise was the only one to have the same crystal structure as natural turquoise.

Imitation turquoise has been made with pyrite inclusions and apparent veins of matrix. The stones are plastic-coated gibbsite, a clay-like aluminium hydroxide. Hot point tests inside the drill hole produce a characteristic plastic odour. No absorption line could be seen at 432 nm; the RI was 1.55. Stones had a sugary, speckled appearance.

Dyed magnesite has been used as a turquoise simulant. It has RI readings of near 1.51 and 1.70, the high birefringence of 0.19 suggesting a carbonate. The SG was close to 3.0. Magnesite will effervesce with warm HCl.

Blocks of azurite-malachite formed by compressing and impregnating chalky porous azur-malachite nodules have been offered as cabochon material. The material is tough, shows good colour and polishes well. A hydraulic ram compresses the nodules into a dense block which is then stabilized by a pore-filling technique similar to that used in stabilizing turquoise.

Gilson makes a coral imitation in various shades of reddish pink. The chief constituent is calcite from a French source, but some ingredients not found in natural coral are present. The SG is 2.44 against the natural 2.6–2.7.

Both imitation and natural have a hardness of about 3.5 and both will effervesce with acids. The RI of the imitation is about 1.55 against 1.49–1.65 for natural coral. Gilson coral shows a brecciated structure while natural coral shows a wood-like graining. Gilson coral gives a reddish brown streak; natural coral gives only a whitish mark.

Coral has been imitated by barium sulphate with a plastic binder. A 21.26 carat orange-red sawn slab gave an RI of near 1.58 with an SG of approximately 2.33. The hardness was estimated at 2.5–3. Whitish pink veining could be seen under magnification.

Dyed marble has been offered as a coral imitation, the deception easily yielding to a drop of HCl. A spot RI showed high birefringence. Instead of the usual coral structure a sugary, granular texture was observed.

Ceramic alumina has been used in the manufacture of cameos. The hardness of a white on blue piece was over 8 and the blue colour could have been due to cobalt. The RI was vague, between 1.75 and 1.76. Later the blue was indeed proved to be due to cobalt.

27

Plastics

Maggie Campbell Pedersen

Plastics have been used to simulate gem materials – especially those of organic origin – since their invention in the mid-nineteenth century.

The word 'plastic' tends to be used generically to describe certain man-made materials, though in its true sense it refers more to a physical state than to a certain type of material. A plastic material is one that can be shaped by a process (usually heat and pressure) and retain that shape afterwards.

Horn and tortoiseshell are natural plastics, and have been heated, shaped and moulded for centuries. Shellac (a resin based on the secretion of the lac insect) and some tree resins could also be called natural plastics.

Most plastics are organic high polymers. These are compounds, rich in carbon, made up of long chains of repeating chemical units or molecules linked together end to end, which can slide against each other when treated in some way. (Hence the name polymer: poly = many, mer = units.)

Sometimes heating produces an extra chemical reaction whereby the molecules in one chain cross-link with molecules in another chain in a 3-dimensional pattern, producing a 'thermoset' material in which the shaping process cannot be repeated. Other plastic materials – the 'thermoplastics' – do not cross-link during processing and can be repeatedly re-heated and re-shaped.

Most plastics are amorphous as their molecules are randomly arranged, but in some the molecular structure can become partially organized and the plastic is termed 'semi-crystalline'. The degree of crystallinity is not enough to confuse results on a polariscope. The hardness, SG and RI of plastics vary according to the polymer type and grade, but the SG is seldom less than 1.08 and so cannot be confused with amber – polystyrene is a notable exception with an SG that may be as low as 1.05. Most plastics pare under a sharp knife.

There are hundreds of different plastics on the market today, and most occur in many different grades and qualities, all manufactured to suit specific uses. Usually they contain additives to change their properties, some examples being dyes to give colour, fillers to give bulk or strength, or chemicals to make them flame-retardant.

In gemmology there are typically five uses for plastics:

1. To impregnate gem materials to stabilize them (for example red soft coral, which, in its natural state, is rough, brittle and has a dull lustre).
2. To cap or back thin pieces of inexpensive gem material to stabilize them and give them more bulk (for example abalone shell).
3. To back translucent or semi-translucent gem materials to change or enhance their colour.
4. To simulate a gem material.
5. As a raw material in its own right. Plastic jewellery is usually colourful and can vary in quality from the inexpensive to stunning designer pieces.

Most plastics are light and feel warm to the touch. This makes them a poor simulant for minerals but excellent for most of the gem materials of organic origin.

The gemmologist is concerned with only a few of today's multitude of plastics. Of the organic gem materials, amber, tortoiseshell and mother-of-pearl are still widely copied in plastics, and ivory copied to a limited extent.

Of more interest to the collector or the gemmologist are the early, man-made, plastics which were used to copy all the organics with varying success. The first of these combined natural polymers with chemicals – an example is vulcanite. Also called ebonite or hard rubber, vulcanite was made by curing natural rubber in the presence of sulphur – a process called 'vulcanization'. It was patented in England in 1843 and again the following year in America.

Vulcanite is an opaque thermosetting material of dark colour. Its early uses were for items such as electrical insulation and car tyres. Later it was coloured black by the addition of soot, and mass produced and moulded as a simulant of hand-carved jet jewellery. The resulting jewellery was very black, could take an intricate impression from a mould, was warm to the touch and light – as is jet. However, unlike jet, vulcanite loses its high lustre and deep black colour with time and fades to a khaki hue due to its sulphur content.

Vulcanite is the only jet simulant to give a brown streak (or brown powder when scratched with a sharp instrument). When rubbed vigorously between thumb and fingers it can give off a smell of rubber and sulphur, but this is not a totally reliable test as some pieces are lacquered to preserve their colour and lustre. When tested with a hot point, vulcanite gives off a smell of burning rubber and sulphur.

Gutta percha is a totally natural polymer, made from a particular species of rubber, but it was not used in the jewellery trade. However, the name is often used generically – especially in the US – to cover all black materials used for jewellery or decorative items, including shellac and vulcanite. *Bois durci* is another black material that was based on albumen (from ox blood or egg whites) and mixed with a filler. It was moulded using large, heavy presses and was therefore not suitable for use as jewellery, but was used for larger decorative items such as candle sticks or plaques.

The real breakthrough in imitating organic gem materials came with the invention of a material based on cellulose from wood or cotton, chemically modified with nitric acid and plasticized with camphor. Produced originally as 'Parkesine' in England in 1862, the material was finally patented in America in 1870, where it was sold as 'Xylonite' but became known by its more common name *celluloid*.

It was a highly inflammable semi-synthetic polymer, but it was versatile as it could be produced in pale colours and in opacities varying from totally opaque to totally transparent. It was therefore useful when imitating ivory, coral and especially tortoiseshell.

Celluloid can degrade with time, depending partly upon its storage conditions. The process is irreversible and the item finally disintegrates completely. During the process it may turn sticky, and give off a smell of camphor. Celluloid should *never* be tested with a hot point due to the risk of combustion.

Cellulose acetate was a safer alternative to cellulose nitrate, and was developed in 1894, using acetic acid with the cellulose instead of nitric acid. It was first used commercially some 30 years after its discovery. As with cellulose nitrate, cellulose acetate could be produced in varying colours and opacities. It can also degrade, giving off a vinegary smell. The celluloids have SGs varying from 1.29 to 1.80.

Casein was first manufactured in Germany in 1899 under the trade name Galalith. It was made from the milk protein waste remaining from butter and cream production, and mixed with formaldehyde to prevent putrifaction. The material was easy to dye, both throughout and on the surface, and had a high lustre and a silky feel. It was used to imitate most of the organic gem materials, including mother-of-pearl which involved the addition of fish scales. It was also used as a turquoise simulant.

Casein has an SG in the range of 1.32–1.39. A hot point test may give an aroma of burnt milk. There is still a small amount of casein in production today.

Bakelite was the first totally synthetic plastic. It took years to perfect, but was finally patented in 1907 in USA. It is a mixture of phenol and formaldehyde, and is usually made with some form of filler which renders it opaque. Phenol formaldehyde can also be produced as a clear material, which is moulded from a syrup rather than from a powder ('cast phenolic'). In this form it has been a very successful amber simulant.

'Bakelite' has become the generic term for the various forms of phenol formaldehyde, all of which have higher SGs than amber, and so will sink in saturated salt solution. When abraded, Bakelite tends to give a yellow streak, but this can be masked by dyes present in the material. Phenol formaldehyde will emit a smell of carbolic when touched with a hot point. It is a tough and hard-wearing plastic and is still in production today.

There are various modern plastics encountered in the gem and jewellery trade, for example the *epoxies*, which have various uses such as stabilizing brittle materials. They are also cast in intricate moulds. *Polystyrene* is an inexpensive material to make, can be produced in any colour or opacity, and is used in cheap jewellery such as beads. It has a low SG and is therefore sometimes used to imitate amber. Burning a scraping of the material will give a pungent smell and thick, black smoke.

Polyester is used for more expensive hand-finished items of jewellery, and can be produced in all colours and opacities.

The most commonly encountered plastics in jewellery today are possibly the *acrylics*, also called 'Perspex' or 'Lucite'. Acrylics came onto the market in 1936, and have been used as a substitute for glass in large or vulnerable areas such as roofing, as they have exceptional clarity and almost total light transmission. They are hard, rigid, bright, can be dyed any colour and are extremely resistant to weathering and to the adverse effects of UV light in daylight. In jewellery they are used to imitate amber, or are used in the production of inexpensive plastic jewellery. Early 'Perspex' or 'Lucite' pieces are considered collectors' items, along with other early plastics. The material is also used in modern designer pieces where its versatility can be put to great effect. Acrylics have a very high lustre. When tapped against a hard surface, they make a characteristic 'clink' sound.

28

Pearls

Grahame Brown

Pearls are rounded concretions of calcium carbonate and an organic matrix that are secreted by cells of the outer mantle of many species of saltwater and freshwater molluscs. While any mollusc[1] (see *Figure 28.1*) is theoretically capable of producing a pearl, nacre[2] (see *Figure 28.2*) – forming molluscs principally, but not exclusively, secrete those pearls that have long been used for personal regal adornment.

Pearls form from the same materials and by the same processes which secrete the shells of their host bivalves, gastropods and cephalopods. Both shells and their pearls have a composite structure of calcium carbonate (aragonite and/or calcite), an organic matrix (essentially of the protein conchins, formerly conchiolin and carbohydrate) and water. Although the organic content of pearls is usually less than 1%, it is this minor component that determines both the crystallographic properties and orientation of the calcium carbonate that is secreted by the mollusc, and the body colour of the pearl.

In order to understand how pearls form, one must first understand the basic biology, gross anatomy and microscopic structure of those nacre-producing bivalve molluscs that are commonly termed either saltwater 'pearl oysters' or freshwater 'pearl mussels'.

Introductory Taxonomy

A 'pearl oyster' is a mollusc that belongs to the class Bivalvia. Bivalves have a transversely flattened soft body that lies protected between two hinged calcified valves or shells. Nacreous marine 'pearl oysters' belong to two genera of the family Pteriidae, the genera Pinctada and Pteria; while nacreous freshwater 'pearl mussels' belong to the family Unionidae.

Some common species of mollusc that are considered to be pearl-producing include:

- *Pinctada maxima* (Jameson, 1901), the silver- or gold-lipped pearl oyster.
- *Pinctada margaritifera* (Linnaeus, 1758), the black-lipped pearl oyster.
- *Pteria (=Magnavicula) penguin* (Röding, 1798), the black-winged pearl oyster or *mabé*.
- *Magaritifera margaritifera* (Linnaeus, 1758), the European freshwater pearly mussel.
- *Haliotis iris* (Martyn, 1784), the New Zealand paua shell.
- *Strombus gigas* (Linnaeus, 1758), the pink conch of the Caribbean.

Gross Anatomy of a Nacreous Bivalve

When viewed in longitudinal section, a nacreous bivalve mollusc has two concave nacre-lined valves that are joined posteriorly by a hinge and a central ligament. These valves enclose the body and the adductor muscle of the mollusc. An epithelium (skin) lined mantle covers the mollusc's body and then extends forwards into two flattened sheets that intimately line both valves to form a pallial cavity into which the mollusc's two sickle-shaped oxygen filtering gills project from the body of the mollusc. The valves of each bivalve are closed by the contraction of its powerful adductor muscle and are opened by passive relaxation of the mollusc's posterior ligament. A conical muscular foot (which allows some mobility in young bivalves) and secreted fibrous byssus which attaches the shell to its substrate complete the gross structure of a bivalve mollusc.

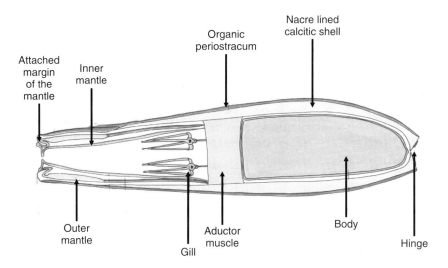

Figure 28.1   Gross structure of a bivalve mollusc (After Doumenge, 1990, p. 16)

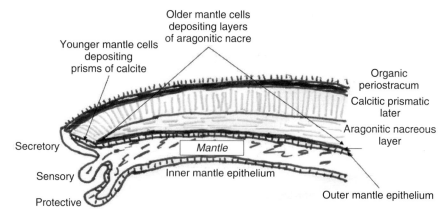

Figure 28.2   Detailed structure of a cross-section of the valve (shell) and mantle of a nacreous bivalve

The mantle covering the body of the nacreous mollusc is attached directly to the inside of each valve by a series of muscles that leave a curved scar, on the nacre lining the valve, termed the 'pallial line'. From the pallial line the mantle stretches outwards lining the inner surface of each valve and be attached peripherally to the rim of each valve. This mantle tissue lining each valve consists of two layers of epithelial (skin) cells supported by underlying connective tissue. It is the secretory epithelial cells of the outer mantle which are responsible for the sequential secretion of the shell's outer tough organic protective covering (periostracum), its strengthening calcitic prismatic layer, and its inner lining of microscopically thin layers of nacre. These tissues are secreted sequentially by specialized cells located on the outer fold of peripheral edge of the mantle. The outer fold also isolates the outer cells of the mantle, and their secretions, from the direct influence of the aqueous environment in which the bivalve lives.

Depending on their age, epithelial cells of the outer mantle of nacreous molluscs rhythmically secrete organic matrices of very specific chemical composition into which single crystals of calcite, and later aragonite, are deposited to form the complex layered structure of the nacreous molluscan shell. In contrast, the inner mantle cells that line the pallial cavity have no secretory function.

Nacreous bivalve shells grow in a radial manner from the centre of their hinge to their outer rim. The external surface of each valve is covered by a tough organic periostracum, and the valve is formed layers of prismatic calcite that are lined internal by smooth layers of aragonitic nacre. In some species of Pinctada, a lip of contrasting colour surrounds its commonly silver-grey coloured nacre. Hence, these shells are termed 'gold-lipped', 'silver-lipped', or 'black-lipped' 'pearl oysters'.

Microscopic Structure of a Nacreous Valve

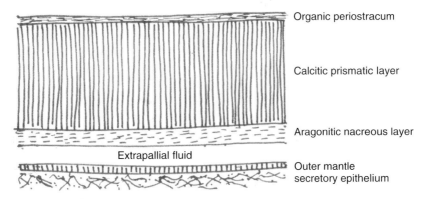

Organic periostracum

Calcitic prismatic layer

Aragonitic nacreous layer

Extrapallial fluid

Outer mantle
secretory epithelium

*Figure 28.3 Microscopic structure of a cross-section of the valve (shell) and mantle
of a nacreous bivalve*

When a cross-section of a nacre-lined valve is examined at low to
moderate magnification, it will be seen to be composed of three distinct
layers. The outer layer is a dark leathery organic layer (periostracum)
that is composed of the keratinous scleroprotein conchin. The periostracum is secreted by the youngest cells of the outer mantle. Being a
protective structure, the periostracum does not increase in thickness
once it has been formed. The central layer is known as the prismatic
layer. This consists of polygonal prisms of calcite that are oriented
at right angles to the surface of the shell. These prisms of fibrous
calcite, which are secreted into previously secreted organic prism
sheaths, have a uniform shape from species to species – but do vary
in size, depending on species. The prismatic layer is secreted by the
second youngest of the peripheral epithelial cells covering the outer
surface of the mantle. The internal nacreous layer of the shell consists
of rhythmically deposited mircoscopically thin parallel layers of
tile-like crystals of aragonite, deposited into very thin layers of
honeycomb-like organic matrix. The nacreous layer has two distinct
components: an outer (fibrous) layer that is deposited against the
prismatic layer, and an inner layer that is deposited between the
fibrous outer nacreous layer and the external surface of the mantle.
Cells responsible for secreting both the organic matrix and crystals of
aragonitic nacre are the oldest epithelial cells. Nacre continues to be
deposited on the inside of the shell throughout the life of the nacreous
bivalve.

The key element in the formation of a molluscan shell is the very thin film of fluid (extrapallial fluid) that maintains intimate contact between the secretory cells of the outer mantle and the inner surface of the shell. It has been hypothesized that the cells of the outer mantle form shell structure by extracting calcium and carbonate from seawater and also secreting specific molecules of organic matrix into the extrapallial fluid. These proto-shell components then diffuse through the extrapallial fluid to be incorporated onto the existing inner surface of the shell in the following sequence:

1. Attachment of organic molecules onto the existing shell surface;
2. Polymerization of these organic molecules into a three-dimensional lattice termed an 'organic matrix'; and
3. Deposition of calcium carbonate into the respective organic matrices as single crystals of either calcite or aragonite – the chemistry, crystallography and orientation of which are determined by the specific chemistry of the mollusc's already-deposited organic matrix.

Depending on species, the inner nacreous lining of molluscan shells, which essentially determines the type of pearl secreted by a particular mollusc, can be formed from several types of calcified tissue. This inner lining is commonly formed from sheet nacre, which has a brick wall structure and is found in all species of pearl 'oyster' and pearl 'mussel'. Less common is columnar nacre, which is composed of tabular crystals of aragonite of uniform size that are stacked vertically with coinciding centres. This highly iridescent form of nacre lines the shells of univalve abalones. Rarest is crossed lamellar structures that consist of thin parallel laths/rods of aragonitic calcium carbonate that have directions that alternate between adjacent circumferential lamellae. These form the non-nacreous flame-patterned inner linings of univalve pink conch shells.

The precise structure of the platelets of aragonite that form nacre was the subject of an investigation by Gutmannsbauer and Hänni (1994). These researchers revealed that, irrespective of species, the overlapping tile-like crystals that form mother-of-pearl, or nacre, consist of single crystals of aragonite. This finding contradicted the long-held belief that these crystals were twinned aragonitic trillings. The single crystals of aragonite that form nacre were shown to have a pseudohexagonal tabular shape, a thickness of 400–1500 nm and a diameter of ~5000 nm (5 μm). This research also reconfirmed that the c-axes of all crystals of nacreous aragonite were oriented at right angles to the inner surface of the valve of the mollusc.

When the structure of a nacreous natural pearl is studied under the high magnification possible with an electron microscope (EM), the concentric

Single crystal/s
of aragonite

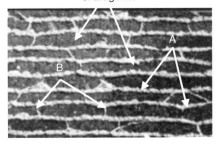
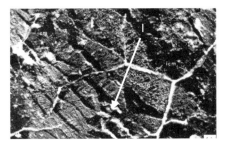

Figure 28.4   The EM structure of sheet nacre. Note (LHS) the interlamellar matrix (A) that lies between rhythmically deposited layers of crystals, and the intercrystalline matrix (B) that is found between adjacent crystals deposited in the same layer. Membranous intracrystalline matrix (I) that occurs within single crystals is illustrated on the RHS electron micrograph

layers of sheet nacre that form the pearl are seen to have a decidedly 'brick wall' structure. This wall is formed from 'bricks' (single crystals) of aragonite that have been cemented together with an organic matrix or 'mortar' of organic conchin. The conchin matrix into which the individual crystals of aragonite are deposited has a decidedly honey-comb structure, and three recognizable components: interlamellar matrix (A) that lies between rhythmically deposited layers of aragonite crystals; intercrystalline matrix (B) that is found between adjacent crystals deposited in the same layer; and membranous intracrystalline matrix (I) that occurs within individual crystals. Overall, the interlamellar organic matrix has about one-fifth the thickness of that of an individual aragonite crystal.

Natural Pearls

How Natural Pearls Form

Any outer mantle cells of a mollusc, which are capable of secreting a shell, also are capable of secreting pearls, for pearls are formed by the same physiological processes and by the same materials as the shells of their host mollusc.

Natural pearls form as a consequence of an irritant – a boring parasite, a tumor-like growth or physical irritant such as a sharp fragment of shell – that stimulates secretory cells of the outer mantle to deposit shell struc-ture(s) to cover and neutralize the irritant.

If the cells of the outer mantle completely coat the irritant to form a cell-lined pearl sac, then a whole pearl will be formed by progressively deposited layers of circumferental nacre. As the formed pearl sac is lined by cells capable of secreting calcified shell structures, these cells will then progressively form and secrete shell constituents (organic matrix, calcite and then aragonite) around the irritant as the epithelial cells lining the pearl sac mature with increasing age. Consequently, if a natural whole pearl is sectioned transversely, on examination the natural pearl would be seen to have an 'onion-like' structure of thin concentric alternating layers of aragonitic nacre and organic matrix that surround a small hollow cavity in the centre of the pearl. This central cavity represents the residue of the original irritant that first caused the pearl to form, and the conchin that was initially deposited around this irritant. In some natural pearls, an intermediate layer of prismatic calcite, of variable thickness, may be found underlying the nacre.

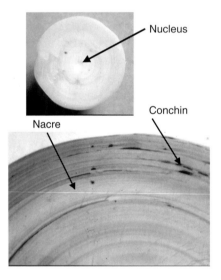

Figure 28.5 Cross-section of a natural nacreous pearl

If, on the other hand, the irritant is trapped in the extrapallial fluid, at or near the inner lining of the shell, then the cells of the outer mantle will progressively cover the irritant with layers of nacre and so create a blister pearl. It is not uncommon to find a degenerating organic material within the hemisphere of nacre that forms a natural blister pearl.

While any mollusc has potential for producing pearls, only a few species (mostly of nacreous bivalves) produce natural pearls that have commercial relevance.

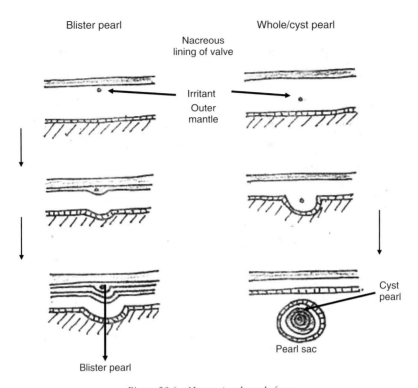

Figure 28.6   How natural pearls form

The Appearance of a Nacreous Pearl

A nacreous natural pearl has a quite complex appearance, for while its body colour reflects, in general, one of the colours displayed by the nacre of the mollusc from which it was derived, it is the subtle additions to this basic body colour that determines the overall appearance, appeal, quality and value of a pearl. It is generally considered that the aesthetic appeal of a pearl is controlled by the intermingling of such visual features such as its body colour, its lustre, any surface iridescence that is superimposed over its body colour, and the presence of any subtle subsurface irides-cence that is termed *orient*.

The Causes of a Pearl's Colours

The cause(s) of colour of nacre is still subject to some debate. While it is generally accepted that the body colour (black, silver, gold, etc.) of a pearl is determined by the presence of organic pigments, such as carotenoids, porphrys, melanins, etc., which are associated with the organic matrix into which particular species of mollusc secrete their nacre, other factors are undoubtedly involved.

For example, Liu, Shigley and Hurwit (1999) and Liu, Hurwit and Tian (2003) investigated the cause of the iridescent colours produced by shell from both the black-lipped pearl oyster *P. margaritifera* and the gold-lipped pearl oyster *P. maxima*. Using a combination of electron microscopy and diffraction generated by an argon ion laser, these researchers clearly established that the strong iridescent colours of *P. margaritifera* shell were caused by diffraction of incident white light at a fine-scale, parallel, smooth and even groove structure of ~3.38 μm width that was the equivalent of a 296 grooves/millimetre two-dimensional reflection diffraction grating. In contrast, the moderately strong iridescence of *P. maxima* nacre was caused by a reflection grating structure of 156 grooves/millimetre density. These iridescences were very directional as no iridescence could be observed along the direction of the groove. As diffused light could not produce iridescent colours from this shell, these authors also suggested that it is diffused light that is responsible for the white body colour displayed by most pearls.

In 2002, Liu, Hurwit and Shigley (2002) also investigated the cause of iridescent colours in nacre of the abalone *Haliotis rufescens*. These investigators found that the iridescence of this nacre was due to its columnar structure that was formed from a two-dimensional slightly rounded square array of slightly convex tiles of aragonite of ~5 μm size. This array of tiles diffracted incident light so that the diffracted wavelengths were visible in any direction of observation. As the maximum diffraction efficiency of this two-dimensional square grating occurred in the blue-green, these were the dominant iridescent wavelengths displayed by this nacre. However, this dominant wavelength can change as the angle of viewing is changed.

The Pearl's Lustre

The lustre of a pearl is determined by the amount and quality of white light reflected from its surface. The smoother the nacre and the more spherical the pearl, the sharper will be the back-reflection from its surface, and therefore the higher the lustre of the pearl. In contrast, pitted, poor quality nacre will have poor lustre due to the scattering of incident white light at the surface of the pearl.

The Pearl's Orient

A nacreous pearl's *orient* represents the subtle, attractive iridescent wavelengths that are reflected from microscopically thin and even subsurface nacreous layers of the pearl. This visual effect is the result of iridescence which is generated by some combination of diffraction and/or interference of white light at very thin, even thickness circumferential lamellae of nacre below the surface of the pearl. This somewhat subtle iridescence is intensified by the thinness and evenness of the circumferential nacreous lamellae that form the pearl, for the thinner the lamellae, with respect to the wavelength of visible light, the more spectral colours will be produced.

External Surface of a Pearl

When the thin circumferential lamellae of aragonite crystals intersect the external surface of a nacreous pearl, they give the surface its identifying appearance of swirling somewhat parallel serrations. This 'finger print'-like pattern, which is generated by the edges of aragonite platelets as successive extremely thin lamellae of nacre outcrop on the surface of the pearl, identifies nacre when viewed with moderate magnification.

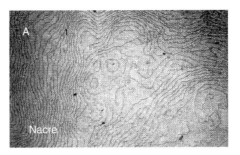

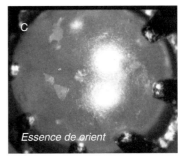

Figure 28.7   Identifying hand lens of the external surfaces of a nacreous pearl (A), a pink conch pearl (B) and an essence-de-orient coated imitation pearl (C)

In contrast, non-nacreous pearls, such as pink conch pearls, have a porcellanous, sometimes 'flame'-patterned external surface, while man-made nacre or *essence-de-orient* has a 'blotting paper' texture which may not completely cover the external surface of the imitation pearl's underlying bead.

Sources of Natural Pearls

The pearl's colour (due to pigments in the pearl's organic matrix and/or thickness and evenness of its constituent layers of aragonite crystals), lustre (due to light reflection from the smoothness or perfection of the pearl's external surface) and orient (due to its subsurface iridescence and reflectivity) are determined by the colour, lustre and orient of the inner nacreous or non-nacreous lining of the mollusc's shell in which the pearl grew. For example:

- The oriental pearl 'oyster' *Pinctada radiata/imbricata* from the Persian Gulf and the Gulf of Mannar yields mostly smallish cream to yellowish pearls that display subtle orient.
- The silver- or gold-lipped pearl 'oyster' (*P. maxima*) from North-West Australia yields nacreous pearls of white, silver and yellow to golden hue that also display orient.
- The black-lipped (*P. margaritifera*) pearl 'oyster', from Pacific Oceania, yields grey to black pearls that also can display a spectacular contrasting iridescence.
- *Pteria (=Magnavicula) penguin*, the black-winged pearl oyster of the tropical waters, yield distinctively iridescent pearls that have a characteristic somewhat metallic (brown to greyish) body colour. *Pteria sterna*, the rainbow wing shell from the Gulf of California, also yields pearls with highly iridescent black to dark brown nacre.
- Freshwater (*Unio*) mussels from Scottish rivers, and the Mississippi River, yield pastel-hued highly lustrous pearls with limited orient.
- The univalve *Haliotis*, or abalone, produces mostly baroque nacreous pearls that display strikingly iridescent nacre.
- The pink conch (*S. gigas*) of the Caribbean yields pink to orange 'flame'-patterned pearls that have a non-nacreous porcellanous surfaces.
- The giant clam (*Tridacna gigas*) of tropical waters yields white porcellanous (non-nacreous) pearls that have little economic value.

Some Natural Pearls

Today, natural pearls are rare, very expensive and are not the basis of any commercial fishery. Traditional and still occasional sources of natural pearls include:

- Saltwater pearls from historical pearling beds in the Persian Gulf, Gulf of Mannar between India and Sri Lanka, off Hepu on the coast of China, waters to the north of Australia, the islands of Oceania, the Gulf of Mexico, and waters surrounding Pearl Island off the coast of Venezuela;

- Freshwater pearls from Scottish and European rivers, and the Mississippi River and its tributaries in the USA;
- Rare baroque saltwater abalone pearls from Californian and New Zealand waters; and
- Non-nacreous pink pearls from the pink conch of the Caribbean.

Natural nacreous pearls occur rarely as whole pearls, and more commonly as hemispherical blister pearls. Whole natural pearls are rarely spherical, and the thinness and evenness of their circumferential layers of nacre determine the strength and quality of iridescence displayed by a particular pearl. As an example, compare the appearance and structure of moderately iridescent Pinctada (silver-lipped pearl 'oyster') nacre with that of highly iridescent Haliotis or abalone pearl nacre.

Traditionally, nacreous natural pearls have been valued by the factors of rarity, size or weight, perfection of shape, colour, lustre and orient, and freedom from surface imperfections. The weight of natural pearls has been calculated in units, termed the 'pearl grain' that weighs 50 mg.

Pinctada and Unio Pearls

Whole natural nacreous Pinctada ('pearl oyster') and Unio ('pearl mussel') pearls are formed from onion-like concentric layers (lamellae) of tile-like single crystals of aragonite that have been deposited into an organic matrix of very specific chemical composition. Some natural nacreous pearls may also have a core of radiating prismatic calcite, and/or a central cavity of degenerated organic matter. Degenerating organic cores, and their evolved gases, are thought to be the cause of colour in some greyish blue coloured natural pearls.

Natural blister pearls have a mostly hemispherical form, and may have a partly hollow base that is partly filled with residual degenerated organic material.

Natural Pinctada and Unio pearls vary in hue from white to silver, yellow to golden, black and various pastel hues – depending on the colour of the nacre of the particular species of bivalve mollusc (either saltwater 'pearl oyster' or freshwater 'pearl mussel') in which the pearls formed and grew. These pearls have an average chemical composition of 85–90% calcium carbonate, 10–13% organic material (conchin, carbohydrates and pigments) and about 2–4% water. Other elements found commonly in natural pearls include Mn^{2+}, which occurs in higher concentrations in freshwater pearls than saltwater pearls, and Sr^{2+} that replaces Ca^+_2 in the calcium carbonate of the shell. Minor variations in the chemical composition of pearls generally reflect differences in the chemistry of the waters in which the pearls grew. Natural Pinctada and Unio pearls also have a hardness between $2\frac{1}{2}$ and $3\frac{1}{2}$ on Mohs' scale, an SG that ranges from 2.68 to 2.74 for saltwater pearls and from 2.66 to 2.78 for freshwater pearls, and they may yield a vague 'spot' RI of ~1.54 on the gem refractometer.

Abalone Pearls

Of the more than fifty living species of abalone, many larger-sized highly nacreous species, such as *H. rufescens*, the red abalone of California, *H. ruber*, and the New Zealand *paua*, will naturally secrete both blister and quite baroque whole pearls. While natural blister pearls are relatively common on the nacre of the univalve abalone, according to Howarth (1988) most free abalone pearls will be found principally in older molluscs, in one of three locations. Hollow, horn- or shark's tooth-shaped pearls may form in pearl sacs located in the gonad (reproductive organ) of the abalone; almost solid crescent shaped baroque pearls can be found associated with the marginal mantle around the periphery of the shell; and small solid rounded pearls can be located in the body of the abalone.

Natural abalone pearls display the following identifying features (Brown, 1985):

- Rounded to baroque shapes.
- Highly iridescent, multi-coloured nacre that is seldom completely smooth and often included by brownish conchin.
- Concentric lamellar structures that are readily displayed on an X-radiograph.
- Internal voids, particularly in horn- or tooth-shaped pearls, that cause the SG of these pearls to range from 2.61 to 2.70. These internal voids are readily seen on an X-radiograph.
- A distinctive greenish fluorescence to LWUV irradiation that modifies to a dull dusky green colour when examined with SWUV wavelengths.

The distribution of the multi-coloured rainbow-like diffraction hues, which characterize the nacre of abalone pearls, is determined by the genetics of the particular species of Haliotis, the very specific (dominant blue-green) wavelengths that are diffracted back from this nacre, and by the curvatures inherent in the surfaces of these oft-baroque pearls. Depending on the angle of observation, common hues displayed by abalone nacre include pink, orange, green, blue, lavender, silver and gold.

Conch Pearls

Conch pearls are non-nacreous calcareous concretions that are produced by the Queen or pink conch, *S. gigas*, the habitat of which ranges from Bermuda to the Caribbean. The pink conch is a univalve mollusc that is fished mainly for its meat. It has been estimated that flame-patterned non-nacreous conch pearls of white to yellowish brown, orange and attractive pink to red body colour are recovered at the rate of one pearl from every 10 000 harvested conchs (Fritsch and Misiorowski, 1987). These pearls have a mixed but mainly aragonitic chemical composition, and in thin section display a dark organic-rich core that is surrounded by circumferentially deposited thin calcific growth lamellae that have a crossed-lamellar structure. The pink to red colours of conch pearls are caused by the presence of organic carotenoid pigments, and their flame patterns are associated with the outcropping of crossed-lamellar structures on the surface of these pearls.

Conch pearls have the following gemmological characteristics:

- They are usually small in size (2–3 mm in diameter) and only rarely (~10%) reach a size, shape and colour suitable for use in jewellery.
- Their most desirable shape is symmetrically ovoid to roundish.
- Colours range from whitish to yellow, brown, and the most desired pastel pink to red (Quarenghi and Sicaguato, 1997).
- The porcellanous surface of these pearls may be decorated with an attractively silvery 'flame' pattern, which has been attributed to 'oriented crystal growth'.
- The pearls are tough, have a hardness of 4–5 on Mohs' scale, have an SG of ~2.85, and a 'spot' RI of ~1.51.

Pearls from Edible Oysters

When discovering about pearls from unlikely sources, the reader occasionally is informed of some lucky diner who finds their fortune while enjoying a meal of oysters. Usually, the financial aspirations of the finder are quickly dashed, for pearls from edible oysters are non-nacreous, usually quite unattractive in colour, and have little intrinsic value. Some species of edible oyster, such as the common oyster *Ostrea edulis*, yield pearls extremely rarely. In contrast, the white hammer oyster, *Malleus albus*, from the Indo-Pacific region produces pearls much more frequently.

Pearls from edible oysters are usually brownish to opaque white-coloured, non-nacreous, have an SG of 2.20 or slightly more, and are formed from alternating concentric layers of radiolucent conchin and radiopaque calcite. Bolman (1941) classified these pearls as bring conchin pearls.

Coconut Pearls

The acquisitive collector of unusual pearls, who diligently searches the backstreets and bazaars of south-east Asia and/or the Indonesian Archipelago for rare additions to his or her collection, could be offered a rare treat – the coconut pearl. With some secrecy and drama, a vendor may offer for examination an ovoid to roundish, whitish, non-nacreous mass of up to 1.5 cm diameter. Usually the potential purchaser is regaled with a suitably colourful story of how this treasure had been acquired by the vendor.

According to the 5th edition of Webster's *Gems* coconut pearls may be either rounded natural pearls that have been harvested from 'The clam of Singapore', *Hippopus hippopus*, or will be found as round nut-like concretions in the coconut that are recovered from 'blind' coconuts in which the coconut pearls represent an aberrant growth of the coconut's embryo.

With respect to Webster's first category of coconut pearls, this description should be amended for, as purchased, coconut pearls may be either natural clam pearls, or more likely they will be man-made pearls, of Buginese origin, that have been deliberately manufactured by turning, carving and polishing the thick-walled shell of a clam.

It seems quite likely that Indonesia may be the prime source of the quite effective imitations of the coconut pearls that from time to time are reported in the gemmological literature. As manufactured, these imitation coconut pearls are whitish to greyish pear-shaped porcellanous masses that are frequently longitudinal ridged along their length. Brown *et al.* (1988) revealed that the shell-like origin of these impostors can be readily identified by determining their:

• SG that lies within the range 2.70–2.85
• 'Spot' RI of 1.53
• Parallel banding that is often visible by transillumination and
• Easy-to-see evidence of surface scratches and grooves that have been created by man.

The so-called 'Philippino black pearl' is a related imitation pearl. These pearls, which are sold to tourists as 'genuine Philippino black pearls', are in fact polished worked pieces of black shell. The ever-present growth banding of these shells identifies this at times quite convincing impostor.

Cultured Pearls

Cultured pearls can be defined simply as pearls that have been produced from host molluscs with the assistance of man.

Varieties of commercially cultured pearls that are presently being farmed worldwide include:

• Whole, half- and *keshi*[3] bead nucleated akoya pearls that have been cultivated in the akoya 'pearl oyster' *P. imbricata*.

- Whole, half-, and *keshi* bead nucleated cultured white, silver, cream or golden South Sea pearls that have been cultivated in the silver- or gold-lipped 'pearl oyster' *P. maxima*.
- Whole, half- and *keshi* bead nucleated cultured black South Sea pearls that have been cultivated in the black-lipped 'pearl oyster' *P. margaritifera*.
- Whole freshwater cultured pearls that have been cultivated following mantle tissue grafting or bead nucleation of the Chinese 'pearl mussels', *Cristaria plicata* and *Hyriopsis cumingii*, and various other species of freshwater mussel from the USA.
- Half-pearls cultivated in a range of molluscs that include several species of abalone (*H. iris*, *H. rufescens*), several species of wing shell (*P. penguin*, *P. sterna*), and several species of freshwater mussel from the Mississippi River.

Brief History of Pearl Cultivation

Historical evidence suggests that thirteenth-century Chinese likely discovered the basic skills of pearl cultivation. The end result of these early experiments was the production of nacre-coated Buddhas on the inside of the shell on the Chinese native freshwater mussel, *C. plicata*. It has been suggested that these devotional objects were produced by gluing metal images of the Buddha between the shell and mantle of the freshwater mussel that is commonly found in Chinese rivers.

The first European experiments in pearl cultivation have been attributed to the eighteenth-century Swedish naturalist Carl von Linne, more commonly known as Linnaeus. He bored a small hole through the end of the shell of the European freshwater mussel, and inserted through that hole a small fragment of limestone that was held in place by a silver wire. His results were not encouraging. Later, during the 1760s, Linnaeus did produce cultured pearls when he inserted part-drilled MOP beads that were held on a T-shaped metal holder through holes drilled through the shells of mussels.

Yet another little known pioneer cultivator of pearls was the entrepreneurial American, La Place Bostwick, who reportedly cultured pearls in several species of both salt- and freshwater molluscs.

Further developments in pearl cultivation technology occurred in the late nineteenth century, when Kokicho Mikimoto experimented with, perfected and patented a process for producing semi-spherical blister pearls by cementing a bead of MOP to the valve of the Japanese pearl oyster *P. imbricata*. However, Mikimoto's long search for the secrets behind successful culture of spherical pearls was less successful. Indeed, Mikimoto's patent for spherical pearl culture, which was granted in 1914, was not based on his personal research. Instead, it was based on the work of a Japanese dentist, Otokichi Kuwabra. It is indeed ironic that while Mikimoto became the world's major producer and promoter of spherical bead nucleated cultured pearls, he did not invent the presently used commercial technology for culturing bead nucleated cultured pearls.

Technology for the cultivation of round (whole) bead nucleated saltwater pearls is jointly claimed to have been invented during the early years of the twentieth century by two Japanese inventors: Nishikawa, an inexperienced marine biologist, and Mise, a carpenter. However, presently available evidence suggests that the claimants could have obtained knowledge of this technology from William Saville-Kent – the English-born marine biologist who first cultivated blister pearls and more arguably round pearls in *P. maxima* of the waters surrounding Thursday Island many years before Nishikawa and Mise made their patent applications. The story of the patent battles between Nishikawa and Mise make fascinating reading. Readers interested in obtaining more details about this controversy should consult O'Sullivan *et al.*'s paper '*The old and the new of Australian pearl production*' presented to World Aquaculture '98 in Las Vegas, USA, during February 1998.

Also during the early 1890s, another Australian – a Mr G.S. Streeter, who was unrelated to the famous pearler W.E. Streeter – grew what have been claimed to be Australia's first cultured pearls in Roebuck Bay, just to the south of Broome. Streeter was reputed to have cultured half-pearls by drilling small holes through the shell of *P. maxima* and then inserting small studs of MOP, which were held in place by a piece of thread pulled through the hole in the shell, to act as nuclei. This attempt was not an economic success, but photographs of Streeter's pearl culture facility still exist.

Bead Nucleated *Akoya* Cultured Pearls

Traditional Japanese Culture of Round *Akoya* Pearls

Technology required for the cultivation of bead nucleated whole saltwater pearls is jointly claimed to have been discovered in Japan during the early years of the twentieth century by Nishikawa and Mise. Irrespective of the contentious contribution made by Nishikawa and Mise, it was Mikimoto – the ex-noodle pedlar – who deservedly is renowned for actively promoting and popularizing world acceptance the round bead nucleated seawater cultured pearl, which is well known in the trade as the *akoya* pearl. This cultured pearl first made significant inroads into the world's pearl markets during the early years of the 1920s.

For the first 50 years of its use, Japanese implantation and pearl culturing technology remained both a trade and a national secret. As a consequence, the Japanese-produced *akoya* bead nucleated cultured pearl dominated the world pearl market, and Japanese technology was the only technology used by trained Japanese technicians to create bead nucleated cultured pearls. These pearls were cultivated in the Japanese bivalve *Pinctada imbricata (=martensi, fucata)*, using aragonitic bead implants manufactured from freshwater Mississippi mussel shell and a graft tissue implant from an unspecified region of the mantle of a 'sacrifice' *P. imbricata* pearl 'oyster'.

Traditionally, <8 mm diameter round *akoya* cultured pearls were culti-vated in the oft narrow, funnel shaped, relatively shallow, sheltered, even temperature bays on the east coast of the Japanese main island of Honshu at Ago Bay and the nearby bays of Gokasho, Matoya, Kagaimura, at Mie to the south of Tokyo, and on the island of Tsushima off the western coast of Japan.

Surprisingly, the only detailed description of traditional Japanese pearl culturing technology was published, in 1949, as *Report* No. 122 from the Natural Resources Section of the General Headquarters for the Supreme Commander for the Allied Powers. This important report, which was written by Dr A.R. Cahn, had the title *Pearl Culture in Japan*.

According to Cahn (1949), the Japanese produced their bead nucleated cultured pearls in the following sequence:

Provision of 'Mother Oysters'

Traditionally, native *akoya* 'mother oysters' were collected by female divers, known as *amas*, who free-dived to depths of 10 m to recover up to 10 shells per dive. Their catch was sorted by hand into two categories: 2–3-year-old 'oysters' that were sold to pearl farmers who sowed these into in shallow water to continue their growth to 3–4 years of age; and 3–4-year-old 'oysters' that were taken to pearl farms where they were allowed to recuperate from their capture by being placed in shallow water with a rocky bottom.

From 1924 onwards, the diminishing supply of natural 'mother oysters' was based on an ever increasing extent on the organized collec-tion of spat (free swimming larvae) following the spawning of *akoya* 'oysters' that occurred from July to September each year. Spat collectors used ranged from suspended bundles of birch branches to vertically suspended bunches of shells, and Mikimoto-patented darkened wire mesh covered spat collecting cages – for Mikimoto's research had revealed that spat developed an aversion to light just before they wanted to settle and attach to a firm substrate and begin to grow. Up to

50 spat collecting cages were suspended at depths of 6 m from individual bamboo-framed rafts that were moored in known spawning waters. Strategically located large drums were used to support each floating spat collecting complex of 5–10 individual rafts that were lashed together. Spat collecting cages remained in place until November, when their content of ~10 000 spat of an 0.5 cm average length were removed and transferred to rearing cages.

Following collection, the spat were removed to compartmentalized rearing cages that were of similar design to that of the spat collecting cages, but were covered with finer mesh to exclude predators. These cages were distributed either over the sea bottom in sheltered waters, or suspended from rafts where the juveniles were allowed to grow until they were 1 year old. The juvenile 'pearl oysters', which averaged 2.4 cm in diameter, were then sown on a shallow (3–5 m) rocky substrate on the sea bottom to mature and grow for a further two years.

In contrast, modern akoya culture technology is based on the grow-out of hatchery bred spat to produce healthy 'mother oysters' that are suitable for implantation.

During June, July and August, mature (3-year-old) 'oysters' were collected by women divers and brought to floating cleaning barges which had been towed to the collecting areas by launches. Here the 'oysters' were cleaned, before being examined and sorted for health and maturity. Suitably mature 'oysters' were then placed in culture cages suspended from rafts for acclimatization in shallow water for about 10 days to allow the 'oysters' to recover from the shock of their removal and cleaning, and to allow them to adjust physiologically to live in shallow water conditions.

Preparation for Operation

Following acclimatization, the 'mother oysters' were prepared for implantation by being encouraged to gape using one of several traditional methods that included immersion in stagnant sea water, immersion in running sea water or being dumped dry on the wharf for about half an hour. Once the oysters had gaped, the blades of opening forceps were inserted between the valves that were then gently forced apart about 1 cm to allow a tapered bamboo wedge to be inserted. The oyster was then ready for implantation of the nucleus and graft by the seeding technician. To ensure success, this operation had to take place within two hours of the oyster being pegged.

While 'keyers' were kept busy pegging oysters, the implant technicians were occupied organizing their equipment at their work stations. Each technician had his or her own set of instruments that

consisted of: a pair of straight surgical scissors for cutting an unspecified strip of mantle tissue from a 'sacrifice oyster', a scalpel for trimming the tissue graft, two softwood discs on which the graft tissue was
trimmed and cut, a pair of fine pointed tissue tweezers, a white porcelain vessel, a beaker of clean seawater, and a supply of good-quality
muslin for cleaning the graft tissue, a custom-designed workbench, an
oyster clamp for holding the pegged oyster during the operation, two
porcelain vessels, a beaker of seawater, a wooden instrument rest,
cloth for wiping and cleaning instruments, a shallow wooden tray for
holding oysters, a wooden tray for holding wedges removed after the
operation, sponges, various shallow cups for holding nuclei of various
size, and a suite of custom-designed surgical instruments that
included scalpels of specific design, forceps, a spatula, tissue retrac-
tors, and lifters of various design for managing incised tissue and
inserting both the graft and the nucleus.

Preparation of graft tissues required the sacrifice of a mature 'oyster',
which was known to have good-quality nacre, to provide a strip of tissue
of 7 × 70 mm approximate dimensions from the outer edge of its mantle.
Following smoothing, cleaning and trimming of this strip of mantle on
the wet graft trimming block, the remaining 2–3 mm wide strip of outer
mantle epithelium was diced into ~2–3 mm segments suitable for insertion with the bead implant. The size of the graft was determined by the
diameter of the bead to be implanted, as generally the graft should cover
about one-half to one-third the circumference of the implanted bead.
Another important factor was that the graft had to be cut and oriented in
such a manner that, when implanted, the nacre-producing cells of the
outer mantle must face the implanted bead.

A series of perfectly round, perfectly smooth nacreous Mississippi
freshwater mussel shell beads of up to 6 mm diameter were required to
complete the implant. Traditionally, these beads have been manufactured
from the thick shells of species of Mississippi River freshwater mussels
that include the genera *Amblema*, *Aquadrula*, *Pleurobema* and *Megalonais*.
The advantages of these beads are that they are aragonitic – so aragonitic
nacre can be efficiently deposited on the surface of the bead – and they
do have a hardness and SG similar to that of nacre. Traditionally, beads
used for implanting *akoya* 'oysters' were prepared in 0.3 mm increments
to yield beads of 1.2–6.6 mm diameter.

The Operation

Holding the pegged oyster in a shell clamp with its right valve uppermost, the implant technician then used specially designed instruments to
perform the necessary surgery. The sequential steps involved in this

operation were: following successful grafting and implanting, oysters were returned by using a spatula to retract the mantle to expose the body and foot of the 'oyster'; using a hook of a retractor to hold, extend and prevent movement of the very muscular foot of the 'oyster' during the operation; making a small incision into the foot of the 'oyster' before dissecting a narrow channel into the large gonad (sex gland) of the 'oyster'; inserting the graft tissue into its prepared bed with a probe-like graft lifter; placing a nucleus of appropriate diameter into the prepared channel and against the nacre-producing cells of the graft using a cup-tipped nucleus lifter; and finally closing the wound by smoothing the incised tissues together before the retractor was removed from the foot of the 'oyster', the peg removed from between its valves, and the 'oyster' was removed from the shell clamp.

A well-trained technician could graft and implant 25–40 oysters per hour using this technique.

Convalescence

Those 'operated oysters' that had been successfully grafted and implanted were returned to seawater-filled holding trays, and then to hinged wire frame culture cages that held 50–60 'oysters' in four compartments. The seeded 'oysters' were allowed to convalesce from the trauma of the operation by being suspended horizontally at depths of 2–3 m from rafts anchored in sheltered waters located near the seeding facility. Here the 'operated oysters' remained undisturbed for 4–6 weeks before they were lifted and inspected for health. All healthy 'oysters' were then transferred to culture cages attached to permanent culture rafts that were towed by barges to carefully monitored protected waters some distance away from the onshore facilities in which they were seeded.

Pearl Growth

On the culture rafts 'operated oysters' were allowed to grow until the implanted beads were coated with an adequate thickness of nacre. This essential time of cultivation occurred under carefully monitored conditions at depths of 2–3 m, with each raft supporting 60 cages that contained a total of 3000–3500 seeded 'oysters'. Except for periodic cleaning (at least three times per year), the seeded 'oysters' remained undisturbed for a period of cultivation of 3–4 years (*but today from 6–18 months*). During periodic cleaning, all encrusting marine growths were removed; the 'oysters' were checked for health; and they were tested (by X-ray) for nucleus retention and rate of pearl growth.

Following periodic cleaning, 'operated oysters' were again returned to clean cages and returned, once more, to the sea.

Harvest

After cultivation had been completed, harvesting of pearls usually began late in the year, just before winter. It was generally believed that choosing this time for harvesting was because exposure to the cold water of early winter caused the operated oysters to secrete a thin layer of translucent nacre over the nacre already deposited – thus improving the overall quality of the pearl's nacre.

The oysters were usually sacrificed at harvest, and any cultured round and *keshi* pearls removed from their tissues. Very rarely, a limited number of healthy young oysters may have had their pearls surgically removed and a new larger bead implanted. This, however, was not common, for *akoya* 'oysters' have a relatively short life span of about 7–8 years.

Processing

Harvested round *akoya* pearls were then prepared for marketing in a number of sequential steps. These steps included: washing and tumbling the pearls in a mild abrasive to remove organic debris (slime) and to smooth and polish the nacre; sorting the pearls for shape and the removal of all unmarketable pearls; grading the pearls into size categories by sieving; lot sorting the pearls into parcels of 100 pearls of approximately the same quality and size; sorting of these lots into three major categories of quality – A-grade (spherical or near-spherical pearls of good colour and lustre), B-grade (baroque pearls of good colour and lustre, and spherical and near-spherical pearls of medium colour and lustre) and C-grade (poor-quality pearls that had limited ornamental value); and regrading the graded pearls into each of three subcategories (Aa, Ab, Ac/Ba, Bb, Bc/Ca, Cb, Cc) based on the criteria of form, colour, lustre and surface perfection.

If the pearls were required for necklaces, the carefully graded pearls were then sent to the drilling room where they were: examined carefully and individually for surface blemishes; marked to indicate where drilling would remove a blemish; drilled; bleached in warm hydrogen peroxide to remove unacceptable natural colours and produce a uniformly white pearl; dyed with a suitable alcohol-based aniline dye to provide uniform market-acceptable colours (*this step in the processing of akoya pearls is still not admitted by the Japanese akoya pearl industry*); and graded, matched and strung into strands. Strands were then inspected, graded for quality, weighed, one hundred necklaces of similar quality were then tied together, again weighed, and were then ready for the market.

Modern Culture of Round Akoyas

Ago Bay, in Ise, Mie Prefecture, is the traditional home for the cultivation of *akoya* pearls of 4–8 mm diameter. In this comparatively narrow bay, which is not self-cleansing, over 1000 pearl farms presently operate from shore bases and also use outer bays for maturing pearls. The modern Ago Bay pearl cultivation industry is based on spat reared stock that is allowed to grow for 2–3 years before being operated on to yield bead nucleated cultured pearls. Over the last decade or so, decreasing nucleus retention rates (55–65% v. 25% today) and economic pressures to obtain a marketable crop of pearls has led to the use of shorter and shorter cultivation times as little as six-month duration.

Today, the yearly work schedule used to cultivate *akoyas* in Ago Bay comprises:

January	Harvest previous year's pearls
February	Preparation of oysters for seeding
March to May	Seeding
June	X-raying for nucleus retention
	Locating seeded oysters in 24–28 pocket panels in a suitably sheltered bay
August to September	Periodic mechanical cleaning of seeded oysters
	Check water temperature
	Remove dead oysters from panels
October	Buying 'baby' oysters
	Routine cleaning and maintenance
November	Nurturing 'baby' oysters
	Routine cleaning
	Continuous checking for cold currents
December	Routine cleaning and checking

Quality Assurance of Akoyas

As of January 1999, the Japan Pearl Exporters Association (JPEA) replaced the Japanese government-run pearl inspection system (Japan Pearl Inspection Office) that has been operating since 1952. Presently, *akoya* pearls (strands and bags of individual pearls) that pass examination by JPEA will receive a quality assurance tag. Inspection Offices in Kobe and

Tokyo will administer the JPEA's Cultured Pearl Quality Inspection and Tag System by applying the following criteria to akoyas whether the pearls are of Japanese, Chinese or other origin.

Nacre Thickness

All pearls will be inspected for thickness of nacre. Pearls with nacre so thin that the core nuclei can be seen through the nacre will be rejected.

Lustre and Clarity

Pearls with low intensity of lustre will be rejected. Pearls that are strongly permeated by calcite and organic matter, resulting in muddy grey, brown or blue colours, will be rejected.

Surface Blemishes

Pearls that exhibit excessive amounts of blemishes, such as bumps and pits will be rejected.

Nacre Damage

Pearls that exhibit visible cracks in the nacre or on the nuclei will be rejected.

Pearls that display any signs of nacre peeling or chipping will be rejected.

Processing

Pearls exhibiting signs of damage from bleaching or dyeing or are judged to be unstable in terms of permanency of appearance and quality will be rejected.

Pearls that display characteristics of uneven dye methods or overdyeing resulting in excess dye residue on the surface of a pearl will be rejected.

Marketing *Akoyas*

Akoya pearls are sold by weight, with units of the momme of 3.75 g and the kan of 1000 momme being the Japanese and world standard for wholesale trading in *akoya* pearls. At the retail level, these pearls are sold by the strand, or by the gram or carat weight for individual pearls.

Structure of *Akoya* Bead Nucleated Cultured Pearls

Round *akoya* cultured pearls may visually resemble natural pearls, but their structure is quite different in that they consist of a central aragonitic shell bead that is coated with <1 mm of cultivated nacre. A thin layer of (usually bleached and dyed) organic conchin separates the bead from

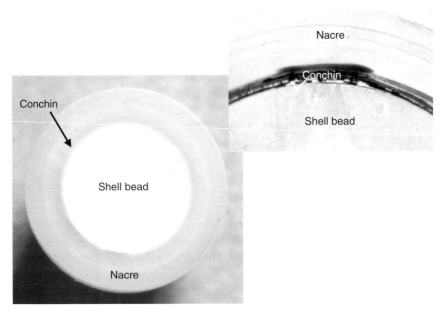

Figure 28.8   Cross-section of a bead nucleated cultured akoya pearl

its surrounding nacre. In some bead-nucleated cultured pearls, calcite is deposited against the bead before the grafted cells become mature enough to produce aragonitic nacre.

Unfortunately, a lot of modern Japanese *akoyas* are produced so rapidly that their bead actually shines through their very thin layer of cultivated nacre. This all too frequent modern occurrence increases the friability of these pearls and lowers their value significantly.

It is a fact that *akoyas* seldom exceed a diameter of 8 mm, and that the thickness of nacre surrounding the bead of modern *akoya* is <<0.5 mm. Due to the presence of their shell beads, *akoyas* usually display a higher SG (~2.70) than natural pearls (~2.67).

Japanese *Keshi* Pearls

The term *keshi* has a Japanese origin, for it is used to describe the 'poppy seed' pearls that were first discovered by Japanese cultivators as by-products of the *akoya* bead nucleated pearl cultivation industry. These small, <5 mm, mostly 'rice grain'-shaped pearls may form naturally, cut commonly form in pearl sacs from which the mollusc has expelled the implanted bead. They are recovered at harvest and after drilling are strung into multi-strand necklaces that closely resemble necklaces of natural 'seed' pearls.

Japanese *Akoya* Cultured Half-Pearls

Akoya 'mother oysters', deemed to be unsuitable for round pearl production, or healthy 'mother oysters' that have had their pearl(s) surgically removed, can be used for the production of hemispherical bead nucleated cultured *akoya* half-pearls.

In this process, hemispheres of appropriate size and shape, which have been made from talc or MOP (in the past) or polymer (at present), are glued to nacre lining the valves of the *akoya*, and the implanted 'oyster' is then returned for cultivation for a further traditional cultivation period of 1–2 years. Over that time, the implant becomes coated with a thin hemisphere of nacre of <0.5 mm thickness. After the implanted 'oyster' has been harvested, the cultured blister is cut from the shell, its contained nucleus removed, and the inner surface of the blister is cleaned, bleached and over recent years painted an appropriate colour. The cavity in the hemispherical blister is then filled with a suitably pigmented polymer, and a flat, polished base of MOP is cemented to the hemisphere of nacre to complete maufacture of the half-pearl.

Older *akoya* half-pearls commonly had rounded rather than flat bases, and deliberately they were made heavier to heft by including a disused MOP bead in the resin with which the hemisphere and base of the half-pearl were cemented together.

Over recent years, production of half-pearls from some Japanese manufacturers includes half-pearls in a range of rather vivid non-traditional colours, and half-pearls of square to rectangular shape.

It is indeed unfortunate that the trade uses the term *mabé* to describe all bead nucleated cultured half-pearls – for the true *mabé* is natural pearl or bead nucleated half-pearl cultivated in the *Pteria (=Magnavicula) penguin* otherwise known as the wing-shell, black-wing pearl 'oyster', or penguin wing pearl 'oyster'. These very attractive (and expensive) cultured half-pearls display a highly iridescent metallic lustre and a light brownish to greyish somewhat metallic body colour that is quite unique and totally unlike the appearance of *akoya* half-pearls.

Other Producers of *Akoya* Pearls

People's Republic of China

Due to the increasing pollution of Japan's pearl culturing waters, excessive mortality of stock over recent decades, and economic trends that have tended to force pearl farmers to use ever-decreasing cultivation times, the quality of the Japanese *akoya* pearl continues to decrease. As a consequence, China is rapidly overhauling Japan as a major supplier of small size round bead nucleated saltwater cultured pearls. Presently, most Chinese *akoya* farms are small in scale, and mostly family run.

Commercial cultivation of Chinese seawater pearls began at the end of 1950s. In 1958, the first pearl culturing facility for seawater pearls was set up in Hepu, in Guangxi province, a province that once was the most famous for producing natural seawater pearls. Over the following 30 years, researchers have successfully developed technologies for cultivating the Chinese akoya *Pinctada chemmitzii* (Philippi, 1849). Subsequently, the annual production of Chinese seawater pearls (*akoyas* and to a limited extent white South Sea pearls) has reached ~35 tons, which makes China one of the major providers of cultured bead nucleated seawater pearls to world markets.

In China, *akoya* pearl culture occurs in the protected and previously unpolluted bays of the coastlines southern provinces of Guangxi to the Vietnamese border, Gaungdong, and offshore northern Hainan Island. However, production from Hainan is comparatively small. To date, pearl culture has focussed largely on Chinese species of *akoya*, the *P. chemmitzii*, which are produced by spat culture. After 15 months of growth, Chinese *akoya* 'pearl oysters' attain a size of 70 mm and are then suitable for implantation. Following seeding, the seeded 'oysters' are returned to the farm for a further 11 months cultivation before they are sacrificed and the pearls harvested. Survival rates from spat to 70 mm size and from seeding to harvest are reported to be approximately 35 and 60%, respectively.

Akoya pearl farming methods, presently being employed in China, are rather simple, and cultivation practices vary in quality and are relatively uncontrolled. For example, rather than cleaning the oysters 4–5 months after implantation, some farmers simply harvest pearls with those pearls having an extremely thin coating of poor-quality nacre. The end result of this short sightedness has been the flooding of the market with Chinese *akoyas* of inferior quality and low value. Standards of cultivation, however, are steadily improving, as is the quality of Chinese enhancement (bleaching and dyeing) processes. Consequently, better quality Chinese *akoyas* now are likely to have thicker nacre, better quality nacre, and are larger and better coloured than those pearls produced, say, five years ago.

Presently, the cultivation of Chinese seawater pearls occurs either in state-run, collective and individual pearl farms, among which the individual pearl cultivation farms are most prominent. Most Chinese-cultivated akoyas are sold to Japanese interests, who process the pearls for subsequent export from Japan.

One interesting feature of the Chinese akoya is that they often have a discernable flat spot on their external surface – possibly caused by the lack of 'turning' of the 'mother oysters' after implantation.

Vietnamese Akoyas

Vietnam is likely to become the next commercial source of *akoya* pearls, with several pearl cultivation farms operating in Haipong, North Vietnam, near the Vietnamese–Chinese border, and at Nha Trang in south-eastern Vietnam just to the north of Ho Chi Minh City.

Bead Nucleated South Sea Cultured Pearls

For over 50 years, larger nacreous saltwater bivalves such as the silver- or gold-lipped 'pearl oyster' (*P. maxima*) and the smaller black-lipped 'pearl oyster' (*P. margaritifera*) also have been implanted (using basic *akoya* technology) to yield large diameter (<10 mm) bead nucleated cultured pearls. These bead nucleated cultured pearls are known in the trade as white or black South Sea pearls. The common colours displayed by white South Sea pearls include pearls of white, silver, cream to yellow and golden colour. The colours of black South Sea pearls range from silver, brownish to golden and grey to black. The major producers of white South Sea pearls, in value order, are Australia, Indonesia, the Philippines and Myanmar. French Polynesia is the major producer of black South Sea pearls, with the Cook Islands and other small Pacific nations being only minor producers of these pearls.

White South Sea Pearls

World production of white South Sea pearls for 2003 increased to 7.5 tons, from an estimated 5 tons in 2002. Major producers for that year were Indonesia (3.75 tons), Australia (2.4 tons), Philippines (1.2 tons) and Myanmar (0.15 tons). This is the first time that Indonesian production has outstripped Australian production. However, the value of Australian white South Sea pearls far outstrips that of pearls produced from either Indonesia or the Philippines.

Australian White South Sea Pearls

Australia is the world's largest producer of white South Sea pearls – that is, bead nucleated cultured pearls of more than 10 mm diameter that are cultivated in the gold- or silver-lipped 'pearl oyster' *P. maxima*. Australian white South Sea pearls have an average diameter of 12–14 mm; however, diameters may range from 10 mm to an exceptional 20 mm plus. While the <8 mm diameter Japanese *akoya* cultured pearls may have a coating of nacre of <0.4 mm thickness, an average 15 mm Australian South Sea pearl has nacre thickness of 4 mm, and a distinctive lustre and orient to match.

Other essential products of Australia's white cultured pearl industry include half-pearls up to 25 mm diameter that are available in a range of shapes, and an intriguing range of designer-ready baroque accidents of the culturing process, the *keshi*.

Australia's pearl culturing industry commenced in 1965 when Pearls Proprietary Ltd established a pearl culture farm at Kuri Bay, some 420 km north of Broome. By 1963 this farm was producing 60% of the world's quality white South Sea pearls. Presently, the high-quality products from pearl farms operating in Western Australia to the north and south of Broome, and in the Northern Territory around the Coburg Peninsula and on the islands to the north-east of the Gove Peninsula dominate the burgeoning Australian white cultured pearl industry. A few Queensland-based pearl cultivators, who operate in Torres Strait at several locations down the eastern seaboard of Cape York as far south as Cooktown, are presently only minor contributors to Australia's overall production of white South Sea pearls.

In Western Australia – the largest producer of Australian white South Sea pearls – the pearl culturing industry is based on 16 pearling licensees, who occupy 94 pearl farm leases that cover a total area of 184 square nautical miles. Pearling traditionally has been managed as a 'gauntlet' fishery that allows predetermined quotas of *P. maxima* of 120–160 mm size to be harvested annually. Larger size *P. maxima* are left as breeding stock.

The Western Australian pearl culturing industry is based on a government-predetermined annual quota of shell suitable for implantation. Since 1992 pearl farmers have had the option of supplementing their supplies with 2-year-old hatchery-bred shells if supplies of natural shells do not reach quotas. Presently, the industry is based on an allowable annual quota of 500 000 'wild' and 350 000 hatchery-bred stock.

Because areas suitable for pearl cultivation are limited, new leases are virtually impossible to obtain, and the annual quota of wild oysters is unlikely to increase. Therefore, the quality rather than the quantity of Australian white South Sea pearls should continue to improve over the next decade or so.

Modern Australian Pearl Culture Technology for Round White South Sea Pearls

According to O'Sullivan *et al.* (1995), in Western Australian waters, a licensee's annual quota of 'wild shell', of >120 mm dorso-ventral shell length, is collected by teams of up to six divers from the sandy offshore shell beds that stretch from Exmouth Gulf in the south to the Lacepede Islands in the north of Western Australia. Up to 10 dives per day are undertaken by these divers.

To collect shell, hookah-equipped divers are towed underwater behind boats that drift with the current over predesignated areas of the known shell beds. Once collected, the 'wild' shell is raised to the surface, cleaned, inspected for appropriate size and then stored in 'kangaroo' net panels of 6 or 8 shells at 'dump sites' that are located on the sea bed of the area from which the shell was harvested. At these 'dump sites' the net panels are either neatly stacked on the sea bottom, or tied onto longlines that are suspended a couple of metres above the seabed.

In the past, 'wild' shell was transported from these 'dump sites' to near-shore farms, where seeding was performed. This caused unacceptability mortality in the transported 'wild shell'. So today, most West Australian cultivators significantly reduce mortality rates, which used to occur as a consequence of this transfer, by seeding their 'wild' shell (in mid-year) at the collecting grounds – using large modern ships that are equipped with surgically clean laboratory and operating facilities.

Onboard ship, mostly Japanese-trained technicians seed each 'wild' shell, in the traditional manner, with a round nucleus and a small piece of mantle tissue from a 'sacrifice oyster'. The nuclei, which are manufactured from thick shell of the freshwater Mississippi mussel, can have diameters of 6.6–14 mm – depending on the size of the shell to be implanted, and whether or not the shell had been previously implanted. During the season, implant technicians each operate on 550–600 shells per day, using strictly sterile conditions and a dexterous surgical technique.

Recent advances in implant technology, to improve the quality of harvested pearls and decrease post-implant mortality, include: transfer technology for mother 'oysters' that is as stress free as possible; other improvements involve use of high-grade aragonitic nuclei that are perfectly spherical, smoothly polished and are free of any softer banding that is likely to create 'circles'; application of strict aseptic techniques during the seeding operation; the use of proprietary antibiotic impregnated nuclei to minimize post-implant infection and increase the rate of nucleus retention, and artificial Bironite[TM4] nuclei as a replacement for increasingly scarce and expensive shell nuclei.

Following seeding, shells are returned to net panels on bottom longlines moored at the original 'dump' sites from which they came prior to seeding. Here they remain for up to three months, during which time they are inspected and turned regularly by divers to ensure that an even envelope of nacre-secreting cells forms in the pearl sac develops around the implanted nucleus. After this stage has been completed, the net panels are retrieved (during the warm

October–December months) and the seeded shell are very carefully transported by boat or seaplane to pearl farm grow-out sites that are located in well protected coastal bays and inlets along the remote coastline of north-western West Australia, and the Northern Territory's Coburg Peninsula.

During grow-out, the implanted shell may be held in net panels and suspended either from well-protected less expensive to operate surface longlines (e.g. Western Australia and Northern Territory) or from bottom posts and longlines that are located about 13 m down. Here the shell is less vulnerable to cyclonic conditions, and does require continuing maintenance by underwater divers (e.g. Western Australia). Net panels can be held either on individual posts or on underwater longline fences. During the two-year pearl cultivation period of grow-out, implanted shells are cleaned every two to four weeks, either by underwater divers or by boat-mounted high pressure cleaning machines. Four to six months after seeding, X-rays are used to check each shell for nucleus retention and pearl formation. Shells that have rejected nuclei are held until the following year. At that time they are re-seeded if they are healthy and have little scarring from the previous attempted seeding. Re-seeding will not occur if shells have commenced forming *keshi* pearls as a consequence of the tissue implant grafting into the incision and forming a pearl-producing pearl sac.

Pearls are harvested in the cooler winter months, after about two years of cultivation. This time is deliberately chosen, for in the winter months nacre secretion is slower, more uniform and the nacre displays its best lustre. Also the lower temperatures seem to reduce stress and consequent mortality in the shells.

Each day's harvest is initially sorted by shipboard implantation technicians into saleable bead nucleated pearls, rejects and *keshis*. The reason for initial sorting is quite simple, for the financial reward for implant technicians is often tied to the productivity and quality of pearls recovered from shells that he or she has previously seeded. The saleable pearls are then graded into categories of size, shape and colour before being transported to the company's headquarters. Here the pearls will be where they are carefully cleaned of residual mucus or salt, before the better quality are gently tumbled with a mild abrasive, such as cooking salt, to remove any adherent organic material (dried mucus) and thus optimize their lustre and orient. Lower grade pearls are only tumbled in finely ground pumice to enhance smoothness, for South Sea pearls are claimed not to be colour-enhanced in any way by bleaching and/or dyeing. Following cleaning, the pearls are sized, classified and graded into specific categories ready for sale to world markets.

The yearly work schedule on a typical Australian white South Sea pearl farm is detailed below:

Typical Work Schedule for an Australian Pearl Farm

January   Prepare for wild shell collection, organize dive crews, fishing gear, paperwork and licence fees.

February   Fishing for 20 000 wild shell begins, linked to tide charts. (Note: Tides in the area can vary by 10 m per day.)

March   Collected shell is 'dumped' on the seabed on site owned or leased by the company and allowed to rest. Maintenance of dumped shells, turning and cleaning them. X-ray shells seeded last year to check if implanted nuclei have been rejected. Oysters that reject nuclei are usually re-seeded.

April   Water temperature begins to drop as winter approaches, rest period for the shells.

May   Ongoing farm work, turning and cleaning previous two-year's seeded oysters kept suspended in wire panels in the water column.

June   Prepare for operating on oysters to implant nuclei. (Note: Some technicians may come from overseas, and some companies have boats fitted as mobile laboratories so seeding can be done on the pearling grounds.) Seeding and harvesting begin.

July   Normal operating time for pearls, seeding new oyster, re-seeding those that have rejected nuclei. Oysters that produce acceptable pearls are also re-seeded.

August   Harvest of previous year's seeded shell continues, and then a two-month turning program follows operations. The oysters are turned over to encourage production of round pearls.

September   Turning operated shell.

October   Turning, cleaning and change of areas.

November   Transportation of operated shell to grow-out areas.

December   Dump and clean gear.

The Australian industry currently estimates that its overall mortality rate for seeded shell, over the two-year cultivation period, is a mere 5%. That is, 95% of successfully seeded 'pearl oysters' will produce a pearl. When to this very small mortality rate is added the 4% of molluscs that must be sacrificed to yield the *siabou* (nacre-secreting graft) tissue, these statistics clearly highlight the remarkable success of Australian white South Sea pearl cultivators.

The yield of quality pearls usually determines that only 45–50% of shell that contained a saleable pearl will be reseeded – this time with a nucleus of the same size as that of the pearl that was removed. The reason for less shell being seeded a second time, irrespective of producing a saleable first crop pearl, is simply that as the reseeded pearl is likely to be of lesser quality than the original pearl only shells capable of producing a good-quality first crop pearls are reseeded. The shell is then cultivated for a further two years to produce second crop pearls. Of the ~15% of implanted shells that remain suitable for implantation with a third nucleus, these are capable of yielding 17–19+ mm pearls. Indeed, experimentally, up to four pearls have been produced from a single shell before it was considered unsuitable for round pearl production.

Australian White South Sea Keshi Pearls

It has been estimated that less than 10% of seeded *P. maxima* reject their implanted bead nucleus and continue to secrete a seedless pearl in the pearl sac formed from grafted mantle tissue that was inserted surgically into the mollusc's gonad. The baroque shapes, attractive silver, white and cream colours, and variable size of the Australian *keshi* pearls make these attractive pearls most desirable to jewellery designers. Precise details of the annual production of *keshis* in the Australian industry are not available, but it is interesting to note that for some years annual data released by the Australian Bureau of Statistics had revealed that at least 15% of Australian pearl exports are of non-bead nucleated *P. maxima* pearls.

Australian White South Sea Cultured Half-Pearls

Once a pearl 'oyster' is deemed to be incapable of producing further round pearls, it can be used for half-pearl production. Commonly, half-pearl culture is also used to generate quick income for a newly established pearl farm. The implant and manufacturing techniques used to produce white South Sea half-pearls is similar to that used in *akoya* technology, with cultivation time for half-pearls commonly being 10–12 months. Following sacrifice of the implanted pearl 'oyster' to recover its cultured blisters, adductor muscles from the molluscs are dried and exported to south-east Asia as culinary delicacies. The remaining shells are also sold overseas as a source of MOP.

Other Producers of White South Sea Pearls

Other major cultivators of white South Sea pearls include Indonesia, the Philippines and recently Myanmar.

Indonesia

Pearl cultivation activities are widespread throughout the Indonesian Archipelago. Aru Island, only 500 km north of Darwin, is Indonesia's major producer of cultured pearls. Most farms are located in Maluku (the Moluccas), West and East Tenggara, Central, South, East, Southwest and North Sulawesi, East Timor, Lampung, Bali, Java and Irian Jaya. The number of pearl farms operating in Indonesia is unknown; but five of the largest farms hold 100 000–250 000 implanted 'oysters', and about 15 farms seed 50 000 pearl 'oysters' annually. About 25 (mostly joint-venture) companies each produce about 40 kg of pearls per year. Due to a general shortage of wild stock, Indonesian pearl cultivators must rely heavily on hatchery stock for seeding.

   Due to warmer water conditions, Indonesian cultured pearls tend to have warmer hues than Australian South Sea pearls. At least 50% of annual Indonesian production is of pearls of cream and yellow colour. Best-quality Indonesian nacre is thick (compared to *akoya* nacre) and has a distinctive light pink creamy colour with a good lustre that tends to hide ever present minor surface imperfections. For economic reasons the majority of Indonesian pearls are in the 8–13 mm range. These pearls are obtained by implanting the oysters with a small (e.g. 6–7 mm) bead, then cultivating them for a mere 12 months.

The Philippines

Pearl culturing began in the Philippines with Sukeo Fujita's failed venture at Zamboanga in 1916. After the Second World War, a further attempt by K. Tashima at culturing pearls at Zamboanga also failed. Early commercial pearl farming projects were begun in the Philippines during the early 1960s. These were centred on Mindanao, the large island in the south of the Philippine Archipelago. However, other farms soon sprang at Zamboanga and several islands (Guiuan, Suridao, Cebu, Bohol) to the north of Mindanao, and off both ends of Palawan in the Sulu Sea to the east of Mindanao. Unfortunately, for a variety of reasons, most of these ventures failed. However, by 1997, over 20 pearl farms were operating in eight different regions throughout the Philippines to produce 225 300 kg of pearls annually. By the early 1980s, larger farms with sound financial backing were established and began operating on the Palawan Island (e.g. Jewlmer's Bugsuk farm in southern Palawan). These farms relied on an annual catch of 300 000 'wild oysters'. By the end of that decade stock was becoming scarce, for divers were required to recover the oysters from depths of 40–50 m.

   Although live shells were traditionally dived for by the Badjaoes – a semi-nomadic people from the Sulu archipelago – by the late 1990s, six of the Philippine's largest producers had mastered hatchery technology; so hatchery cultivation of spat now forms the backbone of the Philippines cultured pearl industry.

The Philippines has 37 operational pearl farms, with Palawan and Busuanga to the north, the major pearl-producing areas that cultivate 70% of the country's South Sea pearls. Most modern Philippine farms are concentrated around Busuanga in northern Palawan and Bugsuk in southern Palawan.

As the Philippine's coastal waters are generally remote and tropical, pearl 'oysters' can be seeded and pearls harvested 3–4 times per year – particularly during wet season that lasts from June to November. Cultivation times of 18–24 months produce pearls of 9–16 mm diameter that have colours that range from white to silver, cream, champagne and most desired golden. Generally speaking, cultured pearls from the Philippines display much the same characteristics as Indonesian cultured pearls.

While the majority of Philippine cultured pearl production consists of pearls of cream to golden colour, ongoing University-based research is revealing ways of increasing the yield of white South Sea pearls from the gold-lipped species of *P. maxima* used. Also, by careful selection of gold-lipped pearl oysters, with respect to the choice of mantle tissue suitable for grafting, some farmers are achieving a 40% return of gold pearls from their crop.

Myanmar

The remote tropical waters of the Andaman Sea, around the islands of Myanmar's Mergui (now Taninthayi) Archipelago, have traditionally hosted large populations of subspecies of both gold-lipped and *akoya* pearl 'oysters'. In the past, the nomadic Salang people fished these waters for MOP and the reward of occasional natural yellowish rosé hued pearls that had world renowned lustre and orient. During the late nineteenth to early twentieth century, both naked diving and hard hat diving were used to recover natural pearls and pearl shell from the waters surrounding the Mergui Archipelago's almost 900 islands. However, over-exploitation of this natural resource almost wiped out the much-valued natural Burmese pearl.

Pearl culturing began in colonial Burma in 1945 with a joint venture between the then government and the Japan-based Takishima Pearl Company. The Burma Pearl Syndicate began pearl cultivation, using Japanese technology and expertise, at what became known as Pearl Island (formerly Sir J. Malcolm Island) in the Mergui Archipelago. However, as the Japanese partner refused to transfer any knowledge with respect to pearl cultivation to the Burmese, the joint venture was nationalized in 1963 and replaced by a government-controlled self-taught Burmese pearl culture that marketed any pearls produced in Yangon at periodic Gems, Jade and Pearl Emporia, run by the Myanmar Gems Corporation.

Since October 1988, when the government of Myanmar began to encourage foreign investment, several joint venture pearl cultivation enterprises have been established. Some seeding is performed by Japanese and other foreign technicians, while others are indigenous citizens of Myanmar. Myanmar's modern pearl cultivation industry, which is spread over islands such as Sir J. Malcolm or 'Pearl' Island, Owen Island, Ravenshaw Island, Domel Island and Bentinck Island, is a small but growing industry based on 'mother oysters' produced by spat culture. The seeded 'oysters' are grown out in kangaroo nets suspended from longlines located just offshore due to the depth of water that surrounds these islands.

A recently published report indicated that Myanmar's production of white South Sea pearls ranges from 12 to 17 mm in diameter. The average yield is claimed to be 10% rounds, 20% baroques and 70% pear shapes. Silver nacred pearls account for 40% of total production, while yellow to golden pearls account for the remaining 60%. An estimated 40% of pearls cultivated in the waters of Myanmar are claimed to have 'clean' surfaces.

Black South Sea Pearls

Black South Sea pearls are the product of the black-lipped pearl-oyster – known to scientists as the *Pinctada margaritifera* (Linnaeus, 1758). Various subspecies of *P. margaritifera*, and its close relatives, are distributed widely throughout tropical Indo-Pacific waters from the Persian Gulf and the Red Sea to the Gulf of California, and from Japan to the Southern islands of the Pacific. Of these, the large, dark nacre-lined subspecies *P. margaritifera cumingii* (Reeve, 1857) is found in five archipelagos that stretch from the Marquesas to the Austral Islands, and from the Leeward Islands to the Gambier Islands. These bivalves occur in greatest abundance in the lagoons of the 81 islands or so that form both the Taumotou Atolls and Gambier Archipelago.

In the Pacific Ocean, commercial pearl culturing operations are centred on French Polynesia and the Cook Islands, with French Polynesia producing 93–95% of all black South Sea pearls. Small-scale black pearl culturing industries are currently being established by several smaller island nations in Oceania.

Culturing of black South Sea pearls began in French Polynesia in 1961, when at the suggestion of Jean Domard, then Head of the Farming and Fisheries Department of French Polynesia, experiments in pearl culture, based on Japanese technology, commenced at Bora Bora. Following a successful harvest in 1963, this experimental pearl culture continued and was joined by a private pearl farm set up in Manihi by the Rosenthal brothers. Cultured pearl production really commenced following the establishment of the pearl farms of Jean-Claude Brouillet's on South Marutea, and Robert Wan on the Gambier Islands. Following a successful joint auction, conducted by these producers and the international pearl merchant Salvadore Assael, the authenticity of what became known commercially as the Tahiti black pearl became established – particularly after it was recognized as distinct gemmological entity by the Gemological Institute of America in 1976.

From 1978, under the auspices of the Fisheries Department, the flourishing newly established cooperative pearl farms were grouped together as GIE *Poe Rava Nui*. Advances in techniques of spat collection, devised by M. Mizuno, were progressively transferred to members of this cooperative during the 1980s. This lead to a boom in the production and export of the cultured black pearl. As a consequence, since 1983 the cultured pearl has been French Polynesia's premier export.

Private pearl farms in French Polynesia belong to one of three organizations: Poe O Tahiti Nui, which represents 42 farmers, including 15 large Producers, GIE *Poe Rava Nui*, which includes about 450 family farms and cooperatives, and the *Syndicat Professionnel des Perles de Tahiti* (SPPP), which has as its Chairman French Polynesia's largest producer, Robert Wan, whose private company produces 50% of all pearls cultivated in French Polynesia.

Modern French Polynesian Pearl Culture Technology for Round Black South Sea Pearls

The cultivation of black South Sea pearls in French Polynesia is primarily conducted in carefully selected areas of the large tropical to sub-tropical lagoons that lie within atolls. As lagoons usually have only one entrance–exit, they are very sensitive to environmental pollution that potentially can rapidly destroy any established pearl culture. Pearl culturing in French Polynesia is conducted in several distinct phases. These phases include the collection of 'wild' spat, growth of spat to the juveline stage, growth of juveniles to maturity, seeding and the grow-out of seeded 'oysters' to yield pearls.

The French Polynesian culturing process begins with the collection of 'wild' spat from lagoons. Following fertilization of ova by sperm in open water, the developing lava drift with the current for about a month while their shells continuously grow and the lava convert into spat. Under natural conditions, the floating spat eventually sink and either settle by successfully attaching themselves to a suitable hard substrate, such as coral, or die and be buried in the sand at the bottom of the lagoon. Artificial spat collection takes advantage of this settling stage by providing an artificial substrate (spat collector) on which the spat can easily settle. These spat collectors, which are usually manufactured from rolled-up dark polythene shade screen, are suspended from subsurface long-lines some 3–5 m below the surface of areas of lagoon across which spat are known to drift. Best settling of spat occurs during the change of seasons, especially at the beginning of the southern summer. Settled spats are collected and removed from the spat collectors at the end of each year. They are then sorted before being placed in fine mesh polythene bags, which are attached to longlines at depths of 5–7 m in deeper waters of the lagoon, where the spat are allowed to grow to year-old juveniles. Following a year's growth, the juvenile shell is once more sorted and then placed in net panels or other suitable baskets to undertake a further two year's growth before the pearl 'oysters' are suitable for seeding. While growing from to spat to juvenile to maturity, individual pearl 'oysters' are kept separated from each other and are cleaned on a regular basis, at least every 3–6 months.

Seeding is performed, at onshore facilities located on the side of the lagoon, by a mix of Japanese and indigenous technicians who use conventional Japanese seeding technology. Following seeding and recuperation, the seeded pearl oysters are then relocated to another area of the lagoon to grow-out for a further period of 12–24 months while suspended below the surface of the lagoon. Cultivation times do vary, for some farmers, concentrating on producing small-size pearls of <10 mm diameter, usually limit their cultivation times to 12–14 months. In contrast farmers interested in producing 12–14 mm pearls use cultivation times of 24 months.

At harvest, on average, 40 pearls are obtained for every 100 seeded pearl oysters. Of these 40 pearls: 3 (7.5%) will be unsaleable, 6 (15%) will be round or semi-round, 20 (50%) will be semi-baroque and 11 (27.5%) will be baroque.

Once harvested, the cultured black pearls of French Polynesia, which may range in colour from pale grey to silver and iridescent brownish, purplish and black, are graded by size, shape and quality. Although the sizes of French Polynesian black South Sea pearls may range from 8 to 16 mm, most pearls have diameters of 9–12 mm. The five categories used to describe the shape of the pearls include round, semi-round, semi-baroque, baroque and ringed. The final factor to be considered in grading these pearls is quality, which represents a mix of colour, lustre and surface perfection. Four grades (A, B, C, D) are used to assess quality of French Polynesian black South Sea pearls.

French Polynesian Cultured Half-Pearls

As in Australia, and elsewhere, French Polynesia has a comparatively small, specialized black bead nucleated half-pearl production, which is an economically viable addition to round pearl culturing operations.

Black South Sea Pearls from the Cook Islands

The Cook Islands consist of a group of 15 islands that are spread over a 2 million square kilometre area of the Pacific Ocean between Samoa to the west and French Polynesia to the east. All commercial pearl farming is conducted on Penrhyn and Manihiki atolls in the northern group of islands. Future production is likely from three additional sites that include the atolls of Pukapuka, Rakahanga and Suwarrow.

Pearl cultivation commenced in the Cook Islands in the early 1970s, when an Australian was granted an exclusive license to cultivate pearls in Manahiki lagoon. Due to strong resistance from the Manahiki Island Council the operation closed in 1982. Families of Cook Islanders commenced pearl farming operations shortly thereafter and currently produce ~5% of world production of black pearls.

Pearl cultivation techniques used by Cook Islanders are modified versions of those used in French Polynesia.

Golden South Sea Pearls

Indonesia and the Philippines are the major suppliers of golden-hued bead-nucleated South Sea cultured pearls using *P. maxima*. Individual farmers usually select the golden pearls from the crops of South Sea pearls they cultivate and sell these to specialist pearl merchants. High-quality golden South Sea pearls have diameters from 9 to 14 mm, display gold-coloured nacre with a dark orange to red tint and are highly valued. Other commercial sources of golden South Sea pearls are the island of Anami, between Okinawa and Japan, which is the source of cool water greenish copper-coloured gold pearls, and Australia, which is a source of yellow to light golden pearls. Old sources of golden pearls, such as Myanmar and Malaysia, are no longer of any economic significance.

Bead nucleated South Sea pearls of yellow to golden hue are relatively rare, and quite expensive to purchase today. Over the years, the production of natural coloured golden South Sea pearls has decreased in number, size and saturation of golden hue largely because farms in Indonesia, the Philippines and to a lesser extent Australia are using more laboratory-bred white-lipped spat to increase their production of much more readily saleable white South Sea pearls. The end result of this situation is that this laboratory-bred stock are only capable of producing 10–40% of cream to golden pearls, of which a mere 5–10% are likely to have a good golden hue. This contrasts strongly with the past (particularly in Australia) when 'wild' yellow-lipped stock was routinely implanted to yield pearls. This culture yielded crops made up of 60–90% of cream, yellow and golden pearls.

Colour-Treated Golden Pearls

The lack of saleable golden-hued pearls has led to several treatments being applied to off-coloured South Sea pearls to yield this currently fashionable colour. Colour enhancements include bleaching and dyeing that can be detected by examining the drill hole in the pearl for the presence of yellow to golden dye between the shell bead and surrounding nacre of these treated pearls; and a still secret, possibly heat-based, treatment that modifies undrilled pearls to a golden colour. Elen (2001) recommended the use of three key observations for identifying this colour enhancement. First, natural coloured yellow to golden pearls usually have an even distribution of colour, while treated golden pearls can display small spots of darker colour concentrated within blemishes on the surface of the colour-enhanced pearl. Secondly, while yellow- to golden-coloured nacre tends to fluoresce an even light yellow, light brown or greenish yellow under LWUV, heat-treated yellow to golden pearls of comparatively even golden body colour displayed an uneven LWUV fluorescence in moderate to strong light yellow, very light blue or light brownish orange colours. Thirdly, when the reflectance spectrum of these pearls was examined in the UV region, natural coloured yellow to golden nacre displayed absorption maxima at 350–365 nm and 420–435 nm that increased in intensity with the saturation of golden colour. In contrast, any absorption of treated golden pearls in this region was, at best, very weak and commonly absent.

Other Cultured Half-Pearls

True Mabé Cultured Half-Pearls

To Japanese of the late nineteenth and early twentieth centuries, the black-lipped (penguin or mabé) wing shell *Pteria penguin* (Röding, 1798) was renowned for its rare natural pearls. These pearls usually had baroque shapes, diameters of 2–13 mm and displayed silvery-white and sometimes golden colours. As the shells of this bivalve were lined with high-quality nacre that displayed a distinct rainbow-like iridescence superimposed on its somewhat metallic silver to golden hues, during the first decades of the twentieth century, pioneering Japanese cultivators became interested in applying recently developed Japanese pearl culture technologies to this mollusc.

Records reveal that around 1908 Sokichi Igaya, and others, unsuccessfully attempted to culture both whole and half-pearls in the mabé pearl oyster in the waters off the island Yuikojima to the south of Japan. Over the next 40 years some minor commercial success was achieved and small quantities of cultured mabé half-pearls were exported to Europe. After the Second World War, some limited success in this culture was achieved, but much interest was stimulated in 1954 when research began into the artificial breeding of suitable 'mother' shells. Research and development of suitable spat culture techniques for *P. penguin* was finally successful, and by 1970 the Tasaki Shinju Co. Ltd had established a working spat cultivation facility, known as the Anami Breeding Centre, which was capable of producing up to 10 000 juvenile shell per year. By 1975 the Tasaki Mabé Pearl Company had been established and the company began to market mabé half-pearls. Other companies followed suit, but Tasaki has remained the leader in the field by producing commercial crops of mabé three-quarter pearls from the early 1980s, and then the first cultured whole pearls in 1985.

Japanese production of cultured mabé half-pearls in *P. penguin* takes 5–6 years: with 3–4 years required to produce suitable 'mother' shells, and a further $1\frac{1}{2}$–2 years required for grow-out following implantation of the hemispherical nuclei. The product consists of half-pearls commonly of 14–15 mm diameter, but with diameters that can range upwards from 10 to 22 mm. According to Muramatsu (1985), the colour of mabé pearl nacre can be silvery light-pink, silvery dark rosé, blue or gold. Superimposed on these attractive body colours is the unique iridescence that characterizes mabé nacre.

Today, the status of Japan's mabé pearl cultivation industry is difficult to determine, due to the reluctance of Japanese cultivators and their related grower's associations to supply any relevant information about the industry.

Australian Mabé Cultured Half-Pearls

Interest in the cultivation of mabé half-pearls in Australian species *P. penguin* is in its infancy, with this interest currently centred in Western Australia at Houtman Abrolhos (off Geraldton), Shark Bay (near Denham), and the waters off Port Hedland, and at Orpheus Island, near Cardwell in far North Queensland. The reason for this interest lies in the fact that the aragonitic nacre lining the shells of Australian species of

P. penguin has comparatively narrow arc of attractive brownish to golden-coloured iridescent nacre that is located between the narrow peripheral black calcitic rim of the shell and the dominant silver to platinum coloured iridescent nacre that lines most of the shell. By carefully gluing polymer hemispheres of appropriate size over selected areas of nacre under the mantle of this bivalve, cultured hemispherical blisters of silver to platinum to brown to golden colour can be produced from this mollusc.

Culture is based on harvested spat that are allowed to grown on the farm for 12 months until they attain a dorso-ventral length of 120 mm (Brown, 2001). Implantation occurs in the months of January and February, when water temperatures are at their highest and the molluscs are most robust. Five hollow, hemispherical polymer implants are glued to the valves of each mollusc, three on one valve and two on the other. Care must be taken that the implants are so positioned that the mollusc can close its valves.

After implant, each *P. penguin* is drilled through its ear (laterally to the hinge) to facilitate its suspension from drop lines that are attached to offshore longlines. Usually up to 10 implanted shells are suspended in water of 3–4 m depth for a grow-out period of 18 months. Minimal cleaning and maintenance is required until harvest. In contrast, Western Australia mabé farmers use traditional framed pocket nets to hold their implanted molluscs during grow-out.

Harvest of the seeded *P. penguin* occurs in the winter months of July–August, when water temperatures are lowest and nacre production is slow but of its best quality. At harvest the implanted molluscs are sacrificed, the valves of each mollusc separated, and any cultured blister pearls are either sawn or drilled from the shells. Following the grinding of each blister into an appropriate shape, its hemispherical implant is removed, the interior of the blister is cleaned with hydrochloric acid, and the manufacture of the mabé pearl is completed by filling the hollow blister with a suitable polymer and gluing a base of flat, polished *P. penguin* nacre to the blister. A final trimming and polishing step assures a well-polished, symmetrical product.

These cultured mabé half-pearls can be readily identified by the unique iridescence that is superimposed on the grey to platinum, and brown to golden body colours of these pearls.

Cultured Abalone Half-Pearls

Attempts at culturing pearls in the univalve abalone, to yield a more predictable and therefore more commercial product, have been in progress since Louis Boutan's pioneering experiments in France from the late nineteenth century to the mid-1920s, and La Place Bostwick's American experiments that he described in rather glowing terms in 1936. By the mid-1950s, the Japanese researcher Dr Kan Uno had developed culture technology suitable for growing half-pearls in several species of abalone. To achieve this, he drilled an opening through the abalone shell to allow a nucleus to be positioned under the mantle and cemented into place. A few years later, the Korean father-and-son team of Ki-Sun and Won-Ho Cho were culturing half-pearls and *keshi* pearls in Korean species of abalone. Most recently, after years of research, Canadian Professor Pete Fankboner devised and patented a screw pegging process to generate an internal convexity in the abalone shell which this haemophilic univalve gastropod could cover with nacre. Today, half-pearls are being cultivated commercially by applying this patented process to red abalone *H. rufescens* on several farms in California (USA).

New Zealand is the home of two successful abalone pearl culture operations that utilize the New Zealand paua, *Haliotis iris* (Martyn, 1784). These commercial operations produce half-pearls of up to 20 mm diameter that display a range of green to blue colours that have pinkish to golden overtones.

According to Hutchins (2004), abalone pearl culture in New Zealand is based on a government-controlled annual quota of 'wild' abalone that are harvested by professional scuba divers and then transferred to farms located in protected, pollution free waters at several locations around the south island of New Zealand. The abalone are seeded, with either a single large or several smaller flat-domed polymer implants of various shape, by secret technology that avoids damage to the body and mantle of the abalone, for these univalves are haemophiliacs. Seeding takes throughout the year, with the grow-out period of 18–36 months taking place in the dark (for abalone will only secrete nacre in the dark) either in large black polymer barrels deliberately held at sea or in onshore black lined tanks through which fresh seawater is continuously circulated. Once a week the implanted abalone and containers in which they live are thoroughly cleaned, and the abalone are fed a proprietary mix of fresh seaweed several times a week. Harvest between May and December yields cultured blisters of round, pear and oval shape, up to 20 mm diameter (average 10–14 mm), and with attractive blue, green to gold, red and violet hues in their nacre. These cultured blisters are manufactured into half-pearls by cutting the blisters from the shell, trimming the blister into a marketable shape, cleaning the inside of the blister to remove excess conchin, and filling the blister with a polymer of appropriate colour before a base of polished abalone nacre is cemented to the filled blister.

Mantle Grafted (Non-Nucleated) Freshwater Cultured Pearls

Biwa Pearls from Lake Biwa

Although the Biwa pearl, the first non-nucleated mantle tissue grafted freshwater cultured pearls, entered the marketplace during the mid-1960s, the Chinese were culturing nacre-coated Buddahs in indigenous freshwater mussels from at least the thirteenth century AD onwards.

Japanese research into the culturing of freshwater pearls began in the early 1920s with attempts at culturing pearls in Japan's indigenous mussel, the *Hyriopsis schelegi*. (Martens, 1861) or *ike-cho-gai*. During 1928, Fujita, an early associate and friend of both Nishikawa and Mikimoto, first attempted the culture of whole freshwater pearls in Lake Biwa, which is Japan's largest freshwater lake located about 10 km north of Kyoto. This attempt was unsuccessful. Through the 1930s, Fugita and his co-worker Yoshida continued experiments and achieved limited success in culturing poor-quality bead nucleated pearls in Lake Biwa's largest freshwater mussel, the *H. schelegi*. However, it was their discovery that accidentally implanted mantle tissue could initiate pearl growth in this mussel that led to the commercialization of non-nucleated pearl production. From 1945, Seiishiro Udo, and his technician Keisaburo Sakiyoshi, began to experiment with freshwater pearl culture with this local mussel in a $2\frac{1}{2}$ acre, 1–2 m deep, bamboo-fenced retaining pen in the 25-acre Hirako reservoir – a dammed arm of Lake Biwa that was located near the village of Shina. Their endeavours were ultimately successful and during the early 1960s the Shinko Pearl Company became the first commercial producers of what became to be known as Biwa pearls.

H. schlegeli is a large, slow-growing phytoplankton-feeding mussel that requires 6–7 years to reach maturity and reaches a length of 130 mm (13 cm). Its external surface has a dark greenish brown colour, while the internal surfaces of its valves are coated with high-quality lustrous nacre that may be quite discoloured where it underlies the edge of the mantle. Depending on the season of the year, these mussels could lie either partly or fully embedded in the sand and mud on the floor of the lake.

Other early secondary sources of non-nucleated freshwater cultured pearls included Lake Kasumiga in Ibaraki Prefecture to the east of Tokyo and Gifu near Lake Biwa to the north of Nagoya. By 1980, Japanese production of tissue grafted freshwater pearls reached a peak of 6.3 tons. Soon thereafter, production began to rapidly decline due to catastrophic environmental contamination of Lake Biwa. By 1998, the production of Japanese mantle grafted Biwa freshwater pearls had dropped to 214 kg.

Crowningshield (1962) revealed that Biwa pearls were cultured in the following sequence. From October to April local fishermen gathered molluscs suitable for implant with a bottom trawl fitted with a rake that was dragged across the muddy bottom of Lake Biwa. Mature mussels were sold to pearl farms, where the mussels were held in wire baskets suspended ~1 m below the surface of retaining pens that had been constructed in the reservoir. When ready for tissue grafting, the mussels were removed from their retaining pens and were brought to the laboratory dry. After the mussels had gaped, their valves were gently opened by insertion of forceps and the mussels were either pegged with a wooden wedge or held open with the forceps. A 'sacrifice' mussel that had good-quality nacre was chosen, and avoiding the deeply pigmented edge of the mantle scissors were used to cut a strip of mantle tissue 40 × 5 mm from outer surface of the mantle that was associated with quality nacre. This strip of tissue was then placed on a wooden block for cleaning, trimmed to remove non-nacre-producing epithelial cells of the inner mantle, and then cut into 4 × 4 mm squares. This graft tissue was stored in oxygenated water that contained antibiotics to reduce the possibility of transmitting infection during the grafting procedure. The mussel to be implanted was then positioned on a specially designed wooded operating rack, valves uppermost, and its valves were gently opened with shell forceps. Multiple small incisions were then made into the outer mantle of each valve, and a small piece of 'sacrifice' mantle tissue is placed into each incision. Then multiple incisions were pressed closed, the shell opening forceps removed and the grafted mussel allowed to recuperate in fresh, oxygenated water. Depending on the shape and original location of the implanted graft tissue, multiple small baroque pearls of various colour and shape could be produced after 18–36 months cultivation.

After the grafting operation, the mussels were placed in plastic-coated wire baskets that were suspended from stakes about 1–2 m below the surface of the reservoir. Here they were left to convalesce from the trauma of grafting. After this period, the implanted mussels were inspected for health and then moved to cultivation areas where they were suspended at depths of 2–3 m from suspension wires strung between poles that were sunk vertically into the mud on the bottom of the reservoir. At the end of an optimal period of culture of three years, any pearls that had grown were carefully harvested by incision, followed by their removal with tweezers. About 60% of implanted mussels survived, with ~90% of implanted mussels growing their full quota of 10 or more first crop pearls. The average yield was 40% oval pearls, 30% baroque pearls and <30% rounded pearls. As the life span of the *H. schlegeli* is about 13 years (twice that of *P. imbricata*), by returning the mussels for further periods of cultivation, their already formed pearl sacs could grow second crop pearls (40–50% roundish), and indeed a third crop of freshwater pearls that were generally of poor colour and lustre.

Following harvest, the Biwa pearls were sorted: for shape into centres, flats, grains, crosses, doublets/triplets, irregulars, sticks and dragons; for colour into the white, pink and orange group, and the fancy coloured group of lilac, wine, green and blue-coloured pearls; and for at least three grades of nacre quality before being either drilled and strung for necklaces or sold as harvested. Off-coloured Biwa pearls were usually value-enhanced by routine bleaching and dyeing.

Cultured Biwa pearls did not have a central bead, but typically contained an X-radiolucent cavity that could vary in size from large in baroque first crop pearls to rather slit like in rounder second crop pearls. The original Biwa pearls had lustrous slightly crinkly surfaces and seldom exceeded 7 mm in size. Their colours were natural and when exposed to X-rays they displayed a strong yellow-green fluorescence and phosphorescence due to the manganese content of freshwater. Usually they did not fluoresce when exposed to LWUV as they are inert under LWUV.

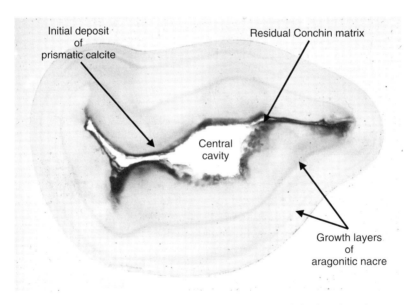

Figure 28.9   Cross-section of a Biwa mantle tissue grafted cultured pearl

It was indeed unfortunate that within a few years pollution virtually wiped out the freshwater pearl culture at Lake Biwa. Subsequent attempts at cultivating spat-bred *H. schlegeli* in artificial freshwater ponds has had some success, but returning implanted mussels to the polluted Japanese freshwater lakes has not been particularly successful. As a consequence China has replaced Japan the world's major producer of freshwater mantle grafted cultured pearls.

Biwa Pearls from Lake Kasumigaura

The culture of mantle grafted freshwater pearls also occurred in Lake Kasumigaura, near Tokyo. As *H. schlegeli* was not native to Lake Kasumigaura, it was first transplanted from Lake Biwa during the 1930s. By the 1950s a significant population had been established, and the commercial cultivation of Biwa pearls commenced in 1963, with an estimated 180 000 mussels being harvested and implanted during that year.

Initially the mussels were grafted with twenty-one pieces of mantle tissue: five in the body and eight on each side of the mantle. Due to the technical skill required to successfully use this technique, it was soon replaced by what is now the standard implantation technique of grafting forty pieces of mantle tissue in two rows of ten on each side of the mantle. The first harvest two year later (1965) yielded two-thirds of an ounce of pearls per mussel.

Within four years, that is 1966, the stock of mussels in Lake Kasumigaura had been depleted, and water quality also started to deteriorate. From 1968 until commercial cultivation virtually ceased in the early 1980s, cultivators bred their own stock mussels and continued to struggle unsuccessfully against increasing pollution of the lake and the stagnation of its water.

Chinese Freshwater Pearls

The Chinese have a long history of successful freshwater pearl culture, for in the thirteenth century the Chinese were culturing nacre-coated Buddhas in river mussels by gluing tin, lead and wax images of Buddha to their shells and then allowing the mussels to coat these with nacre. Kunz and Stevenson (1908) even records that China then had a considerable industry for making 'culture' pearls by implanting mussels indigenous to the rivers of Chenkiang Province with beads of MOP or porcelain. He claimed that after cultivation, and following trimming, near-spherical cultured pearls were produced.

The Chinese began to cultivate freshwater pearls in response to the rapid demise of the Japanese industry. In the mid-1960s, an engineer, Xie Shaoke, discovered how pearls could be produced from the indigenous freshwater mussels of Tai Lake. He experimented into the development of appropriate cultivation technology for freshwater pearls, and his first commercial crop of freshwater pearls, cultivated with Biwa technology, was harvested in China in 1968. Although the scale of production was not very large, its success had a profound influence to the whole industry; for from that time provinces to the south of the Yangtze River became the major areas for cultivation of Chinese freshwater pearls.

This newly developed technology was rapidly distributed by the government to many farmers throughout China. From the end of 1970s, substantial demands for Chinese freshwater cultured pearls in international markets caused the price of these cultured pearls rise sharply. As a consequence, no longer could natural mussels be used to satisfy the increasing demands of pearl cultivators. At that time, a successful technique for breeding pearl mussels was discovered and this provided strong support for the rapidly growing Chinese freshwater pearl cultivation industry. From the 1980s onwards, the output of freshwater cultured pearls, especially those cultivated in *Cristaria plicata*, increased rapidly.

By the early 1980s, the cultivation of freshwater cultured pearls became very popular in China, with the industry becoming well established in quite a few provinces on the southern banks of the Yangtze River. The small cultured pearls were exported almost exclusively to Japan, where they were processed and exported to major markets in Germany, Switzerland, the Middle East and the US. In an attempt at avoiding competition between cultivators, the Chinese government instituted an Export License system in an attempt at controlling ever-expanding production.

However, by the end of 1980s, the prices obtained for freshwater cultured pearls, mostly cultured in *Cristaria plicata*, dropped dramatically. The basic reason for this decline in return was that far too many pearls of poor quality were being produced. A rapid decline (depression) in the mostly family-based Chinese freshwater pearl cultivation industry followed.

During the 1990s, many large commercial pearl culturing enterprises that had strong financial and technical support appeared in the Chinese pearl cultivation industry. These new farms used the slow-growing *Hyriopsis cumingii* as their 'mother' mussels and instigated many innovations in laboratory-based mussel breeding, methods of mantle grafting and techniques of pearl cultivation. As a consequence, these new entrants into the industry overcame previous difficulties and also achieved many technical and financial advances. The scale of Chinese freshwater pearl cultivation rapidly expanded and the output of pearls of high quality increased remarkably. New, large pearl trading centres were established, and thousands of sales people shuttled actively between these new pearl farms and the markets.

By the end of 1992, the government-controlled export licence system for pearls had been abolished and huge amounts of freshwater cultured pearls of much better quality and greater roundness entered international markets. Overproduction once more saturated the market and in a mere two years Chinese freshwater pearl market again fell into crisis. From this time onwards, control of output and overall improvements in quality have became major challenges for the Chinese freshwater cultured pearl industry. For example, as a consequence of the increases in price of pearls that began during the mid-1990s, and lasted until 2001, major freshwater pearl producers in China expanded their production into large-sized farms of up to 200 Ha area in the hope of making greater profits. Unfortunately, the large farm concept failed due to factors such as difficulties associated with maintaining optimal nutrient content in large bodies of water, rapid spread of diseases, increasingly high mortality rates of implanted mussels and poor quality of the harvested pearls. As a consequence, producers now restrict their farms to 3–6 Ha, and balance increased production costs against higher returns for harvested pearls.

Presently, China produces freshwater pearls in a huge range of exotic shapes and colours – mainly along the Chang Jiang (Yangtse River) in the lake and pond areas of Zhejiang and Jiangsu provinces in particular, but also in Anhui, Hubei, Jiangxi and Hunan Provinces to the west, and Fujian Province to the south. However, while Zhejian province remains China's major production base for freshwater cultured pearls, deteriorating conditions have led to the replacement of Anhai and Hubei provinces, which lie on the northern bank of the Yangtse River, by Jiangxi and Hunan provinces, which lie on the oppostite bank of the river.

Two species of freshwater mussels are used in the Chinese freshwater pearl cultivation process: *Cristaria plicata* (Leach, 1815), the Buddha pearl or river mussel (*he bang* in Chinese), which yields yellowish pearls of low nacre quality that have a high growth rate; and *Hyriopsis cumingii* (Lea, 1852), the triangle mussel (*san jiao* or *san jiao bang* in Chinese), which yields good quality pearls with a soft lustre but a slow growth rate.

Chinese mantle tissue grafted freshwater cultured pearls were initially produced by simple Biwa technology to tissue-graft mature Chinese freshwater mussels of the species *C. plicata*. The technique involved mantle grafting 2–3-year-old mussels. Up to 40 pieces of outer mantle epithelium, of 2 × 2 mm size, from a 2–3-year-old 'sacrifice' mussel, were inserted into pockets cut into the inner mantle on each side of the mussel. The tissue grafted mussels were then returned to freshwater ponds that ranged in size from a duck pond to water reservoirs and lakes. After a cultivation period of a further 12–18 months, a harvest of pearls of 4–7 mm size was obtained. In the years of low market prices, from 1996 to 1998, some farmers kept their crops longer in the water, awaiting an improvement in market price. Because of the closeness of the inserted tissues, some pearls grew into each other and merged forming what is commonly considered as 'peanut pearls'.

Following harvest the pearls were graded, and in some cases processed before being sold in China's major wholesale markets for cultured freshwater pearls at Weitang and Zhuji. Weitang, now the minor market of the two, is located to the north-west of Shanghai on the south bank of the Yangtse River, while the newer Zhui market is located to the south-west of Shanghai, and about 18 km from Zhuji City.

Significant numbers of Chinese freshwater pearls need to be processed soon after harvest as a value-degrading whitish film of colloidal hexagonal calcium carbonate mineral, vaterite, soon covers their external surfaces.

Processing of Chinese tissue-grafted freshwater pearls usually involves the sequential steps of:

1. Sorting (grading) for size, shape, colour and perfection of surface.
2. Drilling to remove flaws, allow access for cleaning and bleaching.
3. Bleaching to remove dark spots and lighten colour. Agents used include bleaching agents (e.g. hydrogen peroxide), solvents and surface-active agents.
4. Whitening with fluorescent brightening agents.
5. Colour enhancement by dyeing and/or irradiation.
6. Washing to remove toxic by-products of colour enhancement.
7. Polishing in tumblers filled with wax and bamboo, walnut shells or sheepskin.

During the 1990s, the Chinese freshwater pearl industry entered a new phase. The first method of mantle grafting was used to 1985, when the industry produced small baroque pearls with wrinkled surfaces, known as the 'rice crispies' in the trade, by mantle grafting *C. plicata*. In the second phase, which began in the early 1990s, modified grafting and culturing techniques were used on a different species of mussel, the *Hyriopsis cumingii*, or triangle mussel, when it was mantle grafted to produce larger, rounder pearls that had a smooth coating of nacre. The new techniques of grafting and cultivation originated from farms around Zhuji in Zhejiang Province, a major pearl producing and supplying centre in China. Stocks of *H. cumingii*, essential for this new technique of cultivation, are mostly produced both by spat culture from spat obtained from mussels of 'wild' origin.

Farms for the cultivation of second phase freshwater pearls are located mainly in provinces with large areas of freshwater in downstream provinces on the Yangtze River. Breeding of baby mussels mainly occurs in Zhejiang, Jiangsu, Hubei, Hunan, Anhui and Jiangxi provinces, while the growing of mature mussels is mainly concentrated in Zhuji city of

Zhejiang province and in Suzhou city of Jiangsu province. The processing and marketing of second phase freshwater cultured pearls is concentrated in Shangxiahu town of Zhuji city and Weitang town of Suzhou city.

Culturing techniques used to produce second phase freshwater pearls involve some modifications to the traditional Chinese technique. First, a unique spat culture process, that involves parasitism[5] of a freshwater fish such as the yellowhead or bighead carp, is used to produce immature mussels that are then grown to maturity in nutriment-rich water in bamboo baskets that can hold 50–1000 young mussels and protect them from predators. Second, 1-year-old mussels of 7–10 cm in length are mantle grafted with small (2 × 2 mm) pieces of outer mantle epithelium from areas of the mantle of younger and smaller mussels that display good-quality nacre. Third, the grafting technician picks up a piece of graft mantle with a pointed tool and then makes a small incision from left to right between the inner and outer epithelia of the mantle with a sharp curved probe. The piece of graft tissue is then inserted into the incision and the pointed instrument holding the graft is removed while the incision is closed by pressure. This process is then repeated until up to 40 grafts have been inserted from upper to lower and left to right. The number of pieces of tissue grafted into particular mussels depends on the size and health of a given mussel. Fourth, grafted mussels are allowed to recover in gently flowing, fresh clear water for 2–4 weeks. After a month, the grafted mussels are inspected for health before being placed in individual plastic nets or five or six mussels are placed in a bamboo basket or nylon net that are hung at 0.5–1 m intervals from longlines that are buoyed by plastic bottles. They are then allowed to grow-out for from 1 to 3–5 years in a freshwater environment to which appropriate nutriments are regularly added. Fifth, unlike the harvest of first phase pearls that occurred during the cooler winter months – usually between October and February – second phase pearls are harvested at any time of the year for the triangle mussel produces nacre more slowly, resulting in a smoother, higher quality nacre that displays a good lustre. Indeed, some larger cultivators are harvesting 100–150 000 pearls six times per year.

One interesting facet of modern Chinese pearl cultivation is that smaller-scale farmers often cannot afford to cultivate mantle tissue grafted mussels for up to seven years; the average time needed to obtain an economic yield of large diameter and/or rounded pearls. As the approximate cost of cultivation per mussel are about US$0.25 for one year of cultivation, US$2.50 for a two-year cultivated mussel, and up to US$7.50 for a four-year pearl-bearing mussel, small-scale pearl farmers may rear their mussels for 2–3 years, then sell them to larger-scale farmers for further cultivation.

The yield of Chinese freshwater pearls is governed by several well-known factors:

1. Chinese freshwater pearls require longer cultivation times to yield larger pearls. For example, <3 mm take 1–2 years, 3–5 mm take 2–3 years, 5–7 mm take 3–4 years, 7–9 mm take 5–8 years, and >9 mm take 8–10 years.
2. Round freshwater cultured pearls, even if cultured using modern technology, are very, very rare. To understand the rarity of a perfect, round 9 mm mantle tissue grafted freshwater pearl, one must consider that out of 600 tons of Chinese freshwater cultured pearls production of 8 mm or larger in diameter, only 3% will be considered round. Of these, only 5% will be considered top quality. Overall, these large size, round pearls represent a mere 0.015% of a 600 ton annual production. The yield of 9 mm and 10 mm pearls of perfect colour and shape is even smaller, for these pearls are extremely rare and their cultivation times are extremely long and may involve several discrete periods of cultivation on different farms to maximize their growth rate.
3. Out of the total production of Chinese freshwater cultured pearls, completely round mantle-grafted cultured pearls are found in the following percentages at harvest: 3–4 mm (10.0%), 5–6 mm (5.0%), 7–8 mm (1.0%), >8 mm (0.01%).

The natural colours of these freshwater pearls are orange, lavender, pink and white. Yang *et al.* (2004) suggest that these colours are produced by a combination of metallic trace elements (Zn,Mn,Mg,V) and organic pigments known as carotenoids.

Further clues to the culturing techniques used to produce round non-nucleated freshwater pearls are obtained when cross-sections of these pearls are examined under magnification.

It is very difficult to detect any residual soft tissue in the centre of the pearls, but its presence has been detected by Akamatsu *et al.* (2003) using direct X-radiography. However, what can be observed are 2–3 abrupt changes in growth (and often colour). These indicate that changed growth conditions would result due to resale of the grafted mussels to other cultivators to allow further years of cultivation.

The appearance in some pearls of distinct nacreous cores also suggests that previously harvested pearls may have been used as a nucleus to obtain preferred larger, rounder pearls.

If the structure of a tissue grafted (non-nucleated) cultured freshwater pearl (FWP) is compared with that of a bead-nucleated cultured saltwater pearl, the seawater pearl will have only 10% of its cross-sectional area covered with nacre while content of nacre of the Chinese non-nucleated cultured pearl averages over 99.9%.

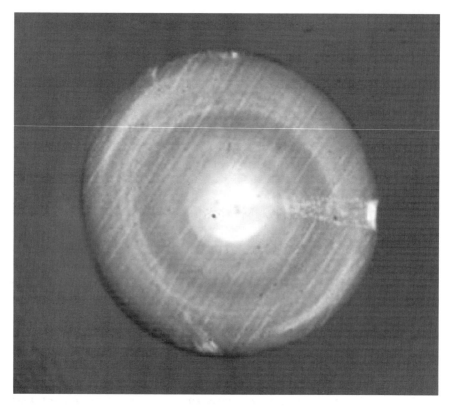

Figure 28.10   Multiple growth layers present on the cross-section of a Chinese round freshwater cultured pearl

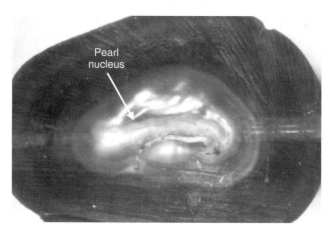

Figure 28.11   Cross-section of a drilled, black dyed nucleated Chinese freshwater cultured pearl

Round Nucleated Freshwater Pearls

Japanese Kasumiga(ura) Pearls

These relatively new bead nucleated freshwater pearls are being cultivated in artificial ponds and in Lake Kasumiga to the north-east of Tokyo in a hybrid mussel produced by crossing the Lake Biwa mussel (H. *schlegeli*) and the Chinese triangle mussel (*Hyriopsis cumingii*). This hydrid mussel is claimed to be pollution resistant and able to tolerate the surgical insertion of a round bead nucleus.

Precise details of how these cultured pearls are produced remains a commercial secret. Culture of these pearls occurs primarily in artificial freshwater ponds on the banks of Lake Kasumiga and in some rare unpolluted areas of the lake. From May to July, 1-year-old 'mother' hybrid mussels are implanted with a MOP bead and a graft of carefully chosen outer mantle epithelium. Due to the multihued colours displayed by this hybrid mussel's nacre, particular care is taken that the mantle tissue chosen for grafting purposes overlies areas of nacre in the 'sacrifice' mussel that have the most desirable natural colour(s) and best-quality nacre. Dilenburger (2004) has suggested that the seeding operation requires a high level of skill, as the implant must be placed in the direct proximity of important organs such as the liver. It has been claimed that pearls are harvested, between December and January, after a growth period of four years, and when the shells have attained a width of ~25 cm in width. Harvest time is around December and January of each year.

The first small crop of Kasumiga pearls was harvested in 1993, with the first pearls being released onto world markets in 1997.

Kasumiga(ura) pearls are bead nucleated, of large size (>8 mm but possibly reaching 13–14 mm in diameter), of round/rounded/baroque shape, and are produced in a range of attractive natural colours that include off-white, peach to orange and lavender to violet hues. More unusual natural colours, such as the valuable green-gold colour, that occur as a result of 'sub-ideal placement' are found predominantly on baroque pearls. The long-term stability of these natural colours has been questioned.

An examination of a strand of 9–13 mm diameter pearls by the SSEF Lab in 2000 revealed that the pearls had been nucleated with previously drilled beads, that is, the beads had two drill holes per pearl; implanted beads had diameters of 7–8 mm; and beads were surrounded by an average of 0.2 mm nacre.

Little else is known about these newest of cultured pearls, other than they are bead-nucleated, production is very limited, they come in a range of attractive colours and are very expensive to purchase.

Figure 28.12 A tomogram of an undrilled Japanese Kasumiga(ura) bead nucleated cultured pearl, showing that it was nucleated with a drilled MOP bead

Chinese Round Bead Nucleated Freshwater Pearls

Since 1991, Chinese cultivators also have succeeded in producing 6–12 mm round bead nucleated cultured pearls from freshwater mussels. The cultivation of Chinese freshwater bead nucleated pearls is centred around Chenghai city in Guangdong province and in Jiangxi province. The diameters of some freshwater bead nucleated cultured pearls may even attain 17 mm.

The round pearls are being produced by a variety of mostly secret techniques that include bead nucleation and a graft of outer mantle epithelium into the body of the mussel, nucleation with a bead of paraffin wax and a graft of outer mantle tissue, reimplantation of a larger bead into the pearl sac from which a pearl has been surgically removed, and pearl implantation in which a previously cultured freshwater pearl is possible rounded before being implanted into a 'mother' mussel together with a graft of outer mantle epithelium.

Other Chinese Cultured Freshwater Pearls

Other varieties of freshwater pearls presently being cultured in China include nucleated pearls that display coin, tablet or flat triangular shapes due to the specially shaped wafers of MOP with which the pearls have been nucleated; tissue grafted pearls that are available in a wide variety of designer-friendly shapes; the so-called 'regenerated' pearls that are small second crop *keshi* pearls that grow in pearl sacs from which previously cultivated pearls have been surgically removed; dumbbell-shaped freshwater pearls consist of rounded bead and tissue nucleated pearls that are fused (grown) together and often of differing colour; and most recently cultured half-pearls.

Freshwater Cultured Pearls from the USA

The family company of America's pioneer freshwater pearl cultivator, the late John Latendresse, commercially produces three types of cultured freshwater pearls. These include hemispherical blister pearls, bead nucleated whole pearls of various unique shape and *keshis*.

To produce these cultured pearls, hookah-equipped divers collect mature 'mother' mussels of prespecified size from the Mississippi River and its tributaries. After collection, the mussels are cleaned, placed into the pockets of 'kangaroo' nets that hold up to 18 shells. Mussels suitable for seeding are then transported to the pearl farm where, following inspection and sorting, the mussels are suspended from rafts made from sealed polythene pipes that are so arranged as to leave sufficient space between both mussels and nets to allow the bivalves to recuperate. At the farm, 'mother shells' are kept separated from 'sacrifice' mussels. Nucleation occurs at the farm using American-trained technicians and American implant technology based on traditional Japanese methods devised by Latendresse and his Japanese-born wife. Implant technicians seed at least two species of *Unio* mussels with MOP nuclei of specific shape, and graft tissue, to obtain cultured pearls of drop, coin, bar, oval and navette shape. Following recuperation under the controlled conditions of laboratory-based ponds, the implanted mussels are returned to the pearl farm for periods ranging from $1\frac{1}{2}$ (for blister pearls) to 3–5 years (for whole pearls) – depending on the nature of the implant. Fancy shape cyst pearls, cultivated in the body of the mussel, are recovered by killing the mussel at harvest. After several decades of experiment, the mortality rates for American freshwater cultured pearls have been reduced to less that 5%.

American freshwater cultured pearls are predominantly silver-white in colour, and are not colour-enhanced by dyeing. The nacre of these pearls is seldom perfectly smooth, and the shapes of the pearls are quite unique.

Imitation Pearls

Imitations of pearl are many and varied. The most common imitations are manufactured from glass, plastic or shell spheres that have been coated either internally or externally with an imitation nacre known as *essence-d'orient*. Other effective imitations of pearls consist of worked pieces of thick MOP of various colour, or composite imitations such as the *coque de perle* and *Osmeña* pearls, which are the most effective imitations of blister pearls.

Essence-Coated Imitations

Manufacture of imitation of pearls possibly followed the Spanish plunder of pearling beds off Venezuela during the late fifteenth and early sixteenth centuries, when Spain's rapidly decreasing supplies of natural pearls were supplemented by Venetian-manufactured hollow beads of pearlescent glass that were filled with wax to give the imitations some heft. Next pearl imitations to arrive on the market were known as 'Roman pearls'. These imitations were manufactured from hollow beads of colourless glass that were first painted inside with a pearlescent varnish, and then filled with wax.

According to Robert Tardieu (1994), author of *Langeac: Au temps ou l'on enfilait des perles*, the technique for manufacturing artificial pearls, by coating hollow glass spheres internally with *l'essence de orient*, was discovered at the end of the seventeenth century, during the reign of French King Henry V, by a Parisian rosary maker known as M. Jacquin. This artisan discovered *essence-d'orient*, accidentally, when he noticed that water containing the scales of the bleak – a small fish from the Seine – had reflections that resembled those from the surface of a pearl. His discovery that the fine film (of guanine crystals) that coated the scales of this fish caused this iridescence led him to create an imitation nacre varnish, or *essence-d'orient*, that was based on these recovered crystals. As a consequence, Paris became the centre for the manufacture of glass imitation pearls. Interestingly, while the hollow glass balls were coated with *essence* in Paris, the glass balls (*bulles*) were hand-blown at Langeac – a small traditional lace-making town some 35 km to the east of Le Puy, in the Haute Loire of central France. These quite effective imitations, known as Roman pearls, were often used to decorate the *dentelle*, the hand-made lace for which Langeac is equally famous.

Chikayama (1997) revealed that Japanese began copying the manufacture of the 'Roman pearl' imitation in 1913 at the glass-making area to the south of the Osaka Prefecture. A few years later the Japanese were manufacturing other imitations of pearl. These included essence-coated shell beads from Chinese river mussels (1917), which proved to be too expensive to market, and solid alabaster glass beads that were coated externally with *essence* (1919). Techniques for 'pin dopping' and 'string dipping' were invented in the early 1920s to increase the production of imitation pearls. The next Japanese invention was the production of *essence*-coated plastic beads during the 1960s. With time, essence-coated plastic was used to imitate both round saltwater and baroque freshwater pearls. These imitations were cheap and their lightness made them easy to identify. As fish scale essence became more difficult to obtain, this was replaced by a potentially toxic modern basic lead carbonate varnish, and in recent times a titanium dioxide based varnish. Over the last couple of decades, these have been the *essences* used to coat shell beads, which today have returned once more to become the premier pearl imitation – the shell-based pearl imitation.

For over 40 years, some of the better quality (solid) glass bead imitations of pearls have been marketed as Majorica pearls™. Since 1965 these have been manufactured on the Spanish island of Majorca, and marketed from Barcelona. Pough (1965) revealed that on this island glass beads were traditionally hand made on clay-slurry-coated wires from long rods of white glass of secret composition. Following their removal, these hand-made beads were sieved for size and checked for quality. Depending on quality the beads were then either strung on wires or else mounted individually on wires set vertically into a wooden block. The beads were then dipped into pans of *essence*, and the coating allowed to dry. This step is repeated at least twice. More expensive imitations are coated with natural essence and are polished with a chamois cloth between dippings. Once completed, the pearls are coated with a colourless protective layer before being removed from their pins or wire, graded and strung into necklaces. Although now automated, this process is still used to produce Majorica pearls™ today.

Polished Shell Imitations

Other surprisingly effective imitations of pearl consist of shell beads cut and polished from either thick nacreous or non-nacreous shell of various colour. Such imitations are sold to tourists as valuable black pearls from the Philippines, and in Indonesia are marketed as 'natural trochus pearls' (they are indeed cut from the columella of trochus shells) or 'coconut pearls'. In the USA, 'French River pearls' are manufactured from the baroque hinges of thick freshwater mussels such as the thick mucket.

Fortunately, all of these fakes are readily identified by longitudinal oriented striations, readily visible to transillumination, which characterize the shell origin of this imitation.

Composite Imitations

For some years oval greyish segments of highly iridescent nacre from the nautilus shell have been used in the manufacture of the quite readily identifiable greyish to bluish pearl imitation termed the *coque de perle*. These highly lustrous imitation pearls consist of cement-filled bluish grey dyed oval sections of nautilus shell, together with supporting inner septa, which have been glued to a flat base of polished MOP. Similar quite lustrous imitations, sometimes called *Osmeña* pearls, are manufactured from sections of the nacreous shells of the trochus and other unknown species. Overall, *Osmeña* pearls have similar structural features to the *coque de perle*, but do not display the characteristic dyed colour and identifying septa of the nautilus, when examined by direct X-ray.

The Identification of Pearls

Identification of a pearl, or pearls, seldom requires use of conventional gemmological testing techniques. Instead, the gemmologist must use a combination of observed visual features, and some specialized techniques of identification such as direct X-radiography, and if available X-ray diffraction, to identify the type of pearl.

Recognition of value enhancement in pearls is of increasing significance to gemmologists. While some treatments, such as: pearl 'skinning', routine colour enhancement of drilled *akoya* cultured pearls by bleaching and dyeing, bleaching of pearls to produce uniformly white pearls, and surface waxing to enhance the lustre of pearls, are considered to be acceptable trade practices, other value enhancements of pearls, such as dyeing and/or irradiation of pearls to enhance their colour, now need to be recognized and disclosed.

The 'Tooth Test'

In spite of what some non-gemmologist jewellers believe, an 'educated eyeball' and the gritty feel of nacre on teeth (due to overlapping platelets of nacre covering the external surface of a pearl) will not discriminate natural from cultured pearls. However, to the pearl expert who does not wear dentures, the absence of 'grittiness' could suggest one's teeth are examining an imitation pearl.

Recommended Identification Protocol

To identify an individual pearl or a strand of pearls, the following steps should be followed:

Step 1. A general examination of the pearl(s)
Step 2. A detailed examination of the pearl(s) external surface(s), drilled hole(s), and transillumination of the pearl(s)
Step 3. Direct X-radiography
Step 4. Other laboratory-based tests that could include X-ray luminescence, cathodoluminescence, specialized spectroscopy such as UV-visible, X-ray fluorescence, Fourier IR and FTIR, Raman spectroscopies, X-ray diffraction and perhaps examination with an endoscope if a working instrument is still in existance
Step 5. Detection of value enhancement.

Step 1: General Examination

Strands of pearls should be examined initially by first holding the strand taunt between both hands, and then examining the necklace as a whole against a neutral coloured background of light grey or matt white colour using fluorescent white overhead illumination from an articulated double bar desk lamp. Similar conditions of lighting should be used when examining a single pearl. Use of low magnification, such as a ×4 head loupe, may assist this examination.

The following observations can be used to assist the discrimination between natural and cultured pearls.

- Due to the rarity of natural pearls of uniform size and shape, strands of natural pearls are almost always graduated. So, if a strand consists of spherical pearls of very uniform size and shape, it likely contains bead nucleated cultured pearls (Plate 32).
- For the same reason, uniformity of colour matching usually will be much better in a cultured pearl or imitation pearl necklace than in a natural pearl necklace.

  This observation can often be confirmed when strands of pearls are examined under LWUV. While strands of bleached akoyas, white South Sea pearls and white freshwater pearls usually display a uniformly strong milky bluish white fluorescence, strands of natural pearls usually contain pearls that fluoresce with variable intensities, often in yellowish to greenish to tan colours. Strands of natural coloured black pearls commonly will display a brownish to reddish LWUV fluorescence of variable intensity, while strands of dyed and/or irradiated cultured pearls commonly are inert to LWUV, or may display fluorescence to LWUV of different colour but will be of relatively uniform intensity.

- 'Circles' and 'fish tails' are characteristic shapes of bead nucleated cultured pearls.
- If a strand of pearls or a single pearl 'blinks', when rotated and transilluminated at arm's length, the presence of a bead within the pearl should be suspected.
- Usually, cultured bead nucleated South Sea pearls, recent Chinese round freshwater pearls, bead nucleated cultured freshwater pearls from lake Kasumiga in Japan and rare natural pearls are the only round pearls likely to have diameters in excess of 8 mm.

Step 2: Detailed Visual Examination

First, the external surface of pearl(s), and any drilled channels in those pearl(s), must be thoroughly cleaned of adherent debris and grease by washing the pearls in luke-warm soapy water with a soft cloth and soft bristle toothbrush. Once clean and dry, the identification of most pearls can be simply accomplished with the 10× hand lens and/or low power binocular microscope and good fibre-optic illumination, by examining:

The Nacre of the Pearl, noting the presence of:

- A thumb print-like pattern due of serrations caused by the edges of overlapping lamellae of platelets of aragonitic nacre intersecting the external surface of both natural and cultured nacreous pearls. This hand lens appearance contrasts strongly with the smooth 'blotting paper' texture of the *essence de orient* coating on imitation pearls, the frequent loss of part of this coating from underlying glass, plastic or shell beads that form the base of these imitations, or the smooth glassy surface of a hollow glass sphere coated internally with *essence d'orient*.

    Benerjee and Habermann (2000) recommends that examination of the surface of nacre at higher magnifications can prove useful for discriminating between the nacre of Chinese tissue grafted pearls and natural pearls, for the freshwater cultured pearls display finely granular nacre that is covered with fine parallel ridges that run straight across the surface of the nacre.
- The smooth, often scratched worked surface of a polished shell bead imitation pearl. Transillumination usually readily reveals the parallel growth banding of the shell from which the bead imitation was manufactured, irrespective of the colour of that shell.

- Learned differences in smoothness, lustre and orient of the nacre of akoya, South Sea, and freshwater cultured pearls.
- Accumulations of dye on the surface, or under the surface of dyed mantle grafted freshwater pearls and *akoya* bead nucleated saltwater cultured pearls.

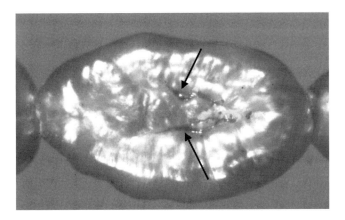

Figure 28.13   Dye infilling surface defects in a dyed Biwa mantle tissue grafted freshwater cultured pearl

- The porcellanous 'flame'-patterned surface of a non-nacreous pearl, as exemplified by the subtle 'flames' that decorate the external surface of white, pink, orange and brownish conch pearls.
- Distinctive planes of junction on natural blister pearls, cultured half- or three quarter pearls, mabé pearls and the composite imitations known as *coque-de-perles* and Osmeña pearls.

Secondly, the drill hole should be examined with either: transmitted illumination of variable intensity and magnification from a 10× hand lens or a low power binocular microscope; incident illumination and magnification from a 10× hand lens or a low power binocular microscope; or transilluminated miniature fibre-optic illumination of variable intensity and magnification from a 10× hand lens or a low power binocular microscope, noting:

- The concentric, thin 'onion'-like alternating layers of aragonite and dark-coloured conchin that form a natural pearl. Note, however, that the colour of the walls of the drill hole in a natural pearl commonly grades downwards from yellowish (peripherally) to a darker brown towards the centre of the pearl.

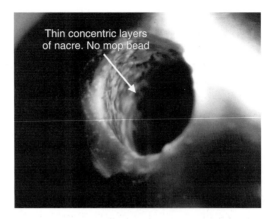

Figure 28.14 Identifying appearance of the drilled drill hole in a natural black saltwater pearl

- The presence of a greyish or white central shell bead, and thin layer of brownish or other coloured organic conchin between this bead and its surrounding circumferential layer of nacre, in a bead nucleated cultured pearl.

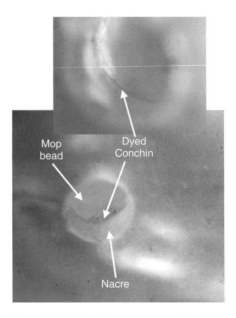

Figure 28.15 Identifying appearance of the drilled hole in a bead nucleated saltwater cultured pearl

With respect to this observation, it should be remembered that as bead nucleated cultured *akoya* pearls are routinely bleached and dyed, the thin layer of conchin between the bead and nacre of this pearl is usually dyed the same colour as that displayed by the body colour of the pearl. Coloured South Sea pearls also should be checked for this feature, for although these cultured pearls used to be dyed (only rarely), the frequency of this treatment is increasing.

The observed thickness of nacre also can be used to provide some indication of the type of pearl the gemmologist could be examining, as the thickness of nacre surrounding the bead of a South Sea pearl is ~1 mm, while <0.5 mm of nacre usually surrounds the beads of most akoya pearls of modern production.

The stringing channels of black pearls require careful examination, for the pearl being examined could be either a natural coloured black pearls, a natural coloured black *keshi* pearl, a bead nucleated black South Sea cultured pearl, a black dyed bead nucleated cultured pearl, a black and/or irradiated dyed mantle grafted freshwater cultured pearls or an irradiated bead nucleated or non-nucleated cultured pearls. These black pearls can be distriminated with relative ease, by careful examination of the pearl's drilled stringing channel, and noting that:

– Natural black pearls will have an 'onion-like' structure of alternating thin darkly coloured lamellae of aragonite and conchin.
– Natural black keshis will have a similar structure to natural black pearls, but an irregular cavity may be detected in the centre of the pearl.
– Bead nucleated cultured pearls have a white to greyish MOP bead, usually a thin layer of black conchin, and a circumferential black nacre of up to 1 mm thickness.
– Black dyed bead nucleated cultured pearls will have a dyed black bead and black nacre. Black dye may also be found adhering to the surface of stringing channels in the pearls. Sometimes the black dye also stains the thread on which the black dyed pearls have been strung.
– Black-dyed mantle-grafted freshwater cultured pearls may have black dye adhering to surface defects, and in the drill hole and staining the thread.
– Irradiated black bead nucleated pearls are characterized by having a dark-coloured bead and lighter-coloured white to greyish nacre surrounding that bead.
– Very iridescent irradiated black freshwater pearls will display an external surface on which the thumb-print pattern of nacre loses its definition.

• The partial absence of the usually thin layer of artificial *essence d'orient* that surrounds the stringing channel of the glass (solid or hollow), plastic or shell bead of an *essence d'orient* coated imitation

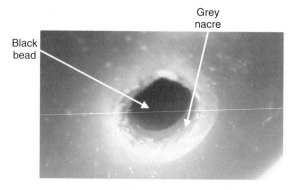

Figure 28.16 Identifying appearance of the drilled stringing channel in an irradiated bead nucleated saltwater cultured pearl

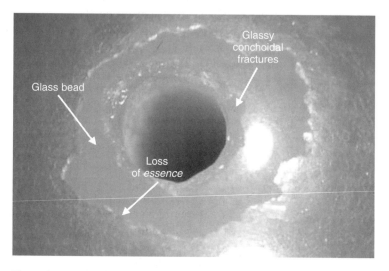

Figure 28.17 Identifying appearance of the stringing channel in essence coated solid glass bead imitation pearl

pearl. In addition, glassy conchoidal fractures may be observed surrounding the stringing channel of glass imitation pearls, while the periphery of the stringing channel of plastic imitation pearls will often display 'mould marks' and will peel when judiciously pared with a small, very sharp blade.

- The presence of a detectible central cavity in the wrinkled nacre surrounding the stringing channel of mantle grafted cultured freshwater pearl that are often known as a Biwa or Chinese 'rice crispie' pearls.
- The presence of two or three alternating growth layers in a round >8 mm Chinese mantle grafted cultured freshwater pearl.

Thirdly, *Transilluminate the pearl*: transillumination of a pearl, centrally located over an adjustable diaphragm, can usually reveal the striped (layered) structure of the shell bead that is centrally located in a bead nucleated saltwater cultured pearl, or the vague shadow cast of a large irregular central cavity in a non-nucleated (freshwater) cultured pearl. Remember, however, due to the absorption characteristics of black nacre and dyed black shell beads, transillumination will prove to be of little use for examining bead nucleated black pearls.

Step 3: Direct X-radiography

Few working gemmologists will have access to facilities for X-radiographing pearls to determine their identity; for by law X-radiography is a technique of identification only available in well-equipped large research or gem testing laboratories that have suitably qualified staff. Therefore most working gemmologists will have to rely on a detailed visual examination, as specified in steps 1, 2 and perhaps refer the pearl(s) to a laboratory to provide answers to questions such as:

Is the pearl natural, cultured or imitation?
If the pearl is cultured, what type is it?

Those gemmologists and/or scientists who are legally qualified and licensed to use X-rays for diagnostic or research purposes must have access to at least a 10 mA, 70–90 kV X-ray unit that produces a collimated beam of X-rays; a dark room equipped with controlled development and fixing facilities for X-rays; and a supply of fine grain X-ray film that will record a pearl's internal structure with maximum definition. Lead foil masks, or immersion in a non-staining contrast medium should be used to minimize fogging of the developed film by scatter of X-rays by the pearl. Pearl(s) should be examined with a bracket of exposures ranging from under-exposure to over-exposure to ensure that both the outer and inner structures of the pearl are recorded in diagnostic detail on the film. Each pearl should be radiographed in at least two predetermined orientations at right angles to each other. Exposed X-rays must be developed and fixed to the manufacturer's precise specifications, and importantly all processed radiographs should be examined dry, with 5–10× magnification, with X-ray viewers equipped with transmitted white light of variable intensity.

According to Kennedy (1998), direct X-radiography has two uses in pearl identification. First, direct X-radiography can be used to provide a general survey (overall impression) of the composition of a pearl necklace – particularly if the pearls have been strung in such a manner that the stringing holes are not accessible to visual examination. Secondly, to ensure an accurate identification of each pearl in a necklace the pearls must be unstrung, and individual X-radiographs used to determine the identifying radiographic structures of each pearl. Also, individual X-radiographs should be used to identify any pearl that is undrilled, pegged or set into jewellery.

When pearls are examined by direct X-radiographs, the resulting negative images on the films (white for calcified tissues and black for soft tissues) do provide a permanent record of the identifying structural features of the majority of pearls. For example:

- Natural pearls display thin arcs and rings of radiolucent black conchin within the white X-radiopaque image that represents a pearl's nacre. Sometimes natural pearls display a small central X-radiopaque cavity.

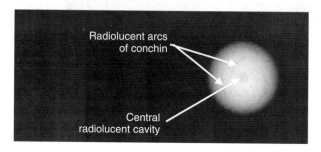

Figure 28.18   X-radiographic features of a natural pearl

- *Keshi* pearls have the same structure as natural pearls, but they may display a central radiolucent cavity of irregular outline and of variable size.

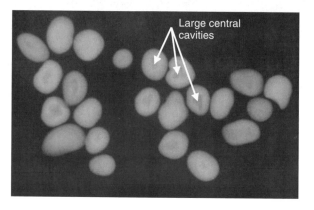

Figure 28.19   X-radiographic features of natural keshi pearls

- Bead nucleated cultured pearls display, from the inside out, a dense white structureless central X-radiopaque bead, a thin to thick black circumferential layer of radiolucent conchin, and comparatively thin (for the *akoya* pearl) to thick (for the South Sea pearl) slightly less X-radiopaque external layer of nacre.

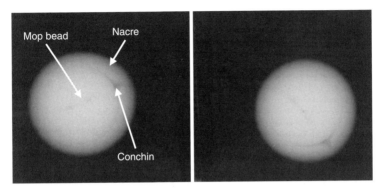

Figure 28.20   X-radiographic features of a white South Sea pearl that has a thick layer of nacre

- Non-round mantle grafted Biwa-type cultured pearls usually display a relatively large central radiolucent cavity of quite variable size and dimensions.

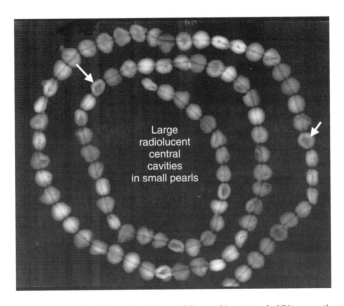

Figure 28.21   X-radiographic features of the pearl in a strand of Biwa mantle grafted freshwater cultured pearls

- Rounded Chinese freshwater cultured pearls may contain a rather flattened central X-radiolucent cavity of variable size and dimensions if they have been mantle grafted. In contrast, if the pearls have been bead nucleated they will display the same X-radiographic features as those of the *akoya* and South Sea pearls. Occasionally the X-radiograph of rounded Chinese freshwater pearls will indicate that they have been formed from a radiopaque central off-round possibly pearl nucleus, a circumferential layer of X-radiolucent conchin and an external layer of X-radiopaque nacre.
- Cultured half-pearls are shown to consist of a thin X-radiopaque hemisphere of nacre, a flat X-radiopaque base of polished shell, and cavity filling of adhesive polymer that is usually X-radiolucent.
- Imitation pearls display a variety of appearances on a X-radiograph. Polymer bead imitations give no image as the material is X-radiolucent. In contrast, solid glass beads and shell beads provide a dense white structureless image. Hollow glass beads have a characteristic image that consists of a thin X-radiopaque rim surrounding a large X-radiolucent central cavity.

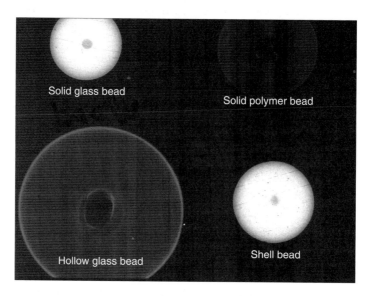

Figure 28.22   X-radiographic features of various imitation pearls

In summary of this step in the identification of pearls it must be stressed that direct X-radiography will not necessarily reveal the identity of some pearls, such as South Sea pearls with thick layers of nacre, or natural pearls that display no X-radiolucent arcs or central concentrations of radiolucent conchin. Also, the interpretation of identifying features recorded on direct X-radiographs of pearls is very much a learned and practised skill.

Step 4: Other Tests

Those pearls that can not be identified either by observation of identifying features or by direct radiography should be submitted to a recognized pearl identification laboratory for more sophisticated testing by, for example, X-ray fluorescence analysis, cathodoluminescence, X-ray diffraction, microchemical analysis by electron microprobe, Raman spectroscopy or, perhaps, endoscopy.

X-ray Luminescence of Pearls

In the laboratory, whether a pearl has a saltwater or freshwater origin can be confirmed or denied by testing the pearl's luminescence to high-voltage X-radiation.

When the shell bead of a cultured pearl is irradiated with X-rays it will fluoresce, and then phosphoresce a distinctive yellow-green colour. This X-ray luminescence response is typical of the freshwater mussel shell from which the beads of cultured pearls have been fashioned. This characteristic luminescence is due to a high concentration of manganese (Mn^{2+}) in freshwater shell.

For the same reason, natural freshwater pearls, and undyed and unirradiated mantle grafted freshwater cultured pearls also will fluoresce and phosphoresce a similar bright yellow-green colour under X-ray excitation. In contrast, as natural saltwater pearls, and the nacreous 'skin' of pearls cultured in a saltwater marine environment have a relatively low manganese content, this renders these pearls inert or only very weakly fluorescent to X-radiation.

Unfortunately, the precise identity of a pearl, which fluoresces and phosphoresces a garish bright yellow-green colour under X-ray, must remain in doubt since the pearl could be either a natural freshwater pearl, a tissue grafted freshwater cultured pearl or a bead nucleated cultured pearl with relatively thin nacre. Fortunately, on a bead nucleated cultured pearl, in which the X-ray induced luminescence has to shine through the layer of non-fluorescing nacre that surrounds the shell bead of the pearl, this subdues the intensity of colour observed by an amount proportional to the thickness of its nacre. Therefore, if a pearl in question fluoresces a yellow-green colour of medium intensity, it could be either a natural pearl or bead nucleated cultured pearl. However, according to Kennedy (1998), if a yellow-green phosphorescence follows the fluorescence, this indicates that the pearl is more likely to be a bead nucleated cultured pearl. Confirmation that the luminescence is emanating from the shell nucleus can be obtained by angling the drill channel to face the window of the X-ray machine. Then, the yellow-green colour will be observed to be more intense within the drill channel. However, if the 'skin' of nacre of a nucleated cultured pearl is thick enough, it is possible that no fluorescence let alone phosphorescence will be observed.

Cathodoluminescence

Cathodoluminescence (CL), luminescence emitted in response to bombardment with high-speed electrons, can be used to discriminate saltwater from freshwater pearls, and also to discriminate natural freshwater pearls from Chinese tissue grafted cultured pearls. This discrimination is based on the presence or absence of Mn^{2+} in the nacre of the pearls, and differences in luminescent intensity of natural freshwater and Chinese tissue grafted cultured pearls.

Banerjee and Habermann (2000) have revealed that Mn^{2+} activated CL spectra of freshwater pearls and shell is characterized by a strong peak at 566 nm (in the green) and a weaker peak at 420 nm (in the blue) related to biological aragonite and a peak at 640 nm (in the orange) attributed to biological calcite. Non destructively irradiation of a pearl's surface in a cold cathode CL-microscope will reveal that the intensity of 566 nm CL is much higher in freshwater pearls than tissue grafted freshwater cultured pearls due to the basic fact that the surface of natural freshwater pearls contain more Mn^{2+} than the surface of freshwater cultured pearls. Saltwater pearls do not contain Mn^{2+} and do not luminesce when exposed to CL.

UV-VIS Spectroscopy

Research by Elen (2001, 2002) has revealed that the natural colour of pearls can be confirmed by the combined use of UV-VIS (ultraviolet-visible) reflectance spectroscopy and LWUV fluorescence.

With respect to natural coloured pearls derived from the nacre of the black-lipped pearl 'oyster' *P. margaritifera*: its white nacre displayed a strong light yellowish LWUV fluorescence, and no identifying absorption features; its black to grey nacre displayed a most identifying absorption at 700 nm, other absorptions at 495 and 405 nm, and a LWUV fluorescence that ranged from reddish to reddish brown to brownish of weak to moderate intensity; its yellow nacre could display the 700 nm absorption, often accompanied by a 495 nm absorption that identifies

P. margaritifera nacre. Elen further suggested that if visible absorptions are not present, the observed presence of an absorption feature between 330 and 385 nm in UV, accompanied by a light yellow/greenish yellow/greenish brown/light brown LWUV fluorescence will identify *P. margaritifera* as the source of the nacre.

With respect to treated black nacre it is important to remember that this usually does not fluoresce when exposed to LWUV.

In contrast, with respect to pearls from the nacre of the silver-lipped 'pearl oyster' *P. maxima*: its white nacre from *P. maxima* generally displayed a moderate to strong light blue to light yellow LWUV fluorescence and no identifying absorption features. However, its yellow to golden nacre displayed a characteristic increase in absorption between 330 and 385 nm as the tone of saturation of the yellow to golden colour increased. This was associated with an even distribution of body colour and a LWUV fluorescence of moderate intensity that changed from light blue/light yellow to light brown, greenish yellow, greenish brown or brown as the colour of the nacre darkened.

In contrast, treated (possibly heated) yellow to golden nacre did not display the 330–385 nm absorption of yellow *P. maxima* nacre, and its LWUV fluorescence tended to be be patchy, in spite of the fact that the distribution of treated colour could be either even or display small spots of concentrated colour.

X-ray Diffraction of Pearls

If grave doubts exist about whether a bead is present in a pearl, then the X-ray diffraction technique (or more specifically the lauegram method) can be used to identify whether the pearl is natural or a bead nucleated cultured pearl. This identification is based on differentiating the X-ray diffracting properties between the layered structure of the shell bead (theoretically thick aragonitic nacre from the Mississippi River mussel) within a bead nucleated cultured pearl, and the concentric symmetrical aragonitic nacreous structure of a natural pearl.

In practice, all X-ray diffraction patterns obtained on the negative X-ray film have a large central circle 'blackened' by the X-rays passing centrally through the pearl. This black spot, in theory, should be surrounded by recognizable, distinctive patterns of spots created by the X-rays that are diffracted further outwards by the crystal structure of the pearl.

If a very fine collimated beam of X-rays is passed through the precise centre of a bead nucleated cultured pearl, perpendicular (at right angles) to the layers of its shell bead, the X-rays will encounter crystals with the same symmetry as displayed by a natural pearl. The pattern obtained on the film which will vary from a hexagonal outline to a hexagonal arrangement of spots is caused by the hexagonal symmetry down the *c*-axis of the aragonite platelets that form nacre. The major spots are situated at apices of the pattern, giving rise to the description of the pattern as a '6-spot pattern'. If a second lauegram is taken at right angles to the first direction, the X-rays will now be travelling parallel to the layers of the shell bead. A different crystal symmetry, perpendicular to the principal axis of the aragonite platelets, will be encountered. This results in a squarer or more rectangular pattern with four internal spots within the outline that reflect the 2-fold symmetry of the aragonite crystal structure in this direction. Hence the description of these being '4-spot patterns'.

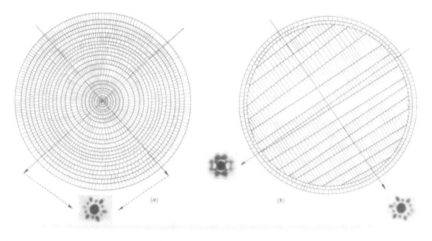

Figure 28.23   Identifying X-ray diffraction patterns for (LHS) a natural pearl and (RHS) a bead nucleated cultured pearl

In contrast, if the pearl is natural, an identical '6-spot pattern' would have been obtained at right angles to the orientation at which the first '6-spot pattern' was obtained. Simply put, two hexagonal diffraction patterns taken from directions perpendicular to each other, in the absence of any contrary evidence, indicates that the pearl is natural.

With any pearl, great care must be taken to ensure that the X-ray beam passes precisely through the centre of growth of the pearl. The reason for this precision is that in thick-skinned bead nucleated cultured pearls a hexagonal pattern also may be obtained in both directions. The reason for this is that the thick nacreous 'skin' of this cultured pearl produces an anomalous '6-spot pattern'. However, if the direction of the X-ray beam can be made to precisely parallel the layers of its bead nucleus, a discriminating superimposed '4-spot' pattern should possibly be discernible within the overall hexagonal pattern.

The Endoscope

The endoscope was an instrument that was invented, in late 1926, to facilitate examination of the stringing channels in drilled pearls. The endoscope was designed to facilitate discrimination of natural pearls from the then newly marketed Japanese bead nucleated cultured *akoya* pearls. Although this instrument is very rarely used today, its principle of operation should be understood by gemmologists.

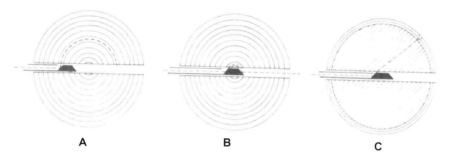

Figure 28.24   Paths of light from the endoscope needle through the pearl in a natural pearl (A&B), and a bead nucleated cultured pearl (C)

The endoscope utilizes a very thin hollow metallic needle to penetrate the stringing hole of a pearl, so that a strong narrow beam of white light that is delivered through the needle to a 45° pyramidal mirror located at its end is reflected vertically into the pearl being examined. As the needle of the endoscope is inserted into the stringing channel of the pearl and directed towards the centre of the pearl, the reflected narrow beam of light will be either transmitted around a hemispherical layers of nacre in a natural pearl and returned into the stringing channel, so that a flash of light can be viewed in an eyepiece located opposite the inserted needle; or, the beam of light will be transmitted to the surface of the pearl along the flat layers that form the bead in a bead nucleated cultured pearl. In this situation, an identifying spot of light then will be observed on the external surface of that cultured pearl.

With respect to the use of the endoscope, it is important to remember that as natural pearls, *keshis* and tissue grafted cultured freshwater pearls are all formed from alternating circumferential layers of aragonitic nacre and organic conchin, each will give the same response when examined with the endoscope.

Step 5: Detection of Value Enhancement

The fifth step in the identification of a pearl is to establish whether or not it has been value-enhanced by techniques that are presently considered to be either acceptable or unacceptable to the trade, trade regulatory bodies and of course the buying public. Here the gemmologist must make the oft-difficult distinction between what are considered to be acceptable and unacceptable trade practices with respect to the value enhancement of pearls.

It is generally agreed that accepted trade practices for the value enhancement of pearls include:

- Traditional 'peeling' of natural pearls.
- Tumbling of newly harvested South Sea pearls in proprietary formulations designed to remove residual residues and give the pearl's nacre a marketable lustre.
- The routine bleaching and dyeing of *akoyas* before they are strung.
- Whitening pearls by bleaching their nacre.
- Enhancing the lustre of pearls by waxing their external surfaces.

However, many questions still remain unanswered, with respect to how should the gemmologist deal with many other value enhancements for pearls that are being applied with increasing frequency today? In today's pearl market, how acceptable is the use of:

- Dyeing technology, to yield commercially acceptable black, golden or other attractive colours in South Sea and mantle grafted freshwater cultured pearls?
- Gamma irradiation, and associated dyeing technology to darken and make much more iridescent the colour of *akoya*, South Sea, and mantle grafted freshwater cultured pearls?
- Heat treatment to induce golden hues in pearls?
- Mechanical buffing and polishing to improve both the shape and lustre of the pearl's nacre?
- Various surface coatings to enhance the lustre of pearls?

The following guidelines for identifying value-enhanced pearls are therefore offered for consideration and use.

Bleached and Dyed Pearls

Colour enhancement (by bleaching followed by dye impregnation) cannot be detected readily on the external surface of undrilled whole pearls. If a pearl is undrilled, laboratory analysis by, for example, IR spectroscopy or Raman spectroscopy, will be required to definitively identify the presence of any colour-enhancing dye.

In contrast, colour enhancement, by dyeing, is detected relatively easily – once bead nucleated cultured pearls, or tissue grafted freshwater cultured pearls have been drilled. Examination of stringing channels or pegging channels in drilled, bleached and dyed pearls will generally reveal: dye concentrated in the organic conchin layer between the shell bead and the outer nacreous layer of dyed bead nucleated cultured akoya and South Sea cultured pearls; dye staining the walls of drill channels in all bleached and dyed pearls; and dye stained thread in some strung dyed pearls.

With the exception of drilled silver nitrate treated black bead nucleated *akoya* cultured pearls, which have an identifying X-radiographic appearance due to the radiopacity of particles of silver impregnating the conchin layer surrounding their bead; and dyed rough surfaced tissue grafted freshwater Biwa or Chinese freshwater cultured pearls, which often have dye adherent to their external surface, strings discoloured by dye, and dye adherent to the walls of the drill channels in the pearls.

Irradiated Pearls

The nacre of γ-irradiated bead nucleated cultured pearls, such as *akoyas*, tends to display greyish rather than black hues. If the drilled stringing channels in these greyish irradiated *akoyas* are examined, they will be shown to have a greyish to black shell bead that is covered by a comparatively thin layer of white nacre. These colour-enhanced pearls also are inert to LWUV irradiation. Other pearls, the colours of which are commercially value-enhanced by γ or other forms of irradiation such as electron irradiation include off-coloured bead nucleated white South Sea pearls, and lower colour grade mantle grafted freshwater pearls. Dark greyish South Sea bead nucleated pearls that owe their colour to irradiation are not common. However, they do display the same identifying features as γ-irradiated *akoyas*.

Electron-irradiated Chinese tissue grafted cultured pearls range in colour from greyish to iridescent black, depending on the radiation dose. These pearls are coloured throughout, and are inert to LWUV irradiation. Remember, however, black is not a natural colour that is found in Chinese mantle nucleated freshwater cultured pearls.

Recently, the pearl market has been inundated with black mantle grafted freshwater cultured pearls, of up to 10 mm diameter, that have nacre that is highly iridescent. These pearls are now known to have been colour-enhanced by a still secret Chinese process that allegedly involves a combination of dyeing with an unspecified silver salt followed by irradiation. These treated black pearls have the following identifying features: they are of unusually iridescent nacre; they change colour from black to dark brown when examined in incandescent light; they show a loss of definition in the 'thumb print' pattern on their nacre; and their nacre is inert to LWUV irradiation.

Bleached Pearls

Both natural and cultured pearls are bleached (with hydrogen peroxide, or perhaps sodium hypochlorite) to lighten overly dark hues and decrease the visibility of brownish patches of conchin in their nacre. With the exception of the routine bleaching followed by dyeing of *akoya* bead nucleated cultured pearls, how prevalent is the use of bleaching among pearl producers is unknown. However, some industry insiders suggest that up to 90% of all white pearls have been bleached prior to their sale.

As guidelines with respect to detecting bleached pearls have not been published, gemmologists should carefully examine the external surfaces of suspect pearls, at magnifications of 10–40×, to reveal tell-tale visual evidence that the nacre of the pearls has been bleached. Increased surface roughness, due to chemical dissolution of margins of aragonitic platelets exposed on the surface of nacre, and dissolution of the organic matrix into which individual aragonite platelets have been deposited, will identify the bleaching of nacre. Examination of drilled stringing channels in bleached bead nucleated pearls also will reveal that the conchin layer between bead and nacre has been decolourized, and in some circumstances either partly or completely dissolved by the bleaching agent. An additional observation worth making is to examine the white nacre under LWUV. Bleached nacre fluoresces uniformly a strong blue-white colour, while natural coloured nacre fluoresces in range bluish to greenish hues of quite variable intensity.

Treated Golden Pearls

Golden pearls created by subjecting undrilled pearls to a secret possible heating process first appeared in markets in 1993. These treated golden pearls are claimed to be colour stable and that they can be polished without loss of their induced colour.

Elen (2001) has suggested that the possibly heat treated golden pearls can be discriminated from pearls with natural coloured yellow to golden-coloured nacre by the absence of absorption between 330 and 385 nm in their VIS reflectance absorption spectrum, the presence of concentrations of colour in surface defects on these pearls, and an uneven surface fluoresce to LWUV not associated with the distribution of colour on the surface of the pearls.

Waxed Pearls

Lower-quality pearls often have their lustres enhanced by the simple process of tumbling in a mild abrasive to remove adherent external deposits, followed by the waxing of their external surfaces to enhance the lustre of the nacre. Fortunately, the presence of this greasy coating on waxed pearls can be detected readily by 10× hand lens examination, or by the mark left behind when a fingernail is drawn across the surface of a waxed pearl.

Polymer-Coated Pearls

Both white and black South Sea cultured pearls have the lustre enhanced by deliberately coating the pearls with thin films of colourless polymer. It is understood that this treatment is performed in Japan, with thicker layers being applied to pearls of poorer lustre.

Fortunately, polymer-coated pearls can be detected with the naked eye by closely examining the lustre of the pearl, for polymer-coated pearls display a rather muted orient beneath the polymer film, and on examination the gemmologist will find visual evidence of a thin coating of polymer and perhaps some residual polymer attached to areas on the surface of the pearl. Evidence of wear and scratching of the coating could also possibly be observed. It also has been reported that polymer-coated white pearls glow pinkish on the top and bluish on the side; while polymer-coated black pearls glow purplish on the top and green on the side.

Surface Polishing

'Pearl skinning' has been used traditionally to improve the value of natural pearls by removing unattractive layers of nacre from the pearl by hand and then trimming and polishing the surface. Today, it is common practice to tumble-polish cultured pearls in mild abrasives to remove adherent deposits, and enhance the lustre of the pearl's nacre by the mild polishing action of that abrasive. However, when the presence of obvious scratches and grooves on the external surface of pearls reveals that they have been ground and polished to improve their shape and/or lustre, a decision must be made whether or not it is appropriate that surface-polished pearls should be identified on any gemmological report.

Faceted Pearls

Faceted pearls are now an acceptable fashion accessory, with some individual pearls being covered with more than 150 triangular facets. This treatment is easy to detect and this value enhancement is presently being applied to both white and black South Sea pearls as well as mantle grafted freshwater cultured pearls of various hue.

References

Akamatsu, S., Li, T.Z., Moses, T.M. and Scarratt, K. (2003) The current status of Chinese freshwater cultured pearls. *Gems & Gemology*. 37(2), 96–113.

Banerjee, A. and Habermann, D. (2000) Identification of Chinese freshwater pearls using Mn^{2+} activated cathodoluminescence. *Carbonates & Evaporites*. 15(2) 138–148.

Bolman, J. (1941) *The Mystery of the Pearl*. Brill: Lriden.

Brown, G. (1985) The abalone and its pearls. *Australian Gemmologist*. 15(11), 400–403.

Brown, G. (2001) North Queensland mabé pearls. *Gemmologie*. 30(4), 193–198.

Brown, G., Kelly, S.M.B. and Snow, J. (1988) A coconut pearl? *Australian Gemmologist*. 16(10), 361–362.

Cahn, A.R. (1949) *Pearl Culture in Japan*. Report No. 122 from the Natural Resources Section of the General Headquarters for the Supreme Commander for the Allied Powers; Tokyo. 92 pp.

Chikayama, A. (1997) Japanese imitation pearls. Proceedings of the XXVI International Gemmological Conference, Idar Oberstein.

Crowningshield, R. (1962) Fresh-water cultured pearls. *Gems & Gemology*. 10(9), 259–273.

Dilenburger, B. (2004) The Kasumigaura pearl. *Australian Gemmologist*. 22(4), 156–161.

Elen, S. (2002) Identification of yellow cultured pearls from the black-lipped oyster *Pinctada margaritifera*. *Gems & Gemology*. 38, 66–72.

Elen, S. (2001) Spectral reflectance and fluorescence characteristics of natural colour and heat-treated 'golden' South Sea cultured pearls. *Gems & Gemology*. 37, 114–123.

Fritsch, E. and Misiorowski, E.B. (1987) The history and gemology of the pink conch. *Gems & Gemology*. 23, 208–221.

Gutmannsbauer, W. and Hänni, H. (1994) Structural and chemical investigations on shells and pearls of nacre-forming salt- and freshwater bivalve molluscs. *Journal of Gemmology*. 24(4), 235–240.

Howarth, P.C. (1988) *The Abalone Book*. Naturegraph Publishers: Happy Camp: California.

Hutchins, P. (2004) Culturing abalone half-pearls. *Australian Gemmologist*. 22(1), 10–20.

Kennedy, S.J. (1998) Pearl identification. *Australian Gemmologist*. 20, 2–19.

Kunz, G. and Stevenson, C.H. (1908) *The Book of the Pearl*. The Century Co: New York.

Liu, Y., Shigley, J. and Hurwit, K.N. (1999) Iridescence colour of a shell of the mollusc Pinctada margaritifera caused by diffraction. *Optics Express*. 4(5), 177–182.

Liu, Y., Hurwit, K.H. and Shigley, J. (2002) Iridescence of a shell of the abalone Haliotis rufescens. Caused by diffraction. *Journal of Gemmology*. 28(1), 1.5.

Liu, Y., Hurwit, K.M. and Shigley, J. (2003) Relationship between the groove density of the grating structure and the strength of iridescence in mollusc shells. *Australian gemmologist*. 21(10), 405–407.

Muramatsu, M. (1985) *Mabé peral. In Pearls of the world*. Les Joyaux Special Edition. pp. 79–86.

O'Sullivan, D., Cropp, D. and Bunter, O. (1995) Pearls of Australia: An overview of pearl production techniques in Australia. *Australian Gemmologist*. 19(3), 155–161.

Pough, F.H. (1965) Mallorca and imitation pearls. *Gems & Gemology*. 11(5), 273–280.

Quarenghi, M. and Sicaguato, R. (1997) *Conch Pearls*. Bel Eclat; Tokyo.

Webster, R. (1994) *Gems: Their Sources, Decriptions and Identification*. 5th ed., Butterworths-Heinemann: Oxford.

Tardieu, R. (1994) *Langeac: Au temps ou l'on enfilait des perles*. (Self published in French).

Yang, M.Y., Guo, S.G. and Shi, L.Y. (2004) Study on the compositions and colouring mechanisms of freshwater cultured pearls. *Journal of Gems & Gemmology*. 6(2), 10–13 (In Chinese).

Notes

1. Molluscs are soft-bodied non-segmented invertebrate (without backbone) animals that typically have a hard external shell which may be either single (univalve) for the abalone and pink conch, or bivalve for many species of 'pearl oyster' and 'pearl mussel'.

2. Nacre, or mother-of-pearl (MOP), is formed from iridescent layers of flattened crystals of aragonitic calcium carbonate, which that have been deposited rhythmically in progressively deposited microscopically thin layers of organic matrix and that line the shells of certain species of both fresh- and saltwater molluscs.

3. *Keshi* is a Japanese term used to describe 'small' pearls that may be of either natural or cultured origin. Natural *keshi* pearls are those small pearls found in 'wild' shell previously untouched by human hands.

Cultured *keshi* pearls occur as a consequence of some malfunction of the implant process in which a small pearl rather than a bead nucleated cultured pearl is produced as intended by the pearl farmer. As it is virtually impossible to discriminate natural from cultured *keshi* pearls, all are simply termed 'keshi' pearls.

4. Bironite is a man-made Australian-produced nucleus that is manufactured from sintered natural dolomite. This dolomitic ceramic is stable, takes a high, smooth polish, is well tolerated when implanted, allows nacre to be intimately deposited on its surface, and can be drilled with minimal risk of fracture. Several Australian and Indonesian pearl cultivators are presently trailing this nucleus.

5. Eggs fertilized from male and female broodstock from different geographical sources metamorphose into floating parasitic *glochidia* that must attach themselves to the fins or gills of freshwater fish, such as the yellowhead or bighead carp, in order to obtain nourishment and grow into young mussels – a metamorphosis that takes from 5 to 14 days depending on water temperature. At this stage the young transparent mussels fall from the fish and are harvested.

29

The Biological Gem Materials

Grahame Brown

The biological gem materials, traditionally termed the 'organic gem materials' or 'organic gems', consist of a group of attractive, relatively rare and mostly durable jewellery and decorative materials that have an animal or plant origin. As man's use of biological gem materials continues to increase, the traditional organic gem materials of amber, coral, ivory, jet, tortoiseshell and pearl (which are described in a separate chapter) have been expanded to include an ever increasing number of products of either terrestrial, marine and avian origin, as summarized in *Table 29.1*.

While the standard gemmological properties of most biological gem material have been determined, seldom are these needed to identity the material. Indeed, as the natural shape, softness, porosity and potential friability of biological gem materials rarely permits their manufacture into conventionally shaped gemstones, identification of most of these materials, their value enhancements and their imitations must be determined by observing those characteristic hand lens and/or microscope features that reflect either the unique structure of each natural material, the physical nature of their value enhancements or their method of manufacture.

For convenience, the common biological gem materials, their enhancements and their imitations, and their identifying features will be described in the order of their perceived importance to gemmologists.

Amber

Amber is a generic term used to describe those moderately hard fossilized plant resins (natural polymers) that have been derived from the protective exudates of several species of ancient conifers and flowering trees. Ambers range in age from less than a million to more than 300 million years old. Amber is an ancient biological gem material, for European

Table 29.1
Modern biological gem materials

Animal origin			Plant origin	
Terrestrial	*Avian*	*Marine*	*Terrestrial*	*Marine*
Ivory & Teeth	Hornbill casque	Pearl	Amber, Copal resin & other solid plant resins	Vegetable coral – Black – Golden
Bone & Antler	Claws & Beaks	Shells	Vegetable ivory	
Calcific concretions	Feathers	Mother-of-Pearl	Seeds, Nuts, Fruit skin, Gourds	
Horns		Operculum	Wood	
Hoofs		Calcific coral – Precious – Reef-building	Jet/Coal	
Claws		Tortoiseshell	Gutta-percha	
Hair		Ivory & Teeth	Vulcanite	
Skin & Leather Insect exoskeletons		Chitinous claws Skin	Lacquer	
Fossil nacre				

prehistoric (Palaeolithic) man is known to have stored unworked pieces of amber in caves in Hautes Pyrenées some 15 000 years ago. The earliest known examples of worked amber include beads from Gough's cave in southern England, dated 11 000–9000 BC, and a pendant depicting four angular human figures from a bog in West Zealand (Denmark) that has been dated 7000 BC.

How Amber Forms

Trees exudate resin as a defensive mechanism against injury, disease and attack by insects and fungi. The exuded resin can infill internal fractures in the wood of the tree, form pockets of resin in the wood, infill surface-reaching wounds in the tree, form deposits of resin under the bark of the tree, or run down the external surface of the bark.

The process of transforming exuded resin into amber (*amberization*) is incompletely understood. Basically the process involves the protracted application of heat and pressure, under oxygen-free (anaerobic) conditions, to convert soft, sticky volatile-rich organic plant resin into a solid natural plastic. These changes involve, first, polymerization of the organic matrix of the resin to produce copal (resin); then, evaporation of residual volatiles from the copal to produce amber, as illustrated in *Figure 29.1*.

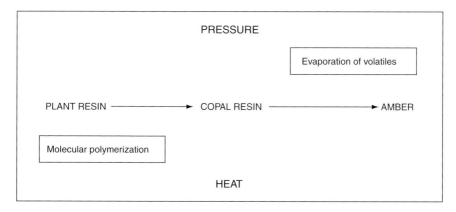

Figure 29.1   How amber forms (After Platt)

Available evidence suggests that both Baltic and Dominican ambers achieved their anaerobic environments by protracted immersion under seawater. The length of time needed to complete the transformation of plant resin to copal to amber varies with conditions of burial, but can take from 2 to 10 million years to complete.

What is Amber?

Chemically, amber is a hydrocarbon (organic compound of carbon, oxygen and hydrogen); but the chemical compositions of ambers from various sources worldwide are quite variable. For example, early chemical analyses revealed that Baltic amber could be discriminated from other ambers on the basis of its content of 3–8% succinic acid. This discovery of the succinic acid content of Baltic amber led to the now somewhat irrelevant subdivision of ambers into *succinites* and succinite-free *retinites*. The chemical compositions, colours and other physical properties of ambers are known to be influenced by factors such as the species of tree that bled the original plant resin, the conditions of burial for the plant resin and the length of time the resin remained buried. Based on his identification of pieces of bark including Baltic amber, in 1863 the German botanist Goeppert named the extinct conifer *Pinites succinifera* as the source of Baltic amber. Subsequent research has revealed that this ancient pine tree never existed. Instead, present evidence suggests that amber can be derived from two distinctly different varieties of resin-producing trees: temperate gymnosperms (plant whose seeds are found in cones = conifers), such as pine trees, and tropical angiosperms (flowering trees), whose seeds are enclosed in an ovary to form a fruit. Amber has been found in both primary and secondary sedimentary deposits that range in age from the Carboniferous (<300 million years) to Quaternary (0–2 million years).

Varieties of Amber

Amber is recovered from primary and secondary sedimentary deposits worldwide. Significant commercial deposits of amber occur in Europe, in and around the Baltic Sea and in the Dominican Republic that forms eastern two-thirds of the Caribbean island of Hispañola.

Baltic amber is believed to have formed from resin exuded from the Baltic Amber Forest that originally covered most of what is now Norway, Sweden, Finland and Eastern Russia. It has been suggested that ancient rivers deposited proto-amber in their deltas. Here the soft resin began its long metamorphosis into copal, then amber. Subsequent glacial activity eroded these deposits and redeposited their amber content around the southern coast of what is now the Baltic Sea – particularly in the Samland Peninsula of today's Poland. Historically, Baltic amber has been recovered as either:

- *Sea amber*, or *scoopstone*, which is washed onshore and fished from the sea with 'amber-catchers' off the coastlines of Denmark, Germany, Poland, Russia, the Baltic Republics (Lithuania, Latvia, Estonia) and Sweden; or
- *Pit amber*, which is mined by open cut and hydraulic methods from *blue earth* – a 1–9 m-thick amber-bearing level of 25–40 million years age that lies about 25–40 m below large areas of the Polish coastline from the north-west side of Gdansk (Danzig) Bay to the Karlingrad Oblast of eastern Russia. Amber-bearing *blue earth* also extends northwards into Lithuania, Latvia and Estonia.

Dominican amber, a 15–35 million years succinic acid-free *retinite*, is derived from an extinct species of resin-producing algarroba tree, the *Hymenaea protera*. Amber is found as fractured fragments in layers of lignite or carbonaceous clay in three areas of the Dominican Republic:

1. Within the Ciabo Valley of the Cordillera Septentrional to the north and north-east of the Dominican Republic's capital of Santiago de los Caballeros. Here amber occurs at an elevation of 300–1300 m.
2. At Cotui near the centre of the island, where the amber has an estimated age of some 15–17 million years.
3. The Eastern Area that is located north-east of the southern coastal port of Santo Domingo.

Other ambers of historical but minor commercial interest include *burmite* from the Hukawung Valley (Myanmar), *simeite* from the Simeto River (Sicily), *rumanite* from Buzau (Romania) and *chiapas* amber from the Andean region of Mexico's southernmost state.

Gemmological Properties

For the gemmologist, amber's properties of interest are summarized in *Table 29.2*.

Table 29.2
The gemmological properties of amber

Colour	Ranges from colourless to yellow, brown, red and black. Green- and blue-coloured ambers are rare. The colour of some Dominican amber is enhanced by this amber's bluish fluorescence in daylight.
Hardness	Variable, 1–3 on Mohs' scale. Some amber can be scratched with a sharp fingernail.
Fracture	Conchoidal.
Sectility	Chips when tested with a sharp blade.
Specific Gravity	1.05–1.10 (Floats on a saturated salt solution).
Refractive Index (distant vision)	1.54 (S.R.) Caution: Refractometer contact liquid may soften the surface of some amber.
Diaphaneity	Ranges from transparent to opaque, depending on size, number and distribution of included gas 'bubbles'.
Lustre	Resinous, but sometimes waxy.
Luminescence	Variable, but a weak bluish white fluorescence to LWUV<SWUV is the most common.
Thermal properties	Thermoplastic, softening at 150–200 °C Will yield a slightly acrid pine-like odour when a hot point is applied judiciously to an inconspicuous area of the amber.
Electrical properties	Triboelectric, becoming negatively charged and capable of attracting small pieces of paper.
Solubility	The surface of amber is unaffected when a drop of a highly volatile organic solvent, such as ether, chloroform or alcohol, is applied to its surface and allowed to evaporate.
Characteristic inclusions	The presence of rare, aesthetic inclusions is of the major reasons why amber has been a long-desired and valued gem material. Amber may be included by: 'bubbles' (some two-phase), flow swirls, remnant vegetable matter, insects, very rarely small animals and minerals such as particles of quartz sand.

Identification

Although amber has characteristic gemmological properties, accurate determination of some of amber's key identifying properties (SG, RI) is often not feasible. Consequently, the gemmologist must identify amber, particularly set ambers, by a combination of the following carefully applied observations:

1. Amber will float on a saturated salt solution that has an SG of ~1.10.
2. The surface of amber is not softened when a small drop of volatile organic solvent (e.g. ether, chloroform or alcohol) is applied to a surface and allowed to evaporate.
3. Amber will chip (not peel) when an inconspicuous edge (e.g. periphery of the stringing hole of a bead) is tested, under magnification, with a small very sharp blade such as a surgical scalpel blade.

It is important to remember that the potentially destructive hot point test has little practical application in the identification of carved and polished amber. However, if this test is applied to rough amber, then the sharp tip of a pointed wire should be heated until it glows red. This tip should be allowed to cool, slightly, before it is applied judiciously to the surface of the suspect amber to create a 'puff of smoke'. When this smoke is smelt, amber will yield a slightly acrid 'piney' aroma while many plastics will yield an acrid 'burnt plastic' aroma.

Due to amber's comparative rarity, and increasing value in the market-place, it is commonly value-enhanced as well as imitated.

Value-Enhanced Amber

Amber is commonly value-enhanced either by dyeing, heat treatment, clarification or reconstruction.

Dyed Amber

Being a hydrocarbon, amber can be readily impregnated by solvent-based organic dyes. Unfortunately these dyes cannot be identified by conventional gem-testing techniques. However, the observed presence of reflective disc-shaped 'sun spangle' inclusions in a brightly coloured amber will be suggestive that the amber has (at least) been heated and cooled by man. These inclusions are always induced in amber that has been softened by heat and then cooled by man (Figure 29.2).

Sunspangle inclusion

Figure 29.2   Red-dyed amber. Note the identifying induced 'sunspangle' inclusion (arrowed)

Heat-Treated Amber

Heat treatment of amber in the presence of oxygen mimics Nature's method of creating darkened reddish brown *antiqued* or *burnt amber*. This treatment cannot be detected by the gemmologist, unless 'sun spangle' inclusions have been induced in the 'burnt' amber.

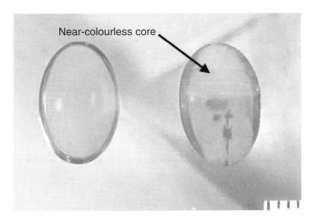

Near-colourless core

Figure 29.3   Coated amber with its underlying colourless amber
revealed where the coating has been ground off

A commercial heat treatment is also being used to coat light-coloured ambers with a thin, darker coloured layer of orange-brown coloured amber (Figure 29.3). These *coated ambers* can be detected by using the 10× hand lens or low-power gemmological microscope to locate the distinct junction between this treated amber's thin brownish coating and its pale-coloured core. The presence of layers of minute surface-migrating gas 'bubbles', and randomly oriented 'sun spangle' inclusions will assist identification of this heat treatment.

Clarified Amber

'Bubble'-included *cloudy ambers* are clarified (enhanced in diaphaneity) by slowly heating them in an oil of similar RI – such as rapeseed (canola) or linseed oil. The oil penetrates the amber and infills its internal 'bubbles'. On cooling, the presence of the solidified oil, infilling 'bubbles', makes the amber transparent. Organic dyes may be added to the clarifying oil to enhance the colour of the clarified amber (Figure 29.4). Following clarification, slow cooling of the amorphous heated amber is essential to minimize the occurrence of identifying circular stress fractures ('sun spangles' or 'nasturtium leaves') in *clarified amber*.

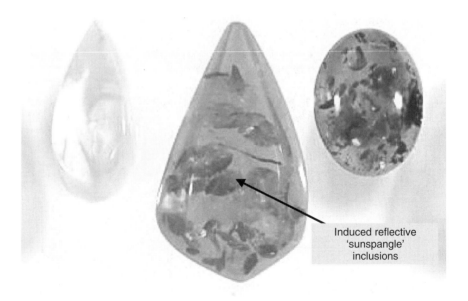

Induced reflective 'sunspangle' inclusions

Figure 29.4   Untreated 'cloudy' amber (LHS), with clarified amber included by 'sunspangle' inclusions (Centre & LHS)

Reconstructed Amber

To create *reconstructed or pressed* amber, which is also termed *ambroid* or *Spiller imitations*, heat (~180 °C) and pressure (120 000 psi) are applied to either carefully cleaned small pieces of reject amber or ground-up amber off-cuts that are held in a large metal pressure vessel. Following careful cooling, this treatment creates a fused mass of amber that commonly is used for the manufacture of sizeable objects such as cigarette and cigar holders, and beads. In a variant of this treatment, powdered amber is softened under heat and pressure before being extruded into rods of homogeneous reconstructed amber that are slowly cooled to minimize the formation of 'sun spangles'.

Various forms of reconstructed amber can be identified by careful 10× hand lens or low-power microscope examination, looking for (Figure 29.5):

1. A cloudy to swirling texture, or an irregular colour distribution, due to the fusion of many small pieces of amber of varying colour.
2. Stretched and distorted 'bubbles' of entrapped air between the fused fragments of amber.
3. A patchy extinction and/or display of a mosaic of strain interference colours when reconstructed amber is examined in polarized light, or is rotated between crossed polars.

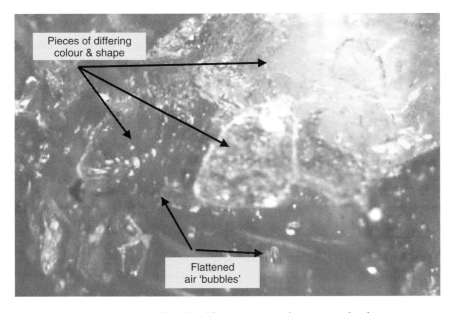

Pieces of differing colour & shape

Flattened air 'bubbles'

Figure 29.5   Identifying hand-lens appearance of reconstructed amber

Amber Imitations

Amber is imitated by a range of natural and man-made materials that include non-fossil plant resins such as copal resin and kauri gum, man-made polymers (plastics), glass and composite materials.

Copal Resin

Copal resin is a solidified, volatile, rich, diaphanous, brittle, colourless, yellow, brown or rarely reddish plant resin that is derived from various tropical angiosperms. Copal resin represents an intermediate stage in the conversion of fresh plant resin to amber. Copal resin is found in New Zealand (kauri gum), the Philippines and Indonesia (manila), Malaysia (dammar), east and central Africa (congo) and Colombia where the copal resin is commonly sold as Colombian amber.

Copal resins make excellent imitations of amber. However, with the exception of their slightly lower hardness of 2, slightly lower SG of 1.03–1.09, increased solubility in volatile hydrocarbons, increased volatility that yields a 'pine'-like odour when copal is firmly rubbed with the heel of the hand, and increased shallow polygonal surface crazing that is restricted to the outer 2 mm of the copal, these effective amber imitations have the appearance, gemmological properties, sectility (chips on peeling) and general pattern of inclusions virtually identical to those of amber.

However, the increased solubility of copal resins to volatile organic solvents, which is a potentially destructive test, can be used to effectively discriminate copal resins from amber, as:

1. The surface of copal resin (and not amber) will become sticky following the judicious application of a very small drop of volatile organic solvent (e.g. ether, chloroform, acetone, alcohol) to an inconspicuous area of the suspect resin, and then allowing this drop to evaporate; and
2. A yellow stain will be deposited on the surface of an alcohol-moistened, rough textured white cloth or tissue, when an inconspicuous surface of the copal resin is rubbed briskly against this surface. Amber will not produce such a stain.

Gemmologists should remember that potentially destructive solubility tests must be applied judiciously and under magnification to inconspicuous areas of objects being tested, and with the owner's permission.

Man-Made Polymers

For almost a century, a large range of man-made polymers (plastics) that include cellulose nitrate, cellulose acetate, formaldehyde-casein, phenol formaldehyde or *bakelite*, polymethyl methacrylate or *perspex*, polystyrene, polyester casting resin and the epoxy resins have been used effectively to imitate amber.

With respect to the discrimination of these polymer imitations from amber and copal resin, as most polymer imitations of amber have SGs greater than that of amber (1.08) and copal (1.03–1.08), they will readily sink in a salt-saturated solution of water. In addition, polymers used to imitate amber will peel, rather than chip when tested at an inconspicuous edge with a small sharp blade.

Polystyrene is the only exception to this generalization, for as it has an SG of 1.03 it also will float on a saturated salt solution. However, polystyrene can be effectively distinguished from amber and copal resin, as:

1. It peels when pared with a sharp knife, as do all other plastics.
2. It is extremely soluble to applied volatile organic solvents.

Glass

Glass imitations of amber pose few problems of identification, for amber-coloured glasses are much heavier to heft than amber, are colder to touch, and are likely to show other glass-like characteristics such as included gas 'bubbles' and glassy conchoidal fractures.

Composites

Composite imitations of amber include modern insects embedded in amber-coloured polymers such as polyester casting resin, copal resin, or rarely amber, and polybern – a man-made composite imitation that consists of fragments of amber embedded in polyester resin.

The modern insect that has been embedded in amber-coloured polyester resin can be readily identified by observing that:

1. The yellowish casting polymer will peel under a sharp blade rather than chip.
2. The casting resin will have an RI that is different from that of amber.
3. To the experienced eye the included insect will be modern; it will have been cast without a struggle; and will not display 'bubble/s' of exuded gas or have a white coating due to decomposition of its body tissues.

Insect-in-amber or animal-in-amber imitations are more likely to be manufactured from copal resin for copal melts at <150 °C and it melts rather than burns. These fakes will contain large, mostly single, well-positioned modern insects or small animals that show no sign of any struggle to avoid entrapment, and importantly no evidence of decompositional changes are found from ancient insects or animals trapped in amber. The copal resin

Figure 29.6   Polybern imitation of amber

surrounding the insect or animal will display the solubility and presence of volatiles that characterize copal resin.

Polybern, which is commonly a Polish product, will display evidence of moulding (mould marks and trapped external air 'bubbles'); it will peel under a judiciously applied small sharp blade; and will be seen to consist of two distinct components – differently coloured and differently shaped fragments of amber that have been embedded in a man-made polymer of both differing colour and RI (Figure 29.6).

Coral

Corals are polyp-bearing marine animals that live in colonies, belong to the phylum Coelenterata, and are found throughout the world's seas and oceans from freezing polar regions to equatorial reefs and at all depths from the intertidal zone to the bottoms of the deepest oceanic trenches. The word *coral*, which is used to describe the secreted skeletons of the corals' polyps, is derived from the ancient Greek word for coral, *korallion*. Historically, this word was used to describe the precious red coral from the Mediterranean that is now described scientifically as *Corallium rubrum*.

For thousands of years, *C. rubrum* from the Mediterranean Sea has been used for the fabrication of items of jewellery and decoration. Along with amber, this solid red coral may have been used as currency, or traded, since Palaeolithic man first roamed the earth. The corals used by man for jewellery and decorative purposes are either *precious coral*, which is derived from the solid calcitic axis and branches of several species of corallium genus, or *non-precious corals*, or jewellery corals, which are derived of the axes and branches of various species of calcitic and organic corals other than those belonging to the genus *Corallium*.

Precious Coral

What is Precious Coral?

Precious (red) *C. rubrum* is an ancient gem material that was and is harvested from relatively shallow, sunlight-protected reefs that are found in the Mediterranean Sea off the coasts of Italy, Sardinia, southern France and Spain, and north Africa. Other species of corallium corals are found

Figure 29.7 A colony of Corallium rubrum *coral*

particularly in the northern Pacific Ocean. To date, worldwide, twenty-five species of corallium coral have been identified, and of these six species are commonly traded for use in jewellery.

Tree-like red to pink corallium corals are branching and may be either fan-like or bush shaped (Figure 29.7). Unlike reef-building corals the polyps of *C. rubrum* live on the external surface of the solid calcitic endoskeleton of this coral. The axis and branches of this coral comprise fused particles of calcium carbonate and are commonly coloured red to pink by the presence of a carotenoid pigment. These corals are very slow growing, having growth rates of ~1 mm per year in Mediterranean Sea, and up to 8 mm per year for Pacific Ocean species. These corals also are relatively long lived – attaining ages of up to 75 years and heights of up to 1 m.

Sources of Precious Coral

For over a thousand years, colonies of *C. rubrum* have been randomly dredged by ship-towed, heavy, wooden, cross-shaped *inangas* that were lowered to depths of 10–20 m to entangle and remove small colonies of this coral from light-protected caves and reefs in the Mediterranean Sea. This traditional method of harvest used to provide sufficient precious coral to satisfy world demand. Although this historic fishery is still operational, today's precious coral is selectively and, more efficiently, harvested by SCUBA divers from caves at considerable depth.

By the late nineteenth century, the demand for precious coral was such that new deposits of red *C. japonicum* and *C. eliatus* were discovered and exhaustively fished by Japanese fishermen from beds scattered south-wards from Japan and the Philippines. Unfortunately, within half a century, overfishing rapidly depleted these deposits. Today, the Pacific coral fishery still survives, for in the early 1960s beds of very deep water pink *C. secundum* were discovered, at depths exceeding 1000 m, off the coast of the Hawaiian Island of Oahu. For a decade this difficult-to-access resource was harvested by Hawaiian interests using the miniature deep-water submarine *Alvin*. Since this time Japanese exploration has revealed a series of beds of corallium corals occupying the tops of the Emperor Seamounts that stretch from the Hawaiian Islands to north-west towards Japan. Presently, these deposits are being harvested, at depths of 400 m, by Japanese fishermen who destructively deep-water dredge the tops of these seamounts.

Today, raw precious coral – harvested from both the Mediterranean Sea and the Pacific Ocean – is marketed internationally and manufactured into carvings, beads of various shape and cabochons. Coral processing occurs in three locations: Torre del Greco (Italy), Japan and Taiwan. The processing of raw coral into items of jewellery and decorative objects is a relatively simple operation. For example, once the external organic coenenchyme has been stripped from *corallium* coral, the squat branching coral is sectioned, examined for imperfections and graded. The graded segments or whole coral trees are then pre-formed, carved and polished into items of jewellery or decoration.

Gemmological Properties

The gemmological properties of precious corallium coral are given in *Table 29.3*:

Table 29.3
The gemmological properties of precious corallium coral

Chemical composition	Disordered magnesian calcite with a Mg^{2+} content of ~10%, with minor calcium sulphate, phosphates, silica and iron oxide. The reddish hues of corallium corals are caused by the presence of organic pigments, known as *trans* carotenoids, in the organic matrix of the coral.
Colour	White, pink, red, brownish red, orange. Deep blood-red and pink colours are the most valued.
Hardness	$3\frac{1}{2}$–4 (Mohs' scale)
Fracture	Flat to hackly
Sectility	Readily carved but quite brittle.
Specific Gravity	2.65 (cf. Quartz)
Refractive Index (distant vision)	Difficult to obtain but should lie somewhere between 1.48 and 1.65 when determined by the birefringence 'spot method' RI reading method.
Diaphaneity	Opaque
Lustre	Waxy on polished surfaces
Luminescence	Mostly inert, but a weak purplish red LWUV fluorescence may be visible on lighter coloured corals.
Solubility	Readily effervesces when a drop of hydrochloric acid is applied to its surface.
Characteristic inclusions	Parallel longitudinal striations extending from central canal onto external surface of axis and branches.

Identification

Precious coral, irrespective of its colour or origin, can be readily identified in hand specimen by (Figure 29.8):

1. Its solid texture, with extremely few, small cavities (other than clusters of pin point holes that are the remnants of the coral's central canal) being visible on the polished external surface of the coral.

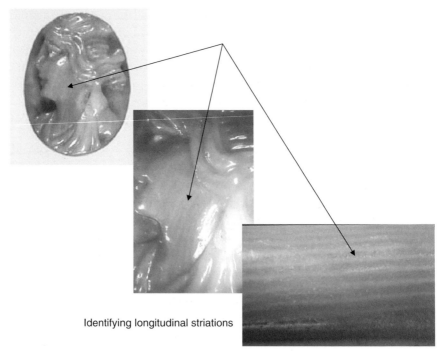

Identifying longitudinal striations

Figure 29.8   Identifying hand-lens features of precious red coral

2. The presence of parallel longitudinal striations on the surface of the axis and branches of the coral.

Value-Enhanced Precious Coral

Pale-coloured corallium coral is being value-enhanced commercially by impregnating the coral with dye. While the presence of red dye may be detected in the stringing holes, in surface defects or on the string of necklaces of dyed coral beads, often the presence of dye is difficult to detect by simple observation. In the laboratory, IR spectroscopy and Raman spectroscopy can be used to identify foreign dye molecules used to colour-enhance precious coral.

Precious Coral Imitations

Precious corallium coral is effectively imitated by several look-alike materials that include glass, plastic, dyed chalcedony, dyed reef build-ing coral, dyed and polymer impregnated reef building coral, so-called 'Sponge Coral™', dyed bamboo coral and Gilson's ceramic coral imitation.

All of these imitations are simple to distinguish from precious coral, for not one of its imitations will display the solid, pit-free, longitudinally striated surface that identifies precious coral. In addition, glass, porcelain and dyed chalcedony imitations will feel cold; plastic imitations will heft lightly and feel warm and any striations on their surfaces will not be parallel; while dyed and/or polymer-impregnated reef-building corals will display evidence of dye, and easily visible cavities in their external surfaces that may or may not be filled with 'bubble'-included colourless polymer.

Sponge coral™ is manufactured by dyeing red the solid, though porous Pacific Ocean coral *Melithaea ocracea*. This allegedly natural coloured red coral is identified by its porous surface, the lighter hued reticulated pattern that is readily visible on its polished surfaces, and the presence of red pigment in defects, in stringing holes and on the polished surface of the coral.

Red-dyed bamboo coral does display longitudinal striations on its external surface. However, the presence of remnants of the black coral, and dye infilling defects in the surface of this coral, will readily detect this imitation (Figure 29.9).

Although the colour of the Gilson™ (now Kyocera™) coral imitation is a good look-alike for that of precious coral, this imitation is a ceramic (Figure 29.10). Consequently, the surface of this imitation of precious coral

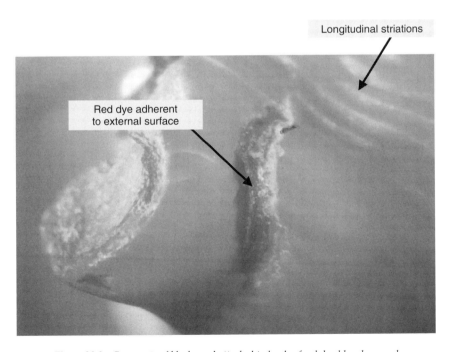

Longitudinal striations

Red dye adherent to external surface

Figure 29.9   Remnants of black coral attached to beads of red dyed bamboo coral

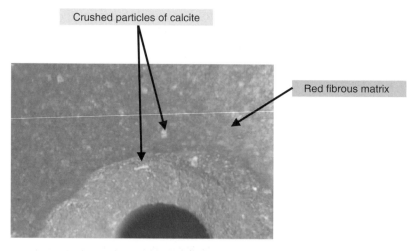

Crushed particles of calcite

Red fibrous matrix

Figure 29.10   Hand lens characteristics of Gilson™ coral imitation

displays has the appearance of terrazzo floor when examined with the hand lens or with a low-power microscope. Additionally, the surface of this imitation is not longitudinally striated, for it consists of small angular particles of crushed white calcite that are held in an appropriately coloured, somewhat fibrous matrix of unknown composition.

Jewellery Corals

The corals commonly used for jewellery purposes include the organic corals *black coral* and *gold coral*, and *bamboo coral*, which is a coral of mixed calcitic-organic composition.

Black Coral

Antipatharian, or thorny black corals, are solid keratinous or horny corals that have flexible endoskeletons. Commercial black coral is obtained from the *Antipathes grandis* and *A. dichotoma* – tree-like thorny organic or vegetable corals that grow as large bushes of up to 2 m height in tropical reef waters that are accessible to SCUBA divers. Stripped of their soft polyp-bearing coenenchyme, the structure of the axis and branches of a black coral closely resembles that of a terrestrial tree. But, the external surfaces of antipatharian black or 'thorny' coral are rough, for they are covered by fine, radially arrayed spines. Shaped and polished surfaces of this thermoplastic black coral take a very high polish. Although appearing opaque and black in incident light, thin sections of this coral are translucent and brownish red coloured.

Black (thorny) antipatharian coral has the following properties of interest to gemmologists. See *Table 29.4* for details.

Table 29.4
The gemmological properties of thorny black antipatharian coral

Chemical composition	Organic, its principal constituent being the scleroprotein antipathin.
Colour	Black, but brownish red in thin sections.
Hardness	1–2 (Mohs' scale)
Fracture	Splintery
Sectility	Readily carved
Specific Gravity	1.35
Refractive Index (distant vision)	1.56
Diaphaneity	Opaque in thick sections, translucent in thin sections.
Lustre	Resinous
Luminescence (UV)	Inert
Thermal properties	Thermoplastic, softening at 100–150 °C. An applied hot point yields a salty 'burnt hair' aroma.
Solubility	Insoluble in acids, alkalis and organic solvents
Characteristic features	Radially arrayed spines

Figure 29.11   (a) Spiny (thorny) axis and branches of black coral (b) Thorns (spines) identifying antipatharian black coral

Thorny black coral is readily identified in hand specimen by (Figure 29.11):

1. Its tree-like structure in cross-section.
2. Observation of remnants of radial spines that are often visible when polished surfaces of this black coral are examined in tangential illumination and with magnification. White light transmits readily through thin superficial layers of this coral to give these layers a waxy brownish red colour against which the presence of this coral's identifying spines is immediately obvious.

Gemmologists are warned that any trade in black antipatharian coral is prohibited under internationally agreed-to CITES regulations.

Gold Coral

The *gold corals* consist of several species of deep water coral that have gold- to golden-coloured keratinous endoskeletons. For some decades, *Gerardia sp.* has been commercially harvested from the deep (300–400 m) waters off Hawaii and the line of seamounts that lie to the north-west. The flattened, branched colonies of this coral, that may reach a height of 250 cm, are the major commercial source of gold coral.

Small tree-like colonies of Gerardia, and other related species, yield a beautiful, lustrous gold-coloured organic coral that has the following gemmological properties and features (*Table 29.5*).

Table 29.5
The gemmological properties of gold coral

Chemical composition	An organic scleroprotein of undetermined composition.
Colour	Gold to brownish with a slightly dimpled and/or ridged surface.
Hardness	1–2 (Mohs' scale)
Fracture	Splintery
Sectility	Readily carved, ground and polished
Specific Gravity	1.40
Refractive Index (distant vision)	1.58
Diaphaneity	Opaque, but brownish in thin sections.
Lustre	Resinous, verging on chatoyant
Luminescence (UV)	Inert
Thermal properties	Thermoplastic, softening between 150–180 °C. An applied hot point yields the smell of 'burnt hair'.
Characteristic inclusions	External surfaces of this coral may be either dimpled or finely ridged.

Gold coral is usually readily identified, by 10× hand-lens examination, by noting (Figure 29.12):

1. The tree-like structure visible on cross-section of branches.
2. The presence of a characteristic dimpled or longitudinally ridged golden surface on this coral.

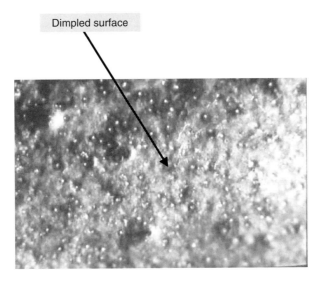

Figure 29.12   Identifying dimples in the external surface of Gerardia golden coral

Poorer-quality gold coral is value-enhanced by impregnating its shrinkage fractures either with whitish opaque paraffin wax or solidified colourless polymer. Wax impregnation is identified by observing the presence of low melting point wax either on the surface of or infilling defects within the treated coral.

Polymer-impregnated gold coral can be readily identified, in hand specimen, by observation of:

1. A coating of colourless polymer on the surface of the gold coral that gives an iridescent sheen to the surface of the coral.
2. The presence of 'bubble'-included colourless polymer infilling shrinkage defects in the coral, and dimples on the external surface of the coral.

An imitation of gold coral is also being produced in the Philippines by bleaching black thorny coral in a warm hydrogen peroxide solution. The hydrogen peroxide denatures the antipathin of black coral, thus

Figure 29.13 Identifying features of bleached black coral imitating gold coral

bleaching the outer layers of this organic coral to a beautiful golden colour (Figure 29.13). Colour-modified golden coral is readily discriminated from gold coral by the obvious presence (to 10× hand-lens examination) of the remnants of radial spines from the original antipatharian coral on the external surface of this treated coral.

Bamboo Coral

Bamboo coral is yet another coral, with considerable jewellery potential, that grows with precious pink coral and gold coral in the deep waters off Oahu in the Hawaiian Islands, off the Alaskan coastline, and in the deep waters to the south of Tasmania.

The Hawaiian bamboo coral, *Lepidisis olapa*, is a bioluminescent, flexible, whip coral that has a single left or right spirally twisted axis of 2–3 m length. The coral obtains its flexibility, and its name of bamboo coral, from a structure that consists of slightly curved calcitic internodes of white coral of 3–7 cm length that alternate and are permanently attached to short (5 mm) connecting nodes of black organic coral. Interestingly, the white coral component of bamboo coral has the properties and visual characteristics of *corallium* coral, while its black coral component has the properties of *antipatharian* coral.

This attractive jewellery coral is identified, in hand specimen, by its alternating structure of short brownish keratinous (black coral) nodes and longer calcitic internodes (Figure 29.14).

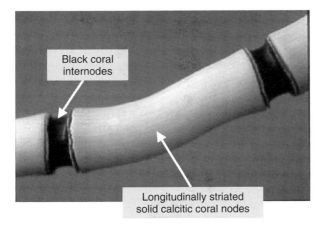

Black coral
internodes

Longitudinally striated
solid calcitic coral nodes

Figure 29.14   Identifying hand lens features of bamboo coral

Examination of a specimen of bamboo coral, with the 10× hand lens, will reveal:

1. Longitudinal striations along its solid white calcitic coral internodes that closely resemble those found on corallium coral.
2. Black coral nodes that are firmly attached to the white coral internodes.
3. The concentric lamellar structure of the keratinous internodes.

When the internodes of bamboo coral are dyed red or pink, this colour-enhanced coral can be difficult to discriminate from precious corallium coral – unless evidence of residual dye and/or wax is observed.

Ivory

Ivory is a long-used biological gem material, for early man used ivory to carve a 32 000-year-old dot-inscribed ivory pendant, which was recovered from caves at Le Conte in France's Dordogne Valley; a 26 000-year-old articulated male figure that was carved from mammoth ivory was discovered near Brno in the Czech Republic; and the 25 000-year-old Venus of Brassempouy was recovered from a Paleolithic cave near Brassempouy in France. Ivory, certainly, has been used as a biological gem material for many thousands of years.

What is Ivory?

Ivory is obtained from the yellowish, elastic, moderately hard dentine of continuously growing teeth (tusks) that belong to certain species of mammal such as both the recent and fossil elephant, the walrus, the hippopotamus, the sperm whale, the narwhal and some species of ungulates such as the boar and the warthog. Tusks have various shapes, but they are formed from similar calcified tissues. These highly modified teeth have a central pulp cavity that is surrounded initially by circumferential layers of dentine and then a thinner layer of cementum that attaches the tusk to its bony socket via connecting soft tissue periodontal ligament. In the tusks of some young elephants, dentine that erupts into the mouth may be covered with hard white enamel to protect tip of the tusk against wear and damage.

Dentine (ivory) consists of 25% organic material and water, and 75% inorganic apatite. At the microscopic level dentine consists of a mass of microscopic collagenous tubules, that radiate outwards, from the pulp cavity to the cementum, through an organic matrix that is impregnated with near-submicroscopic oriented crystals of the carbonate-hydroxyapatite mineral dahllite [$Ca_5(PO_4)_3(F,Cl,OH)$]. As the size, shape, and distribution, and orientation of these dentinal tubules vary with various species of tusk-bearing mammals, this variability produces the visual features that assist identification of the species from which a particular ivory was sourced.

Elephant ivory, either recent or fossil origin, is recognized to be the principal ivory of commerce. This ivory is derived from the two upwardly curving, continuously growing incisor teeth (tusks) of either the Asian elephant (*Elephas maximus*), the African elephant (*Loxodonta africana*) or the now-extinct wooly mammoth (*Mammuthus primigenius*).

Determination of the gemmological properties of an ivory will not greatly assist the identification of its source animal, for all ivories are whitish to creamy yellow coloured, have a hardness of $2\frac{1}{2}$–3 on Mohs' scale, have a smooth to splintery fracture, have SG range from 1.7 to 1.9, will give a 'spot' RI of 1.50, are readily carved and polished to a high lustre, have a resinous-waxy lustre, are opaque and will commonly fluoresce bluish white under LWUV.

Identification of Elephant Ivory

As a consequence, the identification of the principal ivory of commerce, elephant ivory, and its discrimination from its common imitations, must rely heavily on the fact that this ivory has identifying hand-lens features. Over a century and half ago, the English anatomist Sir Richard Owen recognized that elephant ivory could be positively identified by the 'decussating engine turned' pattern visible on the surface of cross sections of the tusk (Figure 29.15). As a consequence, all the gemmologist has to do, to identify elephant ivory, is to turn the unknown ivory, and

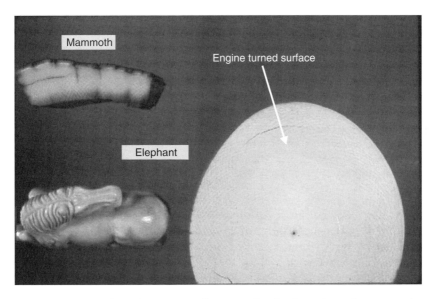

Figure 29.15 Identifying 'engine turned' pattern that is observed on the cross-section of tusks of elephants

closely examine its surface in either reflected or transmitted light, until the pattern illustrated below is noted.

Espinosa and Mann, from the US Fish & Wildlife Forensic Laboratory, also have recommended that that by measuring the angle of intersection of the outer intersections of so-called of 'Schreger lines' (Owen's 'decussating engine turned' pattern) in elephant dentine that lie closest to the outer cementum of the tusk, that modern elephant ivory, which displays with acute intersections acute of less than 90° can be discriminated from mammoth ivory that has intersections at 115° or more.

Gemmologists are warned that any trade in both African and Asian ivory is prohibited under internationally agreed-to CITES regulations.

Identification of Other Commercial Ivories

As the gemmologist may be required to identify ivories, other than elephant ivory, the following guidelines are offered to facilitate the identification of ivory derived from the teeth of the sperm whale (*Physeter catodon*), and the tusks of the walrus (*Odobenus rosmarus*), the teeth and tusks of the hippopotamus (*Hippopotamus amphibious*) and the long, spiral tusk of narwhal (*Monodon monoceros*). The identifying features for these ivories can be found by careful examination of the cross-sectional surfaces of these tusks and teeth.

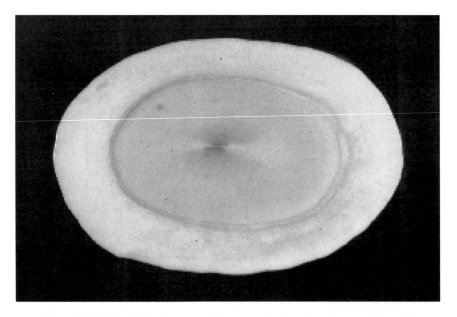

Figure 29.16   Identifying features on the cross-sectional surface of sperm whale tooth

When the cross-sectional surface of a conical, curved sperm whale tooth is examined, it will have an oval outline and will display a 'bull's-eye' pattern of concentrically deposited thin lamellae of dark yellow to brownish primary dentine that is surrounded by a layer of lighter coloured cementum. A stellate remnant the original pulp cavity of the tooth may be located centrally (Figure 29.16).

The cross-sectional surface of the large curved tusk of the walrus also has a flattened, somewhat irregular oval outline (Figure 29.17). The centre of each tusk uniquely is filled with a core of 'rice bubble'-like dark secondary osteodentine. This core is surrounded by lighter coloured, silky, radially arrayed fibrous primary dentine, and a thin outer layer of lighter coloured cementum. Fine longitudinal cracks (radial cracks on the cross-section) often penetrate from cementum into dentine along the length of the tusk.

Hippopotamus ivory is derived from the curved upper and lower tusks and peg-shaped incisor teeth of these animals (Figure 29.18). Unprocessed, the hippo's curved upper tusks are rounded, triangular in cross-section and have a deep groove that extends down the inner curved surface of the tusk. The hippo's large lower tusks are triangular in cross-section, while its incisors have a squarish to round cross-section. When cross-sectional surfaces of these teeth and tusks are examined, they display a core of finely banded, wavy layers of light yellowish primary

Figure 29.17   Identifying features on the cross-sectional surface
of a walrus tusk

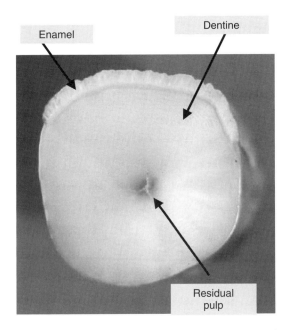

Figure 29.18   Identifying features on the cross-sectional
surface of the incisor tooth of a hippopotamus

dentine that surround central yellow-brown secondary dentine and/or a remnant pulp cavity. This dentine is coated on some surfaces by a thick coating of ribbed enamel, and on the remaining surfaces by thin outer coating of light yellow cementum. Prominent rhythmic growth layers in the primary dentine give the polished surface of hippopotamus ivory its identifying 'crinkly' or 'watered silk' appearance when examined in reflected light.

Narwhal ivory is very difficult to identify in hand specimen, unless the tusk is uncarved or the large central pulp cavity of this tusk is visible. In cross-section the tusk is rounded, but has peripheral indentations due to the spiral twisting that characterizes the narwhal tusk (Figure 29.19). The large pulp chamber of this tusk is surrounded by yellowish primary dentine that displays prominent concentric growth rings. A clearly defined whitish transition zone separates the dentine from surrounding brownish radially fibrous cementum. The cementum of this tusk frequently displays longitudinal cracks that follow depressed areas in the spiral pattern. When these structural features are not visible in one carving made from this rare ivory, other advanced laboratory tests, such as examining a thin peeling

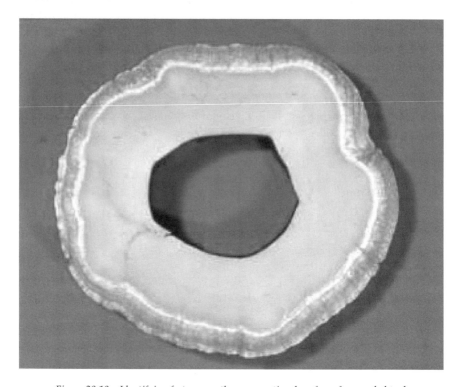

Figure 29.19 Identifying features on the cross-sectional surface of a narwhal tusk

under high magnification or Raman spectroscopy will be required to identify this ivory.

Value-Enhanced Ivory

Having identified the source of an ivory, the gemmologist should remain cautious, for some fossil elephant and walrus ivories are commercially value-enhanced by impregnating these somewhat porous ivories with colourless polymer or wax. This treatment, which enhances the colour, lustre and durability of the fossil ivory, can be detected by careful hand-lens examination that will reveal the presence of colourless ('bubble'-included) polymer, or opaque greyish wax, either coating the surface of or impregnating surface-reaching fractures and defects in value-enhanced fossil ivories.

Also, due to high content of organic material in ivory it is relatively easily stained. As a consequence both elephant and walrus ivory can be stained with either monochrome or polychrome pigments, or 'antiqued' to a brownish colour. Such coloured ivories are relatively common in the markets of India and South-east Asia, but also will be found on antique ivories from China. As ivory's natural colour is yellowish to pale brown-ish, any differently hued ivory should be suspected of being stained by man. However, determining the precise nature of the staining material will require use of advanced instrumentation such as IR or Raman microspec-troscopy that is not commonly available to the trade gemmologist.

Gemmologists are warned that while commercial trade in sperm whale ivory is subject to prohibitions under CITES regulations, native people are allowed to harvest and trade walrus ivory and narwhal ivory. Presently, there are no restrictions on the international trade in hippopotamus ivory.

Ivory Imitations

Ivory's popularity inevitably has led to this material being imitated by many look-alike materials that include glass, plastic, bone and vegetable ivory. Fortunately, identification of ivory imitations requires little more than careful 10× hand-lens examination.

Glass is identified by the combination of its coldness to touch, the observed presence of glassy conchoidal fractures, included gas 'bubbles' and mould marks that may completely or partially surround the imitation.

Polymer imitations of ivory are identified by their lightness to heft, unless deliberately filled, the presence of both positive or negative 'bubbles' trapped in fine detail on the external surface of any moulding, the discovery of mould marks around the periphery of a moulding, and the emission of acrid aroma from an inconspicuous surface when judiciously tested with a hot point.

Bone, particularly the solid cortical bone that covers the large long bones of animals such as the camel, the buffalo or the ox, and the bony antlers of the both deer and the moose, can be identified by this material's intrinsic porosity (Figure 29.20). Hand lens or low-power microscope examination of the surface of bone usually will readily reveal bone's identifying porosity, and thus enable its ready discrimination from ivory that displays no porosity. Rarely, the use of high magnification may be required to observe the holes and channels in bone, termed 'Haversian canals', through which bone receives its essential blood supply.

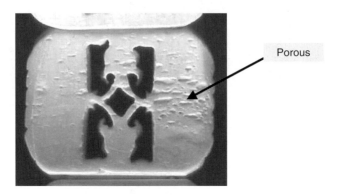

Porous

Figure 29.20   The identifying porosity of solid cortical bone

The whiteness of bone, as opposed to yellowness of ivory, is another visual clue that can be used to identify bone, unless the bone has been deliberately stained.

To decrease the visibility of bone porosities, some bone is either deliberately heavily stained, or else impregnated with suitably pigmented polymer or wax. These impregnators, usually, can be recognized by careful hand-lens observation.

Vegetable ivory, a long-used and effective imitation of elephant ivory, is derived from the large-size dried nuts of several species of palm tree. Once hardened, these nuts are readily carved, easily dyed, and their polished surfaces have a waxy lustre. For many years vegetable ivory has been used to produce miniature carvings, and some early buttons.

The corozo or tagua palm (*Phytelephas macrocarpa*), from Central America and northern South America, is the common commercial source of vegetable ivory (Figure 29.21). Vegetable ivory from this large ivory-coloured nut has a hardness of $2\frac{1}{2}$, an SG of 1.40–1.43, and give a 'spot' RI of 1.54.

Figure 29.21   Tagua palm nuts, the major commercial source of vegetable ivory

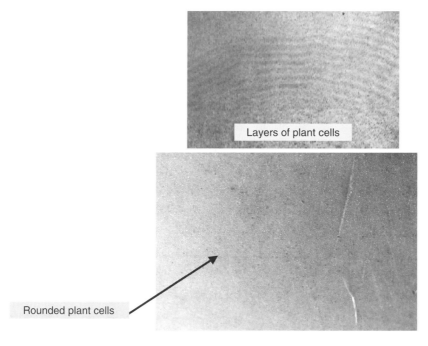

Layers of plant cells

Rounded plant cells

Figure 29.22   Hand-lens characteristics of vegetable ivory

Other sources of vegetable ivory include the hollow northern and central African doum or gingerbread palm (*Hyphaene thebaica*), the real fan palm of southern Africa (*Hyphanae petersania*) and the *Metroloxyn* or sago palm of the Pacific.

Examination of the polished surface tagua nut vegetable ivory, at moderate magnifications, may reveal the rounded outlines of the rhythmically deposited layers of plant cells that form this nut (Figure 29.22). If these plant cells are sectioned longitudinally, they will display a rounded cylindrical shape. If, however, the cells are cross-sectioned the cells will have a circular outline. Visually, the surface of vegetable ivory closely resembles that of the cut surface of an apple.

Tortoiseshell

Tortoiseshell is not actually a shell but a horny, organic substance that covers the skeleton of the hawksbill turtle (*Eretmochelys imbracata*). The dorsal surface or carapace of the hawksbill turtle is made up of 13 plates, or schutes, each of which is composed of keratin. Keratin is a fibrous protein similar to that forming the horns and hooves of cattle, the fingernails of humans, and the claws of animals and birds. These plates of tortoiseshell have rich, warm, yellow, translucent colour and are attractively marbled and spotted with spots and patches of contrasting reddish, chestnut-brown colour. The ventral surface, or plastron of the hawksbill turtle, is covered by 12 plates that have a uniformly pale translucent amber-yellow colour and show no mottling or streaks. This tortoiseshell is termed *blond tortoiseshell* or *yellow-belly*. It is the finest, most highly prized and valued form of tortoiseshell.

Common tortoiseshell has a translucent yellowish body colour that is attractively patterned by irregularly dispersed reddish brown patches that are formed from aggregates of individual rounded particles of the blackish brown biological pigment, melanin.

Today, tortoiseshell is little exploited, for like all sea turtles, the hawksbill turtle is now a protected animal that is on the edge of extinction. When the hawksbill turtle was declared an endangered species in 1975, international trading in post-1975 tortoiseshell was declared to be both illegal and ecologically unacceptable. In 1993, the World Wide Fund for Nature confirmed the listing of the hawksbill turtles listing on Appendix 1 of CITES (the Convention on International Trade in Endangered Species). By 1994 the USA and 115 other countries had banned the import and export of sea turtle products, yet the pressures on sea turtles are not abating.

Gemmologists must be aware that all trade in tortoiseshell is banned in Australia. This ban is a consequence of Australia ratifying CITES regulations that govern trade in endangered animals.

Identification

The identification of tortoisehell is simple, for hand-lens examination of tortoiseshell readily reveals the presence of aggregates of small brownish to blackish melanin pigment particles that form the irregular patches that pattern the yellowish keratinous matrix of the tortoiseshell (Figure 29.23). As confirmatory evidence of identification, a judiciously applied hot point test will yield the distinctive aroma of burnt hair from the surface of tortoiseshell.

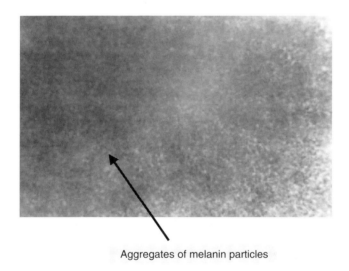

Aggregates of melanin particles

Figure 29.23   Hand-lens characteristics of tortoiseshell

Gemmological Properties

The gemmological properties of tortoiseshell are of little importance to its identification, but for completeness these are given below in *Table 28.6.*

Tortoiseshell is imitated commonly either by yellowish brown polymers or keratinous horn.

Polymer imitations of tortoise-shell are easy to detect for they are not included by aggregates of melanin pigment, when these materials are examined with the 10× hand lens. Instead, brownish swirls of colour, and included gas 'bubbles', readily identify the common polymer imitations of tortoiseshell.

Table 29.6
The gemmological properties of tortoiseshell

Chemical composition	A keratoprotein included by aggregated granules of the pigment melanin
Colour	Yellowish (body) colour patterned brownish red by aggregated masses of the black biological pigment melanin. *Blond tortoise-shell* lacks pigment particles.
Hardness	$2\frac{1}{2}$ (Mohs' scale)
Fracture	Smooth to finely fibrous
Sectility	Readily peeled and carved with a sharp blade.
Specific Gravity	1.30
Refractive Index (distant vision)	1.55
Diaphaneity	Translucent
Lustre	Greasy to resinous
Luminescence	Blue-white fluorescence under LWUV.
Thermal properties	Thermoplastic at about 100 °C. An applied hot point yields the distinctive smell of 'burnt hair'.
Characteristic inclusions	Small masses of brownish black melanin granules.

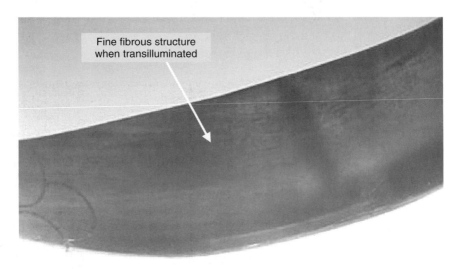

Fine fibrous structure
when transilluminated

Figure 29.24   The fine parallel fibrous structure that identifies bovine horn

The keratinous horns of ruminants, such as the buffalo and other bovines such as the bull, are yet another organic material that could be an effective imitation of tortoiseshell, particularly pigmentless blond tortoiseshell (Figure 29.24). However, this quite common keratinous gem material is readily identified, when transilluminated, by its structure that consists of a solid mass of fine parallel arrayed fused hairs. In addition, a judiciously applied hot point test will yield the distinctive aroma of burnt hair and confirm that the material under test is indeed keratinous.

Jet

Jet is an ancient gemstone that man has used both as a talisman and a jewel for over four thousand years. This opaque black biological gem material is found in several locations around the world. The principal commercial source of jet has long been the Whitby district of North Yorkshire (England). Other sources include Spain's Asturias Province, France's Aude Département, Württemburg in Germany and the Henry Mountains of Utah (USA). Of these deposits only the British and Spanish deposits have been worked commercially.

What is Jet?

Jet is an opaque black coalified fossilized drift wood from the conifer-like, 180 million years old, *Araucaria* or monkey puzzle tree that has been dried and fractured following death of the tree; thus causing some of its woody structure (Figure 29.25). This wood was subsequently transported downriver by floods, during which sand grains entered shrinkage cracks

Figure 29.25   Specimens of rough and processed jet

and fractures in the wood, before becoming waterlogged and sinking into the black mud of the seafloor.

Here accumulated sediments (now surrounding shales) crushed and flattened the wood before it was converted into jet under seawater by the temperature and pressure created by deep anaerobic (oxygen-free) burial. As the jetonisation process progressed, the woody tissues were progressively replaced by carbon that migrated from the outside inwards along the medullary rays of the original wood. In some wood the centre of the wood also became silicified first, forming *cored jet*.

Jet is so effectively fossilized (by carbon replacement) that a thin section can readily reveal precise structural details of the crushed original wood that was the source of this essentially Victorian organic gem material.

Jet is chemically related to brown coal, or lignite, as it has a chemical composition of carbon (75.2%), oxygen (12.5%), hydrogen (7.0%), sulphur (4.6%) and nitrogen (0.7%). However, jet differs from look-alike black anthracitic coal in that it is solid and tough. In contrast, anthracitic bituminous black coal is brittle due to its possession of two directions of distinct jointing.

In the past, gem-quality jet was mined from what is now the North York Moors National Park in North Yorkshire. This park is located near Whitby, a port on the north-east coast of England. At this locality, jet occurs sporadically as thin seams and nodules in a specific strata of rock, termed 'jet rock', that is formed from a series of black laminated shales of about 7–9 m thickness. Best jet occurs in random pockets in the upper 3 m of the 'jet rock'. Landmarks used by miners to locate jet-bearing rock included the fossil ammonite *Harpoceras exeratum*, which is unique to 'jet rock', and a limestone band known as Top Jet Dogger, which lies immediately above the jet rock and weathers into large discs known locally as 'mermaids' dining tables'. When freshly mined, Whitby jet has a decidedly sulphurous oily smell. Within 'jet rock', jet occurs as isolated masses that include 'plank' jet, which occurs as elongated masses in the plane of bedding of the shales, and spindle-shaped 'cored jet' that has an irregular silica core.

Once mined, the jet was sorted and graded, prior to its manufacture into jewellery, into 'hard jet', which is suitable for carving, grinding and polishing, and 'soft jet' that is so brittle that it tends to crack shortly after it has been worked or subjected to heat.

Gemmological Properties

Whitby jet displays characteristics of interest to the gemmologist, as illustrated in *Table 29.7*.

Table 29.7
The gemmological properties of jet

Chemical composition	Carbon with minor impurities.
Colour	Black
Hardness	4 (Mohs' scale)
Fracture	Sharply defined, lustrous conchoidal
Sectility	Readily carved, ground and polished, but brittle unless pared with a very sharp blade.
Streak	Dark brown
Specific Gravity	1.30–1.40 (mean 1.33)
Refractive Index (distant vision)	Very indistinct at ~1.66
Diaphaneity	Opaque, but translucent (reddish brown) in very thin sections.
Lustre	Velvety to waxy
Luminescence	Inert to UV irradiation
Thermal properties	An applied hot point yields the smell of burning coal (cf. the aroma from a steam locomotive).
Characteristic inclusions	None

Identification

Jet is a difficult material to identify by conventional gemmological testing. In fact, objects carved from jet are commonly identified by a process of elimination, based on the knowledge that if all possibilities are systematically eliminated then what remains following this process of elimination must be jet. However, if feasible, following this process of elimination, the identification of jets should be confirmed by conventional gem testing.

So, applying this process of elimination to the identification of jet:

1. Polymer imitations of jet are distinguished by their lightness to heft, inability to yield a streak, the presence of mould marks around the periphery of the moulding, the presence of gas 'bubbles' either included in the material or breaking its surface, by the material being softened by an applied hot point test and perhaps emitting an aroma of burnt plastic.
2. Black glass imitations of jet are discriminated by their intrinsic coolness compared to the comparative warmth of jet, and by observation of glassy conchoidal fractures, the presence of mould marks around the periphery of the moulding, the presence of included gas 'bubbles' if light can be transmitted through thin sections of the glass, the presence in very thin sections of a swirling distribution of colour, and inability of glass to yield a streak.

3. Obsidian, a rare but visually effective imitation of jet, is distinguished by its glassy conchoidal fracture, its SG of 2.3–2.5 v ~1.33 for jet, its single RI of 1.50 v ~1.66 for jet, the intrinsic coldness of obsidian compared to the comparative warmth of jet, and no streak.

4. Black onyx, a possible jet imitation, is distinguished by its SG of 2.6 v ~1.33 for jet, its RI of 1.53–1.54, with a form birefringence of 0.004, and no streak.

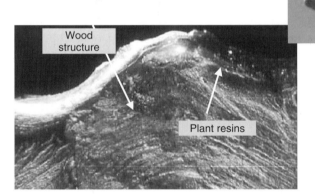

Figure 29.26   Identifying features of bog oak

5. Bog oak is an Irish imitation of jet that is based on modern wood that has been stained dark brown by tannins as a consequence of being buried in a peat bog (Figure 29.26). This dark-coloured wood is commonly carved into objects that are associated with distinctly Irish themes. Bog oak can be identified readily in hand specimen, as this jet imitation has a dark brown colour rather than the black colour of jet, its polished surfaces have a poor lustre and may display evidence of the grain of the wood and occasional remnants of plant resin between the fibres of the wood. Bog oak yields a very pale brown streak.

6. Anthracite coal, an uncommon imitation of jet from the southern states of the USA, is identified by the brilliant lustre of its polished

surfaces and the presence of a characteristic presence of an uneven fracture surface due to anthracite's possession of two distinct sets of joints. Antracite also will readily mark a sheet of white paper with a black streak.

7. Moulded vulcanite and moulded gutta-percha are two quite challenging imitations of Victorian jet jewellery.

Vulcanite is a black rubbery material that was discovered by Robert Goodyear, the inventor of the rubber car tyre. Goodyear manufactured vulcanite by heating sulphur-impregnated natural rubber under considerable pressure. When moulded and subsequently polished, this tough, lightweight, thermosetting material had a colour and lustre similar to that of jet. These features made vulcanite ideal for the production of imitation jet cameos and simple items of jewellery that did not contain undercuts.

In hand specimen vulcanite can be identified, and distinguished from jet, by its light heft, the presence of residual mould marks around the periphery of the object, its lack of wear as vulcanite is tougher than jet, its lack of a conchoidal fracture, and importantly, the change in colour from black to khaki to yellowish brown that occurs when vulcanite is exposed to sunlight for over time (Figure 29.27). This colour change

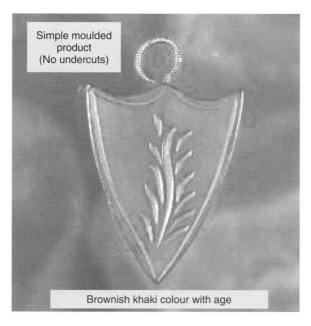

Figure 29.27   Vulcanite displaying its degraded acidic khaki-coloured surface

*Figure 29.28   Cameos moulded from gutta-percha. Note the loss of the black paint that exposes
the granular moulded surface of the underlying gutta-percha*

occurs as the sulphur content of vulcanite breaks down liberating acid
that can be detected by litmus paper. Streak is of little use for detecting
vulcanite from jet, for both yield a brown streak.

In contrast, gutta-percha is a thermoplastic material that is derived
from the coagulated sap from macerated leaves of the mazer wood
(gutta-percha) tree of South-east Asia (Figure 29.28). Gutta-percha, or
GP, is so soft that it is easily scratched by the fingernail. When
unfilled, GP is a greyish to brownish natural unvulcanized rubber that
must first be softened and moulded at ~60 °C before having its external surfaces painted black to imitate the colour of jet. The soft, rather
granular surface of unpainted, moulded gutta-percha does not take
a good polish. These features, which are readily observed in hand
specimen, will readily distinguish gutta-percha from both vulcanite
and jet.

If, following this process of elimination, the gemmologist is left with a
lustrous, black, opaque carving that displays conchoidal fracture – then it
is most likely to have been carved from jet. With care, the gemmologist
may confirm the identity of the material by a judiciously applied hot
point test. Jet will not melt, but will burn and emit the characteristic
odour of burning coal.

Shell

Marine shells are the source of a range of jewellery and decorative items that the gemmologist can be requested to identify. Shell-based materials of particular interest to gemmologists include nacre (MOP), the fossil nacre that is marketed as ammolite, shell cat's-eyes and shell cameos.

Mother-of-Pearl

Shell is the hard, predominantly calcium carbonate-containing external protective coating that is secreted by the soft tissue mantle of most molluscs. When the inner lining of a shell is formed from regularly secreted parallel layers of platelets of aragonite, that have been deposited as single crystals in a 'honeycomb' organic matrix, this iridescent shell is known as nacre, mother-of-pearl or MOP.

Nacre derives its iridescence from the light-diffracting properties of its parallel layers of single platelet-like crystals of aragonite as they outcrop on the surface of the nacre. This iridescence can be either subtle and directional, as displayed by MOP, or strong and visible in all directions of observation, as displayed by the nacre of the abalone (Plate 33). For detailed explanations of how the iridescence of shell is generated, interested gemmologists should consult papers published by GIA researchers Liu, Shigley and Hurwit (1999), Liu, Hurwit and Shigley (2002), and Liu, Hurwit and Tian (2003).

Commercial sources of MOP include the nacre of bivalves such as: *Pinctada maxima* (the gold- or silver-lipped pearl oyster), *P. margaritifera* (the black-lipped pearl oyster) and of course the gastropod trochus shell (*Tectus niloticus*). The abalone (*Haliotis iris, rufescens*, etc.) are sources of nacre that display strong iridescence.

With particular respect to the nacre of *H. iris*, the paua shell, please remember that most of today's commercial paua shell has been dyed to further enhance its colour. So, always check out all paua shell for evidence of dye between its layers of nacre.

Ammolite

Ammolite™ is the trade name for a Canadian biological gem material that was first described by Dowling in 1913. Ammolite consists of the crushed and little altered remains of 71 million year old fossil ammonites, *Placenticeras meeki* and *P. intecalare* (Figure 29.29). These fossil ammonites are mined from randomly scattered deposits in the Bearpaw Formation near Letherbridge southern Alberta, Canada.

Fractures in the nacre
infilled with sediments

Figure 29.29   Ammolite, fossilised ammonite nacre

Here the fossilized ammonites occur at a depth of about 15 m in iron-stone concretions that are associated with beds of shale. In Canada, gem-quality Ammolite is found in the area between Alberta's Red Deer River and the Milk River.

The highly iridescent nacre, found in these fossil ammonites, has been little altered by burial. Their mineral component consisting of 96% aragonite, with traces of strontium, titanium and copper and iron. The organic content of the fossil nacre also has remained little altered; for it contains about 5% conchin and less than 1% water. Ammolite, as the material is termed commercially, has a hardness of 4, an SG of 2.80–3.05, and RI of 1.52–1.67.

Red, green and orange iridescent colours are commonly displayed by ammolite™, while blues and violets are quite rarely seen in this organic gem. Ammolite™ is often cut with its shale retained backing to give the material some durability. It also may be fabricated into triplets. Lower grades of ammolite may be stabilized by impregnation with plastic.

Ammolite is readily identified in hand specimen, by this fossil nacre's shell-like iridescence, and the presence of intersecting linear crush fractures that are often infilled with fine-grained greyish sedimentary shale or reddish brown ironstone.

Operculum

Operculum, or the shell cat's-eye, consists of the small 'door', or operculum, that is used by some univalve molluscs to protect their body parts against predation (Figure 29.30). Opercula used for jewellery purposes are derived from a snail-like sea univalve mollusc, the *Turbo petholatus*. The turbo's operculum consists of a circular low-domed calcitic mass that has a convex

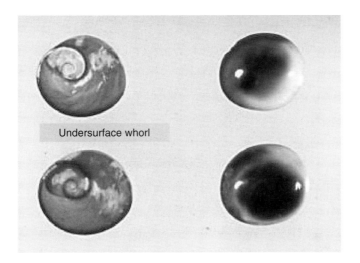

Undersurface whorl

Figure 29.30   Identifying features of the shell cat's-eye

green, yellow and brown patterned porcellanous external surface, and a flattish white base that has an incised spiral pattern. The shell cat's-eye is not chatoyant; it is simply decorated by an external eye-like pattern. The polished surface of the shell cat's-eye displays a reasonably good lustre.

Shell Cameos

Shell cameos are intricately carved items of jewellery and decoration that have been crafted in Italy since the sixteenth century. Major shells from which shell cameos are carved include the bullmouth or red helmet (*Cypraecassis rufa*), the black helmet (*C. madagascaris*) and the pink conch (*Strombus gigas*). As each of these shells is formed from layers of fibrous aragonite crystals, of contrasting colour, and differing fibre orientation, the external surface of these shells can be carved back to yield a white subject against a contrasting coloured background. For example: the red helmet shell yields a white subject on red to orange-red background, the black helmet yields a white subject on dark brown to black background, while the pink conch yields a white subject on pink background.

According to the traditional shell carvers of Torre del Greco, shells used for carving shell cameos include:

- *C. madagascarensis*, the '*sardonica*', '*sardonyx shell*', the black helmet shell of the Atlantic and Indian Oceans, and sea to the north of the Antilles.
- *C. rufa*, the '*corniola*' or red or bullmouth helmet shell from Madagascar and Zanzibar.

- *C. cornuta*, the *'orange shell'* from the Indian Ocean.
- *S. gigas*, the pink conch from the Caribbean.
- *Cypraea tigris* and *Porcellina tigrina*, cowrie shells from the South China Sea.

Identification

As structural studies have indicated that the cameo shells were formed from alternately coloured layers of fibrous aragonite crystals, the constituent fibrous crystals of which are oriented at right angles to each other in helmet shells, and parallel to each other in the pink conch, the following observations can be used to determine the source shell of a shell cameo.

1. Hand-lens examination of the two-colour carved surface of a helmet shell cameo should reveal that each coloured layer appears

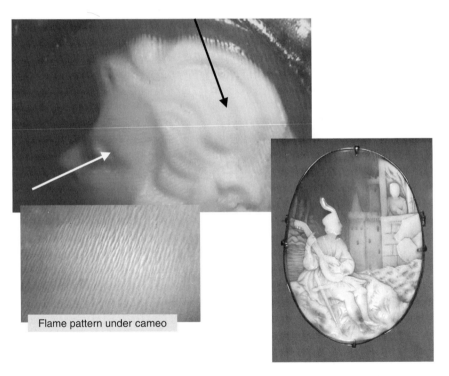

Flame pattern under cameo

Figure 29.31   Identifying features of a helmet shell cameo. Arrows show striations run at 90° to each other

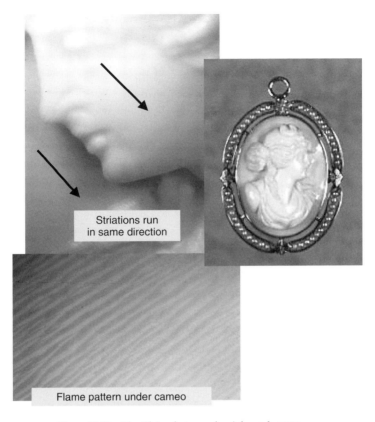

Striations run
in same direction

Flame pattern under cameo

Figure 29.32   Identifying features of a pink conch cameo

to be longitudinally striated – with the direction of striation in alternating coloured layers being oriented at right angles to each other (Figuer 29.31).

2. Hand-lens examination, of the two-colour carved surface of a pink conch shell cameo, should reveal that each coloured layer appears to be longitudinally striated, with the direction of striation in alternating coloured layers being parallel to each other (Figuer 29.32).

3. An additional indicator, for the source shell of a particular shell cameo, will be obtained by examining the non-nacreous concave undersurface of the cameo that represents concave inner surface of the shell, for each cameo shell has its own distinctively coloured 'flame' pattern. That is, red to orange flames characterize the red helmet cameo, brownish to black flames characterize the black helmet cameo and pink 'flames' will be found on the pink conch cameo.

References and Further Reading

Biological Gem Materials

Campbell Pedersen, M. (2002) *Gem and Ornamental Materials of Organic Origin*. Elsevier Butterworths Heinemann: Oxford.

Amber

Brown, G. and Snow, J. (1988) Is it amber? *Australian Gemmologist*. 16, 409–16.
Crowningshield, R. (1993) Gem Trade Lab Notes: Treated ambers. *Gems & Gemology*. 29, 122–23.
Fraquet, H. (1987) *Amber*. Butterworths: London.
Hutchins, P. and Brown, G. (1996) A new amber treatment. *Australian Gemmologist*. 19, 302.
Platt, G. (2004) http://www.gplatt.demon.co.uk.
Ross, A. (1998) *Amber: The Natural Time Capsule*. The Natural history Museum: London.

Coral

Brown, G. (1976) Two new precious corals from Hawaii. *Australian Gemmologist*. 12, 371–77.
Brown, G. (1977) Black coral, true or false? *Australian Gemmologist*. 13, 35–41.
Brown, G. (1979) Gold corals – Some thoughts on their discrimination. *Gems & Gemology*. 16, 240–44.
Brown, G. (1988) Bamboo coral: A new coral from Hawaii. *Australian Gemmologist*. 16, 449–54.
Gimenez, R.G. *et al.* (1996) Australian coelenterate corals. *Revue de Gemmologie*. No. 126, 99. 13–16.
Grigg, R.W. and Brown, G. (1991) Tasmanian gem corals. *Australian Gemmologist*. 17, 399–404.
Liverino, B. (1989) *Red Coral: Jewel of the Sea*. Edizioni Analisi: Bologna.

Ivory

Brown, G. and Moule, A.J. (1977) The structural characteristics of elephant ivory. *Australian Gemmologist*. 13, 13–17.
Brown, G. and Moule, A.J. (1997) The structural characteristics of various ivories. *Australian Gemmologist*. 13, 47–60.
Edwards, H.G.M. and Farwell, D.W. (1995) *Spectrochimica Acta*. Part A. 51, 2073.

Edwards, H.G.M., Farwell, D.W., Holder, J.M. and Lawson, E.E. (1997) *Journal of Molecular Structure*. 435, 49–58.

Edwards, H.G.M., Farwell, D.W., Holder, J.M. and Lawson, E.E. (1997) *Spectrochimica Acta*. Part A. 53, 2403–409.

Espinosa, E.O. and Mann, M.-J. (1993) The history and significance of Schreger pattern in proboscidean ivory characterization. *Journal of the American Institute of Conservation Online*. 12(3), 241–48.

Espinosa, E.O. and Mann, M.-J. (1999) *Identification Guide for Ivory and Ivory Substitutes*. CITES & US Fish and Wildlife Service.

Owen, R. (1856) The ivory and teeth of commerce. *Journal of the Society of Arts*. 213, 65–73.

Penniman, T.K. (1984) Pictures of ivory and other animal teeth, bone and antler. Occasional paper of Technology, Pitt Rivers Museum, University of Oxford.

Rolandi, V. (1999) Characterisation of recent and fossil ivory. *Australian Gemmologist*. 20(7), 266–76.

Schabilion, S. (1983) *All in a Nutshell*. Keystone Comedy: Flora, Missippi.

Jet

Brown, G. (1993) Vulcanite or gutta-percha: The debate continues. *Canadian Gemmologist*. 14, 40–43.

Marshall, C.E. (1974) The jet age. *Australian Gemmologist*. 12, 42–55.

Muller, H. (1987) *Jet*. Butterworths: London.

Shell

Barnson, D. (2000) *Ammolite 2*. Barnson: Selkirk.

Brown, G. (1986) The gemmology of the shell cameo. *Australian Gemmologist*. 16, 153–161.

Brown, G. (1988) Paua shell: New Zealand's distinctive organic gem. *Australian Gemmologist*. 16(10), 367–70.

Liu, Y., Shigley, J. and Hurwit, K. (1999) Iridescent color of the shell of the mollusc Pinctada margaritifera caused by diffraction. Optics Express. 4(5), 177–182, http://epubs.osal.org/oearchive/source/8639.htm.

Liu, Y., Hurwit, K. and Shigley, J. (2002) *Journal of Gemmology*. 28(1), 1–5.

Liu, Y., Hurwit, K. and Tian, L. (2003) Relationship between groove density of the grating structure and the strength of iridescence pf mollusc shells. *Australian Gemmologist*. 21(10), 495–07.

30

The Fashioning of Gemstones

Geoffrey Dominy

Although the origins of gem cutting are unknown, it is thought that as early as 4800 BC people used reeds such as bamboo, or conch-like shells, to fashion ancient seals. Later, by crudely rubbing stones against each other, they learned not only to fashion tools and weapons but also to transform brightly coloured pebbles into gems that could be worn for personal adornment. As the years progressed, they developed techniques that led to the arrival of the cabochon cut and intricate engravings using both transparent and opaque materials. Later, during the tenth and eleventh centuries, Europeans engaged in the art of faceting gemstones which eventually led to the establishment of the first gem cutters guild in Paris in 1290. A century later, diamond cutting and polishing were established in Nuremburg, Germany.

During the fifteenth century, great advances were made in diamond cutting by a Belgian named Louis de Berquem, who announced that he had discovered a method of cutting diamonds. This made possible the eventual development of the round brilliant cut, which was originally credited to a Venetian cutter named Vicenzio Peruzzi during the latter part of the seventeenth century. Over the years, this underwent a gradual transformation until 1914 when Marcel Tolkowsky published his treatise on the use of certain angles and proportions to maximize brilliance.

Without question, the greatest advances, in both diamond and coloured gemstone cutting, have occurred in the latter part of the twentieth century. Modern equipment has permitted the development of new techniques which have, in turn, led to the advent of new and exciting cutting styles. Today, owing to the establishment of numerous organizations and trade journals, there is a tremendous interest in gemstone cutting, not only from a commercial aspect but also from an amateur standpoint.

Styles of Cutting

Cabochon Cut

The earliest form of cutting simply gave a curved surface to the stone, and was known as the cabochon cut. There are four styles used today (*Figure 30.1*): the simple cabochon which has a convex (dome-shaped)

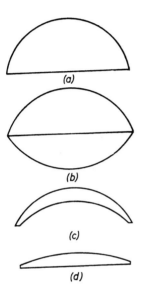

(a)

(b)

(c)

(d)

Figure 30.1   The cabochon cuts: (a) simple cabochon, (b) double cabochon, (c) hollow cabochon, (d) tallow-topped cabochon

upper surface and flat base; the double cabochon in which the base is also convex, but less steep than the top; the hollow cabochon in which the underside is hollowed out to form a concave depression making a concavo-convex form; a low-domed simple cabochon known as a tallow top owing to its resemblance to a drop of candle grease.

   In stones cut *en cabochon*, the outline of the setting edge may be circular, oval, polygonal or of a fancy shape; a common fancy outline is a heart shape. The cut is used for those stones which are translucent or opaque, such as turquoise and jade, and for stones showing optical effects such as asterism and chatoyancy, which must be cut so as to bring out to the full the star or cat's-eye effects. The play of colour in opal and the adularescence of moonstone are best shown by this style of cutting. Deep colour almandines are often cut as hollow cabochons in order that the depth of colour may be lightened, and in some cases the hollowed back is foiled to make the stone appear more brilliant. Such hollow cabochons of almandine garnet are called carbuncles, a name used since ancient times for cabochon-cut red stones, but now usually restricted to the almandine garnet.

Very often, heavily included commercial-grade sapphires are cut in the double cabochon style, and such stones are sometimes set with the flatter side uppermost as this gives a better colour and lustre. Star corundum and chrysoberyl cat's-eyes are set with the steeper dome to the front while the curved back of the stone is usually left unpolished. *Figure 30.2* shows a star sapphire cut in the cabochon style.

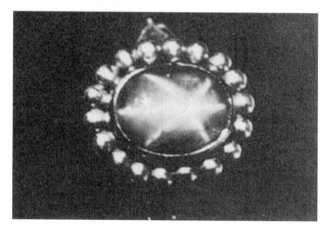

Figure 30.2   A large star sapphire cut in the cabochon style

Rose Cut

One of the first styles of cutting to employ flat facets was the rose cut or rosette, a name derived from a fanciful resemblance in the arrangement of the facets to the petals of an opening rosebud. The origin of the style is not known with any surety but it was developed sometime before the sixteenth century. The rose cut consists of an upper portion only, the underside being just a single large flat base. The upper portion takes the form of a pyramid with three-sided facets meeting at the apex at a more or less steep angle. The rose cut is mostly used for diamonds owing to its economy in material, allowing small cleavage fragments and macles to be used up. In Victorian jewellery the then popular bright blood-red pyrope garnets from Bohemia were also cut in this style.

The most common form of rose cut is the round rose. In its most usual form the facets are arranged in multiples of six and in two groups, the upper six facets constituting the crown or star while the lower facets are the cross facets or, as they are often known, the teeth or dentelle (*Figure 30.3*). (The facet names of a later development towards the brilliant cut stone are shown in *Figure 30.4*.) Small diamond chips and flattish crystals may have only three near-triangular facets.

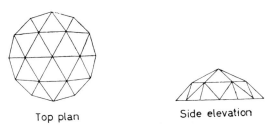

Top plan   Side elevation

Figure 30.3   The round rose style of cutting

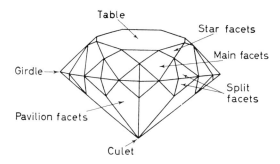

Figure 30.4   The facet names of the brilliant-cut type stone

In the Dutch or crowned rose the height of the pyramid from the base is usually equal to half the diameter of the stone, and the distance from the base to the crown is three-fifths of the total height; the diameter of the crown is three-quarters that of the base, which may be round or oval. The Brabant or Antwerp rose is similar to the Dutch rose except that the pyramid is more flattened. The distribution of the facets is the same in each of these three types. The cross rose and the rose recoupée are modifications of the facets and are shown in *Figures 30.5* and *30.6*.

The rosette has a crown of the normal six triangular facets, but the dentelle of the round rose is replaced by six rectangular facets; it is a simpler form. The double rose may be described as two roses joined together base to base, but in certain cases the heights of each may not be equal, and if the top half of the double rose is excessively elongated a drop-shaped form, termed a briolette (*Figure 30.7*), is produced, but such a stone can be faceted with rectangular facets in place of the triangular facets.

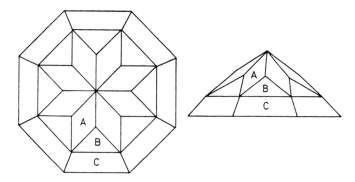

Figure 30.5   The cross rose style of cutting

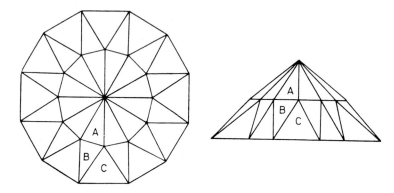

Figure 30.6   The rose recoupée style of cutting

Figure 30.7   The briolette

Figure 30.8   The table cut

Table Cut

What was probably the first advance from the mere polishing of the dia-mond octahedron – termed 'point stones' – was the table cut (*Figure 30.8*). This cut was produced by truncating one of the corners of a diamond octahedron until a face equal to about half the width of the stone was secured. The opposite corner was then removed in the same way leaving a face about a quarter the width of the stone. The whole of the faces were then polished. The table cut, a style extensively used by the ancient Indian cutters, may have many variations both in the depth of the stone and in the outline of the setting edge. Such table-cut diamonds are usually encountered only in old Indian jewellery, but a similar cut was used by the Nuremberg diamond polishers in 1373. One modification of this cut is made very thin and acts as a window or coverglass for small miniatures set in rings. Such stones are usually known as portrait stones.

Trap Cut

The table cut may well be said to have been the forerunner of the trap cut which is so popularly used at the present time for important diamonds and coloured gems. This cut, sometimes called the step cut, has the large table facet surrounded by a series of strip-like rectangular facets which increase in steepness towards the setting edge (the girdle); the lower half of the stone (the pavilion) has similarly arranged rectangular facets decreasing in steepness towards the basal facet at the point (*Figure 30.9*).

The outline of such stone can take many forms. They may be square, rectangular, triangular, kite-shaped, keystone-shaped and lozenge-shaped, and other polygonal forms. The rectangular form with the corners truncated, producing an eight-sided outline, is often called the emerald cut on account of its extensive use in cutting emerald (*Figure 30.10*). Small trap-

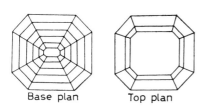

Base plan   Top plan

Figure 30.9   The trap or step cut

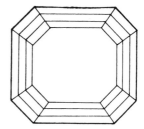

Figure 30.10   The emerald cut

cut stones of a long rectangular shape are known as baguettes and are extensively used today as accent stones especially in pieces of designer jewellery. The term *calibré* is applied to small trap-cut stones, usually of square outline, which are cut to special sizes to fit channelled settings; in fact they are tailored to fit, and that is literally the meaning of the name. Some of these small calibré stones, although trap cut, do sometimes deviate from the true definition in that the outline may be curved, forcing curvature on some of the facets, but these facets are still strip-like. A common modification of the trap cut is the cross cut, or scissors cut as it is sometimes called. Here the side facets consist of four triangular facets corresponding to each of the rectangular facets of the regular trap cut (*Figure 30.11*). The double scissors cut as a double set of these triangular facets on each side facet. A square trap-cut stone is called a *carré*.

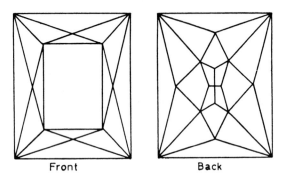

| Front | Back |

Figure 30.11   The scissors cut or cross cut

Brilliant Cut for Diamonds

The most common cut for diamond is the brilliant cut. When correctly cut it not only produces the best balance of fire and brilliancy, but also enables the best use to be made of the octahedral form of the diamond crystals. The style has become so universally used for diamond that the name 'brilliant' is now synonymous for diamond.

The standard brilliant cut (*Figure 30.12*) consists of 58 facets, 33 on the top or crown of the stone, and 25 on the base or pavilion. The crown has a large eight-sided central facet called the table which is surrounded by 8 triangular star facets. These in turn are surrounded by 8 quadrilateral facets called bezels, of which 4 are called quoins and the other 4 templets; the significance of these latter names will be more apparent when the methods of cutting diamonds are discussed. Surrounding the bezels and meeting the edge of the girdle, or setting edge of the stone, are 16 triangular facets which are known as upper girdle facets.

The back portion of the stone, or pavilion as it is called, has 8 long five-sided facets called the pavilions which are symmetrical with the 8 bezels on the crown of the stone. Corresponding to those on the front of the stone there are 16 lower girdle facets which join up with the pavilions. At the point of the stone where the pavilions meet at the bottom there is

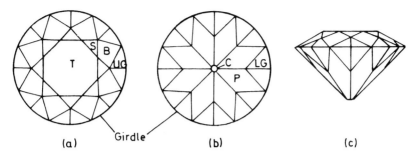

Figure 30.12   The standard brilliant cut consists of (a) the crown of the stone, comprising an octagonal table facet T, surrounded by 8 triangular star facets S, 8 quadrilateral bezel or kite facets B, and 16 triangular upper girdle facets UG, making 33 facets in all, and (b) the base or pavilion of the stone, consisting of a very small octagonal culet C, surrounded by 8 five-sided pavilion facets P, and 16 triangular lower girdle facets LG, making 25 pavilion facets and a total of 58 facets for the complete brilliant. It should be noted that many of these facets have alternative names, but those given represent modern practice

usually ground on a small facet parallel to the table facet of the crown. This culet as it is called, put on to prevent splintering, is sometimes omitted in small stones and is microscopic in most modern brilliants.

In order to ensure maximum brilliance in a diamond (which has a critical angle of 24° 26'), the back facets need to be at an angle of approximately 41° to the plane of the girdle, and the angle between the side facets of the crown and the plane of the girdle needs to be between 35° and 37°. As the angle of the octahedron is 54° 44' a considerable wastage of material – some 50–60% – occurs in cutting such a stone from a diamond crystal; even when the crystals are sawn the loss is great (*Figure 30.13*).

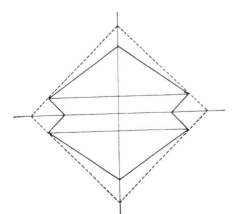

Figure 30.13   Even with a sawn diamond crystal the loss of material when cut into the brilliant style is great

Modifications of the Brilliant Cut

There are many modifications of the brilliant cut, both in outline and in the number of facets. Marquise (*Figure 30.14*) is the name used when the outline is boat-shaped, although navette is used in a similar connection. A pear-shaped outline is called a pendeloque (*Figure 30.15*), and an outline which tends to be squarish with rounded corners is termed 'cushion-shaped' (*Figure 30.16*). Many of the old Brazilian diamonds were cut in

Figure 30.14   The marquise or navette

Figure 30.15   The pendeloque

Figure 30.16   The cushion-shaped brilliant

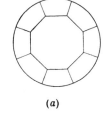

(*a*)   (*b*)

Figure 30.17   (a) The eight cut, (b) the Swiss cut

this fashion to retain weight even though it resulted in a reduction in overall brilliance. For small diamonds the single cut or eight cut is often used, and consists of 8 facets surrounding the table and 8 pavilion facets below the girdle (*Figure 30.17a*). The Swiss cut has 17 facets on the crown (including the table) and a further 17 facets on the pavilion (if there is a culet) for a total of 34 facets (*Figure 30.17b*). The zircon cut (used extensively for zircon) is characterized by a second set of facets surrounding the culet (*Figure 30.18*).

For large stones a greater number of facets than the 58 of a standard brilliant may be put on, but they will still conform to the general symmetry. More elaborate modifications of the brilliant cut are the king cut which has a twelve-sided table with 48 surrounding facets to form the crown of the stone, and 37 facets including the culet on the pavilion, for a total of 86 facets in all (*Figure 30.19*). The magna cut has a ten-sided table facet surrounded by 60 facets on the crown and 41 facets on the pavilion, making 102 facets in all (*Figure 30.20*).

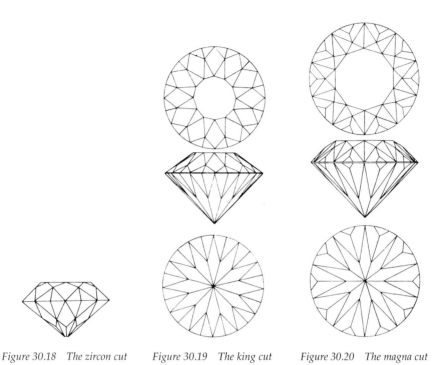

Figure 30.18   The zircon cut   *Figure 30.19   The king cut*   *Figure 30.20   The magna cut*

New Diamond Cuts

During the 24th World Diamond Congress in Singapore (August 1989), Nicholas Oppenheimer, chairman of the Central Selling Organization, presented five new diamond cuts (Zinnia, Sunflower, Dahlia, Marigold and Fire Rose) (*Figure 30.21*). Created by Gabi Tolkowsky, after preliminary studies by the CSO, these new designs were developed to better utilise rough that was either irregularly shaped, heavily included, small or strongly coloured. Adaptable to sizes ranging from mêlée to several carats, the new cuts not only improved yield but also, in many cases, enhanced either brilliance, clarity or colour. Compared with the more conventional cuts, the new designs represent a radical departure in terms of their angles and proportions (*Table 30.1*). As can be seen in the table, variants can still exist within each cut depending on the shape of the rough.

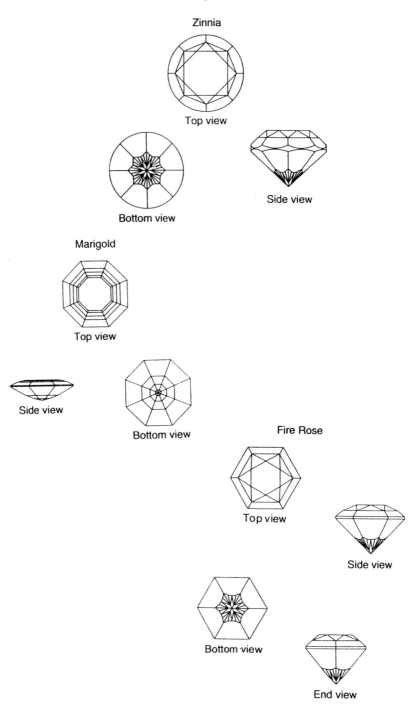

Figure 30.21   The Tolkowsky Flower Cuts (courtesy of De Beers)

Sunflower

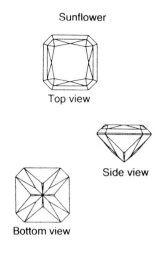

Top view

Side view

Bottom view

Dahlia

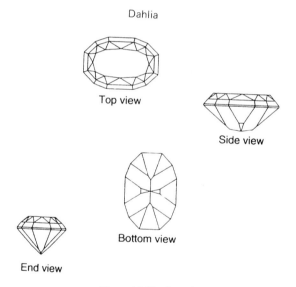

Top view

Side view

End view

Bottom view

Figure 30.21   (cont.)

In addition to these new cuts, several other cuts have been developed over the last 25 years. The Radiant Cut (designed and patented by Henry Grossbard) consists of 70 facets and combines a step-cut crown with a brilliant-cut pavilion for added brilliance. The Trillion Cut (designed by Leon Finker and patented by L F Industries) is triangular in shape and consists of 50 facets. The Trilliant Radiant Cut resembles the Trillion but is usually cut

Table 30.1
Proportion comparison chart for new diamond cuts

Shape	Table (%)	Crown height (%)	Girdle thickness	Pavilion depth (%)
Round Brilliant	57.5	14.6	Medium	43.1
Zinnia	52	16	Thick to very thick	46
Sunflower	53–58	17.5–24	Thin to thick	42.5–51.5
Dahlia	56	15	Medium to thick	49
Marigold	51	12.5	Very thick	35
Fire Rose	47–62	15–20	Thin	45–51

Based on an analysis conducted by GIA personnel: other specimens may vary.

with more rounded sides for additional weight retention. The Quadrillion Cut (patented and trademarked by Ambar Diamonds) is square-shaped and consists of 49 facets. The Princess Cut is similar in appearance to the Quadrillion but is usually cut with steeper crowns, thicker girdles or deeper pavilions to provide greater yield from the rough.

Perfection in cutting, which is sometimes called the 'make' of the stone, is much more important in the case of diamonds than in other stones, for faults and imperfections adversely affect the value of the diamond. Such deficiencies can include (1) unsymmetrical outlines or large culets, (2) spread tables, (3) overthick or irregular girdles, (4) naturals or extra facets, (5) polishing marks, and (6) bearding, which is often caused during the bruting process and resembles a series of fine lines extending in from the girdle.

Other Cuts

In addition to the cuts already discussed, there are a multitude of other cuts used for coloured gemstones. Today, it would seem that there is no limit to the variety of cuts available. Each new issue of a trade journal brings new and exciting cutting styles that are sometimes merely variations on an existing theme. One of the newest cuts to emerge in the marketplace is the fantasy cut, which is used extensively for amethyst, citrine and blue topaz. Characterised by its irregular form, it is favoured particularly by designers who wish to create unique one-of-a-kind pieces of jewellery. Other favoured cuts include the mixed cut, which consists of a brilliant-cut crown and a step-cut pavilion; the Barion cut (developed by Basil Watermeyer), which was designed to improve the brilliance of square and rectangular stones and consists of 62 facets; the Portuguese cut, which can have up to 177 facets; and the cushion cut, which is a cross between an oval and a rectangle.

Intaglios and Cameos

Gemstones are also carved. Heavily included, translucent and opaque gemstones may be carved in the shape of leaves, flowers or stylized designs. Intaglios (*Figure 30.22*) are incised carvings used mainly for seals, while cameos (*Figures 30.23* and *30.24*) have the carving raised

Figure 30.22   Intaglio carved in an amethyst

Figure 30.23   Shell cameo

Figure 30.24   A fine stone cameo (white on a greyish brown background), being a carved copy by R. Hahn of Idar-Oberstein of the Tazza Farnese (actual size about 200 mm)

above the surface. In both cameos and intaglios a layered material having two distinct colours is often used, in particular banded agates and shells of the giant conch and the helmet shell.

Mosaics

The fitting together of small pieces of coloured ornamental stones to form a pattern or picture is called a mosaic or inlay and may be either of two types. The first of these is the Florentine mosaic, which is sometimes known under the names *intarsia* or *pietra dura* (*Figure 30.25*). The pattern is made up of suitably shaped pieces of coloured stone – marbles, coral, malachite, opal, lapis lazuli and turquoise are often used – cemented into a recess cut into, usually, a black marble slab which forms the background. Parquetry is somewhat similar except that the pieces of stone are cut in geometrical shapes and then set in a metal mount. Some Scotch agate brooches are built on this style. The second type, the Roman or Byzantine mosaic (*Figures 30.26* and *30.27*), is made not of natural minerals but of small pieces of coloured glass rods of uniform length cemented in an upright position to a recess in, usually, a glass frame. The patterns depicted are often those of ancient ruins.

Figure 30.25   A Florentine (intarsia or pietra dura) mosaic

Figure 30.26   Roman mosaics

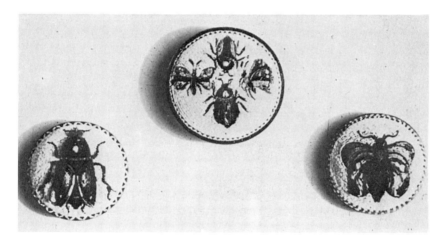

Figure 30.27   Roman mosaic

Gemstone Cutting

Cutting Considerations

Before any cutting can commence, a thorough examination of the rough must take place, since it is the decisions made during this inspection that will ultimately determine the value of the finished stone. Although the objectives of the cutter will vary according to the gemstone being cut and the order book, there are some fundamental points that need to be considered.

Weight Retention

The most important factor is the amount of weight that will be lost during the cutting process. If this exceeds the acceptable level (50–60% in diamonds and up to 75% in coloured gemstones), this will result in either (1) a per carat price which is too high, or (2) a reduction in profits in order to remain competitive.

Factors that influence the amount of weight loss include the following:

Shape

The shape that the cutter chooses (which is largely dependent on the original outline of the rough) will ultimately dictate how much weight is lost or recovered. When all other factors are considered, the shape chosen must result in the greatest amount of weight retention in order to maximize profit potential.

Calibration

Since manufacturers tend to produce mounts with settings that are designed to accommodate calibrated stones (i.e. 5 mm × 3 mm, 6 mm × 4 mm, 8 mm × 6 mm), most cutters will normally choose to sacrifice weight in order to produce finished goods that will fit into these ready-made settings.

Rather than penalize the cutter, it is generally accepted that calibrated stones will sell for a higher price per carat to compensate for the additional weight loss.

Brilliance

In order to achieve maximum brilliance, a cutter must adhere to certain angles (*Table 30.2*) to ensure that the majority of incoming rays (entering through the crown) strike the pavilion facets at an angle greater than the critical angle. The critical angle is given by the following formula:

$$\text{sine (critical angle)} = \frac{\text{RI surrounding medium}}{\text{RI gemstone}}$$

As the surrounding medium is normally air with an RI of 1,

$$\text{sine (critical angle)} = \frac{1}{\text{RI gemstone}}$$

Table 30.2
Gem facet angles (degrees)

Gemstone	Bezel	Star	Upper girdle	Pavilion main	Lower girdle
Andalusite	43	28	47–49	39	41
Apatite	43	28	47–49	39	41
Beryl	42	27	46–48	43	45
Chrysoberyl	37	22	41–43	42	44
Corundum	37	22	41–43	42	44
Diamond	35	20	39–41	41	43
Feldspar	42	27	46–48	43	45
Garnet (demantoid)	43	28	45–49	40	42
Garnet	37	22	41–43	42	44
Iolite	42	27	46–48	43	45
Opal	41	26	45–47	45	47
Peridot	43	28	47–49	39	41
Quartz (coloured)	42	27	46–49	43	45
Quartz (colourless)	45	30	47–49	41	43
Spinel	37	22	41–43	42	44
Spodumene	43	28	47–49	39	41
Topaz	43	28	47–49	39	41
Tourmaline	43	28	47–49	39	41
Zircon (high)	35	20	39–41	41	43
Zircon (low)	37	22	41–43	42	44

Source: Edward J. Soukup, *Facet Cutter's Handbook*.

It is equally important that the rays reflected from the pavilion facets meet the crown facets at an angle less than the critical angle, otherwise they will not be returned to the eye. Failure to achieve this will result in a higher yield at the expense of overall brilliance. To encourage cutters to prioritize brilliance, well-cut stones usually sell at a premium. Generally speaking, the more expensive the rough, the less likely the cutter is to follow these guidelines. In this case, a compromise is normally sought between brilliance and weight retention. In diamonds and lighter-coloured gemstones, cut is far more important since it can make a stone more appealing and saleable. In darker-coloured gemstones (where colour is more of a factor), adjustments (such as shallower pavilions) are sometimes made to ensure that the stone does not appear overly dark. It is also important that stones are within the design parameters of jewellery which is practical to wear.

Position of Inclusions

In some cases, the position of certain inclusions may dictate that a different shape be chosen. In certain gemstones (aquamarine, blue topaz, kunzite and tanzanite) the presence of inclusions has a detrimental effect on value, whereas in others (ruby, sapphire, emerald and pink tourmaline) there is a greater tolerance. Therefore a cutter may opt to cut a smaller aquamarine that is free of inclusions rather than a larger, more included stone when considering overall value. Similarly, a piece of rough that is well suited for an emerald cut may be fashioned as a pear shape in order to remove an inclusion that could present problems from a durability standpoint.

Optical Phenomena

Stones with the potential for optical phenomena (such as asterism, chatoyancy or labradorescence) must be oriented correctly otherwise the finished stone will not display the optical effect to best advantage, and this can result in a reduction in value.

Cleavage Planes

Since it is difficult to produce a good polish parallel to a cleavage plane, a cutter must orient the stone so that this is avoided. At the same time, care must be taken during the cutting process to avoid undue pressure on stones possessing cleavage planes since this can cause a stone to break.

Colour

Since colour accounts for 50–60% of the overall value in a coloured gemstone, every effort must be made to orient the stone so that the most desirable colour is seen in a face-up position (through the table). In some cases, however, the presence of two or more colours (pleochroism) seen at the same time (e.g. in andalusite) is highly desirable, and this must be accomplished even if it compromises some of the other value factors.

Diamonds

Methods of Dividing Stones

In the case of large rough diamonds, and in any diamond containing inclusions, advantage may be taken of the perfect octahedral cleavage of diamond to separate the stone into smaller pieces. This is done so that pieces of a more workable size may be obtained, and to remove included parts, so that the maximum yield and purity are assured.

The separation of diamonds by the use of cleavage is the job of the cleaver who, with the aid of a sharp-edged piece of diamond set in a holder, cuts a small groove or 'kerf' in the cleavage direction (*Figure 30.28*). For this operation the stone to be cleaved is set into a holder, and when

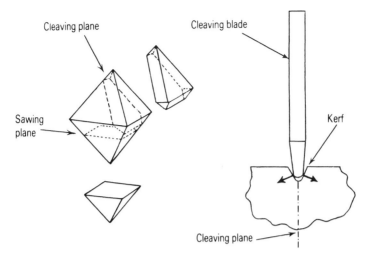

Figure 30.28

the groove has been cut to a sufficient depth, this holder with the stone mounted on it is placed in a lead support attached to the bench. A blunt steel knife is then inserted into the groove with the blade parallel to the cleavage direction. A sharp steady blow is then struck on the back of the blade, when the stone, if correctly grooved and the blade set right, will split into two pieces (*Figure 30.29*).

Figure 30.29   The cleaving blade in position ready for the blow which will cause the diamond to split in two (courtesy of De Beers)

A more modern method (invented in 1905) for separating diamonds into two pieces is by sawing the stones. Well-formed octahedral crystals are commonly sawn in two (*Figure 30.28*). The sawing is carried out by the use of a rapidly revolving diamond-charged phosphor-bronze disc, and the stone to be sawn is first marked out in Indian ink to show the direction in which it is to be sawn. The stone is then set between two metal 'dops', which are fixed to a counterpoised movable arm on the sawing machine. The diamond is suspended above the diamond saw, and by manipulating adjusting screws it can be moved laterally for centring over the saw. The phosphor-bronze disc is about 75 mm in diameter and has a thickness of 0.06 mm, and is run at a speed of approximately 6000 rpm. The edge of the saw is coated with diamond powder mixed with olive oil applied by a hand roller, which not only forces the diamond powder into the edge of the disc but also spreads this edge to give a 'set' so that the saw will not bind in the cut (*Figure 30.30*).

Figure 30.30   Charging the sawing disc with diamond powder (by courtesy of Shell Petroleum Company Limited)

The diamond is then gently lowered on to the edge of the revolving saw, starting the cut at a corner of the crystal (*Figure 30.31*). It is also necessary to ensure that the sawing will be across the 'grain', that is across the traces of the cleavage edges. While the cutting process is proceeding the saw needs constant replenishment with the diamond powder and olive oil, for if the disc runs dry it is likely to tear and cause the stone to break. One operator, the sawyer, deals with this and may look after a battery of as many as 40 saws (*Figure 30.32*). It may take up to 8 hours to cut a crystal weighing 1 ct. The cut, in the case of well-formed octahedral crystals, is made through the thickest part so as to produce two equal halves.

Figure 30.31   A diamond crystal, cemented in a double-sided dop, is lowered under pressure on to the phosphor-bronze saw blade (courtesy of Monnikendam)

Figure 30.32   A double bank of machines attended by the diamond sawyer (courtesy of Monnikendam)

Grinding

The next process is to grind the stone into a truncated double cone (*Figure 30.33*). Previously carried out by a laborious method of hand grinding by working two diamonds one against the other, grinding is now done mechanically. This grinding, termed 'bruting' or 'girdling', is carried out by mounting one stone on the headstock of a special lathe. Another diamond, selected for its sharp edges, is mounted at the end of a long holder which the bruter or grinder holds firmly under his arm. He grinds one stone against the other until both are of the required shape,

Figure 30.33   The bruter rondisting a sawn octahedral diamond. A corner of the other section of the sawn stone is clamped in a dop stick and used as the cutting tool (courtesy of Monnikendam).

a double cone with one end, where the table facet will be placed, flatter than the other. The roughly rounded edge, called the rondist, will become the girdle in the case of brilliant-cut stones, for bruting is not done in the case of emerald-cut stones and rarely for baguette shapes.

From the bruter the stone goes to the polishers who grind on the facets to complete the finished product. As a stone cannot be polished except across a grain it is necessary for these directions to be known (*Figures 30.34 and 30.35*). A certain nomenclature is used in describing the grain. Stones which have been sawn are said to be 'four-point', for the table facet will be parallel to a cube face – a face of fourfold symmetry – and will have two directions of cleavage traces. 'Three-point' stones are those which have the table facet parallel to an octahedral face, a face of trigonal symmetry. Along this direction the stone can only be cleaved and cannot be sawn; further, polishing such a surface is difficult and slow. Only cleavages and spinel twins, the so-called 'macles', are in this category. Macles are disliked by the diamond polisher owing to the twin plane, or naat, which gives the stone a cross grain. 'Two-point' stones have the table facet parallel to a dodecahedral face and give two directions for polishing.

Technique of Faceting and Polishing

Polishing operations are usually divided into two groups and it is usual for the craftsmen to specialize on one group. These are the cross cutter (kruisworker) who cuts the table facet and four main side facets, and the corresponding four back facets and the culet. Care must be taken at this stage to ensure the correctness of the angle of these facets from the girdle,

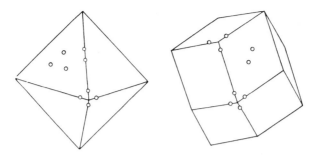

Figure 30.34   Situations of the 'point' on (left) a diamond octahedron
and (right) a rhombic dodecahedron

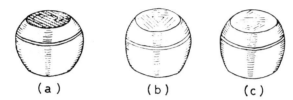

(a)   (b)   (c)

Figure 30.35   The grinding directions on the tables of bruted
stones: (a) four point, (b) three point, (c) two point

a gauge being used to check them. From the cross cutter the stone goes to
the brillianteerer who puts on all the remaining facets.

The cutting of the facets and the polishing are carried out in one operation. This is performed on a revolving horizontal 'lap' or 'scaife', a cast iron
lap about 12–18 mm in thickness and having a diameter of about 300 mm.
The scaife is most carefully balanced so as to run perfectly true on the spin-
dle, and the scaife, after the scouring of the surface with sandstone blocks,
is charged with diamond powder and olive oil. The speed of rotation used
is between 2400 and 3000 rpm.

The diamond to be polished is mounted in a 'dop' which consists of a
brass cup with a soft copper tail. The dop is filled with soft solder into
which the diamond to be polished is mounted whilst the solder is still
pasty. In modern practice a mechanical dop, which came out between the
two world wars, is extensively used (*Figure 30.36*), but for some cuts the
older solder dop is still better. The copper tail of the dop is gripped in a
movable bench clamp called a 'tang' which, with its two feet on the bench
and the dop set with the diamond resting on the scaife, forms a tripod.

One or two stops fixed to the bench, which is called the mill, prevent
the tang being carried round by the motion of the lap, and, in order to
give added pressure, lead weights are loaded on to the arm of the tang.
The operator can feel whether he is polishing the stone in the correct
direction for the grain. Moving the angle of the tang in relation to the

Figure 30.36   Mechanical clamp-type dops holding diamonds on a scaife. The dops are mounted in the end of tangs, which are weighted to achieve the necessary polishing pressure (courtesy of De Beers)

scaife will enable the best polishing position to be found. By bending the copper stalk of the dop most of the side facets can be polished on, and the stone may only need resetting when the rear facets are to be polished. Constant inspection is made as the work of polishing proceeds, as all facets must be symmetrical and of uniform size for the group. On completion the stone is cleaned in acid and methylated spirits, after which it is passed to the sorting rooms for grading into size and quality.

Coloured Gemstones

The methods used by the lapidary for the cutting of stones other than diamond, although similar in many respects, differ in that there are two distinct processes: a definite cutting where the facets are ground on leaving a finish like ground glass, and a final polishing of the facets.

Cabochons

Once the pattern or shape has been decided, a template is used to scribe the desired outline on to the rough. In order to save time, a diamond-charged trim saw is often used to remove any unwanted material. The preformed stone is then secured on to a dopping stick (usually made of wooden dowelling measuring anywhere from 6.35 mm to 22.25 mm in diameter) using special dopping wax. The dopping process is very important because unless the stone is centred and secured properly, problems can occur during the grinding process which will affect the eventual symmetry of the finished stone. After the stone is gently heated, the wax (usually in stick form) is melted over an alcohol lamp and dabbed on to the end of the dopping stick until a sufficient amount of wax has been collected. The heated stone is then carefully placed on to the wax (with the template markings still visible), centred and secured by moulding the wax around the base of the stone. Once the wax has cooled, the corners of the stone are rounded using a 100 to 120 grit wheel. Since additional material will be removed during subsequent polishing, care must be taken not to grind too closely to the template outline otherwise the finished stone will be smaller than originally planned. Using 200 to 230 grit wheels, wheel marks and deep scratches can then be removed by either dry or wet sanding. Generally speaking, there is a preference for wet sanding, especially when dealing with heat-sensitive stones since it reduces friction. Final sanding is carried out using a finer grit (320 to 400) until the stone acquires a semi-gloss or satiny finish. For stones of greater hardness (8 or more) a finer grit (600) is used. The final polish is achieved by using a variety of polishing compounds including tin oxide, cerium oxide, chrome oxide, Linde A and levigated alumina. Once all grinding and polishing has been completed, the stone is then removed from the dopping stick (by gently heating the wax) and cleaned using denatured alcohol. At the discretion of the cutter, the back of the stone will then be either sanded or polished.

Faceted Stones

Once the cutter has made a preliminary examination, the rough is normally trimmed to remove any unwanted material or inclusions that could cause problems during the cutting process. Once this has been accomplished, the trimmed rough is secured to a dopping stick (which is generally two-thirds the size of the stone) using a special faceting wax. During the faceting of a gemstone, this procedure will be performed twice, once for the crown and once for the pavilion. As with cabochons, correctly dopping the rough is very important since it will not only ensure that the finished stone is symmetrical but also help to avoid problems in trying to realign facets should the stone become loose during the grinding and/or polishing process.

Once the stone has been properly secured, the cutter can begin to cut the stone into the desired shape. Prior to the 1960s, the jamb-peg method (described in earlier editions) was preferred. During the last 30 years, however, a variety of electrically powered faceting machines (*Figure 30.37*) has emerged in the marketplace which allow today's lapidaries to achieve a higher level of precision and accuracy.

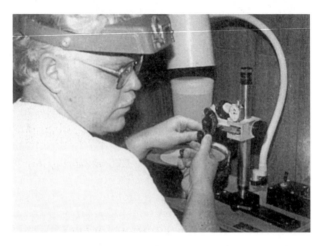

Figure 30.37   An Alpha-Taurus electrically powered faceting machine (courtesy Victor Sawatzky)

Today's faceting machines consist of a faceting head (*Figure 30.38*) and a master lap (*Figure 30.39*). The faceting head consists of a central rod (which is normally secured by a rack-and-pinion assembly), a

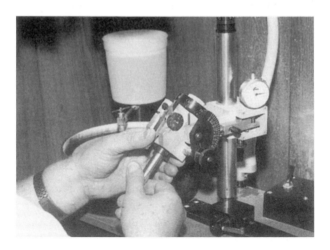

Figure 30.38   The Alpha-Taurus faceting head showing index gear and protractor (courtesy Victor Sawatzky)

Figure 30.39   The Alpha-Taurus faceting head and master lap (courtesy Victor Sawatzky)

sleeve assembly and a dop arm. The dop arm is regulated by an index gear and gear trigger which is located at the right end of the arm. A pointer, located towards the centre of the arm, is used to measure the angle of elevation which can be ascertained by using the protractor attached to the sleeve assembly. A slot, which is cut into the base of the machine, allows the operator to adjust the lateral movement of the faceting head. The master lap (which remains on at all times) is usually made of machined iron or aluminium. In order to achieve the best results, it is imperative that the master lap be absolutely levelled otherwise it will be virtually impossible to cut a gemstone correctly. Surrounding the master lap is a drip pan which can be removed for easy cleaning.

Since it is important that none of the cutting and polishing laps becomes contaminated, many cutters will use a series of laps (invariably pre-charged) tailored to meet their own specific needs. Preforming and faceting are carried out on a cutting lap which is made of either copper, solid steel or aluminium in a variety of grit sizes ranging from 180 up to 3000. Final polishing is performed (using a variety of polishing compounds as noted above) on a polishing lap made of either plastic, tin, tin-typemetal, typemetal, finely charged diamond tin, wax or even ceramic.

Although no standard procedure is used in the faceting of gemstones, many cutters agree that it is wise to keep detailed records of the index readings used so that if a stone has to be recut, the task can be accomplished more easily.

Gemstone Carving

The carving and engraving of gemstones (*Figure 30.40*) is carried out by the use of fine burrs like dentists' drills. These may be made of carborundum or more commonly of hardened steel, and are charged with diamond

Figure 30.40   A magnificent carving in rock crystal by the German carver Martin Seitz

dust in olive oil. The worker has before him a number of these burrs of different shapes and sizes. Such tools are driven from a fixed headstock by electric motor, a suitable burr being screwed into the chuck for the work in hand. The art of gem carving embraces not only the cutting of cameos and intaglios but quite large carvings in hardstone. The hollows of agate cups and ashtrays are cut by a similar headstock fitting with a larger burr called a cupping tool.

Tumbled Gemstones

Tumbled gemstones can be produced by using either a rotary tumbler (*Figure 30.41*) or a new vibrating tumbler. By using first an abrasive agent and then a polishing compound, it is possible to process large

Figure 30.41   Machine for tumbling gems

quantities of baroque-shaped gems at the same time. Although the procedure takes considerable time, it does provide an economical means of processing polished gemstones. Again, it is important to thoroughly clean the machines between stages to ensure that they do not become contaminated.

Today, tumblers come in a variety of shapes and sizes and are geared to both the hobbyist and the commercial operator. When operating a tumbler, it is wise (1) to thoroughly clean the material prior to placing it in the machine, (2) to operate the tumbler in a cool place, and (3) to check the contents daily in order to avoid a potentially dangerous gas buildup which many believe is caused by the chemical breakdown of the silicon carbide in the grit when combined with water.

A Word About Polishing

The older conception of the polish of a material was that it was simply one of superfine grinding so that the hills and the dales of the uneven surface were ground finer until any scratches thereon were too small to be visible. The work of Beilby during the early part of the twentieth century showed that this was not so, except in a small number of cases.

Beilby found evidence that in polishing there is an actual flow of the solid surface with the formation of a liquid-like layer. This layer, termed 'the Beilby layer', lies like a coat of varnish over the underlying scratches. That the scratches are only varnished over by the layer may be demonstrated by etching with acid when they are revealed again.

In 1937 Finch experimented with the new technique of electron diffraction, which reveals the structure of the surface layer of a solid. He substantiated and extended Beilby's findings. Briefly put, Finch found that there are four different types of polish surface in minerals. They are as follows:

1. No Beilby layer as the melting point of the material is too high for the necessary local fusion. In this case, therefore, the polishing is just a fine abrasion. Such is the polish surface of diamond.
2. A Beilby layer forms but immediately recrystallizes in conformity with the underlying structure. This is what occurs when corundum and quartz are polished.
3. The Beilby layer forms but remains amorphous except parallel to important crystal planes, or on long heating. Calcite and kyanite are examples.
4. In this case the Beilby layer remains amorphous on all surfaces. This occurs with zircon and spinel.

More recently, Huddlestone discovered more evidence of the existence of the Beilby layer by the use of Nomarski interference contrast techniques.

Beilby's contention that in general polishing involved surface flow of the polished material has often been attacked, but the account of his meticulous experiments given in his book *Aggregation and Flow of Solids* (London, 1921) and the later confirmation of his theory by modern methods by Finch (*Science Progress*, April 1937) are convincing and worthy of mention here despite continuing controversy.

The Carat Weight

The price of a finished gemstone depends upon the quality, the perfection of cutting and the weight. The unit of weight used almost exclusively for gemstones is the metric carat of 200 mg (0.2 g), a unit which is of comparatively recent derivation.

A carat weight has been used for the weighing of gold and gemstones since ancient times, but the derivation of the weight is obscure. Seeds which have a fairly constant weight were used as 'weights'. Indeed the English grain (gr) was derived from the use of a seed of wheat taken from a well-ripened ear, and the carat weight was said to have been derived from the weight of a seed of the carob or locust tree, whose seeds are remarkably constant in weight whether taken from the centre or the ends of the pod. The seeds of the carob tree (*Ceratonia siliqua*) are chocolate brown and are of a flattened pear-shaped form. According to Spencer the seeds have an average weight of 0.197 g. The same author refers to the Greek weight ceratium, and also the Roman siliqua, as being equivalent to $\frac{1}{144}$ oz, or $3\frac{1}{3}$ gr, which is only slightly more than the $3\frac{1}{6}$ gr of the old English carat.

Also claimed as the origin of the carat weight are the orange kidney-shaped seeds with a black spot at one end obtained from the coral tree (*Erythrina corallodendron*). These have an average weight of 0.197 g, but are not so consistent as the seeds of the carob tree. Kuara, the African name for one species of the coral tree, has been suggested as the origin of the name carat, but the most likely origin of the carat, in both name and weight, is the *Ceralonia siliqua*.

The carat weight varied considerably in different parts of the world. At one time this difference was from 0.1885 to 0.2135 g. The value of 0.204304 g was defined for the carat by the UK Standards Department of the Board of Trade and reported in the proceedings of the Weights and Measures Acts of 1878, 1888 and 1889, but the old English carat was never a legal weight.

The international aspect of trading in gemstones called for an international standard. In 1871 an attempt was made by Paris gem merchants to standardize the carat at 0.205 g and this was approved in 1877, but contrary to their hopes it did not gain international recognition. In 1907 Paris again took the lead in a second attempt to obtain world standardization. The Comité International des Poids et Mesures proposed a metric carat of 200 mg, a weight approximately 2.5% less than the old English carat weight. This proposal was accepted by the Quatrième Conférence Générale des Poids et Mesures in Paris, and it approached through diplomatic channels those other countries who would be interested. Eventually practically all countries legalized the position and the metric carat is now universally standard.

In England, the inclusion of the metric carat, as a unit of weight, was put through by an Order of Council on 1 April 1914. The metric carat can also be described in terms of points (100 to a carat) or grains (4 to a carat).

The only true way of determining carat weight is by using a precision scale (encased in a glass case to avoid draughts) and accurate to three decimal places, since a gemstone can only be sold as a carat when the third decimal place is 9 (i.e. 0.999 ct). When estimating the weight of a mounted stone, use is made of either a stencil gauge with circular or cushion-shaped holes having diameters corresponding to the girdle diameters of correctly proportioned diamonds, or a more expensive caliper gauge such as the Moe (*Figure 30.42*) or Leveridge gauge (*Figure 30.43*). Except for the Moe group, such instruments measure the girdle diameter and the depth of the stone from table to culet, usually in millimetres. By using accompanying tables or weight estimation formulae (*Table 30.3*), a close approximation, often less than 5% in error, may be obtained. In order to compensate for proportional deficiencies, weight correction factors must be used to ensure reasonable accuracy.

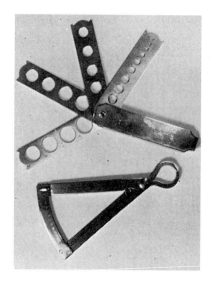

Figure 30.42   Diamond gauges: top, stencil gauge; bottom, caliper gauge

Figure 30.43   Leveridge's caliper gauge

Table 30.3
Weight estimation formulae

Diamonds

Round brilliant: average diameter × average diameter × depth × 0.0061 × weight correction
factor

Oval brilliant: average diameter × average diameter × depth × 0.0062 × weight correction
factor

Heart-shaped brilliant: length × width × depth × 0.0059 × weight correction factor

Triangular brilliant: length × width × depth × 0.0057 × weight correction factor

Baguette: length × width × depth × 0.00915

Tapered baguette: length × width × depth × 0.00915
(width is the average of the two parallel sides)

Emerald cut: length × width × depth × adjustment factor × weight correction factor

Length to width ratio	*Adjustment factor*
1.00:1.00	0.0080
1.50:1.00	0.0092
2.00:1.00	0.0100
2.50:1.00	0.0106

Marquise brilliant: length × width × depth × adjustment factor × weight correction factor

Length to width ratio	*Adjustment factor*
1.50:1.00	0.00565
2.00:1.00	0.00580
2.50:1.00	0.00585
3.00:1.00	0.00595

Pear-shaped brilliant: length × width × depth × adjustment factor × weight correction factor

Length to width ratio	*Adjustment factor*
1.25:1.00	0.00615
1.50:1.00	0.00600
1.66:1.00	0.00590
2.00:1.00	0.00575

Weight correction factor varies from 1 to 12% and is dependent on diameter and girdle
thickness

Coloured gemstones

Faceted
Add 2 to 6% depending on bulge factor

Round brilliant: diameter × diameter × depth × SG × 0.0018

Oval faceted: diameter × diameter × depth × SG × 0.0020 (diameter = length + width × 0.50)

Emerald cut: length × width × depth × SG × 0.0025

Rectangular: length × width × depth × SG × 0.0026

Square: length × width × depth × SG × 0.0023

Marquise: length × width × depth × SG × 0.0016

Pear shape: length × width × depth × SG × 0.00175

Cabochons
General: length × width × depth × SG × 0.0026

Sources: GIA Diamond Grading Lab Manual; The Guide Pricelist and Reference Manual.

A modern electronic balance is shown in *Figure 30.44*.

Less expensive stones (including flame-fusion synthetics and imitations) are normally sold either by the gram or by the piece (based on measurements).

Figure 30.44   An electronic balance for weighing gemstones

Handling Stones

For the handling of gemstones, a variety of tongs are now available. These include spring-loaded, retractable prong and locking types in a full range of shapes, sizes and prices (*Figure 30.45*). Since the springiness and the width of the tongs can vary greatly, it is important to select a pair that is most suitable for everyday use.

Miniature scoops or shovels are also popular, especially when dealing with groups of small stones. Stone papers, with liners to prevent stones piercing through, are a popular way of storing or transporting

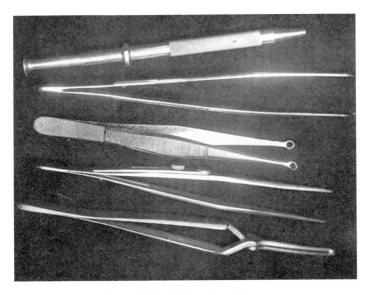

Figure 30.45   Various types of tongs

gemstones. Very often, especially when handling expensive or brittle stones, a fold of lint or cotton wool is used to prevent stones from rubbing against each other and causing damage. When this occurs, the effect is called 'paper wear' and can result, in extreme cases, in the stones having to be recut.

31

Value-Enhanced Gemstones and Gem Materials

Grahame Brown

The artificial treatment of gem materials, such as rough, faceted or cabochoned gemstones, rough and fashioned ornamental materials, and rough and fashioned organic gem materials, to modify their natural appearance and ultimate value is currently accepted to be legitimate trade practice, provided any induced changes are stable under normal conditions of display, wear, and storage; no deceptive trade practices are used to market the value-enhanced materials and appropriate disclosure of the treatment is made at the point of sale of the value-enhanced material.

As colour and clarity enhancements are now routinely applied, to increase the value of most gem materials, ethical gem merchants, jewellers and gemmologists must now decide precisely how much of a gemstone's known history of value enhancement should or should not be disclosed to its owner or potential purchaser.

International organizations associated with the gemstone and jewellery industry have long recognized the existence of the ethical dilemma posed by value enhancement. As a consequence, bodies such as The International Jewellery Confederation (CIBJO), The International Colored Gemstone Association (ICA) and The American Gem Trade Association (AGTA) have attempted to provide very specific guidelines for disclosure of known and/or detected value enhancements in gemstones and gem materials that are being marketed, or require to be identified and/or valued.

Gemmologists wishing to become familiar with current guidelines, with respect to the disclosure of value enhancement, should seek guidance from such world accepted publications such as the CIBJO Blue Book, up-to-date copies of which are available online at www. cibjo.com.

Irrespective of the guidance offered by these international organizations, gemmologists examining gemstones on a day-to-day basis still need practically applicable answers to such questions as:

- Which gem materials are presently being value enhanced?
- Which value-enhanced gem materials can be identified by the trade gemmologist who does not have access to sophisticated laboratory tests?
- What is the short- and long-term stability of particular value enhancements, under normal conditions of wear, storage or display?
- Precisely which value enhancements should or should not be disclosed to customers or clients?

Unfortunately most information about value enhancement, its identification and its disclosure is scattered throughout rather voluminous literature. So, in an attempt at updating available knowledge on commonly available value-enhanced gem/ornamental/organic materials, these will be classified by treatment method, described, and then methods of identification available to the trade gemmologist will be presented, before guidance will be offered with respect to the disclosure of detected value-enhancements.

From the practical viewpoint five basic technologies have been and are being used to commercially enhance the value of gemstones, gem and ornamental materials and organic gem materials. These basic technologies are:

Coating
Lasering
Dyeing and Impregnation
Heat Treatment
Irradiation and Annealing.

Application of Surface Coatings

Foil Backs and Mirror Backs

For many centuries, coloured and/or reflective coatings have been deliberately applied to the pavilion facets of glass imitations and paler coloured gemstones to enhance the colour, brilliance and overall appearance of these materials. As these coatings are potentially deceptive, all antique jewellery should be carefully examined for either coated or mirror-backed stones.

Fortunately, simple hand lens examination of pavilion facets, either directly or obliquely through crown facets, will identify:

- A foil back, by the immediately visible coloured metallic foil attached to its pavilion facets.
- A mirror back, by the highly reflective 'mirror' that has either been applied to or is located on the mount below pavilion facets. Older mirror backs may also display a distinctive iridescent tarnish on their 'mirrors'.
- A coated glass, such as Aurora stone™, that has selected facets coated with a highly reflective iridescent film.

Foil-backs, mirror-backs and coated glasses *must be disclosed*.

Somewhat higher magnifications, and specialized illumination, will be required to detect such modern coated gemstones as coated diamonds, polymer-coated coloured stones, gemstones and other materials coated with microscopically thin films of metal or metallic oxide(s), coated beryl crystals imitating emerald, diamonds coated with ion-implanted thin films of proprietary composition, perhaps diamonds coated with films of CVD synthetic diamond, and value enhancements as simple as acrylic spray-coated porous gem materials.

As each of these coatings is intended to deceive they *must be routinely disclosed*.

Coated Diamonds

A coated diamond is an off-coloured diamond that has had its colour enhanced by applying a suitable blue-coloured film to selected areas of its pavilion facets. Coated diamonds of varying sophistication have been marketed for many decades. Examples of this treatment include techniques for:

- Whitening yellowish body coloured diamonds by applying small symmetrically distributed spots of bluish purple indelible ink to pavilion facets, usually just below the girdle.
- Completely coating pavilion facets with a suitably coloured varnish, such as coloured nail polish.
- Vacuum sputtering a thin film of adhesive bluish anti-reflective magnesium fluoride lens coating onto areas of the pavilion facets of a yellowish diamond.

Fortunately, as both indelible ink and coloured nail polish coatings are readily removed by solvents or mild abrasives, a thorough cleaning often will remove both the coating and its induced colour enhancement. But, while small spots of indelible pencil may be difficult to detect by hand lens examination or microscope examination, nail polish coated

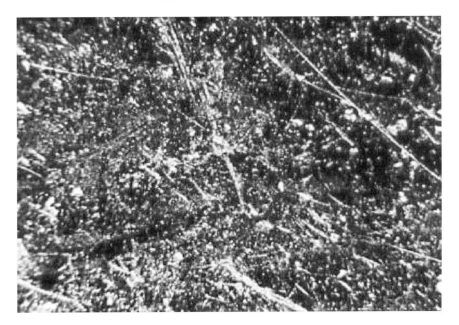

Figure 31.1   Scratched 'bubble' included, iridescent surface patch of magnesium fluoride located under the girdle of a coated diamond (×40, tangentialillumination)

diamonds should easily be detected by routine examination of all facets of a suspect diamond.

Commercial services for colour enhancing diamond have been available since the early 1950s. These now rarely used diamond coating services use vacuum sputtering technology to coat carefully selected areas on pavilion facets of yellowish diamonds with a very thin, strongly adhesive, bluish (complementary coloured) film.

White light entering the crown facets of a coated yellowish diamond will first be selectively absorbed, leaving predominantly yellowish wavelengths to refract towards pavilion facets. At the pavilion facets, these yellow wavelengths will interact with the complementary coloured blue coating on these facets to generate greyish white wavelengths – some of which will be internally reflected back towards the crown facets. To the inexperienced eye, the colour of a blue-coated diamond will appear enhanced, by becoming more greyish to greenish white.

Gemmologists needing to learn how to identify these quite rare coated diamonds should study the seminal paper that investigated this treatment (Miles 1992–93). Miles recommended use of a combination of diffused transmitted and overhead (tangential) illumination, and slight rotary stone movement, to examine both the external (pavilion) surface and opposite facets of a suspect diamond for evidence of :

1. Slight patches of yellowish residue, on the surface, which change to a bluish colour as the diamond is slightly rotated.
2. A dark line just below a facet.
3. An iridescent, scratched, 'bubble' – included patch on a pavilion facet or girdle.

As this coating seldom covers all pavilion facets, visible patches of a deliberately applied coating may appear to the gemmologist as iridescent 'splotches', 'splotches' with a slightly granular texture, 'bubble' included patches, or 'brush strokes'.

While the coatings used to colour-enhance yellowish diamonds may be removed by abrasives, boiling in solvents such as concentrated sulphuric acid, immersion in an ultrasonic cleaner, or the application of heat from a blow torch; such drastic treatments will return the colour of the diamond to its original state. Consequently, such treatments are not recommended as suitable for use as tests by gemmologists.

Coloured Polymer-Coated Gemstones

Polymer-coated gemstones first appeared in Bangkok in the late 1980s. Examples reported included red polymer coated faceted quartz imitating ruby, green polymer coated pale jadeite imitating Imperial jadeite, and red polymer coated pale grey star ruby imitating star ruby. These polymer-coated gemstones had the following identifying features:

1. Spherical gas 'bubbles' included in the thin coloured polymer coating.
2. Surface defects in the polymer coating.
3. Colour 'swirls' in the polymer coating.
4. The polymer coating could be softened by a judiciously applied hot point.

Thin Film Coated Gemstones

Over the last decade, colourless gemstones have been coated (vacuum sputtered) with microscopically thin films of pure metals or metallic oxides to induce unnatural colours in either crystals or faceted stones.

For example, the bluish hue of Aqua aura™ quartz and topaz is caused by the selective transmission of blue wavelengths by the thin film of vacuum sputtered pure gold that coats these treated gemstones. Although the presence of this thin film of gold does not modify the gemmological

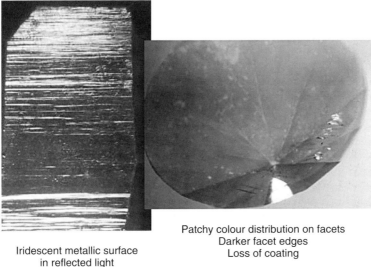

Iridescent metallic surface
in reflected light

Patchy colour distribution on facets
Darker facet edges
Loss of coating

*Figure 31.2   Identifying features of greenish blue AquaAura™ quartz that is coated
with a microscopically-thin layer of pure gold*

properties (SG, RI) of either the quartz or the topaz, the coating does display identifying iridescence when examined with tangential illumination.

Kammerling *et al.* (1990) revealed that when greenish blue Aqua aura treated quartz and topaz gems were examined with the gemmological microscope, in diffused direct transmitted light, they displayed:

1. A coppery surface iridescence in tangential illumination.
2. Diffused dark outlines of some facet junctions.
3. A patchy blue colour distribution on some facets.
4. White facet junctions, irregular white abrasions and surface pits, where the treatment either did not 'take', or had been abraded away.

A few years ago, Golay Buchel (Thailand) begun marketing Tavalite™, a colour-enhanced faceted stone that consisted of either white topaz or cubic zirconia that had been coated by vacuum sputtering with a coating of a proprietary metallic oxide or oxides that had a thickness near to that of visible light. Differential surface absorption and reflection by this thin coating produced faceted stones of either purple, yellow, light blue, green or red colour that changes colour with any change in viewing angle.

Coated quartz crystals were described by Bank *et al.* (1999). According to these authors, the coating of colourless quartz was performed in a vacuum chamber using a high voltage (440 V) applied to the specimens. A positive electrical charge was applied to the stone to be treated, so that it formed the anode in the system. Various metals were then evaporated at the negatively charged cathode and the metallic ions were attracted across the vacuum to be deposited as a thin metal film onto the quartz. Coating with titanium oxide produces an opaque, metallic layer, which appears to be rainbow-coloured. These treated rock crystal clusters are sold as Rainbow quartz™ in the trade. Using, respectively, silver and platinum produce transparent bluish and reddish coatings. Purplish layers are obtained by using platinum and gold. Rock crystal coated with a thin coating of gold have an aquamarine colour and is sold as Aqua aura™. It is important to remember that the presence of these thin metallic films has no influence on either the SG or RI.

Blue, bluish green and green surface-diffused topaz have been produced by heat diffusion of a cobalt-rich or cobalt and nickel-rich powders to induce respectively blue and green colours in previously colourless topaz. Hodgkinson (1998) revealed that the colour of these treated topaz, which is uniform to the naked eye, is confined to a thin layer on or just below the surface of the treated colourless topaz. When examined (immersed) under magnification the diffused layer has a patchy colour distribution. Chipped facets reveal underlying colourless topaz. This coating may produce a negative reading with the gem refractometer. When examined with the spectroscope, cobalt absorption bands at 560, 590 and 640 nm are visible, and this colour-enhanced topaz commonly gives a red response when examined with the Chelsea filter.

Green-Coated Beryl

Coating is also used to create visually effective imitations of emerald crystals. Although green-coated beryl crystals should be simple to detect, this is not always so. Several very experienced emerald dealers have been hoodwinked into purchasing very expensive coated beryl fake emeralds.

To create these most impressive imitation emerald crystals, pale-coloured beryl crystals may be either coated with green polymer or green paint. Other variants of the green-coated beryl, imitating emerald, include drilled crytstals that are filled with green polymer or some other transparent green adhesive material, and colourless beryl crystals that have been deliberately fractured and rejoined with a strong green coloured adhesive before being deceptively glued into position onto a suitable natural matrix.

Detection of imitation emerald crystals requires careful hand lens examination, to detect:

1. Defects in an external green coating.
2. 'Bubbles' in an external green coating.

Polymer coated

Composite crystal

Fracture filled with green glue

Figure 31.3   Various imitation emerald crystals

3. Solubility of an external green coating when wiped with a swab moistened in an organic solvent, such as nail polish remover.
4. Surface reaching fractures filled with green pigment.
5. Pigment-filled, 'bubble' – included cores.
6. Adhesive-filled planes of junction between the crystal and its matrix.

Coating with DLC and Synthetic Diamond

In the Gem News feature of the Summer 1991 *Gems & Gemology* the possibility was raised that faceted softer coloured stones could be coated, by chemical vapour deposition plasma jet technology, with films of approximately 0.08 µm thickness of transparent amorphous Diamond-Like Carbon or DLC. It would appear that this treatment has not yet become a commercial reality. However, Hammer and Schmetzer (2000) reported that faceted synthetic moissanites were being coated by a 1–2 µm thick layer of diamond-like carbon (DLC) using patented CVD technology. If these coated synthetic moissanites are simply tested by commonly used diamond probes, which measure thermal inertia or the rate at which a fixed amount of heat is dispersed, then a diamond response will be obtained from a coated diamond imitation. In contrast, careful examination for double refraction will reveal the identity of the synthetic moissanite that underlies the synthetic diamond coating.

In 1991 Fritsch introduced gemmologists to the possibility that single crystal diamond films could be grown epitaxially on natural diamond surfaces by CVD at low pressures. These early single diamond crystal films had a critical thickness ($<2\ \mu m$) at which the deposited diamond turns polycrystalline. At the experimental level, colourless diamonds were coated with a thin single crystal film of blue type IIb synthetic diamond. The diamonds turned dark bluish grey and became highly electroconductive due to the 200 ppm boron content of the coating. Fortunately the thin film of type IIb synthetic diamond did not cover facet junctions; so the facet junctions gave the appearance of unusual white lines.

Recently, following the announcement in August 2003 that Apollo Diamond Inc. of Boston USA had developed a CVD synthesis process to grow synthetic diamonds for jewellery purposes, it would seem likely that in the foreseeable future this technology could be used for such purposes as repairing surface defects in natural diamonds, or perhaps modifying their colours by applying a suitable coloured film of synthetic diamond. The identifying features of such treated diamond are yet to be reported in the literature.

Ion-Implanted Coloured Diamonds

InColor Enhanced Diamonds™, of Heath (Ohio), are producing colour-enhanced green and black diamonds by ion-implantation of the surface of off-coloured diamonds with a 1–2 MeV linear accelerator that uses ion feeds of proprietary composition. Only a few minutes of treatment are required to produce green diamonds, while the production of black diamonds requires up to several hours of treatment. No residual radio-activity can be detected following this treatment. The identifying features of these treated diamonds, by Moses *et al.* (2000), included a bright iridescence observed on the pavilion facets of green treated diamonds, and the shallow depth of colour, mottled appearance, metallic lustre and brownish yellow colour in reflected light of the black-treated diamonds.

Spray Coating

Gemmologists should also remember that technology as simple as acrylic sprays are being used to enhance the surface lustre of somewhat porous gem materials, such as lapis lazuli and perhaps turquoise, or heavily fractured materials such as jadeite. Acrylic-coated gemstones may be detected by simple hand lens examination, and the solubility of these coatings to organic solvents.

Lasering

For several decades laser technology has been used to drill narrow channels down to dark inclusions in diamond that are then dissolved out with hot hydrofluoric acid.

Traditional lasered diamonds require little more than hand lens or low-power dark-field microscope examination to reveal the relatively straight whitish channels and vacated inclusion cavities that identify this treatment.

A new laser treatment for diamond has recently appeared the market. In this treatment the carefully directed energy of several thin laser beams is used to expand a cleavage associated with a near-surface dark mineral inclusion or to develop a cleavage down to that inclusion. As a result, artificial 'feathers' (cleavages), which have quite similar appearances to naturally occurring cracks or cleavages, are created to the surface of a nearby facet. Lateral movements of the laser then widen out the channel to provide direct access of the HF leaching solution. When examined with magnification the channel looks like a cleavage, except for unnatural worm-like channels or a series of spiky steps used to widen the cleavage. The identifying features of this new laser technique were first described by McClure *et al.* (2000). These researchers emphasized that identification of this treatment is dependent on a thorough microscopic examination looking for the presence of mirror-like transparent surface-reaching that contain reflective, transparent 'feathers' that contain unnatural irregular channels that connect the acid leached inclusion to the surface of a facet. As only faceted diamonds are lasered, no 'drag' marks will be visible on polishing lines that cross the facet from which the feather exits.

Irrespective of the technique of lasering used to enhance the clarity of a diamond, gemmologists must remember that the visibility of laser channels can be significantly decreased by infilling these channels with high RI glass.

All lasered diamonds *must be disclosed*.

Dyeing and Impregnation

Porous gem materials either may be dyed to enhance their colour, or strengthened and colour-enhanced by impregnation Dyes used to colour-enhance gem materials are usually based on colourfast organic or inorganic pigments. In contrast, colourless liquid impregnators are used to in-fill pore spaces and surface-reaching fractures in pale-coloured and friable gem materials. These impregnators may be deposited under vacuum, before they are solidified either chemically or thermally.

Dyed Chalcedony

For many centuries the cryptocrystalline quartzes (agate and chalcedony) and quartz tiger's eye have been routinely dyed with inorganic pigments. As induced colours of dyed agate, carnelian-sard, onyx, green chalcedony, blue chalcedony and red/blue/brown/green tiger's eye are quite colour stable, and do not influence the value of the material, such treatments *are accepted by the trade without qualification.*

Dyed and Impregnated Opal

Precious opal is another relatively porous variety of quartz that may be either dyed, to enhance its *play-of-colour*, or impregnated to increase its durability as well as enhancing its *play-of-colour* if a dye is added to the impregnator. Internally fractured opal is quite porous and many (mostly unsuccessful) attempts have been made to enhance the clarity of precious opal by impregnating its internal fractures with various silica-based solutions. Claims have been made that undetectable impregnation of cracked and crazed Australian precious opal has been accomplished. As no independent description of this possible value-enhancing treatment of opal has been published, little is known of how effective it may be, or how or whether this treatment could be detected.

Since the mid-1950s porous cream to grey-coloured Andamooka sand-stone matrix opal and, most recently, an opalized conglomerate known locally as 'coloured concrete', have been impregnated by particulate carbon to create at least two types of carbonized Andamooka matrix opal. The starting material for this black opal look-alike is a fine-grained, porous, opal-cemented sandstone and/or opal-cemented conglomerate from Andamooka (South Australia) that is included by pinpoints and patches of precious opal. Traditionally, colour-enhanced carbonised Andamooka matrix opal has been manufactured by impregnating prepolished cabs of Andamooka matrix opal with a saturated sugar solution; then, reducing the sugar to carbon by soaking the cabochons in warm sulphuric acid. The more recently used opalized conglomerate is carbon impregnated by impregnation with an unspecified organic material followed by heating under reducing conditions at 500 °C. Once the sandstone-based, treated Andamooka matrix opal is lightly polished, the superficially distributed carbon impregnator allows this black-dyed opal to be used to imitate pin-fire Lightning Ridge black opal. In contrast, although the depth of penetration of the carbonization is much greater in the treated conglomerate, this treated opal is much more porous and does not take a good polish. Consequently, this form of black-dyed Andamooka matrix opal is used for polymer-coated specimens and in doublets and triplets.

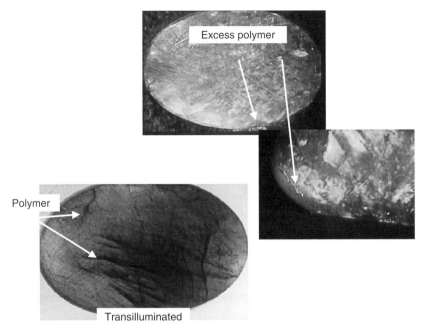

Figure 31.4   Identifying features of polymer-impregnated opal

Fortunately, detection of carbonized Andamooka matrix opal is relatively simple; for hand lens examination, and an educated eye, will reveal pin points and small patches of precious opal that are strongly contrasted against the opaque, carbon impregnated granular sandstone and conglomerate matrix of this colour-enhanced opal. Brown (1991) and Keeling and Townsend (1996) have authored useful descriptions of the two major varieties of Andamooka-treated matrix opal.

Since the early 1990s, polymer-coated carbonized (black-dyed) Andamooka matrix opal first appeared in the marketplace. This soft, porous Andamooka 'concrete' opal (sandstone opal) is being carbonized and plastic-coated to imitate black opal. Several years later, precipitated silver nitrate also has been used experimentally to colour-enhance Andamooka matrix opal to create silver nitrate dyed Andamooka matrix opal.

Gemmologists are reminded that all carbonized matrix opals *must be disclosed*.

Plastic impregnated Brazilian precious opal is yet another colour-enhanced opal that *must be disclosed*. This opal originally had a pale yellowish potch colour, rather similar to that of some South Australian opal. As a considerable amount of Brazilian opal is heavily fractured, this opal is being value-enhanced, by vacuum impregnation with appropriately pigmented thermosetting polymers (plastics), to yield opal with white, black, brown, blue and orange potch colours. While G.I.A. researchers Fritsch and Stockton (1987) suggested that infra-red spectroscopy should be used to identify the polymer impregnator in colour-modified opals, transillumination of a polymer impregnated opal, with diffused light from a light box, should clearly highlight any polymer-filled internal fractures in Brazilian opal.

Dyed Gemstones

Before leaving the consideration of colour-enhanced quartzes, mention should be made of quench-crackled crystalline quartz – a very ancient source of emerald and ruby imitations, and 'Swiss or German lapis' – a blue-dyed jasper imitation of lapis. Transillumination with diffused white light will readily detect the dye infilling induced fractures in quench-crackled crystalline quartz and beryl.

Dyeing porous microcrystalline jasper, with potassium ferrocyanide and ferrous sulphate, creates an unnatural coloured lapis lazuli imitation. This imitation, which historically has been termed 'Swiss or German lapis', displays neither the brassy 'specks' of pyrite, nor the white veins of calcite that characterize lapis lazuli. Additionally, the 'Swiss lapis' imitation inevitably will display glistening microcrystals of quartz within microscopic cavities in the jasper.

Each of these imitations *requires disclosure*.

Effective imitations of emerald or ruby can be made economically by quenching a heated, faceted colourless quartz, beryl or pale-coloured corundum in a suitably coloured dye solution. The dye-filled surface-reaching fractures, giving these imitations their colour, are most effectively visualized by examination, with the gemmological microscope, unimmersed (because some dyes are quite soluble, even in water), using diffuse transmitted light. Always remember . . . dye-filled fractures in gemstones must be traced to the surface of the gem before dye impregnation can be positively proven.

Schmetzer *et al.* (1992) described the gemmological properties of a new ruby imitation, red dyed? quench-crackled natural colourless to pale-coloured sapphire. These fakes are apparently being sold in India as heat-treated Indian rubies. They *must be disclosed*.

Identifying features of these fake rubies include:

1. An absence of the 694 nm fluorescent doublet of ruby.
2. Inertness to LWUV, except for dye-filled fractures that fluoresced an intense yellow-orange.

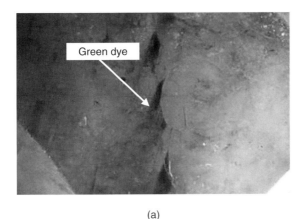

(a)

(b)

Figure 31.5   Green dye infilling surface-reaching fractures in quench-crackled pale aquamarine that imitates emerald (×24)

3. Inertness to SWUV.
4. Red dye filled surface-reaching fractures.
5. No response to rubbing the surface of the fake with an acetone-moistened cotton swab.

Value-Enhanced Jade

Jade is another valuable gem material that is commonly colour- and value-enhanced by bleaching, dyeing and/or impregnation. Although the external surfaces of jades are routinely waxed, this trade-accepted treatment *does not require disclosure*. However, any jadeite that has been colour-enhanced by any combination of bleaching, dyeing and polymer impregnation *must be disclosed*.

Jadeite, particularly jadeite of well-coloured or attractively patterned Imperial green and/or mauve hue, is gem material much favoured by Chinese *aficionados*. Over the last 50 years, the ever increasing cost of obtaining *ying yu* (hard green jade), from difficult-to-access Burmese (Myanmar) deposits, has led to the development of a range of technologies that aim to enhance the value of less valuable jadeite by submitting this comparatively low cost material to various combinations of bleaching, dyeing and/or impregnation.

The following treated jades are traded in the marketplace.

A-Jade

For centuries carved jades have had their difficult-to-polish surfaces deliberately coated with a thin layer of bee's, paraffin, or other solid wax. The purpose of these thin coatings is to enhance the lustre of polished surfaces of the jadeite, as well as infilling surface-reaching fractures and pits.

It is generally accepted that this waxy coating of otherwise untreated jadeite (A-jade) *does not need disclousre* . . . providing significant amounts of the wax have not been used to deeply impregnate previously bleached jadeite.

Gemmologists should remember that, as most carved and polished jades have been 'waxed' to enhance the jade's lustre, the presence of this coating could yield a confusing 'spot' RI of about 1.45, instead of the jadeite's identifying spot RI of 1.66 or 1.67.

Readily scratched wax coatings on jadeite are readily detected by 10× hand lens examination. Their presence can be confirmed as the wax readily melts under an applied hot point.

Less commonly, some type-A jadeite has been coated with a colourless polymer to enhance its lustre and infill surface-reaching fractures and voids. Such thin, often discontinuous coatings can be softened by an applied hot point, yielding a 'burnt plastic' aroma. Polymer-coated A-jade *should be disclosed*.

B-Jade

Since the mid-1980s, a new value-enhanced green jadeite, and more recently value-enhanced mauve and other multi-coloured jadeites, have begun to be commercially marketed.

In the case of the now quite common bleached and polymer-impregnated green jadeite, the original the material was a green streaked jadeite with a strong yellowish, brownish or greyish body colour. Following prolonged soaking in strong acids, such as hydrochloric and nitric acid (sour plum juice), discolouring iron compounds infilling sur-face-reaching fractures and grain boundaries are leached out. Subsequent impregnation of the bleached jadeite with a suitable colourless polymer converts it into a bright green translucent jadeite that now may also have a white body colour.

B-jade will display one or more of the following features:

1. Enhanced visibility and some dissolution of grain boundaries will be observed when the surface of the jade is examined under magnifica-tion due to acid. A good description of this feature was provided by Tay (2001).
2. Evidence of colourless polymer on the surface of the jade or infilling surface-reaching fractures, and depressions in the surface of the jade.
3. A superficial bluish white to yellowish green luminescence to LWUV.

In 1993 Hodgkinson recommended that untreated, polymer-impregnated and dyed jadeite could be effectively detected by applying 1–3 droplets of concentrated hydrochloric acid to a flat surface of the jadeite. When the acid is applied to the surface of natural coloured jadeite, the acid will diffuse, via common surface-reaching fractures in the jadeite, to reappear some distance from its site of application. This re-emerging acid will stain a white tissue brown due to the presence of iron removed from the jadeite by the capillary activity of the acid. In contrast, if the acid is applied to the surface of polymer-impregnated jadeite, the acid will sit on the surface until it evaporates. Dyed jadeite responds to this test by 'sweating' the dye profusely into an aureole that surrounds the spot on which the acid was originally applied. According to Hodgkinson (1993), waxing the surface of jadeite does not appear to influence the action of acid on the jadeite.

Presently, these gemmological tests should be considered indicative, for they may not positively identify this value enhancement. Consequently, Fritsch (1994) clearly warned . . . 'Although in a number of cases type-B jadeite can be identified using classical gemmological techniques, one should emphasize that pieces of high value should be submitted to a properly equipped laboratory' . . . for to date, the only definitive test for identifying bleached polymer-impregnated jadeite is the demonstration of 'a very intense group of peaks around 2900 cm^{-1}, and accompanying features in the mid- and near-infrared' that identifies the particular poly-mer used as the impregnator in the value enhancement.

C-Jade

As good-quality green jadeite has become increasingly rare and costly over the last three or four decades, inexpensive and unattractive pale greyish to yellowish jades are commonly colour-enhanced by dyeing them with a mixture of either blue and yellow aniline dyes, or a solution of a suitable green Cr^{3+} salt.

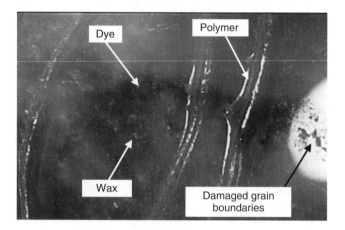

Figure 31.6   Value enhanced A + B + C jade

The resulting green-dyed jadeite can be detected by conventional gemmological testing, by:

1. Transilluminating the jadeite with strong diffused light and examination of the interior of the jade with the gemmological microscope, to detect green dye filled internal fractures that penetrate to the surface of the jade.
2. Examining the jadeite with a Chelsea filter, for some green-dyed jadeites appear orange-red, while natural coloured green jadeite remain green.
3. Examining with the hand spectroscope should reveal a broad dye band in the red from 630 to 670 nm, rather than the three shaded absorptions at 630, 660 and 690 nm that identify naturally coloured chromian green jadeite.

Increasing quantities of violet, lavender and lilac (and even brown) dyed jadeite have been released into the world's markets over the last decade. Fortunately, dye can be detected in the surface-reaching fractures of C-jades.

Mixed Treatments

Wu (1999) described mottled emerald green and brown jades that had been value-enhanced by bleaching, dyeing and polymer impregnation. This jade was a mixed B- + C- jade.

When examined with the microscope both green and brown dyes were noted concentrated in surface-reaching fractures, and in spaces surrounding grains. This jade did not display the dye line (630–670 nm) absorption that is so often stated to identify green-dyed jade. Neither did the brownish patches give any response either to the Chelsea filter or the hand-held spectroscope.

Value-Enhanced Turquoise and Lapis Lazuli

Impregnation and/or dyeing technologies also are almost routinely applied to improve the marketability of such porous and friable ornamental gem materials as American turquoise, and Chilean lapis lazuli. To this end, both turquoise and lapis may be colour-enhanced by aniline dyes, as well as being 'stabilized' by impregnation with colourless paraffin, bee's wax, aqueous silica or colourless polymer. While dyed turquoise and dyed lapis *must be disclosed*, when turquoise and lapis are impregnated with colourless wax or polymer this treatment *need not be disclosed* for it is considered to be normal trade practice.

Fortunately, blue-dyed turquoise and lapis are readily detected by looking for accumulations of dye in surface defects, surface-reaching fractures, in and around the stringing holes of beads, and staining included minerals such as calcite. Judicious wiping of the surface of the material with an acetone-moistened swab – noting the presence of a blue dye on the surface of the swab – will confirm dyeing of these materials. Be warned, however, for coatings of wax that are applied to hide defects in a polished surface may have to be detected and removed before any solubility test for dyeing can be applied. Also, some recent dyes, used to colour-enhance lapis, are very resistant to the solvent effect of acetone. Such dyed lapis requires brisk rubbing of its surface, with 10% hydrochloric acid, to yield an identifying dark blue stain.

Zachery-treated turquoise has been available in the marketplace for over a decade. This treated turquoise, which is produced by a still-secret treatment, enhances both the porosity and colour of the turquoise. Unfortunately, this enhancement *cannot be detected by standard gemmological testing*. A useful description of the properties and features of Zachery-treated turquoise was provided by Fritsch *et al.* (1999).

'Oiled' Emerald

Since antiquity, the valuable gemstone, emerald, has been routinely 'oiled' to improve both its colour and its clarity. Over the centuries, all-too-frequent surface-reaching fractures in emerald have been impregnated with a variety of materials that include mineral (Joban) oil

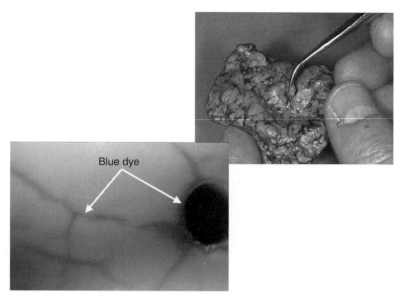

Figure 31.7   Identifying features of dyed turquoise

(RI = 1.48), paraffin wax (RI = 1.52), thermosetting colourless cedar wood oil (RI = 1.50/1.51) or Canada balsam (1.53/1.55), synthetic chemically set polymer fracture sealers such as the epoxy resins Opticon™ (1.55 – unset, 1.58 – set), Araldite (1.57 – unset) and UV cured polymethacrylates such as EpoTek (1.55). The basic requirements for each of these fracture fillings is that they be colourless and have an RI in the same range as that of emerald.

Ultrasonic cleaning, vacuum impregnation and judicious heating ensure that the chosen sealer completely infills all fractures, and if set, they can be removed only by prolonged immersion in a suitable solvent.

Fortunately, the presence of 'oiled' fractures in emerald can be readily detected by examining surface-reaching fractures for the presence of colourless to yellowish resinous residues. The visibility of the set oils, to low-power microscope examination, can be enhanced if the contents of the fractures are illuminated by strong transmitted diffused light. Colourless polymers (particularly unset plastics), impregnating surface-reaching fractures in emerald, can be very difficult to detect. However, most polymers, either set or unset, and some 'oils' do fluoresce when exposed to LWUV. 'Oiled' emeralds can be further distinguished from plastic-filled emeralds, for the oil will 'sweat' onto the surface when the emerald is exposed to an increase in temperature from a hot point held just above the point of exit of a surface-reaching fracture.

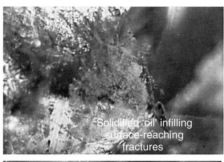

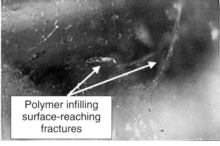

Figure 31.8   'Oil' and colourless polymer infilling
surface-reaching fractures in an 'oiled' emerald

According to Hänni (1998), of the Swiss Gemmological Lab, an 80–90% success rate at identifying organic filling materials in emerald can be achieved using a combination of visual observation, infrared spectroscopy and a micro Raman spectrophotometer. But, difficulties in identification do arise when the filling materials are mixed, decompose due to ageing or chemical attack, or a lesser known filling material has been used to fracture-fill an emerald.

Hänni recommended use of the following three-step protocol to identify 'oiled' emeralds.

Step 1: Visual Examination

The presence of fluorescing 'oiled' surface-reaching fractures will be quickly revealed by examining the emerald with LWUV, using appropriate precautions. Then, a microscope equipped with dark field illumination or fibre-optic lighting then should be used to examine the content of 'oiled' surface-reaching fractures.

Traditional fracture-filling materials have poor durability, particularly as 'oils' dry out rapidly. This allows air to enter the fractures so that the borders between the yellowish to greyish solidified 'oil' and air within the fracture become visible. In contrast, polymer infilling substances tend to develop finer dendritic patterns when drying out or shrinking due to polymerization. As the polymers decompose they appear milky white and are difficult to remove. The presence of trapped 'bubbles' of air within fracture fillings will confirm the presence of a foreign material within the fracture.

As 'oils' are more soluble than polymers, acetone may be used to identify fissures containing 'oil'. When a drop of acetone is applied to an 'oiled' fissure for about five seconds the 'oil' will dissolve. Evaporation of the acetone from the fissures should be observed under magnification. In contrast, a solid polymer fracture filler, such as set Opticon™, is much more stable and will be unaffected by the acetone test.

While examining a suspect filled fracture in an emerald under magnification, in either incident or dark-field illumination, an identifying 'flash effect' may be observed when the fracture filling has an RI above 1.55.

Step 2: Examination by IR Spectrophotometer

Positive identification of the impregnator is the second step in identifying an 'oiled' emerald. This identification is possible in a laboratory that is equipped for Fourier transform IR spectrometer (FTIR). When used either in its transmission or diffused reflectance mode, the IR spectrophotometer will produce a complete spectrum which will reveal all the substances in the sample.

One of the major problems associated with the detection of the IR absorptions of organic materials in emeralds is the fact that emerald also absorbs these wavelengths over large areas of the spectrum. As a consequence, absorptions from organic materials can only be identified where emeralds leave open some 'windows'. Two ranges within the IR spectrum are useful for identifying fracture fillings in emerald. The first is between 1200 and 1700 cm^{-1} and the second is between 2700 and 3200 cm^{-1}. For emerald Hänni recommended that the lower range be used. The major reason given for this choice is that the fluorescence of the emeralds to wavelengths used in this instrument coincide with the peaks of the organic substances in the higher range.

Step 3: Examination by Micro Raman Spectrophotometry

The microscope-mounted Raman spectroscope is yet another instrument that can be used to identify emerald treatment. The microscope is equipped with a laser light source of 514 nm wavelength, a spectrometer and binocular optics for targeting small areas within a gemstone. The filled fissure should be examined, in reflected light and at 50× magnification, with a very small area within the filled fracture centred in the viewing area by a cross-hair eyepiece. When the instrument is switched from the viewing mode to the analysing mode, the laser beam, which is only a few microns in diameter, is directed onto the sample at the centre of the cross hairs and so generates a spectrum. Reference spectra of filling substances (cedar wood oil, 'palm oil', Epon 828, etc.), which have been stored in a computer, are then used to identify peaks and intensities in the recorded Raman spectrum of the emerald and so identify the filling material.

At the present time neither 'oiled' nor colourless plastic-impregnated emeralds *require disclosure*; but *an explanatory statement should be appended to descriptions* of all polymer-impregnated emeralds.

In contrast, the modern technique of oiling faceted emeralds, as well as emerald rough, with green coloured oil, is deliberately intended to deceive, and grossly enhance value. As a consequence, all agree that green-dyed emeralds *must be disclosed*. Fortunately, deep green coloured oil is very easy to detect by microscopic examination.

Clarity-Enhanced Diamonds

For well over a decade diamond's surface-reaching cleavages, laser drill channels and defects on the surface in faceted diamonds have been filled with high RI glass to enhance clarity. This technique was devised by Zvi Yehuda, of Ramat Gan, Israel, and released commercially in 1987. A high RI transparent glass, containing Pb, Cl, O and variable amounts of Bi, is introduced into surface-reaching defects in diamond under conditions of high pressure and temperatures of around 400 °C.

According to Koivula *et al.* (1989) and Shigley *et al.* (2000) this fracture-filling of diamond can be identified by using the gemmological microscope to reveal:

1. The presence of flow structures, entrapped 'bubbles', a crackled web-like texture and a light brown colour in the fracture infilling.
2. An observed 'flash effect', from yellowish orange to blue, or pinkish purple to yellowish green, as the filled fracture is slightly rotated from edge on viewing position – and examined at low magnification and using dark-field or lateral illumination.
3. Glass filled surface-reaching fractures.
4. Whitish areas where the fracture filling reaches the surface.

In addition, glass infilled cleavages in a diamond can readily be detected by the filler's distinct opacity to X-rays.

Yehuda is no longer the only commercial fracture infiller of diamond, for from early 1990s, Yehuda's dominance of the market for clarity-enhanced diamonds has been challenged by the release of clarity-enhanced diamonds

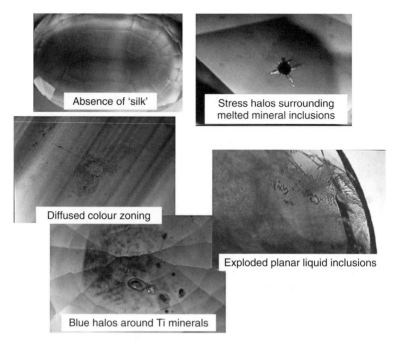

Absence of 'silk'

Stress halos surrounding melted mineral inclusions

Diffused colour zoning

Exploded planar liquid inclusions

Blue halos around Ti minerals

Figure 31.9   Identifying features on conventionally heat-treated sapphire and ruby

by other treaters such as Koss & Shechter, Goldman-Oved and relatively recently the Oved Diamond Company.

All fracture-filled diamonds display the same identifying features as the original Yehuda-treated diamonds, and *must be disclosed*.

Fracture-Filled Coloured Stones

Due to the ease with which surface-reaching fractures can be cleaned of discolouring residues, a range of heavily fractured coloured stones is being colour- and clarity-enhanced by deliberate impregnation of their clean surface-reaching fractures with colourless polymers, glass that has an elevated lead content, as well as glassy residues that remain after the commercial heat treatment of ruby and sapphire. Examples of fracture-filled coloured stones include polymer-filled sapphires and tourmalines, glass-filled emeralds and rubies, and of course flux glass filled rubies and sapphires.

Key identifying features of fracture-filled coloured stones include:

1. Observed differences in lustre between a facet on the coloured gemstone (high lustre), and the colourless glass or polymer infilling a surface-

reaching fracture that exits the facet (lower lustre), when the external surfaces of the stone are examined with tangential illumination.
2. The observed presence of included 'bubbles' within the infilling glass or polymer.

When identified, these fracture-filled coloured gemstones *must be disclosed*.

Be warned, however, for residual 'flux' used in the commercial heat treatment of ruby and sapphire can enter surface-reaching fractures, partly 'heal' these fractures and mimic the appearance of 'healing' fracture that are common both in natural and flux-grown ruby and sapphire.

How flux 'healed' fractures in ruby and sapphire *should be disclosed is still subject to consideration.*

Treated Pearls

One of the dilemmas of disclosure for the gemmologist concerns the routine bleaching and dying of cultured pearls to obtain uniformly hued pearls for stringing into necklaces.

It is a well-known fact that virtually all Japanese *akoya* (bead nucleated) cultured pearls, some poorly coloured South Sea bead nucleated cultured pearls, and considerable numbers of Chinese non-nucleated freshwater cultured pearls are routinely drilled, bleached in warm hydrogen peroxide, and then dyed with suitable dyes to enhance and unify their induced colours. This colour enhancement usually can be detected simply by examining the stringing hole of the pearl (with the hand lens or low-power binocular microscope and fibre-optic illumination) for evidence of dye accumulated in the drill channel, between the shell bead and the surrounding nacre of the cultured pearl, or staining the shell bead in the centre of the pearl (Plate xx).

While *no disclosure* is required for dyed akoya pearls, other dyed bead nucleated cultured pearls *must be disclosed*. In addition a warning should be given that the colours of aniline-dyed cultured pearls are not particularly stable, either to heat or organic solvents.

The problem of disclosure becomes a little more complicated when grey to black pearls have to be examined; for considerable price differentials do exist between: natural black pearls, black bead nucleated cultured pearls, cultured bead nucleated pearls that have been dyed black either by silver nitrate or aniline dye, cultured bead nucleated pearls that have had greyish to blackish colours induced by high-energy irradiation, and cultured non-nucleated pearls described commercially as being 'laser-treated'.

With the notable exception of so-called 'laser-treated' non-nucleated cultured pearls, the practical discrimination of black pearls is not too difficult, when the stringing holes of these grey to black pearls are examined with either the 10× hand lens or low power microscope and fibre-optic illumination, for:

1. Natural black pearls have an 'onion-like' structure of alternating concentric lamellae of nacre and black organic conchiolin. Natural nacre commonly fluoresces a brownish red colour of variable intensity when examined under LWUV.
2. Black bead nucleated pearls have a central greyish white shell bead that is surrounded by a thin layer of dark-coloured conchiolin and a variable thickness of black nacre that also fluoresces brownish red colour of variable intensity in LWUV.
3. Both the nacre and shell bead of black-dyed bead nucleated cultured pearls will be dyed black. Also, black-dyed nacre is inert to LWUV.
   Older silver nitrate dyed black pearls can be discriminated from black aniline-dyed pearls by observing X-radiopaque deposits of particulate silver between the nacre shell bead of the silver nitrate dyed pearl.
4. Greyish to black bead nucleated cultured pearls, that have been colour-enhanced by irradiation, will display a dark nucleus (caused by irradiation of the Mn content of the freshwater shell bead) and greyish white nacre (that has not been colour-enhanced by irradiation). Irradiated nacre is also inert to LWUV.

Both black-dyed bead nucleated cultured pearls and irradiated cultured pearls, *must be disclosed*.

So-called 'laser-treated' grey to black non-nucleated pearls, which are actually either irradiated or dyed and irradiated cultured pearls, usually have a distinctively iridescent lustre. The nacreous surface of these pearls has a distinctively waxy characteristic, and the nacre is inert to LWUV. When detected 'Laser-treated' black pearls *must be disclosed*.

Also *requiring disclosure* are bead nucleated South Sea pearls that have been colour-enhanced to a golden colour by a presently unspecified proprietary treatment. According to Elen (2001), these colour-enhanced pearls can be recognized by the presence of concentrations of dark colour in surface defects on the pearls. These concentrations of colour are also reflected in the partly LWUV fluorescence of this treated nacre.

As price differentials do exist between naturally coloured and dyed Biwa or Chinese freshwater non-nucleated cultured pearls, the dyeing of freshwater pearls *should be disclosed*, however the incidence of dyeing of Chinese freshwater cultured pearls is so extensive that *their disclosure status is yet to be determined*. However, dyed non-nucleated cultured pearls can be identified by the observing:

1. Their unnatural colours.
2. Dye accumulated on the external surface, in or around the stringing hole or in the drill channel of the dyed pearl.
3. Dye staining the string of a necklace.

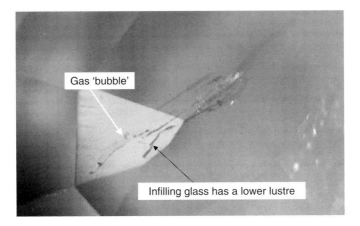

Figure 31.10   Solid 'bubble' included flux glass infilling surface-reaching fractures (×40)

Other Dyed Gem Materials

Other gem materials, that may be either dyed and/or impregnated, and as a consequence must be identified and *disclosed*, include, for example:

- Blue-dyed howlite-imitating turquoise.
- Dyed marble, dolomite, magnesite imitating many ornamental materials.
- Dyed quench-crackled marble.
- Cr^{3+}-dyed chalcedony-imitating chrysoprase.
- Plastic-impregnated and/or dyed reef-building coral-imitating precious coral.
- Dyed amber . . . virtually impossible to detect, except for the observation of 'sun spangle' inclusions.
- Dyed and/or waxed ivory.
- Dyed/plastic impregnated bone.

Heat Treatment

The application of carefully controlled heat, to a gemstone or gem material, may induce colour changes in the material by either changing the valence of transitional elements, by oxidizing or reducing their colour-creating ions, modifying the hydration state of colour-causing inorganic compounds, oxidizing or reducing coloured organic molecules, moving light absorbing unpaired electrons from and between atoms and molecules, or diffusing ionic colour-causing impurities.

Trade-Accepted Heat Treatments

Over the years, simple heat treatment regimes have been devised to produce such commercially acceptable permanent colour changes in gems and gem materials as converting carnelian to sard, brown tiger's eye to red tiger's eye, or pale agate to brown banded agate, transforming unaesthetic gemstone colours into more marketable hues, for example, amethyst to citrine, yellow-green beryl to aquamarine, brownish zircon to colourless or blue zircon, brownish yellow topaz to pink topaz, and abolishing unacceptable pleochroic colours, thus creating: purplish blue tanzanite from khaki green zoisite, strong hued pastel-coloured tourmalines, andalusite with a strong red-green pleochroism, and more.

As each of the above heat treatments creates attractive colour stable materials, and are commercially acceptable. Therefore trade-accepted heat treatments *do not require either identification or disclosure.*

Heat-Treated Amber

Red to reddish brown 'burnt' amber is manufactured from pale-coloured amber by slowly heat treating it under oxidizing conditions. Currently there is no known test that will identify reddish amber that has been artificially heat treated to enhance its value, so this value enhancement *cannot be disclosed.*

Clarified amber is manufactured by heating cloudy amber in rapeseed oil that has a similar RI to that of amber. Oil infills fractures and air bubbles within the amber. However, as the clarified amber cools, disc-like stress fractures or 'sun spangles' are induced throughout. These highly decorative inclusions characterize and identify clarified amber. This value enhancement of amber *must be disclosed.*

In contrast, coated amber, pale-coloured amber coated by a thin brownish orange coating, manufactured by slowly heating cloudy amber from 50–200 °C over a 4–5 hour period, raises some disclosure problems. While this colour enhancement cannot be positively identified as being man-induced, Hutchins and Brown (1996) suggest the observed presence of:

1. A deep brown 'skin' coating a core of near-colourless amber.
2. 'Sun spangle' inclusions in the core.

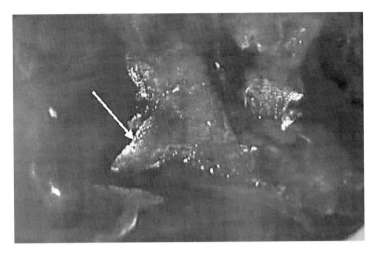

Figure 31.11   Glass-filled surface-reaching fracture (35×)

3. Surface migrating masses of small gas 'bubbles'.
4. A unique orange fluorescence (LWUV > SWUV).

These identify the value-enhanced form of amber. When detected, coated amber *must be disclosed*.

Heat-Treated Ruby and Sapphire

For many decades the marketing of vast amounts of commercially heat-treated sapphires and rubies has complicated problems of disclosure associated with heat-treated gemstones.

Conventional Heat Treatment

Initially, unsophisticated methods were employed in South-East Asia to treat corundum. Heat treatment has been used to remove silk, to develop asterism, to lighten dark blue hues, to induce blue colour in whitish yellow Sri Lanka *gueda* sapphire, to produce valuable reddish, blue, yellow-brown and green hues from poorly coloured corundum, to remove strong colour zoning from both natural and synthetic corundum, and to induce natural-looking 'fingerprint' inclusions in synthetic corundum.

 As these heat treated colour and clarity changes in sapphire and ruby are stable, they significantly increase the value of rough. Consequently, heat treatment is now routinely applied to virtually all rough sapphire and ruby before it is faceted.

Although *no disclosure* is currently required for heat-treated ruby and sapphire, gemmologists should be aware that this value enhancement does stress the rough. Consequently, a significant risk factor (of fracture) exists in all heat-treated natural rubies and sapphires. That is, heat-treated corundum is potentially much more friable than unheat treated corundum.

This is a factor that is well worth remembering, for Kammerling *et al.* (1990) suggest that heat-treated ruby and sapphire can be identified using the gemmological microscope to identify its:

1. Remarkable absence of 'silk' or the presence of 'blobs' in three directions.
2. Colour banding that appears to have diffused.
3. Diffused blue colour surrounding concentrations of titanium.
3. Stress fractures surrounding somewhat melted mineral inclusions.
4. Exploded 'liquid' inclusions.
5. A chalky greenish blue surface fluorescence to UV.
6. Rare 'pock' marked, or multi-planed girdles.
7. Lack of the 'natural' 450 nm absorption.

Diffusion Coating

More concerning, with regard to heat-treated sapphire, was the appearance, first in the early 1980s, of diffusion-treated sapphires, and then in the late 1980s/early 1990s, of deep diffusion-treated sapphires. The properties and features of these colour-enhanced sapphires have been well described and illustrated by Kane *et al.* (1990).

In this very specific heat-treatment technique, high temperatures are applied for long periods of time to faceted pale-coloured corundums that have been either coated with a paste of or enclosed in a powder of synthetic blue sapphire. The end result of this heat-treatment regime is that a very thin layer (<0.5 mm) of synthetic sapphire is fused to, and diffused into, the surface of the natural corundum, artificially creating either a blue sapphire with superficial colour, or a sapphire or ruby with superficial asterism; additional titanium is added to the original powder, and the diffusion-coated sapphire is subsequently annealed.

There is little argument that diffusion heat treatment (coating) of sapphire is a technique that is intended to deceive, and to enhance the value of inferior-coloured sapphire. So this value enhancement *must be detected and disclosed.*

Consequently, gemmologists must be competent to detect this form of heat treatment, ideally using diffused transmitted light, and immersion of the suspect sapphire in di-iodo-methane with an RI of 1.74, to identify (with the gemmological microscope or suitable magnification), in addition to the characteristic features of heat-treated corundum:

Patchy colour distribution
of facets

Accentuated
facet edges

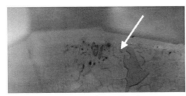

'Bleeding' of coating into
surface reaching fractures

Pock marked facets & girdles

Figure 31.12   Identifying features of diffusion-coated blue sapphire

1. The thinness (up to 0.4 mm) of the coating of synthetic blue sapphire that has been heat-diffused onto the surface of the pale faceted reject *gueda*, or pale sapphire core of this imitation.
2. A localized surface blotchiness, when the corundum is examined immersed in di-iodo-methane (an immersion fluid having an RI (1.74) – virtually identical to that of sapphire (1.76–1.768)). This blotchiness is caused by the removal of some of the diffused coating during a routine repolishing step that inevitably follows diffusion coating.
3. 'Bleeding' of the coating into defects in surface of the natural corundum that has been diffusion-coated.
4. Accentuated facet edges, due to the presence of a double layer of synthetic corundum on these edges.
5. A patchy surface fluorescence due to the loss of some of the coating during repolishing.
6. 'Pock' marked girdles, and multi-faceted girdles, that are more likely to be seen on diffusion-coated sapphires, than sapphires that have been simply heat-treated.

In late 1992, Richard Pollack, of United Radiant Applications, a major producer of diffusion-treated sapphire, claimed to have mastered the technique of creating diffusion-treated ruby by coating faceted cores of pale yellow sapphire with a heat-diffused coating of synthetic ruby. Whether these ruby imitations are economic propositions is a moot point. As a consequence of examining 27 samples of Pollack's product, McClure *et al.* (1993) identified the following diagnostic features for red diffusion-treated corundum:

1. Uneven distribution of facet-to-facet colour, and a dark ring around the girdle, when examined immersed in diffused transmitted light.
2. Colour concentrated at facet edges, when examined with diffused transmitted illumination.
3. Spherical voids, within the diffused layer, usually surrounded by concentration of colour.
4. Dense concentrations of small whitish inclusions just under the surface of the diffused layer.
5. Colour emphasizing the shape and margins of the stone when compared with immersed natural or synthetic rubies.
6. Chalky white to yellow surface fluorescence to SWUV of weak to moderate intensity.
7. Anomalous RIs e.g. multiple readings on the same facet, higher readings than normally expected with ruby, and readings above the scale of the conventional refractometer.
8. Atypical dichroism – one colour being an unusual distinct brownish yellow.

Be-Diffused Corundums

Commercial enhancement of the colours of sapphire and ruby, by the diffusion of foreign elements into the corundum by the external application of near melting point temperatures in a furnace that has a controlled atmosphere, is a technique that has been routinely used for many decades. Indeed, the trade-accepted process of heat-treating corundum in a reducing (hydrogen rich and oxygen poor) atmosphere, to lighten blue colours of sapphire and remove the undesirable brownish secondary colours in some rubies, is based on the diffusion of hydrogen, which has the lowest atomic number of any element, into the sapphire or ruby.

According to Coldham (2002), since late 2001 a new method of colour enhancement, which involves the diffusion of beryllium into pink sapphire by high temperature to create attractive orange sapphires, first appeared undisclosed in the Thai gemstone market. Following suspicions raised by several international gemmologists, co-operative research soon revealed that these beryllium-diffused padparadscha-coloured sapphires were in fact pink sapphires that owed their orange colour to a rim of sapphire that had an induced yellow colour. The divalent beryllium ion, which has an atomic number of 3 to 1 for hydrogen, was soon discovered to be the main chromophore that created a stable yellow beryllium-trapped hole colour centre in the sapphire's crystal structure, after that small ion was diffused inwards by the external application of near melting point heat to the corundum in an oxygen-rich atmosphere.

Initial reports indicated that the diffusion of beryllium into these originally pink sapphires could be detected by first immersing the suspect orange sapphire in a fluid of comparable RI, for example di-iodo-methane that has an RI of 1.742. Subsequent examination of the immersed sapphire, in diffused transmitted light and with magnification, allowed observation of a tell-tale circumferential rim of yellow sapphire that followed the facet outline of the sapphire and surrounded its original pink core. French researchers Fritsch *et al.* (2003) revealed that this rim of induced colour also can be revealed when a Be-diffused sapphire is examined in LWUV. Unfortunately, this identifying observation has become less useful following discovery that the beryllium-diffused yellow colour could be induced throughout the whole sapphire, either by treating smaller stones or by increasing the duration of heat treatment.

The problem of beryllium-diffused corundums has been further complicated as a consequence of Thai and other non-Thai heat treaters using high heat to diffuse beryllium and other light elements, such as lithium (atomic number of 2), to create other colour-enhanced corundums that include, for example, yellow to golden sapphires from near-colourless Sri Lankan sapphire or Australian-type green sapphire, bright red ruby from previously purplish to brown ruby, and lighter blue sapphire from dark blue sapphire. Emmett *et al.* (2003) have written a most comprehensive paper in which all aspects of Be-diffusion of corundum are reviewed in detail.

The following observations can be used to identify or raise suspicion with respect to the identity of Be-treated corundums:

1. Melted or partly melted mineral inclusions (zircon in particular) that are sometimes surrounded by disc-like stress halos that also could display distinctive 'fern-like' recrystallization features.
2. Residual flux glass partly infilling surface-reaching fractures and other defects.
3. Surface-reaching fractures and böhmite-coated parting planes that have been partly 'healed' (at depth) by a mixture of flux glass and recrystallized corundum.
4. Blue 'halos' surrounding included titanium-containing minerals.
5. Surface alteration of some facets and areas of the girdle that were partly coated with recrystallized corundum of blocky to platy habit that is the result of an interaction between excess alumina and molten flux.

In addition, Fritsch *et al.* (2003) have suggested that as the orangy fluorescence of natural coloured Sri Lanka sapphire usually disappears following high-temperature heat treatment, the presence of strong orange luminescence in a heat-treated sapphire likely confirms that the sapphire has been Be-diffused.

With respect to the microscope observations, described above, it is important to remember that the presence of one or more of these features only indicates that the gem corundum has been heat treated. The observed presence of these features in no way identifies beryllium lattice (bulk) diffusion as the cause of any induced colour. In addition, over recent months the diagnostic usefulness of these features has further decreased, as available evidence suggests that the features attributed to high-temperature heat treatment are decreasing in severity as it is thought that recent products are being treated at lower temperatures possible in association with increased pressures.

Unfortunately, the positive detection of Be-diffusion is only possible by complex and expensive chemical analysis, so its detection by conventional gem testing is unlikely.

Once detected, however, Be-diffused sapphires and rubies *must be disclosed*.

Verneuil Synthetics with Induced 'Fingerprints'

Flux-healed quench-crackled Verneuil synthetic ruby and sapphire with induced 'fingerprints', that closely resemble Burmese ruby and Sri Lankan sapphire, are being manufactured as by-products of the routine heat treatment of ruby and sapphire in Chanthaburi, Thailand. This enhanced synthetic is produced by first heating Verneuil boules and then plunging them into water. After cobbing to yield small facetable fragments, pieces of synthetic are heat treated in an unspecified colourless flux for three days. For the first two days of the heat treatment process a kerosene-fired kiln is used. Then an acetylene-fired kiln is used to complete the treatment on the third day.

According to Free *et al.* (1999) identifying inclusions in these treated synthetics include:

1. Curved colour banding that was finely developed in the synthetic ruby, but wider and more diffuse in the synthetic blue sapphire.
2. A chequer-board pattern of fractures that have been induced in the synthetic ruby and sapphire by the quench-crackling process.
3. Surface-reaching fractures that have been partly 'healed' by infilling solid colourless to dark-coloured flux. Occasional included gas 'bubbles' can be noted in solidified flux fracture fillings.

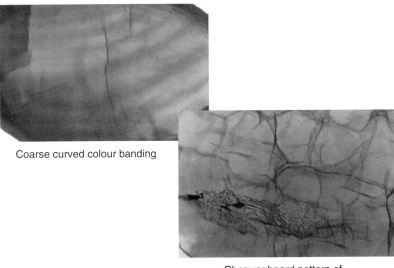

Coarse curved colour banding

Chequer board pattern of
flux filled induced fractures

Figure 31.13   Rough green sapphire from the Anakie sapphire field, Central Queensland,
before (LHS) and after (RHS) diffusion with beryllium (T. Coldham photo)

4. Solidified flux with a quite granular texture that can occasionally be observed when a flux-filled quench-crackled fracture reaches the surface.

This treated synthetic *must be identified and disclosed*.

High-Pressure and High-Temperature Treated Diamond

Pegasus™ diamond, renamed the Monarch™ diamond, and most recently the Beltaire™ diamond, are Lazar Kaplan International's commercial name for brown type IIa nitrogen-poor diamonds that have been whitened by a proprietary GE high-pressure and high-temperature (HPHT) treatment. Recently, small quantities of blue and pink HPHT type IIa diamonds also have begun to be marketed.

High-pressure and high-temperature treated diamonds are, presently, almost impossible to positively identify by normally available gem testing methods, but at least LK International does inscribe the girdles of their HPHT treated diamonds with the identifying inscription 'GE POL'.

Fritsch *et al.* (2001) recommend that suspect HPHT-treated diamonds should first be tested for transparency to SWUV to confirm that they are type IIa diamonds. These researchers further suggest that possible indicators of HPHT treatment, that could be observed with the microscope, but are not readily visible with 10× hand lens examination, include:

1. Whitish or hazy internal graining that ranges in visibility from subtle to obvious.
2. Surface-reaching fractures and cleavages that appear to be partly 'healed' in a manner similar to that observable in heat-treated ruby and sapphire.
3. Surface-reaching fractures and cleavages that appear to have a frosted or granular appearance close to the surface; but change to a glassier appearance deeper within the fracture or cleavage.
4. Black graphitic residues in some surface-reaching fractures and cleavages.
5. Included crystals surrounded by roughly spherical black patches of graphite and an outer halo of tiny outwards radiating fractures.
6. Melted, possibly sulphide, inclusions.

Also, when examined between crossed polars some Pegasus™ diamonds displayed crosshatched, banded and mottled patterns of strain birefringence of moderate to strong intensity.

How HPHT-treated diamonds can be disclosed, if their identifying inscription is removed, is a problem that can only be resolved by advanced Raman fluorescence spectroscopy that is only available in major international gem testing laboratories.

High-Pressure and High-Temperature Treated Yellowish Green Diamond

In 1997 the HRD Institute of Gemmology in Antwerp's Van Bockstael reported the presence in Belgium of several parcels of greenish yellow to brownish type Ia natural diamonds that displayed bright emission lines at 513 and 518 nm, absorption lines at 415 nm (N3), 503 nm (H3), a broad absorption band between 465 and 494 nm, and 982 nm (>1400 °C annealing-induced H2 centre), and whose unusual colours were thought to be due to high-temperature hyperbaric (high pressure) heating of mostly Argyle-type rough in most likely the Russian BARS apparatus. By late 1999, the NovaDiamond Corporation of Provo, Utah, had also introduced and patented a unique HPHT process that converted brown-coloured type Ia diamonds into diamonds that have a vivid yellow-green color. The NovaDiamond™ HPHT process used temperatures of approximately 2000 °C and pressures of 50–60 kilobars to the diamonds. Temperatures below 2000 °C turned the diamonds yellow, while temperatures over 2000 °C turn them yellow-green or dark green colour. Too much heat will convert the diamond to graphite. The NovaDiamond HPHT process is rapid for it can process ten crystals in half-an-hour. NovaDiamond claims to be committed to full disclosure of this treatment, so their treated diamonds will be inscribed with the NovaDiamond logo and a unique serial number, and a detailed certificate will be provided with each diamond. Presently HPHT-treated yellowish green diamonds are being produced commercially in the USA, Russia, Korea and possibly China.

Characteristic features reported for these yellowish green HTHP-treated natural diamonds by van Bockstael (1997) and Reinitz and Moses (1997) include:

1. Greenish yellow to yellowish green induced colours.
2. A distinct greenish overcast (transmission fluorescence) to the strong yellow body colour that is readily visible in daylight and transmitted white light.
3. A strong greenish LWUV fluorescence that was related to brownish planar octahedral growth lines.
4. Strong graining and brownish colour zoning.
5. Induced fractures under the table and in the girdle area.
6. Corrosion (burn) marks and pits, particularly on facets, girdle and surface-reaching fractures and cleavages.
7. Line absorptions at 415 and 503 nm, an absorption band from 465 to 494 nm, and bright emission lines at 513 and 518 nm when examined with the hand spectroscope.

Yellowish green HTHP-treated diamonds *should be detected and disclosed*.

Irradiation and Annealing

High energy Electro Magnetic Radiations, or EMRs, and high-energy particulate irradiation, are used to modify gemstone colours, by artificially creating colour centres in gem materials that would otherwise be colourless or poorly coloured. Radiation-induced colour centres are defects in the gem material's crystal structure where unpaired electrons can be energized above their ground state by absorbing specific energies (wavelengths) from incident light. Colour centres vary in complexity from single atomic or electron vacancies to displaced interstitial ions and electrons, or complex centres involving various atomic substitutions. Subsequent annealing (prolonged heating at high temperature) of radiation-induced colour centres returns some of the radiation-displaced atomic fragments to their previous sites: thus repairing some of the radiation damage and modifying the irradiation-induced colour.

Several types of radiation are commercially used to colour-modify gems and gem materials. These energetic atomic particles and EMRs include charged atomic particles such as high-speed, negatively charged electrons, generated in a linear accelerator or betatron, and high-speed, positively charged protons, alpha particles (helium nuclei) or deuterons generated in linear accelerators or cyclotrons, that only penetrate the gemstone they are irradiating to a very limited depth; uncharged particles such as neutrons from an atomic pile that totally penetrate the gemstone; and high-energy gamma rays that may totally penetrate some gem materials, depending on the atomic weights of the atoms forming the gemstone.

The colour changes (colour centres) that may be induced in gemstones by irradiation include:

Beryl	colourless → yellow beryl, blue → green, pale pink → deep blue Maxixe beryl.
Corundum	colourless → yellow, pink → padparadscha (orange).
Diamond	yellow-brown → green and with subsequent heat treatment yellow, brown, orange, pink.
Fluorite	colourless → range of colours.
Pearl	light coloured → grey, brown, blue, black.
Quartz	colourless, yellow, pale green → smoky, amethyst.
Scapolite	colourless, light pink, blue → blue, red, lavender, purple.
Spodumene	colourless, pink → orange, yellow, green, strong pink.
Topaz	colourless → green-brown . . . with heat treatment blue. yellow → yellow-brown.
Zircon	colourless → brown, red.

Diamond

Diamond is the gem most commonly permanently colour-modified by the technology of irradiation and annealing. Due to the significant price differential, between naturally coloured and artificially coloured fancies, *disclosure of this treatment is mandatory*.

According to Nassau (1984), commercial colour modifiers of diamond use two forms of irradiation. Today electron irradiation in a linear accelerator, Van der Graaff generator, or betatron is not commonly used. This expensive and now not-much-used method of irradiating diamond only induces skin-deep (<1 mm) colours in diamond by displacing carbon atoms from near its surface. The dark greenish to bluish brown colours induced in diamond by this irradiation are subsequently converted, by annealing, into yellow-greens, yellows, browns, blues, and rare oranges and reds depending on the type of the diamond initially irradiated.

Useful descriptions of the identification criteria that gemmologists can use to identify irradiated-annealed diamonds can be found in Collins (1982) and Nassau (1984).

So-called 'cyclotroned diamonds' are identified by:

1. Their 'skin deep' colour distribution as evidenced by observation of a halo around the girdle of table-treated diamonds or an 'umbrella' around the girdle in culet-treated diamonds. In emerald cuts this 'umbrella' is represented by strong colour bands parallel to the keel of the treated diamond.
2. The presence of the 594 nm absorption in their visible absorption spectrum.
3. The absence of electro-conductivity on blue-coloured 'cyclotroned' diamonds.

Neutron irradiation in an atomic pile is a relatively inexpensive method of inducing dull greenish brown colours in yellowish off-coloured type Ia diamonds by displacing carbon atoms throughout the diamond. Subsequent annealing produces predominantly yellow, brown and orange-brown diamonds. However, by careful selection of the type of diamond to be irradiated, for example type Ib diamond, rare colours such as pinks mauves and reds can be produced by pile treatment followed by annealing.

Pile-treated diamonds can be identified by:

1. The completeness of their colour modification.
2. The presence of a 594 nm absorption in the orange region of their visible spectrum that is induced by the annealing step.

There are, however, some practical problems to solve before the 594 nm absorption spectra in diamond can be observed by the gemmologist, for this absorption will be virtually impossible for the average gemmologist to see, unless the diamond is illuminated by a cool fibre-optic light source, and is either precooled in a refrigerator, or by a spray refrigerant, dry ice or liquid nitrogen.

Woods and Collins (1986) also have revealed that the annealing-induced 594 nm absorption in irradiated diamond can be removed by prolonged high-temperature annealing at about 800 °C. If this occurs then the annealed yellow diamonds can be identified only by IR spectroscopy detecting the presence of high-temperature annealing-induced 1936 nm (H1c) and/or 2024 nm (H1b) absorptions in their IR spectra. Unfortunately, trade gemmologists who do not have access to an IR spectrophotometer cannot make these observations.

Irradiated-Annealed Pink Diamond

Colour-treated pink diamonds are relatively rare, usually strongly coloured, and inevitably small in size (<0.25 ct) for the starting material required to produce this colour artificially is yellow type Ib diamond. Irradiation followed by annealing produces identifying absorption lines at 595 and 637 nm. Other features in the spectrum of irradiated-annealed pink diamond include discrete absorption lines at 575 and 595 nm (induced by annealing), and yellow and orange emission lines at 596 and 638 nm respectively, which are responsible for the strong orange fluorescence (SWUV > LWUV) of these colour-enhanced diamonds.

In yellow body coloured mixed type Ia and Ib diamonds, the induced pink colour of the 637 nm absorption often results in a mauve to brownish purple-coloured diamond. The final colour attained by an irradiated and annealed type Ia + Ib diamond will depend on two factors, the strength of its yellow colour before irradiation, and the strength of irradiation.

Colour-Enhanced Synthetic Diamond

Moses *et al.* (1993) described the properties of two colour-treated synthetic red diamonds. These brownish orangy-red diamonds displayed the cub-octahedral colour zoning, graining, green-orange (radiation-annealing induced) fluorescence (SWUV = or > LWUV) and metallic Fe-Ni inclusions of type Ib synthetic diamond of Russian manufacture. In addition, the visible absorption spectrum of these diamonds displayed numerous sharp bands (attributed to Ni) between 400 and 800 nm; with several between 500 and 660 nm being visible with the hand-held spectroscope. In addition, the presence of the 595 nm absorption, as well as the H1a and H1b absorptions in the IR, identified these red diamonds as being created by irradiation followed by high pressure and temperature annealing of type Ib synthetic diamond.

The spectroscopy of coloured diamond is a very specialized skill, that must be performed on a cooled diamond, with the assistance of a spectrophotometer, rather than the gemmologist's rather primitive spectroscope. All coloured diamonds should be submitted to a commercial laboratory for the determination of whether or not their colours are either natural or artificial.

Other Irradiated Gemstones

Other common gemstones, that are commercially colour-modified by irradiation and annealing include blue topaz, synthetic amethyst, bead nucleated cultured pearls, pink to red tourmaline and maxixe sapphire-blue beryl.

Blue Topaz

Various shades of topaz can be produced commercially by the use of four separate regimes of irradiation on colourless topaz followed by an essential annealing step. For example, gamma rays from a cobalt-60 source will create a light blue topaz marketed as Cobalt Blue following heat treatment. High-speed electrons from a linear accelerator will yield an attractive sky-blue topaz following annealing; following annealing high-energy neutrons from a nuclear reactor will yield a deep blue topaz known as London Blue; while topaz irradiated first in a nuclear reactor then by a particle accelerator will yield mid-blue topaz after annealing that has been variously named Super Blue, Swiss Blue, American Blue, Californian Blue or Electra Blue.

It is important to remember that irradiation of colourless topaz must be followed by an essential annealing step, to remove unacceptable yellow to brown radiation-induced colours from the induced blue colour.

Residual radiation is seldom a problem with gamma-irradiated topaz if the treated topaz is stored for a sufficient time to allow its induced radioactivity to decay to safe levels. In the USA, in particular, importers of irradiated blue topaz must have these checked, for dangerous levels of residual radioactivity, by accredited laboratories, before they are able to be sold.

Rossman (1981) clearly indicated that conventional gemmological testing is unable to detect whether or not a blue topaz has had its colour induced by man, but the presumption must be made that the majority of blue topaz in today's market has been produced by irradiation and annealing. As these is no difference in value between natural coloured blue topaz and irradiated-annealed blue topaz, *disclosure is not required*.

Synthetic Amethyst

Synthetic amethyst is also commercially produced by irradiation and annealing of iron containing hydrothermally grown synthetic quartz. To manufacture synthetic amethyst, quartz containing a small percentage of charge-compensated Fe^{3+} replacing Si^{4+} is hydrothermally grown. When this synthetic quartz is irradiated by gamma rays, an amethystine colour centre is created.

If the gemmologist wishes to discriminate synthetic from natural amethyst, then this may be achieved by:

1. Observation of characteristic inclusions, such as segments of the seed plate, and the presence of greyish 'bread crumb' inclusions that characterize the hydrothermal growth process.
2. Checking for the lack of Brazil twinning, when the amethyst is examined parallel to its *c*-axis, between crossed polars.
3. Submitting the amethyst for IR spectrographic examination.

Because of the great difficulty and cost of identifying synthetic amethyst this *does not require specific disclosure*.

Irradiated Black Pearls

Whitish cream to yellowish bead nucleated cultured pearls can be colour modified by gamma-irradiation and irradiation in a particle accelerator to create greyish irradiated 'black' pearls. The induced colour of these pearls is caused by the irradiation darkening the shell beads of these pearls, through its interaction with the manganese content of this freshwater shell. Simple observation of a greyish pearl having a black nucleus and whitish nacre will confirm this colour modification. Irradiated black pearls *must be disclosed*.

Pink Tourmaline

Pale pink tourmalines may be irradiated in a gamma cell to yield strongly hued pink to red tourmalines. These radiation-induced hues are quite colour stable; however, care must be exerzised to ensure that they are not radioactive when sold. These irradiation-induced coloured tourmalines *cannot be disclosed*, as this treatment cannot be detected by conventional gemmological testing, unless the treated tourmalines are radioactive.

Some Perceived Problems

Irradiation-annealing technology, for colour-modifying gemstones and gem materials, is not without its problems, for some of the induced colours are not stable and tend to fade when exposed to strong light and heat. Examples of these colour-unstable irradiated-annealed gemstones include dark blue maxixe beryl, brownish orange topaz, and yellow-brown Sri Lankan sapphire. Suspect colour fading gemstones must be subjected to a simple fade test and then *disclosed*.

Irradiation also may induce dangerous levels of residual radioactivity in gemstones such as orange spodumene, some purplish tourmaline, chrysoberyl cat's-eyes and some blue topaz. Perhaps gemmologists of the future may consider radiation detectors, such as Geiger counters, which may become essential gem-testing instruments.

With regard to a radiation safety standard for irradiated-annealed gemstones Ashbaugh (1992) suggested that a gemstone emitting 2 nanocuries of radioactivity per gram can be judged to be safe, and virtually harmless to wear, as this amount of residual radiation it emits is only a very small fraction of general background radiation.

References and Further Reading

Ashbaugh, C. (1992) Gamma-ray spectroscopy to measure radioactivity in gemstones. *Gems & Gemology*. 28(2), 104–11.

Bank, H., Henn, U. and Milisenda, U. (1999) Treated rock crystal. *Gemmologie*. 48(3), 123–24.

Brown, G. (1991) Treated Andamooka matrix opal. *Gems & Gemology*. 27(2), 100–07.

Coldham, T. (2002) Orange sapphires or just lemons? *Australian Gemmologist*. 21(7), 288–93.

Collins, A.T. (1982) Colour centres in diamond. *Journal of Gemmology*. 18(1), 37–75.

Elen, S. (2001) Spectral reflectance and fluorescence characteristics of natural-color and heat-treated 'golden' South Sea cultured pearls. *Gems & Gemology*. 37(2), 114–23.

Emmett, J., Scarratt, K., McClure, S., Moses, T., Douhtit, T., Hughes, R., Novak, S., Shigley, J., Bordelon, O. and Kane, R. (2003) Beryllium diffusion of ruby and sapphire. *Gems & Gemology*. 39(2), 84–135.

Free, J., Free, I., Brown, G. and Linton, T. (1999) Verneuil synthetic corundums with induced fingerprints. *Australian Gemmologist*. 20(8), 342–46.

Fritsch, E. and Stockton, C. (1987) Infrared spectroscopy in gem identification. *Gems & Gemology*. 23(1), 18–26.

Fritsch, E. (1991) Bluish grey synthetic diamond thin films grown on faceted diamonds. *Gems & Gemology*. 27(2), 118–19.

Fritsch, E., Wu, S.-T.T., Moses, T., McClure, S.F. and Moon, M. (1992) Identification of bleached and polymer-impregnated jadeite. *Gems & Gemology*. 28, 176–87.

Fritsch, E. (1994) Type-B jadeite: new polymers and estimating amount of wax. *Jewellery News Asia*. No. 123, 106–10.

Fritsch, E., McClure, S., Ostrooumov, M., Andres, Y., Moses, T., Koivula, J. and Kammerling, R. (1999) The identification of Zachery-treated turquoise. *Gems & Gemology*. 35(1), 4–16.

Fritsch, E., Chalain, J.-P., and Hänni, H. (2001) Identification of GR POL™ diamonds. *Australian Gemmologist*. 21(4), 172–77.

Fritsch, E., Chalain, J.-P., Hänni, H., Devouard, B., Chazot, G., Giuliani, G., Schwarz, D., Rollion-Bard, C., Garnier, V., Barda, S., Ohnenstetter, D., Notari, F. and Maitrallet, P. (2003) Le nouveau traitement produisant des couleurs orange à jaune dans les saphirs. *Revuw de Gemmologie* A.F.G. No. 147, 11–20.

Hammer, V.M.F. and Schmetzer, K. (2000, December issue) Synthetic moissanite determination remains the challenge, *GZ* (European Jeweller), 108–09.

Hänni, H. (1998) Fracture filling in emeralds and its detection with laboratory methods. *ICA Gazette.* July–August, 5–7.

Hodgkinson, A. (1993) Gemstone enhancement – Detection of polymer-treated jadeite. *Journal of Gemmology.* 23(7), 415–17.

Hodgkinson, A. (1998) Scottish gem lab news. *Australian Gemmologist.* 20(4), 154–58.

Hutchins, P. and Brown, G. (1996) An amber treatment. *Australian Gemmologist.* 19(7), 302.

Kammerling, R., Koivula, J., and Kane, R. (1990) Gemstone enhancement and its detection in the 1980s. *Gems & Gemology.* 26(1), 32–49.

Kane, R., Kammerling, R., Koivula, J., Shigley, J. and Fritsch, E. (1990) The identification of blue diffusuin-treated sapphires. *Gems & Gemology.* 26(2), 115–33.

Keeling, J. and Townsend, I.J. (1996) Dyed opalised sandstone and conglomerate: A new gem product from Andamooka. *Australian Gemmologist.* 19(5), 226–31.

Koivula, J., Kammerling, R., Fritsch, E., Fryer, C., Hargett, D. and Kane, R. (1989) Data on filled diamonds: Identification and durability. *Gems & Gemology.* 25(2), 68–83.

McClure, S., Kammerling, R. and Fritsch, E. (1993) Update on diffusion-treated corundum: Red and other colours. *Gems & Gemology.* 29(100), 16–29e.

McClure, S., King, J.M., Koivula, J. and Moses, T. (2000) A new lasering technique for diamond. *Gems and Gemology.* 36(2), 138–46.

Miles, E. (1992–93) Diamond-coating techniques and methods of detection. *Gems & Gemology.* 10(12), 355–64.

Moses, T., Reinitz, I., Fritsch, E. and Shigley, J. (1993) Two treated-color synthetic red fiamonds seen in the trade. *Gems & Gemology.* 29(3), 182–90.

Moses, T., Reinitz, I., Koivula, J., Buerki, P. and McClure, S. (2000) Update on new InColor treated black and green diamonds. *The Loupe.* 9(4), 16–19.

Nassau, K. (1984) *Gemstone Enhancement.* Butterworths: London.

Reinitz, I. and Moses, T. (1997) Treated-color yellow diamonds with green graining. *Gems & Gemology.* 33(2), 136.

Reinitz, I., Koivula, J., Elen, S., Buerki, P. and Muhlmeister, S. (2000) With evidence of a new diamond treatment. *Gems & Gemology.* 36(1), 61–62.

Rossman, G. (1981) Color in gems: The new technologies. *Gems & Gemology.* 17(2), 60–71.

Schmetzer, K., Hänni, H., Jegge, E. and Schupp, F.-J. (1992) Dyed natural corundum as a ruby imitation. *Gems & Gemology.* 28(2), 112–15.

Shigley, J., McCluer, S., Koivual, J. and Moses, T. (2000) New filling material for diamond from Oved Diamond Company: A preliminary study. *Gems & Gemology*. 36(2), 147–53.

Tay, T.S. (2001) Disclosure – Gemstones v synthetics: Does the jewellery industry care? *Australian Gemmologist*. 21(2), 67–75.

van Bockstael, M. (1997) A new treatment. *Antwerp Facets* (December). 49–51.

Woods, G. and Collins, A. (1986) New developments in spectroscopic methods for detecting artificially coloured diamonds. *Journal of Gemmology*. 20(20), 75–82.

Wu, S.-T. (1999) The identification of B + C jade. *Journal of the Hong Kong Gemmological Association*. 20, 27–28.

Identification Tables

Compiled by
Christine Woodward

Note: Many less common materials are not included. The tables remain largely unaltered from the previous edition of *Gems*.

Table 1
Refractive indices, optical characters, hardnesses and specific gravities

Refractive indices are the most important testing factor, so the values given for these are printed in ascending order. The specific gravity values, rather less used in gem testing, in most cases increase with increasing refractive index and fall within limits into defined groups, and therefore may be readily found in the table. Where constants are outside the given range, or may not easily be found within that range, they are printed in italics; the materials are then repeated in the appropriate places, with the contents in ordinary type. (Non-gem materials with SGs less than 1 are not repeated.) The word 'to' between figures implies variation in range, while '–' indicates the greatest and least readings in doubly refractive stones. 'I' means isotropic, 'U' uniaxial, and 'B' biaxial.

Glass imitation stones commonly have refractive indices which lie between 1.50 and 1.70.

Table 1

Refractive index	Birefringence	Optic sign	Optical character	Hardness	Specific gravity	Name
1.3					**1.0**	
1.378–1.390	0.012	Positive	U	5 to 6	3.15	Sellaite
1.4						
1.54	—	—	I	2 to 2.5	1.03 to 1.10	Amber
1.54	—	—	I	2	1.03 to 1.10	Copal resin
1.59	—	—	I	2.25	1.05	Polystyrene

Table 1 (cont.)

Refractive index	Birefringence	Optic sign	Optical character	Hardness	Specific gravity	Name
1.53 (mean)	—	—	—	2	1.00 to 2.00	Meerschaum
1.64 to 1.68	—	—	I	2.5 to 4	1.10 to 1.40, usually 1.30 to 1.35	Jet
1.56	—	—	—	3.5	1.34	Black coral
1.434	—	—	I	4	3.18	Fluorite
1.44 to 1.46	—	—	I	5.5 to 6	1.98 to 2.20	Opal
1.435 to 1.455					1.97 to 2.06	Fire opal
1.60 to 1.63	—	—	I	2	1.15 to 1.20	Vulcanite
1.50	—	—	I	2	1.18 to 1.19	Perspex
1.46	—	—	I	6	2.21	Silica glass
1.460–1.485	0.025	—	B	3.5	2.71	Creedite
1.464–1.500 to 1.468–1.507	0.031 to 0.043	Negative	U	3.5	1.88 to 1.90	Thaumasite
1.61 to 1.66	—	—	—	2.25	1.25 to 2.00, usually 1.26 to 1.30	Bakelite
1.48	—	—	I	5.5 to 6	2.15 to 2.35, usually 2.25 to 2.30	Sodalite
1.487	—	—	—	5 to 5.5	2.22 to 2.29	Analcime
1.480–1.493	0.013	Positive	B	5.5	2.20 to 2.25	Natrolite
1.48 to 1.54	—	—	I	5.5	2.34 to 2.39	Moldavite
1.48 to 1.51	—	—	—	5	2.33 to 2.42	Obsidian
1.486–1.658	—	—	—	3	2.71	Calcite (marble)
1.54	—	—	—	2.5	1.26 to 1.35	Tortoise-shell and horn
1.490 to 1.505	—	—	I	1.5	1.26 to 1.80, usually 1.29 to 1.40	Cellulose acetate plastic
1.55 to 1.56	—	—	I	2.25	1.32 to 1.34	Casein (protein plastic)
1.495 to 1.510	—	—	I	1.5	1.36 to 1.80, usually 1.36 to 1.42	Celluloid
1.54	—	—	—	2.5	1.38 to 1.42	Vegetable ivory
1.49 to 1.53	—	—	I	5	2.36 to 2.51	Tektites
1.490–1.650	0.160	Positive	B	2.5	2.23	Whewellite
1.492–1.504	0.009	Positive	B	4.5	2.19 to 2.22	Yugawaralite
1.496	—	—	I	5.5 to 6.5	2.4	Haüyne
1.496–1.502	0.006	Positive	U	6	2.36 to 2.57	Tugtupite
1.56	—	—	—	2.5	1.60 to 1.85	Deer horn

Table 1 (*cont.*)

Refractive index	Birefringence	Optic sign	Optical character	Hardness	Specific gravity	Name
1.54	—	—	—	2.25 to 2.5	1.70 to 1.98 2.00 1.85 to 1.98	Dentine ivory Elephant Hippopotamus walrus and narwhal
1.5					**2.0**	
1.45	—	—	I	5.5 to 6	2.00 (mean)	Opal
1.464–1.500 to *1.468–1.507*	0.031 to 0.043	Negative	U	3.5	1.88 to 1.90	Thaumasite
1.502–1.681	0.179	Negative	U	3.5 to 4	2.8 to 2.9	Dolomite
1.492–1.504	0.009	Positive	B	4.5	2.19 to 2.22	Yugawaralite
1.54	—	—	—	2.5	2.00	Bone
1.50 (mean)	0.011	Positive	U	2 to 4	2.00 to 2.45	Chrysocolla
1.48	—	—	I	5.5 to 6	2.15 to 2.35, usually 2.25 to 2.30	Sodalite
1.50 (mean)	—	—	I	5.5	2.70 to 2.90	Lapis lazuli
1.50	—	—	I	2	*1.18 to 1.19*	Perspex
1.495–1.505	—	—	—	1.5	*1.29 to 1.40*	Cellulose acetate
1.487	—	—	I	5 to 5.5	2.20 to 2.29	Analcime
1.480–1.493	0.013	Positive	B	5.5	2.20 to 2.25	Natrolite
1.46	—	—	I	6	2.21	Silica glass
1.512–1.523	0.011	Negative	B	5 to 5.5	2.21	Scolecite
1.50 (mean)	0.001	Positive	B	5	2.29	Mesolite
—	—	—	—	4 approx. unglazed	2.3	Porcelain
1.490–1.650	0.160	Positive	B	2.5	2.23	Whewellite
1.48 to 1.54	—	—	I	5.5	2.34 to 2.39	Moldavite
1.48 to 1.51	—	—	I	5	2.33 to 2.42	Obsidian
1.496–1.502	0.006	Positive	U	6	2.36 to 2.57	Tugtupite
1.504–1.516	0.012	Positive	B	6 to 6.5	2.39 to 2.46	Petalite
1.495–1.510	—	—	I	1.5	*1.36 to 1.42*	Celluloid
1.496	—	—	I	5.5 to 6.5	2.4	Haüyne
1.51 (mean)	0.020	Positive	B	1	*1.65 to 2.00*	Ulexite
1.51 (mean)	0.001	Positive	U	5.5 to 6	2.45 to 2.50	Leucite
1.512–1.523	0.011	Negative	B	5 to 5.5	*2.21*	Scolecite
1.517 to 1.525	—	—	I	6.5	2.85 to 2.94	Pollucite
1.51 (mean)	0.022	Negative	U	6	2.42 to 2.50	Cancrinite
1.515–1.717	0.202	Negative	U	3.5 to 4.5	*3.00 to 3.12*	Magnesite
1.518–1.524	0.006	Negative	B	—	2.57 to 2.58	Sanidine
1.49 to 1.53	—	—	I	5	2.36 to 2.51	Tektites
1.52–1.53	0.010	Positive	B	1.5 to 2	2.30 to 2.33	Gypsum (alabaster)

Table 1 (*cont.*)

Refractive index	Birefringence	Optic sign	Optical character	Hardness	Specific gravity	Name
1.52–1.55	0.027	Negative	U	2.5	2.15 to 2.22	Stichtite
1.52–1.54	0.028	Positive	B	5 to 5.5	2.3 to 2.4	Thomsonite
1.52–1.53 to 1.53–1.54	0.008	Negative	B	6	2.56 to 2.59	Orthoclase feldspar (moonstone)
1.522–1.527	0.005	Negative	B	6	2.56	Yellow orthoclase feldspar
1.522–1.530	0.008	Negative	B	6.5	2.56 to 2.58	Microcline feldspar
1.529–1.532 to 1.548–1.551	0.003	Negative	U	5.5 to 6	2.46 to 2.61	Milarite
1.530–*1.685*	0.155	Negative	B	3.5 to 4	2.93 to 2.95	Aragonite
1.553–*1.625* to 1.559–*1.631*	0.072	Positive	B	7.5	2.35	Hambergite
1.586–*1.614*	0.028	Positive	B	4.5	2.42	Colemanite
1.535–1.537	0.002	Pos/Neg	U	4.5 to 5	2.30 to 2.50	Apophyllite
—	—	—	—	6 to 6.5	approx. 2.5	Nevada wonderstone
1.530 to 1.539	0.004 (av.)	Negative	B	7	2.58 to 2.62	Chalcedony
1.53 (mean)	—	—	—	2	*1.00* to 2.00	Meerschaum
1.53–1.54 to 1.54–1.55	0.008 to 0.012	Negative	B	7	2.57 to 2.66, usually 2.57 to 2.61	Iolite
1.530–1.556	0.026	Negative	B	3.5 to 4	2.8 to *3.30*, usually 2.80 to 2.9	Lepidolite
1.532–*1.680*	0.148	Negative	B	3.5	*4.27 to 4.35*	Witherite
1.54	—	—	—	3.5	*1.34*	Black coral
1.54	—	—	—	2.5	*1.38 to 1.42*	Vegetable ivory
1.54	—	—	—	2.5	2.00	Bone
1.54	—	—	—	2.5 to 2.75	*1.70 to 1.98*	Dentine ivory
1.54	—	—	I	2 to 2.5	*1.03 to 1.10*	Amber
1.54	—	—	I	2	*1.03 to 1.08*	Copal resin
1.54	0.004	Negative	U	5.5 to 6	2.55 to 2.65	Nepheline
—	—	—	—	3.5	2.6 to 2.7	Coral
1.54 approx.	—	—	—	7	2.58 to 2.91	Jasper
1.54 approx.	0.011	Positive	B	6 to 6.5	2.62	Albite feldspar
1.518–1.524	0.006	Negative	B	—	2.57 to 2.58	Sanidine
1.54 approx.	—	—	B	1 to 2	2.7 to 2.8	Soapstone
—	—	—	—	3.5 to 4	2.20 to 2.78	Pearl
					2.69 to 2.73	Fine pearl
					2.40 to 2.65	Blue pearl
					2.67 to 2.78	Australian pearl
					2.70 to 2.78	Cultured pearl

Table 1 (*cont.*)

Refractive index	Birefringence	Optic sign	Optical character	Hardness	Specific gravity	Name
					2.20 to 2.66	Black clam pearl
1.542–1.549	0.007	Negative	B	6 to 6.5	2.62 to 2.65	Oligoclase feldspar (sunstone)
1.544–1.553	0.009	Positive	U	7	2.65	Quartz
1.544–1.560	0.016	Negative	U	6	2.63	Scapolite (pink and white)
1.548–1.568	0.020	Negative	U	6	2.71	Scapolite (yellow)
1.460–1.485	0.025	—	B	3.5	2.71	Creedite
1.55	—	—	—	2.5	*1.26 to 1.35*	Tortoise-shell and horn
1.550–1.559	0.009	Positive	B	5 to 6	2.54 to 2.68	Charoite
1.553–1.562	0.009	Negative	B	5.5 to 6	2.80 to 2.85	Beryllonite
1.553–*1.625* to 1.559–*1.631*	0.072	Positive	B	7.5	2.35	Hambergite
1.55 to 1.56	—	—	I	2.25	*1.32 to 1.34*	Casein plastic
1.55–*1.60*	—	—	—	1.5	2.8	Pyrophyllite
1.56	—	—	—	2.5	*1.60 to 1.85*	Deer horn
1.56 (mean)	—	Negative	B	5	2.4 to 2.6	Variscite
1.560–1.568 to 1.565–1.573	0.008	Positive	B	6 to 6.5	2.68 to 2.69	Labradorite feldspar (yellow)
1.56	—	Negative	B	2.5 4	2.5 to 2.7 2.58 to 2.62	Serpentine (bowenite)
1.558–1.563	0.003	Negative	U	7.5	2.65	Synthetic emerald (flux melt)
1.56–1.57 to 1.59–1.60	0.006 to 0.007	Negative	U	7.5	2.69 to 2.75	Emerald
1.571–1.577	0.006				2.69	Colombia (Chivor)
1.578–1.584	0.006				2.71	Colombia (Muzo)
1.578–1.585	0.007				2.74	Tanzania
1.566–1.571	0.005				2.69	Brazil (some)
1.581–1.588	0.007				2.74	Russia and Brazil
1.586–1.593	0.007				2.75	Transvaal
1.586–1.593	0.007				2.75	Zimbabwe
1.585–1.595	0.007 to 0.010				2.73 to 2.74	India
1.584–1.591	0.007				2.74	Habachtal
1.588–1.595	0.006				2.76	Pakistan
1.583–1.590	0.007				2.74	Zambia
1.567–1.576	0.009	Negative	B	6 to 6.5	2.73	Bytownite feldspar

Table 1 (*cont.*)

Refractive index	Birefringence	Optic sign	Optical character	Hardness	Specific gravity	Name
—	—	—	—	2.5	2.60 to 2.85	Pinite
1.57	0.003	Pos/Neg	B	2.5	2.60 to 2.85	Pseudophite
1.57	—	Negative	B	2.5	2.61	Serpentine (williamsite)
1.57 approx.	—	—	—	3 to 4	2.7 approx.	Ophicalcite (Connemara marble)
1.486–1.658	0.172	Negative	U	3	2.71 / 2.5 to 2.6	Calcite / Marble
1.570–1.575 to 1.580–1.586	0.005 to 0.006	Negative	U	7.5	2.68 to 2.73	Aquamarine; yellow and some pink beryls
1.571–1.614	0.043	Positive	B	3 to 3.5	2.90 to 2.98	Anhydrite
1.574–1.588	0.014	Positive	B	5	2.7	Augelite
1.57 to 1.58	—	Negative	B	3.5 to 4	2.5 to 2.7	Bastite
1.57–1.63	—	—	—	5	*3.00 to 3.10*	Odontolite
1.50	—	—	I	5.5	2.70 to 2.90	Lapis lazuli
1.58 (mean)	—	—	—	3	2.80 to 2.99	Verdite
1.580–1.627	0.047	Positive	B	2	2.6	Vivianite
1.58–1.59 to 1.59–1.60	0.008 to 0.009	Negative	U	7.5	2.80 to 2.90	Some white and pink beryls
1.582–1.627	0.028	Positive	U	6	2.73 to 2.90	Catapleiite
1.583–1.593	0.010	Positive	B	5 to 6	2.7	Xonotlite
1.586–1.614	0.028	Positive	B	4.5	2.42	Colemanite
1.59 (mean)	0.020	Negative	B	3.25	2.58	Howlite
—	—	—	—	Varies	2.7 to 2.8	Kakortokite
1.59	—	—	I	2.25	*1.05*	Polystyrene
1.598–1.602 to 1.610–1.611	0.004	Positive	U	5 to 6	2.8 to *3.1*	Eudialyte
1.607–1.610	0.004	Negative	U	5.5 to 6.5	2.74	Sugilite
1.595 (mean)	0.009	Positive	U	5	2.81	Wardite
1.593–1.612	0.019	Negative	U	3.5	*3.00*	Melinophane
1.595–1.616	0.021	Positive	B	3.5	*3.1*	Phosphophyllite
1.595–1.633	0.038	Positive	B	5 to 6	2.62 to 2.88	Pectolite
1.606–1.608	0.002	Negative	U	6 to 7	2.76 to 2.90	Sogdianite
1.597	—	—	I	6 to 6.5	*3.28*	Ekanite
—	—	—	—	3.5 to 4	approx. 2.8	Mollusc shell
—	—	—	—	3.5 to 4	2.84 to 2.89	Conch pearl
1.61–1.64	0.030	Positive	B	6	2.80 to 2.95, usually 2.88 to 2.94	Prehnite
1.61–1.65	0.040	Positive	B	5.5 to 6	2.6 to 2.85 / approx. 2.6	Turquoise / American

Table 1 (*cont.*)

Refractive index	Birefringence	Optic sign	Optical character	Hardness	Specific gravity	Name
					approx. 2.8	Persian and Egyptian
1.603–1.623	0.020	Positive	B	5.5	2.980 to 2.995	Brazilianite
1.502–1.681	0.179	Negative	U	3.5 to 4	2.8 to 2.9	Dolomite
1.616–1.631	0.015	Negative	B	4.5 to 5	2.8 to 2.9	Wollastonite
1.517 to 1.525	—	—	I	6.5	2.85 to 2.94	Pollucite
—	—	—	—	6 to 7	2.85 to 3.20	Unakite
1.530–1.685	0.155	Negative	B	3.5 to 4	2.93 to 2.95	Aragonite
1.66 to 1.67	—	—	I	7	2.96	Boracite
1.654–1.670	0.016	Positive	U	7.5 to 8	2.95 to 2.97	Phenakite
1.60–1.62	0.02	Negative	B	5.5 to 6	2.98	Tremolite
1.625–1.669	0.044	Negative	B	5	2.90–3.00	Datolite
1.62 (mean)	0.027	Negative	B	6.5	2.90 to 3.02	Nephrite
1.6					**3.0**	
1.720–1.735	0.015	Positive	U	9	3.00 to 3.02	Bromellite
1.597	—	—	I	6 to 6.5	3.28	Ekanite
1.598–1.602 to 1.610–1.611	0.004	Positive	U	5 to 6	2.8 to 3.1	Eudialyte
1.606–1.608	0.002	Negative	U	6 to 7	2.76 to 2.90	Sogdianite
1.620–1.642	0.022	Negative	B	5 to 5.5	3.05	Actinolite
1.378–1.390	0.012	Positive	U	5 to 6	3.15	Sellaite
1.60 (mean)	—	Positive	B	6.5	3.1 to 3.2	Chondrodite
1.58–1.59 to 1.59–1.60	0.008 to 0.009	Negative	U	7.5	2.80 to 2.90	Some white and pink beryls
1.602–1.639	0.037	Negative	B	7.5	3.0	Grandidierite
1.603–1.623	0.020	Positive	B	5.5	2.980 to 2.995	Brazilianite
1.607–1.610	0.004	Negative	U	5.5 to 6.5	2.74	Sugilite
1.593–1.614	0.019	Negative	U	3.5	3.00	Melinophane
1.586–1.614	0.028	Positive	B	4.5	2.42	Colemanite
1.665–1.684	0.019	Positive	B	7 to 8	3.09	Lawsonite
1.595–1.616	0.021	Positive	B	3.5	3.1	Phosphophyllite
1.60–1.62	0.02	Negative	B	5.5 to 6	2.98	Tremolite
1.60 to 1.63	—	—	—	2	1.15 to 1.20	Vulcanite (hard rubber)
1.60–1.63 to 1.62–1.65, mean 1.62	0.027	Negative	B	6.5	2.90 to 3.02	Nephrite
1.57 to 1.63	—	—	—	5	3.00 to 3.10	Odontolite
1.595–1.633	0.038	Positive	B	5 to 6	2.62 to 2.88	Pectolite
1.600–1.820	0.220	Negative	U	4	3.45 to 3.70, usually 3.50 to 3.65	Rhodochrosite
1.61 (mean)	0.030	Positive	B	5	3.00	Herderite
1.612–1.633	0.021	Positive	B	5	3.44	Hemimorphite

Table 1 (*cont.*)

Refractive index	Birefringence	Optic sign	Optical character	Hardness	Specific gravity	Name
1.610–1.638	0.008 to 0.010	Positive	B	8	3.53 to 3.56	Topaz
1.571–1.614	0.043	Positive	B	3 to 3.5	*2.90 to 2.98*	Anhydrite
1.616–1.634 to 1.630–1.652	0.014 to 0.022	Negative	U	7 to 7.5	3.00 to 3.12	Tourmaline
1.553–1.625 to *1.559–1.631*	0.072	Positive	B	7.5	*2.35*	Hambergite
1.616–1.631	0.015	Negative	B	4.5 to 5	*2.8 to 2.9*	Wollastonite
1.61–1.64	0.030	Positive	B	6	2.8 to 2.95, usually 2.88 to 2.94	Prehnite
1.61–1.65	0.040	Positive	B	5.5 to 6	2.6 to 2.85	Turquoise
1.611–1.637	0.026	Positive	B	6	3.01 to 3.03	Amblygonite
1.615–1.645	0.036	Negative	B	5.5	3.1	Lazulite
1.70 (mean)	—	—	B	5 to 6	3.1 to 3.5	Chlorastrolite
1.61 to 1.66	—	—	—	2.5	*1.25 to 1.30 (higher if filled)*	Bakelite
1.620–1.642	0.022	Negative	B	5 to 5.5	3.05	Actinolite
—	—	—	—	6 to 7	2.85 to 3.20	Unakite
1.623–1.651 to 1.632–1.662	0.024 to 0.037	Positive	B	6 to 6.5	3.18	Clinohumite
1.625–1.669	0.044	Negative	B	5	2.9 to 3.00	Datolite
1.625–1.635	0.010	Positive	B	3.5	3.97 to 3.99	Celestine
1.62–*1.85*	0.230	Negative	U	5	*4.30 to 4.35*	Smithsonite
1.582–1.627	0.028	Positive	U	6	*2.73 to 2.90*	Catapleiite
1.629–1.664	0.035	Negative	U	4 to 5	3.06 to 3.19	Friedelite
1.630–1.636	0.006	Negative	B	7	3.00	Danburite
1.63–1.64	0.002 to 0.004	Negative	U	5	3.17 to 3.23	Apatite
1.634–1.644 to 1.641–1.648	0.007 to 0.011	Negative	B	7.25	3.15 to 3.17	Andalusite
1.700–1.706	0.006	Negative	B	6	3.09 to 3.12	Thulite (zoisite)
1.434	—	—	I	4	3.18	Fluorite
1.633–*1.873*	0.240	Negative	U	3.5 to 4	3.83 to 3.88	Siderite
1.636–1.648	0.012	Positive	B	3	*4.3 to 4.6*	Barite
2.65–2.69	0.043	Positive	U	9.25	3.17	Carborundum
1.74 to 1.81	—	Neg/Pos	B	4	3.22	Scorodite
1.639–1.648	0.009	Negative	U	6.5 to 7	3.29–3.31	Jeremejevite
1.639–1.667	0.026	Negative	B	5	3.23	Väyrynenite
1.64 to 1.68	—	—	—	2.5 to 4	usually *1.20 to 1.30*	Jet
1.640–1.645 to 1.668–1.680	0.028 to 0.035	Negative	B	4.5	3.1	Eosphorite
1.644–1.697 to 1.658–*1.709*	0.053	Positive	U	5	3.28 to 3.35	Dioptase

Table 1 (*cont.*)

Refractive index	Birefringence	Optic sign	Optical character	Hardness	Specific gravity	Name
1.55–1.70	—	—	—	6.5	2.7 to 3.4	Saussurite
—	—	—	—	5 to 5.5	3.25	Smaragdite
1.486–1.658	0.172	Negative	U	3	*2.71*	Calcite
1.490–1.650	0.160	Positive	B	2.5	*2.23*	Whewellite
1.652–1.672	0.020	Positive	B	*7.5*	3.10	Euclase
1.650–1.689 to 1.653–1.693	0.040	Positive	B	3 to 4	3.19	Ludlamite
1.654–1.667, mean 1.66	0.013	Positive	B	7	*3.30 to 3.36*	Jadeite
1.654–1.670	0.016	Positive	U	7.5 to 8	2.95 to 2.97	Phenkite
1.654–1.689	0.036	Positive	B	6.5	*3.34*	Peridot
1.656–1.668	0.012	Negative	B	6.5 to 7	3.18	Magnesioaxinite
1.658–1.668	0.010	Positive	B	5.5	3.25	Enstatite (colourless)
1.658–1.678	0.020	Positive	B	6.5 to 7	3.25	Fibrolite
1.65–*1.90*	0.25	Negative	B	4	3.74 to 3.95, usually 3.8	Malachite
1.66 to 1.67	—	—	I	7	*2.96*	Boracite
1.660–1.675	0.015	Positive	B	7	3.17 to 3.19	Spodumene
1.660–1.688	0.028	Positive	B	4.5 to 5	3.32	Serandite
1.662–1.712	0.050	Negative	B	5	*3.94 to 4.07*	Durangite
1.663–1.673	0.010	Positive	B	5.5	3.26 to 3.28	Enstatite
1.665–1.678 to 1.668–1.680	0.013	Negative	B	6.5	3.28 to 3.35	Kornerupine
1.665–1.684	0.019	Positive	B	7 to 8	3.09	Lawsonite
1.67–1.70	0.030	Positive	B	5.5	3.29 3.23	Diopside Violane
1.67–1.71	0.038	Negative	B	6.5	3.47 to 3.49	Sinhalite
1.675–1.685	0.010	Negative	B	7	3.27 to 3.29	Axinite
1.68 (mean)	—	—	B	7	3.26 to 3.41	Dumortierite (massive)
1.668–1.685 to 1.673–1.688	0.011	Negative	B	8 to 8.5	3.26 to 3.41	Dumortierite (transparent)
1.660–1.688	0.028	Positive	B	4.5 to 5	3.32	Serandite
1.702–1.750	0.048	Positive	B	6.5 to 7	3.30 to 3.35	Diaspore
1.532–1.680	0.148	Negative	B	3.5	*4.27 to 4.35*	Witherite
1.67–1.68 to 1.71–1.73	0.010 to 0.020	Negative	B	5 to 6	3.4 to 3.5	Hypersthene
1.689–1.695 to 1.694–*1.702*	0.006 to 0.008	Positive	B	4 to 5	3.42 to 3.52	Triphylite
1.69	—	—	I	8	3.4	Rhodizite
1.69–1.72	0.028	Positive	U	5.5	3.89 *to 4.18*	Willemite
1.692–1.700	0.008	Positive	B	6	3.35	Zoisite
1.700–1.712	0.005 (mean)	Neg/Pos	U	6.5	3.32 to 3.47 3.25 to 3.32	Idocrase Californite

Table 1 (*cont.*)

Refractive index	Birefringence	Optic sign	Optical character	Hardness	Specific gravity	Name
1.72	—	—	I	7.25	3.36 to 3.55	Massive grossular garnet
1.736–1.770	0.034	Negative	B	6.5	3.25 to 3.50, usually 3.4	Epidote
1.72–1.73	0.010	Positive	B	6 to 6.5	3.37 approx.	Clinozoisite
1.73	0.010	Positive	B	6	3.40 to 3.70, usually 3.6 to 3.7	Rhodonite
1.612–1.633	0.021	Positive	B	5	3.44	Hemimorphite
1.700–1.743	0.006	Positive	B	6.5 to 7	3.42 to 3.52	Serendibite
1.705–1.711	0.006	Positive	B	7.5	3.40 to 3.58	Sapphirine
1.732–1.744	0.010	Positive	B	7	3.49	Chambersite
2.42	—	—	I	10	3.52	Diamond
1.885–1.990 to 1.915–2.050	0.105 to 0.135	Positive	B	5.5	3.52 to 3.54	Sphene
1.72 usually	—	—	I	8	3.58 to 3.61	Spinel
1.737	—	—	I	6	3.55 to 3.60	Periclase (synthetic)
1.718–1.722	0.004	Negative	U	8	3.60 to 3.61	Taaffeite
1.724 to 1.729	—	—	I	8	3.61 to 3.65	Spinel (synthetic)
1.77 to 1.80	—	—	I	8	3.63 to 3.90	Spinel (Ceylonite)
1.742 to 1.748	—	—	I	7.25	3.65	Hessonite garnet
1.757–1.804	0.047	Positive	U	6.5	3.65 to 3.68	Benitoite
1.715–1.732	0.017	Negative	B	5 to 7 directional	3.65 to 3.68	Kyanite
1.739–1.750 to 1.747–1.762	0.011 to 0.015	Positive	B	7 to 7.5	3.65 to 3.78	Staurolite
1.73 to 1.75	—	—	I	7.25	3.65 to 3.70	Pyrope garnet
1.75 to 1.78	—	—	I	7.5	3.70 to 3.95	Pyrope/ almandine series garnet
1.75–1.76	0.008 to 0.010	Positive	B	8.5	3.71 to 3.72	Chrysoberyl
1.65–1.90	0.25	Negative	B	4	3.74 to 3.95, usually 3.8	Malachite
1.74 (mean)	—	—	I	8	3.75 (mean)	Gahnospinel
1.736–1.753	0.017	Positive	B	5 to 6	3.76 (mean)	Pyroxmangite
1.87	—	—	I	7.25	3.77	Uvarovite garnet
1.73–1.84	0.110	Positive	B	3.5 to 4	3.77 to 3.89	Azurite
1.752–1.815	—	—	—	6	3.8	Planchéite

Table 1 (*cont.*)

Refractive index	Birefringence	Optic sign	Optical character	Hardness	Specific gravity	Name
1.89	—	—	I	6.5	3.82 to 3.85	Demantoid garnet
1.69–1.72	0.028	Positive	U	5.5	3.89 to 4.18	Willemite
2.493–2.554	0.061	Negative	U	5.5 to 6	3.82 to 3.95	Anatase
1.633–1.873	0.240	Negative	U	3.5 to 4	3.83 to 3.88	Siderite
2.583–2.705	0.122	Positive	B	5.5 to 6	3.87 to 4.08	Brookite
1.760–1.768 to 1.770–1.779	0.008 to 0.010	Negative	U	9	3.90 to 4.00	Corundum
1.78 to 1.81	—	—	I	6	3.90 to 4.10	Zircon (low type)
1.720–1.746	0.026	Negative	B	4.5 to 5	3.95	Hodgkinsonite
1.78 to 1.81	—	—	I	7.5	3.95 to 4.20	Almandine garnet
1.625–1.635	0.010	Positive	B	3.5	3.97 to 3.99	Celestine
1.662–1.712	0.050	Negative	B	5	3.97 to 4.07	Durangite
1.671–1.772	0.101	Positive	U	4.5	4.33	Parisite
1.675–1.735	0.060	Positive	B	5	3.98 to 4.04	Legrandite
1.7					**4.0**	
1.689–1.695 to 1.694–1.702	0.006 to 0.008	Positive	B	4 to 5	3.42 to 3.52	Triphylite
1.704 approx.	—	—	—	5	4.00	Arandisite
1.70 approx.	—	—	—	5 to 5.5	3.25	Smaragdite
1.70 (mean)	—	—	B	5 to 6	3.1 to 3.5	Chlorastrolite
1.700–1.706	0.006	Negative	B	6	3.09 to 3.12	Thulite (zoisite)
1.700–1.743	0.006	Positive	B	6.5 to 7	3.42 to 3.52	Serendibite
1.705–1.711	0.006	Neg/Pos	B	7.5	3.40 to 3.58	Sapphirine
1.702–1.712	0.005	Neg/Pos	U	6.5	3.32 to 3.47	Idocrase
					3.25 to 3.32	Californite
1.702–1.750	0.048	Positive	B	6.5 to 7	3.30 to 3.35	Diaspore
1.662–1.712	0.050	Negative	B	5	3.97 to 4.07	Durangite
1.64–1.69 to 1.66–1.71	0.053	Positive	U	5	3.28 to 3.35	Dioptase
1.67–1.70	0.030	Positive	B	5.5	3.29	Diopside
1.67–1.71	0.038	Negative	B	6.5	3.47 to 3.49	Sinhalite
1.710–1.759 to 1.742–1.773	0.031 to 0.050	Positive	B	3.5	4.30 to 4.68	Adamite
1.715–1.732	0.017	Negative	B	5 to 7 directional	3.65 to 3.68	Kyanite
1.67–1.68 to 1.71–1.73	0.010 to 0.020	Negative	B	5 to 6	3.4 to 3.5	Hypersthene
1.69–1.72	0.028	Positive	U	5.5	3.89 to 4.18	Willemite
1.718–1.722	0.004	Negative	U	8	3.60 to 3.61	Taaffeite
1.714 to 1.736	—	—	I	8	3.58 to 3.64	Spinel
1.724 to 1.729	—	—	I	8	3.61 to 3.65	Synthetic spinel

Table 1 (*cont.*)

Refractive index	Birefringence	Optic sign	Optical character	Hardness	Specific gravity	Name
1.72	—	—	I	7.25	3.36–3.55	Massive grossular garnet
1.72–1.73	0.010	Positive	B	6 to 6.5	3.37	Clinozoisite
1.720–1.735	0.015	Positive	U	9	3.00 to 3.02	Bromellite
1.720–1.746	0.026	Negative	B	4.5 to 5	3.95	Hodgkinsonite
1.73	0.010	Positive	B	6	3.6 to 3.7	Rhodonite
1.675–1.735	0.060	Positive	B	5	3.98–4.04	Legrandite
1.73 to 1.75	—	—	I	7.25	3.65 to 3.80	Pyrope garnet
1.732–1.744	0.010	Positive	B	7	3.49	Chambersite
1.736–1.753	0.017	Positive	B	5 to 6	3.76 (*mean*)	Pyroxmangite
1.737	—	—	I	5	3.55 to 3.60	Periclase
1.73–1.84	0.110	Positive	B	3.5 to 4	3.77 to 3.89	Azurite
1.739–1.750 to 1.747–1.762	0.011 to 0.015	Positive	B	7 to 7.5	3.65 to 3.78	Staurolite
1.736–1.770	0.034	Negative	B	6.5	3.25 to 3.50, usually 3.4	Epidote
1.74 (*mean*)	—	—	I	8	3.75 (*mean*)	Gahnospinel
1.74 to 1.81	—	Neg/Pos	B	4	3.22	Scorodite
1.742 to 1.748	—	—	I	7.25	3.65	Hessonite garnet
1.75–1.76	—	Positive	B	8.5	3.71 to 3.72	Chrysoberyl
1.75 to 1.78	— to 0.010	—	I	7.5	3.80 to 3.95	Almandine/ pyrope garnet
1.752–1.815	—	—	—	6	3.8	Planchéite
1.757–1.804	0.047	Positive	U	6.5	3.65 to 3.68	Benitoite
1.760–1.768 to 1.770–1.779	0.008 to 0.010	Negative	U	9	3.90 to 4.00	Corundum
1.77 to 1.80	—	—	I	8	3.63 to 3.90	Spinel (Ceylonite)
1.78 to 1.81	—	—	I	7.5	3.95 to 4.20	Almandine garnet
1.787–1.816	0.029	Negative	U	7.5	4.01	Painite
1.78 to 1.81	—	—	I	6	3.90 to 4.10	Zircon (low type)
2.40	—	—	B	6	4.05	Calcium titanate
1.79 to 1.81	—	—	I	7.25	4.12 to 4.20	Spessartine garnet
1.795–1.845	0.05	Positive	B	6	5.2	Monazite
2.583–2.705	0.122	Positive	B	5.5 to 6	3.87 to 4.08	Brookite
2.368 to 2.371	—	—	I	3.5 to 4	4.08 to 4.10	Zinc blende
2.1	—	—	I	5.5	4.1 to 4.9	Chromite
2.21	—	—	B	5 to 6	4.1 to 6.2	Samarskite
2.62–2.90	0.287	Positive	U	6 to 6.5	4.2	Rutile (natural and synthetic)
—	—	—	—	3.5 to 4	4.2 approx.	Chalcopyrite
1.532–1.680	0.148	Negative	B	3.5	4.27 to 4.35	Witherite
1.62–1.85	0.23	Negative	U	5	4.30 to 4.35	Smithsonite

Table 1 (*cont.*)

Refractive index	Birefringence	Optic sign	Optical character	Hardness	Specific gravity	Name
1.636–1.648	0.012	Positive	B	3	4.3 to 4.6	Barite
1.671–1.772	0.010	Positive	U	4.5	4.33	Parisite
1.710–1.759 to 1.742–1.773	0.031 to 0.050	Positive	B	3.5	4.30 to 4.68	Adamite
1.974–1.984	0.010	Positive	U	3.5	4.34	Powellite
1.95–1.99	0.020	Positive	B	4.5	4.35	Bayldonite
—	—	—	—	7	4.35 approx.	Psilomelane
1.80	—	—	—	7.5 to 8	4.40	Gahnite
1.83	—	—	—	8 to 8.5	4.57 to 4.60	Yttrium aluminium garnet
2.21–2.30	0.090	—	—	6	4.64	Lithium niobate
1.925–1.984	0.059	Positive	U	7 to 7.5	4.67 to 4.70	Zircon (high type)
2.06 to 2.26	—	—	—	6.5	4.7 to 5.0	Euxenite
2.505–2.529	0.024	Positive	U	3.5	4.83 to 4.90	Greenockite
1.92	—	—	I	7.5 to 8	4.84	Yttrium oxide
—	—	—	—	6 to 6.5	4.85 to 4.90	Marcasite
—	—	—	—	3	4.9 to 5.4	Bornite
—	—	—	—	6.5	4.95 to 5.10	Pyrite
2.94–3.22	—	—	U	6.5	4.95 to 5.16	Hematite
1.8					**5.0**	
—	—	—	—	3.5 to 4	5.0	Pentlandite
2.030–2.050	0.020	Negative	U	3.5	5.05	Boléite
1.80	—	—	I	7.5 to 8	*4.40*	Gahnite
1.77–1.80	—	—	I	8	*3.63 to 3.90*	Spinel (Ceylonite)
1.757–1.804	0.047	Positive	U	6.5	*3.65 to 3.68*	Benitoite
1.804–2.078	0.274	Negative	B	3 to 3.5	*6.46 to 6.57*	Cerussite
1.74 to 1.81	—	Neg/Pos	B	4	*3.22*	Scorodite
1.78 to 1.81	—	—	I	7.5	*3.95 to 4.20*	Almandine garnet
1.752–1.815	—	—	—	6	*3.8*	Planchéite
1.787–1.816	0.029	Negative	U	7.5	*4.01*	Painite
1.79 to 1.81	—	—	I	7.25	*4.12 to 4.20*	Spessartine garnet
1.79 to 1.81	—	—	I	6	*3.90 to 4.10*	Zircon (low type)
1.60–1.82	0.220	Negative	U	4	*3.50 to 3.65*	Rhodochrosite
1.83	—	—	—	8 to 8.5	*4.57 to 4.60*	Yttrium aluminium garnet
1.73–1.84	0.110	Positive	B	3.5 to 4	*3.77 to 3.89*	Azurite
1.62–1.85	0.23	Negative	U	5	*4.30 to 4.35*	Smithsonite
1.87	—	—	I	7.5	*3.77*	Uvarovite garnet
1.633–1.873	0.240	Negative	U	3.5 to 4.5	*3.83 to 3.88*	Siderite
1.877–1.894	0.017	Positive	B	2.75 to 3	*6.30 to 6.39*	Anglesite

Table 1 (cont.)

Refractive index	Birefringence	Optic sign	Optical character	Hardness	Specific gravity	Name
1.88–1.99	0.105	Positive	B	5.5	3.52 to 3.54	Sphene
1.89	—	—	I	6.5	3.82 to 3.85	Demantoid garnet
2.41	—	—	I	6	5.13	Strontium titanate
—	—	—	I	6	5.17	Magnetite
1.795–1.845	0.05	Positive	B	6	5.2	Monazite
1.92	—	—	—	6.5	5.30	'Yttralox'
1.938–1.955	0.017	?	B	8	5.35	Rhombic yttrium aluminate
—	—	—	—	3 to 3.5	5.3 to 5.65	Millerite
1.93	—	—	I	5.5	5.5	Microlite
—	—	—	—	6	5.5 to 5.7	Magnetoplumbite
2.7 to 3.0	0.300	Negative	U	2 to 2.5	5.57 to 5.64	Proustite
2.013–2.029	0.016	Positive	U	4 to 4.5	5.66	Zincite
2.85	—	—	I	3.5 to 4	5.86 to 6.15	Cuprite
2.40	—	—	—	6 to 6.5	5.90	Barium titanate
1.918–1.934	0.016	Positive	U	4.5 to 5	5.9 to 6.1	Scheelite
2.31–2.66	0.35	Positive	B	2.5 to 3	5.9 to 6.1	Crocoite
2.09 to 2.18	—	—	I	8 to 8.5	5.95 to 6.06	Cubic zirconia
1.9					**6.0**	
1.91–2.05	0.135	Positive	B	5.5	3.52 to 3.54	Sphene
1.65–1.90	0.25	Negative	B	4	3.74 to 3.95, usually 3.8	Malachite
2.06	0.100	—	U	7	6.00	Simpsonite
1.918–1.934	0.016	Positive	U	4.5 to 5	5.9 to 6.1	Scheelite
1.92	—	—	I	7.5 to 8	4.84	Yttrium oxide
1.92	—	—	—	6.5	5.30	'Yttralox'
1.925–1.984	0.059	Positive	U	7 to 7.5	4.67 to 4.70	Zircon (high type)
1.93	—	—	I	5.5	5.5	Microlite
1.938–1.955	0.017	?	B	8	5.35	Rhombic yttrium aluminate
1.95–1.99	0.020	Positive	B	4.5	4.35	Bayldonite
1.974–1.984	0.010	Positive	U	3.5	4.34	Powellite
2.85	—	—	I	3.5 to 4	5.86 to 6.15	Cuprite
2.36–2.66	0.35	Positive	B	2.5 to 3	5.9 to 6.1	Crocoite
—	—	—	I	5.5	6.0 to 6.3	Cobaltite
2.114–2.140	0.026	Positive	U	3	6.2	Phosgenite
2.21	—	—	B	5 to 6	4.1 to 6.2	Samarskite
1.877–1.894	0.017	Positive	B	2.75 to 3	6.30 to 6.39	Anglesite
1.804–2.078	0.274	Negative	B	3 to 3.5	6.45 to 6.57	Cerussite
2.304–2.402	0.098	Negative	U	2.75 to 3	6.7 to 7	Wulfenite
2.003–2.101	0.098	Positive	U	6.5	6.8 to 7	Cassiterite

Table 1 (*cont.*)

Refractive index	Birefringence	Optic sign	Optical character	Hardness	Specific gravity	Name
2.0 and over					**7.0**	
2.003–2.101	0.098	Positive	U	6.5	6.95	Cassiterite
2.013–2.029	0.016	Positive	U	4 to 4.5	5.66	Zincite
1.915–2.050	0.135	Positive	B	5.5	3.52 to 3.54	Sphene
1.92–2.03	—	—	I	6.5	7.05	Gadolinium gallium garnet
2.030–2.050	0.020	Negative	U	3.5	5.05	Boléite
2.06	0.100	—	U	7	6.00	Simpsonite
1.804–2.078	0.274	Negative	B	3 to 3.5	6.45 to 6.57	Cerussite
2.06 to 2.26	—	—	—	6.5	4.7 to 5.0	Euxenite
2.09 to 2.18	—	—	I	8 to 8.5	5.98 to 6.06	Cubic zirconia
2.1	—	—	I	5.5	4.1 to 4.9	Chromite
2.114–2.140	0.026	Positive	U	3	6.2	Phosgenite
2.120–2.135	0.015	Negative	U	3.5	7.0 to 7.25	Mimetite
2.14–2.22 to 2.19–2.34	0.008 to 0.015	—	B	5.5–6	7.73–7.97	Manganot-antalite
2.21	—	—	B	5 to 6	4.1 to 6.2	Samarskite
2.21–2.30	0.90	—	—	6	4.64	Lithium niobate
2.304–2.402	0.098	Negative	U	2.5 to 3	6.5 to 7	Wulfenite
2.31–2.66	0.35	Positive	B	2.5 to 3	5.9 to 6.1	Crocoite
2.39–2.46	0.07	Positive	B	5.5 to 6	7.46	Stibiotantalite
2.368–2.371	—	—	I	3.5 to 4	4.08 to 4.10	Zinc blende
—	—	—	—	4.5 to 5	7.54	Breithauptite
2.40	—	—	B	6	4.05	Calcium titanate
2.40	—	—	—	6 to 6.5	5.90	Barium titanate
2.41	—	—	I	6	5.13	Strontium titanate
2.42	—	—	I	10	3.52	Diamond
2.493–2.554	0.061	Negative	U	5.5 to 6	3.82 to 3.95	Anatase
2.505–2.529	0.024	Positive	U	3.5	4.83	Greenockite
2.583–2.705	0.122	Positive	B	5.5 to 6	3.87 to 4.08	Brookite
2.62–2.90	0.287	Positive	U	6 to 6.5	4.2	Rutile
2.65–2.69	0.043	Positive	U	9.25	3.17	Carborundum
2.7–3.0	0.30	Negative	U	2 to 2.5	5.57 to 5.64	Proustite
2.85	—	—	I	3.5 to 4	5.85 to 6.15	Cuprite
2.94–3.22	—	—	U	6.5	4.95 to 5.16	Hematite
—	—	—	U	5 to 5.5	7.33 to 7.67	Niccolite
—	—	—	—	3.5	7.2 to 7.9	Domeykite
—	—	—	—	3.5	8.38	Algodonite

Table of OPTICAL PROPERTIES

The optical properties of gem materials

Single refracting or isotropic	{	Amorphous substances and cubic crystals }	One index of refraction, n
Doubly refracting or anisotropic	{	Hexagonal } Trigonal } Uniaxial Tetragonal	Two indices of refraction, ω and ϵ; positive, ω less than ϵ; negative, ϵ less than ω
		Rhombic } Monoclinic } Biaxial Triclinic }	Three indices of refraction, α, β and γ; positive, $\alpha\beta$—γ; negative, α—$\beta\gamma^*$

*When the intermediate index of refraction β is nearer to α, the optic sign is positive; when nearer to γ, the optic sign is negative.

Table 2
Refractive indices

Readers are referred to Table 1 and to the main text for birefringence values, optical character and sign, and for further information on the range of RI values in minerals listed below. Liquids are listed in italics. (I) signifies the RI range of an isotropic mineral or other material.

Refractive index liquids listed in this table should be used with great care. Always use in a well-ventilated area or, for prolonged use, a fume cupboard.

Water	1.33	Mesolite	1.50 (mean)
Sellaite	1.37–1.39	Petalite	1.50–1.51
Fluorite	1.43	Leucite	1.50–1.51
Opal	1.45	Dolomite	1.50–1.68
Silica glass	1.46	Ulexite	1.51 (mean)
Creedite	1.46–1.48	Cancrinite	1.51 (mean)
Thaumasite	1.46–1.51	Magnesite	1.51–1.70 (mean)
Sodalite	1.48	Pollucite	1.51–1.52 (I)
Analcime	1.48–1.49 (I)	Scolecite	1.51–1.52
Natrolite	1.48–1.49	Sanidine feldspar	1.51–1.52
Moldavite and other tektites	1.48–1.54 (I)	Thomsonite	1.51–1.54
		Orthoclase feldspar	1.51–1.53
Obsidian	1.48–1.51 (I)	Microcline feldspar	1.51–1.53
Calcite	1.48–1.65	Albite feldspar	1.52–1.54
Whewellite	1.48–1.65	Gypsum	1.52–1.53
Celluloid	1.49–1.51 (I)	Milarite	1.52–1.55
Tugtupite	1.49–1.50	Stichtite	1.53 (mean)
Yugawaralite	1.49–1.50	Meerschaum	1.53 (mean)
Haüyne	1.49–1.50 (I)	Chalcedony	1.53–1.54
Chrysocolla (Webster)	1.50 (mean)	Oligoclase feldspar	1.53–1.54
or (Ahrem)	*c.*1.59 (mean)	Iolite	1.53–1.55
silicified	as for quartz	Casein	1.53–1.56 (I)
Lapis lazuli	1.50 (mean)	Witherite	1.53–1.67
Perspex	1.50	Apophyllite	1.53–1.54

Table 2 *(cont.)*

Aragonite	1.53–1.68	Sugilite	1.60–1.61
Nepheline	1.54 (mean)	Brazilianite	1.60–1.62
Bone	1.54	Grandidierite	1.60–1.63
Vegetable ivory	1.54	Vulcanite	1.60–1.63 (I)
Ivory (dentine)	1.54	Tremolite	1.60–1.64
Amber	1.54	Rhodochrosite	1.60–1.82
Copal resin	1.54	Topaz (white, blue,	1.61–1.62
Jasper	1.54	yellow, brown)	
Horn	1.54	Hemimorphite	1.61–1.63
Lepidolite	1.55 (mean)	Prehnite massive	1.61–1.64
Scapolite	1.54–1.58		1.63 (mean)
Quartz	1.54–1.55	Pargasite	1.61–1.63
Tortoiseshell	1.55	Amblygonite	1.61–1.63
Soapstone (steatite)	1.55 (mean)	Lazulite	1.61–1.64
Beryllonite	1.55–1.56	Bakelite	1.61–1.66 (I)
Hambergite	1.55–1.63	Nephrite	1.62 (mean)
Polyester resin	1.55–1.56 (I)	Turquoise	1.62 (mean)
Pyrophyllite	1.60 (mean)	Wollastonite	1.63 (mean)
Deerhorn	1.56	Celestine	1.62–1.63
Black coral	1.56	Tourmaline	1.62–1.64 (mean)
Variscite	1.56 (mean)	Actinolite	1.62–1.65
Bowenite serpentine	1.56 (mean)	Friedelite	1.62–1.66
Labradorite feldspar	1.56–1.57	Datolite	1.62–1.67
Synthetic emerald	1.55–1.56	Clinohumite	1.62–1.67
(some flux melt)		Smithsonite	1.62–1.84
Natural emerald and	1.566–1.590	Danburite	1.63–1.64
hydrothermal synthetics	(see Table 1)	Topaz (red, pink,	1.63–1.64
Aquamarine	1.56–1.59	yellow, orange)	
Anhydrite	1.57–1.61	Apatite	1.63–1.64
Pseudophite	1.57 (mean)	Barite	1.63–1.64
Williamsite serpentine	1.57 (mean)	Jeremejevite	1.63–1.64
Ophicalcite	1.57 (mean)	Väyrynenite	1.63–1.66
Bastite	1.57–1.58	Siderite	1.63–1.87
Bytownite feldspar	1.57–1.58	Andalusite	1.64–1.65
Augelite	1.57–1.58	Eosphorite	1.64–1.68
Odontolite	1.57–1.63 (I)	Magnesioaxinite	1.65–1.66
Verdite	1.58 (mean)	Euclase	1.65–1.67
Xonotlite	1.58–1.59	Phenakite	1.65–1.67
Colemanite	1.58–1.61	Boracite	1.65–1.67
Catapleiite	1.58–1.62	Enstatite	1.65–1.68
Howlite	1.59 (mean)	Fibrolite	1.65–1.68
Polystyrene	1.59	Peridot	1.65–1.69
Wardite	1.59 (mean)	Ludlamite	1.65–1.69
Ekanite	1.59	Jadeite	1.66 (mean)
Eudialyte	1.59–1.60	Spodumene	1.66–1.68
Melinophane	1.59–1.61	Durangite (Webster)	1.66–1.71
Herderite	1.59–1.62	or (Gunther)	1.63–1.68
Phosphophyllite	1.59–1.62	Dioptase	1.65–1.70 (mean)
Pectolite	1.60 (mean)	Lawsonite	1.66–1.68
Chrondrodite	1.60 (mean)	Sinhalite	1.66–1.71
Sogdianite	1.60 (mean)	Kornerupine	1.66–1.69

Table 2 *(cont.)*

Kaersutite	1.67–1.77	Spessartine garnet	1.79–1.81 (I)
Diopside	1.67–1.70	Gahnite	1.80
Serandite	1.67–1.70	Monazite	1.80–1.85
Axinite	1.67–1.69	Cerussite	1.80–2.07
Legrandite	1.67–1.73	Yttrium aluminate (YAG)	1.83
Parisite	1.67–1.77	Malachite	1.85 (mean)
Dumortierite silicified	1.68 (mean) as for	Uvarovite garnet	1.87
	quartz	Anglesite	1.87–1.89
Triphylite	1.68–1.70	Demantoid garnet	1.89
Rhodizite	1.69	Sphene	1.90–2.02
Willemite	1.69–1.72	Scheelite	1.92–1.93
Zoisite var. Tanzanite	1.69–1.70	'Yttralox'	1.92
var. Thulite	1.70 (mean)	Yttrium oxide	1.92
Arandisite	1.70 (mean)	Zircon (high type)	1.92–1.98
Smaragdite	1.70 (approx.)	Microlite	1.93
Chlorastrolite	1.70 (mean)	Rhombic yttrium	1.93–1.95
Hypersthene	1.70–1.71 (mean)	aluminate	
Idocrase (yellow)	1.70–1.71	Bayldonite	1.95–1.99
Serendibite	1.70–1.74	Gadolinium gallium	1.92–2.03 (I)
Sapphirine	1.70–1.71	garnet (GGG)	
Diaspore	1.70–1.75	Simpsonite	1.97–2.04
Spinel	1.71–1.73 (I)	Cassiterite	2.00–2.10
Kyanite	1.71–1.73	Zincite	2.01–2.02
Taaffeite	1.71–1.72	Boléite	2.03–2.05
	(up to 1.74)	Euxenite	2.06–2.24
Hydrogrossular garnet	1.72–1.74 (I)	Chromite	2.08–2.16 (I)
Bromellite	1.72–1.73	Cubic zirconia	2.09–2.18 (I)
Clinozoisite	1.72–1.73	Phosgenite	2.11–2.14
Hodgkinsonite	1.72–1.74	Mimetite	2.12–2.14
Spinel (synthetic)	1.72	Manganotantalite	2.14–2.22
Pyroxmangite	1.73–1.75	Samarskite	2.20
Chambersite	1.73–1.74	Lithium niobate	2.21–2.30
Rhodonite	1.73 (mean)	KTN	2.27
Pyrope garnet	1.73–1.76 (I)	Crocoite	2.31–2.66
Azurite	1.73–1.84	Wulfenite	2.28–2.40
Epidote	1.73–1.77 (mean)	Stibiotantalite	2.37–2.45
Periclase	1.73	Zinc blende (sphalerite)	2.37–2.43 (I)
Gahnospinel	1.72–1.75 (I)	Barium titanite	2.40
Hessonite garnet	1.72–1.74 (I)	Calcium titanate	2.40 (mean)
Staurolite	1.74–1.75 (mean)	Strontium titanate	2.40–2.42 (I)
Chrysoberyl	1.74–1.75	Diamond	2.42
Shattuckite	1.75 (mean)	Anatase	2.49–2.55
Benitoite	1.75–1.80	Greenockite	2.50–2.52
Almandine garnet	1.76–1.81 (I)	Brookite	2.58–2.71
Corundum	1.76–1.77	Rutile	2.62–2.90
Pleonaste (Ceylonite)	1.77–1.78 (I)	Silicon carbide	2.65–2.69
Scorodite	1.78–1.81	(carborundum)	
Painite	1.78–1.81	Proustite	2.79–3.08
Zircon (low type)	1.78–1.85 almost	Cuprite	2.84
	isotropic	Hematite	2.94–3.22

Table 3
Specific gravities

Meerschaum	1.00–2.00	Bastite	2.5–2.7
Amber	1.03–1.10	Nepheline	2.55–2.65
Copal resin	1.03–1.10	Yellow orthoclase feldspar	2.56
Polystyrene	1.05	Microcline feldspar	2.56–2.58
Vulcanite	1.15–1.20	Moonstone feldspar	2.56–2.59
Perspex	1.18–1.19	Iolite	2.57–2.66
Jet	1.20–1.30	Sanidine feldspar	2.57–2.58
Bakelite	1.25–1.30	Howlite	2.58
Tortoise-shell and horn	1.26–1.35	Chalcedony	2.58–2.62
Cellulose acetate	1.29–1.40	Bowenite serpentine	2.58–2.62
Casein	1.32–1.34	Jasper	2.58–2.91
Black coral	1.34	Coral	2.6–2.7
Celluloid	1.36–1.42	Vivianite	2.6
Vegetable ivory	1.38–1.42	Marble	2.6–2.7
Deer horn	1.60–1.85	Pinite	2.60–2.85
Ivory (dentine)	1.70–1.98	Pseudophite	2.60–2.85
Thaumasite	1.88–1.90	Turquoise	2.60–2.85
Opal	1.98–2.20	Williamsite serpentine	2.61
Ulexite	1.99	Albite feldspar	2.62
Bone	2.00	Oligoclase feldspar	2.62–2.65
Chrysocolla	2.00–2.45	Pectolite	2.62–2.84
Stichtite	2.15–2.22	Scapolite	2.63–2.71
Yugawaralite	2.19–2.22	Quartz	2.65
Natrolite	2.20–2.25	Synthetic emerald (some)	2.66–2.68
Analcime	2.20–2.29	Pearl	2.67–2.78
Black clam pearls	2.20–2.66	Charoite	2.68
Silica glass	2.21	Labradorite feldspar	2.68–2.69
Whewellite	2.23	Beryl	2.68–2.90
Sodalite	2.25–2.30	Natural emerald	2.69–2.76
Mesolite	2.29	Augelite	2.7
Porcelain	2.30	Ophicalcite	2.7
Alabaster	2.30–2.33	Xonotlite	2.7
Thomsonite	2.3–2.4	Lepidolite	2.7–2.8
Apophyllite	2.3–2.5	Kakortokite	2.7–2.8
Obsidian	2.33–2.42	Lapis lazuli	2.70–2.90
Moldavite	2.34–2.39	Calcite	2.71
Hambergite	2.35	Creedite	2.71
Tektites	2.36–2.51	Bytownite feldspar	2.73
Tugtupite	2.36–2.58	Sugilite	2.74
Petalite	2.39–2.46	Catapleiite	2.73–2.90
Haüyne	2.4	Sogdianite	2.76–2.90
Milarite	2.4–2.6	Pyrophyllite	2.8
Blue pearls	2.40–2.65	Mollusc shell	2.8
Variscite	2.40–2.60	Beryllonite	2.80–2.85
Colemanite	2.42	Dolomite	2.8–2.9
Cancrinite	2.42–2.50	Lepidolite	2.8–2.9
Leucite	2.45–2.50	Wollastonite	2.8–2.9
Rhyolite	2.5 (approx.)	Verdite	2.80–2.99
(Nevada wonderstone)		Eudialyte	2.8–3.1

Table 3 (*cont.*)

Wardite	2.81	Epidote	3.25–3.50
Conch shell	2.84–2.89	Enstatite	3.26–3.28
Pollucite	2.85–2.94	Dumortierite	3.26–3.41
Unakite	2.85–3.20	Axinite	3.27–3.29
Prehnite	2.88–2.94	Ekanite	3.28
Datolite	2.90–3.00	Jeremejevite	3.29–3.31
Nephrite	2.90–3.02	Dioptase	3.28–3.35
Anhydrite	2.91–2.98	Kornerupine	3.28–3.35
Aragonite	2.93–2.95	Diaspore	3.30–3.35
Phenakite	2.95–2.97	Jadeite	3.30–3.36
Boracite	2.96	Idocrase	3.32–3.47
Tremolite	2.98–3.03	Peridot	3.34
Brazilianite	2.980–2.995	Zoisite	3.35
Melinophane	3.00	Hydrogrossular garnet	3.36–3.55
Herderite	3.00	Clinozoisite	3.37
Danburite	3.00	Rhodizite	3.4
Grandidierite	3.00	Sapphirine	3.40–3.58
Odontolite	3.00–3.10	Rhodonite	3.40–3.70
Tourmaline	3.00–3.12	Hypersthene	3.4–3.5
Bromellite	3.00–3.02	Serendibite	3.42–3.52
Saussurite	3.0–3.4	Sinhalite	3.47–3.48
Amblygonite	3.01–3.03	Chambersite	3.5
Hornblende	3.02–3.45	Rhodochrosite	3.50–3.65
Actinolite	3.05	Triphylite	3.51
Friedelite	3.05–3.07	Diamond	3.52
Lawsonite	3.09	Sphene	3.52–3.54
Thulite	3.09–3.12	Topaz	3.53–3.56
Eosphorite	3.1	Spinel (natural)	3.58–3.61
Phosphophyllite	3.1	Periclase	3.59
Lazulite	3.1	Taaffeite	3.60–3.61
Ludlamite	3.1	Strontium titanate (Fabulite)	5.13
Euclase	3.1	Magnetite	5.17
Chondrodite	3.1–3.2	Rhombic yttrium aluminate	5.35
Chlorastrolite	3.1–3.5	'Yttralox'	5.30
Sellaite	3.15	Millerite	5.3–5.65
Andalusite	3.15–3.17	Boléite	5.4
Carborundum	3.17	Microlite	5.5
Spodurnene	3.17–3.19	Magnetoplumbite	5.5–5.7
Apatite	3.17–3.23	Cubic zirconia	5.54–6.06
Magnesioaxinite	3.178–3.190	Proustite	5.57–5.64
Gedrite	3.18–3.37	Zincite	5.66
Clinohumite	3.18	Cuprite	5.85–6.15
Fluorite	3.18	Barium titanate	5.90
Kaersutite	3.20–3.28	Crocoite	5.9–6.1
Scorodite	3.22	Spinel (synthetic)	3.61–3.65
Väyrynenite	3.23	Pyroxmangite	3.61–3.80
Diopside	3.23–3.29	Ceylonite	3.63–3.90
Fibrolite	3.25	Hessonite garnet	3.65
Smaragdite	3.25	Benitoite	3.65–3.68
Californite	3.25–3.32	Kyanite	3.65–3.68

Table 3 (*cont.*)

Pyrope garnet	3.65–3.70	Adamite	4.30–4.68
Staurolite	3.65–3.78	Powellite	4.34
Almandine/pyrope garnets	3.70–3.95	Bayldonite	4.35
Chrysoberyl	3.71–3.72	Psilomelane	4.35
Malachite	3.74–3.95	Gahnite	4.40
Gahnospinel	3.75–4.06	Yttrium aluminate (YAG)	4.57–4.60
Uvarovite garnet	3.77	Lithium niobate	4.64
Azurite	3.77–3.89	Zircon (high type)	4.67–4.70
Shattuckite	3.8	Euxenite	4.7–5.0
Demantoid garnet	3.82–3.85	Greenockite	4.83
Anatase	3.82–3.95	Yttrium oxide	4.84–4.90
Siderite	3.83–3.88	Marcasite	4.85–4.90
Brookite	3.87–4.08	Bornite	4.9–5.4
Willemite	3.89–4.18	Pyrite	4.95–5.10
Corundum	3.98–4.00	Hematite	4.95–5.16
Zircon (low type)	3.90–4.10	Pentlandite	5.0
Hodgkinsonite	3.95	Simpsonite	6.00
Almandine garnet	3.95–4.20	Cobaltite	6.0–6.3
Celestine	3.97–3.99	Phosgenite	6.2
Durangite	3.97–4.07	Anglesite	6.30–6.39
Legrandite	3.98–4.04	KTN	6.43
Painite	4.01	Cerussite	6.45–6.57
Calcium titanate	4.05	Wulfenite	6.7–7.0
Zinc blende (sphalerite)	4.08–4.10	Cassiterite	6.8–7.0
Chromite	4.1–4.9	Mimetite	7.0–7.2
Samarskite	4.1–6.2	Gadolinium gallium garnet	7.05
Spessartine garnet	4.12–4.20		
Rutile	4.2	Domeykite	7.2–7.9
Chalcopyrite	4.2	Niccolite	7.33–7.67
Witherite	4.27–4.35	Stibiotantalite	7.46
Smithsonite	4.30–4.35	Breithauptite	7.54
Barite	4.3–4.6	Algodonite	8.00–8.39

Table 4
Colour dispersions of gemstones

The values given are the differences between the refractive indices of the stones for the Fraunhofer B line, a red ray at 687.0 nm, and the G line, a blue ray at 430.8 nm. Except where stated the values given are for the ordinary ray in uniaxial stones and the alpha direction in biaxial stones, for as indicated, the dispersion is in some cases considerably different for the different rays in doubly refractive stones.

Rutile	0.28–0.30	Lithium niobate	0.13
Anatase	0.213[†]	Cassiterite	0.071
	0.259[‡]	Cubic zirconia	0.058–0.066
Strontium titanate	0.19	Demantoid garnet	0.057
Zinc blende	0.156	Sphene	0.051

Table 4 (*cont.*)

Yttrium oxide	0.050	Iolite	0.017
Gadolinium gallium garnet	0.038–0.045	Spodumene	0.017
Anglesite	0.044	Tourmaline	0.017
Diamond	0.044	Danburite	0.017
Benitoite	0.044	Crown glass	0.016*
Flint glass	0.041*	Datolite	0.016
Zircon	0.039	Euclase	0.016
Dioptase	0.028†	Andalusite	0.016
	0.036‡	Fibrolite	0.015
Epidote	0.030	Chrysoberyl	0.015
Yttrium aluminate (YAG)	0.028	Phenakite	0.015
Hessonite garnet	0.027	Hambergite	0.015
Spessartine garnet	0.027	Smithsonite	0.014†
Willemite	0.027		0.031‡
Almandine garnet	0.027	Topaz	0.014
Scheelite	0.026	Brazilianite	0.014
Staurolite	0.023	Beryl	0.014
Pyrope garnet	0.022	Quartz	0.013
Kyanite	0.020	Apatite	0.013
Spinel (natural and synthetic)	0.020	Feldspar	0.012
Peridot	0.020	Pollucite	0.012
Taaffeite	0.019	Beryllonite	0.010
Idocrase	0.019	Cancrinite	0.010
Sinhalite	0.018	Leucite	0.010
Kornerupine	0.018	Silica glass	0.010
Sodalite	0.018	Calcite	0.008‡
Corundum	0.018		0.017†
Scapolite	0.017	Fluorite	0.007

*These are mean values for glasses of these two types. The dispersion of glasses can vary considerably.
†Ordinary ray.
‡Extraordinary ray.

Table 5
Mohs' scale

1	Talc	6	Orthoclase feldspar
2	Gypsum	7	Rock crystal
3	Calcite	8	Topaz
4	Fluorite	9	Corundum
5	Apatite	10	Diamond

Fingernail is about 2.5; copper coin is about 3; window glass is about 5.5; knife blade is about 6; steel file is about 6.5.

Table 6
Gemstones in order of their hardness according to Mohs' scale

10	{ Diamond Borazon*	6.5–7	{ Diaspore Jeremejevite Serendibite	
9.5	Boron carbide*			
9.25	Carborundum*		{ Benitoite Cassiterite	
9	{ Corundum Bromellite		Chalcedony Chondrodite	
8.5	{ Chrysoberyl Yttrium aluminate (YAG)		Demantoid garnet Epidote	
8–8.5	Cubic zirconia	6.5	Euxenite Hematite	
8	{ Rhodizite Spinel Taaffeite Topaz		Idocrase Kornerupine Nephrite Peridot	
7.5–8	{ Gahnite Phenakite Yttrium oxide		Pollucite Sinhalite 'Yttralox' Zircon (low type)	
7.5	{ Almandine garnet Andalusite Beryl Euclase Fibrolite Grandidierite Hambergite Sapphirine Lawsonite Painite Uvarovite garnet	6–7 6–6.5	Sogdianite { Ekanite Barium titanate Clinohumite Magnetite Marcasite Microcline feldspar Petalite Plagioclase feldspar Pyrite Rutile	
7.25	{ Hessonite garnet Pyrope garnet Rhodolite garnet Spessartine garnet		{ Amblygonite Calcium titanate Cancrinite	
7–7.5	{ Iolite Staurolite Tourmaline Zircon		Catapleite Columbite KTN Lithium niobate	
7	{ Axinite Boracite Chambersite Danburite Dumortierite Jadeite Quartz Simpsonite Spodumene	6	Orthoclase feldspar Periclase Prehnite Rhodonite Scapolite Silica glass Strontium titanate Tugtupite Xonotlite	
6.5–7.5	Sugilite		Zoisite	

Table 6 (*cont.*)

Hardness	Minerals	Hardness	Minerals
5.5–6.5	Gadolinium gallium garnet Haüyne Opal Psilomelane Sugilite	5	Apatite Augelite Beryllonite Dioptase Durangite Herderite Legrandite Mesolite Obsidian Odontolite Pectolite Smithsonite Variscite Wardite
5.5–6	Anatase Brookite Gedrite Hornblende Leucite Melinophane Nepheline Pyroxmangite Sodalite Stibiotantalite Tremolite Turquoise	4.5–5	Apophyllite Breithauptite Scheelite Wollastonite
5.5	Brazilianite Chromite Cobaltite Diopside Enstatite Lazulite Lazurite (lapis-lazuli) Microlite Milarite Moldavite Natrolite Smaltite Sphene Willemite	4.5	Bayldonite Colemanite Eosphorite Hodgkinsonite Pseudomalachite Yugawaralite
		4–5	Bowenite serpentine Friedelite Triphylite
		4–4.5	Zincite
5–7	Kyanite (varies with direction)	4	Cuprite Fluorite Malachite Rhodochrosite Scorodite
5–6	Charoite Chlorastrolite Eudialyte Glass (normally) Hypersthene Kaersutite Samarskite Sellaite	3.5–4	Algodonite Aragonite Azurite Bastite Chalcopyrite Dolomite Domeykite Magnesite Marble Pentlandite Siderite Zinc blende
5–5.5	Actinolite Analcime Datolite Niccolite Scolecite Thomsonite		

Table 6 (*cont.*)

3.5	Boléite Celestine Cerussite Coral Creedite Greenockite Howlite Lepidolite Millerite Mimetite Phosphophyllite Powellite Thaumasite Witherite	2.5–3.5 2.5–3 2.5 2–4	Pearl { Crocoite { Wulfenite { Albertite Ivory Proustite Pseudophite Stichtite Tortoise-shell Vegetable ivory Whewellite Chrysocolla
3–4	Ludlamite	2–3	{ Bone { Plastics
3	Anglesite Anhydrite Barite Bornite Phosgenite Verdite	2–2.5 2 1.5–2	{ Amber Meerschaum { Alabaster (gypsum) Vivianite { Gedanite Pyrophyllite Soapstone
2.5–4	{ Jet { Serpentine	1	Ulexite

*Synthetically produced abrasives.

Table 7
Gemstones in order of their crystal systems

Cubic

Analcime	Garnet	Sodalite
Barium tantalate	Haüyne	Spinel
Boracite	Lazurite (lapis-lazuli)	Strontium titanate
Bornite	Leucite	Yttrium aluminate
Chromite	Magnetite	Yttrium oxide (YAG)
Cobaltite	Microlite	Zinc blende (sphalerite)
Cubic zirconia	Pentlandite	
Cuprite	Periclase	
Diamond	Pollucite	
Fluorite	Pyrite	
Gadolinium gallium garnet (GGG)	Rhodizite	
Gahnite	Smaltite	

Tetragonal

Anatase	Leucite (pseudocubic)	Scheelite
Apophyllite	Melinophane	Sellaite
Boléite	Phosgenite	Tugtupite
Cassiterite	Powellite	Wardite
Chalcopyrite	Rutile	Wulfenite
Idocrase	Scapolite	Zircon

Table 7 (*cont.*)

Hexagonal

Algodonite	Greenockite	Painite
Apatite	Jeremejevite	Simpsonite
Benitoite	Magnetoplumbite	Sogdianite
Beryl	Milarite	Sugilite
Breithauptite	Mimetite	Taaffeite
Bromellite	Nepheline	Thaumasite
Cancrinite	Niccolite	Zincite

Trigonal

Calcite (marble)	Hematite	Rhodochrosite
Corundum	Magnesite	Siderite
Dioptase	Millerite	Smithsonite
Dolomite	Phenakite	Stichtite
Eudialyte	Proustite	Tourmaline
Friedelite	Quartz	Willemite

Orthorhombic

Adamite	Diaspore	Pyrophyllite
Andalusite	Dumortierite	Rhombic yttrium aluminate
Anglesite	Enstatite	
Anhydrite	Eosphorite	Samarskite
Aragonite	Gedrite	Scorodite
Barite	Grandidierite	Sillimanite (*fibrolite*)
Bronzite	Hambergite	Sinhalite
Brookite	Hypersthene	Staurolite
Calcium titanate	Iolite	Stibiotantalite
Celestine	Jeremejevite	Thomsonite
Cerussite	Kornerupine	Topaz
Chambersite	Lawsonite	Triphylite
Chlorastrolite	Marcasite	Variscite
Chrysoberyl	Natrolite	Witherite
Columbite	Olivine (*peridot*)	Zoisite
Danburite	Prehnite	

Monoclinic

Actinolite	Herderite	Pyrophyllite
Augelite	Hodgkinsonite	Sanidine
Azurite	Hornblende	Sapphirine
Bayldonite	Howlite	Scolecite
Beryllonite	Jadeite	Serandite
Brazilianite	Kaersutite	Serpentine
Catapleite	Lazulite	Shattuckite
Charoite	Legrandite	Sphene
Chondrodite	Lepidolite	Spodumene
Clinohumite	Ludlamite	Spurrite
Colemanite	Malachite	Talc (soapstone)
Creedite	Meerschaum (?)	Tremolite
Crocoite	Mesolite	Ulexite

Table 7 (*cont.*)

Datolite	Nephrite	Vivianite
Diopside	Orthoclase feldspar	Whewellite
Durangite	Pectolite	Wollastonite
Epidote	Petalite	Xonotlite
Euclase	Phosphophyllite	Yugawaralite
Gypsum	Pseudomalachite	

Triclinic

Amblygonite	Plagioclase feldspar	Serendibite
Axinite	Pectolite	Turquoise
Kyanite	Pyroxmangite	
Microcline feldspar	Rhodonite	

Amorphous

Amber	Glass	Opal
Billitonite	Ivory	Plastics
Chrysocolla (?)	Jet	Silica glass
Copal resin	Moldavite	Tortoiseshell
Coral (black & golden)	Obsidian	Vegetable ivory
Ekanite	Odontolite	

Table 8
Gem materials according to colour

Colourless	*Colourless (cont.)*
Adamite (Tr–Op)	Dolomite (Tr)
Amblygonite (Tr)	Enstatite (Tr)
Analcime (Tr)	Euclase (Tr)
Anglesite (Tr)	Feldspar var; orthoclase (Tr)
Apatite (Tr)	Fluorite (Tr)
Apophyllite (Tr)	Gadolinium gallium garnet (Tr)
Aragonite (Tr)	Hambergite (Tr)
Augelite (Tr)	Jeremejevite (Tr)
Barite (Tr)	KTN (Tr)
Barium titanate (Tr)	Lawsonite (Tr)
Beryl (Tr)	Leucite (Tr)
Beryllonite (Tr)	Lithium niobate (Tr)
Bromellite (Tr)	Magnesite (Tr)
Calcium titanate (Tr)	Periclase (Tr)
Catapleite (Tr–Tl)	Petalite (Tr)
Celestine (Tr)	Phenakite (Tr)
Chrysoberyl (Tr)	Pollucite (Tr)
Colemanite (Tr)	Quartz (Tr)
Corundum (Tr)	Rutile (synthetic) (Tr)
Cubic zirconia (Tr)	Sanidine (Tr)
Danburite (Tr)	Scolecite (Tr)
Datolite (Tr)	Sellaite (Tr)
Diamond (Tr)	Spinel (synthetic) (Tr)
Diopside (Tr)	Spodumene (Tr)

Table 8 (*cont.*)

Colourless (cont.)
  Strontium titanate (Tr)
  Thaumasite (Tr)
  Topaz (Tr)
  Tourmaline (Tr)
  Vivianite (Tr)
  Whewellite (Tr)
  Yttrium aluminate (Tr)
  Yttrium oxide (Tr)
  Yugawaralite (Tr)
  Zircon (Tr)

Yellow and orange
  Adamite (Tr–Op)
  Amber (Tr and Tl)
  Amblygonite (Tr)
  Anglesite (Tr)
  Apatite (Tr)
  Aragonite (Tl)
  Axinite (Tr)
  Barite (Tr)
  Beryl (Tr)
  Beryllonite (Tr)
  Brazilianite (Tr)
  Calcite (Tr)
  Cancrinite (Tl–Op)
  Cassiterite (Tr)
  Catapleite (Tr–Tl)
  Chalcedony (Tl–Op)
  Chrondrodite (Tr)
  Chrysoberyl (Tr)
  Clinohumite (Tr–Op)
  Copal resin (Tr)
  Corundum (Tr)
  Danburite (Tr)
  Datolite (Tl–Op)
  Diamond (Tr)
  Dolomite (Tl)
  Eosphorite (Tr)
  Eudialyte (Tr–Op)
  Feldspar var; orthoclase (Tr), var;
    sunstone (Tr–Tl)
  Fluorite (Tr–Tl)
  Friedelite (Tr–Tl)
  Garnet vars; hessonite, spessartine and
    topazolite (Tr)
  Greenockite (Tr)
  Idocrase (Tr)
  Jadeite (Tl)
  Jeremejevite (Tr)
  Legrandite (Tr)

Yellow and orange (cont.)
  Marble (Tl–Op)
  Melinophane (Tr)
  Millerite (Tl)
  Mimetite (Tl)
  Opal var; fire opal (Tr–Tl)
  Ophicalcite (Tl–Op)
  Phenakite (Tr)
  Phosgenite (Tl)
  Quartz (Tr)
  Rutile (synthetic) (Tr)
  Scapolite (Tr)
  Scheelite (Tr–Tl)
  Serandite (Tr–Tl)
  Serendibite (Tr)
  Silica glass (Tr)
  Simpsonite (Tr)
  Sinhalite (Tr)
  Smithsonite (Op)
  Soapstone (Op)
  Sphene (Tr)
  Spinel (Tr)
  Spodumene (Tr)
  Staurolite (Tr–Op)
  Stibiotantalite (Tr)
  Thomsonite (Op)
  Topaz (Tr)
  Tourmaline (Tr)
  Triphylite (Tr–Tl)
  Willemite (Tr)
  Wulfenite (Tr–Tl)
  Zinc blende (Tr)
  Zircon (Tr)

Brown
  Amber (Tr–Tl)
  Anatase (Tr)
  Andalusite (Tr)
  Axinite (Tr)
  Barite (Tr–Op)
  Beryl (Tr)
  Brookite (Tl)
  Cassiterite (Tr)
  Chalcedony (Tl–Op)
  Chondrondite (Tr–Tl)
  Chrysoberyl (Tr)
  Corundum (Tr)
  Datolite (Tl–Op)
  Diamond (Tr)
  Diaspore (Tr–Tl)
  Dolomite (Tl–Op)

Table 8 (*cont.*)

Brown (cont.)
Eosphorite (Tr)
Epidote (Tr)
Fibrolite (Op)
Fluorite (Tr)
Garnet var; hessonite and spessartine (Tr)
Hornblende (Tr)
Hypersthene (Tl–Op)
Idocrase (Tr)
Jadeite (Tl–Op)
Kornerupine (Tr)
Marble (Op)
Microlite (Tr–Tl)
Moldavite (Tr)
Nepheline (Tr–Tl)
Obsidian (Tr–Tl)
Peridot (Tr)
Powellite (Tr)
Prehnite (Tl)
Quartz (Tr)
Rhodochrosite (Tl–Op)
Rutile (synthetic) (Tr)
Sapphirine (Tr)
Scheelite (Tr)
Serpentine (Op)
Siderite (Tl)
Sinhalite (Tr)
Soapstone (Op)
Sphene (Tr)
Staurolite (Op)
Stibiotantalite (Tr)
Thomsonite (Op)
Topaz (Tr)
Tourmaline (Tr)
Tugtupite (Tl–Op)
Willemite (Tr)
Zinc blende (Tr)
Zircon (Tr)

Red and pink
Adamite (Tr–Op)
Amber (Tr–Tl)
Andalusite (Tr)
Barite (Tr)
Beryl (Tr)
Cassiterite (Tr)
Chalcedony (Tl–Op)
Chambersite (Tr)
Chrondrodite (Tr–Tl)
Coral (Tl–Op)

Red and pink (cont.)
Corundum (Tr)
Crocoite (Tr–Tl)
Cuprite (Tr)
Danburite (Tr)
Datolite (Tl–Op)
Diamond (Tr)
Dolomite (Tl–Op)
Durangite (Tr–Tl)
Eosphorite (Tr)
Eudialyte (Tr–Op)
Fluorite (Tr)
Friedelite (Tr–Tl)
Garnet var; pyrope and almandine (Tr)
Hodgkinsonite (Tr)
Jadeite (Tl–Op)
Marble (Op)
Microlite (Tr–Tl)
Nepheline (Tr–Tl)
Opal (fire opal) (Tr–Tl)
Painite (Tr)
Pearl (Tl–Op)
Phenakite (Tr)
Powellite (Tl–Op)
Pyroxmangite (Tr–Op)
Quartz var; rose quartz (Tr–Tl)
Rhodizite (Tr)
Rhodochrosite (Tl–Op)
Rhodonite (Tl–Op)
Rutile (Tr)
Sapphirine (Tr)
Scapolite (Tr–Tl)
Serandite (Tr–Tl)
Smithsonite (Tl)
Spinel (Tr)
Spodumene (Tr)
Staurolite (Op)
Stichtite (Op)
Thomsonite (Op)
Thulite (Op)
Topaz (Tr)
Tourmaline (Tr)
Tremolite (Tr)
Tugtupite (Tr–Op)
Vayrynenite (Tr)
Wulfenite (Tl)
Xonotlite (Tr–Op)
Zincite (Tr)
Zircon (Tr)
Zoisite (Tl–Op)

Table 8 *(cont.)*

Purple and violet
  Adamite (Tr–Op)
  Apatite (Tr)
  Axinite (Tr)
  Beryl (Tr)
  Charoite (Tl–Op)
  Corundum (Tr)
  Creedite (faint) (Tl)
  Diopside var; violane (Op)
  Dumortierite (Op)
  Fibrolite (Tr)
  Fluorite (Tr–Tl)
  Garnet var; almandine and rhodolite (Tr)
  Jadeite (Tl)
  Lepidolite (Op)
  Marble (Op)
  Quartz (Tr–Tl)
  Scapolite (Tr–Tl)
  Sogdianite (Tr–Op)
  Spinel (Tr)
  Spodumene (Tr)
  Spurrite (Tl–Op)
  Stichtite (Op)
  Sugilite (Tl–Op)
  Taaffeite (Tr)
  Tourmaline (Tr)
  Zircon (Tr)

Blue
  Anhydrite (Tr–Tl)
  Apatite (Tr–Tl)
  Anatase (Tr–Tl)
  Anglesite (Tr)
  Azurite (Op)
  Barite (Tr–Tl)
  Benitoite (Tr)
  Beryl (Tr)
  Boleite (Tl–Op)
  Catapleite (Tr–Tl)
  Celestine (Tr)
  Cerussite (Tl)
  Chalcedony (Tl–Op)
  Chambersite (Tr)
  Chrysocolla (Tl–Op)
  Corundum (Tr)
  Diamond (Tr)
  Diopside var; violane (Op)
  Dumortierite (Op)
  Euclase (Tr)
  Fibrolite (Tr)
  Fluorite (Tr)

Blue (cont.)
  Grandidierite (Tr)
  Hauyne (Tl–Op)
  Idocrase var; cyprine (Tl–Op)
  Iolite (Tr)
  Jadeite (Tl–Op)
  Jeremejevite (Tr)
  Kyanite (Tr)
  Lapis lazuli (Op)
  Lawsonite (pale) (Tr)
  Lazulite (Op)
  Odontolite (Op)
  Pearl (Op)
  Pectolite (Tl)
  Quartz var; hawk's eye (Op)
  Rutile (synthetic) (Tr)
  Sapphirine (Tr)
  Scapolite (Tr–Tl)
  Scorodite (Tr–Tl)
  Serendibite (Tr)
  Shattuckite (Op)
  Smithsonite (Tl–Op)
  Sodalite (Op)
  Spinel (Tr)
  Topaz (Tr)
  Tourmaline (Tr)
  Turquoise (Op)
  Vivianite (Tr)
  Wardite (Op)
  Zircon (Tr)
  Zoisite (Tr)

Green
  Actinolite (Tr)
  Adamite (Tr–Op)
  Andalusite (Tr)
  Anglesite (Tr)
  Apatite (Tr)
  Bayldonite (Tl–Op)
  Bastite (Op)
  Beryl (Tr)
  Boracite (Tr)
  Cerussite (Tl)
  Chalcedony (Ti–Op)
  Chlorastrolite (Op)
  Chrysoberyl (Tr)
  Chrysocolla (Tl–Op)
  Corundum (Tr)
  Datolite (Tr–Op)
  Diamond (Tr)
  Diopside (Tr)

Table 8 (*cont.*)

Green (cont.)
Dioptase (Tr–Tl)
Dolomite (Tr–Op)
Ekanite (Tl)
Enstatite (Tr)
Epidote (Tr)
Euclase (Tr)
Feldspar var; microcline (Tr–Op)
Fibrolite (Tr)
Fluorite (Tr–Tl)
Gahnite (Tr)
Garnet var; demantoid and uvarovite (Tr)
Garnet var; massive grossular (Tl–Op)
Grandidierite (Tr)
Idocrase (Tr) var; californite (Tl–Op)
Jadeite (Tl–Op)
Kornerupine (Tr)
Kyanite (Tr)
Ludlamite (Tr)
Malachite (Op)
Marble (Op)
Milarite (Tr)
Moldavite (Tr)
Nepheline (Tr–Tl)
Nephrite (Tl–Op)
Ophicalcite (Tl–Op)
Peridot (Tr)
Phosphophyllite (Tr)
Prehnite (Tl–Op)
Pyrophyllite (Op)
Pseudomalachite (Op)
Pseudophite (Op)
Quartz (Tr)
Rhodizite (Tr)
Sapphirine (Tr)
Serpentine (Tl–Op)
Smaragdite (Op)
Smithsonite (Tl–Op)
Soapstone (Op)
Sphene (Tr)
Spodumene (Tr)
Thomsonite (Op)
Topaz (rare) (Tr)
Tourmaline (Tr)
Tremolite (Tl)
Turquoise (Op)
Variscite (Op)
Verdite (Op)
Willemite (Tr–Op)
Wulfenite (Tl)
Yttrium aluminate (Tr)

Green (cont.)
Zircon (Tr)
Zoisite (Tr)

White
Alabaster (Tl–Op)
Anhydrite (Tr–Tl)
Amber (Tl–Op)
Barite (Tl)
Bone (Op)
Cerussite (Tl)
Chalcedony (Tl–Op)
Colemanite (Tl)
Coral (Tl–Op)
Creedite (Tl)
Datolite (Tl–Op)
Dolomite (Tr)
Gypsum var; satin spar (Tl)
Howlite (Tl–Op)
Ivory (Tl–Op)
Jadeite (Tl–Op)
Marble (Op)
Meerschaum (Op)
Mesolite (Tl)
Nephrite (Tl–Op)
Opal (Tl–Op)
Pearl (Tl–Op)
Pectolite (Tl)
Pyrophyllite (Tl–Op)
Quartz var; milky (Tl–Op)
Scapolite (Tr–Tl)
Scolecite (Tl–Op)
Thaumasite (Tl–Op)
Thomsonite (Op)
Ulexite (Tl)
Witherite (Tl–Op)
Wollastonite (Tl–Op)
Wulfenite (Tl–Op)
Yugawaralite (Tl)

Grey
Apophyllite (Tl)
Cerussite (Tl)
Chalcedony (Tl)
Diamond (Tr–Tl)
Feldspar var; labradorite (Op)
Herderite (Tl–Op)
Marble (Op)
Obsidian (Tl–Op)
Pectolite (Tl–Op)
Pyrophyllite (Tl–Op)

Table 8 (*cont.*)

Black	Metallic colours (cont.)
Amber (Op)	Copper red
Anatase (Op)	Pale copper red
Cassiterite (Op)	Brownish-black
Chalcedony (stained) (Op)	Black
Coral (Op)	Black
Diamond (Op)	Black
Garnet (Op)	Black
Jadeite (Op)	Iron black
Jet (Op)	Brilliant black
Kaersutite (Op)	Brilliant black
Marble (Op)	Cobaltite
Obsidian (Op)	Algodonite
Opal (with play of colour) (Tl–Op)	Domeykite
Pearl (Op)	Smaltite
Quartz var; morion (Op)	Pyrite
Rutile (Op)	Pentlandite
Spinel (Op)	Chalcopyrite
Tourmaline var; schorl (Op)	Marcasite
	Breithauptite
Metallic colours	Niccolite
White	Euxenite
Tin white to steel grey	Columbite
Tin white to steel grey	Tantalite
Tin white to steel grey	Magnetite
Brass yellow	Magneto-plumbite
Bronze yellow	Chromite
Golden yellow	Hematite
Grey yellow	Psilomelane

Tr – transparent; Tl – translucent; Op – opaque.

Table 9
The pleochroic colours of the principal gemstones

Actinolite	Yellow; dark green
Anatase (distinct)	Pale blue or yellowish; dark blue or orange
Andalusite (strong)	Yellow; green; red
Apatite (weak except in Myanmar stones)	
Yellow (asparagus stone)	Golden-yellow; greenish yellow
Blue green (moroxite)	Pale yellow; sapphire-blue
Axinite (strong)	Violet; brown; green
Benitoite (strong)	Colourless; greenish to indigo-blue
Beryl (distinct)	
Green (emerald)	Yellowish green; bluish green
Greenish blue (aquamarine)	Colourless to pale yellowish green; pale bluish green
Blue (aquamarine)	Colourless; sky-blue
Pink (morganite)	Pale rose; bluish rose

Table 9 (*cont.*)

Yellow (heliodor)	Pale yellowish green; pale bluish green
Violet	Violet; colourless
Red	Purplish red; orange red
Chrysoberyl (strong in deep colours)	
Yellow	Colourless; pale yellow; lemon-yellow
(Cat's-eye)	Reddish yellow; greenish yellow; green
Green (alexandrite)	
Natural light	Emerald-green; yellowish; columbine-red
Artificial light	Emerald-green; reddish yellow; red
Corundum (strong)	
Red (ruby)	Pale yellowish red; deep red
Blue (sapphire)	Pale greenish blue; deep blue
Green	Yellowish green; green
Violet	Yellowish red; violet-red
Yellow	Imperceptible
Corundum synthetic (alexandrite type)	
In natural light	Pale brownish green; deep mauve
In artificial light	Brownish yellow; deep mauve
Other synthetic corundums show dichroic tints in general agreement with the natural corundums of similar colour	
Danburite (very weak)	Pale yellow; very pale yellow; pale yellow
Dioptase (weak)	Dark green; light green
Dumortierite	Blue-black; blue; colourless. Also black; brown; reddish brown
Enstatite (distinct)	Green; yellowish green; brownish green
Epidote (strong)	Green; yellowish green; yellow
Euclase (distinct to strong)	Colourless; pale green; green
Deep blue	Azure blue; Prussian blue; greenish blue
Fibrolite (distinct)	Colourless; pale yellow; sapphire-blue
Grandidierite	Colourless; dark blue-green; dark green
Iolite (strong)	Pale blue; pale yellow; dark violet-blue
Kornerupine (strong)	
Yellowish green	Yellowish; brown; greenish
Chromian	Emerald-green; bluish grey; reddish purple
Kyanite (distinct)	Pale blue; blue; dark blue
Peridot (distinct)	Yellow-green; green
Quartz (weak)	
Yellow (citrine)	Yellow; slightly paler yellow
Violet (amethyst)	Purple; reddish purple
Smoky quartz	Brown; reddish brown
Rose quartz	Pink; pale pink
Heat-treated amethyst 'topaz' colour	Imperceptible
Scapolite (distinct)	
Pink and violet	Dark blue to lavender blue; colourless to violet
Colourless and pale yellow	Colourless to pale yellow; yellow
Sinhalite (distinct)	Pale brown; greenish brown; dark brown
Sphene (titanite) (strong)	Colourless; yellow; reddish yellow
Spodumene (strong)	
Pink (kunzite)	Colourless; pink; violet
Green (hiddenite)	Bluish green; grass-green; yellowish green
Yellow	Pale yellow; deep yellow; yellow

Table 9 (*cont.*)

Staurolite (distinct)	Red; brown; yellow
Zincian	Green; yellow; red
Topaz (distinct)	
Yellow	Honey-yellow; straw-yellow; pinkish yellow
Blue	Colourless; pale pink; blue
Pink	Colourless; very pale pink; pink
Green	Pale green; bluish green; colourless
Tourmaline (strong)	
Red (rubellite)	Pink; dark red
Blue (indicolite)	Light blue; dark blue
Green	Pale green; dark green
Brown	Yellowish brown; deep brown
Wide colour range gives rise to great variation in pleochroic colours.	
Zircon (weak except in blue)	
Red	Columbine-red; clove-brown
Blue	Colourless; sky-blue
Green	Brownish green; green
Yellow	Brownish yellow; honey-yellow
Brown	Yellowish brown; reddish brown
Zoisite	
Blue (tanzanite)	Blue; purple; sage green
Pink (thulite)	Dark pink; pink; yellow

Table 10
Colour filter

A table of the effects seen when the stone is viewed through a dichromatic filter, such as the Chelsea colour filter, which transmits a band of green and a band of red light. The residual colour will be more intense the stronger the body colour of the stone. The best results are seen when the stone is illuminated with light from a tungsten electric lamp (desk lamp) and the filter held close to the eye.

Green stones		*Green stones (cont.)*	
Alexandrite	red	Glass (pastes)	green
Aquamarine	distinctively green	Hiddenite	slight pink
		Jadeite	green
Aventurine quartz	reddish	Peridot	green
Chrome chalcedony	red	Sapphire	green
Chrysoprase	green	Soudé emerald	green
Demantoid garnet	reddish	(The old type soudé emerald may show red)	
Emerald	pink to red		
(some emeralds from South Africa and India may not show a red hue, but remain greenish)		Stained bowenite	red
		Stained chalcedony	red
		Stained jadeite	red
Enstatite	green	Synthetic corundum	red
Fluorite	reddish	(Alexandrite colour)	

800

Table 10 (*cont.*)

Green stones (cont.)

Synthetic emerald (most)	strong red
Synthetic sapphire	red
Synthetic spinel	red
(Some old types may show green)	
Tourmaline	green
(Certain anomalous green tourmalines have been found to show red)	
Zircon	reddish

Red stones

Garnets	dark red (no fluorescence)
Garnet-topped doublets	dark red (no fluorescence)
Glass (pastes)	dark red (no fluorescence)
Ruby (natural and synthetic)	strong fluorescent red
Spinel	fluorescent red
Spinel (synthetic)	fluorescent red
(The pink synthetic spinel does not show a red colour through the filter)	

Blue stones

Aquamarine	distinctive green
Garnet-topped doublets	greenish blue

Blue stones (cont.)

Glass (pastes)	
Dark blue	red
Light blue	greenish
Lapis lazuli	weak brownish red
Sapphire	blackish green
(The blue sapphire which shows a purple colour under artificial light usually shows red under the filter)	
Sodalite	slightly brownish
Spinel	reddish
Spinel (natural) coloured by cobalt	red
'Swiss lapis'	greenish blue
Synthetic sapphire	dark greenish blue
(The natural and synthetic sapphire are indistinguishable under the colour filter)	
Synthetic spinel	
Dark blue	red
Light blue	orange
'Zircon' colour	orange to red
'Lapis lazuli colour'	bright red
Zircon	greenish

Purple stones

Amethyst	reddish
Violet sapphire	bright red

Table 11
The light spectrum (nm)

Ultra-violet	100–390
Violet	390–430
Blue	439–490
Blue-green	490–510
Green	510–550
Yellow-green	550–575
Yellow	575–590
Orange	590–630
Orange-red	630–650
Red	650–700
Deep red	700–780
Infra-red	780–1 000 000

Table 12
The major Fraunhofer lines (nm)

A	oxygen	759–762
B	oxygen	687–688
C	hydrogen	656.3
D1	sodium	589.6
D2	sodium	589.0
E	iron	527.0
F	hydrogen	486.1
G	iron	430.8
H	calcium	396.8
K	calcium	393.4

Table 13
The principal emission lines useful for calibration (nm)

Barium	588.1 (yellow) 577.8 (yellow) 553.6 (yellow-green) 551.9 (yellow-green) 542.5 (green) 513.7 (green) 493.7 (blue-green) 487.4 (blue) 455.4 (blue)	Barium lines are better produced by arc than by bunsen flame
Calcium	616.2 (orange) 559.0 (yellow-green) 422.7 (violet)	Calcium lines can usually be obtained by inserting a calcium compound in the bunsen flame
Copper	521.8 (green) 515.3 (green) 510.6 (green)	Copper lines are best produced by arc excitation
Lithium	670.8 (red) 610.3 (orange)	Given by lithium compound in bunsen flame. Red line is strong, orange line is weak
Mercury	623.4 (orange) 615.2 (orange) 579.0 (yellow) 577.0 (yellow) 546.1 (green) 435.9 (blue) 434.8 (blue) 407.8 (violet) 404.7 (violet)	The mercury lines are best obtained from a mercury vapour electric discharge lamp. The lines may be observed underlying the continuous spectrum of a fluorescent lighting tube
Potassium	766.9 (deep red) 766.5 (deep red) 404.7 (violet) 404.4 (violet)	Can be induced by potassium compound in bunsen flame
Sodium	589.0 (yellow) 589.6 (yellow)	Sodium compound in bunsen flame
Strontium	687.0 (red) 606.0 (orange) 460.7 (blue) 407.8 (violet)	Strontium compound in bunsen flame
Thallium	535.0 (green)	Thallium compound (Clerici's solution) in bunsen flame

Table 14
Interconversion of wavelength (nm), wave numbers (cm^{-1}), and electron volts

A straight edge placed horizontally across the diagram, using the two outer scales as guides, will give interconversion between the different energies.

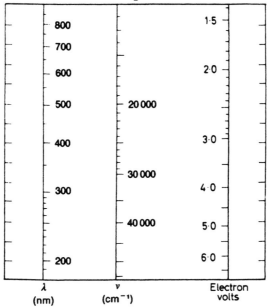

| λ
(nm) | ν
(cm^{-1}) | Electron
volts |

1 electron volt (eV) equals 8066 cm^{-1} (ν);
cm^{-1} equals number of waves in a centimetre

Table 15
The obsorption spectra of gemstones

The following spectra are drawn as seen through a prism-tpe spectroscope.

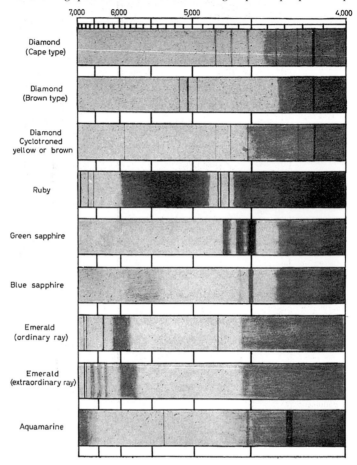

Spectrum plate 1

Table 15 (*cont.*)

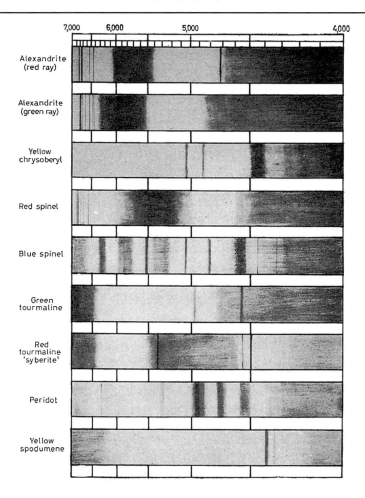

Spectrum plate 2

Table 15 (*cont.*)

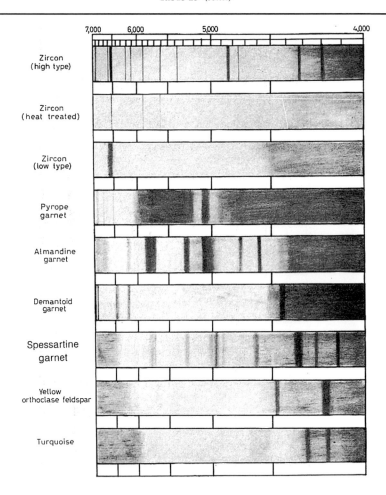

Spectrum plate 3

Table 15 (*cont.*)

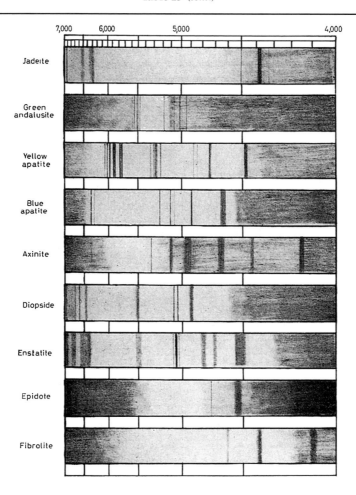

Spectrum plate 4

Table 15 (*cont.*)

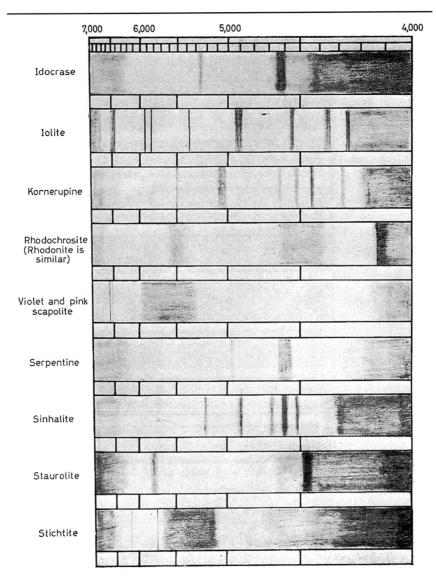

Spectrum plate 5

Table 15 (*cont.*)

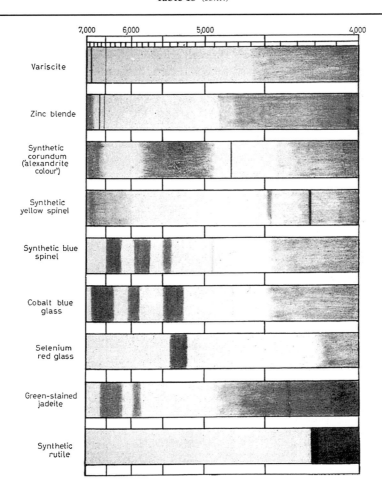

Spectrum plate 6

Table 16
The fluorescent colours of the principal gemstones and their simulants

The following table gives the more common luminescent colours shown by gem materials under each of the three types of radiation – long-wave ultra-violet, short-wave ultra-violet and X-rays. As it is extremely difficult to analyse the shades of colour in the various groups, the wide division into the six colours of the Newtonian spectrum, with the addition of white, has therefore been adopted. It must be understood that under *red* the shade may vary from deep plum-red, through crimson, to rose and pink; and *blue*, from a rich dark blue to a bright whitish blue. The lilac-coloured glow is listed under violet. In the case of synthetic emerald the red glow is much stronger when the stones are irradiated with the additional long-wave ultra-violet and near violet wavelengths present in the fluorescent-type ultra-violet lamp. In many gem species, fluorescence varies in intensity from specimen to specimen.

Colour	Long-wave ultra-violet light (365.0 nm)	Short-wave ultra-violet light (253.7 nm)	X-rays (mean 0.1 nm)
White	Ivory	Synthetic white spinel	Diamond
	Casein	Paste (some)	Scapolite
	Copal resin	Ivory	Synthetic white spinel (some)
	Amber	Casein	
	Tortoise-shell	Copal resin	Cubic zirconia
	Opal	Amber	
	Jadeite (some)	Tortoise-shell	
		Opal	
		Jadeite (some)	
Red	Synthetic ruby	Natural and synthetic ruby	Natural and synthetic ruby (synthetic shows phosphorescence)
	Natural ruby		
	Natural and synthetic pink sapphire	Natural and synthetic pink sapphire	
	Natural and synthetic red spinel	Natural and synthetic red spinel	Natural and synthetic pink sapphire
	Synthetic orange sapphire	Synthetic orange sapphire	Natural and synthetic red spinel
	Natural and synthetic alexandrite	Calcite (some)	Synthetic orange sapphire
	Synthetic (alexandrite) sapphirre	Natural and synthetic alexandrite	Sri Lanka blue sapphire
	Synthetic green sapphire	Synthetic (alexandrite) sapphire	Synthetic blue spinel (some)
	Synthetic blue spinel	Synthetic (alexandrite) sapphire	Synthetic green spinel
	Sri Lanka blue sapphire	Synthetic emerald	Synthetic emerald
	Synthetic emerald	Sri Lanka blue sapphire	Morganite
	Emerald (some)	Diamond	Natural emerald
	Calcite (some)	Natural emerald (some)	Thulite
	Kyanite	Fire opal (brownish red)	Blue apatite
	Fire opal (brownish red)	Synthetic blue spinel	Yellow apatite
		Synthetic green spinel	
		Scapolite	

Table 16 (*cont.*)

Colour	Long-wave ultra-violet light (365.0 nm)	Short-wave ultra-violet light (253.7 nm)	X-rays (mean 0.1 nm)
	Synthetic green spinel Diamond Green grossular Benitoite		

Orange	Kunzite
	
	Sri Lanka yellow
	  sapphire
	Scapolite
	Diamond
	Sri Lanka blue sapphire
	Synthetic (alexandrite)
	  sapphire
	Natural white sapphire
	Synthetic green sapphire
	Synthetic orange
	  sapphire
	Natural mauve spinel
	  (some)
	Topaz (some)
	Sodalite (orange spots)
	Lapis lazuli (orange
	  spots)
	Coral, light red and
	  orange
		Sri Lanka yellow
		  sapphire
		Scapolite
		Diamond
		Sri Lanka blue sapphire
		Synthetic orange
		  sapphire
		Synthetic green
		  sapphire (brownish
		  orange)
		Synthetic (alexandrite)
		  sapphire
		Synthetic pale blue
		  spinel
		Synthetic emerald
		  (early Gilson)
		GGG
		Jadeite dyed
		  lavender (some)
		Synthetic emeralds
		  (early Gilson)
		YAG (some)
			Kunzite
			
			Sri Lanka yellow sapphire
			Calcite (including marble)
			Scapolite
			Synthetic white sapphire
			Synthetic (alexandrite)
			  sapphire
			Synthetic green sapphire
			Natural white sapphire
			Massive green grossular
			  ('Transvaal jade')
			Petalite
			Pale yellow sinhalite
			Topaz (some)

Yellow	Diamond Amber Apatite Zircon Topaz Fire opal (brownish) Paste Kornerupine (some) Scapolite White opal Oiled emerald Synthetic pale green spinel GGG Yellowish green YAG Cubic zirconia (some)	Diamond Zircon Amber Green apatite Violet apatite White opal Paste (some) Yellowish green YAG Cubic zirconia (some) Yellow apatite Green apatite Violet apatite	Diamond Zircon Cassiterite Cultured pearl Freshwater pearl Rhodizite Diopside Lapis lazuli Sodalite (yellow spots) YAG
Green	Synthetic yellow spinel Synthetic yellow-green	Synthetic yellow and green spinels	Synthetic yellow spinel Synthetic yellow-green

Table 16 (*cont.*)

Colour	Long-wave ultra-violet light (365.0 nm)	Short-wave ultra-violet light (253.7 nm)	X-rays (mean 0.1 nm)
	spinel	Willemite	spinel
	Willemite	Opal (some)	Synthetic white spinel
	Opal (some)	Synthetic green spinel	Diamond
	Diamond	Diamond	Paste
	Pale blue natural	Amber	Opal (some)
	spinels (some)	Paste	Natural spinels (some)
	Emerald (some)	Natural and cultured	Taaffeite
	Apatite	pearl (some)	Rhodizite
	Paste (some)	Synthetic blue sapphire	Amblygonite
	Amber	(some)	Topaz
	Synthetic white spinel	Cubic zirconia	Synthetic blue
	and sapphire (very	Emerald (some)	Sapphire (some)
	weakly)		
	Natural and cultured		
	pearl (some)		
Blue	Diamond	Benitoite	Fluorite
	Danburite	Scheelite	Benitoite
	Fluorite (Blue John		
	is inert)	Fluorite	Beryllonite
	Paste	Danburite	Diamond
	Pearl	Diamond	Scheelite
	Casein	Synthetic blue spinel	Synthetic blue sapphire
	Ivory	Synthetic blue sapphire	Kyanite
	Amber	Paste (some)	Phenakite
	Moonstone	Ivory	Synthetic white spinel
	Blue apatite	Amber	(some)
	Natural and cultured	Copal	Synthetic blue spinel
	pearl	Blue apatite	Paste (some)
	Synthetic dark blue	Natural and cultured	Lapis lazuli
	spinel	pearl	Sodalite
		Synthetic dark blue spinel	Topaz
		Synthetic white spinel	Zircon
		Plastic (some)	
Violet	Fluorite (Blue John	Fluorite	Fluorite
	is inert)	Diamond	Danburite
	Diamond	Morganite	Synthetic yellow sapphire
	Apatite	Synthetic pink sapphire	(some)
	Scapolite	Alexandrite-like sapphire	Synthetic white sapphire
	Morganite	Violet apatite	(some)
	Diopside	Yellow apatite	Scapolite
	Alexandrite-like		Pink tourmaline
	sapphire (some)		Zircon
	Synthetic pink sapphire		GGG
	Violet apatite		
	Yellow apatite		

Table 17
Specific gravity correction tables for toluene and 1,2-dibromoethane

Commercial samples of toluene and more particularly of 1,2-dibromoethane may vary appreciably in SG from the figures given below. For really accurate determinations, therefore, it is wise to purchase a fair quantity (say 500 g) which can then be calibrated and will last for years if kept in a stoppered bottle.

A convenient method of calibrating a sample is to carry out a hydrostatic determination with a large piece of pure quartz (say 50 carats) and to work 'backwards', assuming the SG of quartz to be 2.651, according to the formula

$$\text{specific gravity of liquid} = 2.651 \times \frac{\text{loss of weight in liquid}}{\text{weight of stone}}$$

Supposing the SG of a sample of toluene at 11.3 °C is found on calibration to be 0.8734 – a figure corresponding to a temperature of 10.3 °C in the tables – the worker will know that it is necessary to subtract 1 °C from his actual temperature in all future experiments before referring to the table.

(a) SGs of toluene from 5 °C to 25 °C

°C	SG	°C	SG	°C	SG	°C	SG
5.0	0.8787	6.5	0.8772	8.0	0.8757	9.5	0.8742
5.1	0.8786	6.6	0.8771	8.1	0.8756	9.6	0.8741
5.2	0.8785	6.7	0.8770	8.2	0.8755	9.7	0.8740
5.3	0.8784	6.8	0.8769	8.3	0.8754	9.8	0.8739
5.4	0.8783	6.9	0.8768	8.4	0.8753	9.9	0.8738
5.5	0.8782	7.0	0.8767	8.5	0.8752	10.0	0.8737
5.6	0.8781	7.1	0.8766	8.6	0.8751	10.1	0.8736
5.7	0.8780	7.2	0.8765	8.7	0.8750	10.2	0.8735
5.8	0.8779	7.3	0.8764	8.8	0.8749	10.3	0.8734
5.9	0.8778	7.4	0.8763	8.9	0.8748	10.4	0.8733
6.0	0.8777	7.5	0.8762	9.0	0.8747	10.5	0.8732
6.1	0.8776	7.6	0.8761	9.1	0.8746	10.6	0.8731
6.2	0.8775	7.7	0.8760	9.2	0.8745	10.7	0.8730
6.3	0.8774	7.8	0.8759	9.3	0.8744	10.8	0.8729
6.4	0.8773	7.9	0.8758	9.4	0.8743	10.9	0.8728
11.0	0.8727	14.6	0.8691	18.1	0.8656	21.6	0.8621
11.1	0.8726	14.7	0.8690	18.2	0.8655	21.7	0.8620
11.2	0.8725	14.8	0.8689	18.3	0.8654	21.8	0.8619
11.3	0.8724	14.9	0.8688	18.4	0.8653	21.9	0.8618
11.4	0.8723	15.0	0.8687	18.5	0.8652	22.0	0.8617
11.5	0.8722	15.1	0.8686	18.6	0.8651	22.1	0.8616
11.6	0.8721	15.2	0.8685	18.7	0.8650	22.2	0.8615
11.7	0.8720	15.3	0.8684	18.8	0.8649	22.3	0.8614
11.8	0.8719	15.4	0.8683	18.9	0.8648	22.4	0.8613
11.9	0.8718	15.5	0.8682	19.0	0.8647	22.5	0.8612
12.0	0.8717	15.6	0.8681	19.1	0.8646	22.6	0.8611
12.1	0.8716	15.7	0.8680	19.2	0.8645	22.7	0.8610
12.2	0.8715	15.8	0.8679	19.3	0.8644	22.8	0.8609
12.3	0.8714	15.9	0.8678	19.4	0.8643	22.9	0.8608
12.4	0.8713	16.0	0.8677	19.5	0.8642	23.0	0.8607
12.5	0.8712	16.1	0.8676	19.6	0.8641	23.1	0.8606
12.6	0.8711	16.2	0.8675	19.7	0.8640	23.2	0.8605

Table 17 (cont.)

°C	SG	°C	SG	°C	SG	°C	SG
12.7	0.8710	16.3	0.8674	19.8	0.8639	23.3	0.8604
12.8	0.8709	16.4	0.8673	19.9	0.8638	23.4	0.8603
12.9	0.8708	16.5	0.8672	20.0	0.8637	23.5	0.8602
13.0	0.8707	16.6	0.8671	20.1	0.8636	23.6	0.8601
13.1	0.8706	16.7	0.8670	20.2	0.8635	23.7	0.8600
13.2	0.8705	16.8	0.8669	20.3	0.8634	23.8	0.8599
13.3	0.8704	16.9	0.8668	20.4	0.8633	23.9	0.8598
13.4	0.8703	17.0	0.8667	20.5	0.8632	24.0	0.8597
13.5	0.8702	17.1	0.8666	20.6	0.8631	24.1	0.8596
13.6	0.8701	17.2	0.8665	20.7	0.8630	24.2	0.8595
13.7	0.8700	17.3	0.8664	20.8	0.8629	24.3	0.8594
13.8	0.8699	17.4	0.8663	20.9	0.8628	24.4	0.8593
13.9	0.8698	17.5	0.8662	21.0	0.8627	24.5	0.8592
14.0	0.8697	17.6	0.8661	21.1	0.8626	24.6	0.8591
14.1	0.8696	17.7	0.8660	21.2	0.8625	24.7	0.8590
14.2	0.8695	17.8	0.8659	21.3	0.8624	24.8	0.8589
14.3	0.8694	17.9	0.8658	21.4	0.8623	24.9	0.8588
14.4	0.8693	18.0	0.8657	21.5	0.8622	25.0	0.8587
14.5	0.8692						

(b) SGs of 1,2-dibromoethane from 10°C to 20°C

°C	SG	°C	SG	°C	SG	°C	SG
10.0	2.1998	11.1	2.1976	12.2	2.1954	13.3	2.1932
10.1	2.1996	11.2	2.1974	12.3	2.1952	13.4	2.1930
10.2	2.1944	11.3	2.1972	12.4	2.1950	13.5	2.1928
10.3	2.1992	11.4	2.1970	12.5	2.1948	13.6	2.1926
10.4	2.1990	11.5	2.1968	12.6	2.1946	13.7	2.1924
10.5	2.1988	11.6	2.1966	12.7	2.1944	13.8	2.1922
10.6	2.1986	11.7	2.1964	12.8	2.1942	13.9	2.1920
10.7	2.1984	11.8	2.1962	12.9	2.1940	14.0	2.1918
10.8	2.1982	11.9	2.1960	13.0	2.1938	14.1	2.1916
10.9	2.1980	12.0	2.1958	13.1	2.1936	14.2	2.1914
11.0	2.1978	12.1	2.1956	13.2	2.1934	14.3	2.1912
14.4	2.1910	15.9	2.1880	17.3	2.1852	18.7	2.1824
14.5	2.1908	16.0	2.1878	17.4	2.1850	18.8	2.1822
14.6	2.1906	16.1	2.1876	17.5	2.1848	18.9	2.1820
14.7	2.1904	16.2	2.1874	17.6	2.1846	19.0	2.1818
14.8	2.1902	16.3	2.1872	17.7	2.1844	19.1	2.1816
14.9	2.1900	16.4	2.1870	17.8	2.1842	19.2	2.1814
15.0	2.1898	16.5	2.1868	17.9	2.1840	19.3	2.1812
15.1	2.1896	16.6	2.1866	18.0	2.1838	19.4	2.1810
15.2	2.1894	16.7	2.1864	18.1	2.1836	19.5	2.1808
15.3	2.1892	16.8	2.1862	18.2	2.1834	19.6	2.1806
15.4	2.1890	16.9	2.1860	18.3	2.1832	19.7	2.1804
15.5	2.1888	17.0	2.1858	18.4	2.1830	19.8	2.1802
15.6	2.1886	17.1	2.1856	18.5	2.1828	19.9	2.1800
15.7	2.1884	17.2	2.1854	18.6	2.1826	20.0	2.1798
15.8	2.1882						

Table 17 (*cont.*)

(*c*) *The relation between refractive index and SG of Clerici solution and di-iodomethane with suitable dilutants*

If a specimen under test is found to be freely suspended in the liquid, the SG of that liquid may be ascertained from its index of refraction.

Clerici solution: water

Refractive index	SG	Refractive index	SG	Refractive index	SG	Refractive index	SG
1.500	2.584	1.550	3.008	1.600	3.432	1.640	3.770
1.510	2.669	1.560	3.093	1.610	3.517	1.650	3.854
1.520	2.753	1.570	3.178	1.620	3.601	1.660	3.939
1.530	2.838	1.580	3.263	1.630	3.685	1.670	4.023
1.540	2.923	1.590	3.348				

The above table is based on the work of Anderson, Payne and Franklin.

Di-iodomethane: toluene

Refractive index	SG	Refractive index	SG	Refractive index	SG	Refractive index	SG
1.610	2.007	1.650	2.403	1.690	2.809	1.720	3.106
1.620	2.106	1.660	2.502	1.700	2.908	1.730	3.205
1.630	2.205	1.670	2.601	1.710	3.007	1.740	3.304
1.640	2.304	1.680	2.710				

Appendices

Appendix A

Birthstones

Official list issued by the National Association of Goldsmiths of Great Britain and Ireland.

Month	Colour	Official stone
January	Dark red	Garnet
February	Purple	Amethyst
March	Pale blue	Aquamarine
April	White (transparent)	Diamond
May	Bright green	Emerald
June	Cream	Pearl
July	Red	Ruby
August	Pale green	Peridot
September	Deep blue	Sapphire
October	Variegated	Opal
November	Yellow	Topaz
December	Sky-blue	Turquoise

Gemstones for the Days of the Week

Day	Stone
Sunday	Topaz or diamond
Monday	Pearl or crystal
Tuesday	Emerald or ruby
Wednesday	Amethyst or loadstone
Thursday	Cornelian or sapphire
Friday	Emerald or cat's-eye
Saturday	Diamond or turquoise

Emblems of the Twelve Apostles

Apostle	Stone
Andrew	Blue sapphire
Bartholemew	Red cornelian
James	White chalcedony
James-the-less	Topaz
John	Emerald

Apostle	Stone
Matthew	Amethyst
Matthias	Chrysolite
Peter	Jasper
Philip	Sardonyx
Simeon	Pink hyacinth
Thaddeus	Chrysoprase
Thomas	Beryl

Appendix B

Glossary of Unusual Names

This list includes alternative or disused gem names, misnomers and trade names. Misnomers are printed within inverted commas and trade names in italics. These names, many never used, have been retained for their occasional literary and scientific relevance.

Accarbaar (Akabar)	black coral
Accidental pearl	natural pearl
Achirite	Dioptase
Adamantine spar	silky brown sapphire
Adamite	artificial corundum powder
Adelaide ruby	red South African garnet
Adinol	a silicified porphyry or diabase
Aeroides	pale sky-blue aquamarine
Afghan turquoise	dyed magnesite
African emerald	green fluorite
African jade	massive green grossular garnet
African Star Coral	the coral *Allopora nobilis*
Agalmatolite	Steatite; pyrophyllite or pinite
Agaphite	vitreous variety of Persian turquoise
Agstein	Jet
Akori Coral	the coral *Allopora subviolacea*
Alabandine ruby	Almandine garnet
Alalite	Diopside
Alexandrite or Alexandrine	synthetic colour change sapphire or spinel
Alexandrite garnet	misnomer for a colour change garnet
Alaska black diamond	Hematite
Alaska diamond	rock crystal
Alaska jade	Pectolite
Alasmoden pearls	freshwater pearls
Alençon diamond	rock crystal
Aleppo stone	eye agate
Almandine spinel	natural violet spinel
Almandite	synthetic spinel
Almaschite	Romanian amber
Alomite	Sodalite
Aloxite	aluminium oxide powder
Alpine diamond	Pyrite
Alshedite	Sphene
Alumag	synthetic spinel
Alundum	aluminium oxide powder
Amarillo stone	figured chalcedony (Texas)
Amaryl	light green synthetic corundum
Amatrix (Amatrice)	mixture of variscite, chalcedony and quartz
Amause	glass (strass)
Amazon jade	green microcline feldspar
Amberine	yellowish green moss agate

America jade	Californite (idocrase)
American ruby	(1) Pyrope garnet: (2) rose quartz
Amethyst	synthetic purple sapphire
Ametrine	Quartz showing mixed amethyst and citrine
Ammolite	fossilized ammonite shell composed of aragonite
Amourant	synthetic white sapphire top/strontium titanite bottom composite stone
Ampullar pearl	Pearl from the epidermis of the oyster
Ancona ruby	Rose quartz
Andalusite	brown tourmaline
Andesine jade	Andesine feldspar
Angelite	opaque, bluish grey anhydrite
Angelo pearl	imitation pearl comprising plastic-coated mother-of-pearl
Aphrizite	black tourmaline
Apricotine	apricot-coloured garnets or quartz from New Jersey
Apyrite	peach-coloured tourmaline
Aqua Aura	blue, iridescent quartz crystals coated with gold, platinum or silver
Aquagem	light blue synthetic spinel
Aquamarine	synthetic pale blue spinel
Arabian diamond	rock crystal
Arabian magic diamond	synthetic colourless or yellow sapphire
Argillite	a slate-like rock from British Columbia
Arizona ruby	pyrope garnet
Arizona spinel	garnet
Arkansas diamond	rock crystal
Arkansite	transparent brookite
Armenian stone	lapis lazuli
Asparagus stone	yellow-green apatite
Astrilite	lithium niobate
Astryl	synthetic rutile
Atlas pearls	beads of white satin-spar
Australian ruby	garnet
Australite	tektite
Awabi pearl	abalone pearl (Japanese)
Axe stone	nephrite
Aztec stone	smithsonite
Azules opal	water opal with red and green flecks with bluish haze
Azurite	synthetic blue spinel
Azurite	sky-blue smithsonite
Azurlite	pale blue chalcedony
Bacalite	amber from Lower California
Baffa diamond	rock crystal
Bal de Feu	strontium titanate
Balas ruby	red spinel
Bamboo-pearl	Tabasheer
Baroda Gem	foiled back colourless glass
Bastard emerald	peridot
Bayate	ferruginous jasper (Cuba)
BC jade	nephrite from British Columbia
Beccarite	green zircon
Beckerite	succinite

Beekite	agatized coral
Bell pearl	drop pearl
Belomorite	moonstone (Russia)
Bengal amethyst	purple sapphire
Berigem	synthetic greenish yellow spinel
Bernat	polyester resin imitation of amber
Binghamite	Chatoyant quartz with goethite inclusions
Bird's-eye pearl	freshwater pearl with dark rings
Bird's-eye quartz	jasper with colourless spherulites
Bishop's stone	amethyst
Black amber	jet
Black diamond	hematite
Blackfellows buttons	tektites from West and South Australia
Black moonstone	Labradorite feldspar
Blackmorite	reddish yellow common opal from Montana
Blue alexandrite	change colour sapphire
Blue chrysoprase	chalcedony with included chrysocolla
Blue malachite	azurite
Blue opal	lazulite
Blue point pearl	pearls from the *Quadrula undulata*
Blue zircon	synthetic green-blue spinel
Boakite	brecciated green and red jasper
Bobrowka garnet	demantoid garnet
Bohemian chrysolite	moldavite
Bohemian diamond	rock crystal
Bohemian ruby	(1) pyrope garnet; (2) rose quartz
Bohemian topaz	citrine
Boke	rose-coloured coral (Japanese)
Bone turquoise	odontolite
Bornholm diamonds	rock crystal
Bourguignon pearls	wax-filled imitation pearls
Brazilian aquamarine	blue topaz
Brazilian diamond	rock crystal
Brazilian emerald	green tourmaline
Brazilian emerald	synthetic yellowish green spinel
Brazilian peridot	light green tourmaline
Brazilian ruby	red or pink topaz
Brazilian sapphire	blue tourmaline
Briancon diamond	rock crystal
Brighton diamond	rock crystal
Brighton emerald	green glass
Brilliante	synthetic rutile
Bristol diamond	rock crystal
Brilliant-glass	flat diamond used as portrait stone
Brown pearl	conchiolin-rich pearls of low value
Burma sapphire	synthetic blue sapphire
Burnt amethyst	heat-treated yellow quartz
Buxton diamond	rock crystal
Byewater	a colour grade of yellowish diamond
Cabra stone	fluorite
Cacholong	porcellaneous common opal
Calaite (Kalaite)	turquoise
Calao ivory	casque of a hornbill

Calbenite	myrickite
Calcentine	fossilized ammonite shell composed of aragonite
Californian tiger's eye	chatoyant bastite (serpentine)
Californian jade	californite (idocrase)
California moonstone	chalcedony
California onyx	banded stalagmitic calcite and aragonite
California ruby	garnet
California turquoise	variscite
Callainite	near turquoise (Brittany)
Cambay stone	Indian carnelian
Canada moonstone	peristerite feldspar
Canadian jade	nephrite from British Columbia
Canary diamond	yellow diamond
Canary stone	yellow carnelian
Cand (Cann)	fluorite (Cornwall)
Candite	blue spinel
Cape chrysolite	prehnite
Cape emerald	prehnite
Cape ruby	pyrope garnet
Capra Gem	trade name (USA) for synthetic rutile
Carnegiegem	a doublet with synthetic spinel top and strontium titanate base
Carneol	pink-dyed chalcedony
Catalin	phenolic resin plastic
Catalinaite	jasper from Catalina Island
Cathaystone	glass simulating cat's-eye
Catlinite	a clay-like mineral used by early American Indians
Cat's-eye	operculum
Cat's-eye jade	chatoyant tremolite or actinolite
Cat's-eye opal	Harlequin opal showing a streak of light
Cedarite	amber from Manitoba, Canada
Celestial stone	turquoise
Cellon	cellulose acetate plastic
Ceragate	waxy yellow-coloured chalcedony
Cerulene	calcite coloured green and blue by malachite and azurite
Ceylon chrysolite	yellow-green tourmaline
Ceylon diamond	colourless zircon
Ceylon opal	moonstone
Ceylon peridot	yellow-green tourmaline
Ceylon sapphire	synthetic sapphire
Chalchichuitl	Mexican name for jade or other greenstone
Chalcomalachite	malachite-calcite mixture
Chalcomalachite	mixture of malachite, calcite and gypsum
Chameleonite	change colour tourmaline
Champlain marble	massive dolomite from Vermont, USA
Changhua	altered silicic tuff containing cinnabar
Chank pearl	pearl from the *Turbinella scolymas*
Cherry opal	red-coloured common opal
Chicken Blood Stone	altered silicic tuff containing cinnabar
Chicken bone jade	yellowish burned or buried jade
Chicot pearl	natural blister pearl
Chi Ku Pai jade	chicken bone jade
Chinese cat's-eye	shell cat's-eye
Chinese jade	jadeite
Chinese turquoise	mixture of calcite, quartz and soapstone dyed blue

Ch'iung Yü	red jadeite
Chloropal	(1) green common opal (2) an opal-like hydrous silicate of iron
Chlorophane	fluorite which fluoresces on heating
Chlorospinel	green spinel
Chlor-utahlite	variscite
Chrome idocrase	emerald-green idocrase
Chromepidote	chrome-rich epidote from Myanmar, also known as tawmawite
Chrysanthemum stone	a radial aggregate of xenotime and zircon (Japan)
Chrysoberyllus	greenish yellow beryl
Chrysocarmen	red to brown copper-bearing stone with blue or green spots
Chrysolite	undesirable name for yellow-green chrysoberyl or for peridot, etc.
Chrysolithus	yellow beryl
Chrysopal	green common opal
Chrysoprase	green-dyed chalcedony
Chrysoquartz	green aventurine quartz
Cinnabar matrix	quartz with red cinnabar inclusions
Cinnamon stone	Hessonite garnet
Ciro Pearl	imitation pearl
Cirolite	yttrium aluminium garnet (YAG)
Cleiophane	zinc blende
Cloud agate	agate with dark cloud-like markings
Coconut pearl	pearls from the clam of Singapore
Coconut pearl	round concretions found in coconuts: of no value
Collophane	massive fibrous carbonate-hydroxyl apatite
Colorado diamond	transparent smoky quartz
Colorado goldstone	aventurine quartz
Colorado jade	green microcline feldspar
Colorado ruby	pyrope garnet
Colorado topaz	yellow quartz
Coltstone	acrylic resin plastic
Common opal	opal without play of colour (may be coloured)
Comptonite	variety of thomsonite
Congo emerald	dioptase
Contra Luz opal	opal with play of colour produced by transmitted and reflected light
Copper lapis	azurite
Coral agate	agate with a coral-like design, or agate pseudomorphous after coral
Coralline	red-dyed chalcedony
Cornish diamond	rock crystal
Coro Pearl	imitation pearl
Corsican green	Serpentine-bastite with schiller
Corundolite	synthetic white spinel
Cotterite	quartz with inclusions of white clay
Craquelées	crackled rock crystal (fire-stones)
Creolin	brecciated jasper
Creolite	red and white banded jasper (California)
Crispite	quartz or agate with green hair-like needles
Crocidolite opal	opal with included crocidolite (an opal cat's-eye)
Crop pearl	baroque pearl
Cross-grained stones	irregularly shaped and intergrown diamond crystals
Cross stone	(1) staurolite twin crystals (2) chiastolite (andalusite)
Crown Jewels	synthetic sapphire

Crusite	chiastolite
Crystal opal	water opal
Crystolon	carborundum powder
Cymophane	chrysoberyl cat's-eye
Cyst pearl	natural whole pearl
Dallasite	a green and white rock from Vancouver Island, British Columbia
Danburite	synthetic light red corundum
Danburite	synthetic yellow sapphire
Daourite	red tourmaline
Darwin glass	tektite
Dauphiné diamond	rock crystal
Davidsonite	yellow-green beryl
Dear pearl	lustreless natural pearl
Dekorite	phenolic plastic
De la Mar Pearl	imitation pearl
Delatynite	Romanian amber
Delawaritea	aventurine feldspar
Delta pearls	imitation pearls
Demidovite	blue compact chrysocolla
Desert amethyst	solarized glass
Desert glass	obsidian or moldavite
Diagem	strontium titanate
Diakon	acrylic resin plastic
Dialite	strontium titanate/synthetic spinel doublet
Diamanite	yttrium aluminate (YAG)
Diamantine	crystallized boron abrasive powder
Diamolin	yttrium aluminium garnet (YAG)
Diamonair	yttrium aluminium garnet (YAG)
Diamondite	synthetic white sapphire
Diamone	yttrium aluminium garnet (YAG)
Diamonique	yttrium aluminium garnet (YAG)
Diamonte	yttrium aluminium garnet (YAG)
Diamontina	strontium titanate
Diamothyst	synthetic rutile
Diaspore	hydrous aluminium oxide mineral
Dirigem	synthetic green spinel
Disthene	kyanite
Distrene	polystyrene resin plastic
Ditroite	sodalite
Di' Yag	yttrium aluminium garnet (YAG)
Djevalite	cubic zirconia
Dragomite	rock crystal (Galicia)
Duluth agate	Agate from Lake Superior
Durosol	aluminium oxide abrasive powder
Dust pearls	very small seed pearls
Dynagem	strontium titanate
Earth stone	mined amber
Eastenite	chrome enstatite-hypersthene
Ebonite	vulcanized india rubber
Eclogite	pyroxene garnet rock in South African diamond pipes
Edinite	prase
Egeran	idocrase
Egg pearl	natural egg-shaped pearl
Elbaite	pink tourmaline (Elba)

Elco Pearls	imitation pearls
Eldoradoite	blue chalcedony
El Doradoite	yellow quartz (California)
Electric emerald	green glass
Elie ruby	pyrope garnet
Elite Pearls	imitation pearls
Elixirite	banded rhyolite from New Mexico
Ellandra Pearls	imitation pearls
Emeralda	synthetic yellow-green spinel
Emerald malachite	dioptase
Emerald matrix	green fluorite
Emeraldine	dyed green chalcedony
Emeraldite or Emeralite	pale green tourmaline
Emeraldolite	synthetic emerald coating on natural colourless beryl
Emeraudine	dioptase
Emildine (Emilite)	spessartine garnet (South Africa)
Empirite	name for tektites found in Georgia
Endura Emerald	green glass
Enhydros	chalcedony nodules partly filled with water
Epidosite	epidote
Erinide	yellowish green synthetic spinel
Erinoid	casein plastic
Esmeralda	green tourmaline
Essonite	hessonite garnet
Evening emerald	peridot
Eye agate	banded agate with cutting so that bands are concentric
Eye diamond	fish-eye diamond
Eye stone	thomsonite
Fabulite	strontium titanate
Fairburnite	fortification agate (South Dakota)
Fairy stone	staurolite crystal (twinned) or an imitation of same
Falcon's eye	silica pseudomorph of blue crocidolite
Fales	any stone with differently coloured layers
False amethyst	purple fluorite
False chrysolite	moldavite
False emerald	green fluorite
False lapis	lazulite, or dyed jasper
False topaz	citrine or yellow fluorite
Falun Brilliants	lead-glass imitation stones
Fancy pearls	coloured natural nacreous pearls
Fashoda garnet	pyrope garnet
Feather gypsum	satin-spar
Fei-ts'ui	jade
Feldspar-apyre	andalusite (French)
Ferrers (Ferros) Emerald	green glass
Ferrolite	black iron slag proposed for use as a gem
Fianite	Cubic zirconia
Fire jade	a rock consisting mostly of opal, having a tiger's-eye appearance (due to grunerite)
Fire Pearl	billitonite
Flame spinel	orange-red natural spinel
Flash opal	Opal with a single-hued flash of colour
Flèches d'amour	rutilated (sagenitic) quartz
Flinder's diamond	white topaz

Flohmig amber	fatty amber, full of bubbles
Flower agate	chalcedony with flower-like inclusions
Foil back	chatons
Forcherite	orange-yellow opal coloured by orpiment
Fossil pineapple	opal pseudomorph after crystals of glauberite, gaylussite, or gypsum
Fowlerite	rhodonite
Framesite	an aggregate of diamond, bort and carbon from Premier mine
French colour rubies	light red rubies
Frost agate (Frost stone)	agate with white markings
Fuh yu	abalone shell and pearl
Fukien jade	soapstone
Futuran	phenolic resin plastic
Gava gem	synthetic rutile
Galalith	casein plastic
Galliant	a trade name (German) for a diamond simulant, gadolinium gallium garnet (GGG)
Garnet jade	massive green grossular garnet
Gagat	jet (German)
Gemerald	beryl doublet or synthetic emerald coated beryl
Geminair	yttrium aluminium garnet (YAG)
Geneva ruby	reconstructed ruby
German diamond	rock crystal
German lapis	blue-dyed jasper
German Mocoas	imitation moss agate
Gibsonite	pink thomsonite
Gibsonville emerald	greenish quartz
Giguku	jade (Japanese)
Giogetto	black opal
Girasol	(1) fire opal; (2) a type of water opal; (3) moonstone, etcetera
Girasol	trade name for glass used for imitation pearl beads
Girasol pearl	imitation pearl
Girasol sapphire	sapphire cat's-eye
Glass agate	obsidian
Glass lava	obsidian
Glass meteorite	moldavite
Glass opal	hyalite
Glass stone	axinite
Gles or geits	small cleavage cracks in diamond
Goldfluss	aventurine glass
Gold opal	fire opal
Gold quartz	(1) small particles of gold in quartz or quartzite; (2) golden-coloured crystal quartz
Gold sapphire	lapis lazuli
Goldstone	aventurine glass
Gold topaz	golden quartz
Goniobasis agate	agate replacing the fossil gastropod *Goniobasis*
Goodletite	marble forming the matrix of rubies (Myanmar)
Green ear	ear-shaped freshwater pearls
Green garnet	enstatite
Green John	massive green fluorite
Green onyx	stained green chalcedony
Green quartz	fluorite
Green star stone	chlorastrolite

Greenstone	(1) Nephrite; (2) chlorastrolite
Grenalite	staurolite
Griqualandite	crocidolite (tiger's-eye)
Gualdalcanal cat's-eye	operculum
Guarnaccio	yellowish red garnet
Guizhou jadeite	mixture of quartz, green dickite and an organic substance
Gum animé	copal resin
Haida slate	argillite
Hair amethyst	sagenitic amethyst
Hairstone	sagenitic quartz
Hakik	Agate (Indian)
Hammer pearls	baroque (hammer-head-shaped) pearls
Hard mass	imitation gems in hard glass
Hawaiian diamonds	rock crystal
Hawaiite	peridot (Hawaii)
Haystack pearl	natural high-domed button pearl
Heliolite	(1) aventurine feldspar; (2) transparent red labradorite
Heliotrope	bloodstone (quartz)
Hematine	imitation hematite
Hematinon, haematinon	opaque red glass
Hematite garnet	a synthetic iron-rich garnet
Herkimer diamond	rock crystal
Herrerite	blue and green smithsonite
Hinge pearls	elongated baroque pearls from the hinge of the freshwater mussel
Hinjosa topaz	yellow quartz
Höden, hoting	casque of a hornbill
Holstein	fossil wood
Honan jade	soapstone (agalmatolite)
Hope Sapphire	synthetic sapphire-blue spinel
Horatio diamond	rock crystal
Hornbill ivory	casque of a hornbill
Horn coral	black coral
Hot Springs diamond	rock crystal
Hsi jade	clear water or clear black jade
Hsieh jade	ink black jade
Hsiu Yen	green and white jasper
Hungarian cat's-eye	quartz cat's-eye
Hyacinth	(1) orange-brown zircon; (2) hessonite garnet
Hyacinth of Compostella	reddish iron-rich quartz
Hyaline	opalescent milky quartz
Hyalite	clear colourless opal
Hyalithe	red, brown, green or black opaque glass
Iceland agate	obsidian
Iimori jade (*Iimori Stone*)	a type of glass (Japan)
Image stone	agalmatolite
Imperial jade	fine green Chinese jade
Imperial sodden snow jade	white nephrite
Imperial topaz	sherry-coloured topaz
Imperial Yu stone	green aventurine quartz
Inamori (Crescent Vert) Emerald	synthetic emerald
Inanga	grey nephrite (New Zealand)

829

Inca emerald	emerald from Ecuador
Inca stone	pyrites
Indian agate	moss agate
Indian cat's-eye	chrysoberyl cat's-eye
Indian emerald	green-stained crackled quartz
Indian jade	green aventurine quartz
Indian topaz	yellow sapphire
Invelite	phenolic resin plastic
Iolanthite	banded reddish jasper
Irish diamonds	rock crystal
Iron opal	red or yellow common opal
Iserine (Iserite)	black iron mineral used to imitate hematite
Isle of Wight diamond	rock crystal
Isle Royale greenstone	chlorastrolite
Italian chrysolite	idocrase
Italian lapis	stained jasper
Itali	Aztec name for obsidian
Ivorina	casein plastic imitation of ivory
Ivory pearls	spherical ivory beans found in tusk cavities
Ivory turquoise	odontolite
Jacinth	(1) red-brown zircon; (2) hessonite garnet
Jadeolite	green syenite resembling jade, possibly pseudojadeite
Jade Tenace	saussurite
Jadine	Australian chrysoprase
Japan pearl	old name for cultured blister pearl
Japanese coral	dark red coral with white core
Jargoon	colourless or pale-coloured zircon
Jarra Gem	synthetic rutile
Jasp agate	intermediate between jasper and agate
Jaspe fleuri	jasp agate
Jasperine	banded jasper
Jasper jade	green jasper, serpentine, etcetera
Jaspillite	banded hematite and jasper
Jasponyx	banded jasp agate
Jasponal	intermediate jasper and opal
Jet stone	black tourmaline (schorl)
Johannes Gem	synthetic rutile
Johnite	vitreous and scaly turquoise
Jourado Diamond	synthetic colourless spinel
Kahurangi	pale green translucent nephrite
Kalmuck agate (opal)	cacholong
Kandy spinel	almandine garnet
Kan Huang jade	light yellowish jade
Kaolite	moulded imitation cameos, etcetera, in baked clay
Karlsbad Spring stone	banded gypsum used for carvings
Kashgar jade	inferior nephrite
Kawakawa	nephrite (Maori)
Kenneth Lane Jewel	strontium titanate
Kenya Gem	synthetic rutile
Keweenaw agate	agate from Lake Superior
Keystoneite	chalcedony coloured blue by chrysocolla
Khoton jade	inferior nephrite
Kidney stone	nephrite

Kikukwaseki	a radical aggregate of zircon and xenotime (chrysanthemum stone)
Killecrankie diamond	colourless topaz
Kima Gem	synthetic rutile
Kimberlite Gem	synthetic rutile
Kimpi	red or brown jadeite
Kingfisher jade	bluish green jadeite
King Topaz	natural yellow sapphire
Kinradite	orbicular jasper
Kismet Pearls	imitation pearls
Kollin garnet	almandine garnet
Korea jade	bowenite serpentine
Korite	fossilized ammonite shell composed of aragonite
Kyauk-ame	black jadeite
Kyauk-átha	white translucent jadeite
La Beau Pearls	imitation pearls
Labrador moonstone	Labradorite feldspar
Lactoid	Casein plastic
Laguna Pearls	imitation pearls
Lake George diamonds	rock crystal
Lake Superior agate	thomsonite, or correctly agate
Lake Superior fire agate	glass imitation of opal
Lake Superior greenstone	chlorastrolite
Lao Kan C'hing jade	bluish jade
La Paz Pearls	mostly grey and bronze pearls from the Gulf of California; or for coloured pearls from the hammer-head clam
Lapis crucifer	staurolite crystals
Lapis Nevada	thulite-diopside skarn
Lardite	agalmatolite
Larimar	blue pectolite from the Dominican Republic
Laser Gem	a doublet with a synthetic sapphire top and strontium titanate base
La Tausca Pearls	imitation pearls
Lat yay	clouded jadeite
Laurelite	idocrase
Lavendrine	amethyst quartz
Lavernite	synthetic periclase
Lazurfeldspar	bluish orthoclase (Siberia)
Lazurquartz	blue quartz (chalcedony)
Lechosos opal	opal with deep green and red flashes of colour
Leonite	Tibet stone (eosite)
Leuco-sapphire	colourless sapphire
Ligament pearl	hinge pearl
Linde Simulated Diamond	yttrium aluminate (YAG)
Lingah pearl (shell)	pearls (shell) from Persian Gulf
Lintonite	variety of thomsonite
Lithia amethyst	kunzite
Lithia emerald	hiddenite
Lithoxyle (Lithoxylite)	opalized wood
Litoslazuli	massive purple fluorite
Liver opal	menilite (impure opal)
Lluvisnando opal	yellowish water opal with pronounced flames
Love arrows	sagenitic quartz

Love stone	aventurine quartz
Lucinite	variscite
Lusterite	synthetic rutile
Lustigem	strontium titanate
Lux sapphire	iolite
Lynx eye	Labradorite with green flash
Lynx sapphire	Iolite
Macusanite	natural glass from Macusani, Peru
Madeira topaz	brown quartz or brown synthetic sapphire
Magalux	synthetic spinel
Maiden pearl	newly fished pearl
Majorica pearls	imitation pearls
Malacon	glassy brown variety of zircon
Malaya	a variety of East African garnet
Manchurian jade	soapstone
Manganese spar	rhodochrosite
Man Yu	jade of blood-red colour
Maori stone (jade)	nephrite
Marcasite (cut stones)	pyrite
Mari diamond	rock crystal
Mariposite	a foliated rock with bright green streaks of mica
Marmarosch diamond	rock crystal
Marmora diamond	rock crystal
Marvelite	strontium titanate
Marvella Pearls	imitation pearls
Mascot Emerald	Soudé emerald
Mass aqua	hard glass imitation
Mass opal	opal matrix
Matara (Matura) diamond	colourless zircon from Sri Lanka
Matorolite/Matorodite	a suggested name for chrome chalcedony
Maxixe aquamarine	boron-rich deep blue aquamarine (easily fades)
Mayaite	diopside jadeite (Central America)
Mecca stone	cornelian
Medfordite	a type of moss agate
Medina Emerald	green glass
Melanite	black andradite garnet
Melichrysos	yellow zircon
Menilite	banded grey and brown common opal
Meru sapphire	blue zoisite
Meta-jade, *Meta-hsui*	a type of glass (Japan)
Mexican agate	banded calcite or aragonite
Mexican diamond	rock crystal
Mexican jade	green-dyed stalagmitic calcite
Mexican onyx	stalagmitic calcite
Micatite	phenolic resin plastic
Midge stone	moss agate
Milhama Pebbles	jasper pebbles
Milk opal	milky white common opal
Miridis	synthetic rutile
Mixte	composite stone, half real and half imitation
Mock pearl	imitation pearl
Mogok diamond	white topaz
Molochites	green jasper
Montana jet	obsidian

Montana ruby	red garnet
Mont Blanc ruby	rose quartz
Morion	dark brown quartz
Moro	blood-red coral (Japanese)
Moroxite	greenish blue apatite
Mother of emerald	prase
Mountain crystal	rock crystal
Mountain jet	obsidian
Mountain ruby	red garnet
Mozarkite	a chert or flint
Muller's glass	hyalite glassy opal
Muscle pearls	small distorted pearls found near the muscle of the oyster
Mussel egg	freshwater pearl
Mussite	diopside
Mutzschen diamond	rock crystal
Mya yay	best-quality green jadeite
Nassau pearl	pink conch pearl
Nautilus pearl	oval section of Indian nautilus (coq de pearl)
Needle stone	sagenitic quartz
Nerchinsk aquamarine	blue topaz
Nevada diamond	obsidian
Nevada topaz	obsidian
Nevada turquoise	variscite
New Zealand greenstone	nephrite
Nicolo	black or dark brown onyx with thin bluish white layer
Night emerald	peridot
Nigrine	black rutile
Nixonoid	cellulosic plastic
Noble opal	precious opal
Norbide	boron carbide abrasive
Nuummite	iridescent gedrite
Occidental agate	poor-quality agate
Occidental amethyst	real amethyst (violet quartz)
Occidental cat's-eye	quartz cat's-eye
Occidental chalcedony	poor-quality chalcedony
Occidental cornelian	poor-quality cornelian
Occidental diamond	rock crystal
Occidental topaz	citrine
Occidental turquoise	odontolite
Oeil de boeuf	ox-eye labradorite feldspar
Olivene (olivine)	demantoid garnet
Onegite	amethyst with needle-like inclusions
Onyx obsidian	parallel banded obsidian
Onyx opal	banded opal
Opal agate	alternate bands of opal and chalcedony
Opalite	impure common opal, or chert from Lander Co, Nevada
Opalite	plastic imitation of opal
Orange topaz	brownish yellow quartz
Oregon jade	dark green jasper
Oregon moonstone	chalcedony
Oriental agate	good-quality agate
Oriental alabaster	stalagmitic calcite
Oriental almandine	purple-red sapphire
Oriental amethyst	violet sapphire

Oriental aquamarine	aquamarine-coloured sapphire
Oriental cat's-eye	girasol sapphire
Oriental cat's-eye	chrysoberyl cat's-eye
Oriental chalcedony	good-quality chalcedony
Oriental chrysoberyl	yellowish green sapphire
Oriental chrysolite	greenish yellow chrysoberyl or sapphire
Oriental cornelian	deep-coloured cornelian
Oriental emerald	green sapphire
Oriental topaz	yellow sapphire
Orletz	Russian name for rhodolite
Osmenda pearl	same as coq de perle
Owl-eye	eye agate with two similar eyes
Ox-eye	Labradorite feldspar
Ox-eye agate	similar to owl-eye
Ozakite	thomsonite
Pacific cat's-eye	operculum
Padparadscha (h)	orange variety of Sri Lankan sapphire
Pagoda stone (pagodite)	(1) agalmatolite; (2) fossil limestone (3) translucent agate with pagoda-like markings
Pai Yu (Pao Yu)	white jadeite or nephrite
Palmeira topaz	brown synthetic sapphire
Panabase	copper ore, properly tetrahedrite but often copper pyrites in a quartz base
Panama pearls	slate blue to black pearls from the Gulf of Mexico
Pantha	white translucent jadeite
Paphros diamond	rock crystal
Paragon pearls	round pearls of large size
Paragon Pearls	imitation pearls
Paris Pearls	imitation pearls
Paredrite	one of the favas of the Brazilian diamondiferous gravels (TiO_2 plus H_2O)
Passau pearl	freshwater pearl from Central Europe
Paste	glass imitation stones
Pate ce riz	glass imitation of jade
Patona Pearls	imitation pearls
Patricia Pearls	imitation pearls
Pauline Trigere	strontium titanate
Paulite	nearly black hypersthene with coppery inclusions
Peacock stone	malachite
Pearl doublet	cultured blister pearl
Pearl opal	cacholong
Pecos diamonds	rock crystal
Pectolite jade	pectolite
Peganite	variscite
Peiping (Peking) jade	any true jade but usually nephrite
Pelhamine	precious serpentine
Pennsylvania diamond	pyrite
Peredell topaz	greenish topaz
Peridine	green heat-treated quartz
Perigem	synthetic light yellow-green spinel
Petal pearls	distorted flattened pearls
Petoskey stone (agate)	fossil limestone (Michigan)
Phainite	cubic zirconia
Picotite	black spinel

Picrolite	serpentine
Pigeon-blood agate	cornelian
Pilbora Jade	ornamental serpentine
Pingoes d'agoa	colourless water-worn pebbles (Brazil)
Pink moonstone	scapolite
Pinna pearls	pearls from the Pinna mussel
Pin-fire opal	precious opal with play of colour in small patches
Pipe stone	a red siliceous clay (catlinite)
Pistacite	epidote
Pi Yu	jadeite or nephrite with a vegetable green colour
Plexiglas	acrylic resin plastic
Plume agate	moss agate (flower agate)
Point chalcedony	grey chalcedony with red spots
Polka-dot agate	translucent chalcedony with small red, brown or yellow dots
Polyphant stone	serpentine diabase (Cornwall)
Pomegranate ruby	red spinel
Pompadour Pearls	imitation pearls
Poppy stone	orbicular jasper
Porcelainite	metamorphosed baked clay
Potch	miners' term for opal which is colourful but dead
Prasemalachite	chalcedony filled with malachite
Prase opal	nickel-stained green opal
Prismatic moonstone	chalcedony
Prismatic quartz	iolite
Pseudochrysolite	moldavite
Pseudojadeite	probably a green albite feldspar
Pudding stone jade	lighter-coloured nodules of nephrite cemented by darker material
Purpurine	opaque red glass
Pyralin	cellulosic plastic
Pyralmandite	name suggested for the intermediate garnets of the Py/Al series
Pyrandine	alternative to pyralmandite
Pyroemerald	green fluorite
Quartz topaz	citrine
Quebec diamond	rock crystal
Quetzalztli	translucent green jade (Mexico)
Quincite (Quinzite)	(1) pink sepiolite; (2) pink common opal
Radient	synthetic white spinel
Radium diamond	smoky quartz
Rainbow agate (chalcedony)	iridescent agate
Rainbow Diamond	synthetic rutile
Rainbow Magic Diamond	synthetic rutile
Rainbow obsidian	iridescent obsidian
Rainbow quartz	Iris quartz
Raspberry spar	rhodochrosite
Redmanol	phenolic resin plastic
Red Sea pearls	coral beads
Regency Created Emerald	synthetic emerald
Resinoid	phenolic resin plastic
Retinalite	honey yellow serpentine
Rhine diamond	rock crystal
Rhodesian moonstone	bluish white translucent quartz from Mtoko, Zimbabwe

Rhodoid	cellulose acetate plastic
Riband agate	banded agate
Riband (Ribbon) jasper	jasper with stripes of alternating colour
Richlieu Pearls	imitation pearls
Ricolite	banded serpentine
Ripe pearls	pearls with good orient
River agate	water-worn moss agate pebbles
River pearls	freshwater pearls
Rock Crystal	glass
Rock glass	obsidian
Rock (Rocky mountain) ruby	pyrope garnet
Rodingite	grossular-vesuvianite ornamental rock
Rogueite	greenish jasper (Oregon)
Roman Pearls	imitation pearls
Romanzovite	dark brown grossular garnet
Rosaline	thulite
Rosaline	deep pink synthetic sapphire
Rose garnet	xalostocite, or incorrectly for rhodonite
Rose de France	pinkish violet amethyst, or synthetic sapphire of similar colour
Roseki	agalmatolite
Rose Kunzite	synthetic pink sapphire
Roselite (Rosolite)	xalostocite
Rose moonstone	pink scapolite
Rosinca	rhodochrosite
Rossini jewel	strontium titanate
Rosterite	rose-red beryl (Elba)
Rothoffite	yellow to brown andradite garnet
Royal Azel	sugilite
Royalite	purplish red glass
Royal Lavulite	sugilite
Royal topaz	blue topaz
Rozircon	pink synthetic spinel
Rubace (Rubasse)	red-stained crackled quartz
Rubasse	quartz coloured red by scales of iron oxide
Rubicelle	orange-red spinel
Rubolite	red-coloured common opal
Ruby balas	red spinel
Ruby spinel	red spinel
Ruin agate	brecciated agate or agate with patterns of ruins
Sabalite	banded green variscite
Safirina	blue spinel or blue quartz
Safranite (Saffronite)	citrine
Sagenite	quartz with needle-like inclusions
Salamanca topaz	fiery-coloured citrine
San Diego ruby	red tourmaline
Sang-i-yeshan	dark green bowenite serpentine
Saphir d'eau	iolite
Sapphire quartz	blue chalcedony, also blue silicified crocidolite
Sapphire spinel	blue spinel
Sapphirine	correct for Mg.Al silicate mineral. Used also for (1) blue chalcedony; (2) blue spinel (3) blue glass
Sard	translucent brown to reddish chalcedony
Sardium	artificially coloured sard

Sardoine	dark cornelian
Satelite	fibrous serpentine
Saxon chrysolite	topaz
Saxon diamond	topaz
Saxon topaz	yellow quartz
Schaumberg diamond	rock crystal
Schiller spar	bastite
Schmelze	glass
Schnide	blue glassy common opal
Schorl	black tourmaline
Scientific brilliant	synthetic white sapphire
Scientific emerald	green beryl glass
Scientific topaz	synthetic pink corundum
Scotch (Scottish) topaz	yellowish brown quartz
Seed pearls	small pearls less than 0.25 grain
Serra stone	Brazilian agate
Shanghai jade	steatite or talc
Shell marble	lumachella
Siam Aquamarine	blue zircon
Siberian chrysolite	demantoid garnet
Siberian ruby	red tourmaline
Siderite	blue quartz
Siliciophite	chrysotile in common opal
Silver Peak jade	malachite
Sinopal or sinople	reddish aventurine quartz
Sioux Falls jasper	quartzite (South Dakota)
Sira	abrasive aluminium oxide
Slocum Stone	glass imitation of opal
Smaragdite	green zoisite type of rock resembling jade
Smaragdolin	green beryl glass
Smoky topaz	smoky quartz
Soapstone	serpentine
Sobrisky opal	opal from Death Valley, California
Soldered emerald	soudé emerald (composite stone)
Soldier's stone	amethyst
Soochow jade	bowenite serpentine or soapstone
Sorella	strontium titanate
South African jade	massive green grossular garnet
Spalmandite (spandite)	names suggested for intermediate almandine-spessartine garnets
Spanish emerald	green glass
Spanish lazulite	iolite
Spanish topaz	yellowish brown quartz
Sparklite	colourless zircon
Spectrolite	iridescent labradorite from Finland
Sphalerite	zinc blende
Spinach jade	nephrite
Spinel ruby	red spinel
St Stephen's stone	red-spotted white chalcedony
Stantienite	black amber, possibly produced by lightning strike
Starlite	blue zircon
Starolite	star rose quartz doublet
Star-tania	synthetic rutile
Star topaz	yellow star sapphire

Stolberg diamonds	rock crystal
Strawberry pearl	pink-coloured baroque freshwater pearl with a pimply surface
Stremlite	blue zircon
Strongite	synthetic spinel
Styrian jade	pseudophite
Sugar stone	pink datolite (Michigan)
Sun opal	fire opal
Sweetwater agate	fluorescent moss agate from Wyoming
Sweetwater pearls	freshwater pearls
Swiss jade	green-dyed jasper
Swiss lapis	blue-dyed jasper
Synthetic Alexandrite	synthetic corundum or spinel
Synthetic Aquamarine	synthetic corundum or spinel
Synthetic Turquoise	an imitation turquoise
Syntholite	synthetic corundum imitating alexandrite
Syriam (Syrian or Suriam) garnet	almandine garnet
Tabasheer	amorphous opal-like silica found in the joints of some types of bamboo
Taiwan cats'-eye	chatoyant tressolite or actinolite
Takin	engraved emerald (India)
Taltalite	green tourmaline (Brazil)
Tama	jade (Japanese)
Tangiwai(te)	bowenite (New Zealand)
Tania-59	synthetic rutile
Tanzanite	purple-blue zoisite
Tasmanian diamond	rock crystal
Tauridian topaz	blue topaz
Tawmawite	massive chrome-rich epidote
Taxoite	green serpentine (Pennsylvania)
Tecla Pearls	imitation pearls
Tibet stone	mixture of aventurine quartz and quartz porphyry (Eosite)
Tigerite	tiger's-eye
Tirum Gem	synthetic rutile
Titangem	synthetic rutile
Titania Brilliante	synthetic rutile
Titania Midnight Stone	synthetic rutile
Titanium	synthetic rutile
Titanium Rutile	synthetic rutile
Titanstone	synthetic rutile
Tokay lux sapphire	hungarian obsidian
Tomb jade	buried jade which has turned to a red or brown colour
Tooth turquoise	odontolite
Topaz	yellow-brown quartz
Topaz	synthetic yellow sapphire
Topaz cat's-eye	chatoyant yellow sapphire
Topazolite	greenish yellow to yellow andradite garnet
Topaz quartz	brownish yellow quartz
Topaz saffronite	brownish yellow quartz
Tosa coral	Japanese coral
Tourmaline green	synthetic dark green spinel

Trainite	a type of banded variscite
Traversellite	green diopside
Tree stone (tree agate)	mocha stone (moss agate)
Trenton diamond	rock crystal
Triamond	yttrium aluminium garnet (YAG)
Trillium	green or blue-green apatite
Tripletine	emerald-coloured beryl triplet
Tsavorite, Tsavolite	transparent green grossular garnet from Kenya
Tsilaisite	manganese tourmaline
Tube agate	Agate with tubes (often filled) running through it
Turkey fat ore	cadmium-coloured yellow smithsonite
Turquerenite	dyed magnesite
Turritella agate	misnomer for goniobasis agate
Turtle back	(1) chlorastrolite; (2) turquoise or variscite matrix
Turtle-back pearl	oval natural blister pearl with a fairly high dome
Tuxtlite	sodium/magnesium pyroxene, the principal constituent of mayaite
Uigite	chlorastrolite from Skye
Ultralite	red-violet synthetic sapphire
Unionite	pink zoisite (thulite)
Unripe pearls	poor-quality pearls
Ural chrysolite (emerald or olivine)	demantoid garnet
Uralian sapphire	blue tourmaline
Utah turquoise	variscite
Utahlite	variscite
Vabanite	brown-red jasper specked with yellow (California)
Valencianite	adularia feldspar
Vallum diamond	rock crystal
Variolite	dark green orthoclase with lighter-coloured globules
Vashegyrite	an aluminium phosphate mineral, said to be like variscite, but may be yellow or brown
Verde de Corsica	green dialage and labradorite
Verdelite	green tourmaline
Vermeil, Vermeille	orange-red stones, may be zircons, garnets or spinels
Vermilite	cinnabar in opal
Vesuvian garnet	leucite
Vesuvianite jade	californite
Victoria Stone	a type of glass (Japan)
Vienna (Viennese) turquoise	a turquoise imitation
Vigorite	phenolic resin paste
Viluite	idocrase
Vinegar spinel	yellowish orange spinel
Violan(e)	violet massive diopside
Violet stone	iolite
Violite	purple synthetic sapphire
Viridine	green andalusite
Viscoloid	cellulosic plastic
Volcanic chrysolite	idocrase
Volcanic glass	obsidian
Vorobeyevite	pink beryl
Vulcanite	hard black rubber
Walderite	synthetic white sapphire

Wart pearl	baroque pearl
Washita diamond	rock crystal
Water sapphire	Iolite
Water stone	glassy orthoclase, or hyalite opal
Wax agate	yellow to yellowish red agate with waxy lustre
Wax opal	yellowish opal with waxy lustre
Wellington	strontium titanate
White garnet	leucite
White jade	Grossular garnet
Wilconite	purplish red scapolite
Wild pearl	natural pearl
Wilsonite	purplish red scapolite
Wiluite	idocrase
Winchellite	lintonite (thomsonite)
Wing pearl	baroque pearl shaped like a wing
Wisconsin pearls	fine coloured freshwater pearls from the *Unio* of the Mississippi
Wolf's-eye	(1) moonstone; (2) tiger's-eye
Wood agate	agate pseudomorph after wood
Wood opal	Opal pseudomorph after wood
Wood stone	fossil stone
Xihuitl	Turquoise (Aztec)
Xyloid jasper	jasperized wood
Xylonite	cellulosic plastic
Xylopal	opalized wood
Yanolite	violet axinite
Yaqui onyx	onyx marble from Baja, California
Yttralox	an optical ceramic mixture of 90% yttria and 10% thoria; RI = 1.92, dispersion = 0.039, SG = 5.30 and hardness = 6.5; may simulate diamond
Yu	jade
Yui ko lu jade	tomb jade coloured green from bronze objects buried near it
Yu Yen Stone	a massive greenish grey serpentine
Zaba Gem	synthetic rutile
Zabeltitzten diamond	rock crystal
Zarafina	blue spinel or blue chalcedony
Zeasite	wood opal
Zebra stone (jasper)	vari-coloured blue and green tiger's-eye (crocidolite), or limonite with lighter brown layers of shell material
Zenithite	strontium titanite
Zeuxite	green tourmaline (Brazilian)
Zircolite	synthetic white sapphire
Zircon Spinel	synthetic pale blue spinel
Zirctone	synthetic bluish green sapphire
Zonite	chert or jasper (Arizona)
Zonochlorite	near chlorastrolite
Zylonite	cellulosic plastic

Appendix C

Famous Diamonds and Named Large Diamonds

Name	Locality	Weight rough	Weight cut
Abadia do Dourades (light brown)	Brazil	104 Mcts	
Abaete (rose-pink)	Brazil	238 Mcts	
Akbar Shah	India	116 cts	73.60 Mcts
Algeiba Star (yellow)	South Africa?		133.03 Mcts square brilliant
Formerly the Mahjal, recut and renamed			
Al-Nader	?		115.83 Mcts pear
Amsterdam (black)	?	55.85 Mcts	33.74 Mcts pear
Arc	South Africa	381 cts	
Ashberg (amber)	South Africa?		102.48 Mcts cushion
A Steyn	South Africa	141.25 Mcts	
Austrian Yellow	*See Florentine*		
Banjarmasin	Kalimantan		40 cts approx. square
Barkly Breakwater	South Africa	109.25 cts	
Baumgold Brilliant	South Africa	167.25 Mcts	55 Mcts
Baumgold Pears	South Africa	609.25 Mcts	Two pears at 50 Mcts each and 12 others
Bazu	India	104 cts	
Beaumont	South Africa	273 cts	
Beau Sancy	India?		34 cts
Berglen (brown)	South Africa	416.25 Mcts	
Black Diamond of Bahia	Brazil	350 cts	
Black Star of Africa	South Africa?		202 cts
Bob Craig	South Africa	100.50 Mcts	
Bob Gove	South Africa	337 cts	
Brady	South Africa	330 cts	
Braganza	*Probably a topaz of 1680 cts*		
Brazilia (light blue)	Brazil	176.20 Mcts	
Broderick	South Africa	412.50 cts	
Brunswick Blue	?		?6–7 cts
Thought to be part of the Tavernier Blue			
Burgess	South Africa	220 cts	
Cape (canary yellow)	South Africa	167.00 Mcts	
Carmo do Paraneiba (brown)	Brazil	245 Mcts	
Carns	South Africa	107 cts	
Cartier	South Africa		107.7 Mcts pear
Cartier (NY)	South Africa	240.80 Mcts	69.42 Mcts pear and others

Name	Locality	Weight rough	Weight cut

Largest stone is known as the Taylor-Burton

Centenary	South Africa	599 Mcts	273.85 Mcts modified heart
Cent Six	?		106 cts
Cleveland	South Africa	100 + cts	50 cts cushion
Colenso (pale yellow)	South Africa	133.145 Mcts	
Coromandel I	Brazil	180 Mcts	
Coromandel II	Brazil	141 Mcts	
Coromandel III	Brazil	226 Mcts	
Coromandel IV	Brazil	400.65 Mcts	
Cross of Asia (champagne)	?		109.26 Mcts
Cuban Capitol	South Africa	23 cts	
Cullinan	South Africa	3106 Mcts	9 principal stones and 96 others
Cullinan I			530.20 Mcts pear

Also known as the Great Star of Africa. Largest cut stone of fine quality in the world, second largest overall. Set in the sceptre of the British Regalia.

Cullinan II			317.40 Mcts cushion
Cullinan III			94.40 Mcts pear
Cullinan IV			63.65 Mcts square

Cullinans II, III, and IV are also known as the Lesser Stars of Africa

Cullinan V			18.80 Mcts heart
Cullinan VI			11.50 Mcts marquise
Cullinan VII			8.80 Mcts marquise
Cullinan VIII			6.80 Mcts oblong brilliant
Cullinan IX			4.40 Mcts pear
Cumberland	India		32.82 Mcts triangular
Dan Campbell	South Africa	192.50 Mcts	32 Mcts
Darcy Vargas (brown)	Brazil	455 Mcts	
Dary-i Nur (pink)	India		estimated between 175 and 195 Mcts rectangular step cut

Major part of the Great Table

De Beers (yellow)	South Africa	439.86 Mcts	234.5 Mcts cushion
Deepdene (yellow)	?		104.88 Mcts cushion
Dewey	USA	23.75 Mcts	11.15 Mcts
Diario de Minas Geraes	Brazil	375.10 Mcts	
Dresden Green	India		40.07 Mcts
Dresden White	*See Saxon White*		
Dresden Yellow	India		4 brilliants, largest 38 Mcts
Dudley	*See Star of South Africa*		
Du Toit I (yellowish)	South Africa	250 cts	
Du Toit II (yellowish)	South Africa	127 cts	

Name	Locality	Weight rough	Weight cut
Earth Star (brown)	South Africa	248.9 Mcts	111.59 Mcts pear
Edna Star	?		115.0 Mcts emerald cut
Emperor Maximillian	Brazil		41.94 Mcts cushion
English Dresden	Brazil	119.5 cts	78.53 Mcts pear
Eugénie	Brazil?		52.35 Mcts oval brilliant
Eureka	South Africa	21.25 Mcts	10.73 Mcts oval brilliant
Excelsior	South Africa	995.2 Mcts	21 brilliants, largest 69.68 Mcts
Fineberg Jones	South Africa	206.5 cts	
Florentine (light yellow)	India		137.27 Mcts irregular double rose cut
Fly	South Africa	60 cts	
Golconda D	India		47.29 Mcts brilliant
Golconda Doré (yellow)	India	130 Mcts	95.40 Mcts
Golden Door (yellow)	?		104.95 Mcts shield
Golden Hue (yellow)	South Africa?		132.42 Mcts cushion
Governador Valadares	Brazil	108.30 Mcts	
Goyaz	Brazil	600 Mcts	80.00 Mcts and others
Great Brazilian	Brazil		130.00 Mcts
Great Chrysanthemum (brown)	South Africa	198.28 Mcts	104.15 Mcts pear
Great Mogul	India	787 cts	280 cts
Great Star of Africa	*See Cullinan I*		
Great Table	A legendary stone said to be 242 cts – *See Dary-i-Nur*		
Great White	*See Jacob*		
Guinea Star	Guinea	255.1 Mcts	89.01 Mcts
Hanger (pale yellow)	South Africa	123 cts	
Harry Young	South Africa	269.5 cts	
Hope (blue)	India		45.52 Mcts

Probably part of the Tavernier Blue; the Hope is now in the Smithsonian Institution

Name	Locality	Weight rough	Weight cut
Hope of Africa (yellow)	South Africa		151.91 Mcts
Ice Queen	*See Niarchos*		
Idol's Eye	India		70.20 Mcts
Imperial	*See Jacob*		
Incomparable (brownish yellow)	Africa	890 Mcts	407.48 Mcts triolette
Independencia	Brazil	106.82 Mcts	
Indien	India?		250 cts
Iranian I (cape)	South Africa		152.16 Mcts rectangular old brilliant
Iranian II (cape)	South Africa		135.45 Mcts high old cushion

Name	Locality	Weight rough	Weight cut
Iranian III (cape)	South Africa		123.93 Mcts cushion
Iranian IV (yellow)	South Africa		121.90 Mcts multi-faceted octahedron
Iranian V (cape)	South Africa		114.28 Mcts high old cushion
Ituiutaba	Brazil	105 Mcts	
Jacob	South Africa		184.5 Mcts oval
Formerly known as the Victoria, Imperial or Great White			
Jagersfontein	*See Winston*		
Jahan Akbar Shah	*See Akbar Shah*		
Jahangir	India		83.03 Mcts drop-shaped engraved
Jonker	South Africa	726 Mcts	125.65 Mcts largest and 11 others
Jubilee	South Africa	650.8 Mcts	245.35 Mcts brilliant
Julius Pam (yellow)	South Africa	248 cts	123 cts
Juscelino Kubitschak	Brazil	174 Mcts	
Khedive (light yellow)	?		36.61 Mcts rectangular emerald cut
Kimberley (champagne)	South Africa	490 Mcts	55.09 Mcts rectangular emerald cut
Kirti Noor (pink)	India		15 Mcts
Koh-i-Noor or Koh-i-Nur	India	said to be 800 cts	191 Mcts recut to 108.93 Mcts oval brilliant
Kruger	South Africa	200 cts	
La Belle Helene	South Africa	160 Mcts	Two pears 30.38 Mcts and 29.71 Mcts and a marquise 10.5 Mcts
Lesotho B	Lesotho	527 Mcts	
Lesotho C	Lesotho	338 Mcts	
Lesotho Brown	Lesotho	601.25 Mcts	18 stones, largest 71.73 Mcts
Libertador or Liberator	Venezuela	155 cts	4 stones, 3 emerald cuts 39.80 Mcts 18.12 Mcts and 8.93 Mcts; a marquise 1.44 Mcts
Light of Peace	Sierra Leone	435 Mcts	130.27 Mcts pear
Litke	South Africa	205.5 cts	
Mahjal	*See Algeiba Star*		
Marie Antoinette Blue			5.46 Mcts heart
Maria	Russia	106 Mcts	
Matan or *Mattam*	Kalimantan	*Probably a quartz of 367 cts*	
McClean	?		31.26 Mcts cushion
Meister (yellow)	South Africa?		118.05 Mcts cushion

Name	Locality	Weight rough	Weight cut
Minas Gerais	Brazil	172.5 Mcts	
Moon	South Africa?		183 Mcts round brilliant
Moon of Baroda (light yellow)	?		24 Mcts
Moon of the Mountains	India	*Legendary*	121 cts
Mountain of Splendor	India?		135 cts
Nassak	India		89.75 cts recut to 80.3 cts, then to 43.38 Mcts emerald cut
Nawanager	Russia?		148 Mcts
New Star of the South	Brazil	140 cts	
Niarchos	South Africa	426.50 Mcts	128.5 Mcts pear
Rough first called 'Ice Queen' or 'Pretoria'			
Nizam	India		277–340 cts (estimated) partially cut
Nooitgedacht	South Africa	325 Mcts	
Nur ul-Ain (pink)	India		60 Mcts (approx.) oval brilliant
May be part of the Great Table			
Oppenheimer (yellow)	South Africa	253.70 Mcts	
O'Reilly	South Africa	21.25 cts	
Orlov or Orloff	India	about 300 cts	189.6 Mcts rose above, flat, unfaceted below
Orpen-Palmer	South Africa	117.75 cts	
Otto Borgstrom	South Africa	121.50 cts	
Pam	South Africa	115 Mcts	56.6 Mcts
Paragon	Brazil		137.82 Mcts kite
Pasha of Egypt	India		41.06 Mcts recut to 38.19 Mcts, then to 36.22 Mcts
Patos (brown)	Brazil	324 Mcts	
Patrocino	Brazil	120.375 Mcts	
Paul I (pink)	India		13.35 Mcts cushion
Paulo de Frontin	Brazil	49.5 Mcts	
Peace	India		12.25 Mcts heart
Penthièvre (yellow)	?		10 cts oval
Pigott or Pigot	India		48.63 Mcts
Said to have been destroyed			
Pitt	*See Regent diamond*		
Pohl	South Africa	287 Mcts	
Polar Star	India		41.285 Mcts cushion
Porter-Rhodes	South Africa	153.5 Mcts	56.60 Mcts emerald cut, recut to 54.99 Mcts

Name	Locality	Weight rough	Weight cut
Portuguese	South Africa?		127.02 Mcts nearly octagonal emerald cut
Premier Rose	South Africa	353.9 Mcts	137.02 Mcts pear
President Dutra	Brazil	409 Mcts	36 stones
President Vargas	Brazil	726.6 Mcts	29 stones, largest 48.26 Mcts, recut to 44.17 Mcts emerald cut
Punch Jones	USA	34.46 cts	
Queen of Albania	India		49.03 Mcts
Queen of Holland (white, with blue tint)	India?		136.25 Mcts recut to 135.92 Mcts cushion
Raulconda	India		103 cts
Red Cross (yellow)	South Africa	370 cts approx.	205.07 Mcts square brilliant
Regent (white, with blue tinge)	India	410 cts	140.5 Mcts cushion
Regent of Portugal	Brazil	215 cts	
Reitz	*See Jubilee diamond*		
Rojtman (yellow)	South Africa		107.46 Mcts cushion
Sancy	India		55.23 Mcts pear-shaped double rose

The Sancy has been confused with a similarly shaped stone weighing 60.4 Mcts

Name	Locality	Weight rough	Weight cut
Saxon White	India		49.71 cts
Shah	India	95 cts	88.8 Mcts partially cut, engraved
Shah Jahan or Javeri	India		56.71 Mcts table cut
Sierra Leone II	Sierra Leone	115 Mcts	2 stones, 27.14 Mcts and 15.78 Mcts
Soleil D'Or (yellow)	?		105.54 Mcts emerald cut
Southern Cross	Brazil	118 cts	
Spaulding	*See Stewart diamond*		
Spoonmaker or Kasikci	India		86 Mcts rose-cut pear
Stalingrad	Russia	166 Mcts	
Star of Africa	*See Cullinan*		
Star of Arkansas	USA	15.33 Mcts	8.27 Mcts marquise
Star of Denmark	South Africa	105 Mcts	34.29 Mcts
Star of Diamonds	South Africa		107.5 cts

This stone may be the Rojtmann diamond

Name	Locality	Weight rough	Weight cut
Star of Egypt	India		106.75 Mcts emerald cut
Star of Este	India		26.16 Mcts
Star of Minas	Brazil	179.30 Mcts	
Star of Sierra Leone	Sierra Leone	968.90 Mcts	17 stones, largest 53.96 Mcts
Star of South Africa	South Africa	83.50 Mcts	47.69 Mcts pear
Star of the South	Brazil	261.88 Mcts	128.8 Mcts, oval brilliant

Name	Locality	Weight rough	Weight cut
Stewart (light yellow)	South Africa	296 Mcts	123 Mcts brilliant
Sunrise (yellow)	?		100.52 Mcts emerald cut
Tablet of Islam (black)	?		160.18 Mcts emerald cut
Tai Hang Star	South Africa	120 cts	60 cts
Taj-i-mah	India		115.06 Mcts Mogul cut
Taylor-Burton	*Largest stone cut from the Cartier (NY) (see Cartier)*		
Tennant	South Africa	112 cts	68 cts brilliant
Tereschenko (blue)	?		42.92 Mcts pear
Tiffany (yellow)	South Africa	287.42 Mcts	128.51 Mcts cushion
Tiger-eye (amber)	South Africa	178.5 cts	61.5 cts brilliant
Tiros I (brown)	Brazil	354 Mcts	
Tiros II (pink)	Brazil	198 Mcts	
Tiros III	Brazil	182 Mcts	
Tiros IV (brown)	Brazil	173 Mcts	
Unzue Heart (dark blue)	?		30.82 Mcts heart
Erroneously known as the 'Eugenie Blue'			
Vainer Briolette (yellow)	Southern Africa	202.85 Mcts	116.6 Mcts briolette
Van Zyl	South Africa	229.25 cts	
Vargas	*See President Vargas*		
Venter	South Africa	511 cts	
Victoria	*See Jacob*		
Victory	*See Woyie River diamond*		
Vitoria	Brazil	328.34 Mcts	44 stones, largest 30.39 Mcts
Webster kopje	South Africa	124 cts	
Williamson (pink)	Tanzania	54.5 Mcts	23.60 Mcts brilliant
Windsorten	South Africa	104 Mcts	
Winston	South Africa	154.50 Mcts	62.05 Mcts pear
Wittlesbach (blue)	India		35.50 Mcts oval brilliant
Woyie River	Sierra Leone	770 Mcts	30 stones, largest 31.35 Mcts emerald cut

The following large cut diamonds are unnamed:

(brown)	South Africa	755.5 Mcts	545.67 Mcts fire-rose cushion
The largest cut diamond in the world			
(yellow)	?		200.87 Mcts pear
(yellow)	?		180.95 Mcts briolette
(yellow)	?		150 Mcts emerald cut
—	?		141.23 Mcts pear
—	?		118 ct
(yellow)	India		114.28 Mcts briolette
(yellow)	South Africa?		114.03 Mcts cushion
(white)	?		111.82 Mcts heart

Name	Locality	Weight rough	Weight cut
(yellow)	?		108.04 Mcts emerald cut
(brown)	?		107.10 Mcts cushion
(white)	?		106.00 Mcts modified pear
(yellow, treated)	?		104.52 Mcts cushion
—	?		102.65 Mcts antique round
—	?		102.61 Mcts cushion
(light brown)	?		102.42 Mcts pear
(white)	?		101.84 Mcts pear
—	India		101.25 Mcts
(yellow)	?		101.25 Mcts briolette
(white)	?		101.14 Mcts kite

Mcts: metric carats   cts: old carats

Appendix D

Bibliography

Compiled by Michael O'Donoghue

The literature of gemstones is surprisingly large and had never been subjected to scholarly bibliographical techniques until the publication of *Gemology, an Annotated Bibliography*, by John Sinkankas. This two-volume work was published by Scarecrow Press, Metuchen, New Jersey in 1993 and gives full and scholarly details of all gemstone-related books largely but no means exclusively in English, irrespective of date of publication. Many important title-pages are reproduced. In this bibliography the writer has attempted to show something of the development of gemstone study, ranging from early classics to modern scientific books on gem identification. As with any scientific study the bulk of the useful published literature is found in papers in journals rather than in monographs, but considerations of space and the necessity for wholesale citation if this material were to be included have restricted coverage here to separate works. For this edition of *Gems* I have incorporated additional bibliographical material within the main text.

Literature Guides

O'Donoghue, M (1986) *The Literature of Gemstones*. The British Library. Modern bibliography with comments on many items, based on the resources of the British Library.

Gill, J (1978) *Gill's Index to Journals, Articles and Books Relating to Gems and Jewelry*. GIA, Santa Monica. The only extant guide to some at least of the periodical literature.

Sinkankas, J (1976) *Gemstones of North America*. Van Nostrand Reinhold, Princeton. The largest and best area bibliography with considerable material of general gemmological interest. The bibliography is in the second volume.

O'Donoghue, M (1988) *Gemstones*. Chapman and Hall, London. Large bibliography containing many older items.

O'Donoghue, M (1986) *The Literature of Mineralogy*. The British Library. Includes much detail of use to the scientific gemmologist with coverage of abstracting services, maps and dictionaries. Based on the resources of the British Library.

Abstracts

The only way to find out what papers have been published in serious scientific journals is to use hard copy or on-line abstracting services. The best hard copy in *Mineralogical Abstracts*, published by the Mineralogical Society of London from 1922 (covering from 1920). This is now available on-line via the

GeoRef host and includes notices of monographs; it covers all major Western languages. Russian and Slavonic material is found in *Referativnyi Zhurnal. Geologiia*, from 1956. This deals with earth science as a whole; it is not translated and is hard to use. Readers should consult curatorial staff in very large libraries and museums, the only places where this item can be found.

General Earth Science Literature Overview

Anyone beginning a serious academic study of any branch of the earth sciences should be able to find his way round the general literature. The only current guide is the following:

Wood, D, Hardy, J and Harvey, A (eds) (1989) *Information Sources in the Earth Sciences*. Butterworths, London. Includes critical descriptions of bibliographies, indexes and abstracts by M O'Donoghue and of earth science databases by D McKay and M O'Donoghue.

Major Mineralogical Works

Gemmologists will inevitably need to refer to standard mineralogical texts (see O'Donoghue, 1986, under 'Literature Guides'). Some of the more important are listed below; though some date back many years they are still alive even when volumes appear only at very long intervals. Those engaged in gemstone taxonomy (classification and naming) will particularly need to consult mineralogical sources.

The standard detailed classification of minerals is:

Anthony, J W *et al.* (1995–2003) *Handbook of Mineralogy*. 6 vols, Mineral Data Publishing, Tucson, AZ.
Deer, Howie & Zussman (Latest volume, 2004). *Rock-Forming Minerals (multi-volume)*. Geological Society of London, London.
Gains, R *et al.* (1997) *Dana's New Mineralogy* 8th edn, Wiley, New York.

A complementary work, also still theoretically in progress, is:

Hintze, C (1899–) *Handbuch der Mineralogie* (1889–1971) with *Ergänzungsbande und Register*. Revised by Karl Chuboda. Von Veit, Leipzig (and other publishers). An enormous work, by far the finest for locality information even today. Currently extending to 11 volumes in all.

As mineral science proceeds, many species are found to belong to groups. These, with details of chemical composition, crystal system and in some instances colour, can be found conveniently in the various editions of:

Fleischer's latest edition 2004. *Glossary of Mineral Species*. The Mineralogical Record Inc., Tucson. Updated about every five years in monograph and from time to time in *Mineralogical Record*.

General Comprehensive Works

Arem, J (1987) *Color Encyclopedia of Gemstones*. Van Nostrand Reinhold, New York. Alphabetical listing of a very large number of species with accurate scientific comment and the best colour photographs of any extant book on gemstones. First published in 1977.

Cavenago-Bignami, S (1980) *Gemmologia* (in Italian). 4th edn. 3 vols. Hoepli, Milan.

O'Donoghue, M (1988) *Gemstones*. Chapman and Hall, London. A survey of a wide range of gem species from the point of view of the geologist and mineralogist with considerable topical detail appropriate to these disciplines.

In all studies dealing with the natural world, taxonomy (classification and naming) is a continuing duty, and for this reason older surveys which were comprehensive for their time should still be kept to hand. The best of these is:

Bauer, M (1895–6: 1969) *Precious Stones*. Latest reprint. Tuttle, Rutland, Vermont. English translation by LJ Spencer of *Edelsteinkunde*, originally published in parts. The Tuttle reprint has some additional comment. A Dover Press reprint was published in 1968.

Gem Testing

O'Donoghue, M and Joyner L (2003) *Identification of Gemstones*. Butterworth-Heinemann, Oxford.

Schlossmacher, K (1969) *Edelsteine und Perlen*. 5 Aufl. Schweizerbart'sche, Stuttgart. The standard German textbook on gem testing until the 1970s. Illustrations poor by modern standards.

Smirnov, V (1973) *Dragotsennye i tsvetnye kamni poleznoe iskopaemoe*. Nauka, Moscow. A general gem testing work in Russian.

Gemstone Occurrence and Mineralogy

Bancroft, P (1984) *Gem and Crystal Treasures*. Western Enterprises and Mineralogical Record, Fallbrook, CA. Superbly illustrated and accurate survey of major gem and mineral occurrences world-wide.

Kivienko, E (1980) *Prospecting and Evaluation of Deposits of Precious and Economic Stones* (in Russian). Nedra, Moscow.

Scalisi, P and Cook, D (1983) *Classic Mineral Localities of the World: Asia and Australia*. Van Nostrand Reinhold, New York. Includes useful information, otherwise hard to find, on older gemstone deposits. Welcome reprints of crystal diagrams from Goldschmidt (see under 'Crystals').

Keller, P (1990) *Gemstones and their Origins*. Van Nostrand Reinhold, New York. One of the first attempts to place gemstones in their geological context. Very well illustrated.

Van Landingham, S L (1985) *Geology of World Gem Deposits* [collection of papers by various authors]. Van Nestrand, New York.

Crystals

The study of crystals is important for the gemstone prospector and for the student. The all-time most comprehensive survey is still:

Goldschmidt, V (1916–23) *Atlas der Krystallformen*. Carl Winters Universitätsbuchhandlung, Heidelberg. 9 vols. Partly text and diagrams, the latter drawn from journals up to the dates of publication. More than one reprint has been issued in recent years.

Phillips, F (1971) *An Introduction to Crystallography*. 4th edn. Oliver and Boyd, Edinburgh. As lucid as this rather complex topic can be made.

Wight, Q (1993) *The Complete Book of Micromounting* The Mineralogical Record, Tucson. [Recommended for the study of crystals.]

Colour

Nassau, K (1983) *The Physics and Chemistry of Color*. Wiley, New York. Admirably lucid.

Nassau, K (1985) *Gemstone Enhancement*. Butterworths, London. Gemstone colour alteration is the serious topic of the late twentieth century: this treatment is by far the best introduction to this vexed question.

Inclusions

Koivula, J and Gübelin, E (1992) *Photoatlas of Inclusions in Gemstones*. 2nd edn of a work first published in 1986. ABC Edition, Zurich. Gübelin and later Koivula have made the study of inclusions their own. Gübelin's first book, *Inclusions as a Means of Gemstone Identification*, was published by the GIA in 1953 and was followed in 1979 by the 2nd edition of *Internal World of Gemstones*. The current joint work has contributions by Meyer, Roedder, and Stalder on diamond inclusions and genesis, fluid inclusions, and quartz inclusions, respectively.

Chikayama, A (1973) *Gem Identification by the Inclusions*. Tokyo. English title of *Inkūrujon ni yoru hō seki no kanbetsu: kenbikyō shasin to sono nanbetsuhō*. *Illustrated entirely in colour with English headings only.* (In Japanese.)

Gem Testing Instruments

Read, P (1983) *Gemmological Instruments*. 2nd edn. Butterworths, London. The only guide in English from one of the pioneers in the field.

Quick Reference Publications and Tables

Read, P (1988) *Dictionary of Gemmology*. 2nd edn. Butterworths, London. Reference source for principal gem materials, gemmological terms and test instruments.

Sinkankas, J (1972) *Gemstone and Mineral Data Book*. Winchester Press, New York. Later reissued as *Field Collecting Gemstones and Minerals*. The only place to find numerical data, conservation details and much more.

Gunther, B (1981) *Bestimmungstabellen für Edelsteine, synthetische Steine, Imitationen*. Lenzen, Kirschweiler. Virtually in German and English and set out in tabular form.

Textbooks

Read, P (2005) *Gemmology*. Elsevier, Oxford. The only topical text of gemmology at student level, with specimens of examination questions. Update of *Beginner's Guide to Gemmology* (1980).

Synthetic and Imitation Stones

Arem, J (1973) *Man-Made Crystals*. Smithsonian Institution Press, Washington DC. Excellent and well-illustrated account of crystal growth with many gem examples.

Nassau, K (1980) *Gems Made by Man*. Chilton, Radnor, PA. Unrivalled account of the history and current growth techniques used to manufacture gemstones.

O'Donoghue, M (1983) *Identifying Man-Made Gemstones*. NAG Press, London. The standard guide to testing synthetic and imitation gemstones with notes (for 1983) on the alteration of colour. US edition has the title *A Guide to Man-Made Gemstones*.

O'Donoghue, M (1976) *Synthetic Gem Materials*. Worshipful Company of Goldsmiths, London. A literature survey with citations from scientific as well as gemmological literature.

O'Donoghue, M (1997) *Synthetic, Imitation and Treated Gemstones*. Butterworth-Heinemann, Oxford.

O' Donoghue, M (2005) *Artificial Gemstones*. NAG Press, London.

Balitskii, V and Lisitsyna, E (1981) *Sinteticheskie analogi i imitatsii pririodykh dragot-sennykh kamnei* (in Russian). Nedra, Moscow.

Michel, H (1926) *Die kunstlichen Edelsteine*. Leipzig. Very useful account of early gemstone manufacturing methods.

Macinnes, D (1973) *Synthetic Gem and Allied Crystal Manufacture*. Noyes Data Corporation, Park Ridge, NJ. Reproduces patents which are otherwise hard to locate.

Yaverbaum, L (1980) *Synthetic Gems Production*. Noyes Data Corporation, Park Ridge, NJ. Updates Macinnes (see above).

Fashioning

Sinkankas, J (1985) *Gem Cutting*. Van Nostrand Reinhold, New York. The best book on the subject.

Watermeyer, B (1980) *Diamond Cutting*. Purnell, Cape Town. Good and exhaustive account.

Pricing

Sinkankas, J (1968) *Van Nostrand's Standard Catalog of Gems*. Van Nostrand Reinhold, New York. Later issues have a supplementary page updating the text but prices are now out of date; the criteria by which they are reached remain the same, however, and there is a good deal of additional material.

Suwa, Yasukazu (2000–) *Gemstones, Quality and Value*. English and Japanese versions. In progress, Tokyo, published by the author and Sekai Bunka Publishing, Inc.

Gemstones of Particular Localities

North and Central America

Sinkankas, J (1959, 1976) *Gemstones of North America*. Van Nostrand Reinhold, New York. The 1976 volume is additional to the first and partly updates it. Central American locations are included. Outstanding bibliography in the 1976 text.

Myanmar

Iyer, L (1953) *The Geology and Gem-Stones of the Mogok Stone Tract, Burma*. Geological Survey of India, Calcutta. Forms vol. 82 of *Memoirs of the Geological Survey of India*.

Sri Lanka

Gübelin, E (1968) *Die Edelsteine der Insel Ceylon*. Published by author. Luzern.

Pakistan

Kazmi, A and O'Donoghue, M (1990) *Gemstones of Pakistan: Geology and Gemmology*. Gemstone Corporation of Pakistan, Peshawar. The first geological/gemmological survey of a single country with fine colour photographs.

Australia

Chalmers, R (1967) *Australian Rocks, Minerals and Gemstones*. Angus and Robertson, Sydney.

Brazil

Sauer, J (1982) *Brazil, Paradise of Gemstones*. Rough gem crystals beautifully illustrated in colour.

Asia and Australia

See entry for Scalisi and Cook under 'Gemstone Occurrence and Mineralogy'.

Africa

Keller, P (1992) *Gemstones of East Africa*. GIA, Santa Monica.

Individual Gem Species

Diamond

Balfour, I (1992) *Famous Diamonds*. 2nd edn. NAG Press, Colchester.
Bruton, E (1978) *Diamonds*. 2nd edn. NAG Press, London. An adequate survey, useful for beginners and students.
Copeland, L (1960–) *The Diamond Dictionary*. Various edns. GIA, Santa Monica.
Cram, L A *Journey though Colour*. vols 1, 2A and 2B published so far of this intended complete survey of Australian opal. Magnificently illustrated.
Hofer, S C (1989) *Collecting and Classifying Colored Diamonds*, Ashland Press, New York.
Hwang, J W (1995) *Natural Bleach Jadeite Identification* [in Chinese].
Liddicoat, R (1992) *The GIA Diamond Dictionary*. GIA, Santa Monica.
Orlov, Y (1977) *The Mineralogy of the Diamond*. Wiley, New York. Translation of *Mineralogiia alamaza*, Moscow, 1973. Invaluable for its discussion of diamond crystals and of the diamond deposits of the former USSR.
Balfour, I (1992) *Famous Diamonds*. 2nd edn. NAG Press, Colchester. The best up-to-date account of named diamonds, their description and whereabouts.
Wilks, J and Wilks, E (1991) *Properties and Applications of Diamond*. Butterworth-Heinemann, Oxford. The best scientific survey currently in print.
Williams, A (1932) *The Genesis of the Diamond*. Benn, London. 2 vols.
Williams, G (1906) *The Diamond Mines of South Africa*. Buck, New York. First published in 1902. 2 vols. Alpheus F and Gardner F Williams were related, and their two books are the finest study of diamond occurrence and mining extant.
Lenzen, G (1970) *The History of Diamond Production and the Diamond Trade*. Barrie and Jenkins, London. Translation of *Produktions- und Handelsgeschichte des Diamanten*.
Lenzen, G (1983) *Diamonds and Diamond Grading*. Butterworths, London. Translation of *Diamantenkunde*, 1979.
Pagel-Thiessen, V (1973–) *Handbook of Diamond Grading*. Various edns. Frankfurt. Well illustrated.
Bobrievich, A (1957) *Almazy sibiri*. Gosgeoltekhizdat, Moscow.
Bobrievich, A (1959) *Almaznye mestorozh deniya Yakutii*. Gosgeoltekhizdat, Moscow. Two authoritative accounts of the Siberian diamond deposits.
Freire de Andrade, C (1953) *Diamond Deposits in Luanda*. Companhia de Diamantes de Angola, Lisbon. 2 parts.

Beryl

Sinkankas, J (1981) *Emerald and Other Beryls*. Chilton, Radnor, PA. Contains an exhaustive bibliography. A shorter version edited by Read is published by Butterworth. Likely to remain the standard work on beryl for many years.
Vlasov, K and Kutukova, E (1960) *Izumrudnye kopi*. Akademiya Nauk SSSR, Moscow.

Kazmi, A and Snee, L (eds) (1989) *Emeralds of Pakistan: Geology and Genesis.* Geological Survey of Pakistan and Van Nostrand Reinhold, New York. Multi-author coverage in depth and a model of what such a study should be. Each section has its own bibliography.

Jade Minerals

Keverne, R (ed.) (1991) *Jade.* Anness, London. A multi-author coverage of the jade minerals and of the artefacts made from them.

Hansford, S (1968) *Chinese Carved Jades.* Faber, London. Originally published as *Chinese Jade Carving*, 1950. The best scholarly study with a good bibliography.

Nott, S (1936) *Chinese Jade throughout the Ages.* Batsford, London. Reprinted by Tuttle, Rutland, Vermont, 1962. A comprehensive study and good for the time.

Learning, S (1978) *Jade in Canada.* Geological Survey of Canada, Ottawa. Papers of the Geological Survey of Canada, no. 78-19. A good geological/mineralogical study. Coverage is not confined to Canada.

Bulletin of the Friends of Jade. 5004 Ensign St, San Diego, California 92117, USA. Occasional series with good quality papers.

Turquoise

Pogue, J (1915) *The Turquoise.* National Academy of Sciences, Washington DC. *Memoirs of the National Academy of Sciences*, vol. 12, part 2. The best coverage with considerable detail on artefacts. A reissue with a memoir of Pogue was published by Rio Grande Press, Glorieta, New Mexico in 1973.

Menchinskaya, T (1981) *Biryuza* (in Russian). Nedra, Moscow. Very extensive bibliography and authoritative text.

Quartz

O'Donoghue, M (1987) *Quartz.* Butterworths, London. An up-to-date scientific study with an extensive bibliography.

Dake, H (1938) *Quartz Family Minerals.* Whittlesey House, New York. A shorter study than the above but still useful.

Liesegang, R (1915) *Die Achate.* Dresden. Despite its age, the best treatment of agate dyeing.

Opal

Kalokerinos, A (1971) *Australian Precious Opal.* Nelson, Melbourne. Very fine colour photographs and original ideas on varietal nomenclature.

Eyles, W (1964) *The Book of Opals.* Tuttle, Rutland, Vermont. The best coverage of opal finds in the USA.

Leechman, G (1978) *The Opal Book.* Ure Smith, Sydney. Excellent overall coverage. Reprint of 1961 original.

Gunn, J (1971) *An Opal Terminology.* University of Sydney Australian Language Research Centre, Sydney. Occasional paper no. 15.

Wollaston, T (1924) *Opal, the Gem of the Never-Never*. Murby, London. Excellent for the early days of Australian opal.

Cody, A (1991) *Australian Precious Opal: A Guide Book for Professionals*. Andrew Cody, Melbourne. One of the best short studies.

Cram, L (1991) *Beautiful Queensland Opal*. Robert Brown, Buranda, Queensland. The best book on Queensland opal with a short history of the fields.

Garnet

Rouse, J (1986) *Garnet*. Butterworths, London. Surprisingly the only monographic study of the garnet group in English.

Tourmaline

Kuz'min, V (1979) *Turmalin* (in Russian). Nedra, Moscow. Includes a 300-entry bibliography.

Benesch, F (1991) *Der Turmalin: eine Monographie*. 2 durchgesehene und verbesserte Aufl. Urachhaus, Stuttgart. An extremely large-format book with superb colour photographs and paintings. Fine coverage of geochemistry, paragenesis and also of the place of tourmaline in religious literature.

Dietrich, R (1985) *The Tourmaline Group*. Van Nostrand Reinhold, New York. The only major study in English and of a high standard.

Corundum

Hughes, R (1990) *Corundum*. Butterworths, London. The best and almost the only modern coverage of this major species.

Mumme, I (1988) *The World of Sapphire*. Mumme, Port Hacking, NSW. A good and the only coverage of sapphire in English at the time of writing.

Pearl

Kunz, G and Stevenson, C (1908) *The Book of the Pearl*. Macmillan, London. Very large comprehensive study which is still valuable.

Farn, A (1986) *Pearls: Natural, Cultured and Imitation*. Butterworths, London.

Strack, E (1982) *Perlenfibel*. Rühle-Diebener, Stuttgart.

Amber

Fraquet, H (1987) *Amber*. Butterworths, London.

Rice, P (1974) *Amber, the Golden Gem of the Ages*. Van Nostrand Reinhold, New York.

Schlee, D (1980) *Bernstein-Raritäten*. Staatliche Museum für Naturkunde, Stuttgart. Very fine coloured photographs of inclusions in amber.

Larsson, S (1978) *Baltic Amber: A Paleobiological Study*. Scandinavian Science Press, Klampenberg. Entomonograph no. 1.

Jet

Muller, H (1987) *Jet*. Butterworths, London.

Crystal Growth

Gemmologists need to know something of this complex and often unpredictable process. The best modern guide is:

Brice, J (1986) *Crystal Growth Processes*. Blackie, Glasgow. Up-to-date coverage of all commercial and experimental growth methods.

A guide to the literature (which is often hard to find) including conference papers is:

O'Donoghue, M (1988) *Crystal Growth: A Guide to the Literature*. The British Library. Many accounts of the growth of crystals are concealed in conference proceedings, which are set out here, as well as monographs and journals in all languages.

Journals

Association Journals

All students of gemmology should subscribe to at least one of the official journals (i.e. the organ of a major gemmological association). The following is a selection (published quarterly unless indicated otherwise):

ExtraLapis series. Multi-author monographs devoted to particular minerals, many of them gem species. Very fine illustrations in colour and useful diagrams. Weise Verlag, München. In progress.

Journal of Gemmology. Gemmological Association and Gem Testing Laboratory of Great Britain. 27 Greville St, London, EC1N 8TN, UK. Scientific papers of a high standard; very comprehensive coverage of abstracts and reviews in all languages.

Gems and Gemology. Gemological Institute of America. Illustrated entirely in colour; valuable country surveys and notes from GIA gem testing laboratories.

Australian Gemmologist. Gemmological Association of Australia. PO Box 35, South Yarra, Victoria, Australia. Particularly valuable for studies of organic materials and for Gemmology Study Club reports on a very wide variety of materials. (*Wahroongai News*, the organ of the Queensland branch of the GAA, has a very wide informal coverage and is well worth consulting. PO Box 7184, East Brisbane, Queensland, Australia.)

Zeitschrift der Deutschen Gemmologischen Gesellschaft. Deutsche Gemmologische Gesellschaft, Postfach 122260, D-55743 Idar-Oberstein, Germany. Papers of high scientific standard. Irregular.

Revue de Gemmologie. Association Française de Gemmologie, 14 rue Cadet, Paris 75009, France. Emphasis on French jewellery history.

La gemmologia. Istituto Gemmologico Italiano, Viale Gramsci 228, 20099 Sesto S. Giovanni, Milan, Italy. High-quality papers with English abstracts.

Boletin del Instituto Gemológico Español. Instituto Gemológico Español, Victor Hugo 1,3°, 28004 Madrid, Spain. Irregular.

Current Awareness Journals

Gemmologie Aktuell. Published several times a year and reproduced from typewriting. Covers new species, man-made substances etc. Address as for *Zeitschrift* above.

Popular and Trade Journals

Lapidary Journal. PO Box 80937, San Diego, California 92138–0937, USA. A monthly survey with many trade advertisements and with particular reference to amateur collecting and cutting. The best of its kind.

Jewelers' Circular-Keystone. Chilton Company, Chilton Way, Radnor, Pennsylvania 19089, USA. Very good trade and gemstone coverage. Monthly.

Mineralogical Journals

Mineralogical Record. PO Box 35565, Tucson, Arizona 85740, USA. Superb coverage of the mineral world with fully illustrated papers (many of gem interest) successfully bridging the gap between educated amateur and professional.

Mineralogical Abstracts. Mineralogical Society, 41 Queen's Gate, London SW7 5HR, UK. Quarterly. World-wide coverage of all earth science journals with mineral (including gem) content. Now only available on-line.

Map

World Map of Gemstone Deposits. Swiss Gemmological Society and Kummerly and Frey, Bern, Switzerland. In colour.

Jewellery History

Jewellery Studies. Society of Jewellery Historians, c/o Dept of Prehistoric and Romano–British Antiquities, The British Museum, London WC1B 3DG, UK. An occasional journal with papers of a high scholarly standard. Appears irregularly.

Jewellery Pricing

This often vexed topic can be studied conveniently by subscribing to or at least consulting the major saleroom catalogues. These have descriptions of the lots with notes of major stones (Myanmar) rubies and Kashmir sapphires often have certificates of origin) and notes of estimates (before sale) of price for all but the most important and debatable items. In catalogues of the major houses the lots are illustrated in colour and thus form an invaluable record of fashion and of important stones. Long runs of the best sale catalogues are valuable, but they should be completed by notes on the prices actually paid (and of where items are withdrawn for one reason or another). The major catalogues are those of:

Christie's, 8 King St, London SW1Y 6QT, UK
Sotheby's, 34–35 New Bond St, London W1A 2AA, UK.

Both houses hold sales in their home country, in the USA and in major European cities, particularly in Switzerland where the most prestigious sales are held.

Index